| | | |
|---|---|---|
| Boltzmann's constant | $k$ | $1.38 \times 10^{-23}$ J/K |
| Avogadro's number | $N_0$ | $6.02 \times 10^{23}$/mole |
| Elementary charge | $e$ | $1.60 \times 10^{-19}$ C |
| Electron mass | $m_e$ | $9.1 \times 10^{-31}$ kg |
| Proton mass | $m_p$ | $1.67 \times 10^{-27}$ kg |
| Planck's constant | $h$ | $6.63 \times 10^{-34}$ J s |
| Mass of earth | $M_e$ | $5.98 \times 10^{24}$ kg |
| Mass of moon | $M_m$ | $7.36 \times 10^{22}$ kg |
| Mass of sun | $M_s$ | $1.99 \times 10^{30}$ kg |
| Earth-moon distance | | $3.80 \times 10^5$ km |
| Radius of earth | $R_e$ | $6.35 \times 10^3$ km |
| Electrical constants: | $k_0 = 1/(4\pi\epsilon_0)$ | $9.00 \times 10^9$ N m$^2$/C$^2$ |
| | $k_0/c^2 = \mu_0/4\pi$ | $10^{-7}$ N s$^2$/C$^2$ |
| | $\epsilon_0$ | $8.85 \times 10^{-12}$ F/m |

# Physics

JAN 2 6 1992
FEB - 4 1992
NOV 2 6 1992
FEB 2 6 1993
AUG - 4 1993
MAR 7 1994
21
OCT 1 5 1995
NOV 2 6 1995
APR 6 2000

*From the front to rear cover is a span of almost 4000 years in the progress of science. Stonehenge on the front cover was built in 1900 B.C. by the scientists of that period in England. Fermilab on the rear cover was built in A.D. 1970 by scientists in the United States; a view of the central laboratory building is shown. [Stonehenge photo by Preston Lyon; Fermilab photo by Photography Department, Fermi National Accelerator Laboratory.]*

*Enrico Fermi, 1901–1954*

# PHYSICS

## Jay Orear
CORNELL UNIVERSITY

Macmillan Publishing Co., Inc.
NEW YORK

Collier Macmillan Publishers
LONDON

Copyright © 1979, Jay Orear

Printed in the United States of America

All rights reserved. No part of this book may be reproduced or transmitted in any form or by any means, electronic or mechanical, including photocopying, recording, or any information storage and retrieval system, without permission in writing from the Publisher.

Macmillan Publishing Co., Inc.
866 Third Avenue, New York, New York 10022

Collier Macmillan Canada, Ltd.

Library of Congress Cataloging in Publication Data

Orear, Jay.
  Physics.

  Includes index.
  1. Physics.   I. Title.
QC21.2.073        530         77-28180
ISBN 0-02-389460-1

Printing: 1 2 3 4 5 6 7 8    Year: 9 0 1 2 3 4 5

# Preface

This book is intended as a text for a one-year or a one-and-a-half-year introductory physics course for science and engineering students. Prior knowledge of calculus is not needed, but a course based on this book should be concurrent with a calculus course. The mathematical and conceptual level of this book is not intended to exceed that of currently popular texts. However, this book differs from most others in two respects:

1 There is a unified presentation of up-to-date physics.
2 Whenever possible, "laws" of physics are derived from the basic principles; that is, the distinction between the basic principles and what is derivable from them is emphasized.

The relationships among the different fields of physics (as well as science and technology) are made clear. Seemingly independent topics are unified and tied together into one grand overview. When a new "law" of physics, such as magnetic force on a moving charge or equipartition of energy, is introduced, I try to make clear whether it is truly a new fundamental law or can be derived from preceding material. In most cases there are simple derivations that help delineate the logical structure and beautiful unity of what might otherwise seem to be an encyclopedic listing of diverse phenomena and laws. Such so-called laws as Ampere's law, Faraday's law, the equipartition of energy, magnetic force, Ohm's law, slower speed of light in matter, Hooke's law, and Huygens' principle are derived from the more fundamental laws where possible. They are not presented as new, independent laws of physics. I feel obligated to explain the reason "why" whenever possible. If a derivation is too difficult, the reader is at least told that a derivation is possible and plausibility arguments are given. These "secondary laws" make possible further experimental predictions upon which the survival of the basic laws depends.

If all of electromagnetism is to be "derived" from Coulomb's law, a prior understanding of special relativity is helpful. For this reason relativity is presented in Chapters 8 and 9, before electromagnetism (Chapters 15–21). Nevertheless, the chapters on electromagnetism are written in such a way that those wishing to do so may skip Chapters 8 and 9. If relativity is to be deferred, it should not be deferred past Chapter 24.

Since a true understanding of the structure of matter and many other physical phenomena rests upon quantum theory, the basic principles of quantum theory are presented in Chapters 24 and 25. Understanding of the quantum theory is strengthened in the chapters that follow by quantitative applications in atomic physics, solid-state physics, nuclear physics, astrophysics, and particle physics. With this background in quantum mechanics and relativity, it is even possible to make simple calculations of the radius and constituents of neutron stars and black holes (Chapter 30).

Some may worry that such "advanced" topics as neutron stars, black holes, Fermi energy, conservation of parity, quarks, holography, time dilation, and intensity interferometry are too difficult for beginning students. I include them because these new developments capture the imagination of students as they read of them in newspapers and magazines. Students come to college expecting to learn more about such exciting things in their physics course. My classroom experience shows that many of these so-called advanced topics are easier for students to grasp than the implications of Newton's third law.

Another legitimate concern is whether students planning to become engineers should be exposed to these modern ideas. I find that few professors of engineering want the introductory course to be an applied physics course or an engineering course. The introductory physics course for engineering majors may be the students' only chance to see how the different fields tie together. It may be their only chance to learn of new developments and how these relate to other fields of science and technology. Thus there is some effort to relate the study of physics to other fields of science and give attention to the impact of science on society. For example, the world shortage of energy is a central theme that runs throughout the book. Other social, political, economic, and philosophical implications of science are discussed. Not only does this physics course lay the theoretical foundations for a future profession, but it should also contribute to the cultural background expected of a person who will be involved with science and technology. And, as previously implied, an understanding of relativity and quantum mechanics is necessary for proper understanding of most natural phenomena.

Many examples are given, with emphasis upon those of social significance. Although some examples are used to illustrate interesting side applications, most provide instruction in problem-solving technique. Those examples that might be too difficult for a beginning student are labeled with a star. In addition to the worked-out examples distributed throughout the text, there are many problems at the ends of chapters.

These are divided into "Exercises" and "Problems." "Exercises" are merely easier problems; they tend to have fewer and easier steps. Each group is arranged approximately in the order of the presentation in the text. I have made a serious attempt to provide problems corresponding to real-life situations. Thus some of the problems involve topics from earlier chapters. Not only is this closer to real life, but also it provides review and helps develop retention and transfer. Another element of realism is that not all the necessary information is provided in every problem, though the information can be found elsewhere in the text. On the other hand, a few problems contain more information than is necessary.

The text avoids modernized spellings and innovative word usage. This is hard on the author, especially when it comes to the sexism built into the English language. When I use the word *man* or its derivatives, I mean Webster's definition **I** rather than definition **II**. (Webster's **I** is "a member of the human race"; Webster's **II** is "a male human being.") Of course, a more serious problem than the bias built into our centuries-old language is the present-day bias that discourages young women from the study of physics or engineering.

The SI system of units (mks) is used throughout the book. For practical reasons other units are occasionally mentioned. Although the presentation of electromagnetism uses the SI units, most of the electricity equations are displayed in a form easily usable for those who prefer to teach electricity in gaussian (cgs) units.

A uniform color-code system has been adopted for all vectors. Arrows indicating velocity and speed are always grey. Acceleration vectors are open arrows outlined in black. Force vectors are open arrows outlined in color. Electric field and electric lines of force are colored, magnetic field is lighter-colored, and electric currents are black arrows. Resultant vectors are of greater width than their components. These and other graphical conventions should make the illustrations easier to interpret.

For a shorter or less demanding course, sections that can be eliminated are indicated by stars in the Contents.

I am grateful to my colleagues and students at Cornell University for their encouragement and for giving me opportunities to try out much of the material here in teaching the introductory physics course for engineers over the last ten years. My greatest debt is to Enrico Fermi who not only taught me much of the physics presented here but how to approach it and teach it. My main purpose in writing this book is an attempt to present the spirit and excitement of physics in the clear way that Fermi might have done.

Jay Orear

# Contents

## 1 Introduction 1

- 1-1 The Nature of Physics 1
- 1-2 Units 3
- 1-3 Dimensional Analysis 7
- 1-4 Accuracy in Physics 10
- 1-5 Mathematics in Physics 13
- ★1-6 Science and Society 17
- Appendix 1-1. Correct Answers to the Common Mistakes 18

## 2 One-Dimensional Motion 22

- 2-1 Velocity 2
- 2-2 Average Velocity 24
- 2-3 Acceleration 27
- 2-4 Uniformly Accelerated Motion 29

## 3 Two-Dimensional Motion 38

- 3-1 Free-Fall Trajectories 38
- 3-2 Vectors 39
- 3-3 Projectile Motion 45
- 3-4 Uniform Circular Motion 47
- 3-5 Earth Satellites 49

## 4 Dynamics 55

- 4-1 Introduction 55
- 4-2 Definitions 56
- 4-3 Newton's Laws of Motion 57

★ Starred sections may be skipped or deferred.

|   |   |
|---|---|
| 4-4 | Units of Force and Mass   61 |
| ✳ 4-5 | Contact Forces and Friction   62 |
| 4-6 | Problem Solving   65 |
| 4-7 | Atwood's Machine   69 |
| 4-8 | The Conical Pendulum   70 |
| 4-9 | Conservation of Momentum   71 |

## 5 Gravitation   79

| | |
|---|---|
| 5-1 | The Universal Law of Gravitation   79 |
| 5-2 | The Cavendish Experiment   82 |
| 5-3 | Kepler's Laws of Planetary Motion   84 |
| 5-4 | Weight   86 |
| ★5-5 | The Principle of Equivalence   90 |
| ★5-6 | The Gravitational Field Inside a Sphere   91 |

## 6 Work and Energy   97

| | |
|---|---|
| 6-1 | Introduction   97 |
| 6-2 | Work   98 |
| 6-3 | Power   100 |
| 6-4 | The Dot Product   101 |
| ✳ 6-5 | Kinetic Energy   103 |
| ✳ 6-6 | Potential Energy   106 |
| 6-7 | Gravitational Potential Energy   108 |
| 6-8 | Potential Energy of a Spring   110 |

## 7 Conservation of Energy   115

| | |
|---|---|
| 7-1 | Conservation of Mechanical Energy   115 |
| 7-2 | Collisions   120 |
| 7-3 | Conservation of Gravitational Energy   124 |
| 7-4 | Potential Energy Diagrams   127 |
| 7-5 | Conservation of Total Energy   129 |
| ★7-6 | Energy and Biology   133 |
| 7-7 | Energy and the Automobile   134 |
|   | ★Appendix 7-1. Conservation of Energy for N Particles   139 |

## 8 Relativistic Kinematics   144

| | |
|---|---|
| ★8-1 | Introduction   144 |
| ★8-2 | Constancy of the Speed of Light   145 |
| ★8-3 | Time Dilation   150 |

- ★ 8-4 The Lorentz Transformation  153
- ★ 8-5 Simultaneity  157
- ★ 8-6 Doppler Effect of Light  159
- ★ 8-7 The Twin Paradox  161

## 9 Relativistic Dynamics  169

- ★ 9-1 Einstein's Addition of Velocities  169
- ★ 9-2 Definition of Relativistic Momentum  172
- ★ 9-3 Conservation of Momentum and Energy  173
- ★ 9-4 Equivalence of Mass and Energy  175
- ★ 9-5 Kinetic Energy  178
- ★ 9-6 Mass and Force  180
- ★ 9-7 General Relativity  181
  - ★ Appendix 9-1. Momentum–Energy Transformation  184

## 10 Rotational Motion  190

- 10-1 Rotational Kinematics  190
- 10-2 The Vector Cross Product  192
- 10-3 Angular Momentum  193
- ★ 10-4 Rotational Dynamics  195
- ★ 10-5 Center of Mass  200
- ★ 10-6 Rigid Bodies and Moment of Inertia  203
- ★ 10-7 Statics  206
- ★ 10-8 Flywheels  209

## 11 Oscillatory Motion  215

- 11-1 The Harmonic Force  215
- 11-2 The Period of Oscillation  218
- 11-3 The Pendulum  220
- 11-4 Energy of Simple Harmonic Motion  223
- ★ 11-5 Small Oscillations  224
- ★ 11-6 Intensity of Sound  228

## 12 Kinetic Theory  234

- 12-1 Pressure and Hydrostatics  234
- 12-2 The Ideal Gas Law  239
- 12-3 Temperature  241
- 12-4 Equipartition of Energy  244
- 12-5 Kinetic Theory of Heat  246

# 13 Thermodynamics 253

- 13-1 First Law of Thermodynamics 253
- 13-2 Avogadro's Hypothesis 254
- 13-3 Specific Heat 255
- 13-4 Isothermal Expansion 260
- 13-5 Adiabatic Expansion 260
- 13-6 The Gasoline Engine 264

# 14 Second Law of Thermodynamics 270

- 14-1 The Carnot Engine 270
- 14-2 Thermal Pollution 273
- 14-3 Refrigerators and Heat Pumps 273
- 14-4 The Second Law of Thermodynamics 276
- ★14-5 Entropy 279
- ★14-6 Time Reversal 284

# 15 Electrostatic Force 290

- 15-1 Electric Charge 290
- 15-2 Coulomb's Law 292
- 15-3 Electric Field 295
- 15-4 Lines of Flux 297
- 15-5 Gauss' Law 300

# 16 Electrostatics 308

- 16-1 Spherical Charge Distributions 308
- 16-2 Linear Charge Distributions 312
- 16-3 Flat Charge Distributions 313
- 16-4 Electric Potential 317
- 16-5 Capacitance 323
- ★16-6 Dielectrics 327

# 17 Current and Magnetic Force 335

- 17-1 Electric Current 335
- 17-2 Ohm's Law 337
- ★17-3 Direct Current Circuits 341
- 17-4 Magnetic Force—Experimental 346
- 17-5 Derivation of Magnetic Force 348
- 17-6 Magnetic Field 349

| | | |
|---|---|---|
| 17-7 | Magnetic Field Units 353 | |
| ★17-8 | Relativistic Transformation of $\mathcal{B}$ and $\mathbf{E}$ 355 | |
| | ★Appendix 17-1. Transformation of Current and Charge 359 | |

## 18 Magnetic Fields 365

| | |
|---|---|
| 18-1 | Ampere's Law 365 |
| 18-2 | Current Distributions 368 |
| 18-3 | The Biot–Savart Law 372 |
| ★18-4 | Magnetism 376 |
| 18-5 | Maxwell's Equations with Steady Currents 379 |

## 19 Electromagnetic Induction 384

| | |
|---|---|
| 19-1 | Motors and Generators 384 |
| 19-2 | Faraday's Law 388 |
| 19-3 | Lenz's Law 390 |
| 19-4 | Inductance 391 |
| 19-5 | Magnetic Field Energy 394 |
| ★19-6 | Alternating Current Circuits 398 |
| ★19-7 | $RC$ and $RL$ Circuits 405 |
| | ★Appendix 19-1. Loop of Arbitrary Shape 411 |

## 20 Electromagnetic Radiation and Waves 417

| | |
|---|---|
| 20-1 | The Displacement Current 417 |
| 20-2 | Maxwell's Equations in General Form 420 |
| 20-3 | Electromagnetic Radiation 422 |
| 20-4 | Radiation from a Sinusoidal Current Sheet 423 |
| ★20-5 | Nonsinusoidal Current Sources—Fourier Analysis 427 |
| 20-6 | Traveling Waves 431 |
| 20-7 | Energy Transmission by Waves 435 |
| | ★Appendix 20-1. Derivation of Wave Equation 438 |

## 21 Interaction of Radiation with Matter 442

| | |
|---|---|
| 21-1 | Radiated Energy 442 |
| 21-2 | Momentum of the Radiation Field 446 |
| 21-3 | Reflection by a Good Conductor 448 |
| ★21-4 | Interaction of Radiation with a Nonconductor 449 |
| ★21-5 | The Origin of the Index of Refraction 450 |

*21-6 Electromagnetic Radiation in an Ionized Medium 455
*21-7 Radiation by Point Charges 457
Appendix 21-1. Method of Phasors 463
Appendix 21-2. Wave Packets and Group Velocity 464

## 22 Wave Interference 473

22-1 Standing Waves 473
22-2 Interference with Two Point Sources 476
22-3 Interference from Multiple Sources 479
22-4 The Diffraction Grating 481
22-5 Huygens' Principle 484
22-6 Single Slit Diffraction 486
*22-7 Coherence and Incoherence 488

## 23 Optics 496

*23-1 Holography 496
*23-2 Polarization of Light 500
23-3 Diffraction by a Circular Aperture 506
23-4 Optical Instruments and Resolution 508
*23-5 Diffraction Scattering 513
*23-6 Geometrical Optics 515
*Appendix 23-1. Brewster's Law 521

## 24 The Wave Nature of Matter 525

24-1 Classical Versus Modern Physics 525
24-2 The Photoelectric Effect 526
24-3 The Compton Effect 529
24-4 Wave-Particle Duality 530
24-5 The Great Paradox 531
24-6 Electron Diffraction 535

## 25 Quantum Mechanics 541

25-1 Wave Packets 541
25-2 The Uncertainty Principle 543
25-3 Particle in a Box 549
25-4 The Schrödinger Equation 552
25-5 Finite Potential Wells 554
25-6 The Harmonic Oscillator 557

## 26 The Hydrogen Atom 564

- 26-1 Approximate Solution to Hydrogen Atom 564
- 26-2 The Three-Dimensional Schrödinger Equation 566
- 26-3 Exact Solutions for Hydrogen Atom 568
- 26-4 Orbital Angular Momentum 571
- 26-5 Photon Emission 577
- ★26-6 Stimulated Emission 580
- ★26-7 The Bohr Model 582

## 27 Atomic Physics 590

- 27-1 The Pauli Exclusion Principle 590
- 27-2 Multielectron Atoms 592
- 27-3 The Periodic Table of the Elements 596
- ★27-4 X Rays 600
- 27-5 Molecular Binding 601
- ★27-6 Hybridization 604

## 28 Condensed Matter 608

- 28-1 Types of Binding 609
- 28-2 Free Electron Theory of Metals 612
- 28-3 Electrical Conductivity 616
- ★28-4 Band Theory of Solids 619
- ★28-5 Semiconductor Physics 625
- ★28-6 Superfluidity 632
- 28-7 Barrier Penetration 632
  - ★Appendix 28-1. Applications of the p–n Junction (Radio and TV) 636

## 29 Nuclear Physics 642

- 29-1 Nuclear Size 643
- 29-2 The Basic Nucleon-Nucleon Force 648
- 29-3 The Structure of Heavy Nuclei 654
- 29-4 Alpha Decay 660
- 29-5 Gamma and Beta Decay 664
- 29-6 Nuclear Fission 667
- 29-7 Nuclear Fusion 670

# 30 ★Astrophysics 677

- 30-1 The Energy Source of Stars 678
- 30-2 Star Death 680
- 30-3 Black Holes 681
- 30-4 Quantum Mechanical Pressure 683
- 30-5 White Dwarf Stars 684
- 30-6 Neutron Stars 688
- 30-7 Black Hole Critical Mass 693
- 30-8 Summary of Experimental Evidence 693

# 31 Particle Physics 698

- 31-1 The Weak Interaction 699
- 31-2 High Energy Accelerators 702
- 31-3 Antimatter 706
- 31-4 Conservation of Leptons 709
- 31-5 The Hadrons 710
- 31-6 Quarks 716
- 31-7 Nonconservation of Parity 719
- 31-8 Summary of the Conservation Laws 722
- 31-9 Problems for the Future 724

## Appendix A 729
- Physical Constants 729
- Astronomical Constants 730

## Appendix B 731
- Conversion of Units 731
- Electrical Units 731

## Appendix C: Mathematical Appendix 733
- Geometry 733
- Trigonometry 733
- Binomial Expansion 734
- Quadratic Equation 734
- Some Derivatives 734
- Some Indefinite Integrals 734
- Vector Products 734
- The Greek Alphabet 735

**Answers to Odd-Numbered Exercises and Problems** 737

**Index** 745

# Introduction

The main goal of physics is to seek out and understand the basic laws of nature upon which all physical phenomena depend. The history of science reveals a progression to deeper and deeper levels of understanding where, at each successive level, the basic laws or theories become simpler and fewer in number. For example, the number of basic particles and interactions has usually become smaller as time goes on. This historical observation, that the closer we get to the truth, the simpler the basic laws become, was formulated by the philosopher William of Occam in the fourteenth century and is called Occam's razor.

Scientists seek the ultimate truth about the physical world. Whether the progression to deeper levels of truth will continue is unknown. Most scientists have faith that mankind will continue to get closer and closer to the "ultimate truth." To give some idea of where we now stand in this quest, there is a single basic equation $\lambda = h/p$ (along with its physical interpretation) that, when applied to the known elementary particles and the known forces between them, explains in principle atomic physics and chemistry. And, since biology is believed to be a complex of chemical reactions, biology is "explained" as well.

The study of what is most basic leads to the fundamental building blocks of matter or elementary particles, such as protons, electrons, neutrons, and photons. Thus, a major preoccupation of physicists is the study of elementary particles, their properties and their interactions. So far only four basic interactions have been found, and upon these are based all the observed forces and interactions in the universe. These four interactions are summarized in Table 1-1.

If the elementary particles and their interactions are truly basic, they should explain not only the world of the small but also the world of the large. As far as we know the same laws of physics that rule the elementary particles also rule the stars and galaxies. The study of how the basic laws explain the structure of stars and galaxies is also within the domain of

## 1-1 The Nature of Physics

**Table 1-1.** *The four basic interactions upon which all known forces and interactions are based*

| Type | Source | Relative Strength | Range |
|---|---|---|---|
| 1. gravitational | mass | $\sim 10^{-38}$ | long |
| 2. weak | all elementary particles | $\sim 10^{-15}$ | short ($\sim 10^{-15}$ m) |
| 3. electromagnetic | electric charge | $\sim 10^{-2}$ | long |
| 4. nuclear | hadrons (protons, neutrons, mesons) | 1 | short ($\sim 10^{-15}$ m) |

physics (see Chapter 30). All fields of physical and biological science have their roots in physics.

Some of the basic laws of nature are quite contrary to our everyday experience and thus violate common sense. For example, the concepts behind the equation $\lambda = h/p$ can give $2 + 2 = 0$ and also $2 + 2 = 8$! This is illustrated in Figure 1-1, where a beam of electrons is aimed at an opaque barrier that has two holes, A and B, in it. We put a small Geiger counter far behind the barrier and plug up hole B. For this condition the counter counts 2 electrons per second. Now we open hole B and close hole A. Again we get 2 counts per second. We now open both holes together. In this situation we get no counts at all! Not only is the whole less than the sum of its parts but it is even less than either of the parts by itself. If we prefer to count 8 electrons per second, we just move the Geiger counter slightly sideways and we can find a place where the whole is twice as big as the sum of its parts. All this may be hard to accept, but it is true in principle, and quite similar phenomena have been observed in the laboratory (see Figure 24-11 on page 536). This particular phenomenon is due to the wave nature of matter. In Chapter 24 we shall learn that all particles have certain wave properties and that such phenomena then follow naturally.

We can gain additional insight into what physics is by noting what it is not. Astrology, psychokinesis, witchcraft, spiritualism, life after death, miracles, black magic, and telepathy either invoke forces that have never been detected by physicists or violate basic laws of physics. In the same

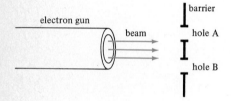

**Figure 1-1.** *Electron gun shooting beam of electrons through holes A and B.*

category are the speculations of Immanuel Velikovsky, which currently enjoy a large following. We quote from the April 1974 *AAAS Bulletin*:

The session with Immanuel Velikovsky and a panel of diverse experts exemplified what science is all about. Velikovsky is the author of the famous theory in which a real cosmic catastrophe supposedly provides an explanation of the common themes in ancient myths of widely diverse and apparently separate peoples. According to Velikovsky, Venus separated itself from Jupiter in historic times, nearly hit us twice during Moses' lifetime, then collided with Mars, after which things settled down considerably.

One can only speculate on the appeal that Velikovsky has for some young people. Certainly he challenges the scientific establishment and he is physically compelling. Perhaps Velikovsky, like other gurus today, offers an easy belief, a mystical, magical relief from an orderly universe whose scientific beauty seems to require higher mathematics for full appreciation. Perhaps a teleological explanation for human existence and history is more comfortable.

Some of Velikovsky's speculations violate the basic conservation laws of energy, momentum, and angular momentum.

## 1-2 Units

Much of physics deals with measurements of physical quantities such as length, time, frequency, velocity, area, volume, mass, density, charge, temperature, and energy. Many of these quantities are interrelated. For example, velocity is length divided by time. Density is mass divided by volume, and volume is a length times a length times a length. Most of the physical quantities are related to length, time, and mass. Some of these relationships are shown in Table 1-2. We will study these physical quantities as they appear later in the book. The basic quantities of length ($l$), time ($t$), and mass ($m$) are called dimensions. Hence velocity has dimensions $l/t$, which we can also write as $lt^{-1}$. In this book both forms, $l/t$ and $lt^{-1}$, will be used.

### Length

The definitions of length, area, and volume are given in Euclidean geometry. There are several standard units of length in use today such as the meter, inch, foot, mile, and centimeter. As of 1978 all nations of the earth, except for Burma, Liberia, Yemen, Brunei, and the United States, officially use, or are converting to use, the metric system. Although the British system of units is still the official system in the United States, American scientists use the metric system almost exclusively; hence, we will make use of the metric system throughout this book. The United

**Table 1-2.** *Some physical quantities in terms of length l, mass m, and time t*

| Quantity | Dimensions |
|---|---|
| area | $l^2$ |
| volume | $l^3$ |
| velocity | $lt^{-1}$ |
| acceleration | $lt^{-2}$ |
| density | $ml^{-3}$ |
| momentum | $mlt^{-1}$ |
| force | $mlt^{-2}$ |
| energy | $ml^2t^{-2}$ |
| frequency | $t^{-1}$ |
| angular momentum | $ml^2t^{-2}$ |
| pressure | $ml^{-1}t^{-2}$ |

**Table 1-3.** *Prefixes for metric units*

| Prefix | Abbreviation | Power of Ten |
|--------|--------------|--------------|
| tera   | T            | $10^{12}$    |
| giga   | G            | $10^{9}$     |
| mega   | M            | $10^{6}$     |
| kilo   | k            | $10^{3}$     |
| centi  | c            | $10^{-2}$    |
| milli  | m            | $10^{-3}$    |
| micro  | $\mu$        | $10^{-6}$    |
| nano   | n            | $10^{-9}$    |
| pico   | p            | $10^{-12}$   |
| femto  | f            | $10^{-15}$   |

States is now in the process of a slow, voluntary conversion to the metric system.

The meter was originally defined in terms of the distance from the north pole to the equator. This distance is very close to 10,000 kilometers (km) or $10^7$ meters (m). Until recently the standard meter of the world was the distance between two scratches on a platinum-alloy bar that is kept at the International Bureau of Weights and Measures in France. Now the standard meter in France has been specified in terms of the number of wavelengths of light of a certain spectral line of the isotope krypton-86. The United States inch is defined as exactly equal to 2.54 centimeters (cm). Within the metric system, conversion to other units of length is simple; one merely adds a prefix indicating the desired power of ten (see Table 1-3).

## Time

Time is a physical concept; thus its definition is related to certain laws of physics. For example, the laws of physics say that to very great accuracy the period of rotation of the earth must be constant. This fact can then be used to define a basic unit of time called the mean solar day. Also the laws of physics say that the period of oscillation of a vibrating slab of crystal in a crystal oscillator should remain constant if the temperature and other external conditions are kept constant. So an electronic crystal oscillator can be made into a very accurate clock. The accurate, battery-operated, digital wristwatches use crystal oscillators.

However, if the rotation of the earth is measured by accurate crystal oscillators, it is found that the earth's speed of rotation is slowing down. This effect is well understood and is due mainly to the tidal forces on the earth. But then, if identical crystal oscillators are compared, it is found that they drift a small amount in frequency. These drift effects are also well understood. Our understanding of the laws of physics leads us to expect that one could get greater accuracy by making use of the vibrational frequency of electrons in an atom. Experiments on atomic clocks agree with the theory. At present the most accurate clocks count up the vibrations of radiation emitted by cesium-133 atoms. The second is now defined as the length of time for $9.19263177 \times 10^9$ vibrations of radiation emitted by cesium-133 atoms.

When we base a concept such as time on the laws of physics, we cannot be sure that these laws are absolutely correct. For example, suppose the speed of light is slowly increasing with time. This would cause a change in some of our standards of length and time. So far there is no experimental evidence that any of the universal physical constants is changing with time, but this does not rule out the possibility of a very slow change beyond the accuracy of present measurements. In the course of this book we shall see that it is not uncommon for a "sacred" law of physics to be overthrown by new experimental data. We must learn to be open minded about our existing "laws" of physics and be prepared to modify them if experimental evidence should ever appear against them. No matter how beautiful and compelling is a given physics theory, it ultimately rests on experimental foundations because physics deals with the physical world. A physics theory will predict new experiments on which to test the theory. If it fails any such test, it must either be modified or abandoned.

### *Mass*

Mass is also a physical concept and must be defined in terms of certain laws of physics. In Chapter 4 we give the modern definition of mass in terms of the law of conservation of momentum. In the metric system the unit of mass was originally defined as that amount of mass contained in 1 cubic centimeter ($cm^3$) of water at a specified temperature and pressure. This amount of mass is called the gram. Thus the density of water is 1 gram per cubic centimeter ($g/cm^3$) by definition. The present world standard of the kilogram (kg) is a particular cylinder of platinum–iridium alloy preserved in a vault in Sèvres, France. For benefit of those who still think in pounds, $1 \text{ kg} = 2.204 \text{ lb}$.

### *mks and cgs*

In physics quantities such as force and energy are usually measured in terms of units based on meters, kilograms, and seconds, or in units based on centimeters, grams, and seconds. The former is called the mks system of units and the latter the cgs system of units. Although conversion from mks to cgs units merely involves the shifting of decimal points, it is important when working problems to convert all units to mks or all to cgs. It is extremely important in problem solving never to mix mks with cgs. In this book we shall follow the recent trend and give preference to mks over cgs. The system of mks units of length, mass, and time when combined with the unit of Kelvin for temperature and the ampere for electric current is called the S.I., which stands for *Système International* (in French).

We use the following abbreviations: m for meter, kg for kilogram, g for gram, s for second, h for hour, K for Kelvin, and A for ampere. Then km would be kilometer, cm is centimeter, $\mu s$ is microsecond, ns is nanosecond, and so on.

Most answers to physics problems contain a number and a unit. We stress that such an answer is incomplete if the unit is not specified. Numerical answers should never be given without explicitly writing down the unit as well. Units have a quantitative size and are an essential part of a numerical answer.

### Conversion of Units

Quite often the units of quantities given as data or asked for as an answer are not all in the most convenient system for solving the given problem. It is also quite common to encounter mixed units, such as miles per hour (mph) for a velocity. If the problem is being worked in the British system, such a velocity must first be converted into feet per second (ft/s). Usually we shall work such problems in the mks system; then we must express all velocities in meters per second (m/s).

As an example we shall convert a velocity of 60 mph to its value in m/s by the method of substitution. For the sake of clarity we shall enclose each newly substituted quantity with parentheses.

$$v = 60 \frac{\text{mi}}{\text{h}} = 60 \times \frac{(1 \text{ mi})}{(1 \text{ h})}$$

Now in place of the quantity (1 mi) we substitute its equivalent in meters $(1.61 \times 10^3 \text{ m})$:

$$v = 60 \times \frac{(1.61 \times 10^3 \text{ m})}{1 \text{ h}}$$

In place of (1 h) in the denominator we substitute $(3.6 \times 10^3 \text{ s})$:

$$v = 60 \times \frac{(1.61 \times 10^3 \text{ m})}{(3.6 \times 10^3 \text{ s})}$$

$$= 26.8 \text{ m/s}$$

An alternate method of conversion is to multiply by 1 where

$$\left( \frac{1.61 \times 10^3 \text{ m}}{1 \text{ mi}} \right) \quad \text{and} \quad \left( \frac{1 \text{ h}}{3600 \text{ s}} \right)$$

both have the value 1:

$$\frac{60 \text{ mi}}{\text{h}} \times 1 \times 1 = 60 \frac{\text{mi}}{\text{h}} \times \left( \frac{1.61 \times 10^3 \text{ m}}{\text{mi}} \right) \times \left( \frac{\text{h}}{3600 \text{ s}} \right)$$

$$= \frac{60 \times 1.61 \times 10^3 \text{ mi} \times \text{m} \times \text{h}}{3600 \text{ h} \times \text{mi} \times \text{s}} = 26.8 \frac{\text{m}}{\text{s}}$$

It is often convenient to use negative exponents rather than putting units in the denominator. Then m/s is written as m s$^{-1}$. Both ways of expressing units will be used in this book.

## 1-3 Dimensional Analysis

In many physics problems it is necessary to start from one or more basic equations and derive a specialized formula. As an example, we would like a specialized formula for the velocity of a car in terms of its acceleration $a$ and distance traveled, $x$, if it starts from rest at constant acceleration. In this case the basic equations are the definitions of velocity and acceleration, which are given in the next chapter. The specialized formula to be derived is $v = \sqrt{2ax}$. Suppose we cannot remember how to do the derivation or that we get stuck at some intermediate point. Fortunately there is a simple and powerful procedure that can be used to derive or check formulas in most cases. The procedure, called dimensional analysis, gives the correct functional form except for a dimensionless proportionality constant. In the preceding example dimensional analysis would give $v \propto \sqrt{ax}$, but it could not give the factor $\sqrt{2}$. (The symbol $\propto$ means "is proportional to.")

The procedure is to set up a generalized relationship:

$$v \propto a^p x^q \tag{1-1}$$

where $p$ and $q$ are unknown exponents. Then one merely checks the dimensions on both sides. The three basic quantities mass, length, and time are called dimensions. The dimensions of velocity are distance per unit time or $l\,t^{-1}$. On the right hand side we have

$$\text{dimensions of } [a^p x^q] = \left(\frac{l}{t^2}\right)^p (l)^q$$

$$= l^{p+q} t^{-2p}$$

Equating the dimensions of both sides gives

$$lt^{-1} = l^{p+q} t^{-2p}$$

Since the exponent of $l$ must be the same on both sides, we have

$$1 = (p + q) \tag{1-2}$$

Equating exponents of $t$ gives

$$-1 = -2p$$

or

$$p = \frac{1}{2}$$

Now substitute $\frac{1}{2}$ for $p$ in Eq. 1-2:

$$1 = \left(\frac{1}{2}\right) + q$$

$$q = \frac{1}{2}$$

Putting $p = \frac{1}{2}$ and $q = \frac{1}{2}$ into Eq. 1-1 gives

$$v \propto a^{1/2} x^{1/2}$$

or

$$v \propto \sqrt{ax}$$

As pointed out in the previous paragraph, this is off by $\sqrt{2}$. Quite often we will be lucky and the proportionality constant will be 1. In the following example the correct proportionality constant happens to be 1.18. (The "star" on an example means that the question is too difficult for a beginning student to answer on his own.)

---

**\*Example 1.** Use dimensional analysis to "derive" a formula for the speed of sound in a gas of density $\rho$.

ANSWER: The only possible variables are the pressure $P$, temperature $T$, and density $\rho$ of the gas. We shall see in Chapter 12 that only two of these variables are independent (for a given density, the pressure is proportional to temperature). Hence

$$v \propto \rho^p P^q$$

The density $\rho$ has units of kg/m$^3$ or $ml^{-3}$ for its dimensions. Pressure is force per unit area and, according to Table 1-2, has dimensions $ml^{-1}t^{-2}$. Now take the dimensions of both sides of the preceding equation:

$$lt^{-1} = (ml^{-3})^p \, (ml^{-1}t^{-2})^q$$

Equating exponents

$$\begin{aligned} \text{of } m: & \quad 0 = p + q \\ \text{of } t: & \quad -1 = -2q \\ \text{of } l: & \quad 1 = -3p - q \end{aligned}$$

The solution is $p = -\frac{1}{2}$ and $q = \frac{1}{2}$. So

$$v \propto \sqrt{\frac{P}{\rho}}$$

The exact answer for a gas like air is

$$v = 1.18 \sqrt{\frac{P}{\rho}}$$

In many kinds of problems it is not necessary to know the proportionality constant. An example of such a problem is to compare the speed of sound in two different gases at the same pressure. Dimensional analysis tells us that the speed of sound goes inversely as the square root of the density.

The astute student will use dimensional analysis to check all derivations and calculations whenever possible. When at an impass on an exam problem, it is better to work it by dimensional analysis than to do nothing at all. Our next example illustrates the power of dimensional analysis. We can apply it to physics situations where we have no knowledge or understanding and still get useful results.

**Example 2.** Approximately how fast must a car weighing 1 metric ton (1000 kg) go so that the force of air resistance is comparable to its weight? (Its weight will be about $10^4$ kg m s$^{-2}$ in mks units.) Assume the cross sectional area of the car is about 2 m² and the density of air is about 1 kg m$^{-3}$. Assume the force of air resistance depends on the cross sectional area $A$, the density $\rho$ of the air that the car is pushing ahead of it, and the velocity $v$ of the car:

$$F_{\text{air resist.}} \propto A^p \rho^q v^r \quad (1\text{-}3)$$

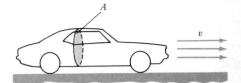

**Figure 1-2.** *Area A is the cross-sectional area of the car.*

ANSWER: According to Table 1-2 the dimensions of force are $mlt^{-2}$; hence

$$mlt^{-2} = (l^2)^p (ml^{-3})^q (lt^{-1})^r$$

Equating dimensions

of $m$:  $\quad 1 = q$
of $l$:  $\quad 1 = 2p - 3q + r$
of $t$:  $\quad -2 = -r$

The solutions are $q = 1, r = 2, p = 1$. Putting these values into Eq. 1-3 gives:

$$F_{\text{air resist.}} \propto A\rho v^2$$

It turns out that this equation is correct to within about a factor of 2 (see page 136). Solving for $v$ gives

$$v \sim \sqrt{\frac{F}{A\rho}}$$

When the force is as large as the car's weight, $F = 10^4$ kg m s$^{-2}$, and we have

$$v \sim \sqrt{\frac{10^4 \text{ kg m s}^{-2}}{(2\text{m}^2)(1 \text{ kg m}^{-3})}} = 70 \text{ m/s} = 155 \text{ mph}$$

The conclusion is that the engine power needed to drive a car at ~155 mph is comparable to the power needed to drive it up a vertical wall (assuming the wheels could hold on). Also we see that this air resistance force goes as the square of the velocity, so the same car at 40 mph will experience only one sixteenth as much resistive force. Clearly one can save on fuel by driving more slowly.

## 1-4 Accuracy in Physics

Physics is sometimes called an exact science. However students who have taken a physics lab might think differently—and in a sense they are right. In general instruments do not give exact measurements. A distance of 5 cm can be read to an accuracy of about 1 part in 100 from a cheap plastic ruler. But, in addition to the limitation of reading accuracy, the plastic expands with temperature; thus there exists what is called a systematic error. Systematic errors can be quite subtle and difficult to estimate in all fields of science. It is a sobering experience to take a physics lab and try to discover and estimate all the systematic errors involved. Professional physicists are continually plagued with the same problem.

One of the most accurate measuring instruments is a frequency meter or scaler, which counts the total number of vibrations of an oscillator. One of the most stable and accurate oscillators is the cesium beam atomic clock, which is controlled by the frequency of a well-defined hyperfine transition in the ground state of the cesium-133 atom. Two such oscillators will keep in step with each other to about 1 part in $10^{12}$. They are so accurate that the second of time has been redefined to be equal to 9,192,631,770 accumulated periods of oscillation of an "ideal" cesium beam atomic clock, as mentioned in Section 1-2.

### Accuracy of the Mean

Accuracy can be improved by repeating the same measurement several times and taking the average. For example, suppose there are $n$ identical atomic clocks all measuring the same unknown time interval where the readings of the clocks are $t_1, t_2, \ldots, t_n$, respectively. (They were turned on and off simultaneously.) Then the best value for the time interval will be the mean

$$\bar{t} = \frac{t_1 + t_2 + \cdots + t_n}{n}$$

If the typical error of a single clock is $\sigma$, then the error of $\bar{t}$ can easily be shown to be $(\sigma/\sqrt{n})$. We can express this result as $t = \bar{t} \pm \sigma/\sqrt{n}$, where $\sigma$ is the typical error of a single measurement. This works for random errors, but not for systematic errors.

### Significant Figures

Suppose in an experiment to measure velocity by means of a ruler and an accurate clock, a body is observed to move 10 cm in "exactly" 3 seconds. Then

$$v = \frac{10 \text{ cm}}{3 \text{ s}} = 3.33333 \text{ cm/s}$$

There is a problem of how many decimal places to use for expressing 10/3 as a decimal. The convention is to use at most one more decimal place than the certainty of the result. Thus, if the 10 cm had been measured to 1% accuracy, the result could be expressed as $v = 3.33 \pm 0.03$ cm/s. Since the true value of $v$ lies somewhere between 3.30 and 3.36 cm/s, the first two 3's are significant figures and the third 3 is somewhat uncertain. It is poor practice to write the result as $v = 3$ cm/s or $v = 3.333$ cm/s. The form $v = 3.33$ cm/s is preferred. To use more decimal places would be not only superfluous but misleading. We would be claiming that our result was better than it really was.

Suppose the velocity $v = 3.33$ cm/s is to be added to another velocity $v' = 4.51$ m/s and that $v'$ is also known to 1% accuracy.

$$\begin{aligned} v &= \phantom{00}3.33 \text{ cm/s} \\ v' &= 451.\phantom{00} \text{ cm/s} \\ \hline v + v' &= 454.33 \text{ cm/s} \end{aligned}$$

Note that, if we quote the answer as given, we are implying that the accuracy of our result is better than 1 part in $10^4$ instead of 1 part in $10^2$. In this case the answer should be quoted as 454 cm/s.

---

**Example 3.** A student repeats measurements of the period of a pendulum with the same stopwatch. His measurements vary among each other by $\frac{1}{10}$ s on the average. How many times must he repeat the measurement in order to determine the period to an accuracy of $\frac{1}{100}$ s?

ANSWER: We use

$$(\text{error of mean}) = \frac{\sigma}{\sqrt{n}}$$

$$(\tfrac{1}{100} \text{ s}) = \frac{(\tfrac{1}{10} \text{ s})}{\sqrt{n}}$$

$$\sqrt{n} = \frac{(\tfrac{1}{10} \text{ s})}{(\tfrac{1}{100} \text{ s})} = 10$$

$$n = 100$$

To improve accuracy by a factor of 10 requires $10^2$ repeated measurements.

---

Unless stated otherwise, quantities in this book will be written using three significant figures, which implies at least 1% accuracy. The problems should be worked to this same accuracy.

## 1-5 Mathematics in Physics

Physics has the reputation of involving higher mathematics of great difficulty. Fortunately this is not the case if we are to put emphasis on the basic laws. Occam's razor seems to work here: the more basic the laws, the simpler they become both conceptually and mathematically. The more difficult mathematics usually enters when one tries to calculate something that is not basic, such as the three-body problem (the motion of three mutually interacting bodies). The three-body problem is not basic because it is really the superposition of three interrelated two-body problems. Several hundred years ago Isaac Newton solved the really basic problem—the orbits of two bodies interacting under an inverse square force. The two-body problem of astronomy can be solved using elementary mathematics (see Chapter 5), but to do an accurate job of the three-body problem requires a large electronic computer.

Most of the physics in this book uses simple algebra, geometry, and a little trigonometry. Differential calculus is slowly introduced in Chapter 2 where we use $\frac{d}{dx}(x^2) = 2x$. Starting in Chapter 6 we use a little integral calculus at the level of $\int x\, dx = \frac{1}{2}x^2$.

In Chapter 11 we first encounter derivatives of sine and cosine functions. Vector analysis is introduced slowly in Chapter 3 with vector products deferred to Chapters 6 and 10. The vector analysis in this book is self-contained. This book may be used as a physics text by students who have not had calculus, but who are taking it concurrently. Even though such students need not know calculus at the start, it is useful for them to have a good working knowledge of first-year high school algebra. The following is a list of common mistakes made by students at this level. The reader should readily recognize these mistakes and be able to supply the correct answers. The correct answers are given in Appendix 1-1. A reader who has trouble with these problems will probably have trouble learning physics.

**Some Common Mistakes**

1. $(a + b)^2 = a^2 + b^2$
2. $\frac{1}{a + b} = \frac{1}{a} + \frac{1}{b}$
3. One half of $10^{-10}$ is $10^{-5}$
4. $\frac{A}{B} + \frac{X}{Y} = \frac{A + X}{B + Y}$
5. 4 divided by $\frac{1}{2}$ is 2
6. $\sqrt{16ab} = 4ab$
7. $\frac{1}{2}$ of $10^{-8} = 5^{-8}$
8. $\frac{10^{-10}}{10^{-5}} = 10^{-15}$
9. $\log AB = \log A \log B$
10. $\sin(A + B) = \sin A + \sin B$

Topics in algebra of particular value are use of exponents, logarithms, simultaneous equations, and the binomial expansion. Use of exponents is reviewed in part in the section on Scientific Notation. As a reminder we state the binomial expansion:

$$(1 + a)^n = 1 + na + \frac{n(n-1)}{1 \cdot 2} a^2 + \frac{n(n-1)(n-2)}{1 \cdot 2 \cdot 3} a^3 + \cdots$$

This can be written more compactly by using the summation sign:

$$(1 + a)^n = \sum_{j=0}^{\infty} \left( \frac{n!}{j!(n-j)!} a^j \right)$$

The summation sign $\Sigma$ means to keep adding the same form of term each time with the value of $j$ increased by one. We give two examples:

$$\sum_{j=1}^{5} (j) = 1 + 2 + 3 + 4 + 5$$

$$\sum_{j=1}^{n} (x_j t_j) = x_1 t_1 + x_2 t_2 + \cdots + x_n t_n$$

A common example of the use of simultaneous equations is the elimination of unwanted variables. We give an example from Chapter 3 that uses three simultaneous equations. It is not necessary to know what the symbols stand for in order to solve this problem. The problem is to express centripetal acceleration $a_c$ in terms of radius $R$ and frequency $f$ starting with the three equations:

$$a_c = \frac{v^2}{R} \tag{1-4}$$

$$v = \frac{2\pi R}{t} \tag{1-5}$$

$$f = \frac{1}{t}$$

Our approach will be to use the method of substitution to eliminate $v$ and then again to eliminate $t$. First substitute Eq. 1-5 for $v$ in Eq. 1-4:

$$a_c = \frac{(2\pi R/t)^2}{R} = \frac{4\pi^2 R}{t^2}$$

Now substitute $(1/f)$ for $t$ in the above:

$$a_c = \frac{4\pi^2 R}{(1/f)^2} = 4\pi^2 f^2 R$$

Note our use of parentheses to indicate that a substitution has just been made.

Another mathematical skill of value is the reading and plotting of curves. Some practice in this is given in the next chapter.

## Scientific Notation

Most of the quantities encountered in physics are numerically either much larger than 1 or much smaller than 1. For convenience, the standard practice is to write any quantity, no matter how large or small, as a number between 1 and 10 (called the mantissa) times the appropriate power of ten. This is called expressing the number in scientific notation. For example, the mass of the electron is $9.11 \times 10^{-31}$ kg. The mantissa is 9.11 and the exponent or power of ten is $-31$. The mass of the sun is $1.99 \times 10^{30}$ kg. We see that masses span a range of $\sim 10^{60}$. Distances in physics cover a similar range in magnitude.

In working numerical problems it is good practice first to do the calculation by hand making a crude one-digit approximation for the mantissa. Then the calculation can be repeated using a slide rule or pocket calculator. Typical pocket calculators with floating decimal cover a range $10^{-8}$ to $10^8$ whereas those capable of scientific notation cover $10^{-99}$ to $10^{99}$. Clearly the latter is more useful for a student of physics. Also it is useful for the pocket calculator to have trigonometric (trig) and logarithmic (log) functions and a separate memory for storage of intermediate results. Such scientific pocket calculators can be purchased for about $20.

Another advantage of scientific notation is that, in multiplying or dividing numbers, the exponents are merely added or subtracted respectively. We make use of the relation $10^a \times 10^b = 10^{(a+b)}$ and $10^a/10^b = 10^{(a-b)}$. The following two examples give a little practice in handling large numbers and conversion of units. In addition they have some social and political content relevant to Section 1-6.

**Example 4.** Electrical power production in the United States averaged over the year is 250 million kilowatts (kW). If this power were to be supplied by solar power, how much area must be covered with solar cells? Assume 10% efficiency for conversion of solar to electrical energy. The average solar power received at midday in the southern United States is $\sim 1$ kW/m².

ANSWER: After converting to electrical power the 1 kW/m² becomes 100 W/m². Averaged over 24 h it is ~25 W/m². Let $P_{total} = 2.5 \times 10^{11}$ W be the total power needed and $A_{total}$ be the total area collecting solar energy. Then

$$\frac{P_{total}}{A_{total}} = 25 \text{ W/m}^2$$

$$A_{total} = \frac{P_{total}}{25 \text{ W/m}^2} = \frac{2.5 \times 10^{11} \text{ W}}{25 \text{ W/m}^2} = 10^{10} \text{ m}^2 \quad (1\text{-}6)$$

To convert to square miles we use

$$1 \text{ mi} = 1.61 \text{ km} = 1.61 \times 10^3 \text{ m}$$

or

$$1 \text{ m} = \frac{1 \text{ mi}}{1.61 \times 10^3}$$

Substitution into Eq. 1-6 gives

$$A_{total} = 10^{10} \left(\frac{1 \text{ mi}}{1.61 \times 10^3}\right)^2 = \frac{10^{10}}{1.61^2 \times 10^6} \text{ mi}^2 = 3.9 \times 10^3 \text{ mi}^2$$

This would fit into a square plot of land only 62 miles on a side. A federal plot of land of about this size exists just north of Las Vegas. Any road map of Nevada shows it as "Nevada Proving Grounds U. S. Atomic Energy Commission, closed to public." Perhaps such land could eventually be used for such purposes.

Whether or not the United States uses this or similar land to supply solar power is a question involving social and political values, economics, ecology, technology, and politics. It has little to do with physics.

In this calculation we have ignored such considerations as energy storage and transmission losses.

**Example 5.** The United States stockpile of nuclear weapons in 1960 was estimated to be ~40 million kilotons equivalent of TNT. (The Hiroshima bomb was equivalent to 15 kilotons of TNT.)

If the entire stockpile were delivered to an enemy country containing 200 cities, how many Hiroshima bombs would this be per city?

ANSWER:

$$\text{bombs/city} = \frac{(40 \times 10^6 \text{ kilotons})}{(15 \text{ kiloton/bomb})(200 \text{ cities})} = \frac{40 \times 10^6}{30 \times 10^2} \frac{\text{bombs}}{\text{city}}$$

$$= 1.3 \times 10^4 \text{ bombs/city}$$

Compared to Hiroshima, this is a factor of $\sim 10^4$ overkill. If the goal is complete destruction of such a city, more than one Hiroshima bomb would be needed. However, some people might regard 13,000 bombs per city as wasteful and dangerous.

In recent years there has been a trend to convert large yield bombs to smaller yield, so that the present stockpile has a significantly lower overall yield than it did in 1960. This trend is offset by an increase in the total number of nuclear bombs.

The two preceding examples are typical of the kind of calculations encountered by scientists. A scientist should be able to do such order of magnitude calculations on the back of an envelope without slide rule or pocket calculator.

## 1-6 Science and Society

The previous two examples illustrate how science and technology have interacted with society in a life and death way. The decision whether to have a factor of 10 or a factor of 10,000 overkill is not made by scientists. The decisions of how many bombs to stockpile and whether to continue testing is not made by scientists or engineers but by governmental leaders who have difficulty understanding powers of ten, much less science itself. Perhaps we and other countries would have fewer bombs if governmental leaders had sufficient training in science and mathematics.

In the 1978 United States Congress there were three scientists in the House and one in the Senate. Congressmen don't know where to turn for technical advice. They have difficulty evaluating different positions on technical matters, such as the economics of solar power, the supersonic transport, ABM (antiballistic missile), MIRV, the B-1 bomber, ozone depletion, cruise missiles, the Trident submarine system, and so on. For this reason the American Association for the Advancement of Science (AAAS) and the American Physical Society in 1973 started a program of Scientist Congressional Fellowships under which real scientists would work for a year or more in Congressional offices. One of the first such Congressional fellows was Jessica Tuchman, a PhD in biophysics from Cal Tech, who summarized the problem: "You can get experts to talk on opposite sides of almost any question, which is what happened with the ABM and SST debates. But because there is no one around to evaluate, you get tremendously cynical people. And that's why we're here—to translate."

To physicists, understanding the world around us is a valuable goal in itself. *Homo sapiens* is the only animal capable of such understanding. Our scientific understanding is a central part of our modern culture and civilization. Those who are intellectually alive cannot help but strive to obtain this scientific understanding.

A secondary and more commonly expressed reason for the study of physics is the usefulness an understanding of science has to a person living in this modern, technological age—the age of automation, pollution, nuclear power, computers, space travel, missiles, and nuclear bombs. Almost every edition of the daily newspaper contains articles that cannot be fully understood without a knowledge of physics. Almost every day there are articles on nuclear weapons, nuclear power, energy conservation, solar or thermonuclear energy, new weapons, pollution control, space travel, UFO's, new scientific discoveries and devices, and so on. One wonders whether people who have little understanding of science are capable of making competent policy decisions on such vital topics. Yet, upon these decisions rest the survival of human civilization as we know it.

## Appendix 1-1  Correct Answers to the Common Mistakes

1  $(a + b)^2 = a^2 + 2ab + b^2$

2  $\dfrac{1}{a + b} = (a + b)^{-1}$

3  $\frac{1}{2}$ of $10^{-10} = \frac{1}{2} \times 10^{-10} = 5 \times 10^{-11}$

4  $\dfrac{A}{B} + \dfrac{X}{Y} = \dfrac{AY + BX}{BY}$

5  $\dfrac{4}{\frac{1}{2}} = 8$

6  $\sqrt{16ab} = 4\sqrt{ab}$

7  $\frac{1}{2}$ of $10^{-8} = \frac{1}{2} \times 10^{-8} = 5 \times 10^{-9}$

8. $\dfrac{10^{-10}}{10^{-5}} = 10^{(-10+5)} = 10^{-5}$

9. $\log AB = \log A + \log B$

10. $\sin(A+B) = \sin A \cos B + \cos A \sin B$

## Exercises

(Exercises are problems that happen to be shorter or simpler.)

1. This problem is called "The Fuller Task" by science educators. They have found that the majority of students from typical American colleges cannot work this problem. In working this problem, assume you do not know the conversion from miles to kilometers.

    The state of Ohio is converting all of its highway mileage signs to a dual English-metric system. We show an example of a sign that you might see as you drive toward Cleveland.

    Assume that the state of Nebraska converts its road signs to the same system. As you drive toward Wahoo you might see our other sign. Compute the number to put in the blank shown on the sign to the right.

2. Another standard problem that science educators find cannot be worked by about half of the typical college students is shown below.

    Triangle A and Triangle B are similar triangles.

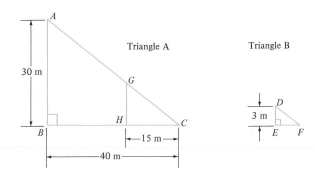

    (a) What is the length of $EF$ in Triangle B?
    (b) What is the length of $GH$ in Triangle A?

3. If $K = \tfrac{1}{2}Mv^2$ and $P = Mv$, what is $K$ in terms of $P$ and $M$?

4. Evaluate $8^{2/3}$, $8^{-2/3}$, and $8^{3/2}$.

5. Show that $\dfrac{1+\beta}{\sqrt{1-\beta^2}} = \sqrt{\dfrac{1+\beta}{1-\beta}}$. What is $\dfrac{1-\beta}{\sqrt{1-\beta^2}}$?

6. If $M = \dfrac{M_0}{\sqrt{1 - v^2/c^2}}$, what is $v$ in terms of $M$, $M_0$, and $c$?

7. A certain brand of soap comes in two sizes, both of the same shape. The larger size is 50% longer than the small size. How much more soap is in the large bar?
8. If an 8-in. pizza costs $1.50, how much should a 12-in. pizza cost?
9. Two grams of $H_2$ gas contains $N_0 = 6.02 \times 10^{23}$ molecules. What is the mass of the hydrogen atom?
10. Expand $\left(1 - \dfrac{v^2}{c^2}\right)^{-1/2}$ by the binomial theorem. Give the first three terms. If $v/c = 0.1$, what is the ratio of the third to the second term?
11. Simplify $\exp\left(-\ln \dfrac{1}{x}\right)$  ($\exp \equiv e = 2.718$.)
12. At what value of $x$ does $y = \exp(-x^2/2)$ drop by a factor of 2?
13. At what value of $t$ does $y = \exp(-t/\tau)$ drop by a factor of 2? Give answer in terms of $\tau$.
14. The edge of a cube is measured to 1% accuracy. What is the accuracy of the volume computed from this measurement?

**Problems**   (The following problems tend to be more difficult than the preceding Exercises.)
15. An object 100 m away subtends an angle of 1°. How high is it? (Work this without using trigonometric functions.)
16. If $v = v_0 + at$ and $x = x_0 + v_0 t + \tfrac{1}{2}at^2$, what is $v$ in terms of $x_0, v_0, a$, and $x$?
17. Using the following information, find $E$ in terms of $e$ and $R$ only.

$$E = \tfrac{1}{2}mv^2 + U \qquad U = -e^2/R \qquad mv^2/R = e^2/R^2$$

18. If $a = v^2/R$, $v = 2\pi R/t$, and $\nu = 1/t$, what is $a$ in terms of $\nu$ and $R$?
19. If $p = M_0 \gamma v$, $E = M_0 \gamma c^2$, and $\gamma = (1 - v^2/c^2)^{-1/2}$; what is $E$ in terms of $M_0$, $p$, and $c$?
20. Plot $y$ versus $x$ when $y = \sin^2 x$ where $x$ is in degrees.
21. A sound wave is transmitted down a metal rod of cross sectional area $A$. The wave is generated by varying a force $\Delta F$ applied to the end. The elongation $\Delta l$ of the rod will be proportional to $\Delta F$ and is given by the formula

$$YA \dfrac{\Delta l}{l} = \Delta F \qquad \text{or} \qquad Y \equiv \dfrac{\Delta F/A}{\Delta l/l}$$

where $Y$ is defined as the Young's modulus (it is a constant for a given metal). Derive a formula for the speed of sound inside the rod in terms of $Y$ and the density $\rho$ using dimensional analysis ($\rho = $ mass/volume).
22. The power delivered by "burning" uranium for a time $t$ in a nuclear reactor is $P = 10^{-3} mc^2/t$ where $m$ is the uranium mass and $c = 3 \times 10^8$ m/s is the speed of light. If $m$ is in kilograms and $t$ in seconds, $P$ will be in watts. The efficiency for converting this energy into electrical power is about 30%. How many grams of uranium fuel must be used per day to supply the average United States electrical power?

23. A typical American car gets 15 miles to the gallon and travels at 55 mph.
    (a) If there are 3800 g of gasoline to the gallon, how many grams are being expended per second?
    (b) If gasoline can deliver 30,000 joules (J) of energy per gram and $1 \text{ W} = 1 \text{ J/s}$, how many watts of power is the car consuming? Is this more than is consumed by the average home?
24. Assume a stockpile of 40 million kilotons of TNT bombs was dropped uniformly over 200 cities of average area 200 km² each. If the bombs did not explode, how thick a layer of bombs would be covering the cities? Assume TNT has a density of 1 g/cm³.
25. In relativity theory $\gamma = m/m_0$ is the ratio of relativistic mass to rest mass. The velocity is $v = c(1 - 1/\gamma^2)^{1/2}$. Use the binomial expansion to get the first two terms.
26. Show that

$$\gamma_+ - \gamma_- = \frac{2\beta_0\beta}{\sqrt{(1-\beta_0^2)(1-\beta^2)}}$$

where
$$\gamma_+ = (1 - \beta_+^2)^{-1/2} \qquad \beta_+ = \frac{\beta_0 + \beta}{1 + \beta_0\beta}$$

$$\gamma_- = (1 - \beta_-^2)^{-1/2} \qquad \beta_- = \frac{\beta_0 - \beta}{1 - \beta_0\beta}$$

27. A pendulum has a period of 2.5 s. Forty measurements are made with stopwatch A, which has a random measuring error of $\sigma = 0.1$ s.
    (a) What is the accuracy of the mean?
    (b) If 10 measurements of the period are made with stopwatch B which has $\sigma = 0.05$ s, what is the accuracy of the mean period obtained using watch B?
    (c) Suppose the 40 measurements in (a) and the 10 measurements in (b) are both made in the same experiment. Now with what accuracy could the pendulum period be quoted?
28. In the preceding problem it is seen that four measurements with watch A are worth one measurement with watch B. What is the accuracy of the mean in part (c) of Problem 27 if 10 measurements are made with each stopwatch?
29. If $\tan \alpha = \dfrac{\sin \theta + \sin 2\theta}{1 + \cos \theta + \cos 2\theta}$, show that $\alpha = \theta$.

    (Hint: $\sin(a + b) = \sin a \cos b + \cos a \sin b$; $\cos(a + b) = \cos a \cos b - \sin a \sin b$.)
30. Prove that $\cos(k + \Delta k)x + \cos(k - \Delta k)x = 2(\cos \Delta kx)(\cos kx)$.
31. Express $\sin 2A$ in terms of $\sin A$.
32. The quantity $C = (A - B^2)$. If $A = (1010 \pm 1)$ and $B = (30.0 \pm 0.1)$, what is the percentage accuracy of $C$? Express $C$ in scientific notation with the proper number of significant figures.
33. Repeat Problem 32 for $A = (920.0 \pm 0.1)$ and $B = (30.00 \pm 0.01)$.

# One-Dimensional Motion

Chapter 2 and Chapter 3 together are a study of what is called kinematics. Kinematics deals with the relationships between the position, velocity, and acceleration of a particle or body under study. In kinematics we are not concerned with where the acceleration or force comes from. The study of forces and how they are produced is called dynamics and is introduced in Chapter 4. In this chapter we shall only deal with motion in a straight line. We shall usualy choose the $x$ axis to be along the straight line under consideration.

**2-1 Velocity**

In this age of automobiles, velocity is a concept learned in childhood. The car speedometer reads the magnitude of the instantaneous velocity in miles per hour (mph) or kilometers per hour (km/h). **Velocity is the time rate of change of distance.**

### *Constant Velocity*

If a car is moving with constant velocity, the distance $x$ traveled in a time $t$ is $x = vt$. If the car was at $x = x_0$ at a time $t_0$, then

$$x - x_0 = v(t - t_0)$$

or

$$v = \frac{x - x_0}{t - t_0} \quad \text{(constant velocity)} \qquad (2\text{-}1)$$

where $v$ is a constant.

This relationship between $x$ and $t$ is plotted in Figure 2-1. Strictly speaking, the term "speed" is used rather than velocity when one is only

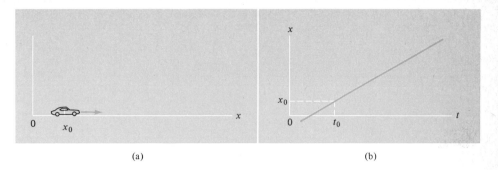

**Figure 2-1.** (*a*) *Car is at $x_0$ at time $t_0$.* (*b*) *Plot of the position of the car versus the time for a car moving at constant velocity.*

speaking of the magnitude or absolute value of the velocity. However, our expression for $v$ can be either positive or negative, where the sign indicates the direction. If $v$ is negative, the motion is toward decreasing values of $x$.

In working practical problems it is useful to have conversions from one set of units to another. Using the substitution method on page 6:

$$\frac{60 \text{ mi}}{1 \text{ h}} = \frac{60(5280 \text{ ft})}{1(3600 \text{ s})}$$
$$= 88 \text{ ft/s}$$

so 60 mph = 88 ft/s. Also 1 mi = 1.61 km, and

$$60 \text{ mph} = 96.6 \text{ km/h}$$
$$= 26.8 \text{ m/s}$$

## Instantaneous Velocity

If the car is speeding up or slowing down, Eq. 2-1 will give the wrong value for the speedometer reading unless a very small value of $x - x_0$ is used. We shall let $\Delta x$ stand for a very small value of $x - x_0$ and $\Delta t$ stand for the small time interval in which the car moved the distance $\Delta x$. Then the instantaneous velocity is the limit of $\Delta x / \Delta t$ as $\Delta t$ approaches zero:

$$v \equiv \lim_{\Delta t \to 0} \left[ \frac{\Delta x}{\Delta t} \right] \qquad (2\text{-}2)$$

This is precisely the definition in calculus of the first derivative of $x$ with respect to $t$. In calculus notation Eq. 2-2 may be rewritten as

$$v \equiv \frac{dx}{dt} \quad \text{(definition of instantaneous velocity)} \qquad (2\text{-}2)$$

(The symbol "$\equiv$" means "is defined as.")

> **Example 1.** Suppose $x$ is increasing as the square of the time; that is, $x = At^2$. What is the instantaneous velocity at a time $t_1$?
>
> ANSWER:
>
> $$v = \frac{d}{dt}(At^2) = A\frac{d(t^2)}{dt} = 2At$$
>
> At the time $t_1$, $v = 2At_1$.

In general the derivative of $t^n$ is

$$\frac{d}{dt}(t^n) = nt^{n-1}$$

Example 1 could be worked without using differential calculus by letting $x_2$ be the position at a later time $t_2$, then

$$\frac{\Delta x}{\Delta t} = \frac{x_2 - x_1}{t_2 - t_1} = \frac{(At_2^2) - (At_1^2)}{t_2 - t_1}$$

For $t_2$ substitute $(t_1 + \Delta t)$:

$$\frac{\Delta x}{\Delta t} = \frac{A(t_1 + \Delta t)^2 - At_1^2}{\Delta t} = \frac{2At_1 \Delta t + A(\Delta t)^2}{\Delta t}$$

$$= 2At_1 + A\,\Delta t$$

Now take the limit as $\Delta t \to 0$ and the second term vanishes. Hence

$$v_1 = 2At_1$$

In light of Example 1 it is seen from Figure 2-2 that the slope of the $x$ versus $t$ curve is the instantaneous velocity. In Figure 2-2 the curve $x = At^2$ is plotted. The colored line has slope $(x_2 - x_1)/(t_2 - t_1)$ which, in the limit of $t_2$ approaching $t_1$, is the slope of the curve.

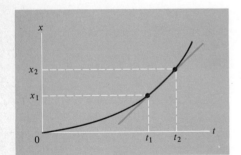

**Figure 2-2.** *Colored line has slope* $= (x_2 - x_1)/(t_2 - t_1)$. *This illustrates that slope of x versus t curve is instantaneous velocity.*

## 2-2 Average Velocity

In this section we shall obtain the formula

$$\bar{v} = \frac{x - x_0}{t} \quad \text{(average velocity)} \qquad (2\text{-}3)$$

for average velocity, where $x - x_0$ is the distance traveled in the time $t$. It would be convenient if we could just define average velocity as $(x - x_0)/t$ and then move on; however, this would be logically incorrect because average is already a well-defined quantity. We must start with the definitions of mathematical average and instantaneous velocity, and then derive a formula for average velocity.

In order to form a mathematical average, one must use a weighting factor for each item to be included in the average. For example, if a car has velocity $v_1$ for a time $t_1$ and velocity $v_2$ for a time $t_2$, the average velocity with respect to time is by definition:

$$\bar{v} \equiv \frac{v_1 t_1 + v_2 t_2}{t_1 + t_2} \quad \text{(definition of weighted average)} \quad (2\text{-}4)$$

If we had used $x_1$ and $x_2$ as the weighting factors instead of $t_1$ and $t_2$, we would have obtained the average velocity with respect to distance. (Velocity averaged with respect to distance is used in hydrodynamics.) In kinematics there is a convention that the term average velocity implies "averaged with respect to time" unless specified otherwise.

**Example 2.** Suppose a car passes through a restricted zone at 20 km/h for 10 km and then travels at 60 km/h for the next 10 km. Will the average velocity be exactly halfway between 20 and 60; that is, will the average velocity be 40 km/h?

ANSWER: Let us first calculate the weighting factors $t_1$ and $t_2$.

$$t_1 = \frac{x_1}{v_1} = \frac{10 \text{ km}}{20 \text{ km/h}} = \tfrac{1}{2} \text{ hr}$$

and

$$t_2 = \frac{x_2}{v_2} = \frac{10 \text{ km}}{60 \text{ km/h}} = \tfrac{1}{6} \text{ h}$$

Now substitute these weighting factors into Eq. 2-4:

$$\bar{v} = \frac{20 \text{ km/h} \, (\tfrac{1}{2} \text{ h}) + 60 \text{ km/h} \, (\tfrac{1}{6} \text{ h})}{(\tfrac{1}{2} \text{ h}) + (\tfrac{1}{6} \text{ h})} = 30 \text{ km/h}$$

Now we shall consider the more general situation where the velocity of a body keeps changing; that is, it is $v_1$ for a short time $t_1$, $v_2$ for time $t_2$, $v_3$ for time $t_3$, and so on. Then the average velocity will be

$$\bar{v} = \frac{v_1 t_1 + v_2 t_2 + \cdots + v_n t_n}{T} \qquad (2\text{-}5)*$$

where $T = t_1 + t_2 + \cdots + t_n$. Note that any term $v_j t_j = x_j$, where $x_j$ is the distance traveled during the time $t_j$. Then Eq. 2-5 becomes

$$\bar{v} = \frac{x_1 + x_2 + \cdots + x_n}{T}$$

The sum $x_1 + x_2 + \cdots + x_n$ is called the net displacement and algebraically it equals $x - x_0$ where $x_0$ is the starting position and $x$ is the position after the time $T$.

This completes our derivation of the formula $\bar{v} = (x - x_0)/T$ for average velocity. We note that a body which travels at a constant speed of 60 km/h and then quickly turns around and returns at the same speed has an average velocity of zero even though its average speed (the average of $|v|$) is 60 km/h.

---

**Example 3.** Suppose a car moving at 60 mph can come to a stop in 5 seconds by slamming on the brakes. Also assume the velocity is decreasing uniformly during the 5 seconds. (Then the average velocity is $\bar{v} = 30$ mph $= 13.4$ m/s.) How much distance is needed in which to make the stop?

ANSWER: We solve Eq. 2-3 for $x - x_0$ and use 13.4 m/s for $\bar{v}$:

$$x - x_0 = \bar{v}t = \left(13.4 \frac{\text{m}}{\text{s}}\right)(5\text{s}) = 67 \text{ m}$$

---

The answer to Example 3 is just about the best most cars can do on a dry road. A driver who knows his physics should expect to go at least ten

---

*Using the summation symbol, $\sum$, Eq. 2-5 could also be written as

$$\bar{v} = \frac{\sum_{j=1}^{n} v_j t_j}{\sum_{j} t_j}$$

For those readers who happen to be familiar with integral calculus Eq. 2-5 can be expressed as

$$\bar{v} = \frac{\int_a^b v \, dt}{t_b - t_a}$$

car lengths before coming to a stop from 60 mph. Already the little physics we have learned is giving important practical results.

---

**Example 4.** A bicyclist encounters a series of hills. Uphill speed is always $v_1$ and downhill speed is always $v_2$. The total distance traveled is $l$, with uphill and downhill portions of equal length. What is the cyclist's average speed?

ANSWER: Let $t_1$ be the total time spent pedaling uphill. Then $t_1 = (l/2)/v_1$ and the downhill time will be $t_2 = (l/2)/v_2$. Substitution into Eq. 2-4 gives

$$\bar{v} = \frac{v_1(l/2v_1) + v_2(l/2v_2)}{l/2v_1 + l/2v_2} = \frac{2}{1/v_1 + 1/v_2} = \frac{2v_1 v_2}{v_1 + v_2}$$

---

## 2-3 Acceleration

We all have a qualitative understanding of acceleration. In a car acceleration is produced by pressing down on the accelerator pedal. The farther the pedal is pushed, the greater the acceleration. While under acceleration, the speed is increasing and the passengers are pushed by the backs of their seats. The amount of push is a quantitative measure of acceleration. Pressing down on the brake pedal gives the same effect—except that now it is a negative acceleration (also called deceleration). **Acceleration is the time rate of change of velocity.**

### *Uniform Acceleration*

By definition a body is moving with uniform or constant acceleration if its velocity is uniformly increasing with time. The acceleration $a$ is constant when

$$v - v_0 = at$$

or

$$a = \frac{v - v_0}{t} \quad \text{(constant acceleration)} \quad (2\text{-}6)$$

where $(v - v_0)$ is the increase in velocity during a time $t$. Acceleration $a$ has units of meters per second squared (m/s²) in the mks system and feet per second squared (ft/s²) in the British system.

## Instantaneous Acceleration

If the acceleration is changing with time, we must measure the change in velocity ($\Delta v$) over a small time interval ($\Delta t$). Then

$$a = \lim_{\Delta t \to 0} \left(\frac{\Delta v}{\Delta t}\right)$$

or

$$a \equiv \frac{dv}{dt} \quad \text{(definition of instantaneous acceleration)} \quad (2\text{-}7)$$

In this chapter and the next we shall be dealing with motion resulting from uniform acceleration. However, in later chapters, when we treat simple harmonic motion and inverse square law forces, the acceleration will vary with position and with time.

## Gravitational Acceleration

It is a remarkable experimental fact that near the surface of the earth *any* object when released will accelerate toward the center of the earth with an acceleration of 9.8 m/s² or 32 ft/s². What is so remarkable is that this is independent of the mass, composition, or velocity of the body. (If there is significant air resistance the acceleration will be less.) This special value of the acceleration is given a special symbol $g$.

$$g \equiv 9.8 \text{ m/s} = 32 \text{ ft/s} \quad \text{(gravitational acceleration)}$$

We shall always consider the symbol $g$ as a positive quantity. Then if the $x$ axis is pointing up, the acceleration will be $a = -g$.

---

**Example 5.** A body has acceleration $g$ for one entire year. What is its final velocity if it starts from rest?

ANSWER: According to Eq. 2-6

$$v = gt = (9.8 \text{ m/s}^2)(3.16 \times 10^7 \text{ s}) = 3.09 \times 10^8 \text{ m/s}$$

---

## Effect of Relativity

Note that in Example 5 we obtained a velocity slightly greater than the speed of light, which is $2.998 \times 10^8$ m/s. A basic principle that we shall study in Chapter 8 is that no body can travel faster than the speed of light.

We are thus forced to conclude that something is wrong with the equation $v = at$. If we wish to be exact, we should be using the equations from Einstein's theory of relativity rather than equations such as 2-6. The equation in relativity theory that corresponds to $v = at$ is

$$v = at \left[1 + \left(\frac{at}{c}\right)^2\right]^{-1/2}$$

Here $c$ is the speed of light and $a$ is a uniform acceleration as measured by an observer on the moving body. Note that for $at$ much larger than $c$, this gives $v \approx c$. And for $at$ much less than $c$, the bracket is essentially equal to 1 and we have $v \approx at$.

Since the modifications due to relativity theory are quite insignificant when dealing with "ordinary" velocities, it is reasonable to continue with our study of classical mechanics using Eq. 2-6, which is a very good approximation to the exact relativistic relation. A detailed discussion of the modifications due to relativity theory is given in Chapters 8 and 9.

## 2-4 Uniformly Accelerated Motion

So far we have a relation that tells us the velocity once we know the acceleration and time. However, often we want to know the position of a body rather than its velocity. We want to obtain an equation for $x$ in terms of $a$, $t$, and initial velocity $v_0$. We can use Eq. 2-3 to obtain an equation for $x$:

$$x = x_0 + \bar{v}t \qquad (2\text{-}8)$$

In uniformly accelerated motion the velocity is uniformly increasing from the value $v_0$ to the value $v$. As seen in Figure 2-3 the average value of the velocity is $\frac{1}{2}(v_0 + v)$, which is the average height of the curve.

$$\bar{v} = \tfrac{1}{2}(v_0 + v)$$

We may substitute $\frac{1}{2}(v_0 + v)$ for $\bar{v}$ in Eq. 2-8 to obtain

$$x = x_0 + \tfrac{1}{2}(v_0 + v)t$$

From Eq. 2-6 we have $v = v_0 + at$. Now substitute $(v_0 + at)$ for $v$ in the preceding equation:

$$x = x_0 + \tfrac{1}{2}[v_0 + (v_0 + at)]t$$
$$x = x_0 + v_0 t + \tfrac{1}{2}at^2 \quad \text{(for constant } a\text{)} \qquad (2\text{-}9)$$

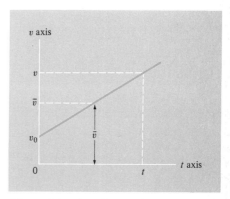

**Figure 2-3.** *Plot of $v$ versus $t$. Average velocity is the height of the curve at the halfway point.*

We see that the position of a body starting from rest increases as the square of the time when undergoing uniform acceleration. As shown in Figure 2-4, a falling body starting from rest falls a distance $x = \frac{1}{2}gt^2$.

Equation 2-9 is plotted in Figure 2-5(a). If we take the time derivative of both sides, we obtain

$$\frac{dx}{dt} = v_0 + at$$

This is plotted as curve (b) and is by definition the velocity $v$. Then if we differentiate again, we obtain $d^2x/dt^2 = a$, which is plotted as curve (c). Since this is $dv/dt$, it is the acceleration.

**Example 6.** Replot Figure 2-5 for the case where $v_0$ is a negative quantity. At what time $t_1$ will the velocity be zero?

ANSWER: The plot for negative $v_0$ is shown in Figure 2-6. In order to get the value of $t$ when $v = 0$ we solve Eq. 2-6 for $t$ and substitute $v = 0$:

$$t = \frac{v - v_0}{a}$$

$$t_1 = \frac{(0) - v_0}{a}$$

$$= -\frac{v_0}{a}$$

We note that $t$ will be a positive quantity when $v_0$ is negative and $a$ is positive.

**Example 7.** One of the ways of rating automobiles is to specify how long it takes to accelerate from 0 to 60 mph. In some cars the limitation is the slipping of the tires rather than the power of the engine. Good tires can give the car an acceleration of about $0.5g$. For such a car, how many seconds would it take to reach 60 mph (88 ft/s) and over what distance?

ANSWER: Since $v_0 = 0$ we have

$$v = at \quad \text{or} \quad t = \frac{v}{a} = \frac{88 \text{ ft/s}}{\frac{1}{2}(32 \text{ ft/s}^2)} = 5.5 \text{ s}$$

$$x = \frac{1}{2}at^2 = \frac{1}{2}(16 \text{ ft/s}^2)(5.5 \text{ s})^2 = 242 \text{ ft}$$

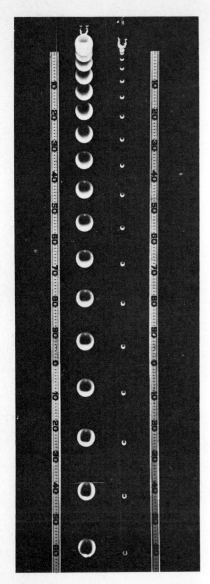

**Figure 2-4.** Stroboscopic photograph of two freely falling balls of unequal mass. Such a strobe photograph is taken by opening up the camera lens and flashing the light every thirtieth of a second. Note that the small mass hits the floor at the same time as the heavy mass. Both balls were released simultaneously with lower surfaces in line. [Courtesy Educational Development Center.]

**Figure 2-5.** (a) A plot of $x = x_0 + v_0 t + \frac{1}{2}at^2$. The derivative of that curve is plotted in (b); that is, curve (b) is the slope of (a). Curve (c) is the slope of (b).

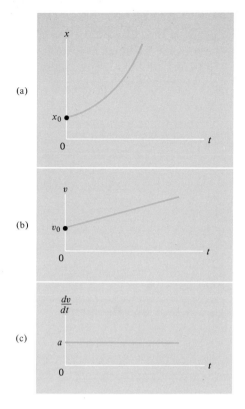

Example 7 is also related to the shortest time and distance in which a car can be stopped. If the maximum deceleration is also $0.5\,g$, $a = -16\text{ ft/s}^2$. We use Eq. 2-6 $v - v_0 = at$, where $v_0 = 88$ ft/s is the initial velocity and $v = 0$ is the final velocity. Then

$$0 - 88 \text{ ft/s} = (-16 \text{ ft/s}^2)t$$
$$t = 5.5 \text{ s}$$

As one should expect, this is the same result as in Example 7. If we took a movie of the car in Example 7 and ran the film backward, we would see a car slowing down in reverse with a deceleration of 16 ft/s. It takes the same time for the car to slow down as to speed up.

By the same reasoning, if one throws a ball straight up in the air it takes just as long to come down as it took to go up. What is the instantaneous acceleration and velocity of the ball just as it reaches its maximum height? The instantaneous velocity is zero, and one might be tempted to say the acceleration must be zero when the velocity is zero. However, no matter what the velocity of the ball, it is continuously decreasing at a rate of $9.8 \text{ m/s}^2$; hence, the instantaneous acceleration is $a = -9.8 \text{ m/s}^2$.

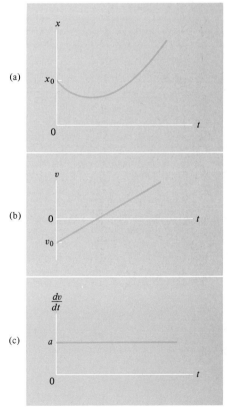

**Figure 2-6.** Same as Figure 2-5 except that $v_0$ is negative rather than positive.

SEC. 2-4/UNIFORMLY ACCELERATED MOTION

**Example 8.** Assume that for comfortable air travel the horizontal component of acceleration should never be greater than 10 m/s (which is close to $g$). What is the shortest possible time in which one could make a trip from New York to Boston (a trip of 280 km)?

ANSWER: Starting from rest in New York, there will be constant acceleration to the halfway point and then constant deceleration for the remaining half of the trip in order for the plane (or rocket) to be able to land in Boston. If $x_1$ is the half-distance, then

$$x_1 = \tfrac{1}{2}at_1^2 \quad \text{where } t_1 \text{ is the half-time}$$

$$t_1 = \sqrt{\frac{2x_1}{a}} = \sqrt{\frac{280 \times 10^3 \text{ m}}{10 \text{ m/s}^2}} = 167 \text{ s}$$

Total time $= 335$ s $= 5.58$ min.

### Velocity-Distance Relation

In Eq. 2-9 we have the distance in terms of time. As seen in Example 9, it is also useful to have a relation between the distance and the velocity without knowing the time. This can be obtained by solving Eq. 2-6 for $t$ and substituting into Eq. 2-9:

$$t = \left(\frac{v - v_0}{a}\right)$$

Substituting into Eq. 2-9

$$x = x_0 + v_0\left(\frac{v - v_0}{a}\right) + \frac{1}{2}a\left(\frac{v - v_0}{a}\right)^2$$

$$= x_0 + \frac{v^2 - v_0^2}{2a}$$

$$v^2 - v_0^2 = 2a(x - x_0) \quad \text{(for constant } a\text{)} \quad (2\text{-}10)$$

**Example 9.** The speed necessary to go into orbit around the earth is 8 km/s. It is also necessary to travel about 200 km in order to get out of the atmosphere. Suppose a rocket can reach this speed after traveling 200 km under constant acceleration. What would be the acceleration?

ANSWER: In Eq. 2-10, $v_0 = 0$ and $x = 2 \times 10^5$ m

$$v^2 = 2ax$$

$$a = \frac{v^2}{2x} = \frac{(8 \times 10^3 \text{ m/s})^2}{2(2 \times 10^5 \text{ m})} = 160 \text{ m/s}^2 = 16.3g$$

This is about the limit of acceleration a well-trained astronaut can take for such a length of time.

There are a few cases where a passenger has fallen out of a high-flying airplane without a parachute and lived to tell about it. Such a passenger could be decelerated by a soft, deep snow bank or the branches of a tree. Assume that the passenger can tolerate $a = 50g$ for such a short time. How thick must be the snow bank, or how high the tree? Fortunately a freely falling human body stops accelerating when the force of the air resistance becomes equal to the force of gravity. This occurs at a velocity of about 120 mph or 53 m/s. If $x$ is the thickness of the snow, we have

$$v^2 - v_0^2 = 2ax$$

$$0 - v_0^2 = -100gx$$

$$x = \frac{v_0^2}{100g} = \frac{(53 \text{ m/s})^2}{100(9.8 \text{ m/s}^2)} = 2.9 \text{ m}$$

One of the well-documented cases of a person surviving without a prachute is reported by R. G. Snyder in volume 131, page 1290, of the *Journal of Military Medicine* (1966):

During one of the battalion drops, from 1,200 feet on a clear, relatively warm day, an observer noted what appeared to be an unsupported bundle falling from one of the C-119 airplanes; no chute deployed from the object. The impact looked like a mortar round exploding in the snow. When the aidmen reached the spot, they found a young paratrooper flat on his back at the bottom of a $3\frac{1}{2}$ foot crater in the snow, which consisted of alternating layers of soft snow and frozen crust. He could talk and did not appear injured.

## Summary

Motion in a straight line with constant velocity is described by

$$x = x_0 + vt$$

The same equation holds for average velocity $\bar{v}$:

$$x = x_0 + \bar{v}t$$

Instantaneous velocity

$$v = \frac{dx}{dt}$$

For uniform acceleration $a$ we have

$$x = x_0 + v_0 t + \tfrac{1}{2}at^2 \qquad v = v_0 + at$$

and also

$$v^2 - v_0^2 = 2a(x - x_0)$$

Instantaneous acceleration

$$a = \frac{dv}{dt} = \frac{d^2x}{dt^2}$$

## Exercises

1. One km/h equals how many ft/s?
2. A car travels at constant velocity $v = -10$ km/s starting at $x = 50$ km when $t = -2$ s.
   (a) Plot $x$ versus $t$.
   (b) At what time will it reach $x = 0$?
3. In Example 4 what is $\bar{v}$ for the bicyclist averaged with respect to distance?
4. A body starts from rest with uniform acceleration. It goes a distance $s$ in a time $T$. What is its instantaneous velocity at time $T$ in terms of $s$ and $T$?
5. A car travels a distance $x_1$ at velocity $v_1$ and then an additional distance $x_2$ at velocity $v_2$. What is the velocity averaged with respect to distance? (Now $x_1$ and $x_2$ are the weighting factors.)
6. At a time $t_1$ a body has a position $x_1$ and velocity $v_1$. At a later time $t_2$ it has a position $x_2$ and a velocity $v_2$.
   (a) What is its average velocity in terms of these quantities?
   (b) What is its average acceleration in terms of these quantities?
7. Given the following graph of the acceleration of a particle moving along the $x$ axis, sketch the graphs of its velocity and its position as functions of time. Assume that at $t = 0$, $x = 0$, and $v = 0$.

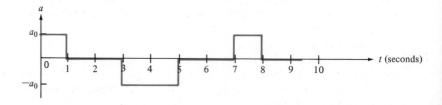

8. Plot $x$ versus $t$ for positive $x_0$ and $v_0$ if the acceleration $a$ is a constant negative quantity.

9. Plot $x$ versus $t$ where $x_0$ and $v_0$ are negative. The acceleration $a$ is a constant negative quantity.
10. In Figure 2-6 extend the curves to negative values of $t$.
11. Repeat Example 8 for a horizontal acceleration equal to $2g$.
12. What is the maximum speed in Example 8?
13. Repeat Example 8 for a distance to the other side of the earth. (Then $x_1$ is one quarter the earth's circumference.)
14. Under the conditions of Example 8, how long would it take to make a round-the-world trip?
15. A steel ball is bouncing up and down on a steel plate with a period of oscillation of 1 s. How high does it bounce?
16. A car traveling 60 mph crashes into a solid wall. This is the same impact as if it had been dropped from what height?
17. A certain U. S. Army rocket fired vertically at constant acceleration attains a speed of 600 mph by the time it reaches 1,000 ft. How many times $g$ is its acceleration?
18. Assume there is a 1-min warning time for an ABM (antiballistic missile) to intercept an incoming missile that will be directly over the ABM at a height of 200 km. If the ABM has an acceleration of $10g$, is the 1-min warning time adequate?
19. Terminal velocity in air for a human body is about 55 m/s. From what height must a body fall in vacuum in order to reach this speed?
20. A rocket is fired with constant acceleration of $16g$. How far will it travel before escape velocity of 11.3 km/s is reached?

## Problems

21. A body starts with initial velocity $v_0$ with uniform acceleration. It goes a distance $x$ in a time $T$. What is its instantaneous velocity at time $T$?
22. If $x = At^n$,

$$\frac{\Delta x}{\Delta t} = A \frac{(t + \Delta t)^n - t^n}{\Delta t}$$

Expand $(t + \Delta t)^n$ using the binomial expansion. For $t = 1$ s and $\Delta t = 0.1$ s, what are the first-, second-, and third-order terms?
23. A right fielder is 200 ft from home plate. Just at the time he throws the ball to home plate, a runner leaves third base and takes 4.5 s to reach home plate. If the maximum height reached by the ball was 64 ft, did the runner make it to home plate in time?
24. A person throws a ball straight up in the air and catches it 2 s later.
    (a) What was the initial speed of the ball?
    (b) How high did the ball rise above the point from which it was thrown?
25. A body moving with a velocity of 10 m/s is uniformly decelerated, coming to rest in a distance of 20 m.
    (a) What is its deceleration?
    (b) How long a time was required for the body to come to rest?
    (c) Plot $v$ versus $t$. Plot $x$ versus $t$.

26. In order to get into orbit an astronaut has to accelerate from zero velocity during the time $T$ to the orbital velocity of 8 km/s. Assume the spacecraft's acceleration during this takeoff period is $4g$ and that it travels straight up. How long does the spacecraft take to reach orbital velocity, and how far does it travel during this time?

27. An earth satellite has an orbit 400 km high. A super-powerful cannon is pointed vertically in an effort to shoot down the satellite. Assume the acceleration due to gravity is constant and ignore air resistance. What must be the muzzle velocity for the cannon shell to just barely reach the satellite? How long would it take the shell to reach the satellite?

28. Suppose a car traveling 60 mph had a head-on collision with a heavy truck also traveling at 60 mph. Assume that during the collision the truck is not slowed down. What is the equivalent height from which the car must be dropped on its nose to sustain comparable damage to its front end?

29. In Example 5 suppose the body travels for $\frac{1}{2}$ year. What is the final velocity? Give both classical and relativistic results.

30. A body starting from rest experiences a uniform acceleration of $g$ for 1 year. According to relativistic mechanics what is its final velocity as observed from the starting point? If it accelerates for 10 years, how close will it get to the speed of light?

31. According to relativity theory, how long would it take the above body to reach 99% the speed of light?

32. A particle starts from rest and undergoes accelerations as plotted in the figure for the first 4 s.
    (a) Plot a graph of $v$ versus $t$ for this particle.
    (b) Plot a graph of distance versus time.
    (c) What will be the maximum velocity during the 4 s?
    (d) How far will the particle go in the 4 s?

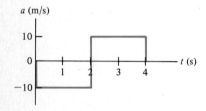

Problem 32

33. An apartment dweller sees a flower pot (originally on a window sill above) pass the 2.0-m-high window of his fifth floor apartment in 0.1 s. The distance between floors is 4.0 m and $g = 9.8$ m/s². From which floor did the flower pot fall?

34. A ball is dropped from a height of 64 ft to a flat surface and bounces to a height of 16 ft.
    (a) What is the velocity of the ball just before it touches the surface?
    (b) How much time elapses from the instant the ball is dropped to its arrival at the top of its bounce?
    (c) What is the velocity of the ball just after it leaves the surface?
    (d) Plot $y$ versus $t$.

35. (This problem uses integral calculus.) A particle starts from rest with constant acceleration $a$ and reaches a velocity $v_1$. Show that $\bar{v}$ averaged over distance is $\frac{2}{3}v_1$. Plot $v$ versus $x$.

36. (This problem uses integral calculus.) Suppose the acceleration $a = At$. Show that $x = A(t^3/6)$ for a body starting from rest.

37. (Integral calculus not necessary.) For Problem 36, what is $v$ as a function of $t$?

38. (This problem uses integral calculus.) Repeat Problem 37 for a body starting from a position $x_0$ with a velocity $v_0$.

39. Because of air resistance, the amount of gasoline used by a car has a term proportional to the square of the velocity. The volume of fuel $V_{gas}$ used to go a distance $x$ at a fixed speed $v$ is given by

$$\frac{V_{gas}}{x} = A + Bv^2$$

Use the data plotted in Figure 7-15 for the Pinto to answer the following:
(a) What is $A$ in liters/km? (A liter is $10^3$ cm$^3$.)
(b) What is $B$ in (liters/km)/(km/h)$^2$?
(c) What is $B$ in (liters s$^2$)/m$^3$?

# Two-Dimensional Motion

In the preceding chapter we discussed motion along a straight line. Whether the line was vertical or horizontal, we called it the $x$ axis. Now we shall treat motion in a plane. Usually it will be a vertical plane. In that case we shall use $x$ for the horizontal coordinate and $y$ for the vertical coordinate. We shall see that two-dimensional motion can be treated as two independent one-dimensional motions.

## 3-1 Free-Fall Trajectories

We shall use Figure 3-1 to show that if a cannon ball is fired at an angle $\theta$ from the horizontal, it will follow the path of a parabola. In view (a) the cannon shoots a white ball straight up with an initial velocity $(v_0)_y$. The camera shutter is left open and a strobe lamp is flashed 10 times per second. According to Eq. 2-9 the vertical position of the ball will be given by

$$y = (v_0)_y t - \tfrac{1}{2}gt^2 \tag{3-1}$$

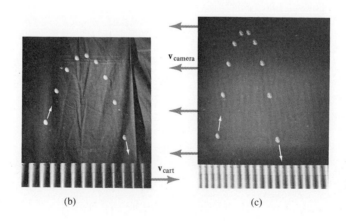

**Figure 3-1.** *Strobe photos of ball fired from vertical cannon. Flash rate is 10 per second. (a) Cannon at rest. (b) Cannon on cart moving to the right. (c) Same as (a) except camera is moving to the left.*

(a)  (b)  (c)

In view (b) the same vertical cannon is moving uniformly to the right when the ball is fired, and in view (c) the situation is the same as in view (a) except that the camera is moving to the left with a velocity $-(v_0)_x$. Then according to an observer moving along with the camera, the horizontal position of the cannon ball will be

$$x = (v_0)_x t \qquad (3\text{-}2)$$

and the vertical position must be the same as given by Eq. 3-1, since we are describing the very same physical situation.

The equation for the path as seen in view (c) can be obtained by solving Eq. 3-2 for $t$ and substituting into Eq. 3-1:

$$y = (v_0)_y \left(\frac{x}{(v_0)_x}\right) - \frac{1}{2}g\left(\frac{x}{(v_0)_x}\right)^2$$

$$= \frac{(v_0)_y}{(v_0)_x} x - \frac{g}{2(v_0)_x^2} x^2 \qquad (3\text{-}3)$$

This is the equation for a parabola.

In view (b) the camera is at rest as in (a) and the cannon, which is on a cart, is moving to the right with constant horizontal velocity $v_{cart}$. Its horizontal position is given by Eq. 3-2 using $v_{cart}$ for $(v_0)_x$. Since the gravitational acceleration of a freely falling body is observed to be independent of the velocity, the vertical position of the ball is still given by Eq. 3-1 and the path must still be a parabola. Figure 3-1 verifies that in either case the shape of the path is a parabola. As will be shown in the next section, the magnitude of the initial velocity in Figure 3-1(b) is $v_0$ where $v_0 = \sqrt{(v_0)_x^2 + (v_0)_y^2}$.

## 3-2 Vectors

Now that we are discussing motion in a plane, we will be adding and subtracting velocities that are not always in the same direction. Starting with the next chapter, we shall be adding and subtracting forces not always in the same direction. Such additions and subtractions can be simplified by using the mathematical definition of a vector. A vector has both magnitude and direction, but no definite position in space. For example, the initial velocity of the ball in Figure 3-2 can be specified by giving its magnitude in m/s and its angle $\theta$ to the $x$ axis. It can also be specified by giving its components $(v_0)_x$ and $(v_0)_y$. The relation between $v_0$ and its components can be obtained using the Pythagorean theorem. Let

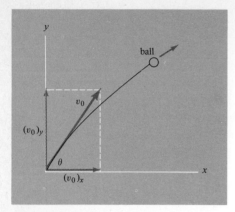

**Figure 3-2.** *Initial velocity $v_0$ and its x and y components.*

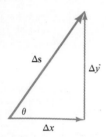

**Figure 3-3.** *Relation between displacement $\Delta s$ and its x and y components.*

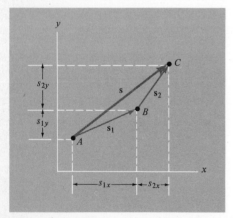

**Figure 3-4.** *Addition of two vectors by placing the tail of the second vector on the tip of the first.*

$\Delta x$ and $\Delta y$ be the horizontal and vertical distances covered by the ball in a time $\Delta t$. Then the straight line distance covered, or displacement, is

$$\Delta s = \sqrt{(\Delta x)^2 + (\Delta y)^2}$$

as shown in Figure 3-3. Dividing both sides by $\Delta t$ gives

$$\frac{\Delta s}{\Delta t} = \sqrt{\left(\frac{\Delta x}{\Delta t}\right)^2 + \left(\frac{\Delta y}{\Delta t}\right)^2}$$

or

$$v = \sqrt{v_x^2 + v_y^2}$$

Also $v_x = v \cos \theta$ and $v_y = v \sin \theta$. In all three dimensions

$$v = \sqrt{v_x^2 + v_y^2 + v_z^2}$$

### Addition of Vectors

To complete the definition of vector we must specify how to add vectors having different directions. The rule for vector addition is contained in the following definition: **A vector is a mathematical quantity that has both magnitude and direction. Each component of the sum of two vectors is equal to the sum of the corresponding components of the two vectors.** (See Figure 3-4.)

An example of how to add two vectors is shown in Figure 3-4. Vector $\mathbf{s}_1$ is the displacement from point $A$ to $B$, and $\mathbf{s}_2$ is the displacement from $B$ to $C$. The resultant displacement from $A$ to $C$ is the vector sum $\mathbf{s}$. We see from Figure 3-4 that

$$s_x = s_{1x} + s_{2x}$$
$$s_y = s_{1y} + s_{2y}$$

and if the vectors are not in the $xy$ plane we also have

$$s_z = s_{1z} + s_{2z}$$

In this book we use boldface $\mathbf{s}$ for a vector and $s$ or $|\mathbf{s}|$ for the magnitude of the vector. The magnitude of a vector is always a positive quantity. The vector equation $\mathbf{s} = \mathbf{s}_1 + \mathbf{s}_2$ is mathematical shorthand for the above three simultaneous equations. Note that if $\mathbf{s}_1$ and $\mathbf{s}_2$ are not parallel, $s < (s_1 + s_2)$ when $\mathbf{s} = \mathbf{s}_1 + \mathbf{s}_2$. In fact it is sometimes possible for $s$ to be less than the magnitude of any of its parts. Such a situation is shown in Figure 3-5.

## Polygon Rule of Addition

In Figure 3-4 we have placed the tail of $s_2$ at the tip of $s_1$. Then we have drawn vector $s$ connecting the first tail with the last tip. Note that $S_x = S_{1x} + S_{2x}$ and $S_y = S_{1y} + S_{2y}$; hence, by definition the vector $s = s_1 + s_2$. The process of placing the next tail on the preceding tip can be repeated and is called the polygon rule of addition of vectors (see Figure 3-5). Since $s_{1x} + s_{2x} = s_{2x} + s_{1x}$, it is clear that $s_1 + s_2 = s_2 + s_1$ and the order of addition does not affect the result.

The negative of a vector has the same magnitude, but is oppositely directed. A vector can be subtracted by adding its negative. If $v = v_2 - v_1$, then $v = v_2 + (-v_1)$. This is illustrated in Figure 3-6.

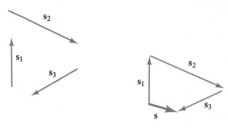

**Figure 3-5.** *The polygon rule of vector addition is used to obtain* $s = s_1 + s_2 + s_3$.

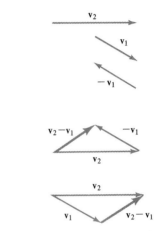

**Figure 3-6.** *In the upper triangle* $-v_1$ *is added to* $v_2$ *to obtain* $v_2 - v_1$. *An alternate method is shown in the lower triangle.*

---

**Example 1.** The vectors $v_1$ and $v_2$ have the same length $v$. The angle between them is $\theta$. What is the magnitude of $v_2 - v_1$?

ANSWER: Let us call $\Delta v = |v_2 - v_1|$. It is the base of the isosceles triangle in Figure 3-7. In either right triangle

$$\sin(\tfrac{1}{2}\theta) = \frac{\tfrac{1}{2}\Delta v}{v}$$

Hence

$$\Delta v = 2v \sin(\tfrac{1}{2}\theta)$$

---

Many of the quantities we shall encounter in physics are vector quantities. Whether or not a physical quantity behaves as a mathematical vector must ultimately be tested by experiment. Some of the vector quantities we shall study are displacement, velocity, acceleration, force, momentum, angular momentum, torque, electric field, magnetic field, and current density.

If a vector is multiplied (or divided) by a number, the resulting quantity is a vector. For example, if a small displacement $\Delta s$ is divided by $\Delta t$, the resulting quantity (a velocity) is a vector:

$$v \equiv \lim_{\Delta t \to 0}\left[\frac{\Delta s}{\Delta t}\right] \quad \text{(definition of velocity)}$$

Similarly, if a vector $\Delta v$ is divided by $\Delta t$, the resulting quantity (an acceleration) is a vector:

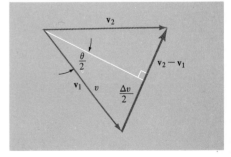

**Figure 3-7.** $v_1$ *and* $v_2$ *are two vectors of equal length. The heavy vector is their difference.*

$$\mathbf{a} \equiv \lim_{\Delta t \to 0} \left[ \frac{\Delta \mathbf{v}}{\Delta t} \right] \quad \text{(definition of acceleration)}$$

We can show that velocities add vectorially by considering the example of a boat sailing in moving water. Let $\Delta \mathbf{S}$ be the displacement of the boat with respect to the water and $\Delta \mathbf{S}_w$ be the displacement of the water with respect to the shore, both in a time $\Delta t$. Then if $\Delta \mathbf{S}'$ is the displacement of the boat with respect to the shore we have $\Delta \mathbf{S}' = \Delta \mathbf{S}_w + \Delta \mathbf{S}$. Now divide both sides by $\Delta t$:

$$\frac{\Delta \mathbf{s}'}{\Delta t} = \frac{\Delta \mathbf{s}_w}{\Delta t} + \frac{\Delta \mathbf{s}}{\Delta t}$$

In the limit as $\Delta t$ approaches zero we have

$$\mathbf{v}' = \mathbf{v}_w + \mathbf{v}$$

In the following example $\mathbf{v}$ is the boat velocity in a frame of reference or coordinate system that is at rest with respect to the water. It is the velocity the pilot or other boaters would measure if they could not see the shore. A person in a different frame of reference, namely, at rest with respect to the shore, would see a different velocity $\mathbf{v}'$ as given by the preceding equation.

**Example 2.** A ferryboat wishes to go straight across a stream that is flowing at 5 km/h to the east as shown in Figure 3-8. The pilot knows his speed with respect to the water is 10 km/h. At what angle must he head the boat and what will be his speed with respect to the shore?

ANSWER: Let $\mathbf{v}$ be his vector velocity with respect to the water and $\mathbf{v}_w$ be the velocity of the water. Then $\mathbf{v}' = \mathbf{v} + \mathbf{v}_w$ is the vector velocity as viewed from the shore (it must point due north). This vector triangle is shown in Figure 3-8(b) and is a 30°-60° right triangle. The pilot must head the boat at 30° to the west of north. The magnitude of $\mathbf{v}'$ is cos 30° times 10 km/h or 8.66 km/h. Note that this resultant vector is smaller in magnitude than the direct sum of its parts; that is, $8.66 \neq 10 + 5$.

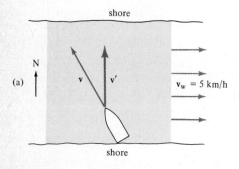

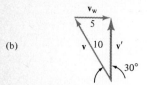

**Figure 3-8.** *In order to cross a 5-km/h river, a 10-km/h boat must "sail" at 30° from the intended direction.*

### Resolution of Vectors

We have been discussing situations where the resultant vector is obtained from its components. But there are also occasions where a single vector is known and its components are sought. A common example is to

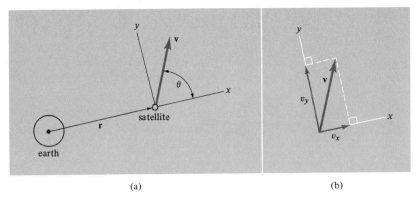

**Figure 3-9.** (a) *A satellite has velocity* **v** *when at a distance* **r** *from the earth.* (b) *Perpendiculars are dropped from the tip of* **v** *in order to establish the x and y components of* **v**.

determine the radial and tangential velocities of a satellite in orbit about the earth as shown in Figure 3-9(a). In order to obtain the radial velocity, we choose the $x$ axis to be along the radial direction $r$. The radial component $v_x$ is obtained by dropping a perpendicular from the end of vector **v** to the $x$ axis as shown in Figure 3-9(b). It will be of length $v_x = v \cos \theta$. The tangential velocity is obtained by dropping a perpendicular onto the $y$ axis. This procedure is also called projecting the vector onto the $y$ axis.

**Example 3.** A mass $m$ is being pulled down on an inclined plane by a gravitational force $\mathbf{F}_g$ as shown in Figure 3-10. What is $F_\parallel$, the component of force that is parallel to the surface?

ANSWER: In Figure 3-10 we have drawn the $x$ axis in the direction parallel to the surface. Then we drop a perpendicular from the tip of $\mathbf{F}_g$ to the $x$ axis. We see from the figure that the force

$$F_\parallel = F_g \sin \alpha$$

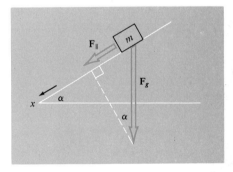

**Figure 3-10.** *Gravitational force* $\mathbf{F}_g$ *is pulling m straight down. The component* $\mathbf{F}_\parallel$ *parallel to incline is obtained by dropping a perpendicular to the surface.*

A practical example of resolution of vectors explains how a sailboat can sail into a wind. In Figure 3-11(a) the sailboat is sailing at an angle of 45° to the wind. Since the component of $\mathbf{v}_{\text{wind}}$ in the direction of the boat's motion is opposed to the motion, we wonder how a sailboat could ever work its way upwind.

The explanation involves resolution of the vector force acting on the sail. For a flat sail, the force $\mathbf{F}_{\text{sail}}$ on the sail due to the wind will push perpendicular to the surface as shown in Figure 3-11(b). Because of the keel (or centerboard) below the hull, the boat can only move along the $x$ axis. We note that the component of force in this direction, $F_x$ is pointed in the direction of motion. (This explanation has invoked some approximations among which was the assumption of a flat sail. In actual practice a

SEC. 3-2/VECTORS 43

**Figure 3-11.** (*a*) *A sailboat is sailing at an angle 45° to the wind.* (*b*) *The component of force on the sail in the direction of motion is* $F_x$. *This force is moving the boat.*

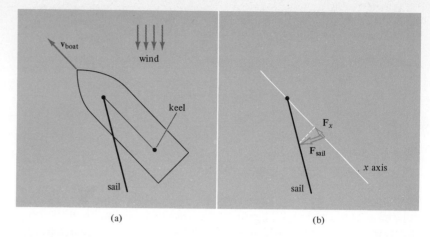

sail bows out and there is an additional air foil effect also tending to move the boat forward.)

### Unit Vectors

A vector **v** could also be specified by listing its three components $(v_x, v_y, v_z)$. The more usual notation in physics texts is

$$\mathbf{v} = \mathbf{i}v_x + \mathbf{j}v_y + \mathbf{k}v_z$$

where **i**, **j**, and **k** are defined as unit vectors along the *x*, *y*, and *z* axes, respectively. The vector **i** has length of one unit and is pointing parallel to the *x* axis, as shown in Figure 3-12.

Mathematicians also have rules for multiplication of vectors. We shall have no need to multiply vectors until Chapter 6. At that time we shall discuss multiplication of vectors.

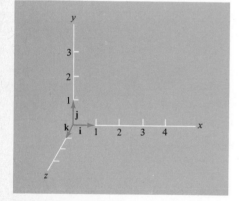

**Figure 3-12.** *The three unit vectors* **i**, **j**, *and* **k**.

**Example 4.** The displacement vector from the origin to the position of a particle (also called the position vector) is specified in terms of three constants as

$$\mathbf{S} = \mathbf{i}a_1 t + \mathbf{j}(a_2 t - a_3 t^2)$$

Find $|v_0|$, **v**, and **a** (the acceleration).

ANSWER: For velocity we use the definition:

$$\mathbf{v} = \frac{d\mathbf{S}}{dt} = \mathbf{i}a_1 + \mathbf{j}(a_2 - 2a_3 t)$$

At $t = 0$:
$$\mathbf{v}_0 = \mathbf{i}a_1 + \mathbf{j}a_2 \qquad |\mathbf{v}_0| = \sqrt{a_1^2 + a_2^2}$$

The acceleration is
$$\mathbf{a} = \frac{d\mathbf{v}}{dt} = \frac{d}{dt}[\mathbf{i}a_1 + \mathbf{j}(a_2 - 2a_3 t)] = \mathbf{j}(-2a_3)$$

The acceleration vector has a constant magnitude of $(2a_3)$ and is pointing down (along the negative $y$ direction). Note that the equation for $\mathbf{S}$ is the equation of a parabola in vector notation.

## 3-3 Projectile Motion

An age-old military problem (since the invention of the slingshot) is how to aim a cannon (or slingshot) given the range $R$ of the target and the muzzle velocity $v_0$ of the projectile. We want to solve for $\theta$ in Figure 3-13. The path of the trajectory is given by Eq. 3-3, if we set $(v_0)_x = v_0 \cos\theta$ and $(v_0)_y = v_0 \sin\theta$. Then

$$y = (\tan\theta)x - \left(\frac{g}{2v_0^2 \cos^2\theta}\right)x^2 \qquad (3-4)$$

In order to obtain the range $R$, set $y = 0$ and $x = R$:

$$0 = (\tan\theta)R - \left(\frac{g}{2v_0^2 \cos^2\theta}\right)R^2$$

$$R = \frac{2v_0^2 \sin\theta \cos\theta}{g} = \frac{v_0^2}{g}\sin 2\theta \qquad (3-5)$$

$$\sin 2\theta = \frac{gR}{v_0^2}$$

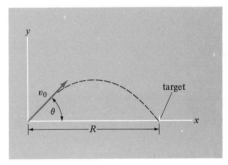

**Figure 3-13.** Path of projectile fired at angle $\theta$ with muzzle velocity $v_0$. Range is $R$.

It is seen that maximum range is achieved when $2\theta = 90°$ or $\theta = 45°$. Example 5 is a modern-day version of this classic problem.

**Example 5.** A SLBM (submarine launched ballistic missile) is fired at our city from a distance of 3000 km. Assume it is detected at firing time. How much warning time is there, and what is the launch velocity $v_0$? Make the assumptions of a flat earth, constant gravitational acceleration, and a firing angle of $45°$ and that the rocket coasts over all but the very beginning of its trajectory.

ANSWER: First we solve Eq. 3-5 for $v_0$.

$$v_0 = \sqrt{gR} \quad \text{when } \theta = 45°$$
$$= \sqrt{9.8 \times 3 \times 10^6} \text{ m/s} = 5.42 \text{ km/s}$$

(This happens to be 68% of the velocity necessary to put the missile into orbit.) The $x$ coordinate of the missile position is given by

$$x = (v_0 \cos \theta)t \quad \text{or} \quad t = \frac{x}{v_0 \cos \theta}$$

The total time $T$ occurs when $x = R$. Therefore

$$T = \frac{R}{v_0 \cos \theta} = \frac{3 \times 10^6}{5.42 \times 10^3 \times 0.707} \text{ s} = 783 \text{ s} = 13 \text{ min}$$

We see from Example 5 that the maximum warning time for an enemy missile attack is about 10 min (which is not enough time for the evacuation of a city).

**Example 6.** This is called the shoot-the-monkey problem. Assume the monkey lets go of the tree shown in Figure 3-14 when the gun is fired. At what angle should the gun be pointed so that the bullet hits the monkey while in free fall? The answer turns out to be independent of the muzzle velocity.

**Figure 3-14.** *The shoot-the-monkey problem. What must $\theta$ be? (a) Just before the gun is fired. (b) At time $t$ after the gun is fired. Both monkey and bullet have dropped the same distance $h_0$ from the straight line.*

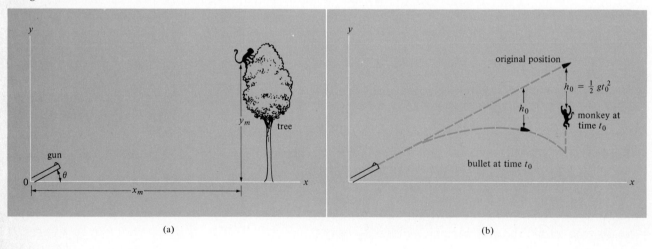

(a)

(b)

CH. 3/TWO-DIMENSIONAL MOTION

ANSWER: Let $x_m$ and $y_m$ be the initial coordinates of the monkey. Let $t_1$ be the time of collision. At collision the height of the monkey is

$$y = y_m - \tfrac{1}{2}gt_1^2$$

and the height of the bullet is

$$y = (v_0 \sin \theta)t_1 - \tfrac{1}{2}gt_1^2$$

Equating these two gives

$$t_1 = \frac{y_m}{v_0 \sin \theta} \qquad (3\text{-}6)$$

The $x$ coordinate of the bullet at time $t_1$ is

$$x_m = (v_0 \cos \theta)t_1$$

Now substitute the right-hand side of Eq. 3-6 for $t_1$. Then

$$x_m = (v_0 \cos \theta)\left(\frac{y_m}{v_0 \sin \theta}\right)$$

or

$$\tan \theta = \frac{y_m}{x_m}$$

We see from this that the gun must be pointed directly at the monkey!

## 3-4 Uniform Circular Motion

We now shall consider a body moving in a circular path of radius $R$ with constant speed $v$. Even though $v$ is constant, the vector **v** is not constant because it is continuously changing direction. The change in **v** is a vector $\Delta \mathbf{v}$, which is nonzero. Hence the vector acceleration, $d\mathbf{v}/dt$, must be nonzero. **This acceleration by virtue of the changing direction of the velocity is called centripetal acceleration, $a_c$.** We shall now show that the magnitude of $a_c$ is $v^2/R$ and that its direction is always toward the center of the circle. In order to calculate the magnitude $a_c$ we must find the velocity difference for two successive positions. Assume the body takes a time $\Delta t$ to go from position 1 to position 2 in Figure 3-15(a). Let $\Delta \mathbf{v} = \mathbf{v}_2 - \mathbf{v}_1$. Then

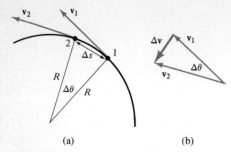

**Figure 3-15.** (*a*) Two successive positions in uniform circular motion. (*b*) The vector difference between the two velocity vectors.

$$\mathbf{a}_c = \lim_{\Delta t \to 0} \left( \frac{\Delta \mathbf{v}}{\Delta t} \right) \qquad (3\text{-}7)$$

Note that the angle $\Delta\theta$ between $\mathbf{v}_1$ and $\mathbf{v}_2$ is the same as $\Delta\theta$ in Figure 3-15(a) (their sides are mutually perpendicular). Thus the triangle in Figure 3-15(b) is similar to the triangle in part (a), and we have

$$\frac{\Delta v}{v} = \frac{\Delta S}{R}$$

or

$$\Delta v = \frac{v}{R} \Delta S$$

where $\Delta S$ is the straight-line distance between positions 1 and 2. Dividing both sides by $\Delta t$ gives

$$\frac{\Delta v}{\Delta t} = \frac{v}{R} \frac{\Delta s}{\Delta t}$$

Now take the limit as $\Delta t \to 0$, then $(\Delta v/\Delta t) \to a_c$ and $(\Delta s/\Delta t) \to v$. We obtain

$$a_c = \frac{v^2}{R} \qquad \text{(centripetal acceleration)} \qquad (3\text{-}8)$$

Note that the vector direction of $\Delta \mathbf{v}$ is perpendicular to $\mathbf{v}$, in the limit as $\Delta t \to 0$ and thus it is pointing toward the center of the circle. We see that centripetal acceleration always points toward the center of the circle.

Often it is more convenient to express the centripetal acceleration in terms of $R$ and $T$, where $T$ is the period of revolution or the time for the particle to make one complete trip around the circle. The particle speed is the distance covered in one revolution divided by $T$:

$$v = \frac{2\pi R}{T}$$

We substitute this for $v$ in Eq. 3-8 and obtain

$$a_c = \frac{(2\pi R/T)^2}{R}$$

$$a_c = \frac{4\pi^2}{T^2} R \qquad (3\text{-}9)$$

Some readers may be familiar with the term centrifugal force or centrifugal acceleration. Such a force or acceleration only occurs when the observer is in a rotating frame of reference (the observer is accelerating). We shall confine our discussions to observers "at rest" or moving with uniform velocity (see definition of inertial frame of reference in Chapter 4) hence we shall never encounter centrifugal acceleration.

---

**Example 7.** What is the centripetal acceleration of a body on the earth's equator due to the earth's rotation?

ANSWER: Here $T = 1$ day $= 8.64 \times 10^4$ s and $R = R_e = 6370$ km. Substitution into Eq. 3-9 gives

$$a_c = \frac{4\pi^2(6.37 \times 10^6)}{(8.64 \times 10^4)^2} \text{ m/s}^2 = 0.034 \text{ m/s}^2$$

This is 0.35% of $g = 9.8$ m/s². If the Earth were a perfect sphere a person on the equator would weigh 0.35% less than when near either pole. This is one reason why it is easier to beat previous athletic records on the equator than in higher latitudes.

---

## 3-5 Earth Satellites

A common question of those who have not studied physics is, "What keeps an earth satellite up?" Once the rocket power is turned off, shouldn't it fall toward the center of the earth with acceleration $g$ as do all other bodies near the surface of the earth? The answer is yes, low-flying earth satellites in orbit have acceleration 9.8 m/s² toward the center of the earth. If they did not, they would "fly" off in a straight line tangent to the earth. Any body moving in a circle automatically has an acceleration $v^2/R$. If it is in circular orbit around the earth, the force to give this acceleration is supplied by gravity and we have

$$g = \frac{v_c^2}{R_e} \qquad (3\text{-}10)$$

where $v_c$ is called the critical orbital velocity and $R_e = 6370$ km is the radius of the earth. Solving Eq. 3-10 for $v_c$ gives

$$v_c = \sqrt{gR_e} \qquad (3\text{-}11)$$
$$= \sqrt{(9.8 \text{ m/s}^2)(6.37 \times 10^6 \text{ m})} = 7.90 \text{ km/s} = 4.9 \text{ mi/s}$$

This is the minimum velocity necessary to put a body into orbit. A model of the first such body to go into orbit is shown in Figure 3-16. The period $T$ or time for one revolution is the circumference of the earth divided by $v_c$:

**Figure 3-16.** *Full-scale model of Sputnik I on display in Moscow. [Courtesy Sovfoto]*

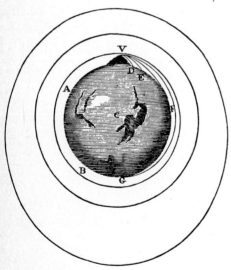

**Figure 3-17.** *Early proposal for an earth satellite.*

$$T = \frac{2\pi R_e}{v_c} = \frac{40{,}000 \text{ km}}{7.9 \text{ km/s}} = 5060 \text{ s} = 84 \text{ min}$$

Our result is in agreement with the well-known orbiting time of the many low-flying earth satellites starting with Sputnik I. Isaac Newton was the first person to make these calculations (about 300 years ago). Figure 3-17 shows a drawing of an earth satellite orbit made by Newton himself. He discussed the firing of a big cannon from a mountain top. If a muzzle velocity of 8 km/s could ever be achieved, he predicted the cannon ball would circle the earth as shown.

It is not necessary to achieve $v_c$ exactly in order to go into orbit. Suppose $v$ were 10% greater than $v_c$, as indicated in Figure 3-18. The acceleration while near the surface of the earth must still be $g$, so we have

$$g = \frac{v^2}{R} \quad \text{or} \quad R = \frac{v^2}{g}$$

where $R$ is the initial radius of curvature of the orbit. In this example $v = 1.1 v_c = 1.1 \sqrt{gR_e}$. Substitution of this value into the above equation gives

$$R = \frac{(1.1\sqrt{gR_e})^2}{g} = 1.21 R_e$$

We see that the initial radius of the orbit would be 21% greater than that of the low-flying satellite in circular orbit. Such a projectile will initially move away from the earth. After traveling for a while, its velocity will have a component pointing away from the center of the earth. The gravitational force opposes this component of motion, with the consequence that $v$ will be reduced so that the projectile will eventually "fall back" toward the earth. As discussed in Chapter 5, the exact path will be an ellipse with one focus at the center of the earth.

If a satellite in circular orbit is an appreciable distance $h$ above the earth's surface we must take into account the experimental fact that the gravitational acceleration decreases inversely as the square of the distance from the center of the earth. See Figure 3-19. The gravitational acceleration at a distance $R_e + h$ from the center of the earth is

$$g' = g \frac{R_e^2}{(R_e + h)^2}$$

(This fact is discussed in Chapter 5.) Equating $g'$ to $v^2/(R_e + h)$ gives

$$\frac{v^2}{R_e + h} = g \frac{R_e^2}{(R_e + h)^2}$$

$$v = \sqrt{gR_e} \sqrt{\frac{R_e}{R_e + h}} = v_c \sqrt{\frac{R_e}{R_e + h}} \quad (3\text{-}12)$$

We see that the velocity is less than the critical orbital velocity.

If a spacecraft is in a high circular orbit and wishes to drop down to a lower orbit, it should turn on its retrorockets pointed in the direction of motion (a force opposing the motion). During the time the retrorockets are firing, the spacecraft will be picking up speed while slowly "falling"

(a)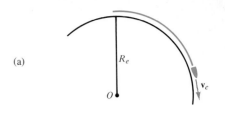

(b)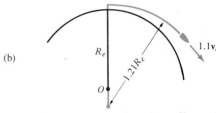

**Figure 3-18.** (a) Orbit of earth satellite with circular velocity $v_c$. (b) The satellite is launched with a 10% higher velocity.

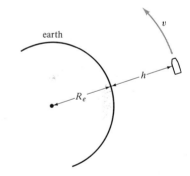

**Figure 3-19.** Satellite in circular orbit at height $h$ above surface of earth.

toward the earth. If a retrorocket was fired from the hood of a car, it would slow down the car; however, Eq. 3-12 shows us that the velocity must increase as $h$ decreases—contrary to common sense. Such maneuvers can be simulated using a computer with proper displays. A game of space war on the computer can be tricky for a novice. If he follows his natural impulse, the spacecraft will tend to do the opposite of what he wants.

## Summary

Vertical and horizontal motions can be treated separately. When there is a uniform vertical acceleration $a_y$, the path is a parabola with

$$x = (v_0 \cos \theta)t \quad \text{and} \quad y = (v_0 \sin \theta)t + \tfrac{1}{2}a_y t^2$$

The range of a projectile is

$$R = \frac{v_0^2}{g} \sin 2\theta$$

Displacement, velocity, and acceleration are vectors. In order to add or subtract vectors one may use the component method or the polygon method. Velocity vectors must be subtracted in order to obtain the vector acceleration. For uniform motion around a circle, this gives the centripetal acceleration, $a_c = v^2/R$. For a low-flying earth satellite $a_c = g$ and $v = \sqrt{gR_e}$.

## Exercises

1. What is the vector **C** in terms of the vectors **A** and **B**? What is the vector **Z** in terms of the vectors **X** and **Y**?
2. If $\mathbf{A} + \mathbf{B} + \mathbf{C} = 0$ and $\mathbf{A} = 2\mathbf{i} + 3\mathbf{j} + 4\mathbf{k}$ and $\mathbf{B} = 5\mathbf{i} + 6\mathbf{j} + 7\mathbf{k}$, then what is **C**? What is the magnitude of **C**? What is the angle between **C** and the $x$ axis?
3. If $|\mathbf{A}| = 3$ m and $|\mathbf{B}| = 2$ m and the angle between them is 30°, what is the projection of **B** onto **A**?
4. In Example 1, what is $\Delta v$ in terms of $v$ and $\theta$ (in radians) when $\theta$ approaches zero? Do not use any trigonometric functions.
5. In Example 3 what is $F_1$, the component of $F_g$ perpendicular to the surface?
6. Repeat Example 5 for 30° firing angle, keeping the range $R$ equal to 3000 km. Find $t$ and $v_0$. Plot $y$ versus $x$. Plot $y$ versus $t$.
7. At what angle must a cannon be fired to achieve half its maximum range?
8. A vector **E** lies along the $y$ axis.
   (a) What is the magnitude of the component of **E** along a $y'$ axis that is at an angle $\theta'$ to the $y$ axis? This component is a vector we shall call **E**'.

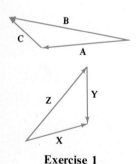

Exercise 1

(b) Now consider a $y''$ axis at an angle of $\theta''$ to the $y'$ axis. What is the magnitude of the component of $\mathbf{E}'$ along the $y''$ axis? We shall call this component $E''$.
(c) What is $E''$ in terms of $E$, $\theta'$, and $\theta''$?
(d) If $(\theta' + \theta'') = 90°$, will $E'' = 0$?
9. In Exercise 8 suppose $\theta' = \theta'' = 45°$. What is $E''$ in terms of $E$? Suppose $\theta' = 30°$ and $\theta'' = 60°$. What is $E'$ and $E''$ in terms of $E$?
10. In Exercise 8, suppose $\theta' = \theta'' = 60°$. What is $E''$? What is its $y$ component? Is it positive or negative?
11. Repeat Example 2 for the case where the ferryboat is traveling at 6 km/h with respect to the water.
12. A particle is moving with constant speed in a circle of radius $R$. Let $f$ be the number of revolutions per second. What is the acceleration of the particle in terms of $f$ and $R$?

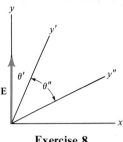

Exercise 8

## Problems

13. Suppose the camera in Figure 3-1(c) is moving with velocity

$$\mathbf{v}_c = -\mathbf{i}\frac{v_c}{\sqrt{2}} - \mathbf{j}\frac{v_c}{\sqrt{2}}$$

and the ball is fired vertically such that

$$y = v_b t - \tfrac{1}{2}gt^2$$

What is the equation for the ball's trajectory as viewed from the camera? Plot this for $v_c = 10$ m/s and $v_b = 20$ m/s.
14. Show that when $\theta = 45°$ in Figure 3-13, the maximum height of the projectile is $\tfrac{1}{4}R$.
15. Suppose the force on the sail in Figure 3-11 is equal to $F_0 \sin \alpha$, where $\alpha$ is the angle between sail and wind. If the angle between keel and wind is $\theta$, what value of $\alpha$ will give maximum speed to the boat?
16. Consider a projectile fired at 30° from the horizontal. The vertical component of the initial velocity is $v_y = 100$ m/s. Ignore air resistance.
    (a) What is the initial velocity?
    (b) Let $T$ be the total time of flight. What is $v_y$ at $t = \tfrac{1}{2}T$? What is the acceleration at this instant of time?
    (c) What is $v_y$ just before $t = T$?
    (d) What is $v_y$ at $t = \tfrac{1}{4}T$?
    (e) Plot $v_y$ versus $t$.
17. What is the maximum height obtained by the projectile in Figure 3-13? Give answer in terms of $v_0$, $\theta$, and $g$.
18. Suppose the bullet in Figure 3-14 hits the monkey just as he reaches the ground. What is $v_0$ in terms of $\theta$ and $x_m$?
19. A sphere is shot at an angle of 30° from the vertical into the air from point $A$. It is acted on only by gravity, not air resistance. After 20 s it lands at a point $B$,

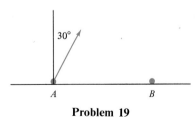

Problem 19

which is at the same vertical height as the point it started. How far above its initial level did the sphere rise?

20. An electron moves with respect to a certain frame of reference with initial position vector

$$\mathbf{r}_0 = x_0\mathbf{i} + z_0\mathbf{k} \quad \text{where } x_0 = 3.0 \text{ m and } z_0 = 1.0 \text{ m}$$

initial velocity

$$\mathbf{v}_0 = v_{0y}\mathbf{j} \quad \text{where } v_{0y} = 2.0 \text{ m/s}$$

and acceleration

$$\mathbf{a}(t) = At\mathbf{j} + B\mathbf{k} \quad \text{where } A = 12.0 \text{ m/s}^3 \text{ and } B = 8.0 \text{ m/s}^2$$

(a) What is the electron's $z$ coordinate at time $t = 0.5$ s?
(b) What is the electron's speed at $t = 1$ s?
(c) What is the angle between the position vector $\mathbf{r}$ and the velocity vector $\mathbf{v}$ at $t = 0$? Is this possible?

21. Consider an airplane whose cruising speed is 300 km/h with respect to the air. It has a scheduled round-trip flight between two points $A$ and $B$, separated by 600 km. Neglect the time to start, stop, and turn around.
(a) How much time will the round trip flight take on a calm day?
(b) How much time will it take on a day when a 60 km/h wind is blowing from $B$ to $A$?
(c) How much time will it take if it is a 60 km/h crosswind?

22. Assume a lunar module is coasting in circular orbit around the moon. If the radius of its orbit is one third the radius of the Earth and the gravitational acceleration of the lunar module is $g/12$ where $g = 9.8$ m/s$^2$, what is the speed compared to that of a low-flying earth satellite?

23. An apparatus is designed to study insects at an acceleration of $100g$. The apparatus consists of a 10-cm rod with insect containers at either end. The rod is rotated about its center.
(a) What will be the insect velocity when its acceleration is $100g$?
(b) What will be the number of revolutions per second?

24. Acceleration due to gravity at a distance $r$ from the center of the earth is $a = g(R_e/r)^2$, where $R_e$ is the radius of the earth. It is desired to have an earth satellite hover over a fixed point on the equator. If the time for one earth rotation is $t_0$, what must be $v$ in terms of $g$, $R_e$, and $t_0$?

25. An earth satellite is in circular orbit at a distance of 240,000 mi from the earth's center. What is its period in days?

26. How high must an earth satellite be in circular orbit in order to make one revolution per day?

27. Show that the period of a satellite at height $h$ above the earth is

$$t = t_c \left(\frac{R_e + h}{R_e}\right)^{3/2} \quad \text{where } t_c \text{ is the period of a low-flying earth satellite}$$

28. The gravitational acceleration on the surface of the moon is $0.14g$ and the moon's radius is $1.74 \times 10^3$ km. How long does it take a lunar module to orbit the moon close to the moon's surface?

# Dynamics

## 4-1 Introduction

One of the main goals of physics is to predict the future (or past) positions and velocities of individual particles interacting with each other. We learned in Chapter 3 that if we know the acceleration of each particle as a function of time, we could, in principle, predict the future position of each particle. As we shall see, in order to know the acceleration, one must know the force on the particle and its mass. Hence this goal of physics reduces in part to the study of forces and what the sources of these forces are.

Fortunately, it turns out that of all the forces observed in nature, there are to our present knowledge only four basic kinds: (1) gravitational, (2) weak, (3) electromagnetic, and (4) nuclear. As we shall see in the next chapter, gravitational force acts on all masses and is produced by mass at a distance. (Mass will be formally defined later in this chapter.) Electromagnetic force acts on charges and currents and is produced by charges and currents. Since atoms contain charged electrons and protons, the forces between atoms are electromagnetic in their origin. Since ordinary matter is made of atoms, most everyday forces, such as the push and pull of a spring or other contact forces, are basically electromagnetic. The electromagnetic force is studied in detail in Chapters 15–21. The nuclear and the weak forces are of short range (they cannot be felt at distances greater than $10^{-14}$ m). It is the nuclear force that holds a nucleus together in spite of the strong electrostatic repulsion between the protons. Nuclear and weak forces are studied in Chapters 29–31.

In order to study the motion of a body produced by a force acting on it, it is not necessary to know what kind of a force it is or where it came from. In this chapter we shall study the effects of forces in general, and in later chapters we shall study the special features of the gravitational, electromagnetic, weak, and nuclear forces. The study of motion produced by forces in general is called dynamics. In dynamics, as opposed to kinematics, we are concerned with real material bodies that have mass, momentum, and energy as well as velocity and acceleration. In Section 4-2 the definitions of mass, force, and momentum are rather briefly stated. More detailed physical interpretations follow in Sections 4-3 and 4-4.

There are several different ways of defining quantities such as mass and force that are mathematically equivalent. We shall use the definitions given in the next section.

## 4-2 Definitions

### Mass

Our definition of mass will be an operational one. First we start with a standard mass of 1 kg. A standard mass of 1 kg (actually 0.99997 kg) can be obtained by taking 1000 cm³ of water at 4°C and atmospheric pressure. This amount of water can be frozen into a block of ice. We can compare an unknown mass $m$ to this standard mass $m_0$ by placing a small compressed spring between them (see Figure 4-1). If the spring is released when the masses are initially at rest they will fly apart in opposite directions with speeds $v$ and $v_0$, respectively. We define the unknown mass $m$ as

$$m \equiv m_0 \frac{v_0}{v} \quad \text{(definition of mass)} \quad (4\text{-}1)$$

### Momentum

We define the momentum of a body as the product of its mass times its vector velocity. We shall use the symbol **P** for momentum:

$$\mathbf{P} \equiv m\mathbf{v} \quad \text{(definition of momentum)} \quad (4\text{-}2)$$

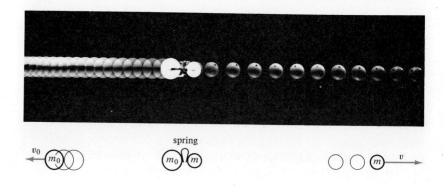

**Figure 4-1.** *Strobe picture of two unequal masses being pushed apart by a spring.* [*Courtesy Educational Development Center*]

## Force

If a single force $\mathbf{F}_1$ is applied to a body of mass $m$, the value of the force is defined as the time rate of change of momentum of the body:

$$\mathbf{F}_1 \equiv \frac{d\mathbf{P}}{dt} \quad \text{(definition of force)} \quad (4\text{-}3a)$$

For a body of constant mass $m$, this is

$$\mathbf{F}_1 = \frac{d(m\mathbf{v})}{dt} = m\frac{d\mathbf{v}}{dt}$$

$$\mathbf{F}_1 = m\mathbf{a} \quad (4\text{-}3b)$$

Equation 4-3b can be used to calibrate a spring scale as shown in Figure 4-2.* The more the spring is stretched, the greater the force and the greater will be the acceleration of the frictionless cart. The spring scale could be calibrated by using a cart of unit mass. The spring scale is pulled until the cart has one unit of acceleration; then a mark of one unit of force would be made at the position of the pointer. The process is repeated for two units of acceleration, which shows where to put the mark for two units of force, and so on.

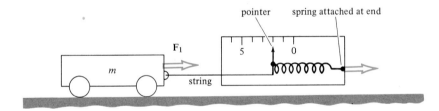

**Figure 4-2.** *Mass $m$ is being pulled by force $\mathbf{F}_1$, which is supplied by spring scale being pulled to the right.*

## 4-3 Newton's Laws of Motion

In order to predict the motion of a body due to forces acting on it, we must have a basic "law" (theory) with which we can make such a prediction. The theory may, or may not, be correct. Only experimental measurements can tell. The basic theory that permits us to predict the motions of bodies consists of three equations, which are called Newton's laws of motion. First we shall briefly state Newton's three laws of motion. This listing will

*We are assuming that the force of the spring is transmitted undiminished via a string to the cart. This assumption will be justified in detail on page 69.

be followed by a discussion of their deeper meaning and significance. These laws were first proposed by Isaac Newton at the end of the seventeenth century.

**Newton's first law**—A body remains in a state of rest or constant velocity (zero acceleration) when left to itself (the net force acting on it is zero). Mathematically this says

$$\mathbf{a} = 0 \quad \text{when} \quad \mathbf{F}_{net} = 0 \quad \text{(Newton's first law)} \tag{4-4}$$

where $\mathbf{F}_{net}$ is the vector sum of all the forces acting on the body.

**Newton's second law**—The time rate of change of momentum of a body is equal to the net force acting on the body. For a body of constant mass this is mass times acceleration.

$$\mathbf{F}_{net} = \frac{d\mathbf{p}}{dt} \quad \text{or} \quad \mathbf{F}_{net} = m\mathbf{a} \quad \text{(Newton's second law)} \tag{4-5}$$

**Newton's third law**—Whenever two bodies interact, the force on the first body due to the second is equal and opposite to the force on the second due to the first.

$$\mathbf{F}_{A \text{ due to } B} = -\mathbf{F}_{B \text{ due to } A} \quad \text{(Newton's third law)}$$

### Discussion of Newton's First Law

The first law states that if $\mathbf{F}_{net}$ is zero, the acceleration $\mathbf{a} = 0$. This seems to be merely a special case of the second law. Even so, it is worth emphasizing because, before the time of Newton, the accepted scientific dogma was based on the teachings of Aristotle. A basic principle in Aristotle's scheme of things was that all bodies must come to rest in the absence of external forces. This seems to check with common everyday experience. We observe that moving bodies, when not being pushed or pulled, usually come to rest rather than continuing on with constant velocity. An automobile will come to rest when the engine is turned off. But according to Newton's first law, the net force on the automobile cannot be zero if it is slowing down. In such a case there are the retarding forces of the air resistance and push of the road against the tires (see Example 1).

Implicit in the first law is an important physical principle: the existence of what is called an inertial frame of reference. Certainly the first law would appear to be violated by an observer who is accelerating. The meaning of the first law is that if a body is free of external forces, there exists a frame of reference where it will be at rest. And, if the body is at rest in one frame of reference, there is a set of frames of reference where the body will have constant velocity. These frames of reference are called inertial systems or inertial frames of reference. A nontrivial prediction of Newton's first law is that, if an observer is in an inertial system as defined by a first body being at rest, then all other bodies that have zero net force on them will also be at rest or have constant velocity.

## Discussion of Newton's Second Law

It is clear that Newton's second law will hold only if the observer is in an inertial system. This is because the right-hand side of the equation $\mathbf{F} = m\mathbf{a}$ would change depending on the acceleration of the observer. We repeat that the form $\mathbf{F}_{net} = m\mathbf{a}$ holds only in the case where $m$ is a constant.

In Newton's time the experiments indicated that $m$ was independent of velocity. However, more recent experiments show that $m$ as defined by Eq. 4-1 does depend on velocity. The experimental result is

$$m(v) = \frac{m_{rest}}{\sqrt{1 - v^2/c^2}}$$

where $m_{rest}$ is the value of mass when at rest and $c = 2.998 \times 10^8$ m/s is the speed of light. (See Chapter 9 on relativity.) Note that for small $v$, $m \approx m_{rest}$; then $m$ can be treated as a constant. For velocities less than 1% the speed of light, we shall treat $m$ as a constant and safely use the equation $\mathbf{F}_{net} = m\mathbf{a}$. (For $v/c = 0.01$, $m = 1.00005 m_{rest}$.) In this book we shall call $m(v)$ relativistic mass. Whenever the term "mass" is used, we shall mean rest mass.

An important point to stress is that the force in Newton's second law is the *net* force. One must take the vector sum of all the forces acting on a given body before applying Newton's second law.

At this point the reader may feel we are using circular reasoning. If force is defined as $\mathbf{F} = m\mathbf{a}$ by Eq. 4-3, then shouldn't the $\mathbf{F} = m\mathbf{a}$ of Newton's second law be true by definition rather than because it is a basic law of nature? First we note that Eqs. 4-3 and 4-5 are not identical; Eq. 4-3 has $\mathbf{F}_1$ (a single force) on the left hand side and Eq. 4-5 has $\mathbf{F}_{net}$. This distinction is very important. It implies that Eq. 4-5 has additional physical content that must be checked by experiment. Equation 4-5 implies the additivity of mass and the vector addition of forces. By additivity of mass we mean that, if two masses $m_A$ and $m_B$ are joined together, the combined object will have mass $m = (m_A + m_B)$ as measured according to Eq. 4-1. This may seem absurdly obvious; however, all speculations about nature must be checked by experiment. There are

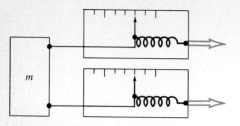

**Figure 4-3.** *Two identical springs. If each spring alone gives an acceleration $a_0$, will the two together give $2a_0$?*

common physical quantities that are not additive, such as the magnitudes of vectors or the addition of volumes. If 1 liter of alcohol is added to 1 liter of water, the combined volume is noticeably less than 2 liters.

One can test for additivity of forces by measuring how much a given spring must be stretched to give a 1-kg mass an acceleration of 1 m/s². This is the mks unit of force and is called the newton (N). Two separate springs could be calibrated, each to give 1 N of force. Then the two springs could both be attached to the same 1-kg mass, and a total force of 2 N would be applied, as shown in Figure 4-3. Again it would seem obvious that the 1-kg mass would have an acceleration of 2 m/s²; however, such a speculation about nature must be carefully tested by experiment. Experiments do show that single forces as defined by Eq. 4-3 add vectorially. We see that the equation $\mathbf{F}_{net} = m\mathbf{a}$ is much more than just a definition and that it implies scalar additivity of mass and vector additivity of force, and this additional content must be tested by experiment.*

### Discussion of Newton's Third Law

Suppose we have a system that consists of $m_A$ and $m_B$ only. Then, as shown in Figure 4-4, the only forces will be $\mathbf{F}_A$ (the force on A due to B) and $\mathbf{F}_B$ (the force on B due to A). These two forces are called interaction forces. To give just three examples, they may be gravitational, electric, or via a contact force (if A and B are in contact). Newton's third law states that in the case of interaction forces between two bodies

$$\mathbf{F}_A = -\mathbf{F}_B$$

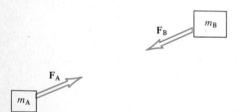

**Figure 4-4.** *Two interacting bodies showing the interaction forces.* $\mathbf{F}_A = -\mathbf{F}_B$.

Note that the two forces in Newton's third law cannot both act on the same body. $\mathbf{F}_B$ is called the reaction force to $\mathbf{F}_A$, and $\mathbf{F}_A$ is called the reaction force to $\mathbf{F}_B$.

As an example consider the three-car toy train pulled by an external force $\mathbf{F}$ in Figure 4-5. Here the interaction forces are transmitted by massless strings. For example, $\mathbf{F}_1(2)$ is the force on $m_1$ due to $m_2$ and $\mathbf{F}_2(1)$ is the force on $m_2$ due to $m_1$. According to Newton's third law, the sum

**Figure 4-5.** *Frictionless train pulled by external force $\mathbf{F}$. All the interaction forces are shown.*

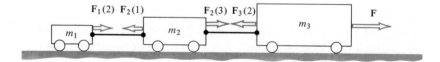

---

*In fact the relation $\mathbf{F}_{net} = m\mathbf{a}$ fails the test of experimental measurement when $m$ is moving near the speed of light; however, the relation $\mathbf{F}_{net} = d\mathbf{P}/dt$ agrees with experiment under all conditions.

$F_2(1) + F_1(2)$ is zero. The acceleration of the train can be found by applying Newton's second law to each of the cars and then adding:

$$F_1(2) = m_1 a$$
$$F_2(1) + F_2(3) = m_2 a$$
$$\underline{F_3(2) + F = m_3 a}$$
$$[F_1(2) + F_2(1)] + [F_2(3) + F_3(2)] + F = (m_1 + m_2 + m_3)a$$
$$F = (m_1 + m_2 + m_3)a$$
$$a = \frac{F}{m_1 + m_2 + m_3}$$

The sums in the square brackets are zero because of Newton's third law.

The discussion of Newton's third law will continue in the Section 4-5, Contact Forces and Friction.

## 4-4 Units of Force and Mass

Historically, mass in the metric system was defined in such a way as to give water a density of 1 g/cm³ when at its maximum density; 1 cm³ of water at 4°C was within the accuracy of measurement equal to 1 g (gram).

The mks unit of force is that force which gives 1 kg an acceleration of 1 m/s. Hence the mks unit of force is 1 kg m/s². This combination is given a special name: the newton (abbreviation N). The cgs unit of force is 1 g cm/s² and is given the special name of dyne.

$$1 \text{ N} = 1 \text{ kg m/s}^2 = 1 \,(10^3 \text{ g})(10^2 \text{ cm})/\text{s}^2 = 10^5 \text{ g cm/s}^2 = 10^5 \text{ dyne}$$

In the British system the same word, pound, is used for force as well as for mass. We recommend the abbreviation lb for pound of mass and lbf for pound of force; 1 lbf is equal to the gravitational force on 1 lb of mass where $g = 9.8$ m/s², and 1 lb of mass is 0.454 kg. Hence, 1 lbf is 0.454 kg times 9.8 m/s or 1 lbf = 4.45 N. Because of the confusion introduced by using the same word for two quite different things and because nearly all nations and all scientists use the metric system, we shall try to avoid using the British units for mass or force.

> **Example 1.** A 1500-kg car is traveling along a level road at 75 mph (33.5 m/s). When the gas pedal is released, the car slows down to 65 mph in 5.0 s. What is the net resistive force (mainly air resistance at this speed). By this simple procedure one could accurately measure the retarding force acting on a car. Students are advised not to perform such an experiment. (The last time the author made this measurement he received a speeding ticket.)

ANSWER: The average acceleration is

$$a = \frac{\Delta v}{\Delta t} = -\frac{10 \text{ mph}}{5 \text{ s}} = -\frac{4.48 \text{ m/s}}{5 \text{ s}} = -0.894 \text{ m/s}^2$$

The average force is $F = ma = (1.5 \times 10^3 \text{ kg})(-0.894 \text{ m/s}^2) = -1.34 \times 10^3 \text{ N}$. This is about 9% of the weight of the car.

**Table 4-1.** *Systems of units for force (we shall use only the top half of this table).*

| | System | Mass | Acceleration | Force |
|---|---|---|---|---|
| | mks | kg | m/s² | 1 N = (1 kg) × (1 m/s²) |
| | cgs | g ($10^{-3}$ kg) | cm/s² | 1 dyne = (1 g) × (1 cm/s²) |
| obsolete { | British | lb (0.454 kg) | ft/s² | 1 poundal = (1 lb) × (1 ft/s²) |
| | British engineering | slug (32 lb) | ft/s² | 1 lbf = (1 slug) × (1 ft/s²) |

## 4-5 Contact Forces and Friction

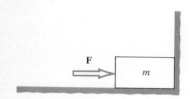

**Figure 4-6.** *Block being pushed against immovable wall.*

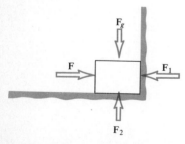

**Figure 4-7.** *All four forces acting on the block shown in Figure 4-6. Forces on wall and floor due to block are not shown.*

When two bodies are pushed into contact, such as a block pushed against a wall or table, there are contact forces present. Not only is there a force on the table due to the block but, according to Newton's third law, there is a force on the block due to the table. The ultimate source of these two forces is the repulsive force between atoms. When the electron clouds of two atoms begin to overlap, there is a repulsive force between them, and as the two atoms are pushed closer together the repulsive force increases. This repulsive force between atoms is electromagnetic in origin and can be very strong compared to gravitational forces. If we push a block against a table, the surface atoms of the block are pushed closer to those of the table until there is a net repulsive force equal and opposite to the applied force. We call such repulsive forces between surfaces, contact forces.

In Figure 4-6 a block of mass $m$ is pushed against a wall with a force **F**. If we blindly applied the equation $\mathbf{F} = m\mathbf{a}$ to the situation in Figure 4-6, we would get the result $\mathbf{a} = \mathbf{F}/m$, which is nonzero. However, the block clearly does not accelerate when **F** is applied. A complete analysis reveals that the atoms in the wall push on the block with a total contact force $\mathbf{F}_1$ that is equal to $(-\mathbf{F})$. The net force is

$$\mathbf{F}_{\text{net}} = \mathbf{F} + \mathbf{F}_1 = \mathbf{F} + (-\mathbf{F}) = 0$$

If gravity is pulling down on the block with a force $\mathbf{F}_g$, there will be a second contact force $\mathbf{F}_2$ pushing up with strength $-\mathbf{F}_g$. Then the net force is the sum of all four forces (see Figure 4-7).

$$\mathbf{F}_{net} = \mathbf{F} + \mathbf{F}_1 + \mathbf{F}_g + \mathbf{F}_2 = \mathbf{F} + (-\mathbf{F}) + \mathbf{F}_g + (-\mathbf{F}_g) = 0$$

It is important in all applications of Newton's second law to calculate the *net* force.

As we proceed further, we shall begin to appreciate the great simplicity and beauty of Newton's laws. However, the correct application of Newton's laws can at times be tricky. Let the following "paradox" serve as a warning.

Consider two blocks, $m_A$ and $m_B$, on a frictionless surface as shown in Figure 4-8. A force $\mathbf{F}$ is applied to block A and transmitted through it to block B. By Newton's third law, block B must exert an equal and opposite force $(-\mathbf{F})$ on block A. Then the net force on A would be the sum of the applied force $\mathbf{F}$ plus the contact force $-\mathbf{F}$ of block B pushing back on A. In this case $\mathbf{F}_{net} = \mathbf{F} + (-\mathbf{F}) = 0$. According to Newton's second law

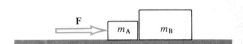

**Figure 4-8.** *Two blocks being pushed along a frictionless surface.*

$$\mathbf{a} = \frac{\mathbf{F}_{net}}{m_A} = 0$$

We are forced to conclude that block A can never move, no matter how large a force $\mathbf{F}$ is applied to it! See if you can find the mistake before reading the next paragraph.

The mistake in the above reasoning is the assumption that the force $\mathbf{F}$ is transmitted through block A and is thus also applied to block B. There is nothing in Newton's laws saying this should be so. Instead, one should assume an arbitrary value $\mathbf{F}'$ for the contact force of A on B. The general procedure for solving most problems in dynamics is to apply Newton's second law to each mass separately. In addition to the force $\mathbf{F}$ acting on $m_A$, there will be the contact force of $m_B$ pushing back, which, according to Newton's third law, is $(-\mathbf{F}')$. The net force on A is then $(\mathbf{F} - \mathbf{F}')$ and Newtons second law gives

$$(\mathbf{F} - \mathbf{F}') = m_A \mathbf{a}$$

For $m_B$, Newton's second law gives

$$\mathbf{F}' = m_B \mathbf{a}$$

The acceleration **a** can be found by adding together the preceding two equations to obtain

$$\mathbf{F} = (m_A + m_B)\mathbf{a} \quad \text{or} \quad \mathbf{a} = \frac{\mathbf{F}}{m_A + m_B}$$

We note that this result could also be obtained by treating the two blocks as one single block of mass $(m_A + m_B)$.

**Figure 4-9.** (*a*) *A sideways force* **F** *applied to block* A. (*b*) *The gravitational force* **F**$_g$ *acting on* A *as well as both components of the contact force (force due to* B *acting on* A). *The vector sum of all forces acting on* A *is zero.* (*c*) *The contact forces due to block* A *that are acting on block* B. *Block* B *is rigidly attached to the ground.*

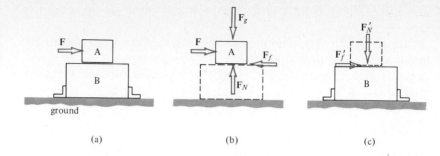

### Friction

So far the contact forces we have discussed have been perpendicular to the surface between the two bodies. We call this a normal force. In addition, there can be a component of contact force along the surface. Such a force parallel to the surface is called a frictional force. For example, consider a block A sitting on top of a block B (Figure 4-9). A small sideways force **F** can be applied to A and it will not move. This means that **F** is opposed by the frictional force **F**$_f$ shown in Figure 4-9(b); that is, **F**$_f$ = −**F**. As **F** is increased there will come a point where block A starts to move. The smoother the surface, the sooner this point will occur. Let us call this critical value of frictional force $(F_f)_s$. (The subscript $s$ stands for "static.") The ratio of $(F_f)_s$ to the normal force $F_N$ shown in Figure 4-9(b) is defined as the coefficient of static friction, $\mu_s$.

$$\mu_s \equiv \frac{(F_f)_s}{F_N} \quad \text{(coefficient of static friction)}$$

It is an experimental fact that for most dry surfaces, $\mu_s$ is approximately independent of $F_N$ and the area of contact.

If $F$ is greater than $(F_f)_s$, the block in Figure 4-9 will move; however, there still will be an opposing frictional force $(F_f)_k$. (The subscript $k$ stands for kinetic.) There is a corresponding coefficient of kinetic friction

$$\mu_k \equiv \frac{(F_f)_k}{F_N} \quad \text{(coefficient of kinetic friction)}$$

For most materials $\mu_k$ is a bit less than $\mu_s$. For dry surfaces it tends to be independent of $F_N$, the contact area, and the velocity.

For smooth wood against smooth wood, $\mu_s \approx \mu_k \approx 0.3$. For rubber tires against concrete, the coefficient of friction can be as large as unity. In many problems involving friction, the coefficient of friction is given. Then the limiting frictional force is obtained by multiplying $\mu$ by $F_N$. We see in the example of Figure 4-9 that $F_N$ is a consequence of the gravitational force $F_g$.

Friction is basically a complicated phenomenon and will not be stressed in this book. A proper explanation would require detailed knowledge of the interactions of the surface atoms, which would involve details of solid state physics and chemistry.

### Weight

In order to calculate the normal force in a typical problem, we must know how to calculate the gravitational force $F_g$. The gravitational force acting on a body is defined as its weight. (This topic is discussed in more detail in Section 5-4.) Since all free bodies near the surface of the earth are observed to have $a_g = g$, Newton's second law gives $\mathbf{F}_g = m\mathbf{a}_g$ or

$$F_g = mg \quad \text{(weight)}$$

**Near the earth's surface, weight = mg.**

## 4-6 Problem Solving

**Example 2.** A car has 25% of its weight supported by each tire. Assume $\mu_s = 0.8$ is the coefficient of static friction between tire and road. (a) What is the minimum stopping time from 60 mph? (Brakes act on all 4 wheels.) (b) What is the fastest start-up from rest to reach 60 mph? (This is a standard measure of automobile performance.)

ANSWER: If the brakes are applied so that the tires don't slip, the net retarding force is $0.8mg$, where $mg$ is the magnitude of the normal force. (If the tires slip, $\mu_k$ should be used rather than $\mu_s$.) If we equate the $F_{net}$ of $0.8mg$ to $ma$, we obtain $a = 0.8g$. The time to stop is

$$t = \frac{v}{a} = \frac{88 \text{ ft/s}}{0.8(32 \text{ ft/s}^2)} = 3.44 \text{ s}$$

Assuming a two-wheel-drive car, where the engine is connected to the rear wheels only, the maximum driving force before tire slippage is one half the above frictional force, which results in an acceleration of half or $0.4g$. The time to reach a velocity of 60 mph is then twice the calculated stopping time or 6.88 s. By using large, soft tires, and putting more than half the weight on the rear wheels, it is possible to do somewhat better than this. Requirements on engine horsepower are given in Chapter 7.

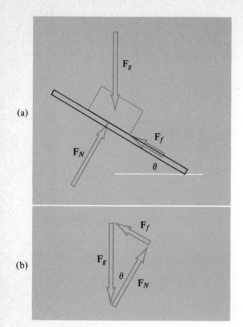

**Figure 4-10.** (a) Block at rest on an incline. (b) The 3 forces acting on the block are added to give $F_{net} = 0$.

**Example 3.** A wooden block is placed on a wooden incline. The angle of inclination is increased until $\theta = 20°$ at which time the block starts to slide and then slowly accelerates. What is $\mu_s$?

ANSWER: In Figure 4-10(a), there are three forces acting on the block such that $\mathbf{F}_g + \mathbf{F}_N + \mathbf{F}_f = 0$. The vector addition is shown in Figure 4-10(b). We see that

$$\tan \theta = \frac{F_f}{F_N}$$

At the maximum angle

$$\tan \theta_{max} = \frac{(F_f)_s}{F_N} = \mu_s$$

$$\tan 20° = \mu_s$$

$$\mu_s = 0.36$$

As soon as the block starts to slide, the frictional force decreases because $(F_f)_k < (F_f)_s$. The net force will be $F_{net} = (F_f)_s - (F_f)_k$.

**Example 4.** If the coefficient of friction between the wheels of a four-wheel-drive car and a steep road is 0.8, what is the steepest incline the car can climb without slipping?

ANSWER: We can use the relation $\tan \theta_{max} = \mu_s$, which was obtained in the previous example:

$$\tan \theta_{max} = 0.8 \quad \text{or} \quad \theta_{max} = 38.6°$$

We see that Jeep-riding on slopes steeper than this would result in a runaway vehicle.

### *Free-Body Diagrams*

In Example 3 and in the problem of the sliding blocks, all the forces acting on the body in question were identified and then added vectorially to form $\mathbf{F}_{net}$. Then this was equated to the mass of the body times its acceleration as if it were a free body acting under the influence of the force

$F_{net}$ alone. A drawing of the body and all the forces acting on it is called a free-body diagram. A useful procedure in solving problems involving forces is as follows:

1. Isolate the body in question.
2. Identify all the forces acting on it including contact and frictional forces.
3. Add these forces vectorially. It is helpful to draw a force diagram to see how the vectors add.
4. Now apply $F_{net} = ma$ to the body in question.
5. If there are still unknown quantities remaining, repeat the process for other bodies in the system.

We shall next apply this free-body diagram approach to four examples: an upside-down roller coaster, acceleration on an inclined plane, Atwood's machine, and the conical pendulum.

## The Roller Coaster

A person of mass $m$ is in a roller coaster doing a loop-the-loop of radius $R$ as shown in Figure 4-11. If the velocity is $v$ when upside down, what is the acceleration, what is the force of the person's body against the seat, and what is the net force acting on the person's body?

ANSWER: The acceleration is by definition $\mathbf{a} = d\mathbf{v}/dt$ no matter how strongly gravity is pulling down. So $a = v^2/R$ as shown on page 48. According to Newton's second law the net force is

$$F_{net} = ma = m\frac{v^2}{R}$$

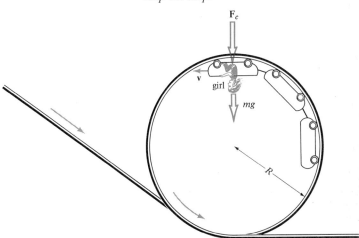

**Figure 4-11.** *Girl and roller coaster are momentarily upside down while doing a loop-the-loop.*

pointing down. This force is supplied by gravity pointing down and by $F_c$, a contact force of the seat on the person, also pointing down. Since $F_{net} = mg + F_c$,

$$m\frac{v^2}{R} = mg + F_c$$

$$F_c = m\left(\frac{v^2}{R} - g\right)$$

According to Newton's third law, this is also the force of the person's body pushing against the seat. (This is by definition the person's apparent weight. When $v^2/R = g$, the person's body is "weightless.")

### The Inclined Plane

We wish to calculate the acceleration of a mass $m$ sliding down a surface inclined at an angle $\theta$ to the horizontal as shown in Figure 4-12. We see in the figure three forces acting on $m$. There is the downward force of gravity, $m\mathbf{g}$. There is a contact force $\mathbf{F}_N$ pushing the mass away from the surface, and there could be a frictional force $\mathbf{F}_f$ opposing the motion down the incline. These three forces are added vectorially in Figure 4-12(b) in order to obtain $\mathbf{F}_{net}$. In this triangle $\mathbf{F}_N$ and $m\mathbf{g}$ are at an angle $\theta$ to each other since they are mutually perpendicular to the sides of the angle $\theta$ in Figure 4-12(a). Hence the far side of the force triangle, which is of magnitude $(F_{net} + F_f)$, is $mg \sin \theta$.

$$F_{net} + F_f = mg \sin \theta$$

Replacing $F_{net}$ with $ma$ gives

$$ma = mg \sin \theta - F_f \qquad (4\text{-}6)$$

In the absence of friction we have

$$a = g \sin \theta \qquad \text{(no friction)} \qquad (4\text{-}7)$$

In the case of friction we replace $F_f$ in Eq. 4-6 with $\mu_k F_N$:

$$ma = mg \sin \theta - \mu_k F_N$$

From the force triangle in Figure 4-12 we see that $F_N = mg \cos \theta$. Making this substitution gives

$$a = g \sin \theta - \mu_k g \cos \theta \qquad (4\text{-}8)$$

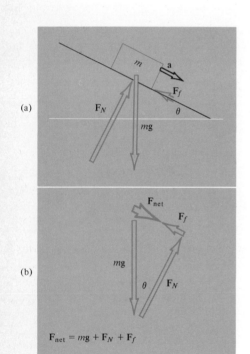

**Figure 4-12.** (a) *Mass m on an inclined plane.* (b) *The three forces acting on m are added to yield* $\mathbf{F}_{net}$.

We see in Eqs. 4-7 and 4-8 that an inclined plane can be used to reduce the acceleration of a body under the influence of gravity.

Suppose the block is sliding, but not accelerating. Then $a = 0$ in Eq. 4-8, and

$$g \sin \theta = \mu_k g \cos \theta$$

or

$$\tan \theta = \mu_k$$

At this particular angle a body would keep sliding at the same speed. We note that this equation is of the same form as the result $\tan \theta = \mu_s$ from Example 3, except that now we have $\mu_k$ rather than $\mu_s$. The distinction is simple. If the block is sliding at constant speed, one must use $\mu_k$; but, if the block is at rest at the maximum angle, one uses $\mu_s$.

We see in these examples of the inclined plane that the contact force $\mathbf{F}_N$ adjusts itself so that $\mathbf{F}_{net}$ is pointing in the direction of the incline.

## 4-7 Atwood's Machine

In mechanics we can encounter a variety of pulley problems where different masses are connected by belts or strings that run over freely rotating pulleys. Usually such belts, strings, and pulleys are considered massless. Then, even though the string is accelerating, the force pulling on one end of the string must equal the force pulling on the other end. For example, in Figure 4-13 the net force has the value $(F_2 - F_1)$ and the string is accelerating to the right. If the string is of mass $m$, we have

$$(F_2 - F_1) = ma$$

But if $m = 0$, we have

$$(F_2 - F_1) = 0 \quad \text{or} \quad F_2 = F_1$$

Figure 4-13. *Forces on a piece of string.*

In Figure 4-14 the force on either mass due to the string is defined as the tension $T$. By Newton's third law this force has the same magnitude as the force of the mass pulling on the string; so by applying the result $F_1 = F_2$ we also have the result $T_1 = T_2$. We see that the tension is the same at both ends of a massless string, so we shall call it $T$. In this pulley problem (called Atwood's machine) we wish to solve for both the acceleration $a$ and the tension $T$. To solve for these two unknowns we need two simultaneous equations. We can get two simultaneous equations by applying Newton's second law to each of the two masses separately; that is, we have two separate free-body diagrams. For $m_1$:

$$F_{1\,net} = T - m_1 g \quad \text{or} \quad m_1 a = T - m_1 g \quad (4\text{-}9)$$

For $m_2$:

$$F_{2\,net} = m_2 g - T \quad \text{or} \quad m_2 a = m_2 g - T$$

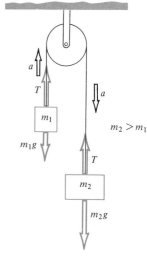

Figure 4-14. *Atwood's machine. $m_1$ moves up and $m_2$ down with acceleration $a$.*

We have assumed a direction for the acceleration $a$ and we take forces in the assumed direction of $a$ as positive. If our guess as to the direction of $a$ were incorrect, $a$ would turn out to be a negative quantity. Now add the two preceding equations:

$$m_1 a + m_2 a = m_2 g - m_1 g \qquad (4\text{-}10)$$

$$a = \frac{m_2 - m_1}{m_2 + m_1} g$$

We see that the acceleration is small when $m_1 \approx m_2$.

The tension is obtained by substituting this expression for $a$ into Eq. 4-9:

$$m_1 \left( \frac{m_2 - m_1}{m_2 + m_1} g \right) = T - m_1 g$$

$$T = \frac{2 m_1 m_2}{m_1 + m_2} g$$

## 4-8 The Conical Pendulum

Figure 4-15 shows a conical pendulum consisting of a mass $m$ in uniform circular motion hanging from a string of length $L$. Hence the acceleration is a centripetal acceleration and $\mathbf{F}_{\text{net}}$ must point toward the center of the circle. Let $v$ be the velocity and $R$ the radius of the circular path. The only forces acting on $m$ are $m\mathbf{g}$ pointing down and the string tension $\mathbf{F}_T$

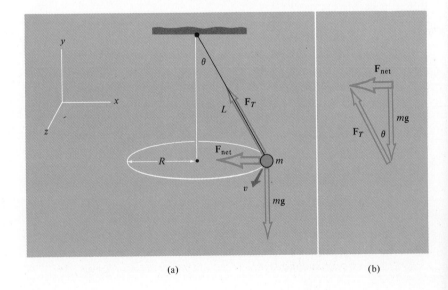

**Figure 4-15.** (a) A conical pendulum consisting of a mass m in uniform circular motion suspended by a string. (b) The vector sum of the forces acting on m.

pointing at an angle $\theta$ from the vertical. These two forces are added vectorially in Figure 4-15(b) to form $\mathbf{F}_{net}$. We see from this force diagram that

$$F_{net} = mg \tan \theta$$

Replacing $F_{net}$ with $ma$ gives

$$ma = mg \tan \theta$$

Now use Eq. 3-9 to replace the centripetal acceleration with $4\pi^2 R/T^2$:

$$m\left(\frac{4\pi^2}{T^2} R\right) = mg \tan \theta$$

Now solve for the period $T$:

$$T = 2\pi \sqrt{\frac{R}{g \tan \theta}}$$

Replacing $R$ with $(L \sin \theta)$ gives

$$T = 2\pi \sqrt{\frac{L}{g} \cos \theta}$$

Note that the period is independent of the mass $m$.
For small $\theta$, $\cos \theta \approx 1$ and

$$T = 2\pi \sqrt{\frac{L}{g}} \quad \text{(for small oscillations)} \quad (4\text{-}11)$$

In this case the period is not only independent of $m$ but also of $\theta$. If we only look at the components of $F_{net}$, $v$, and the position of $m$ in the $xy$ plane, we have the physical equivalent of a simple pendulum swinging back and forth from $x = -R$ to $x = +R$ with the period $T = 2\pi \sqrt{L/g}$. Thus it must be the case that the period of a simple pendulum is also given by Eq. 4-11 when $\theta$ is small.

## 4-9 Conservation of Momentum

In this section we shall derive the law of conservation of momentum from Newton's second and third laws of motion. Later we shall also derive the law of conservation of energy from Newton's laws of motion. Interestingly, it can be done the other way around: Newton's laws of motion can be derived from the laws of conservation of momentum and energy. It is just a matter of choice which to postulate and which to derive. Our presentation here is more traditional and historical.

Actually with advanced mathematics beyond the scope of this book it is possible to derive Newton's laws and the conservation of momentum and energy from the laws of homogeneity of space and time. By the **law of homogeneity of space** we mean that the laws of physics are the same in all positions in space. The **law of homogeneity of time** means that the laws of physics do not change with time. (A corollary is that no physical constants change their value with time.) No matter how plausible such symmetry principles may sound, they must be tested by experimental measurement.

We recall that momentum was defined in Eq. 4-2 as $\mathbf{P} = m\mathbf{v}$. The law of conservation of momentum states that the total momentum of a closed system is constant for all time. By total momentum we mean the vector sum of the momenta of all the particles in the system. By **closed system** we mean there are no external forces acting. The sources of all the forces are within the system itself. If the closed system consists of two interacting particles, $m_A$ and $m_B$, Newton's third law says that $\mathbf{F}_A = -\mathbf{F}_B$ where these forces are as shown in Figure 4-4. Now we use Newton's second law to replace each force with $d\mathbf{P}/dt$:

$$\frac{d\mathbf{P}_A}{dt} = -\frac{d\mathbf{P}_B}{dt}$$

$$\frac{d\mathbf{P}_A}{dt} + \frac{d\mathbf{P}_B}{dt} = 0$$

$$\frac{d}{dt}(\mathbf{P}_A + \mathbf{P}_B) = 0$$

$$(\mathbf{P}_A + \mathbf{P}_B) = \text{a constant} \quad \text{or} \quad \mathbf{P}_{\text{total}} = \text{a constant}$$

We see that the total momentum of the system does not change with time. The above derivation can easily be generalized to cover a system of $n$ particles. As long as there are no external forces we get

$$\sum_j \mathbf{P}_j = \text{constant}$$

or

$$\sum_j \mathbf{p}_j = \sum_j \mathbf{P}_j \quad \text{(conservation of momentum)} \quad (4\text{-}12)$$

where the lower case $p$ stands for the values of momentum at some initial time and the upper case $P$ for the momenta at some later time.

**Example 5.** The two-body explosion. Two masses $m_1$ and $m_2$ are initially at rest with a small spring between them (see Figure 4-1). What is the ratio of velocities after the spring has been released?

ANSWER: According to Eq. 4-12:

$$\mathbf{p}_1 + \mathbf{p}_2 = \mathbf{P}_1 + \mathbf{P}_2$$

where $\mathbf{p}_1 = \mathbf{p}_2 = 0$ is the initial momentum.

$$0 + 0 = \mathbf{P}_1 + \mathbf{P}_2$$

$$\mathbf{P}_1 = -\mathbf{P}_2 \quad \text{or} \quad m_1 \mathbf{v}_1 = -m_2 \mathbf{v}_2$$

$$\frac{v_1}{v_2} = -\frac{m_2}{m_1} \tag{4-13}$$

The minus sign is a reminder that the velocities are in opposite directions.

Note that Eq. 4-13 in this example is the same as Eq. 4-1, which we use as our definition of mass.

**Example 6.** As shown in Figure 4-16 a 3-kg rifle shoots a 10-g bullet with a muzzle velocity of 600 m/s. If the gun is held loosely, what is the recoil velocity of the rifle as it hits the shoulder?

ANSWER: Since both gun and bullet have zero initial momentum, we can use Eq. 4-13 for the ratio of the final velocities. Use subscript g for gun and b for bullet.

$$\frac{v_g}{v_b} = -\frac{m_b}{m_g}$$

$$v_g = -\frac{m_b}{m_g} v_b$$

$$= -\frac{0.01}{3}(-600) \text{ m/s} = 2 \text{ m/s}$$

**Figure 4-16.** *Rifle of mass $m_g$ shown after firing a bullet of mass $m_b$.*

Example 6 illustrates the principle of rocket propulsion. The gun can be considered as a rocket and the bullet a bit of fuel that has been expelled with exhaust velocity $v_b$. The rocket will receive an increment of velocity $v_g$ each time a mass $m_b$ of fuel is ejected.

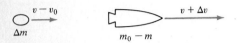

**Figure 4-17.** *Rocket ejects fuel of mass $\Delta m$ with exhaust velocity $v_0$.*

\* **Example 7.** (This example should be skipped by those who have not yet mastered integral calculus.) A rocket consisting of initial mass $m_0$ starts from rest. Fuel of total mass $m$ has been ejected so far and the rocket has reached a velocity $v$. If the exhaust velocity is always $v_0$ relative to the rocket, what is $v$ as a function of $m$?

ANSWER: Figure 4-17 is a view in the lab system where an additional amount of fuel $\Delta m$ has been ejected. (In the lab system $v = 0$ when $m = 0$.) We can get a relation between $\Delta m$ and $\Delta v$ by using the law of conservation of momentum. First we give the mass $m_0 - m$ a velocity increase of $\Delta v$. This is a momentum increase $\Delta P_1 = (m_0 - m)\Delta v$. Now remove a piece $\Delta m$ and reduce its velocity by $v_0$. This is a momentum decrease $\Delta P_2 = (\Delta m)v_0$. In order to conserve momentum these two quantities must be equal:

$$(m_0 - m)\Delta v = (\Delta m)v_0$$

$$\Delta v = v_0 \frac{\Delta m}{m_0 - m} \quad \text{or} \quad dv = v_0 \frac{dm}{m_0 - m}$$

The velocity $v$ can be obtained by integrating both sides:

$$v = v_0 \int_0^m \frac{dm}{m_0 - m}$$

$$v = v_0 \ln\left(\frac{m_0}{m_0 - m}\right) \quad (4\text{-}14)$$

The final velocity is achieved when $m_0 - m$ is the mass of the empty rocket. The ratio $m_0/(m_0 - m)$ can be as large as 10 to 1, giving a final velocity of $v = 2.3 v_0$. In order to achieve the higher velocities necessary to go into orbit, multistage rockets must be used.

## Summary

If the force on a body of mass $m$ is known, the acceleration and future positions and velocities can be determined using Newton's three laws of motion.

Law 1  If $\mathbf{F}_{\text{net}} = 0$, $\mathbf{a} = 0$.

Law 2  $\mathbf{F}_{\text{net}} = \dfrac{d\mathbf{p}}{dt} = m\mathbf{a}$, where momentum $\mathbf{P} \equiv m\mathbf{v}$.

Law 3  Force of $m_a$ on $m_b$ is equal and opposite to force of $m_b$ on $m_a$.

A single force can be defined and determined by measuring the acceleration **a** of a known mass; the value of the single force is m**a**. A force of 1 N gives a mass of 1 kg an acceleration of 1 m/s².

An unknown mass can be compared with a standard mass $m_0$ using conservation of momentum. If they start out at rest with a small spring to push them apart, $m\mathbf{v} = -m_0\mathbf{v}_0$.

If a body slides along a surface there will be a contact force that adjusts itself so that $\mathbf{F}_{net}$ is pointing along the surface.

There may be a component of force along the surface opposing the motion of the body which is called the frictional force. The coefficient of friction $\mu = F_f/F_N$ where $F_N$ is the normal force.

For a body sliding down an inclined plane the normal force (normal component of the contact force), gravitational force, and frictional force add vectorially to give $\mathbf{F}_{net}$.

When applied to a closed system of $n$ particles, the law of conservation of momentum says $\sum m_j \mathbf{v}_j = \sum m_j \mathbf{V}_j$ or $\mathbf{P}_{total}$ = a constant.

## Exercises

1. A tractor with a constant speed of 10 km/h pulls a log with a force of $10^3$ N. The gravitational force on the log is 2000 N. What is the net force on the log?
2. At what speed is the relativistic mass $m(v) = 1.01 m_{rest}$?
3. In Figure 4-5 what is the tension in each string in terms of $m_1, m_2, m_3$, and $F$? What is the net force on each car in terms of these quantities?
4. Suppose the force in Example 1 is proportional to $v^2$. How long would it take to slow down from 65 to 55 mph? (Assume the force is constant over this interval and that it is $(60/70)^2$ times its value at 70 mph.)
5. How many poundals in a lbf? How many poundals in a newton? How many kilograms in a slug?
6. In Figure 4-8 what is the net force on $m_A$ and what is the net force on $m_B$? Give answer in terms of $m_A$, $m_B$, and $F$.
7. In Example 1 the car slows down to 58 mph in the next 5 seconds. What is the average net force on the car during this period?
8. The gun in Example 6 is fired horizontally by a hunter on a frictionless ice pond. If the hunter's mass is 60 kg, what is his speed after firing the gun?
9. What is wrong with the following reasoning? A tractor pulls a plow with a force **F**. By Newton's third law the contact force of the ground on the plow is −**F**. Since the sum of these two forces is zero, the plow cannot move.
10. In Figure 4-10 assume $\theta = 30°$, $\mu_s = 0.4$, and $\mu_k = 0.38$. While the block is sliding down, $\theta$ is decreased until it starts decelerating and comes to rest. What is this value of $\theta$? Then $\theta$ is increased until it starts sliding again. What is this second value of $\theta$?
11. Assume the hydrogen atom consists of an electron of mass $9.1 \times 10^{-31}$ kg traveling in a circle of $10^{-10}$ m diameter around a proton. The force of attraction is $9 \times 10^{-8}$ N. What is the electron's speed? How many revolutions per second does the electron make?

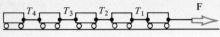

Exercise 12

12. The 5-car toy train shown in the figure is pulled by a child with a force **F** as shown. The mass of each car is $m$.
    (a) What are the tensions $T_1$, $T_2$, $T_3$, and $T_4$ in the strings in terms of $F$ and $m$? Ignore friction.
    (b) What is the acceleration?

13. A pendulum bob is hanging from a string of length $l$ at an angle of 20° from the vertical. It is moving in a horizontal circle.
    (a) If $l = 1$ m, what is its period?
    (b) What is the ratio of its period to that of a simple pendulum of the same length undergoing small oscillations?
    (c) Repeat for 45° instead of 20°.

14. A simple pendulum has a period of 1 s. What is the length of the string?

## Problems

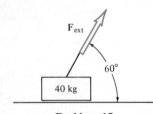

Problem 15

15. A block of 40 kg under the influence of gravity is on a frictionless surface and an external force of 200 N is applied as shown in the figure.
    (a) What is the net force on the block? Give magnitude and direction.
    (b) Suppose $F_{ext} = 800$ N instead of 200 N. Now what is the magnitude of the net force?

16. At what value of $v$ would the person in Figure 4-11 have half his or her apparent weight? Give answer in terms of $g$ and $R$.

17. Four identical blocks each of mass $m$ are pushed along a frictionless table by force $F$. Give answers in terms of $m$ and $F$.
    (a) What is the acceleration of block 4?
    (b) What is the force of block 1 on block 2?

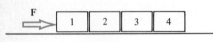

Problem 17

18. In a closed system of three bodies $m_1$, $m_2$, and $m_3$, there are six interaction forces: $F_{12}$, $F_{13}$, $F_{21}$, $F_{23}$, $F_{31}$, $F_{32}$. Prove, using Newton's laws, that $P_1 + P_2 + P_3 = $ a constant.

19. Suppose in Problem 18 the system is not closed; that is, there are three external forces $F_{1\ ext}$, $F_{2\ ext}$, $F_{3\ ext}$ in addition to the six interaction forces. Prove that

$$F_{1\ ext} + F_{2\ ext} + F_{3\ ext} = \frac{d}{dt}(P_1 + P_2 + P_3)$$

20. Suppose $m = m_0(1 - v^2/c^2)^{-1/2}$. Express $F = d(mv)/dt$ in terms of $m_0$ and $v$.

21. Two blocks are tied together with a piece of string and another string is tied to the top block. The blocks are pulled down by the earth's gravity.
    (a) How much force $F$ must be applied to the top string to suspend both blocks at rest?
    (b) How much force $F$ must be applied to the top string to give both blocks an acceleration upward of 2 m/s²? What then is the tension in the string between the two blocks?

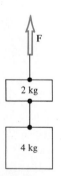

Problem 21

22. $m_1$ and $m_2$ are attached by a string that runs over a frictionless pulley. $m_1$ is held down on the table.
    (a) If $m_1 = 0.1$ kg and $m_2 = 0.3$ kg, what force is required to hold $m_1$ down on the table?

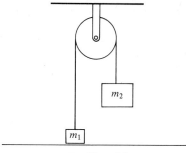

Problem 22

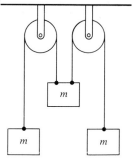

Problem 23

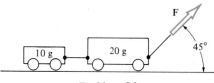

Problem 26

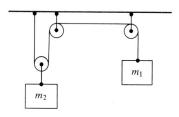

Problem 27

(b) What is the tension in the string when $m_1$ is held down on the table?
(c) What would be the tension if $m_1$ were released?
23. Consider the "double" Atwood's machine with massless strings and pulleys. Ignore friction. Give answers in terms of $m$ and $g$.
   (a) What is the acceleration of the center mass?
   (b) What is the tension in each string?
24. Suppose the driver of the Jeep in Example 4 had forgotten to put it into four-wheel drive. Now at what angle would he start slipping? (Assume 60% of the weight on the rear wheels.)
25. A single force $F$ acts for a time $t_0$ on a mass $m$. What is the increase in momentum of $m$? Give answer in terms of $F$ and $t_0$.
26. A child pulls a frictionless pull toy with a force of $1.4 \times 10^4$ dynes at an angle of 45° as shown.
   (a) What is the acceleration of the toy?
   (b) What is the tension in the string between the cars?
   (c) What is the tension in the string pulled by the child?
   (d) With what force does the floor push the 20-g car?
27. Consider the system of masses and pulleys shown in the figure. Assume a massless string and frictionless pulleys.
   (a) Give the necessary relation between masses $m_1$ and $m_2$ such that the system is in equilibrium and does not move.
   (b) If $m_1 = 6$ kg and $m_2 = 8$ kg, calculate the direction and magnitude of the acceleration of $m_2$.
28. Suppose the angle $\theta$ in Figure 4-12 is increased until the block just starts sliding. Derive a formula for its acceleration in terms of $\mu_k$, $\mu_s$, and $g$.
29. Mass $m$ is moving in a circle in the $xz$ plane. Mass $2m$ is on the axis of rotation (the pulley rotates with mass $m$ but is otherwise stationary). Assume massless pulley and string and no friction. What is the period of rotation of $m$? What is the angle $\theta$?
30. A two-stage rocket has a total mass of 25.5 metric tons before firing. The masses of fuel and shell of each stage are given in the table. After the 20 tons of the first stage fuel is expended, the first stage shell is ejected and the second stage fuel ignited. The exhaust velocity is 1 km/s.
   (a) What is the velocity at the time of separation from the first stage?
   (b) What is the final velocity after all the fuel is used up?

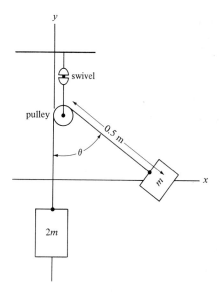

Problem 29

|  | Shell | Fuel |
|---|---|---|
| 1st stage | 2 tons | 20 tons |
| 2nd stage | $\frac{1}{2}$ ton | 3 tons |

(c) Suppose the total 23 tons of fuel and 2.5 tons of shell were combined into an equivalent one-stage rocket. What would be its final velocity?

31. Suppose the two-stage rocket of Problem 30 had a third stage mounted on top. The shell of the third stage is $\frac{1}{5}$ ton and it carries $\frac{4}{5}$ ton of fuel. (Now the total rocket is 26.5 tons.) What is the final velocity of the third stage? Will it reach orbital velocity?

32. A block of wood having a mass of 2 kg is initially at rest on a frictionless horizontal surface. A 5-g bullet having a horizontal speed of 500 cm/s strikes the block and remains imbedded in it. With what final velocity do the block and bullet move after the collision?

33. A father (60 kg) and a daughter (20 kg) are both at rest on a frictionless ice pond. If the father throws a 1 kg ball to his daughter with a horizontal speed of 5 m/s, what is his speed after throwing the ball and what is the daughter's speed after catching it?

34. In Problem 33 assume the ball bounces off the daughter's hand with a velocity of 4 m/s toward the father. Now what is the daughter's speed?

35. A car is going around a banked curve of radius $R$. What should be $\alpha$, the angle of the bank, in terms of $v$, $R$, and $g$? (This is similar to the conical pendulum where the contact force $F_N$ plays the same role as the string force $F_T$.)

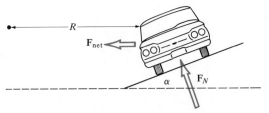

Problem 35. *Car is moving with velocity v pointing into the page.*

36. A road in a small town has a gentle curve of radius $R = 100$ m. The road is not banked. The speed limit is 40 km/h. After a light snow the coefficient of friction for a passenger car is $\mu_s = 0.2$. If a car is traveling at the speed limit, will it skid? If so, at what speed would cars start skidding on this turn?

37. Suppose the curve described in Problem 36 is banked at an angle $\alpha = 10°$? If $\mu_s = 0.1$, at what speed would cars start skidding?

38. A Jeep is slowly descending a 30° incline. It reaches a grassy field where $\mu_s = 0.5$ and $\mu_k = 0.48$. Will it start slipping, and if so, how long does it take to reach a runaway speed of 60 mph (26.8 m/s)?

39. Suppose in Figure 4-12 that $\mu_s = 0.3$ and $\mu_k = 0.2 + Av$, where $A = 2$ s/m.
    (a) A block is released on a 30° incline. What is the initial acceleration?
    (b) What is the limiting velocity?

# Gravitation

Now we shall discuss in more detail one possible source of **F** in the equation $\mathbf{F} = m\mathbf{a}$. One can consider the force **F** as the cause of the acceleration **a**. Everyday examples of forces are the gravitational attraction of the earth on a mass $m$, the attraction of a piece of iron by a magnet, the attraction or repulsion of a magnet by another magnet, the attraction or repulsion of a charged body by another charged body, the force exerted by a spring or rubber band, contact forces, and so on. In this chapter we shall restrict ourselves to gravitational forces.

On a summer day in 1665 while in a contemplative mood, Newton noticed an apple falling to the ground. He asked himself just what was it that made an apple fall to the ground. If there is a force of attraction between the earth and the apple, there must also be a force of attraction between any two masses $m_1$ and $m_2$. Since the force is proportional to the mass of the apple, it must be proportional to each of the two masses $m_1$ and $m_2$ separately; that is, $F \propto m_1 m_2$. (The symbol $\propto$ means "is proportional to.")

Newton also wondered whether the force on the apple would decrease if the apple were moved away from the surface of the earth as in Figure 5-2. He speculated that if it were moved as far away as the moon it should have the same acceleration as the moon. The nature of gravitational force between earth and moon should be the same as between earth and apple.

## 5-1 The Universal Law of Gravitation

**Figure 5-1.** *Cartoon of Newton and apple by N. Mistry.*

> **Example 1.** What is the acceleration of the moon and what is the ratio of the moon's acceleration to the gravitational acceleration at the surface of the earth?
>
> ANSWER: Using Eq. 3-9 for centripetal acceleration, the moon's acceleration is $a = 4\pi^2 r_m/T^2$, where $r_m$ is the distance from earth to moon. The earth to moon distance is $3.86 \times 10^5$ km and the period is $T = 27.3$ days or $2.36 \times 10^6$ s. These values give $a =$

$2.73 \times 10^{-3}$ m/s². Near the surface of the earth the acceleration would be $g = 9.8$ m/s². The ratio is $a/g = 1/3590 \approx (1/60)^2$, which is within errors the same as $R_e^2/r_m^2$.

Newton did the simple calculation in Example 1 and found that the gravitational force on the apple would then be reduced by a factor of $3600 = (60)^2$, which also happens to be the same as the ratio of the squares of the distances. Newton then concluded that the gravitational force between two masses must decrease inversely as the square of the distance between the two masses. He proposed a universal law of gravitational attraction between any two masses:

$$F \propto \frac{m_1 m_2}{r^2}.$$

Capital $G$ is used for the proportionality constant:

$$F = G \frac{m_1 m_2}{r^2} \quad \text{(universal law of gravitation)} \quad (5\text{-}1)$$

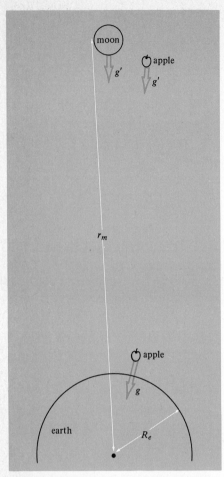

**Figure 5-2.** *Gravitational acceleration of apple decreases when it is moved away from the earth. Moon and apple have same acceleration g' when at same distance.*

**Example 2.** Newton was able to estimate the numerical value of $G$ by assuming the average density of the earth is $\rho = 5 \times 10^3$ kg/m³. (His "guess" for the density happened to be within 10% of the correct value.) What is $G$ in terms of $\rho$, $R_e$, and $g$?

ANSWER: We apply Eq. 5-1 to the force between the earth, $M_e$, and an apple of mass $m$. Then

$$F = G \frac{M_e m}{r^2}$$

If we use $r = R_e$, which is the distance between the center of the earth and the apple, we obtain

$$F = G \frac{M_e m}{R_e^2}$$

According to Newton's second law this must also equal $ma$ where $a = g$:

$$G \frac{M_e m}{R_e^2} = mg$$

$$G = \frac{g R_e^2}{M_e} \quad (5\text{-}2)$$

We use $M_e$ equals the density times the volume: $M_e = \rho(\tfrac{4}{3}\pi R_e^3)$

$$G = \frac{gR_e^2}{\rho\tfrac{4}{3}\pi R_e^3} = \frac{3g}{4\pi\rho R_e}$$

Using $R_e = 6.37 \times 10^6$ m and $\rho = 5 \times 10^3$ kg/m³, this gives $G = 7.35 \times 10^{-11}$ N m² kg⁻², which is 10% higher than the accepted value of $G = 6.67 \times 10^{-11}$ N m² kg⁻².

In comparing the gravitational acceleration at the moon to that at the surface of the earth, Newton was assuming that the earth behaves as if all of its mass is concentrated at the center. Newton guessed that this would probably be the case for a force that varied inversely as the square of the distance. However, he was not able to complete the proof until 20 years later. Presumably this problem was one of the motivations for him to invent integral calculus. The integration is somewhat lengthy and cumbersome and for this reason we shall not present it here. However, in Chapter 16, when we study Gauss' law, we shall present a simple proof that a solid sphere behaves as if all its mass is concentrated at the center. This is not true if one asks what is the gravitational force inside a sphere. If one could dig a hole to the center of the earth the gravitational force would decrease as one approached the center as shown in Section 5-6.

Equation 5-1 is called the universal law of gravitation because the very same law applies to all examples of gravitational force. The same law that explains falling bodies on the earth, also explains the orbits of the planets and comets about the sun and even the motions of huge galaxies of stars in orbit about each other. It permits us to calculate the masses of the earth, sun, and most planets as well as their periods of revolution.

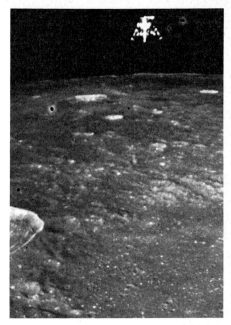

**Figure 5-3.** NASA photo of lunar module in orbit around the moon.

**Example 3.** What is the period of an Apollo lunar module in orbit around the moon just before landing? (See Figure 5-3.)

ANSWER: In the equation $F = ma$ we use $GM_m m/R^2$ for $F$ where $M_m$ is the moon's mass, $R$ is the orbit radius, and $m$ is the lunar module mass. For acceleration $a$ we use $(4\pi^2/T^2)R$. Then

$$G\frac{M_m m}{R^2} = m\left(\frac{4\pi^2}{T^2}R\right)$$

$$T^2 = \frac{4\pi^2}{GM_m}R^3 \tag{5-3}$$

$$T = 2\pi\sqrt{\frac{R^3}{GM_m}}$$

**Figure 5-4.** NASA photo of lunar rover on the surface of the moon.

Using $R \approx 1740$ km, which is the moon's radius, $M_m = 7.35 \times 10^{22}$ kg, and $G = 6.67 \times 10^{-11}$ N m² kg⁻², we get $T = 6.5 \times 10^3$ s or 108 min.

**Example 4.** A synchronous earth satellite is one that hovers over the same point on the equator forever. What is the distance of the synchronous satellites from the center of the earth?

ANSWER: In order for a satellite to hover it must have the same period of revolution as the period of rotation of the earth, which is 24 h. According to the inverse square law, the gravitational acceleration is $g(R_e^2/r^2)$, which must be the same as the centripetal acceleration of the satellite; that is,

$$\frac{4\pi^2 r}{T^2} = g\frac{R_e^2}{r^2} \quad \text{where } r \text{ is distance to satellite.}$$

$$r^3 = \frac{gR_e^2}{4\pi^2}T^2$$

Using $R_e = 6.37 \times 10^6$ m and $T = 24$ h $= 86,400$ s gives $r = 42,000$ km.

## 5-2 The Cavendish Experiment

Newton's evaluation of $G$ involved an educated guess as to the average density of the earth. If the earth had contained a superdense core, as stars do, his result for $G$ would have been greatly in error. Why not determine $G$ independently of the earth's mass by performing a laboratory experiment using two masses $m_1$ and $m_2$ as shown in Figure 5-5? Let $F$ be the force of $m_1$ on $m_2$. Then $F = Gm_1m_2/x^2$ or

$$G = \frac{Fx^2}{m_1m_2}$$

where $x$ is the distance between the centers of the spheres. However, for two 1-kg masses 10 cm apart, $F$ is $6.67 \times 10^{-9}$ N, which is less than $10^{-9}$ of the weight of a 1-kg mass and is too small to detect by conventional means.

Henry Cavendish in 1797 devised a clever scheme for measuring such a small force. He made use of the fact that the force required to twist a long, thin quartz fiber by a few degrees is a very small force of about the same

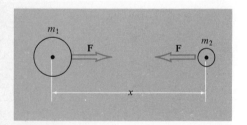

**Figure 5-5.** $F$ is the gravitational force between $m_1$ and $m_2$.

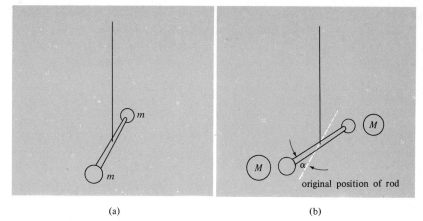

**Figure 5-6.** (a) A rod with small spheres m is hanging by means of a quartz fiber. (b) Two large spheres M are placed near the small spheres and the fiber is twisted by an angle α.

strength as the gravitational force between two lead spheres that almost touch each other. He first calibrated such a quartz fiber. Then he hung two small lead masses mounted on the ends of a light rod, as shown in Figure 5-6(a). Finally, he placed two larger lead spheres near the smaller spheres and measured the angular deflection of the rod, which is the angle α shown in Figure 5-6(b). Refined measurements using Cavendish's method give the value $G = 6.67 \times 10^{-11}$ N m² kg⁻².

## Weighing the Earth

Armed with a reliable value for $G$, Cavendish used it in Eq. 5-2 and solved for $M_e$:

$$M_e = \frac{gR_e^2}{G} \quad (5\text{-}4)$$

His result for the mass of the earth was just as accurate as his measurement of $G$. Not only did Cavendish "weigh" the earth; at the same time, he determined with equal accuracy the mass of the sun, the mass of Jupiter, and the mass of all other planets with observable satellites.

In Figure 5-7, let $M$ be the mass of the sun (or of Jupiter). Let $m$ be the mass of a planet orbiting the sun (or of a satellite orbiting Jupiter). Then $F = GMm/R^2$ and the acceleration is $a = 4\pi^2 R/T^2$. Putting these expressions into $F = ma$ gives

$$G\frac{Mm}{R^2} = m\left(\frac{4\pi^2 R}{T^2}\right)$$

$$M = \frac{4\pi^2 R^3}{GT^3} \quad (5\text{-}5)$$

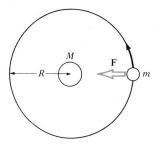

**Figure 5-7.** Body of mass m orbiting about mass M. F is gravitational force.

Thus, if $R$ is the earth to sun distance and $T$ is the period of revolution (1 year), $M$ is the mass of the sun. Similarly, for $R$ we could use the distance from the center of Jupiter to one of its 13 moons, with $T$ as the period of that moon; then Eq. 5-5 gives the mass of Jupiter.

## 5-3 Kepler's Laws of Planetary Motion

Before Newton had postulated his universal law of gravitation, Johannes Kepler found that the motions of the planets could be described by three simple laws. Kepler's laws added strength to the hypothesis of Copernicus that the planets revolved around the sun rather than around the earth.

In 1600 it was religious heresy to suggest that the planets revolve around the sun. In fact, in 1600, Giordano Bruno, an outspoken advocate of the Copernican heliocentric system and a religious dissenter in general, was tried by the Inquisition and burned at the stake. Even the great Galileo was imprisoned, tried by the Inquisition, and made to renounce publicly his beliefs, in spite of the fact that the Pope was supposedly a good personal friend of his.

The dogma of the times holding sacred the teachings of Aristotle and Ptolemy taught that the orbits of the planets were described by complicated motions of circles within circles with the earth at the center. A dozen or so circles of different sizes were needed to describe the orbit of Mars. Johannes Kepler's ambition was to "prove" that Mars and the earth must revolve around the sun. His approach was to find a simple geometrical orbit that would accurately fit all of the vast measurements of the position of Mars. After years of tedious work, he was able to discover three simple laws that very accurately agreed with the data on all the planets. Kepler's laws also apply to satellites revolving about a planet.

**Kepler's first law**—Each planet travels in an elliptical orbit with the sun at a focus of the ellipse.

**Kepler's second law (law of equal areas)**—The line joining the sun and a planet sweeps out equal areas in equal times.

**Kepler's third law**—The cubes of the semimajor axes of any two planetary orbits are to each other as the squares of their periods. For circular orbits

$$\frac{R_1^3}{R_2^3} = \frac{T_1^2}{T_2^2}$$

The semimajor axis is one half of the longest distance across the ellipse.

At the time when Newton was developing his universal law of gravitation, not only did he apply it to falling apples and the moon, but also to the force between sun and planets. He was able to prove that, if and only if the force is an inverse square law, any planetary orbit will be an ellipse with the sun at one focus and that

$$\frac{R_1^3}{R_2^3} = \frac{T_1^2}{T_2^2}$$

for any two planets in circular orbits (for elliptical orbits $R$ is the semi-major axis). He also was able to derive Kepler's law of equal areas from his three laws of motion. The fact that he could derive all three of Kepler's laws, which accurately described the detailed motions of the planets, was taken as a final confirmation of Newtonian dynamics.

Newton's derivations of Kepler's first and third laws are too complicated to repeat here. However, we can present the special case of Kepler's third law, where the two planets are in circular orbits (actually all the planets except Pluto are in nearly circular orbits). If we apply Eq. 5-5 to planet 1 we obtain

$$M = \frac{4\pi^2 R_1^3}{GT_1^2}$$

Applying it to planet 2 gives

$$M = \frac{4\pi^2 R_2^3}{GT_2^2}$$

Equating the two right-hand sides gives

$$\frac{R_1^3}{T_1^2} = \frac{R_2^3}{T_2^2} \quad \text{or} \quad \frac{R_1^3}{R_2^3} = \frac{T_1^2}{T_2^2}$$

Kepler's second law follows directly from the law of conservation of angular momentum. In Chapter 10 we shall derive the conservation of angular momentum from Newton's laws of motion. We shall see that angular momentum $L$ of the planet in Figure 5-8 is

$$L = rmv_\perp \quad \text{(angular momentum)}$$

Then

$$\frac{L}{2m} = \frac{1}{2} rv_\perp$$

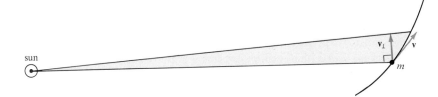

**Figure 5-8.** *Planet m of velocity* **v** *is attracted by the sun.* $v_\perp$ *is the component of* **v** *perpendicular to the line joining it to the sun.*

Note that $\frac{1}{2}rv_\perp$ (the shaded area in Figure 5-8) is approximately the area swept out in 1 second. It is exactly the time rate of area swept; hence,

$$\frac{L}{2m} = \frac{dA}{dt}$$

where $dA/dt$ is the rate that the line joining sun and planet sweeps out the area. According to the law of conservation of angular momentum, the left-hand side is a constant; hence, $dA/dt = $ a constant.

## 5-4 Weight

The weight of a body is not its mass, but it is usually defined as the net gravitational force acting on the body. Near the surface of the earth this will be $mg$ for a body of mass $m$.

**Example 5.** By what factor is an astronaut's weight reduced from that on earth when he is on the moon? Use $M_m/M_e = 0.0123$ and $R_m/R_e = 0.273$.

ANSWER: The weight on the moon is

$$F_m = G\frac{M_m m}{R_m^2}$$

The weight on the earth is

$$F_e = G\frac{M_e m}{R_e^2}$$

The ratio is

$$\frac{F_m}{F_e} = \frac{M_m}{M_e}\left(\frac{R_e}{R_m}\right)^2 = 0.165$$

The astronaut in Figure 5-9 is demonstrating that his weight is one sixth of normal.

This definition of weight can be confusing when applied to an accelerating body. For example, an astronaut in an earth satellite while drifting freely in space considers himself weightless even though gravity is still acting on him. Even the astronaut in Figure 5-9 is weightless until he lands back on the moon. The physiological sensation of weight depends on how hard it is to lift one's arm or head; the pull of one's internal organs against the body frame is proportional to one's weight. We could define physio-

**Figure 5-9.** *NASA photo of astronaut jumping up from the surface of the moon.*

logical weight as proportional to the force of the fluid on the nerve endings in the semicircular canals of the inner ear. We shall now define apparent weight, which is a way of measuring physiological weight.

### Apparent Weight

We define the apparent weight of a body as the reading one would obtain if one were to weigh the body on a spring balance. So, apparent weight is the weight as obtained using bathroom scales. It is the force of the body on the scales. Of course the scales must be oriented so that they are perpendicular to the force of the body. In Figure 5-10 let $\mathbf{F}_w = -\mathbf{j}F_w$ be this force. According to Newton's third law of motion the force of the scales on the man is $+\mathbf{j}F_w$.

Consider a man standing on scales in an elevator accelerating upward. The net force on the man is the $-\mathbf{j}mg$ of gravity pulling down plus the contact force, $\mathbf{j}F_w$, pushing up:

$$\mathbf{F}_{net} = -\mathbf{j}mg + \mathbf{j}F_w$$

Replacing $\mathbf{F}_{net}$ with ($\mathbf{j}ma$) gives

$$\mathbf{j}ma = -\mathbf{j}mg + \mathbf{j}F_w$$
$$F_w = m(g + a)$$

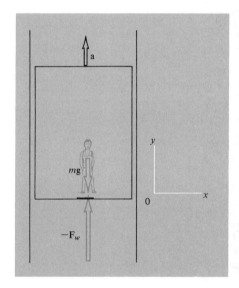

**Figure 5-10.** *Man in an elevator that is accelerating upward. Contact force pushing up on the man is* $(-\mathbf{F}_w)$.

SEC. 5-4/WEIGHT

The apparent weight is

$$\mathbf{F}_w = -\mathbf{j}F_w = -\mathbf{j}m(g + a)$$

We see that the apparent weight is pointing down and is of magnitude $m(g + a)$. (We always use the symbol "g" as a positive quantity.)

We note that if the elevator is decelerating, $\mathbf{F}_{net} = -\mathbf{j}ma$, and then

$$\mathbf{F}_w = -\mathbf{j}m(g - a) \quad \text{(for decelerating elevator)}$$

In terms of the vectors $\mathbf{g}$ and $\mathbf{a}$, the apparent weight is

$$\mathbf{F}_w = m(\mathbf{g} - \mathbf{a}) \tag{5-6}$$

If the elevator were in free fall $\mathbf{a} = \mathbf{g}$ and $\mathbf{F}_w = 0$; that is, the man is weightless. This is the situation with an astronaut in an earth satellite. All spacecrafts are in free fall except during those rare moments when they are using rocket power. "Artificial" gravity can be supplied by spinning the spacecraft (see Example 8).

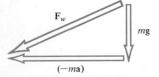

Figure 5-11. *Vector diagram for Example 6.*

**Example 6.** Assume a special car with rocket assistance can accelerate horizontally with $a = 2g$. What is the apparent weight of the driver?

ANSWER: According to Eq. 5-6,

$$\mathbf{F}_w = m\mathbf{g} - m\mathbf{a}.$$

These two vectors, which are at right angles, are subtracted in Figure 5-11. Since the sides of the right triangle are in the ratio of 1 to 2, the hypotenuse is $\sqrt{5}$ times $mg$; that is, $F_w = 2.236mg$.

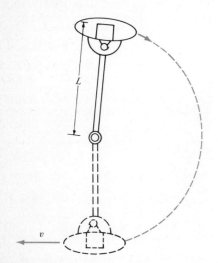

Figure 5-12. *The rocket ride.*

**Example 7.** One of the most exciting amusement park rides this author has experienced is called the rocket. It is essentially a great pendulum that oscillates with larger and larger swings until it reaches the upside down position shown in Figure 5-12. As it swings back down, the "rocket" reaches a maximum velocity $v = 2\sqrt{gL}$ (this is shown in Example 3 of Chapter 7).
(a) What is the acceleration at the bottom?
(b) What then is the net force on the passenger?
(c) What then is the apparent weight of the passenger?

ANSWER: For part (a) the acceleration is $a = v^2/L = 4g$. We can use Newton's second law to get the net force. We merely multiply the acceleration by the mass of the passenger to obtain $F_{\text{net}} = 4mg$. This is also the contact force of the seat, $F_c$, pushing up minus $mg$ of gravity pulling down.

$$F_c - mg = 4mg$$
$$F_c = 5mg$$

By definition the force of the passenger on the seat is the apparent weight. According to Newton's third law, this equals $F_c$ in magnitude. Hence the passenger's apparent weight is $5mg$. Every part of the person's body feels five times heavier than normal.

**Example 8.** Consider a spacecraft consisting of two chambers connected by a tunnel of length 20 m, as shown in Figure 5-13. How many revolutions per second must the spacecraft make in order to supply normal gravity for the passengers?

ANSWER: Let $T$ be the time for one revolution and let $f$ stand for the number of revolutions per second. Then the number of revolutions per second times $T$ in seconds equals 1:

$$fT = 1 \quad \text{or} \quad T = \frac{1}{f}$$

Now substitute $(1/f)$ for $T$ in the equation $a_c = 4\pi^2 R/T^2$:

$$a_c = 4\pi^2 f^2 R$$

$$f = \frac{1}{2\pi}\sqrt{\frac{a_c}{R}}$$

The apparent weight will be $mg$ when $a_c = g$:

$$f = \frac{1}{2\pi}\sqrt{\frac{g}{R}} = \frac{1}{2\pi}\sqrt{\frac{9.8 \text{ m/s}^2}{10 \text{ m}}} = 0.158 \text{ rev/s}$$

So such a spaceship spinning at only 9.5 rev/min will make the passengers who stay 10 m away from the axis of rotation feel at home.

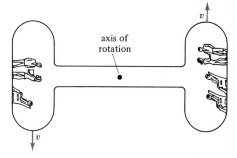

**Figure 5-13.** *Rotating spaceship supplies "artificial" gravity for the passengers. Their apparent weight is the same as on earth.*

## 5-5 The Principle of Equivalence

In Chapter 2 we stated the experimental fact that all bodies near the surface of the earth fall with the same acceleration independent of their mass. This experimental fact led Newton to postulate that gravitational force on a body is proportional to its mass. But just how accurate is this experimental fact? We could invent an alternate hypothesis that the gravitational force is proportional to the number of nucleons (neutrons and protons) in a body instead of the inertial mass as defined on page 56. Then the gravitational force on a helium atom would be exactly 4 times the gravitational force on a hydrogen atom. However the experimentally measured mass of a helium atom is not quite 4 times that of a hydrogen atom. $m_{He}/m_H = 3.9715$. Measurements more accurate than one part in a hundred are required to rule out our alternate hypothesis that gravity is proportional to the number of nucleons. Only experiments can tell which hypothesis is correct.

Strictly speaking, Newton's universal law of gravitation defines the **gravitational mass** of a body. For all we know, so far, gravitational mass may be proportional to the number of nucleons rather than to mass as defined on page 56 (which is also called inertial mass as distinguished from gravitational mass). We shall use the notation $m'$ for gravitational mass. Then the force of gravity between two bodies is $F = Gm'_1 m'_2/r^2$. The mass that appears in the equation $F = ma$ is inertial mass and will be unprimed. An inertial mass $m_1$ in free fall near the earth's surface will have an acceleration $a_1$ where

$$m_1 a_1 = G\frac{M'_e m'_1}{R_e^2} \qquad (5\text{-}7)$$

A mass $m_2$ of a different substance may have a slightly different acceleration $a_2$ where

$$m_2 a_2 = G\frac{M'_e m'_2}{R_e^2} \qquad (5\text{-}8)$$

Dividing Eq. 5-7 by Eq. 5-8 gives

$$\frac{m_1}{m_2}\frac{a_1}{a_2} = \frac{m'_1}{m'_2}$$

We see that if all bodies fall with the same acceleration $a_1 = a_2 = g$, the ratios of the inertial masses will equal the ratios of the gravitational masses. Then if we set inertial mass equal to gravitational mass for one substance, they will be equal for all substances; that is, if $m_1 = m'_1$, then $m_2 = m'_2$.

Newton was able to determine that $a_1 = a_2$ to one part in a thousand. In

1901 a Hungarian physicist, R. Eötvös, made the determination to one part in $10^8$. In 1964 R. Dicke of Princeton University improved upon the Eötvös measurement by a factor of 300. These results strongly suggest that gravitational mass is *exactly* equal to inertial mass for all substances. This is called the principle of equivalence. It is a fundamental law of nature based on experiment as are the other laws.

A consequence of the principle of equivalence is that there is no way to distinguish between acceleration of a laboratory and gravitation. If a physics lab is put into a large elevator and the elevator is accelerated, there is no experiment that can be performed inside the elevator to determine whether the elevator is accelerating or whether the elevator is at rest and instead a new source of gravity had been "turned on." As will be discussed further in Chapter 9, the principle of equivalence is the starting point for Einstein's general theory of relativity.

## 5-6 The Gravitational Field Inside a Sphere

By gravitational field we mean the gravitational acceleration as a function of position. The gravitational field of a hollow spherical shell of mass $m$ and radius $R$ will be $Gm/r^2$ for $r > R$ where $r$ is measured from the center. This is what we mean when we say that a spherical mass behaves as if it is all concentrated at the center. But what will be the gravitational field inside the shell at any arbitrary point as shown in Figure 5-14? First consider the contribution from area $A_1$. It will contribute a force $F_1 \propto A_1/r_1^2$ pulling to the left. Now extend a line from each point on $A_1$ through $P$ to the other side to give an area $A_2$ at a distance $r_2$. Area $A_2$ contributes a force $F_2$ pulling to the right where

$$\frac{F_1}{F_2} = \frac{A_1}{A_2} \times \frac{r_2^2}{r_1^2}$$

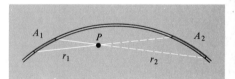

**Figure 5-14.** *Point $P$ inside hollow thin shell. Surface areas $A_1$ and $A_2$ are opposite to each other as seen from point $P$.*

It is easily seen from the geometry of the situation that

$$\frac{A_1}{A_2} = \frac{r_1^2}{r_2^2}$$

This is because the base areas of two similar cones are proportional to the squares of their linear dimensions. If we substitute this into the previous equation we obtain

$$\frac{F_1}{F_2} = \left(\frac{r_1^2}{r_2^2}\right)\frac{r_2^2}{r_1^2} = 1$$

So the two contributions from $A_1$ and $A_2$ exactly cancel each other out.

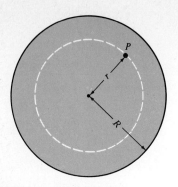

**Figure 5-15.** *Solid sphere with point P at distance r from center. Imaginary spherical surface is drawn through P.*

The area of the entire shell can be covered in this way giving a net force of zero. Hence the gravitational field everywhere inside a hollow sphere is zero. The field inside a hollow sphere with a thick shell is also zero because it can be broken down into a series of concentric thin shells.

In Figure 5-15 we show a solid sphere of radius $R$ where we make an imaginary break at point $P$, at a distance $r$ from the center. According to the previous paragraph the field at point $P$ due to the outer part is zero. Let $m(r)$ be the mass of the inner part. The field due to $m(r)$ is just the field at the surface of a sphere of radius $r$, which is

$$a = G\frac{m(r)}{r^2}$$

The mass of the inner sphere is its density $\rho$ times its volume: $m(r) = \rho\frac{4}{3}\pi r^3$. So the net field at point $P$ is

$$a = G\frac{\frac{4}{3}\pi\rho r^3}{r^2} = \tfrac{4}{3}\pi\rho G r \tag{5-9}$$

We note that the field increases uniformly from the center to the surface. The density is the total mass $m$ divided by the total volume:

$$\rho = \frac{m}{\frac{4}{3}\pi R^3}$$

If this is substituted into Eq. 5-9, one obtains

$$a = G\frac{m}{R^2}\frac{r}{R}$$

Note that this result depends upon the assumption that the sphere is of uniform density. If we assume that the earth is of uniform density and we substitute $(gR_e^2/G)$ for $m$ (see Eq. 5-4), we obtain

$$a = g\frac{r}{R_e} \quad \text{(inside the earth)} \tag{5-10}$$

and

$$a = g\frac{R_e^2}{r^2} \quad \text{(outside the earth)}$$

This is plotted in Figure 5-16.

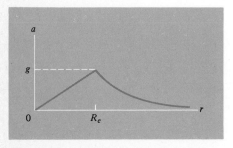

**Figure 5-16.** *Gravitational field of the earth as a function of distance from the center (assuming uniform density). The direction of a is toward the center.*

## Summary

Newton's universal law of gravitation $F = Gm_1m_2/r^2$ applies to all masses. If $m_1$ is the earth, $m_2$ could be an apple or the moon. Or $m_1$ could be the sun and $m_2$ a planet. Then the acceleration of the planet would be $a = Gm_1/r^2$.

If a planet or satellite is in circular orbit its acceleration can be determined from $a_c = 4\pi^2 R/T^2$. Setting this equal to $Gm_1/R^2$ determines $m_1$. By such means the masses of the sun, earth, and satellite-bearing planets were determined. $G$ was originally determined by measuring forces between small spheres in the Cavendish experiment.

Newton with his universal law of gravitation was able to derive Kepler's three laws (which were based on experimental observations). Kepler's laws:

1 A planet follows an elliptical path.
2 Law of equal areas.
3 Cubes of planetary distances (the semimajor axes) go as squares of the periods of revolution.

Apparent weight $\mathbf{F}_w$ is the actual reading one would obtain from bathroom scales. $\mathbf{F}_w = m(\mathbf{g} - \mathbf{a})$.

The gravitational force inside a hollow sphere is zero. Outside the sphere the force is the same as if all the mass were concentrated at the center. If the sphere is of uniform density, the gravitational force inside increases uniformly from the center.

## Exercises

1. A 60-kg student is in an elevator that is accelerating upward with $a_y = 9.8$ m/s². What is the net force in newtons acting on the student?
2. A 60-kg student is in an elevator that is accelerating downward at a rate of 9.8 m/s². What is the net force in newtons acting on the student?
3. Mars is 52% farther from the sun than the earth. Use this information to determine the length of a Martian year.
4. An airplane with constant speed of 300 km/h does part of an outside loop of radius $R$ as shown. What must be the value of $R$ if the passengers are to feel weightless?
5. In Kepler's third law, can we compare the period of the moon to the period of the earth? Can we compare the period of the moon to the period of a moon of Jupiter? Can we compare the periods of all of Jupiter's moons with each other?
6. An elevator starts up from rest at the basement with an initial acceleration of 16 ft/s².
   (a) The passengers' apparent weight will be (increased, decreased, remain the same).

Exercise 4

(b) The period of oscillation of a pendulum in such an elevator would (increase, decrease, remain the same).
(c) When the elevator reaches a velocity of 32 ft/s, it maintains that constant upward velocity. Under this condition a passenger's apparent weight will be (increased, decreased, remain the same) as compared to his weight when at rest.

7. If the moon had twice its present mass but moved in its present orbit, what would be its period of revolution?
8. An elevator starts up with an acceleration of 16 ft/s². What is the apparent weight of a 60-kg man during this acceleration? The elevator then attains a uniform upward velocity of 32 ft/s. Now what is the apparent weight of the man? What would be his apparent weight if the elevator cable broke?
9. A man whose weight is 600 N gets into an elevator at the fiftieth floor of a 100-story building and steps onto some scales. When the elevator begins to move, he sees that the scales read 720 N for 5 s, then 600 N for 20 s, then 480 N for 5 s, after which the elevator is at rest at one end of its track.
   (a) Is he at the top or bottom of the building?
   (b) How high is the building? (This same sort of method can be used to tell an astronaut how far his space capsule has traveled.)
10. The centers of two identical spheres are 1 m apart. What must be the mass of each sphere if the gravitational force of attraction between them is 1 N?
11. At some point between the earth and the moon, the gravitational force on a spaceship due to the earth and moon together is zero. Where is this point? Would the passengers be weightless only when passing this point?
12. What is the value of "$g$" 200 km above the earth's surface?
13. Calculate the mass of the sun using the value of $G$, the earth's distance, and the earth's period.
14. A spaceship travels from the earth toward the sun. At what distance from the earth is the net gravitational force equal to zero?
15. What is the value of the acceleration of the earth about the sun?
16. What is "$g$" at the surface of Mars? Its radius is $3.43 \times 10^6$ m and its density is $3.95 \times 10^3$ kg/m³.
17. Two identical spheres of density $\rho$ are of radius $R$. If they are touching each other, what is the gravitational force between them in terms of $G$, $R$, and $\rho$?
18. In Exercise 17, suppose the spheres are lead and the force is 1 dyne ($10^{-5}$ N). What must be $R$? The density of lead is $11.3 \times 10^3$ kg/m³.
19. Suppose the car in Example 6 accelerates horizontally with $a = g$. What is the driver's apparent weight?
20. What is the apparent weight of the passenger in Example 8 of Chapter 2 when at the halfway point? (At this time the horizontal component of acceleration is zero.)

**Problems**

21. A 30-kg girl is sliding down a rope. Her downward acceleration is $0.1g$.
    (a) What is her apparent weight?
    (b) What is the tension in the rope?
22. In this problem we shall design an amusement park ride capable of producing

weightlessness for short periods of time. Two "rocket" cars are separated by a 20-m arm and rotated with velocity $v$ in a vertical circle as shown. What must be the value for $v$ for the passengers to be weightless when their car reaches the top? What would be their apparent weight when the car reaches the bottom?

23. If the moon's orbital velocity were doubled in such a way as to keep it in a circular orbit, what would be the radius of its new orbit? What would be its new period of revolution?

24. What is the velocity $v$ of an earth satellite in a circular orbit of height $h$ above the earth's surface? Express $v$ in terms of $R$ (the earth's radius), $h$, and $g$. Does the velocity increase or decrease as the satellite is acted upon by a very weak air resistance?

25. At the position of the moon, what is the gravitational force (a) of the earth on a 1-kg mass? (b) of the sun on a 1-kg mass? Do not use $G$ or the mass of the earth or sun.

26. Assume the Viking orbiter circles Mars at a height of 100 km. What is its period of revolution about Mars? The radius of Mars is $3.43 \times 10^6$ m and its average density is $3.95$ g/cm$^3$.

27. Repeat Problem 26 for a lunar orbiter circling the moon at a height of 100 km.

28. Assume our galaxy consists of $10^{11}$ stars of average mass $10^{30}$ kg each. A star at the edge of the galaxy is in a circular orbit of 50,000 light-year radius. What is its velocity and period of revolution? Assume it behaves as if all the mass is concentrated in the center of the galaxy.

29. When suspended beside a mountain, a plumb-bob will be pulled slightly to the side. A mountain of 1 km$^3$ volume and density 2500 kg/m$^3$ is pulling a plumb-bob to the side. Assume the mass of the mountain behaves as if it is concentrated at a point 600 m to the side of the plumb-bob. What will be the angle to the vertical?

30. What must be the period of revolution of the earth for it to fly apart? (Loose objects on the equator would go into circular orbit.) Give answer in terms of $G$, $M_e$, and $R_e$. Also give numerical result.

31. Repeat Problem 30 for the sun.

32. When a star finishes burning its thermonuclear fuel, it undergoes gravitational collapse and contracts. Because of conservation of angular momentum, the quantity $R^2/T$ (where $T$ is the period of rotation) must stay constant. What would be the minimum radius our sun could achieve before it started to fly apart? Its present period of rotation is 27 days.

33. For two planets in circular orbits about the sun, the ratio of the velocities equals the inverse ratio of the radii to some power. What is that power?

34. Suppose a comet is in an elliptical orbit about the sun with semimajor axis $a$ and semiminor axis $b$. For an ellipse the distance from the center to focus is $\sqrt{a^2 - b^2}$. What is the ratio of the velocities $v_2/v_1$ in terms of $a$ and $b$? What is the ratio in terms of the eccentricity $\varepsilon \equiv \sqrt{1 - (b^2/a^2)}$?

35. In Problem 34, what is the ratio $v_3/v_1$?

36. In Example 7 the velocity is $\sqrt{2gL}$ when the "rocket" has fallen halfway down. What then is the apparent weight of a passenger?

37. If the "rocket" in Example 7 is released with the arm in the horizontal position, the velocity will be $\sqrt{2gL}$ when it passes its lowest position. What then is the apparent weight of a passenger?

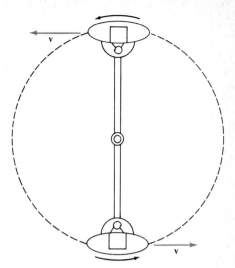

Problem 22

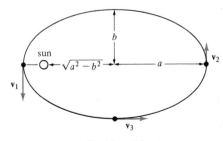

Problem 34

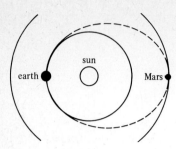

**Problem 41**

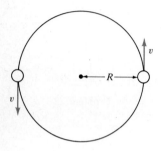

**Problem 42**

38. Assume that the gravitational mass $m'$ of a body is not the same as the inertial mass $m$. Rederive the formula for the period of a conical pendulum in Section 4-8. Show that

$$T = 2\pi \sqrt{\frac{m}{m'} \times \frac{R}{g \tan \theta}}$$

39. Suppose for carbon $m'/m$ is exactly equal to 1, but for lead it is 1.001. Rederive Eq. 4-11 and find the ratio of the periods of small oscillation $T_\text{C}/T_\text{Pb}$ for two identical conical pendulums, one with carbon and the other with lead.

40. Consider a hollow spherical shell of mass $m$ that has an inner radius $R_1$ and outer radius $R_2$. The shell thickness is $R_2 - R_1$. What is the gravitational field inside the shell; that is, when $R_1 < r < R_2$? Give answer in terms of $G, m, R_1,$ and $R_2$. Assume uniform density.

41. A spaceship launched to reach Mars is in an elliptical orbit with major axis equal to the sum of the earth and Mars distances. The spaceship's orbit is the dashed line in the accompanying figure. How long will it take to reach Mars? Mars to sun distance = $2.28 \times 10^{11}$ m.

42. Two stars of equal mass are in circular orbit about their common center as shown.
    (a) What is the net force on either star in terms of $m$, $G$, and $R$?
    (b) Derive a formula for the period of revolution in terms of $m$, $G$, and $R$.

# Work and Energy

## 6-1 Introduction

Energy has become of prime concern to every citizen. The amount of energy easily available on earth is limited, and the limit has just about been reached. Personal wealth is correlated with energy use. For example, the gross national products of nations are almost directly proportional to their energy usage. The production and distribution of a limited commodity in such high demand becomes a prominent social and economic problem with all kinds of technological considerations. It is difficult to make wise and fair decisions without first understanding what energy is, as well as being familiar with the technologies of production and distribution of energy. In the next two chapters we shall study different forms of energy and their conversion from one form to another. We shall study work, kinetic energy, potential energy, heat energy, chemical energy, and power. In later chapters we shall learn about efficiency of converting heat into mechanical and electrical energy. Also the reverse: use of mechanical and electrical energy to extract heat (air conditioning, refrigeration and heat pumps). In later chapters we shall also study electric motors and generators, electromagnetic radiation, nuclear fission and fusion, nuclear reactors, thermonuclear power, and stellar energy.

What is perhaps the most important principle in all of physics will be developed in this and the next chapter—the law of conservation of energy. With it we can place stringent limitations on energy conversion and utilization. It will be a central theme in most of the remaining chapters whether we are talking about mechanics, relativity, gravitation, thermodynamics, electromagnetism, electromagnetic radiation, atomic physics, or nuclear and modern physics. In mechanics the law of conservation of energy will give us a powerful new tool with which to calculate the motion of bodies acting under the influence of different kinds of forces. In many cases it will allow us to bypass Newton's laws of motion and give us a quick easy way for analyzing the motion of a body.

**Figure 6-1.** *Man pulling sled a distance s with a force F.*

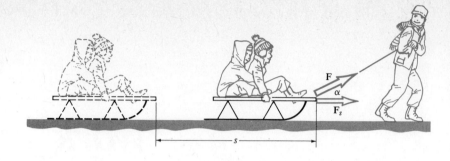

## 6-2 Work

A force acting on a moving body does work on the body. Work has the units of force times distance. Quantitatively **the work done by a force is the component of the force in the direction of motion times the distance moved.** For example, in Figure 6-1 the man pulls the sled with children a distance $s$. He pulls with a constant force **F** on the rope. The work done on the sled by the man is

$$W = F_s s \quad \text{(work done by constant force)}$$

Note that the work is not $Fs$, it is $F_s s$. $F_s$ is the component of **F** in the **s** direction. Since $F_s = F \cos \alpha$, Eq. 6-1 can also be written as

$$W = Fs \cos \alpha \quad \text{(work done by constant force)} \tag{6-1}$$

If the force is not constant we take the force averaged over the distance:

$$W = \bar{F}_s s$$

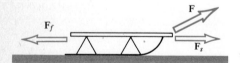

**Figure 6-2.** *Sled showing frictional force $F_f$ and applied force **F**.*

(We note that according to definition this is the integral of $F_s$ over $s$: $W = \int F_s \, ds$.)

Equation 6-1 can be applied to any one of the forces acting on the sled. In addition to the force **F** pulling on the rope, there is a force $F_f$ of friction opposing the motion, as shown in Figure 6-2. The $s$ component of $F_f$ has magnitude equal to $|F_f|$; however, it is negative in sign. Hence $W_f = -|F_f|s$ is the work done by the frictional force.

If the man is walking at a constant pace, the sled is not accelerating and $F_{\text{net}} = 0$. In the horizontal direction $F_{\text{net}} = F_s - F_f = 0$. So in this case the work done by the frictional force is the negative of the work done by the man.

What is the work done by the net force? If $F_{\text{net}} = 0$, it must be zero. If the man starts walking faster, $F_{\text{net}}$ becomes positive; then the work done by $F_{\text{net}}$ would be positive. The sled would speed up and its kinetic energy (defined in Section 6-5) would increase. In Section 6-5 we shall show that the work done by $F_{net}$ is equal to the increase in kinetic energy.

## Energy

In this chapter we shall discuss several forms of energy. Work is our first form of energy. We say that the work done by force $F$ applied to a body or system of bodies increases the energy of the system by that same amount. In the preceding sled example, when $F_{net} = 0$, we see that the applied force increases the energy of the sled while the frictional force is decreasing the energy of the sled by the same amount. Hence the net energy of the sled does not increase. We shall define each form of energy as we come to it. We shall see how it is possible to transform energy from one form into another and that the overall amount of energy will be conserved in a closed system.

## Units

Work and energy have the units of force times distance or newton-meters (N m). The dimensions are $ml^2t^{-2}$. This is such a common unit that it is given the special name "joule" (J). A 100-W light bulb uses 100 J of energy each second. One horsepower (hp) is defined as 746 J of energy per second. In the cgs system the unit dyne-centimeter is given the special name of erg.

$$1 \text{ J} = 1 \text{ N} \times 1 \text{ m} = (10^5 \text{ dyne}) \times (10^2 \text{ cm}) = 10^7 \text{ dyne cm} = 10^7 \text{ erg}$$

Another common unit of energy in atomic and nuclear physics is the electron volt (eV). The conversion is

$$1 \text{ eV} = 1.6 \times 10^{-19} \text{ J} \qquad \text{(definition of electron volt)}$$

---

**Example 1.** In Figure 6-1 $\alpha = 30°$ and the man is walking at a rate of 1.5 m/s. If the man is doing 100 J of work per second, what is $F$? (This would be about $\frac{1}{7}$ horsepower, which, as we shall see, is heavy work for a man.)

ANSWER: In each second $s = 1.5$ m. Using Eq. 6-1:

$$Fs \cos \alpha = W$$

$$F = \frac{W}{s \cos \alpha} = \frac{100 \text{ J}}{(1.5 \text{ m})(0.866)} = 77 \text{ N}$$

> A force of 77 N is equivalent to lifting a mass of 7.9 kg or 17 lb. The rate of doing work in this problem is equivalent to lifting 10 kg about 1 m every second. This is indeed hard work.

## 6-3 Power

If energy ($E$) is being used—that is, transferred from one system to another, or being delivered to a body or system via an external force—then the rate of energy transfer is defined as power. Let $P$ stand for power. Then by definition

$$P \equiv \frac{dE}{dt} \quad \text{(definition of power)} \quad (6\text{-}2)$$

This is the instantaneous rate of energy transfer. In mks units power is joules per second (J/s) with dimensions $ml^2t^{-3}$. The unit J/s is given the special name "watt" (W). A 100-W light bulb uses 100 J/s. In Example 1 the man has a power output of 100 W.

The product (power) × (time) is energy. A common unit of energy is the kilowatt-hour (kW h)

$$1 \text{ kW h} = (10^3 \text{ W}) \times (3600 \text{ s}) = 3.6 \times 10^6 \text{ J}.$$

Typical electrical energy usage in the United States is $1.3 \times 10^{13}$ kW h per day.

Suppose a body is moving with velocity **v** under the influence of force **F**. Then the energy increase due to **F** is

$$dE = F \, ds \cos \alpha$$

$$\frac{dE}{dt} = F \frac{ds}{dt} \cos \alpha$$

$$P = Fv \cos \alpha \quad (6\text{-}3)$$

### Horsepower

Another and earlier unit of power is the horsepower (hp), which is related to the power that can be delivered by a hard-working horse. The unit of power, the horsepower, was established well before the mks system. It turns out that 1 hp equals 746 W.

$$1 \text{ hp} = 746 \text{ W} \quad \text{(definition of horsepower)}$$

We have seen that the work done in Figure 6-1 is

$$W = Fs \cos \alpha$$

where $\alpha$ is the angle between the vector **F** and the displacement vector **s**. We shall now define a quantity called the dot product. It is also called the scalar product, and as we shall see, is a scalar. Consider any two vectors **A** and **B** with angle $\alpha$ between them as in Figure 6-3. Then the dot product is defined as

$$\mathbf{A} \cdot \mathbf{B} \equiv |A| \cdot |B| \cos \alpha \quad \text{(definition of dot product)} \quad (6\text{-}4)$$

## 6-4 The Dot Product

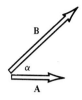

**Figure 6-3.** *Two vectors.*

**The dot product of two vectors is equal to the product of the magnitudes of the two vectors times the cosine of the angle between them.** Whenever two vectors are multiplied together with a dot (not a cross) between them, it is a universal convention that the cosine of the angle between them is part of the product. It is as if the dot stands for $(\cos \alpha)$. For example, the dot product of the two unit vectors in the $y$-direction is $\mathbf{j} \cdot \mathbf{j} = |1| \cdot |1| \cos 0° = 1$. The dot product of $\mathbf{i} \cdot \mathbf{j} = |1| \cdot |1| \cos 90° = 0$. Also note that

$$\mathbf{A} \cdot \mathbf{B} = \mathbf{B} \cdot \mathbf{A} \quad \text{and} \quad \mathbf{A} \cdot (\mathbf{B} + \mathbf{C}) = \mathbf{A} \cdot \mathbf{B} + \mathbf{A} \cdot \mathbf{C}$$

The product

$$\mathbf{A} \cdot \mathbf{B} = (\mathbf{i}A_x + \mathbf{j}A_y + \mathbf{k}A_z) \cdot (\mathbf{i}B_x + \mathbf{j}B_y + \mathbf{k}B_z)$$
$$= A_x B_x + A_y B_y + A_z B_z \quad (6\text{-}5)$$

---

**\*Example 2.** If the vector **A** is a function of $t$, show that

$$A \frac{dA}{dt} = \mathbf{A} \cdot \frac{d\mathbf{A}}{dt} \quad (6\text{-}6)$$

ANSWER:
$$A^2 = A_x^2 + A_y^2 + A_z^2$$

$$\frac{d(A^2)}{dt} = \frac{dA_x^2}{dt} + \frac{dA_y^2}{dt} + \frac{dA_z^2}{dt}$$

$$A \frac{dA}{dt} = A_x \frac{dA_x}{dt} + A_y \frac{dA_y}{dt} + A_z \frac{dA_z}{dt}$$

Let $\mathbf{B} \equiv d\mathbf{A}/dt$, $B_x = dA_x/dt$, and so on. Then according to Eq. 6-5 the right-hand side is equal to $\mathbf{A} \cdot \mathbf{B}$ or $\mathbf{A} \cdot (d\mathbf{A}/dt)$, which completes the proof.

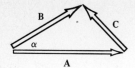

**Figure 6-4**

As an illustration of the power of vector notation, we shall present a quick derivation of the law of cosines. In trigonometry the usual derivation is much longer. In the triangle of Figure 6-4 we can express side **C** in terms of **A** and **B** as $\mathbf{C} = \mathbf{B} - \mathbf{A}$. By squaring both sides, we obtain

$$\mathbf{C} \cdot \mathbf{C} = (\mathbf{B} - \mathbf{A}) \cdot (\mathbf{B} - \mathbf{A})$$
$$C^2 = B^2 - 2\mathbf{A} \cdot \mathbf{B} + A^2$$
$$= A^2 + B^2 - 2AB \cos \alpha$$

We shall now express our definition of work in the dot product notation. If **s** is the displacement of a body, then the work done by a constant force **F** acting on it is

$$W = \mathbf{F} \cdot \mathbf{s}$$

If the force is not constant, the increment of work done in moving an infinitesimal distance $d\mathbf{s}$ is

$$dW = \mathbf{F} \cdot d\mathbf{s} \qquad (6\text{-}7)$$

The work done in moving from point $A$ to point $B$ in Figure 6-5 is

$$W = \sum \mathbf{F}_j \cdot d\mathbf{s}_j = \sum (F_s)_j ds_j$$

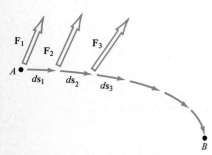

**Figure 6-5.** *Path from A to B is broken down into increments $d\mathbf{s}_j$.*

In the limit of each $ds_j$ approaching zero, this summation is defined as the definite integral of $F_s ds$ from $A$ to $B$ (the integral sign can be thought of as a summation sign):

$$W = \int_A^B F_s \, ds$$

This can also be written as

$$W = \int_A^B \mathbf{F} \cdot d\mathbf{s} \qquad \text{(work done by force } F\text{)} \qquad (6\text{-}8)$$

> **Example 3.** To stretch a spring a distance $x$ requires an applied force of strength $F = kx$. (This linear dependence on $x$ is called Hooke's law.) What is the work done in pulling the spring a distance $x_0$?
>
> Answer: We put $(kx)$ for $F$ in Eq. 6-8 and use $dx$ for $ds$:
>
> $$W = \int_0^{x_0} (kx)\, dx = k \int_0^{x_0} x\, dx = k\left[\frac{x^2}{2}\right]_0^{x_0} = \frac{1}{2} kx_0^2$$

In evaluating the integral we have used $\int x^N \, dx = \frac{1}{N+1} x^{N+1}$.
(Up to this point knowledge of integral calculus has not been necessary.)

---

**Example 4.** A projectile has a velocity $v_A$ parallel to the earth when at a height $h$. It hits the ground at point $B$. What is the work done by the gravitational force?

ANSWER: We need to calculate $W = \int_A^B \mathbf{F} \cdot d\mathbf{s}$ where the angle $\alpha$ between these two vectors is continuously changing. We note that each element of work

$$\mathbf{F} \cdot d\mathbf{s} = mg(ds \cos \alpha).$$

From Figure 6-6 we see that $(ds \cos \alpha) = dy$. We make this substitution and integrate.

$$\int_A^B \mathbf{F} \cdot d\mathbf{s} = \int_A^B (-mg)(dy) = -mg \int_A^B dy = mgh$$

$W = mgh$ is the work done by the gravitational force where $h$ is the initial height.

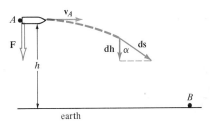

**Figure 6-6.** *Projectile for Example 4.*

## 6-5 Kinetic Energy

We define the kinetic energy of a body of mass $m$ as

$$K \equiv \tfrac{1}{2} m v^2 \quad \text{(definition of kinetic energy)} \tag{6-9}$$

This has the dimensions $ml^2 t^{-2}$, which are the same as those of energy. We shall now show that the kinetic energy of a particle will increase precisely by the work done on it by the *net* force acting on it. The work done by the net force in going from point $A$ to point $B$ is

$$W = \int_A^B \mathbf{F}_{net} \cdot d\mathbf{s}$$

Now replace $\mathbf{F}_{net}$ with $m(d\mathbf{v}/dt)$ and $d\mathbf{s}$ with $(\mathbf{v}\, dt)$:

$$\mathbf{F}_{net} \cdot d\mathbf{s} = \int_A^B \left( m \frac{d\mathbf{v}}{dt} \right) \cdot (\mathbf{v}\, dt)$$

Using Eq. 6-6 we can replace $(d\mathbf{v}/dt) \cdot \mathbf{v}$ with $v(dv/dt)$:

$$\int_A^B \mathbf{F}_{net} \cdot d\mathbf{s} = m \int_A^B \left(v\frac{dv}{dt}\right) dt = m \int_A^B v \left(\frac{dv}{dt} dt\right)$$

The quantity $(dv/dt) \, dt$ equals $dv$ because, for a small time interval $\Delta t$, we have $(\Delta v/\Delta t) \Delta t = \Delta v$. So

$$\int_A^B \mathbf{F}_{net} \cdot d\mathbf{s} = m \int_A^B v(dv)$$

$$= m \left[\tfrac{1}{2} v^2\right]_A^B$$

$$= \tfrac{1}{2} m v_B^2 - \tfrac{1}{2} m v_A^2$$

$$\int_A^B \mathbf{F}_{net} \cdot d\mathbf{s} = K_B - K_A \quad \text{(the work-energy theorem)} \quad (6\text{-}10)$$

This says that the work done by the net force in going from point $A$ to point $B$ equals the kinetic energy at point $B$ minus the kinetic energy at point $A$; that is, the kinetic energy is increased by the amount of work done by the net force. This general relationship between $\mathbf{F}_{net}$ and kinetic energy is called the **work-energy theorem.**

---

**Example 5.** What is the speed $v_B$ of the projectile in Figure 6-6 when it hits the ground at point $B$?

ANSWER: We note that $F = mg$ is the net force. Hence the integral that was calculated in Example 4 is also $\int_A^B \mathbf{F}_{net} \cdot d\mathbf{s}$ and this has the value $(mgh)$. So we can replace the left-hand side of Eq. 6-10 with $(mgh)$:

$$(mgh) = \tfrac{1}{2} m v_B^2 - \tfrac{1}{2} m v_A^2$$

$$v_B^2 = 2gh + v_A^2$$

---

We begin to see advantages in using energy to solve such problems. In Example 5 it was not necessary to calculate the trajectory or the velocity as a function of time.

---

**Example 6.** A 30-m high waterfall flows at the rate of 10 kg of water per second. By what rate does the kinetic energy of the water flowing over the edge increase?

ANSWER: We use $(mgh)$ for the left-hand side of Eq. 6-10:

$$mgh = \Delta K$$

The water flowing over the edge in 1 second gains a kinetic energy of

$$\Delta K = (10 \text{ kg})(9.8 \text{ m/s}^2)(30 \text{ m}) = 2.9 \text{ kJ}$$

If this 2.9 kJ per second could be converted to electrical energy with 100% efficiency, we would have 2.9 kW of electrical power generated.

We see in Example 6 that a respectable waterfall might be able to supply 2 or 3 kW to a house. However, the typical American home is wired for 10 to 20 kW, not 2 or 3! This is a small-scale version of what is becoming one of the world's greatest problems; namely, society has become so power hungry that the earth's conventional power sources can no longer meet the demand. In the United States most of the available hydroelectric power has already been tapped, and it only provides about 4% of the total power demand.

**Example 7.** Initially a mass $m$ is at a height $h$ above the earth and both the mass and the earth are at rest. What is the kinetic energy of the mass compared to the kinetic energy of the earth when they collide?

ANSWER: As seen in Example 4, the work done by gravity on the mass is $W = mgh$ and, according to Eq. 6-10, this will also be the kinetic energy of $m$. In order to get the earth's kinetic energy, we can use the law of conservation of momentum. Since the total momentum of the system is zero, the earth's momentum must be equal and opposite to that of mass $m$:

$$M_e v_e = -mv \quad \text{or} \quad v_e = -\frac{m}{M_e} v$$

$$v_e^2 = \frac{m^2}{M_e^2} v^2$$

$$\tfrac{1}{2} m_e v_e^2 = \frac{m}{M_e} (\tfrac{1}{2} mv^2)$$

$$K_e = \frac{m}{M_e} K$$

In Example 7 we see that the kinetic energy of the earth is $m/M_e$ times that of $m$. For $m = 6$ kg, $(m/M_e) = 10^{-24}$. This is such a small number that we can ignore energy transfer to the earth. However, we cannot ignore momentum transfer to the earth. In Example 7 the earth's momentum has the same value as that of mass $m$. In general when we have a body of mass $m$ interacting with the earth via any kind of force (except friction) we can ignore energy transfer to the earth; the energy will be so small as to be unmeasurable. However, momentum conservation may appear to be violated unless the earth's momentum is taken into account.

## 6-6 Potential Energy

In the next chapter we shall make considerable use of the work-energy theorem, including deriving the law of conservation of energy from it. Since the left-hand side is of the form $\int \mathbf{F} \cdot d\mathbf{s}$, it will be convenient to evaluate this integral for several standard type forces and to give it a special name. The special name is potential energy. (To be precise, the integral is the decrease in potential energy.) Many problems involving energy are greatly simplified if this integral (the potential energy) is calculated in advance. Potential energy can be thought of as energy stored up for future potential use. In many cases it can be converted at will into other useful forms of energy.

We start our presentation by calculating the potential energy between the two masses shown in Figure 6-7, which are interacting via a gravitational or electromagnetic force $F$. Then the change in potential energy of $m_1$ in going from $A$ to $B$ is defined as

$$\Delta U_1 \equiv -\int_A^B \mathbf{F}_1 \cdot d\mathbf{s}_1 \qquad (6\text{-}11)$$

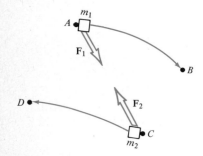

**Figure 6-7.** *A system of two interacting masses $m_1$ and $m_2$.*

The change in potential energy of $m_2$ in going from $C$ to $D$ is

$$\Delta U_2 = -\int_C^D \mathbf{F}_2 \cdot d\mathbf{s}_2$$

$\mathbf{F}_1 = -\mathbf{F}_2$ because of Newton's third law. If $m_2$ is the earth, its displacement will always be so small that $\Delta U_2$ is close to zero. Example 7 shows that $\Delta U_2 / \Delta U_1$ is typically on the order of $10^{-24}$ for the earth. So, if our system consists of a single mass $m$ and the earth, we will use

$$U_B - U_A = -\int_A^B \mathbf{F} \cdot d\mathbf{s} \qquad \text{(change in potential energy)} \quad (6\text{-}12)$$

where **F** is the interaction force between $m$ and the earth. We see that potential energy is defined as the negative work done by the interaction force. **Change in potential energy is the positive work that must be done on a body to move it slowly from $A$ to $B$ in the presence of the interaction force.** (In order to move a body slowly, the applied force must be equal and opposite to the interaction force.)

## Conservative Forces

There is a restriction on the kind of force that can be used in Eq. 6-12. It must be what is called a conservative force. The definition of a conservative force is illustrated in Figure 6-8. If **F** is a conservative force, then

$$\int_A^B \mathbf{F} \cdot d\mathbf{s} \bigg|_{\text{path 1}} = \int_A^B \mathbf{F} \cdot d\mathbf{s} \bigg|_{\text{path 2}} \quad \text{(definition of conservative force)}$$

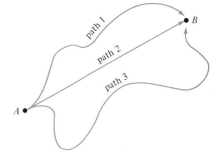

**Figure 6-8.** *Some possible paths between $A$ and $B$. For a conservative force $\int_A^B \mathbf{F} \cdot d\mathbf{s}$ is the same for each path.*

The work done by a conservative force acting on a body is independent of the path taken by the body in moving from any point $A$ to $B$. A mathematically equivalent statement is that $\int \mathbf{F} \cdot d\mathbf{s}$ around any closed path will be zero. Hence, with a conservative force it is not possible to gain (or lose) energy continuously by repeated circuits of the same path.

It turns out that all four of the basic forces that act between elementary particles are conservative. The same must be true of a force that can be reduced to one of the basic forces, such as the force on a mass attached to the end of a stretched spring. If the spring is stretched, the atoms are slightly pulled apart and there is an electrostatic attraction between them proportional to the amount of stretch.

Friction is an example of a nonconservative force. In such a case **F** is always antiparallel to $d\mathbf{s}$ and $\int \mathbf{F} \cdot d\mathbf{s}$ around a closed path would always be negative (the body would continuously lose energy). At this point we might ask how there can be a nonconservative force if all forces are made up of the basic forces and these are all conservative? The answer is that, if we keep track of the potential and kinetic energy of each elementary particle, there are no nonconservative forces. This is called the microscopic approach. However, friction is strictly a macroscopic phenomenon where we can ignore what is happening to the individual particles. A frictional force is due to an average transfer of momentum to the constituent particles (which displays itself as an increase in temperature). Thus, as the kinetic energy of a body experiencing friction is decreasing, the kinetic energy of its particles is increasing (getting warmer). Any force that generates heat is a nonconservative force. As we shall see in Chapter 12, heat is the kinetic and potential energy of the individual particles.

In this section we have learned how to obtain the potential energy once the force is known. Now we shall deal with the reverse problem: how to obtain the force if the potential energy is known. We start with the defining equation, Eq. 6-12, and take points $A$ and $B$ very close together in the $s$ direction.

$$dU = -F_s \, ds$$

If we divide both sides by $ds$, we obtain

$$F_s = -\frac{dU}{ds} \quad (dU \text{ in } s \text{ direction}) \quad (6\text{-}13)$$

For example, if $U$ is known as a function of $x$, $y$, and $z$, then

$$F_x = -\frac{dU}{dx} \qquad F_y = -\frac{dU}{dy} \qquad F_z = -\frac{dU}{dz}$$

for $dU$ in the $x$, $y$, and $z$ directions, respectively.

## 6-7 Gravitational Potential Energy

We wish to determine the potential energy of a mass $m$ at a distance $h$ from the earth's surface as shown in Figure 6-9. In Chapter 5 we learned that Newton's universal law of gravitation says that the force on a mass $m$ at a distance $r$ from the center of the earth is $F = -mg(R_e^2/r^2)$, where $R_e$ is the earth's radius. (The minus sign indicates the direction of the force.) Inserting this expression for $F$ into Eq. 6-12 gives

$$U - U_e = -\int_{R_e}^{r} \left(\frac{-mgR_e^2}{r^2}\right) dr$$

where $U_e$ is the potential energy at the surface of the earth. Now we complete the integration:

$$U - U_e = mgR_e^2 \int_{R_e}^{r} r^{-2} \, dr = mgR_e^2 \left[-\frac{1}{r}\right]_{R_e}^{r}$$

$$U - U_e = mgR_e^2 \left(\frac{1}{R_e} - \frac{1}{r}\right) \quad \begin{array}{l}\text{(gravitational potential} \\ \text{energy at a distance } r \\ \text{from earth's surface)}\end{array} \quad (6\text{-}14)$$

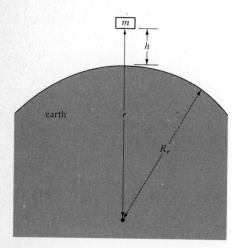

**Figure 6-9.** *Mass m at height h above surface of earth.*

Equation 6-14 describes the work required to move $m$ a distance $h$ above the surface of the earth where $h = (r - R_e)$. We note that the energy required to move $m$ an infinite distance from the earth is

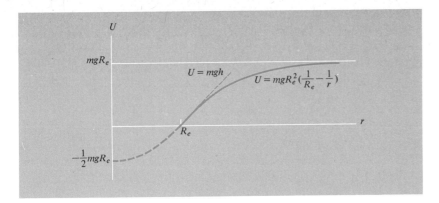

**Figure 6-10.** *Plot of gravitational potential energy measured from the surface of the earth. Dashed line is potential energy inside the earth.*

$$U - U_e = mgR_e^2\left(\frac{1}{R_e} - \frac{1}{\infty}\right) = mgR_e$$

This can be seen in Figure 6-10 where Eq. 6-14 is plotted as a function of $r$. When $m$ is near the surface of the earth, we have $R_e/r \approx 1$ and obtain the following limiting expression

$$U - U_e = mgR_e^2\left(\frac{r - R_e}{R_e r}\right) = mg\frac{R_e}{r}h$$

$$U - U_e \approx mgh \tag{6-15}$$

**Example 8.** What is the gravitational potential energy inside the earth and at the center of the earth?

ANSWER: According to Eq. 5-10, inside the earth $F = -mgr/R_e$. Then

$$U - U_e = -\int_{R_e}^{r}\left(-mg\frac{r}{R_e}\right)dr = \frac{mg}{R_e}\int_{R_e}^{r} r\,dr = \frac{mg}{R_e}\left[\frac{r^2}{2}\right]_{R_e}^{r}$$

$$U - U_e = -\frac{mg}{2R_e}(R_e^2 - r^2)$$

This equation is plotted as the dashed line in Figure 6-10. For $r = 0$ we have

$$U - U_e = -\tfrac{1}{2}mgR_e$$

In Figure 6-10 we have put the surface of the earth at zero potential energy. We could just as well have defined zero potential energy at the center of the earth, or at $r = \infty$, as will be done in the next chapter. The position in space at which we define zero potential energy is arbitrary. As we shall see in the next chapter, what is physically relevant are changes in potential energy.

## 6-8 Potential Energy of a Spring

Figure 6-11 shows an unstretched spring. We place the origin of the $x$ axis at the end of the spring. According to Hooke's law, the conservative force exerted by the spring will be $F = -kx$ where $k$ is called the spring constant. The minus sign indicates that the spring pulls to the left when it is stretched. If the spring is compressed, $x$ is negative and the spring force is pushing to the right. We define $U = 0$ at $x = 0$ and use Eq. 6-12:

$$U = -\int_0^x (-kx)\, dx = k \int_0^x x\, dx$$

$$U = \frac{kx^2}{2} \quad \text{(potential energy of a spring)} \quad (6\text{-}16)$$

Equation 6-16 is plotted in Figure 6-12(b). The corresponding force is shown in Figure 6-12(a).

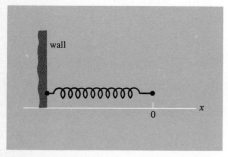

**Figure 6-11.** *Unstretched spring.*

**Figure 6-12.** (*a*) *Force as a function of $x$ for the spring in Figure 6-11.* (*b*) *The corresponding potential energy.*

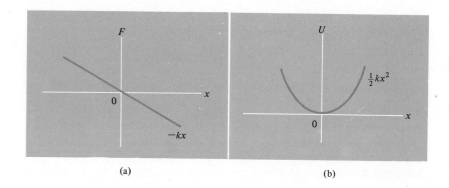

## Summary

Work $W$ done by a force $F$ moving a body from $A$ to $B$ is

$$W \equiv \int_A^B \mathbf{F} \cdot d\mathbf{s}$$

The units are newton meters or joules; 1 N acting on any body does work of 1 J if the body moves 1 m in the direction of the force.

Power is the rate of doing work or transferring energy:

$$P = \frac{dW}{dt}$$

The dot product between two vectors: $\mathbf{A} \cdot \mathbf{B} \equiv AB \cos \alpha$ where $\alpha$ is the angle between them.

Kinetic energy is defined as

$$K \equiv \tfrac{1}{2}mv^2.$$

Work-energy theorem:

$$\int_A^B \mathbf{F}_{net} \cdot d\mathbf{s} = K_B - K_A$$

If $\mathbf{F}$ is a conservative force (or the vector sum of conservative forces) the increase in potential energy is

$$(U_B - U_A) = -\int_A^B \mathbf{F} \cdot d\mathbf{s}.$$

This integral is independent of path for conservative forces. Friction is a nonconservative force. Nonconservative forces acting on a body generate heat energy.

The gravitational potential energy of a mass $m$ acted on by a spherical mass $M$ is

$$U = GMm \left( \frac{1}{R} - \frac{1}{r} \right)$$

where $R$ is the radius and $U = 0$ at $r = R$. The position in space where $U$ is set equal to zero is arbitrary.

## Exercises

1. The present annual energy consumption in the United States is about $7.5 \times 10^{16}$ Btu (1 Btu/h $= 0.293$ W). The average solar energy striking the earth is $\sim 1000$ W/m² when the sun is shining. If the area of the United States is $8 \times 10^6$ km², what is the ratio of present energy consumption to energy input from the sun?
2. A person slowly carries a 10-kg mass a horizontal distance of 5 m. How much work does the person do on the mass?

3. An earth satellite of mass 100 kg is in circular orbit of radius $R = 7000$ km.
   (a) What is the work done on the satellite by the earth's gravitational force in one half orbit?
   (b) Assume the orbit is slightly elliptical and that during the half-orbit it increases its distance by 10 km. Now what is the work done? Is it positive or negative?
4. A 1-ton car starts from rest with a constant acceleration of 100 m/s². How much horsepower must the engine deliver
   (a) at the start?
   (b) After 1 s?
   (c) After 10 s?
   (d) What is the velocity after 10 s?
5. In the United States, annual electric power consumption is $\sim 7$ quads. (1 quad = $10^{15}$ Btu; 1 Btu/hr = 0.293 W). If this were used at a constant rate, how many megawatts (MW) would be used?
6. An unstretched spring is of length $x_0$. It is stretched by a force $F_1$ to a distance $x_1$ where $x_1 > x_0$. Then it is stretched further to $x_2$ where $x_2 > x_1$. How much work was done in pulling the spring from $x_1$ to $x_2$? Give answer in terms of $F_1$, $x_0$, $x_1$, and $x_2$.
7. The potential energy of a body is $U = Ax^2$. What is the force acting on it?
8. Assuming the earth is of uniform density, the potential energy of a mass $m$ on the surface with respect to the center is [positive; negative] and is of magnitude [$GM_e m/R_e$; $\frac{1}{2}mgR_e$; $mgR_e$; none of these]. The potential energy of a mass $m$ on the surface with respect to a point at infinity is [positive; negative] and is of magnitude [$\frac{1}{2}mgR_e$; $mgR_e$; $2mgR_e$; none of these]. The potential energy of a mass $m$ at the center with respect to infinity is of magnitude $mgR_e$ times [$\frac{1}{2}$; 1; 1.5; 2; 3].
9. A bedroom of volume 4 m × 4 m × 2.5 m is sealed tight. Two persons are sleeping in it for 8 h. Will they use up all the oxygen? If not, what percentage will they use? The initial density of oxygen is 0.26 kg/m³. Each person generates 90 W while asleep; 1 g of oxygen is used up for every $10^4$ J of body energy. Is ventilation a necessity?
10. An unstretched spring of length $x_0$ is extended by an additional length $x_1$, at which point the spring force is $F_1$. How much work was done on the spring?
11. A 5-g charged body is moved to the right from point $A$ to point $B$. Assume the body experiences a constant electrostatic force of $2 \times 10^{-5}$ N pointing to the left. How much work must be done to move the body from $A$ to $B$, a distance of 1.5 m? Did its potential energy increase or decrease?
12. A firecracker is moving with a speed of 5 m/s. Suppose it explodes into just two pieces of equal mass. If one of the two pieces has zero speed just after the explosion, what is the ratio of the final kinetic energy to the initial kinetic energy?
13. A 1.2-kg toy train is being pulled with a constant force of $10^{-3}$ N. At first the train speeds up from rest and then reaches a constant velocity $v_0$. If $5 \times 10^{-4}$ J of work is done every second, what is $v_0$?
14. A car is running at 50 hp of mechanical energy. It takes 5 times this amount of fuel energy to produce the mechanical energy in an automobile engine. Suppose this same fuel energy could run an electric generator with 90% efficiency. How many kilowatts of electric energy would be generated? If a

typical home uses 3 kW of electricity on the average, how many homes is this car equivalent to energywise?

## Problems

15. Suppose a force on a particle increases as the square of the distance from the origin ($F = kx^2$). What is the increase in potential energy of the particle as it is moved from $x = 0$ to $x = x_1$?
16. The potential energy of a particle is $U = A/r = A(x^2 + y^2 + z^2)^{1/2}$. What are $F_x$, $F_y$, and $F_z$?
17. Consider a hollow sphere with inner radius $R_1$ and outer radius $R_2$. The potential energy of mass $m$ when at the center is zero.
    (a) What is $U$ of mass $m$ when at a distance $r$ from the center in region I?
    (b) Repeat for region II. The hollow sphere has total mass $M$.
    (c) Repeat for region III.
18. If $U = A/r^2 = A/(x^2 + y^2 + z^2)$, what are $F_x$, $F_y$, and $F_z$?
19. How far must a 1-kg mass drop in order to achieve an increase in kinetic energy of 100 J? How long does it take? Are both answers independent of its initial velocity?
20. A boy pulls a 5-kg sled at constant velocity of 0.5 m/s with a force of 10 N at 30° to the horizontal.
    (a) What is the frictional force in newtons?
    (b) What is the vertical component of force of the ground pushing up on the sled in newtons?
    (c) What is the coefficient of kinetic friction?
    (d) What is the rate of energy loss due to friction?
21. A 1500-kg car is traveling at 32 m/s on a level road. The driver turns off the gas. In 3 s the car slows down to 28 m/s.
    (a) What is the net frictional force acting on the car?
    (b) How many watts must the engine supply to keep the car going at a speed of 30 m/s?
    (c) If gasoline can supply $3 \times 10^7$ J/gal of mechanical energy, how far can the car go on 1 gal at 30 m/s?
22. A ball of mass $m$ is attached to a spring and the spring is attached to a fixed point $P$ as shown. The spring cannot bend. The ball is moving in a circle of radius $R$ (in a horizontal plane) with an angular velocity $\omega$ (rad/s) and the spring constant is $k$. The spring is massless and the plane on which $m$ moves is frictionless.
    (a) What is the tension in the spring at the point where it attaches to $m$ if $m = 1$ kg, $\omega = 1$ rad/s, and $R = 1.0$ m?
    (b) If the relaxed length of the spring is 0.9 m, what is the spring constant $k$?
    (c) If the ball and spring now rotate with $\omega = 2$ rad/s, what is the new radius of the ball's path (radius accurate to 1%)?
    (d) What work has to be done on the ball and spring to increase $\omega$ from 1 rad/s to 2 rad/s?
23. A point mass $M_1$ is at the center of a thin spherical shell that has mass $M_2$ and radius $R$. A mass $m$ is brought in from infinity to a distance $r$.
    (a) What is its potential energy (as measured from infinity) when $r > R$?
    (b) What is its potential energy when $r < R$?

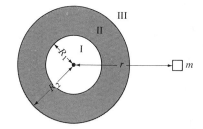

Problem 17

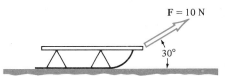

Problem 20

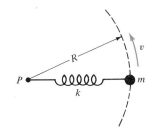

Problem 22

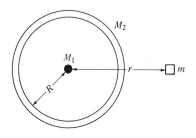

Problem 23

**24.** A solid sphere of density $\rho$ and radius $r$ has a mass $m = \frac{4}{3}\pi r^3 \rho$. If a mass $dm = (4\pi r^2 \rho \, dr)$ is added as a shell of thickness $dr$, the change in potential energy is

$$dU = -G\frac{m\,dm}{r} = -G\frac{(\frac{4}{3}\pi r^3 \rho)(4\pi r^2 \rho \, dr)}{r}.$$

Show that the gravitational potential energy of a solid sphere of radius $R$ and mass $M$ is $U = -\frac{3}{5}GM^2/R$. (Then $\frac{3}{5}GM^2/R$ is the work required to disassemble the sphere and move all mass elements $dm$ to infinity.)

# Conservation of Energy

The law of conservation of energy is a central theme that runs through all of physics and engineering. It puts severe restrictions on the availability of energy and on the conversion of energy from one form to another.

The law of conservation of energy forbids perpetual motion machines of a kind where a closed system continually delivers mechanical energy to the outside, such as the water wheel in Figure 7-1. For ages people have been trying to invent a perpetual motion machine. In spite of the law of conservation of energy, there are still attempts to achieve perpetual motion. Such "nonbelievers" come up with complicated systems of belts, pulleys, swinging and floating weights, and so on. From Figure 7-1 we see that it is easier to achieve perpetual motion on paper than in reality.

## 7-1 Conservation of Mechanical Energy

First we consider the case where the only force acting on a body is a conservative force, $F$. Then the conservative force is the net force, and Eq. 6-10 applies:

$$\int_A^B \mathbf{F} \cdot d\mathbf{s} = K_B - K_A$$

According to the definition of potential energy (Eq. 6-12), the left-hand side is $-(U_B - U_A)$, the negative of the potential energy change. Making this substitution gives

$$-(U_B - U_A) = K_B - K_A$$

$$K_A + U_A = K_B + U_B \quad \text{(conservation of mechanical energy)} \quad (7\text{-}1)$$

**Figure 7-1.** *Waterfall, M. C. Escher, lithograph, 1961.* In the words of the artist: "Falling water keeps a millwheel in motion and subsequently flows along a sloping channel between two towers, zigzagging down to the point where the waterfall begins again. The miller simply needs to add a bucketful of water from time to time, in order to compensate for loss through evaporation." Then Escher goes on to analyze the visual illusion upon which his drawing is based. [*Escher Foundation—Haags Gemeentemuseum, The Hague*]

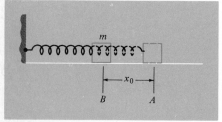

**Figure 7-2.** *Spring is stretched from B to A and then released.*

Equation 7-1 is known as the conservation of mechanical energy, and it applies to a body under the influence of a conservative force whose potential energy is $U$. It says that the sum of kinetic and potential energies of such a body stays constant as long as no other forces are applied.

**Example 1.** A mass $m$ is attached to the end of a massless spring of force constant $k$. The spring is stretched a distance $x_0$. If the mass is released, what will be its maximum velocity? See Figure 7-2.

ANSWER: Before the mass is released, its kinetic energy is zero and its potential energy is $U_A = \frac{1}{2}kx_0^2$ (see Eq. 6-16). Let point $A$ correspond to the stretched position. Then $K_A = 0$ and Eq. 7-1 becomes

$$0 + \tfrac{1}{2}kx_0^2 = \tfrac{1}{2}mv_B^2 + U_B$$

The mass $m$ has its maximum velocity when $U_B = 0$. Then $v_B$ is $v_{max}$, and the preceding equation becomes

$$0 + \tfrac{1}{2}kx_0^2 = \tfrac{1}{2}mv_{max}^2 + 0$$

$$v_{max} = \sqrt{\frac{k}{m}}\, x_0$$

The law of conservation of mechanical energy can be used to solve for final (or initial) velocities in systems where the force as a function of time is complicated or difficult to evaluate. Consider the following two examples.

**Example 2.** A mass $m$ is hanging by a string of length $l$. What velocity $v_0$ must be imparted to it in order for it to just barely reach the top? See Figure 7-3(a).

ANSWER: We shall call the top position point $B$. When passing through point $B$ the centripetal acceleration must be equal to the gravitational acceleration $g$:

$$\frac{v_B^2}{l} = g \quad \text{or} \quad v_B^2 = gl$$

Then $K_B = \tfrac{1}{2}mv_B^2 = \tfrac{1}{2}m(gl)$ and $U_B = mg(2l)$. The sum of these two energies must equal the initial kinetic energy of $\tfrac{1}{2}mv_0^2$. (The initial potential energy is zero.) So

$$\tfrac{1}{2}mv_0^2 = \tfrac{1}{2}mgl + 2mgl$$
$$v_0^2 = 5gl \quad \text{or} \quad v_0 = \sqrt{5gl}$$

**Example 3.** Repeat Example 2 but for the case where $m$ is hanging by a massless rigid rod rather than a string.

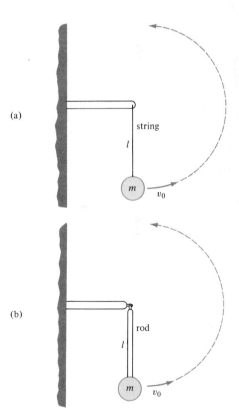

**Figure 7-3.** (a) Mass $m$ hanging by string of length $l$. (b) Mass $m$ hanging by massless rod of length $l$.

ANSWER: Now the rod can support the mass against gravity and $v_B$ can approach zero as $m$ reaches the top. Then $K_B = 0$ when $U_B = 2mgl$ and

$$\tfrac{1}{2}mv_0^2 = 0 + 2mgl \quad \text{or} \quad v_0 = 2\sqrt{gl}$$

Note that if the mass is released from the top position we have the same result, only in reverse. The velocity at the bottom will still be $2\sqrt{gl}$. Hence this result applies to the rocket ride described in Example 7 of Chapter 5.

There can be more than one source of the potential energy of a body. In such a case the potential energy $U$ is the sum of the separate contributions. For example, a mass $m$ hanging from a spring has

$$U = mgy + \tfrac{1}{2}k(y - y_0)^2$$

where $y$ is the height of $m$ and $y_0$ is its height when the spring force is zero. We shall consider such a case in Example 4.

\* **Example 4.** In this example we shall demonstrate that a roped-up rock climber should be able to survive a free fall of any height; that is, the climber's greatest instantaneous acceleration will be less than $25g$. We shall treat a nylon rope as a spring that obeys Hooke's law until it has stretched 25% of its length, at which point it breaks. The climber chooses a rope whose breaking strength is 25 times his or her weight ($F_B = 25mg$).
Since $F_B = k(0.25l)$, the force constant $k$ is

$$k = \frac{F_B}{0.25l} = \frac{25mg}{0.25l} = 100\frac{mg}{l}$$

We suppose the climber falls after having climbed a distance $l$ above the nearest point of support, as shown in Figure 7-4. Then the climber will free fall a distance $2l$. Let $y_{\max}$ be the maximum amount of stretch of the rope, at which point the climber's velocity is zero. The climber's potential energy at this point is

$$U_B = mg(h - l - y_{\max}) + \tfrac{1}{2}ky_{\max}^2$$

Just before the fall the climber's potential energy was

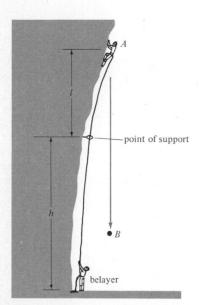

**Figure 7-4.** Climber at A falls a distance $2l$ and then rope stretches a distance $y_{\max}$ to reach point B. Rope is held by belayer so that it binds at point of support.

$$U_A = mg(h + l)$$

Since the kinetic energy is zero at both points $A$ and $B$, we have

$$U_A = U_B$$
$$mg(h + l) = mg(h - l - y_{max}) + \tfrac{1}{2}ky_{max}^2$$

Now replace $k$ with $(100\, mg/l)$:

$$mgl = -mgl - mgy_{max} + \frac{1}{2}\left(100\,\frac{mg}{l}\right)y_{max}^2$$
$$0 = 50y_{max}^2 - ly_{max} - 2l^2$$

The solution is $y_{max} = 0.21l$, which is within the $0.25l$ limit of breaking. We can estimate the maximum acceleration using Newton's second law, where $F_{net}$ is $ky_{max} - mg$

$$ma_{max} = ky_{max} - mg$$

$$a_{max} = \frac{ky_{max}}{m} - g = \frac{(100\, mg/l)(0.21l)}{m} - g$$

$$= 20g$$

We note that this result is independent of the height of the fall. We conclude that if a climber should fall from an overhang, he or she would probably survive—perhaps with some broken ribs. A more serious danger is that the body or head might strike protruding rocks on the way down.

So far we have derived the conservation of mechanical energy for a system consisting of a single body or a single body plus the "immovable" earth. However, the law is more general and holds for all closed, conservative systems of bodies. By closed system we mean no external forces, and by conservative we mean all the interaction forces are conservative and can thus be expressed via a potential energy function.

Let

$$K_t \equiv \sum_{j=1}^{N} K_j$$

be the total kinetic energy of all the particles in the closed system. It is shown in Appendix 7-1 that

$$(K_t)_1 + U_1 = (K_t)_2 + U_2 \quad \text{(conservation of mechanical energy for a closed system)}$$

where $U_1$ and $U_2$ are the potential energies of the system at times $t_1$ and $t_2$, respectively.

The preceding equation says that the sum of the total kinetic plus potential energy at a time $t_1$ equals the total kinetic plus potential energy at a time $t_2$. The symbol $E$ is usually used for this constant value of energy.

$$E = \sum K_j + U = \text{a constant} \quad \text{(conservation of energy for a closed system of particles)} \quad (7\text{-}2)$$

This says that the sum of all the kinetic and potential energies of all the bodies in the solar system (or any other closed system) stays the same no matter what kind of interactions and collisions might take place.

## 7-2 Collisions

When two solid bodies collide, a large contact force quickly builds up at the point of contact. Usually the force is so large that the bodies undergo a momentary compression at the point of contact. The large, but momentary, contact force causes a change in direction and magnitude of the velocity of each body.

### *Impulse*

The magnitude of the contact force during a collision can be estimated by defining a quantity called the impulse **I**. The impulse imparted to a body from time $t_1$ to time $t_2$ is defined as

$$\mathbf{I} \equiv \int_{t_1}^{t_2} \mathbf{F}\, dt \quad \text{(definition of impulse)} \quad (7\text{-}3)$$

The impulse can be related to the change in momentum of the body by replacing **F** with $d\mathbf{P}/dt$:

$$I = \int \left(\frac{d\mathbf{P}}{dt}\right) dt = \int d\mathbf{P} = \mathbf{P}_2 - \mathbf{P}_1$$

Hence

$$\int_1^2 \mathbf{F}\, dt = \mathbf{P}_2 - \mathbf{P}_1 \quad (7\text{-}4)$$

**Example 5.** A 1.5-ton car traveling at 20 m/s collides with a tree. It comes to rest in $3 \times 10^{-2}$ s, which corresponds to a deformation of 30 cm. What is the average force acting on the car during this time?

ANSWER: The average force $F$ times $\Delta t$, the duration of the collision, is the impulse by definition. So

$$\bar{F} \Delta t = P_2 - P_1 = mv$$

$$\bar{F} = \frac{mv}{\Delta t} = \frac{(1.5 \times 10^3)(20)}{3 \times 10^{-2}} \text{ N} = 1.0 \times 10^6 \text{ N}$$

This is about 70 times the weight of the car.

**Example 6.** In the collision described in Example 5 an 80-kg passenger is held by a seat belt that is 5 cm wide and 2 mm thick. If the breaking strength or tensile strength of seatbelt material is $5 \times 10^8$ N/m², will the seat belt break?

ANSWER: The average force on the seat belt during the collision is related to the change of momentum of the passenger by

$$\bar{F} \Delta t = 0 - mv$$

$$\bar{F} = -\frac{mv}{\Delta t} = -\frac{(80)(20)}{3 \times 10^{-2}} \text{ N} = -5.33 \times 10^4 \text{ N}$$

The average force on either side of the seat belt is about half this, or $2.67 \times 10^4$ N.

The cross sectional area is 5 cm times the thickness of 0.2 cm or 1 cm². Hence, the force per unit area is $2.67 \times 10^8$ N/m². This is a bit more than half the breaking strength, so the seat belt will hold.

If, in the collision described in these two examples, no seat belt were used and the passenger's head hit the windshield after it had come to rest, $\Delta t$ for the head-windshield collision would be perhaps a hundred times shorter than the time it took for the car to stop. Since $\bar{F}$ goes as $1/\Delta t$, the head would break.

We conclude our discussion of impulse by noting that in a collision between two bodies, the impulses will be equal and opposite because the forces are equal and opposite in accordance with Newton's third law. Then by Eq. 7-4 the change of momentum of one body will be equal in magnitude to the change in momentum of the other body. This is just another way of stating the law of conservation of momentum.

### Elastic Collisions

When two bodies collide, the collision is either elastic or inelastic. **In an elastic collision the total kinetic energy after the collision is the same as before the collision.** In an inelastic collision, there is a loss of kinetic energy. Discussion of inelastic collisions is deferred to Section 7-5 where the concept of heat energy is introduced.

We shall consider two common types of elastic collisions: (1) head-on collisions and (2) collisions in two dimensions between bodies of equal mass.

Consider the head-on collision illustrated in Figure 7-5. The initial speeds of $m_1$ and $m_2$ are $v_1$ and 0, respectively. The speeds just after the collision are $V_1$ and $V_2$. The problem is to determine $V_1$ and $V_2$ in terms of $v_1$. Since there are two unknowns, we need two simultaneous equations. We use the equality of total kinetic energy and total momentum to supply the two equations.

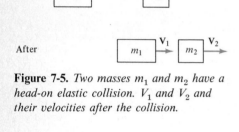

**Figure 7-5.** *Two masses $m_1$ and $m_2$ have a head-on elastic collision. $V_1$ and $V_2$ and their velocities after the collision.*

$$\tfrac{1}{2} m_1 v_1^2 = \tfrac{1}{2} m_1 V_1^2 + \tfrac{1}{2} m_2 V_2^2 \quad \text{(equating kinetic energies)} \quad (7\text{-}5)$$

$$m_1 v_1 = m_1 V_1 + m_2 V_2 \quad \text{(equating momenta)} \quad (7\text{-}6)$$

We obtain $V_2$ by solving Eq. 7-6 for $V_1$ and substituting into Eq. 7-5. The result is

$$V_2 = \frac{2 m_1}{m_1 + m_2} v_1 \quad (7\text{-}7)$$

If Eq. 7-7 is substituted into Eq. 7-6, we obtain

**Figure 7-6.** *If a ball strikes one end, it exchanges velocity via successive collisions until the ball at the other end leaves with the same velocity.*

$$V_1 = \frac{m_1 - m_2}{m_1 + m_2} v_1$$

Note that in the special case of equal masses, $V_1 = 0$ and $V_2 = v_1$; that is, the velocities are exchanged. This principle of velocity exchange is behind the explanation of the device shown in Figure 7-6.

For our second example of elastic collisions we consider the problem in billiards illustrated in Figure 7-7. The goal is to knock the target ball into the side pocket. If the angle to the side pocket is 30° as shown, what should be the angle $\theta$ for deflection of the cueball? We shall ignore friction and english (rotational motion imparted to the balls). As before, we equate the kinetic energies and divide by $\tfrac{1}{2}m$ to get

$$v_1^2 = V_1^2 + V_2^2 \tag{7-8}$$

The law of conservation of momentum gives

$$\mathbf{v}_1 = \mathbf{V}_1 + \mathbf{V}_2$$

Squaring both sides

$$v_1^2 = (\mathbf{V}_1 + \mathbf{V}_2) \cdot (\mathbf{V}_1 + \mathbf{V}_2) = V_1^2 + 2\mathbf{V}_1 \cdot \mathbf{V}_2 + V_2^2$$
$$v_1^2 = V_1^2 + V_2^2 + 2V_1 V_2 \cos \alpha \tag{7-9}$$

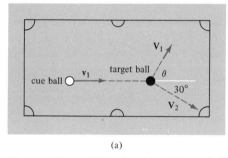

(a)

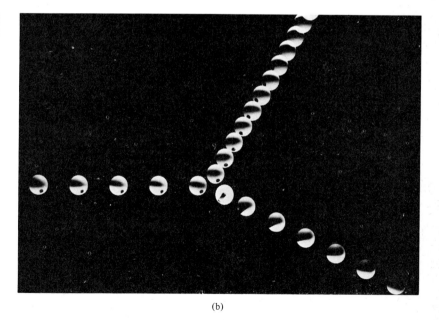

(b)

**Figure 7-7.** (a) Billiards table with cue ball and target ball. What angle $\theta$ must cue ball make in order to knock target ball into side pocket? (b) Strobe photo of a similar collision. [Photo Courtesy Educational Development Center]

where $\alpha$ is the angle between $\mathbf{V}_1$ and $\mathbf{V}_2$. Now subtract Eq. 7-8 from Eq. 7-9 to get

$$\cos \alpha = 0 \quad \text{or} \quad \alpha = 90°$$

We have just proved that in any elastic collision between two equal masses, the two bodies will move off at right angles to each other. Hence the solution to the problem in Figure 7-7 is that $\theta = 60°$.

## 7-3 Conservation of Gravitational Energy

### Velocity of Escape

Suppose a projectile of mass $m$ is fired straight up with velocity $v_1$. How high will it go? Could it escape the earth and go to $r = \infty$? Let its distance from the center of the earth be $r_2$ when at its maximum height. Then its kinetic energy $K_2 = 0$. At all times $K + U$ remains constant so

$$\tfrac{1}{2}mv_1^2 + U_1 = 0 + U_2$$
$$\tfrac{1}{2}mv_1^2 = U_2 - U_1$$

Using Eq. 6-14 for $U$ gives

$$\tfrac{1}{2}mv_1^2 = mgR_e^2 \left( \frac{1}{R_e} - \frac{1}{r_2} \right) \tag{7-10}$$

So

$$r_2 = \left( \frac{1}{R_e} - \frac{v_1^2}{2g} \right)^{-1}$$

is the maximum distance from the center of the earth where $R_e$ is the radius of the earth. We see from Eq. 7-10 that it is possible for $r_2$ to be infinite if $v_1$ is large enough. The velocity such that $m$ just barely reaches an infinite distance is called the velocity of escape, $v_0$. Putting $r_2 = \infty$ into Eq. 7-10 gives

$$\tfrac{1}{2}v_0^2 = gR_e^2 \left( \frac{1}{R_e} - 0 \right)$$
$$v_0 = \sqrt{2}\sqrt{gR_e} \quad \text{(velocity of escape)} \tag{7-11}$$

We recall from Eq. 3-11 that $\sqrt{gR_e} = v_c$ is the critical velocity or the velocity needed to go into a low circular orbit. We see from Eq. 7-11 that the escape velocity is $\sqrt{2}$ times as much. Since $v_c = 8$ km/s, the escape velocity is 11.2 km/s or 7 mi/s. In this discussion we have ignored the gravitational field of the sun (see Example 7).

**Example 7.** What is the velocity needed to escape from the solar system at a distance $R_0 = 93$ million miles from the sun (the earth to sun distance)? Give answer in terms of $R_0$ and $T_0$ (time for earth to orbit the sun).

ANSWER: The velocity of a satellite circling the sun at a distance $R_0$ is $v_c = 2\pi R_0/T_0$. The earth is such a satellite and has $T_0 = 1$ yr $= 3.15 \times 10^7$ s. So

$$v_c = \frac{2\pi \times 93 \times 10^6}{3.15 \times 10^7} \text{ mi/s} = 18.5 \text{ mi/s} = 30 \text{ km/s}$$

The escape velocity is $\sqrt{2}$ times as much or

$$v_0 = \sqrt{2}\frac{2\pi R_0}{t_0} = 42 \text{ km/s}$$

---

**Example 8.** What fraction of escape velocity is needed to send a rocket as far as the moon?

ANSWER: The distance to the moon is 60 earth radii. Substituting $r_2 = 60R_e$ into Eq. 7-10 gives

$$\tfrac{1}{2}v_1^2 = gR_e^2\left(\frac{1}{R_e} - \frac{1}{60R_e}\right)$$

$$v_1^2 = 2gR_e\tfrac{59}{60}$$

$$v_1 = \sqrt{\tfrac{59}{60}}\sqrt{2gR_e} = 0.992v_0$$

More than 99% of escape velocity is needed in order to travel as far as 240,000 mi from the earth.

---

### Energy in Circular Orbit

In Section 6-7 we measured gravitational potential energy from the surface of the earth. This is not very practical when working problems involving other planets or the sun. We need a common position in space from which to measure gravitational potential energy. The convention is to replace $R_e$ in Eq. 6-14 with $\infty$. For a mass $m$ at a distance $r$ from a body of mass $M$, the gravitational potential energy is then

$$U = -\int_\infty^r \mathbf{F} \cdot d\mathbf{s} = -\int_\infty^r \left(-G\frac{Mm}{r^2}\right) dr = GMm \int_\infty^r r^{-2}\, dr = GMm \left[-\frac{1}{r}\right]_\infty^r$$

$$U = -\frac{GMm}{r} \quad \text{(gravitational potential energy measured from infinity)} \quad (7\text{-}12)$$

This is the work required to bring $m$ from infinity to a distance $r$ from $M$. It is also the negative of the work required to move $m$ from position at $r$ to infinity.

For a small mass $m$ in circular orbit of radius $R$ about a large mass $M$, the potential energy is $U = -GMm/R$. The centripetal acceleration is

$$\frac{v^2}{R} = \frac{F}{m} = G\frac{M}{R^2}$$

$$v^2 = G\frac{M}{R} \quad (7\text{-}13)$$

Multiplying both sides by $\tfrac{1}{2}m$ gives

$$\frac{1}{2}mv^2 = \frac{GMm}{2R}$$

This is the kinetic energy. Note that it is one half the potential energy in magnitude. The total mechanical energy is

$$E = K + U = \frac{GMm}{2R} + \left(-\frac{GMm}{R}\right) = -G\frac{Mm}{2R}$$

We note that the total energy $E$ and the kinetic energy have the same magnitude and opposite sign. We shall encounter the same situation when we study the Bohr model of the hydrogen atom.

---

**\*Example 9.** If a rocket is to escape from the solar system, what must be its initial velocity $v_1$ as it leaves the surface of the earth?

ANSWER: In this case the potential energy of the rocket is the sum of potential energy due to the earth alone plus the potential energy due to the sun alone. For the earth alone Eq. 7-12 gives

$$U_e = -G\frac{mM_e}{r_e}$$

where $r_e$ is the distance to the earth. For the sun alone

$$U_s = -G\frac{mM_s}{r_s}$$

where $r_s$ is the distance to the sun.

The initial total potential energy is the sum:

$$U_1 = -G\frac{mM_e}{R_e} - G\frac{mM_s}{R_{es}}$$

where $R_e$ is earth's radius and $R_{es}$ is the earth to sun distance. Let $K_\infty$ and $U_\infty$ be the kinetic and potential energies at infinity. Using $K_1 + U_1 = K_\infty + U_\infty$ gives

$$\tfrac{1}{2}mv_1^2 + \left(-G\frac{mM_e}{R_e} - G\frac{mM_s}{R_{es}}\right) = 0 + 0$$

$$v_1^2 = 2G\frac{M_e}{R_e} + 2G\frac{M_s}{R_{es}} = 2v_{ce}^2 + 2v_{cs}^2$$

using Eq. 7-13, where $v_{ce}$ is circular velocity around earth and $v_{cs}$ is circular velocity around sun. Then, since $v_{0e}^2 = 2v_{ce}^2$ and $v_{0s}^2 = 2v_{cs}^2$,

$$v_1^2 = v_{0e}^2 + v_{0s}^2$$

where $v_{0e}$ is the velocity of escape from the earth alone (Eq. 7-11) and $v_{0s}$ is the velocity of escape from the sun at the position of the earth's orbit (see Example 7). So

$$v_1 = \sqrt{(11.2 \text{ km/s})^2 + (42 \text{ km/s})^2} = 43.5 \text{ km/s}$$

However, the rocket already has 30 km/s of this due to the orbital velocity of the earth. Hence the additional velocity needed is 13.5 km/s.

## 7-4 Potential Energy Diagrams

Since $K + U$ is always constant for a closed system, the value of $K$ at any position can be obtained conveniently from what is called a potential energy diagram. The curve in Figure 7-8 is a plot of $U$ versus $r$ for the gravitational potential energy of two masses $m_1$ and $m_2$. The horizontal line corresponds to the fixed energy of the system $E = K + U$. The value of $K$ can be found easily since $K = E - U$. Graphically $K$ is always the

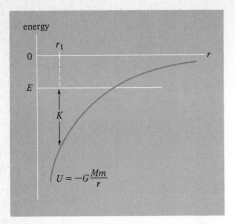

**Figure 7-8.** *Potential energy diagram for gravitational force between masses M and m.*

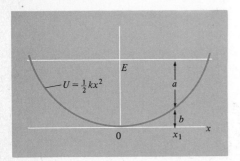

**Figure 7-9.** *Potential energy diagram for a spring having a spring constant k.*

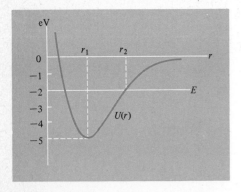

**Figure 7-10.** *Plot of potential energy for two atoms with a distance r between their centers.*

vertical distance between the horizontal line and the curve. This is illustrated in Figure 7-8 for $r = r_1$.

The potential energy diagram for a spring is shown in Figure 7-9. At $x = x_1$ we would like to express $E$, $K$, and $U$ in terms of $a$ and $b$. If $a$ and $b$ are positive quantities, then $E = a + b$, $K = a$, and $U = b$.

There are many instances in physics where the force is complicated and cannot be expressed by a simple analytical function. In many such cases the potential energy curve can be obtained by numerical methods or from computer programs. Then potential energy diagrams are of great value.

**Example 10.** Figure 7-10 is a plot of a typical potential energy $U(r)$ between two atoms in a molecule where $r$ is the distance between their centers.

(a) At $r = r_1$, what are $E$, $K$, and $U$?
(b) At $r = r_2$, what are $E$, $K$, and $U$?
(c) What is the net force at $r_1$?
(d) If $r = r_1$ and the kinetic energy is zero, what then is $E$? (It will be different from that shown in the figure.)

ANSWER: At all values of $r$, $E = -2$ eV, as indicated by the white horizontal line in the figure. The values of $U(r)$ can be read off the curve. At $r = r_1$, $U(r) = -5$ eV. Similarly $U(r_2) = -2$ eV. The kinetic energy is obtained using $K = E - U$.
At $r = r_1$, $K = (-2 \text{ eV}) - (-5 \text{ eV}) = 3$ eV.
At $r = r_2$, $K = (-2 \text{ eV}) - (-2 \text{ eV}) = 0$.
The force can be obtained using Eq. 6-13:

$$F = -\frac{dU}{dr}$$

At $r = r_1$ this derivative is zero; hence, $F(r_1) = 0$.
In part (d) the energy $E$ would be

$$E = K + U = 0 + (-5 \text{ eV}) = -5 \text{ eV}$$

In this case we would have a stable molecule with no atomic motion or vibration.

### Equilibrium

A body is in equilibrium when it is at rest and the net force on it is zero. The net force will be zero when the body is at a local minimum or at a local maximum in the potential energy curve. In the first case the equilibrium is stable, and in the second case it is unstable (a small additional force would set the body in motion).

## 7-5 Conservation of Total Energy

Now we shall go to the more general case where, in addition to a conservative force, which is a function of position only, there may be both a frictional force $F_f$ and an external force $F_{ext}$. Let $\mathbf{F}_c$ stand for the conservative force, and $\mathbf{F}_{ext}$ for the external force that is being applied by some outside agent. The net force is then

$$\mathbf{F}_{net} = \mathbf{F}_c + \mathbf{F}_f + \mathbf{F}_{ext}$$

Application of the work-energy theorem (Eq. 6-10) gives us $\Delta K$, the increase in kinetic energy.

$$\int_A^B (\mathbf{F}_c + \mathbf{F}_f + \mathbf{F}_{ext}) \cdot d\mathbf{s} = \Delta K$$

$$\int_A^B \mathbf{F}_{ext} \cdot d\mathbf{s} = \Delta K + \left(-\int_A^B \mathbf{F}_c \cdot d\mathbf{s}\right) + \int_A^B (-\mathbf{F}_f) \cdot d\mathbf{s}$$

The second term on the right-hand side is by definition $\Delta U$. So

$$\int_A^B \mathbf{F}_{ext} \cdot d\mathbf{s} = \Delta K + \Delta U + \int_A^B \mathbf{F}'_f \cdot d\mathbf{s}$$

where $\mathbf{F}'_f$ is the frictional reaction force, that is, the force of the body pushing against the frictional surfaces. We note that $\int \mathbf{F}'_f \cdot d\mathbf{s}$ is the work done by the body in heating up itself and its surroundings. Physically this heat is the work done to supply additional kinetic and potential energy to the individual particles (atoms and molecules). From a macroscopic point of view this is called the internal energy, $U_{int}$. It is simply the kinetic and potential energies given to the constituent particles that are not included in $K$ and $U$ of the body as a whole. Hence

$$\int_A^B \mathbf{F}_{ext} \cdot d\mathbf{s} = \Delta K + \Delta U + \Delta U_{int} \quad \text{(law of conservation of energy)} \quad (7\text{-}14)$$

Equation 7-14 says that any work done on a body by an external agent equals the increase in kinetic energy plus the increase in potential energy plus the increase in internal energy. All the energy is accounted for—none gets lost. We have an overall conservation of energy.

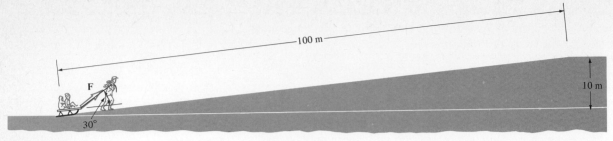

**Figure 7-11**

**Example 11.** Assume sled (plus passengers) has mass of 50 kg. Assume frictional force is 20 N.
(a) How much work must the man do in pulling the sled up a 100-m hill of height 10 m?
(b) If the sled is released from the top, what will be the kinetic energy of the sled when it reaches the bottom? What will be its speed?

ANSWER: (a) Eq. 7-14 says the work done by the man is

$$\int \mathbf{F}_{ext} \cdot d\mathbf{s} = \Delta K + \Delta U + \Delta U_{int}$$
$$= 0 + mgh + |F_f|s$$
$$= 0 + (50)(9.8)(10) + (20)(100) = 6900 \text{ J}$$

(b) Applying Eq. 7-14 for the trip down gives

$$0 = \Delta K + \Delta U + \Delta U_{int} \quad (\text{since } \mathbf{F}_{ext} = 0)$$
$$= \Delta K + (-mgh) + |F_f|s$$
$$\Delta K = mgh - |F_f|s = (50)(9.8)(10) - (20)(100) = 2900 \text{ J}$$
$$\tfrac{1}{2}mv^2 = 2900 \text{ J}$$
$$v = \sqrt{\frac{5800}{50}} \text{ m/s} = 10.8 \text{ m/s}$$

### Inelastic collisions

So far we have only studied elastic collisions where the total kinetic energy after the collision is the same as before the collision. However, when most macroscopic bodies collide, some kinetic energy is lost and converted into heat energy. Let us consider the extreme example where the two bodies stick together after the collision. It could be a billiard ball collision with chewing gum on the target ball, or it could be a head-on collision between a truck and a car as shown in Figure 7-12. If the initial

**Figure 7-12.** *A truck-car collision.*

velocities $v_1$ and $v_2$ are known, we should be able to calculate the final velocity $V$ and the amount of heat energy produced. In any collision, whether elastic or inelastic, total momentum is conserved, so

$$m_1 v_1 + m_2 v_2 = (m_1 + m_2) V$$

$$V = \frac{m_1 v_1 + m_2 v_2}{m_1 + m_2}$$

For example, if $v_1 = -v_2 = 100$ km/h, $m_1 = 15$ tons, and $m_2 = 1.5$ tons,

$$V = \frac{(15 \times 10^3)(100) + (1.5 \times 10^3)(-100)}{16.5 \times 10^3} \text{ km/h} = 81.8 \text{ km/h}$$

The truck only loses about 20% of its velocity, whereas the car reverses direction and is dragged along with the truck. This is one reason why collisions are usually not quite as dangerous in heavier cars.

In the case where the car and truck do not stick and the final velocity of the truck (or car) is known, the final velocity of the car (or truck) can be found using conservation of momentum:

$$m_1 v_1 + m_2 v_2 = m_1 V_1 + m_2 V_2$$

---

**Example 12.** How much kinetic energy is lost in the truck-car collision previously described.

ANSWER:

$$\Delta K = \tfrac{1}{2} m_1 v_1^2 + \tfrac{1}{2} m_2 v_2^2 - \tfrac{1}{2} m V^2$$

$$= \frac{1}{2} \left[ m_1 v_1^2 + m_2 v_2^2 - (m_1 + m_2) \left( \frac{m_1 v_1 + m_2 v_2}{m_1 + m_2} \right)^2 \right]$$

$$= \frac{1}{2} \frac{m_1 m_2}{m_1 + m_2} (v_1 - v_2)^2$$

The relative velocity $(v_1 - v_2) = 200$ km/h $= 55.5$ m/s

$$\Delta K = \frac{1}{2} \frac{(15 \times 10^3)(1.5 \times 10^3)}{16.5 \times 10^3} (55.5)^2 \text{ J} = 2.1 \times 10^6 \text{ J}$$

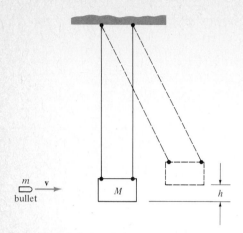

**Figure 7-13.** *The ballistic pendulum. Block rises a distance h after being struck by bullet.*

**Example 13.** The muzzle velocity of a bullet can be determined by using what is called a ballistic pendulum. The bullet is fired into the wooden block shown in Figure 7-13. The block with imbedded bullet climbs a vertical distance h.
(a) What is the muzzle velocity $v$ in terms of $m$, $M$, and $h$?
(b) What fraction of the kinetic energy is lost?

ANSWER: (a) According to the law of conservation of momentum

$$mv + 0 = (m + M)V$$

$$v = \frac{m + M}{m} V \qquad (7\text{-}15)$$

where $V$ is the velocity immediately after the collision. $V$ can be obtained by equating the kinetic energy immediately after the collision to the potential energy $(m + M)gh$ when at maximum height:

$$\tfrac{1}{2}(m + M)V^2 = (m + M)gh$$
$$V = \sqrt{2gh}$$

Hence

$$v = \frac{m + M}{m} \sqrt{2gh}$$

The kinetic energy after the collision is $K' = \tfrac{1}{2}(m + M)V^2$ or, using Eq. 7-15,

$$K' = \frac{1}{2}(m + M)\left(\frac{m}{m + M}v\right)^2$$

$$K' = \frac{m}{m + M}\left(\tfrac{1}{2}mv^2\right) = \frac{m}{m + M}K$$

$$\frac{K'}{K} = \frac{m}{m + M}$$

The fractional energy loss is 1 minus this quantity or $M/(m + M)$. We see that nearly all the kinetic energy is lost in such a collision.

### Internal Energy

As we have seen, internal energy is the additional energy of the individual particles that is not tallied up when the system is viewed macroscopically. If the individual particles change their form (as in a chemical reaction when molecules change from one type to another), the final molecules may have more energy than the initial. Such an increase in microscopic energy must be included in $U_{int}$ when viewing the system macroscopically.

Not only can molecules convert from one type to another, but so can the more elementary atomic nuclei and even elementary particles themselves. Such changes in nuclear energy must also be included in the term $\Delta U_{int}$. As we shall see in Chapter 9, all elementary particles have an intrinsic energy content $m_0 c^2$, where $m_0$ is the mass of the particle when at rest. We shall see that when an elementary particle changes from one type to another, there will be an energy release $(\Delta m) c^2$ where $\Delta m$ is the decrease in total rest mass involved. Such a term must also be included in $\Delta U_{int}$.

Internal energy and heat energy are discussed further in Chapter 12.

## 7-6 Energy and Biology

One common form of potential energy is chemical energy. When atoms combine to form a molecule there is an attractive force between them that does work and releases energy, which usually shows up as heat. In animals the usual source of chemical energy is carbohydrate (various carbon-hydrogen molecules), which combines with oxygen to form $H_2O$ and $CO_2$ with a release of energy. The typical energy release is 20,000 joules per gram of carbohydrate. Almost twice as much chemical energy per gram is contained in animal fat. When carbohydrate fuel is "burned" in muscle cells, about 25% of the energy can be released as mechanical work. A horse can burn fuel at the rate of 2000 W or do mechanical work at the rate of about 500 W over long periods of time. Over shorter periods a horse can do work at a rate of 700 or 800 W. The horsepower of 746 W was based on such measurements.

*Homo sapiens* is a smaller animal and at best can do only 100 W or so of mechanical work per unit time. Even when asleep, a typical adult human burns his fuel at a rate of about 80 W just in order to keep his bodily functions running. This is called the basal metabolic rate. The same amount of power is required to operate an average light bulb. When awake, as in a physics lecture, a student uses ~150 W, the basic 80 W plus about 40 W dissipated in the brain and 15 W in the heart. With moderate exercise, such as cycling 10 mph or swimming 1 mph, a person uses about 500 W. More vigorous exercise, such as basketball, requires about 700 W. With even more vigorous exercise, such as high speed bicycle racing, a human in good shape can exceed 1000 W, but only 100 W or so of this is transmitted outside the body as mechanical energy.

**Example 14.** For how long will 1 lb of fat supply the fuel needed to maintain moderate exercise (500 W)? Another way of stating the question is how long must an overweight person exercise in order to eliminate 1 lb of fat?

ANSWER: Fat contains about 40,000 J/g of fuel. 1 lb is 450 g, which contains $450 \times 40{,}000$ J or $18 \times 10^6$ J. Since $P = E/t$ we have

$$t = \frac{E}{P} = \frac{18 \times 10^6 \text{ J}}{500 \text{ W}} = 3.6 \times 10^4 \text{ s} = 10 \text{ h}$$

10 h of exercise may eliminate 1 lb of fat, but it is usually accompanied by a significant increase in appetite. Another approach to weight reduction would be to stop eating. Then about 0.7 lb of body fat per day would be used just to keep alive.

**Example 15.** How many Calories of food must one eat per day in order to stay alive? One food Calorie ($= 1$ kcal) contains 4180 J of chemical energy.

ANSWER: The minimum power used per day averages between the 80 W while asleep and 150 W while awake. Assume an average of 110 W.

$$E = Pt = (110 \text{ W})(8.6 \times 10^4 \text{ s}) = 9.5 \times 10^6 \text{ J needed per day}$$

This is the energy content of 2260 Calories (kcal) of food.

## 7-7 Energy and the Automobile

In this section we shall apply the basic principles learned thus far in order to estimate the power requirements and fuel consumption of a typical automobile. We shall obtain a basic understanding of the various factors that affect a car's performance. Those who have such a basic understanding are in a better position to optimize performance and conserve fuel.

If there were no friction, a car traveling at constant speed on a level road would use no power. Of course, power is needed to accelerate a car whether starting from rest or passing another car. We shall first calculcate the power or horsepower needed to accelerate a car from rest to 60 mph; then we shall discuss the main source of friction at high speeds and estimate the additional fuel consumption due to high speed travel.

A passenger car is considered to be performing well if it can accelerate from rest to 60 mph in 10 s. This corresponds to a constant acceleration of

$$a = \frac{v}{t} = \frac{88 \text{ ft/s}}{10 \text{ s}} = 8.8 \text{ ft/s}^2 \approx \tfrac{1}{4}g$$

First let us see if there is enough frictional force of pavement against the tires to supply this amount of acceleration. The frictional force $F_f = ma = m(\tfrac{1}{4}g)$. The normal force of the rear tires is $\sim \tfrac{1}{2}mg$. The ratio of $F_f$ to normal force is about $\tfrac{1}{2}$. Hence, the coefficient of friction must be at least 0.5. This is near the maximum coefficient of friction possible for typical passenger tires. Thus, acceleration times significantly shorter than 10 s are not possible for passenger cars. Racing and sports cars with special tires and more "weight" on the rear wheels can do better.

Now we come to the question of how much power the engine must have in order to make full use of this limiting frictional force. For a car of mass $m = 10^3$ kg (1 metric ton) the force must be $F_f = ma = (10^3 \text{ kg})(\tfrac{1}{4}g) \approx 2.5 \times 10^3$ N. As the car is reaching 60 mph, the power delivered by the engine must be (see Eq. 6-3)

$$P = Fv = (2.5 \times 10^3 \text{ N})(27 \text{ m/s}) = 68 \times 10^3 \text{ W} = 90 \text{ hp}$$

So a 1-ton car should have an engine that can deliver 90 hp at 60 mph if it is to be capable of the "ultimate" in performance. To apply more horsepower than this would cause the wheels to spin without any increase in performance. Additional horsepower would be wasteful. Note that $Fv = 0$ just as the car is starting from rest. At this instant of time zero power is needed, and one must take care not to spin the wheels at the start.

### Air Resistance

Now we shall make use of basic principles to estimate how much power is needed to overcome air resistance when the car is traveling at constant speed. As we shall see, the force of air resistance increases as the square of the velocity and thus dominates at high velocities. Other sources of friction, such as friction in the bearings and heat loss in the tires, depend mainly on the number of engine revolutions and are approximately the same per mile independent of the speed. We shall now estimate how much power must be supplied to overcome air resistance at high speeds. The extra fuel needed to supply this power will also be estimated. Most of this extra fuel could be saved by driving at slower speeds.

A small mass of air, $\Delta m$, just in front of the car will be given a kinetic energy $\tfrac{1}{2}(\Delta m)v^2$, where $\Delta m$ is the mass of the air swept out in a time $\Delta t$ and $v$ is the car's velocity. As seen in Figure 7-14,

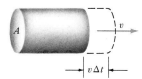

**Figure 7-14.** A column of air is pushed a distance ($v\Delta t$).

$$\Delta m = \rho_{\text{air}} A v\, \Delta t$$

where $\rho_{\text{air}} = 1.3$ kg/m$^3$ is the density of air and $A$ is the effective cross sectional area of the car. The energy loss to the air in the time $\Delta t$ is

$$\Delta E = \tfrac{1}{2}(\Delta m)v^2 = \tfrac{1}{2}(\rho_{\text{air}} A v\, \Delta t)v^2$$

and the power loss due to air resistance is

$$\frac{\Delta E}{\Delta t} = \tfrac{1}{2}\rho A v^3 \qquad (7\text{-}16)$$

Since power $= Fv$ the resistive force or drag is

$$F_{\text{drag}} = \tfrac{1}{2}\rho_{\text{air}} A v^2 \qquad (7\text{-}17)$$

Note that the area $A$ is the effective cross sectional area of the air that moves at or near the speed of the car. Experimental confirmation of Eq.

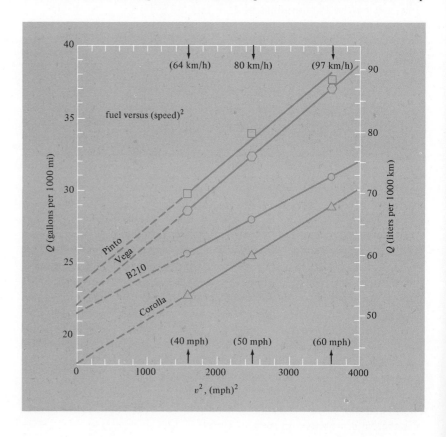

**Figure 7-15.** *The vertical scale is proportional to the force. The fact that the experimental points lie on straight lines shows that $F = K_1 + K_2 v^2$. The term $K_2 v^2$ is due to air resistance.* [From R. E. Barker, Jr.: Am. J. Phys., Jan. 1976]

7-16 is shown in Figure 7-15. Air near the top or sides of the car slips around it and picks up only a small velocity. We shall assume $A \approx 1 \text{ m}^2$ for a typical car. We should not expect the following numerical estimates to be accurate, but they are probably correct to within a factor of 2. For a car at $v = 60$ mph (27 m/s), Eq. 7-16 gives

$$P = \tfrac{1}{2}(1.3)(1)(27)^3 = 12{,}800 \text{ W} = 17 \text{ hp}$$

Note that the power required goes as the cube of the velocity. In order for the car to go at 90 mph, 3.375 times as much power is needed; that is, 57 hp is needed to overcome air resistance at 90 mph.

Now we shall estimate the fuel needed to overcome the air resistance at 60 mph. In a 60-mile trip the energy needed to overcome air resistance is

$$\Delta E = P \Delta t = (12{,}800 \text{ W})(3600 \text{ s}) = 46 \times 10^6 \text{ J}$$

The energy content of gasoline is $31 \times 10^3 \text{ J/cm}^3$. This is $1.2 \times 10^8 \text{ J/gal}$. A good car engine is 25% efficient, so each gallon delivers about $30 \times 10^6$ J of mechanical energy. Hence our car uses about 1.5 gal to go the 60 miles. This is a gas mileage of 40 mi/gal. No car of this cross sectional area could do better than this at 60 mph. The actual gas mileage should be worse than this due to other frictional or resistive losses.

---

**Example 16.** A car is cruising at 60 mph and getting 20 mi/gal. How many kilowatts of power is being used just to keep the car going at constant speed?

ANSWER: The car is using 3 gal/h. Since each gallon has an energy content of $1.2 \times 10^8$ J, this is a power consumption

$$P = \frac{3 \times 1.2 \times 10^8 \text{ J}}{1 \text{ h}} = \frac{3.6 \times 10^8 \text{ J}}{3.6 \times 10^3 \text{ s}} = 10^5 \text{ W} = 100 \text{ kW}$$

We see that one typical car uses up about the same energy as the electricity being used by about 50 homes at any one time. This is why cars are accused of being an energy extravagance.

---

**Example 17.** If a diver makes a belly flop into water, he should not exceed an acceleration $a = 50g$. From what height would a diver experience this large an acceleration if his "belly" has an effective cross sectional area of 0.2 m²?

ANSWER: The velocity upon entering the water is
$$v = \sqrt{2gh}$$
Then according to Eq. 7-17 the drag force is
$$F \approx \tfrac{1}{2}\rho A(2gh)$$
$$ma \approx \rho A g h$$
$$h \approx \frac{m}{\rho A}\frac{a}{g} = 50\frac{m}{\rho A} \quad \text{for } a = 50g$$

For water $\rho = 10^3$ kg/m³. Assume $m = 60$ kg. Then
$$h = 50\,\frac{60 \text{ kg}}{(10^3 \text{ kg/m}^3)(0.2 \text{ m}^2)} = 15 \text{ m}$$

In order to avoid serious injury a higher dive than this should involve a smaller effective area, for example, head first or feet first.

## Summary

If no external forces or frictional forces are acting on a mass, we have conservation of mechanical energy: $(K + U) = $ constant for all positions of the mass. This also holds for a closed system of $N$ masses where $K = \sum_{j=1}^{N} K_j$.

Impulse $\mathbf{I} = \int \mathbf{F}\, dt = \mathbf{P}_2 - \mathbf{P}_1$. In an elastic collision kinetic energy is not lost.

For a mass $m$ just barely escaping from the surface of a spherical mass $M$,
$$\tfrac{1}{2}mv_0^2 + \left(-G\frac{Mm}{R}\right) = 0 + 0$$
or
$$v_0 = \sqrt{\frac{2GM}{R}}$$
is the velocity of escape.

In the presence of an external force $\mathbf{F}_{\text{ext}}$ and friction, the law of conservation of energy is
$$\int_A^B \mathbf{F}_{\text{ext}} \cdot d\mathbf{s} = \Delta K + \Delta U + \Delta U_{\text{int}}$$
where $\Delta U_{\text{int}}$ is the increase of heat, chemical, and rest energy.

Drag force due to a body of effective cross-sectional area $A$ moving through a fluid is $F \approx \frac{1}{2}\rho A v^2$ where $\rho$ is the density of the fluid.

## Appendix 7-1  Conservation of Energy for $N$ Particles

The change in the total potential energy $U$ is by definition:

$$dU = -\sum_j \mathbf{F}_j \cdot d\mathbf{s}_j$$

Now replace each $\mathbf{F}_j$ by $m_j(d\mathbf{v}_j/dt)$ and each $d\mathbf{s}_j$ by $\mathbf{v}_j\, dt$:

$$dU = -\sum m_j \frac{d\mathbf{v}_j}{dt} \cdot \mathbf{v}_j dt$$

Use Eq. 6-3 to replace $(d\mathbf{v}_j/dt)\cdot \mathbf{v}_j$ with $v_j(dv_j/dt)$:

$$dU = -\sum m_j \left(v_j \frac{dv_j}{dt}\right) dt$$

Now replace each $(dv_j/dt)\, dt$ with $dv_j$:

$$dU = -\sum m_j v_j\, dv_j$$

$$= -\sum d\left(\frac{1}{2} m_j v_j^2\right)$$

Hence

$$\Delta U = -\sum \Delta K_j$$

At two different instants of time $t_1$ and $t_2$, $U = U_2 - U_1$ and $\Delta K_j = (K_j)_2 - (K_j)_1$, so

$$U_2 - U_1 = -\sum (K_j)_2 + \sum (K_j)_1$$

$$(K_t)_1 + U_1 = (K_t)_2 + U_2$$

## Exercises

1. In Example 1 what is the velocity of $m$ as it passes through the position $x = \frac{1}{2}x_0$? Give answer in terms of $k$, $m$, and $x_0$.
2. In Example 2 what velocity must be given to $m$ in order for it just barely to reach the same height as its support? Is the result the same for $m$ hanging by a string as for $m$ hanging by a rod?
3. Assume in Example 5 that the passenger's head is 60 cm from the windshield.
   (a) How long will it take from the first contact with the tree until the head hits the windshield?
   (b) Assume the windshield can deform by 3 mm. Assuming no skull deformation, what would be the average deceleration of the head?
4. In Figure 7-5 what is the ratio $K_1'/K_1$ where $K_1'$ is the kinetic energy of $m_1$ after the collision? Give answer in terms of $m_1$ and $m_2$.
5. In Figure 7-5 if $m_2/m_1 = 2$, what is the fractional energy loss of $m_1$ due to the collision?
6. What is the velocity of escape from the moon and from Mars? Numerical values can be obtained from Appendix A.
7. What fraction of escape velocity is needed to send a rocket halfway to the moon?
8. Repeat Example 11 for a frictional force of 30 N.
9. In Example 12 what is the fractional loss of kinetic energy of the system?
10. In Figure 7-13 a 2-g bullet strikes a 2-kg wooden block which rises 10 cm. What is the muzzle velocity of the bullet?
11. A person fasts for 1 week. His average metabolism rate is 100 W. Estimate his weight loss.
12. A 1.5-ton car with 50% of its weight on the rear wheels can accelerate from 0 to 50 mph in 5 s. What is the coefficient of friction for its rear wheels? What power is delivered by the engine just as it is reaching 50 mph?

## Problems

13. In Example 3 let the mass $m$ be released from the top position. Let $\theta$ be the angle between the rod and its initial position. Derive a formula for $v$ in terms of $\theta$.
14. Repeat Example 4 for $F_B = 20mg$. Will the rope break?
15. In Example 4 what should be the breaking strength $F_B$ to give $y_{max} = 0.25l$?
16. In Example 4 assume $l = 20$ m and that there is 2 m of slack in the rope. What will be $y_{max}$ and $a_{max}$?
17. In Example 5 show that the 30-cm deformation and the $3 \times 10^{-2}$ s stopping time are consistent with each other, assuming a constant deceleration.
18. In this problem we shall design an air bag for auto safety. Assume that a human head will escape injury if its deceleration does not exceed $30g$. Assume the worst crash would be a 60 mph (27 m/s) stop in $2 \times 10^{-2}$ s. How thick should the airbag be?
19. In Figure 7-7 assume the target ball has twice the mass of the cue ball. What is $\theta$ when the target ball recoils at a 30° angle as shown?
20. (a) What is the ratio of the energy that must be supplied to a mass $m$ for it to escape from the earth (but not the solar system) to the energy needed for circular orbit?
    (b) Repeat for escape from the solar system as well as the earth.
21. Assume the universe contains only one neutron and one electron. Assume

the electron is in a circular orbit of radius $R$ about the neutron. The only force between them is the gravitational force. Also assume that $m_e vR = 1.05 \times 10^{-34}$ J s.
   (a) What is the electron velocity?
   (b) What is the radius of the orbit?
22. Consider a car of mass 1000 kg. When on a level road it always takes a force of 500 N to push it slowly at a constant velocity.
   (a) What is the frictional force in newtons?
   (b) If a force of 1000 N is applied, what will be the acceleration of the car?
   (c) If the car is parked on the side of a hill and the brakes give out, how far will the car coast before coming to rest if its vertical height on the hill was 10 m above the level ground?
23. Block B is at rest on a frictionless surface. An identical block A is attached to one end of a string of length $R$. Block A is released from the horizontal position and collides with B. The two blocks stick to each other and move together after the impact.
   (a) What is the velocity of the two blocks immediately after impact?
   (b) How far will they rise above the surface?
24. A 1-kg block slides down an inclined plane starting from rest at the top of the plane. The velocity of the block at the bottom of the plane is 100 cm/s.
   (a) What is the work done by this frictional force?
   (b) What is the constant force due to friction?
   (c) If the inclined plane is given a coat of oil and the frictional force is reduced to one tenth of its previous value, what would be the new velocity of the block at the bottom of the plane?
25. In Figure 7-12 assume the car is at rest ($v_2 = 0$). Derive a formula for energy loss of the truck: $\Delta K_1/K_1 = ?$
26. In Figure 7-12 what is the kinetic energy of $m_2$ after the collision?
27. (a) How much work is required to push a 10-g mass up a frictionless inclined plane whose length is 3 m and whose height is 0.5 m?
   (b) Now assume a force of friction between the mass and the inclined plane of 700 dynes. How much work must be done to move the mass up the plane?
   (c) Suppose an applied force of 3000 dynes is used to push the mass up the incline [assume the same frictional force as (b)]. What would be the velocity of the mass when it reached the top of the incline?
28. Use Figure 7-15 to estimate the effective cross sectional area of a Vega. Assume the engine is 20% efficient.
29. Use Figure 7-15 to estimate the net frictional force $F_f$ on a Vega when traveling at low speed. Assume 20% engine efficiency.
30. At what speed would the air resistance losses exceed the frictional losses for a Vega. Use Figure 7-15.
31. In Example 4 assume the climber is of 60-kg mass, $l = 50$ m, and $F_B = 1.1 \times 10^4$ N. What are $y_{max}$ and $a_{max}$?
32. A toy roller coaster works as shown. The car is given a gentle push at position $A$ so that it starts out with essentially zero velocity. It slides down the frictionless track and travels around on the inside of the circular loop of radius $R$. The height $h$ is such that the car can just barely make the trip around the loop without losing contact with the tracks. What is $h$ in terms of $R$? What force does the track exert on the car at point $B$?

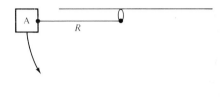
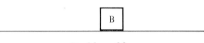

Problem 23

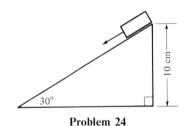

Problem 24

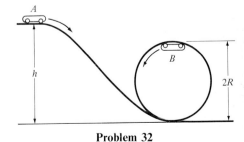

Problem 32

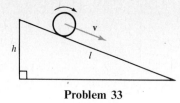
Problem 33

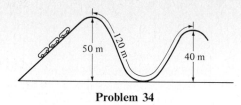
Problem 34

33. Consider a solid sphere starting from rest and rolling down an incline. Assume that the kinetic energy of rotation is always equal to the kinetic energy of translation ($\frac{1}{2}mv^2$) where $v$ is the velocity of the center of mass. It can be shown that the total kinetic energy must always be the sum of the two.
   (a) What will be the total kinetic energy of the sphere when at the bottom? Give answer in terms of $m$, $g$, and $h$.
   (b) What will be $v$ when at the bottom?
   (c) What is the acceleration of the center of mass in terms of $v$ and $l$?

34. A certain roller coaster rises 50 m. If the next rise is 40 m after 120 m of track, what is the maximum permissible value of the frictional force on a 500-kg car? (If $F_f$ were any larger, the car would not be able to reach the top of the second rise. $F_f$ is constant.)

35. How many gallons per hour are needed to keep a Vega going at 60 mph? What is the rate of production of heat energy in kilowatts? Use Figure 7-15.

36. A mass $m$ is hanging by string of length $l$. A second identical mass is sliding along a frictionless surface with velocity $v_0$.
   (a) If the two masses have an elastic collision, to what height $h$ will the first mass rise?
   (b) If the collision is completely inelastic, to what height $h$ will the masses rise?
   (c) How much heat energy is generated in part (b)? Give answer in terms of $m$ and $v_0$.

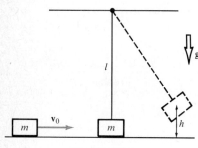
Problem 36

37. A toy gun consists of a ball of mass 10 g resting against a massless spring, with spring constant $k = 400$ N/m, that is compressed 5 cm inside a barrel. A child shoots the gun holding it horizontally 1 m above the ground.
   (a) If the ground is perfectly flat where does the ball land? Ignore air friction.
   (b) Suppose the ball hits the center of a target of mass 40 g hanging from a tree. The ball sticks to the target and they swing up together. How high does the target rise? Treat the target as a point mass on the end of a rigid stick.

38. (a) A bead of mass $m$ ($= 2$ kg) slides without friction or air resistance on a vertical rod as shown in the figure. If the bead is released from rest at position $A$, what will be the bead's velocity after it has dropped distance $d$ ($= 4$ m) to position $B$?
   (b) If a spring having an unstretched length of 3 m and a spring contant of $k$ ($= 22$ N/m) is attached to the bead as shown, what will be the velocity of the bead when it reaches point $B$, after being released from rest at position $A$ as in part (a)? (Note: The bead is heavy enough and the spring is weak enough so that the head does, in fact, reach point $B$!)

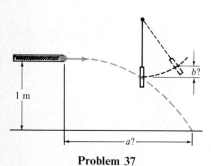
Problem 37

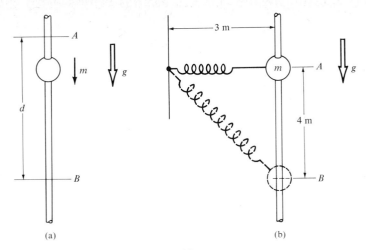

**Problem 38**

39. Assume the truck-car collision in Figure 7-12 is an elastic collision. What is the car velocity after the collision?
40. Assume that the truck-car collision in Figure 7-12 results in a final velocity of 70 km/h for the truck.
    (a) What is the final velocity of the car?
    (b) How much kinetic energy is lost in the collision?

# Relativistic Kinematics

## 8-1 Introduction

Up to now in our presentation of mechanics we have specified that all velocities be much less than the speed of light, which we have denoted as $c$. Now that we have developed the concepts of mechanics, we are ready to explain the reason for the restriction $v \ll c$. Put simply, the reason for such a restriction is that Newtonian mechanics (also called classical mechanics) is not correct. The correct theory is called relativistic mechanics. It is also called the **special theory of relativity.** Newtonian mechanics, it turns out, is an approximation to relativistic mechanics that is very accurate in the region $v \ll c$. In fact, as $v \to 0$, the equations of classical mechanics become exactly correct and, as we have seen, can explain much of the physical world.

One may ask, why bother with relativistic mechanics since most everyday velocities are much less than the speed of light. We list several reasons:

1 A major subject in physics is the study of light, and for light $v = c$.
2 The theory of light is derived from the theory of electricity. Important electrical effects such as magnetic field and electromagnetic induction depend on the speed of light. It is correct to say that the theory of electricity is a relativistic theory. For example, one must first understand relativity theory before one can truly understand magnetism.
3 In the fields of nuclear physics and elementary particle physics, some of the particles do travel near (or at) the speed of light. For example, photons and neutrinos always travel at $v = c$.
4 Modern astronomy has much to do with relativity. Distant galaxies are traveling near the speed of light. New phenomena such as neutron stars, pulsars, and black holes deeply involve relativistic effects.
5 In order to develop our understanding of quantum mechanics we should discuss topics such as photoelectric effect and Compton effect

where it is necessary to know the relativistic relations between energy, mass, and momentum.

6  We shall see that relativity theory blatantly violates common sense and common experience. One's first reaction is that it can't be true. Intellectually and philosophically it is important to give such a situation very close scrutiny. Even today there are still a few intellectuals who do not accept all the consequences of the theory. This is our first example of natural phenomena that openly violate common sense.

7  Most of us have already heard of things such as $E = mc^2$, that no particles or signals can travel faster than the speed of light, the twin paradox, time dilation, the Lorentz contraction, and so on. In this modern technological age these things are becoming part of our culture—at least among the intelligentsia. They should be understood by those who wish to be considered as educated persons.

## 8-2 Constancy of the Speed of Light

The central paradox is that the speed of light must be the same for all observers. Figure 8-1 shows an example of this which violates common sense and is contrary to our experience. Suppose a single pulse (or flash) of light passes by observer A standing on the earth. At the same instant of time observer B goes by with velocity $v_B$ looking at the same pulse of light. According to all we have learned so far, he should see the pulse traveling slower with speed $v'_{pulse} = v_{pulse} - v_B$. However, if one does the experiment, not only does observer A measure $v_{pulse} = c$, where $c = 2.998 \times 10^8$ m/s, but observer B measures $v'_{pulse} = c$ as well, and this for the same pulse at the same instant of time!

Another example would be two observers, one at rest with respect to a distant star and one traveling very fast toward the star, both measuring the speed of light coming from the star. They will both get the same answer, $v_{light} = c$. The main starting point in Einstein's theory of relativity is that the speed of light is always $c = 2.998 \times 10^8$ m/s independent of the velocity of the observer or of the source. Einstein explained this "strange" result by attributing "strange" properties to space and time. He proposed that space, as viewed by a moving observer, "contracts" in the direction of motion by the factor $\sqrt{1 - v^2/c^2}$ and that time as measured by a moving observer "slows down" by the same factor. In fact he "doctored up" space and time in just the right way to give the result $\Delta x'/\Delta t' = c$ for any pulse of light seen by any observer moving with constant velocity ($x'$ and $t'$ are space and time as measured by the moving observer). In the next two sections we shall see quantitatively just how this can be done. For the remainder of this section we shall present the first experimental evidence that the speed of light is independent of both the velocity of the source and the velocity of the observer.

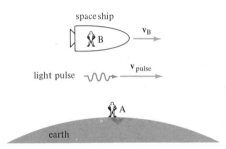

**Figure 8-1.** Observer A on earth and observer B in a spaceship simultaneously measure the speed of the same pulse of light.

## The Michelson–Morley Experiment

Before Einstein's theory of relativity was published in 1905, most physicists reasoned that light waves must have a medium for their transmission just as air is the medium for sound waves. This medium for light waves was called the ether. Then a frame of reference which is at rest with respect to the ether would be a preferred frame of reference. Only in this reference system would $v_{light} = c$. To an observer having velocity $v$ with respect to the ether, the velocity of light would be $(c + v)$ when moving toward the light source. The ether was thought of as a "physical," but massless medium—a difficult concept, to say the least.

Experiments were performed in the 1880's that gave results consistent with the principle that light travels at $v_{light} = c$ independent of the velocity of source or observer, and thus contradicted the ether hypothesis. Since the earth is moving at a velocity $v = 30$ km/s around the sun, supporters of the ether theory reasoned that there must be times of the year when the earth has a velocity of at least 30 km/s with respect to the ether (or the ether has a velocity of 30 km/s with respect to the earth). See Figure 8-2. Then to an observer on earth, light traveling in the same direction as the ether should have a velocity $(c + v)$ with respect to the earth, and light traveling in the opposite direction would have velocity $(c - v)$ where $v$ is at least 30 km/s.

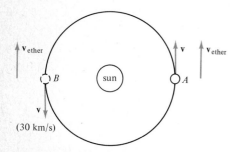

**Figure 8-2.** *Earth's orbit about the sun. If the ether velocity were the same as the earth's velocity at position A, the ether drift would be zero. But at position B the earth would be moving at 60 km/s, with respect to the ether.*

Suppose a rigid arm of length $D$ holds a light source and a mirror with the ether moving as shown in Figure 8-3. Then the time for light to travel from the source to the mirror is $t_1 = D/(c - v)$. The travel time from mirror back to source is $t_2 = D/(c + v)$. The round trip time is

$$t = \frac{D}{c - v} + \frac{D}{c + v} = \frac{2Dc}{c^2 - v^2}$$

$$t = \frac{2D}{c}\left(1 - \frac{v^2}{c^2}\right)^{-1} \tag{8-1}$$

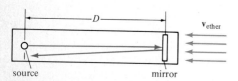

**Figure 8-3.** *Rigid arm holds source and mirror. Ether is moving from right to left with velocity v.*

The time for the light to travel to the mirror and back is given by Eq. 8-1 when the arm is lined up parallel to $v$, the ether velocity. If the arm is rotated 90° so that it is perpendicular to $v$, the light must travel a distance $2D'$ as viewed by an observer at rest with respect to the ether (see Figure 8-4). Then the time to travel to the mirror and back would be

$$t' = \frac{2D'}{c} \quad \text{or} \quad D' = \frac{ct'}{2} \tag{8-2}$$

From the right triangle in Figure 8-4 we have

$$D'^2 = D^2 + \left(\frac{vt'}{2}\right)^2$$

**Figure 8-4.** *Three successive positions of arm moving transverse to ether. In this figure the ether is at rest.*

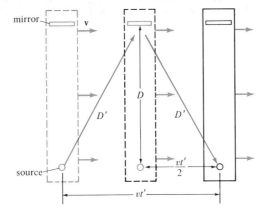

Now by substituting the right-hand side of Eq. 8-2 for $D'$ we have

$$\frac{c^2 t'^2}{4} = D^2 + \frac{v^2 t'^2}{4}$$

Solving this for $t'$ gives

$$t' = \frac{2D}{c}\left(1 - \frac{v^2}{c^2}\right)^{-1/2} \qquad (8\text{-}3)$$

We see that the time difference between parallel and transverse orientations is

$$t - t' = \frac{2D}{c}\left[\left(1 - \frac{v^2}{c^2}\right)^{-1} - \left(1 - \frac{v^2}{c^2}\right)^{-1/2}\right]$$

This can be simplified by using the first two terms of the binomial expansion: $(1 - \varepsilon)^n \approx 1 - n\varepsilon$. Then

$$t - t' = \frac{2D}{c}\left[\frac{v^2}{c^2} - \frac{v^2}{2c^2}\right] = \frac{Dv^2}{c^3} \qquad (8\text{-}4)$$

---

**Example 1.** Suppose the arm is 1 m long. What is the time difference for longitudinal versus transverse orientation when the ether velocity is 30 km/s?

ANSWER:

$$\frac{v}{c} = \frac{30 \text{ km/s}}{3 \times 10^5 \text{ km/s}} = 10^{-4}$$

$$(t - t') = \frac{D(10^{-4})^2}{c} = \frac{D(10^{-8})}{c} = 3.3 \times 10^{-17} \text{ s}$$

In this amount of time light travels about 1/40 of a wavelength of light.

Michelson and Morley realized that they could observe such a small time difference by using an interferometer that already has two arms at 90° to each other. Such an interferometer is shown in Figure 8-5. In the interferometer, light from source S is split by the half-silvered mirror $M_1$. The two light rays are reunited at the screen. If the two light paths are of the same length, there will be a constructive interference at the screen (the two wave amplitudes will add.) The experiment consists of adjusting the mirror positions to give a constructive interference. Then the entire apparatus is rotated 90° by the earth's rotation and the new interference pattern is observed. A change in the time duration of the light paths due to the velocity of the ether should show up as a change in the interference pattern (the two wave amplitudes could subtract and give a reduction in light intensity.) Even a value of $v$ as small as 30 km/s should give a very noticeable effect.

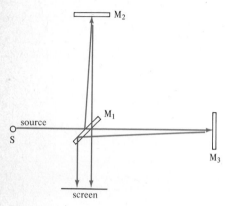

**Figure 8-5.** *The Michelson interferometer. Light from source* S *is split by half silvered mirror* $M_1$ *and reunited at the screen.*

But, after exhaustive trys, Michelson and Morley never obtained any effect at all. One explanation given was that the ether happens by accident to have a velocity of 30 km/s with respect to the solar system and is traveling along with the earth. However, Michelson and Morley repeated their experiment six months later when the earth's velocity was reversed in direction. If indeed the ether theory were true, they should have observed twice the expected effect (see Figure 8-2), but again they obtained a null result.

Another possible explanation was that the earth drags a local region of the ether along with it. However, this would cause the stars to appear to shift in position back and forth each year in a manner different from that which is observed. This explanation was therefore ruled out by astronomical observation.

As a result of these experiments, we conclude that the light emitted by the interferometer source always travels with velocity $c$ with respect to the source and mirrors. A final attempt to explain the null result might be to revise the laws of electricity so that light would always be emitted with velocity $c$ with respect to the source of the electromagnetic waves. This explanation, it turns out, is ruled out on the basis of astronomical observation. If this emission theory of light were true, the motion of a double star should appear distorted and violate Kepler's laws. When one member of the binary is moving toward the earth with velocity $v$, its light would travel the entire distance with velocity $(c + v)$ and arrive early, whereas the light emitted while the star was moving away would arrive late.

One might expect that such a definitive series of experiments would have had a profound influence on Albert Einstein in formulating his solution to the problem. However, the Michelson–Morley experiments were of little concern to Einstein. He was much more bothered about the inconsistencies between the equations of electromagnetic theory and classical mechanics. A favorite thought problem of his was what would happen if one should chase after a light pulse and catch up with it by traveling at $v \simeq c$. So Einstein tackled the problem of what changes in the classical ideas of space and time would be necessary to make the velocity of light appear the same to all observers and to make the equations of electromagnetic theory have the same form for all observers moving at constant velocity with respect to each other.

**The Principle of Relativity**

As we shall see, Einstein's revision of space and time follows directly from two basic principles. The first is the constancy of the speed of light for all observers. (A more general formulation is that there is an ultimate speed of $c = 2.998 \times 10^8$ m/s beyond which no particle can travel. Massless particles such as photons and neutrinos must always travel at $v = c$ for all observers.) In our discussions we have been taking a second basic principle for granted, and that is the principle of relativity which was

**Figure 8-6.** *Albert Einstein (1879–1955).*

first proposed explicitly by Galileo. He proposed that *the laws of physics should be the same for all observers moving with constant velocity relative to each other no matter what the magnitudes and directions of the velocities.* Another way of stating the principle of relativity is that there must be no preferred reference system or no way to determine absolute velocity. Certainly if one closes his eyes while flying on a jet airliner everything feels the same as if it were at rest. The principle of relativity says that there are no physics experiments one can perform inside the jet plane to determine its speed. By inside the jet we mean no communication with the outside.

## 8-3 Time Dilation

We shall start our presentation of the theory of relativity with a simple application of the two principles (constancy of speed of light and principle of relativity) to illustrate why Einstein found it necessary to make changes in the concept of time. We shall apply the two principles to a simple kind of clock called a "light clock." The construction is very simple: just two parallel mirrors a fixed distance $D$ apart, as shown in Figure 8-7(a).

These are idealized clocks with perfectly reflecting mirrors each containing a short pulse of light bouncing back and forth. Let $\tau$ be the time required for a pulse of light to strike the top mirror starting from the bottom. Each time the light strikes a mirror we have a "tick" of the clock. We start with two such identical clocks with their ticks synchronized. The time between ticks is $\tau = D/c$. Now we shall let clock B move to the right with velocity $v$ as in Figure 8-7(b). First we must ask ourselves whether the moving clock will still appear to have the same length as clock A. This question can be answered by imagining a thin paint brush attached to the end of clock B. As it moves past clock A it paints a mark. If the mark is on the end of A, the clocks are the same length. If the mark is below the end, clock B is shortened in length when in motion. Now suppose the paint mark was below the end. Then observer A (who is moving along with clock A) would see that moving light clocks (or any length transverse to the direction of motion) appear shorter; whereas, according to observer B, moving light clocks appear longer. But, according to the principle of relativity, both observers are on an equal basis and both should observe the same effect. This can only happen if both clocks are the same length to both observers.

The remainder of the discussion will be from the point of view of an observer at rest with respect to clock A. Such an observer sees a longer path for the light pulse to travel from one end to the other in clock B. Light pulse B must take the diagonal path in Figure 8-7(b) and, according to the constancy of the speed of light principle, must travel with the same speed as the light pulse in clock A. Hence, light pulse B must take a longer time than light pulse A to reach the top mirror (according to observer A).

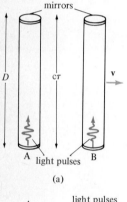
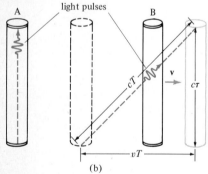

**Figure 8-7.** (a) *Two identical light clocks at $t = 0$. Clock B is moving to the right with velocity $v$.* (b) *The light clocks $\tau$ seconds later as seen by A. Both light pulses have traveled a distance $c\tau$. The pulse in A has reached the end, but the pulse in B still has farther to go.*

Let us call this longer time $T$. The length of the diagonal path is then $cT$. Applying the Pythagorean theorem to Figure 8-7(b) gives

$$(cT)^2 = (vT)^2 + (c\tau)^2$$

Solving for $T$ gives

$$T = \frac{1}{\sqrt{1 - \frac{v^2}{c^2}}} \tau \qquad (8\text{-}5)$$

In relativity theory the factor $(1 - v^2/c^2)^{-1/2}$ occurs so often that it is given the symbol $\gamma$ (gamma).

$$\gamma \equiv \frac{1}{\sqrt{1 - \frac{v^2}{c^2}}} \qquad \text{(definition)}$$

According to the stationary observer, the time between ticks for the moving clock is $T$, which is longer than $\tau$, the time between ticks when either of the clocks is at rest. We must conclude that any observer will observe that moving clocks tick more slowly by the factor $\gamma$ than an identical stationary clock.

In Eq. 8-5 $\tau$ is called the proper time. It is the interval in time between two events as measured by an observer who sees them occur at the same position in space. Then $T$ is the time between the same two events as seen by a moving observer (according to his own clocks).

$$\tau = \frac{1}{\gamma} T \qquad \text{(proper time)} \qquad (8\text{-}6)$$

The proper time of a given clock is time as measured by observers who move along with the clock. An observer moving with respect to a clock will see it measure time intervals $T = \gamma\tau$ according to clocks which are stationary with respect to the observer.

But do light clocks behave this way because of the special nature of light? Should ordinary mechanical clocks whose parts move much slower than the speed of light also slow down by the same factor $\gamma$? Einstein said yes, because it has nothing to do with the nature of the particular clock—it is due to an intrinsic property of time itself. To see this, suppose a light clock and a wristwatch are fastened together, both keeping identical time. Then they are pushed sideways with velocity $v$ and the light clock slows down as it should, but the wristwatch does not. Now we would have a

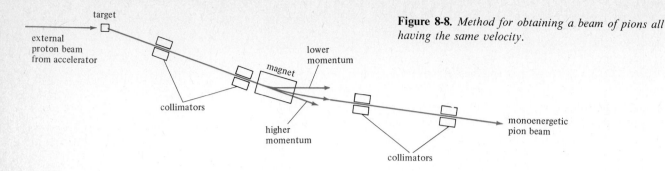

**Figure 8-8.** *Method for obtaining a beam of pions all having the same velocity.*

simple detector of absolute motion—whenever the two clocks agree, they are at rest; whenever the light clock runs slower, everybody knows they are in motion. This, of course, violates the principle of relativity upon which all of our discussion is based.

Since time dilation is a property of time itself, not only are all moving clocks slowed down, but so are all physical processes such as chemical reaction rates slowed down when in motion. Since life consists of a complex of chemical reactions, life also would be slowed down by the same factor. In fact, if biological aging did not slow down at the same rate, we could fasten a biological sample that keeps time (beats of a heart, for example) to a moving light clock and if the heart and clock did not keep the same time we would have a detector of absolute motion, again violating the principle of relativity. Of course, a human being or any other form of life in a fast moving spaceship will not feel or observe any slowing down of life inside the spaceship (see Section 8-7 for further discussion of aging during space travel).

Since all physical processes must be slowed down, so must the half-life of a radioactive sample be slowed down by the same factor. This effect on the half-life has been observed directly to an accuracy of about 1 part in $10^4$ by using a beam of unstable particles moving near the speed of light. The half-life of such particles will be increased by the factor $\gamma$. A very common unstable particle is called the pion (see Chapter 31). The pion has a half-life of about $1.8 \times 10^{-8}$ s and is easily produced by bombarding any material with a beam from a high energy accelerator. A beam of pions, all having the same velocity, can be produced by selecting those trajectories which have the same angle of bend in a magnetic field as shown in Figure 8-8.

> **Example 2.** A beam of pions is produced which has a velocity $v = 0.99c$. (a) By what factor is the lifetime (as measured in the lab) of these pions increased? (b) How long will it take for half of these pions to decay? (c) How far will they travel in this time?

ANSWER: The factor

$$\gamma = \frac{1}{\sqrt{1 - 0.99^2}} = 7.09.$$

The half-life will be 7.09 times longer; that is, $t = 7.09(1.8 \times 10^{-8} \text{ s}) = 12.7 \times 10^{-8}$ s. The distance traveled in this time is $x = vt = 0.99c(12.7 \times 10^{-8} \text{ s}) = 37.9$ m.

Not only has time dilation been observed using microscopic "clocks" in the form of unstable particles but, in 1960, the effect was first observed using what are called Mossbauer clocks. The most stable timekeeping device that can be built with present knowledge is a clock utilizing what is called the Mossbauer effect. Mossbauer "clocks" make use of photons from a radioactive iron isotope imbedded in an iron crystal. Two identical Mossbauer clocks will keep the same time within one part in $10^{16}$. A time shift shows up as an increase in photon counting rate. The amount of shift in clock rate can be measured quantitatively. In this time-dilation experiment, an entire Mossbauer clock was rotated rapidly and found to slow down at the rate $(1 - v^2/c^2)^{-1/2}$ compared to an identical Mossbauer clock at rest.

## 8-4 The Lorentz Transformation

"I pace hardly at all, nevertheless I feel I have come far already."
"You see, my son, that here time transforms into space."
(from *Parsifal*, Transformation Scene, R. Wagner, 1877)

We shall see in this section that a poet's dream has turned out to be close to the truth. The Lorentz transformation (Eq. 8-9) shows that time can transform into space and vice versa. Consider two observers moving with relative velocity $v$. Call them Mr. X and Mr. Prime. Mr. X measures

**Figure 8-9.** *Two observers moving with relative velocity v.*

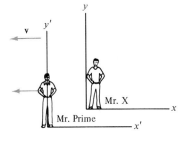

(a) Mr. Prime as seen by Mr. X

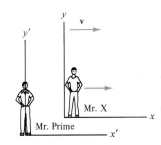

(b) Mr. X as seen by Mr. Prime

events in his $(x, y, z, t)$ coordinate system. The coordinate system used by Mr. Prime we call the primed system as shown in Figure 8-9. In classical mechanics the relation between the two coordinate systems is

$$\begin{cases} x' = x + vt \\ y' = y \\ z' = z \\ t' = t \end{cases}$$

if the two origins are at the same point when the time $t = t' = 0$. With these transformation equations a beam of light traveling to the right with velocity $c$ in the unprimed system will have velocity $(c + v)$ in the primed system. We want to find a different set of coordinate transformation equations so that, if an object travels with $v = c$ in the unprimed system, it will also travel with $v' = c$ in the primed system; that is, if $x = ct$, then $x' = ct'$. The general coordinate transformation is

$$x' = Ax + Bt \qquad (8\text{-}7)$$
$$t' = Et + Fx \qquad (8\text{-}8)$$

where $A$, $B$, $E$, and $F$ can be functions of $v$. (The origins are at the same point when $t = t' = 0$.) We have already seen that $y' = y$, $z' = z$ (the result on page 150 that transverse lengths must appear the same for both observers). In order to solve for the four quantities $A$, $B$, $E$, $F$ we need four relations. First consider a clock fixed at $x = 0$. Let the time between ticks be $\tau$. According to Eq. 8-5, Mr. Prime sees a moving clock where $\gamma\tau$ is the time between ticks. Since Eq. 8-8 must hold for the first tick, $x = 0$, $t = \tau$ and $t' = \gamma\tau$ will satisfy Eq. 8-8:

$$\gamma\tau = E\tau + 0$$

Thus $E = \gamma$.

According to Mr. Prime the clock is moving to the right with velocity $v$; that is, he sees it at $x' = vt'$. Now substitute this as well as $x = 0$ into Eq. 8-7:

$$vt' = 0 + Bt$$

or

$$B = \frac{vt'}{t}$$
$$= v\gamma$$

because, as we have just seen, $t' = \gamma t$.

In order to solve for $A$, place the clock at Mr. Prime's origin. According to the principle of relativity, Mr. X must see the clock move off to the left with velocity $-v$; that is, $x = -vt$ when $x' = 0$. Inserting these values into Eq. 8-7 gives

$$0 = A(-vt) + (v\gamma)t$$

or

$$A = \gamma$$

Equations 8-7 and 8-8 are now of the form

$$\begin{cases} x' = \gamma x + \gamma vt \\ t' = \gamma t + Fx \end{cases}$$

Now we finally make use of the fact that when $x = ct$, $x' = ct'$. Substitute these quantities into the two preceding equations and divide the upper equation by the lower:

$$\frac{ct'}{t'} = \frac{\gamma ct + \gamma vt}{\gamma t + Fct}$$

or

$$c = \frac{\gamma c + \gamma v}{\gamma + cF}$$

Solving for $F$ gives

$$F = \frac{v}{c^2}\gamma$$

Now all four coefficients have been obtained and the final form for Eqs. 8-7 and 8-8 is

$$\begin{cases} x' = \gamma x + \gamma vt \\ t' = \gamma t + \gamma \frac{v}{c^2} x \end{cases} \quad \text{(the Lorentz transformation)} \quad (8\text{-}9)$$

In relativity theory time is sometimes called the fourth dimension. More strictly speaking the quantity $ct$, which has the same units as $x$, $y$, and $z$, behaves as a fourth space coordinate. We see that $ct$ and $x$ can get intermixed depending on the velocity of the observer. In relativity theory $ct$ and $x$ behave in a similar way mathematically.

In Eq. 8-9 we have the primed coordinates in terms of the unprimed. It is also useful to have the unprimed in terms of the primed. This can be done by solving the two simultaneous equations in terms of the two unknowns, $x$ and $t$. A few steps of algebra give the result

$$\begin{cases} x = \gamma x' - \gamma vt' \\ t = \gamma t' - \gamma \frac{v}{c^2} x' \end{cases}$$

Note that these equations are of the same form as Eq. 8-9 except that $v$ is replaced by $-v$. This is just as one would expect since Mr. X sees Mr. Prime moving with velocity $-v$, whereas Mr. Prime sees Mr. X moving with velocity $v$.

### The Lorentz Contraction

Suppose Mr. X tries to measure the length of a meter stick that is at rest in the primed system (its ends are fixed at $x'_1$ and $x'_2$ as shown in Figure 8-10). Then Eq. 8-9 gives

$$x'_2 = \gamma x_2 + \gamma v t_2$$
$$x'_1 = \gamma x_1 + \gamma v t_1$$

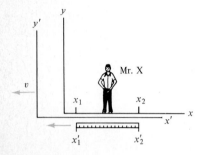

**Figure 8-10.** *Moving meter stick is stationary in primed system.*

Subtracting:

$$x'_2 - x'_1 = \gamma(x_2 - x_1) + \gamma v(t_2 - t_1)$$

In order for Mr. X to make a valid length measurement of a moving object in his frame of reference, he must take care to measure the position of both ends at what he claims is the same instant of time; that is, $t_1 = t_2$. Then the preceding equation becomes

$$x'_2 - x'_1 = \gamma(x_2 - x_1)$$

or

$$(x_2 - x_1) = \frac{1}{\gamma}(x'_2 - x'_1)$$

(length of moving rod) = $\sqrt{1 - v^2/c^2}$ (length of same rod when at rest)

$$l_{\text{moving}} = \sqrt{1 - v^2/c^2}\, l_{\text{rest}} \quad \text{(Lorentz contraction)} \quad (8\text{-}10)$$

If two observers should pass by each other, each holding an identical meter stick in the direction of motion, each observer would "see" the other's meter stick as shortened by the same amount. We put the word "see" in quotation marks because it is important to measure the positions of the two ends simultaneously; whereas if one looked at the two ends by eye there would be a time delay due to the finite travel time of light. To calculate what a photograph of a rapidly moving object would look like is complicated because of the corrections due to the different travel times of the different light paths.

> **Example 3.** A meter stick passes by with a velocity 60% of the speed of light. How long would it appear to be?
>
> ANSWER: According to Eq. 8-10,
>
> $$l = \sqrt{1 - 0.6^2}\,(100 \text{ cm})$$
> $$= \sqrt{0.64}\,(100 \text{ cm}) = 80 \text{ cm}$$

## 8-5 Simultaneity

A good physical reason why one observer thinks that a given meter stick is shorter than another observer thinks it is, is because events simultaneous for one observer are not simultaneous for the other. Remember, in order to measure the length of a meter stick, one must measure the two ends simultaneously.

We shall now demonstrate by using a moving boxcar as an example that two events that are simultaneous to a stationary observer will not be simultaneous for an observer in the moving boxcar. The boxcar is of length $l$ when at rest, as measured by Mr. X, who is standing in the center of the boxcar. (See Figure 8-11.) Suppose at a time $t = t_0$ Mr. X passes by Mr. Prime who is standing beside the railroad track. At this same instant of time, according to Mr. Prime, two lightning bolts strike the ends of the boxcar leaving marks on the railroad track. This provides Mr. Prime an excellent opportunity to measure the length of the boxcar. At his leisure he can measure the distance between the marks and he will find that

$$l' = \sqrt{1 - v^2/c^2}\, l$$

where $l$ is the length when at rest.

An equally amazing result, however, is the fact that Mr. X claims that the lightning bolt on the right struck first. Certainly, as viewed by Mr. Prime, the man in the boxcar is moving toward the light from the right-hand lightning bolt and he will meet this pulse of light first. And if Mr. X's face is illuminated by the right-hand pulse of light first, he reaches it first, no matter who the observer is. But according to Mr. X, the two bolts of lightning were equal distances away, and, if he saw with his own eyes the one from the right occur first, then according to him it must have occurred first. Another observer, Mr. Double Prime, starting at the same position, but moving to the left, would similarly claim that the left-hand bolt struck first.

We conclude that whenever two events occur within the time required for light to travel between them, the order of occurrence is undefined—it

**Figure 8-11.** Mr. Prime sees two lightning bolts strike the ends of the boxcar simultaneously. Mr. X is racing toward the light pulse on the right and meets it first. To him, the right end of his boxcar was hit first.

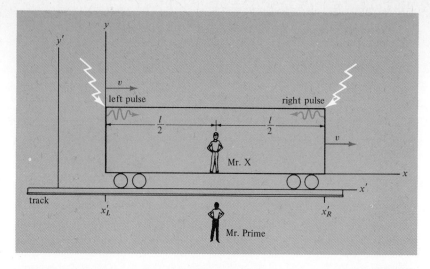

depends on the velocity of the observer. In such cases future events can be made to precede past events by selecting an appropriate moving observer.

**Example 4.** Lightning strikes both ends of a 20-m long car of the Japanese National Railway traveling along the $x$ axis at 200 km/h = 55.6 m/s. According to an observer (Mr. Prime) on the ground, the two ends were hit simultaneously. According to the passengers (in the unprimed system), what was the time difference between the two strikes?

ANSWER: Refer to Figure 8-11. We want to solve for $(t_R - t_L)$. The Lorentz transformation equation for $t'$ is

$$t'_R = \gamma t_R + \gamma \frac{v}{c^2} x_R$$

and

$$t'_L = \gamma t_L + \gamma \frac{v}{c^2} x_L$$

Subtracting:

$$(t'_R - t'_L) = \gamma(t_R - t_L) + \frac{\gamma v}{c^2}(x_R - x_L)$$

According to Mr. Prime, $(t'_R - t'_L) = 0$, and according to any passenger, $(x_R - x_L) = l =$ car length. Substitution into the preceding equation gives

$$(t_R - t_L) = -\frac{v}{c^2} l = -\frac{55.6}{(3 \times 10^8)^2}(20) \text{ s} = -1.24 \times 10^{-14} \text{ s}$$

This value is too small to be measurable. The minus sign means that $t_R$ is smaller than $t_L$; that is, the event at $x_R$ occurred before the event at $x_L$.

## 8-6 Doppler Effect of Light

When an observer moves toward the source of sound the observed frequency of the sound increases and when he moves away the frequency decreases. This frequency shift due to motion is called the Doppler effect. A common example of this is the sound of the horn of an oncoming train. As the train goes by the frequency or pitch of the horn drops. The same sort of phenomenon occurs with light waves. As the source moves toward the observer (or, equivalently, as the observer moves toward the source), the observed frequency of the light increases (the light is "blue shifted"). If the source and observer are moving apart, the observed frequency of the light decreases (it is "red shifted"). Doppler effect of sound is calculated using classical mechanics; however, we must use relativity theory to calculate Doppler effect of light.

In Figure 8-12, B is a source of light received by detector A. A and B are moving away from each other with relative velocity $v$. Assume identical clocks are attached to A and to B and that these clocks both read zero when they pass each other. Suppose a pulse of light is emitted by B when his clock reads $T_B$. We wish to calculate the time $T_A$ when the pulse arrives at A. An observer in the primed reference system sees the stationary clock at A running faster than the moving clock at B. According to A, his clock read

$$t_A = \gamma T_B$$

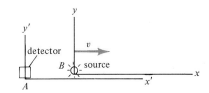

**Figure 8-12.** *Light source B is traveling to the right with velocity v according to observer in primed system.*

when the moving clock B read $T_B$. But we are interested in the time when the light pulse from B arrives at A. It has a travel time in A's frame of $x'/c$. If $T_A$ is the arrival time of the light pulse at A in A's frame:

$$T_A = t_A + \text{(travel time)} = t_A + \frac{x'}{c}$$

Now replace $t_A$ with $(\gamma T_B)$ and eliminate $x'$ by noting that the distance B has traveled in the time $t_A$ is $x' = vt_A = v(\gamma T_B)$. Then

$$T_A = \gamma T_B + \frac{v\gamma T_B}{c}$$

or

$$T_A = \gamma(1 + \beta)T_B \quad \text{where } \beta \equiv \frac{v}{c}$$

The time difference or period of oscillation between two successive pulses is

$$\tau_A = \gamma(1 + \beta)\tau_B \tag{8-11}$$

where $\tau_B$ is the period as measured at the source. The frequency or number of pulses per second is related to period of oscillation by $f = 1/\tau$. Taking the reciprocal of both sides of Eq. 8-11 and using $\gamma = (1 - \beta^2)^{-1/2}$ gives

$$f_A = \frac{\sqrt{1 - \beta^2}}{1 + \beta} f_B$$

or

$$f_A = \sqrt{\frac{1 - \beta}{1 + \beta}} f_B \quad \text{(source receding)} \tag{8-12}$$

where $f_A$ is the number of pulses received per second by A.

The relation is the same whether we are counting ticks of a clock B, or number of oscillations of an oscillator at B, or whether B is a source of light (electromagnetic waves). In the case of light, $f_B$ is the number of wavefronts emitted per second at the source and $f_A$ is the number received per second by an observer moving away in relative motion. Since $f_A < f_B$, the apparent frequency is decreased or "red shifted" when the source is moving away. If the source had been approaching instead, the sign of $\beta = v/c$ would be reversed and the corresponding result would be

$$f_A = \sqrt{\frac{1 + \beta}{1 - \beta}} f_B \quad \text{(source approaching)} \tag{8-13}$$

---

**Example 5.** Red shifts as great as a factor of 3 have been observed in spectral lines emitted by astronomical objects called quasars. What would be the velocity of recession of a quasar with a red shift of a factor of 3?

ANSWER: Using Eq. 8-12 we have

$$f_A = \tfrac{1}{3} f_B \quad \text{or} \quad \tfrac{1}{3} = \sqrt{\frac{1 - \beta}{1 + \beta}}$$

$$1 + \beta = 9(1 - \beta)$$
$$10\beta = 8$$
$$v = 0.8c$$

Distant galaxies and quasars seem to be receding from our galaxy with velocities proportional to the distance. If this linear relationship between distance and velocity holds for the quasar in this example, it would be $12 \times 10^9$ light years away.

## 8-7 The Twin Paradox

Those who follow the space program may have heard that space travelers will not age as fast as their brothers back on earth. But because $v/c \ll 1$ for real space travelers, the effect is so small as to be negligible. However, if a space traveler could travel with the speed of light, he would not age at all. According to an observer on earth, the clocks and all physical processes including life itself on a spaceship of velocity $v$ would be slowed down by a factor $\sqrt{1 - v^2/c^2}$ (see Eq. 8-5).

**Example 6.** Consider two twins, A and B, in the situation shown in Figure 8-13. Twin B takes a round trip space voyage to the star Arcturus at a velocity $v = 0.99c$. According to those of us on earth, Arcturus is 40 light years away. What will be the ages of the two twins when B finishes his trip if they are age 20 at the start of the trip?

ANSWER: According to twin A, the trip would take 1% longer than the 80 years it takes light to travel to Arcturus and back. Thus A would be 20 plus 80.8 or 100.8 years old when B returns. According to twin A the clocks on the spaceship would be running slower by the factor $\sqrt{1 - 0.99^2} = 0.141$. On the spaceship the elapsed time for the trip would then be 0.141 times the 80.8 years of earth time or 11.4 years. Twin B would be 20 plus the 11.4 years or 31.4 years old at the end of the trip. Twin B would then be 69.4 years younger than the twin who stayed behind.

The space traveler does not feel that his time is running slower. In Example 6, twin B sees the distance to Arcturus shortened by the Lorentz

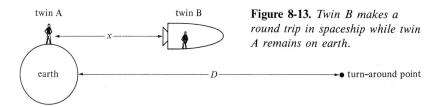

**Figure 8-13.** *Twin B makes a round trip in spaceship while twin A remains on earth.*

contraction. Twin B measures that the distance from earth to Arcturus is $\sqrt{1 - 0.99^2}$ times 40 light years or 5.64 light years. B also sees the earth moving away at the same relative velocity of $v = 0.99c$. So the twin in the spaceship thinks it takes 5.7 years to get to Arcturus or 11.4 years for the round trip. This result checks with the calculations of twin A on earth.

However, we run into an apparent paradox when the space traveler looks back on earth and sees the earth clocks slowed down with respect to his clock. It seems that B should get the result that A is younger, which contradicts our previous reasoning. In fact, if velocity is truly relative, how can we arrive at an asymmetric result at all? According to symmetry should not both twins end up with the same age? At first sight it appears that Einstein's formalism gives rise to a self-contradiction. The paradox is resolved by noting that this problem is inherently asymmetric. The twin on earth always remains in the same inertial reference system, whereas the spaceman changes reference systems in turning around. At the end of this section we shall do the entire calculation from the point of view of the spaceman and still get the result that the earthling ages more, even though the earthling's clocks appear to run slower according to the spaceman.

There has been a long history of controversy concerning the twin paradox (also called the clock paradox). By now nearly all physicists accept the interpretations given here; however, there are still some philosophers, mathematicians, and even one or two physicists of note who claim that the twins actually end up the same physical age. This author is as certain of the slower aging of the space traveler as he is of anything else in physics.

Such time dilation effects are negligible unless the spaceship can achieve a kinetic energy comparable to its rest mass energy. Even the energy release of nuclear fission or fusion is still about 1000 times too low for this. It is impossible in a practical sense for mankind to utilize time dilation in order to make long trips to other stars.

The twin paradox has been verified by several experiments. In one of these, the iron crystal of a Mossbauer clock (see page 153) is heated up and compared with a cold Mossbauer clock. Iron atoms in the heated crystal vibrate back and forth compared to iron atoms in the cold sample, which are almost at rest. Two identical iron nuclei emit radiation of identical frequency when at the same temperature; however, nuclei making the back-and-forth trip (just as twin B) are observed to emit radiation whose average frequency is reduced. This experiment was first performed in 1960 and the fractional frequency change was found to be $\Delta f/f = -2.4 \times 10^{-15}$ for a 1 K temperature increase. This number agrees with the change in $\gamma$ due to the known increase in mean square thermal velocity with temperature.

A second verification was made using large-sized or macroscopic clocks rather than individual iron atoms. The most accurate macroscopic clock in use is the cesium beam atomic clock. In fact, 1 s is now by definition equal to 9,192,631,770 ticks of such a clock. During October 1971 two sets of such clocks were compared. One set was flown around the world in commercial jet airliners. The other set was the reference clocks at the U. S. Naval Observatory. The special theory of relativity predicted that the traveling clocks should have lost $(184 \pm 23)$ ns because of their motion relative to the reference clocks. The observed time loss was $(203 \pm 10)$ ns. We see that the measurement is in agreement with the theory within errors. This experiment is reported in the July 14, 1972, issue of *Science*.

We shall now end this section with a detailed calculation of the space traveler (twin B) observing the slower clocks back on earth. Assume that both B's clock and the earth clock emit a pulse of light with each tick. We want to see if the space traveler also gets the result that more time has elapsed on earth than in his spaceship. This is the crux of the "paradox"; namely, if A's clock is running slower according to B, how can B receive more pulses from the slow clock than from his own fast clock? As we shall see, it is because on the return trip the speeding up of the pulse rate due to the "blue shift" Doppler effect is greater than the "slowing down" effect of time dilation.

We shall now calculate the total number of clock pulses received by the space traveler, both from the clock on the spaceship and from the earth clock. Let $N_B$ be the total number of pulses from B's clock and $N_A$ the total number from the earth clock. The number of pulses from B's clock is

$$N_B = f_0 t_B$$

where $f_0$ is the pulse rate from either clock when at rest and $t_B$ is the total travel time according to B.

$$t_B = \frac{\text{total distance}}{v}$$

As discussed previously, B sees the distance $D$ Lorentz contracted; that is, the round trip distance according to B is $2D/\gamma$. So

$$N_B = f_0 \left(\frac{2D}{\gamma v}\right) \tag{8-14}$$

is the total number of pulses B counts from the spaceship clock.

The number of pulses received by B from the earth clock is

$$N_A = f_1 t_1 + f_2 t_2$$

where $f_1$ and $f_2$ are the observed pulse rates when traveling away from and toward the earth, respectively. The time of each one-way trip is

$$t_1 = t_2 = \frac{1}{2} t_B = \frac{D}{\gamma v}$$

so

$$N_A = (f_1 + f_2) \frac{D}{\gamma v}$$

For $f_1$ and $f_2$ we use the results of Eqs. 8-12 and 8-13:

$$N_A = \left( \sqrt{\frac{1-\beta}{1+\beta}} f_0 + \sqrt{\frac{1+\beta}{1-\beta}} f_0 \right) \frac{D}{\gamma v}$$

$$= \left( \sqrt{\frac{1-\beta}{1+\beta}} + \sqrt{\frac{1+\beta}{1-\beta}} \right) \sqrt{1-\beta^2} \, \frac{D}{v} f_0$$

$$= [(1-\beta) + (1+\beta)] f_0 \frac{D}{v} = \frac{2D}{v} f_0$$

This is the same result that the twin on earth gets while observing the earth clock; hence, the theory is self-consistent. Also twin B has the result in Eq. 8-14 that the elapsed time on the spaceship clock is $\sqrt{1 - v^2/c^2}$ times that of the earth clock. The ratio $N_A/N_B = \gamma$.

In all of the preceding discussions we have been assuming that the turn-around time is much less than the travel time and can thus be ignored. The number of light pulses intercepted by twin B during turnaround will be much smaller than those intercepted during his long trip while at constant velocity.

## Summary

All the results of relativistic kinematics can be derived mathematically starting from the two basic postulates: (1) the principle of relativity (no way to determine absolute motion), and (2) invariance of the speed of light (the speed of light has the same value for all observers).

These two postulates determine the Lorentz transformation equations relating the $x$ and $t$ values of an event as measured by one observer to the $x'$ and $t'$ values of the same event as measured by a second observer where the two observers have a relative velocity $v$ along the $x$ axis:

$$x' = \gamma x + \gamma v t \qquad t' = \gamma t + \gamma \frac{v}{c^2} x \qquad \text{where } \gamma = (1 - v^2/c^2)^{-1/2}$$

It immediately follows from these equations that a moving meter stick appears contracted by the factor $1/\gamma$ (Lorentz contraction) and moving clocks appear to run slower by the factor $\gamma$ (time dilation). If two events that are separated in $x$ by a distance $l$ occur simultaneously for one observer, they will appear separated in time by $\Delta t = -vl/c^2$ to a moving observer (lack of simultaneity).

A moving oscillator will have its observed frequency shifted by the factor $\gamma(1 + \beta)$ where $v = \beta c$ is the velocity of the oscillator (relativistic Doppler effect). Not only do moving clocks slow down by a factor of $\gamma$, but a space traveler who makes a round trip during a time $t$ (on earth) will have aged $(1 - v^2/c^2)^{1/2}$ times $t$ where $v$ is his velocity with respect to the earth (the twin paradox).

## Exercises

1. Suppose the arm in Figure 8-4 is moving through the air at $v = 30$ m/s. The source emits a pulse of sound (not light). $D = 2$ m, $v_{\text{sound}} = 330$ m/s.
   (a) What is the travel time for the pulse to be reflected back to the source?
   (b) Repeat for the arm perpendicular to $v$.
   (c) Repeat part (a) for the time to go from the source to the mirror.
2. Assume the arm in Exercise 1 is stationary. A fan blows air along the arm at a speed of 20 km/h. How much longer does it take the pulse to make the round trip?
3. A Michelson interferometer has 2-m long arms. According to the ether theory, how fast must the ether be moving so that a 90° rotation of the interferometer gives one "fringe shift." ($\Delta t$ is the time for light to travel one wavelength of 0.4 μm.)
4. A massless particle is traveling away from an observer at a speed $u = c$. The observer then chases after it with a speed $v = 0.9c$. What speed does the moving observer measure for the particle?
5. A rod of proper length $l$ moving parallel to its length is measured by an observer to have length $l'$. What is the speed of the rod with respect to the observer?
6. For $v \ll c$, show that the Doppler shift formula gives

$$\Delta f = f_A - f_B \approx -\beta f_B$$

7. Repeat Example 2 for a pion beam of velocity $v = 0.999c$.
8. Distant galaxies and quasars have a red-shift parameter $Z$ defined by $Z \equiv \Delta\lambda/\lambda_0$, where $\lambda_0$ is the wavelength of a given spectral line from a source at rest. $\Delta\lambda$ is the shift in wavelength of this line observed in light from the distant receding source. Use Eq. 8-12 to derive a formula for $v/c$ in terms of $Z$.
9. If the two lightning bolts of Example 4 made marks on the ground, what would be the distance between the marks?
10. A well-known spectral line is observed to have a wavelength 0.5 μm when coming from a distant galaxy. The usual wavelength of this line is 0.4 μm. What is the velocity of recession of the galaxy?
11. Repeat Example 6 for $v = 0.999c$.

12. Let twin B be a space traveler and let twin A stay behind. Twin B travels for 30 years of earth time at a velocity of $v = 0.1c$. How much younger will twin B be than twin A?
13. Eq. 8-13 can be written as

$$\frac{f}{f_0} = \left(\frac{1+\beta}{1-\beta}\right)^{1/2}$$

where $f$ is the Doppler-shifted frequency. What is $df/f_0$ in terms of $\beta$ and $d\beta$?
14. If a clock is flown a distance of 24,000 mi at a speed of 500 mph, how much time will it have lost in nanoseconds with respect to a clock fixed on the earth?
15. Charge density is electric charge per unit volume. If electric charge is independent of the observer's velocity, will the charge density of a charged body appear to increase or decrease for a moving observer? If $\rho_0$ is the charge density at rest, what will be $\rho'/\rho_0$?
16. Assume the edge of the universe is $10^{10}$ light years away as we measure it from earth. Assume a space traveler has a velocity such that $(1 - v^2/c^2)^{-1} = 10^8$. According to the traveler, how many light years away is the edge of the universe?

## Problems

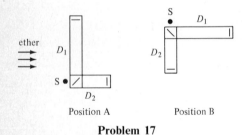

Position A        Position B

**Problem 17**

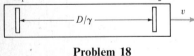

**Problem 18**

17. Suppose the two arms of a Michelson interferometer are of unequal length ($D_1 > D_2$). Let the time difference between round trips in the two arms be $\Delta t_A$ for position A. Then when the interferometer is rotated so that arm 1 is parallel to the ether velocity (position B), the time difference is $\Delta t_B$. Assuming the prerelativity theory of the ether, show that

$$(\Delta t_B - \Delta t_A) \approx \frac{v^2}{c^3}(D_1 + D_2)$$

This is the quantity that Michelson and Morley thought they were measuring.
18. Suppose the light clock in Figure 8-5 is pointed along the direction of motion. Then a "stationary" observer would see its length contracted to be $D/\gamma$. Let $t_1$ be the time it takes for the light pulse to go from $M_1$ to $M_2$ according to the stationary observer. (During this time $M_2$ moves a distance $vt_1$)
    (a) What is the distance traveled by the light pulse in going from $M_1$ to $M_2$ according to the stationary observer?
    (b) Show that $t_1 = D/\gamma(c - v)$.
    (c) How long will it take for the reflected pulse to return to $M_1$ according to the stationary observer?
    (d) Show that the round trip time is $t_1 + t_2 = \gamma\, 2D/c$.
19. Assume the classical ether theory, but also assume that all bodies have a Lorentz contraction in their direction of motion with respect to the ether; that is, the interferometer arm $D_1$ of Problem 17 becomes ($\sqrt{1 - v^2/c^2}\, D_1$) in position B. Repeat Problem 17 under these assumptions to find ($\Delta t_A - \Delta t_B$).
20. Repeat Problem 19, but do not rotate the interferometer by 90°. Then the measured time difference is $\Delta t_A$. Assume the ether velocity is $v$. Now change

the position of the interferometer so that the ether velocity is $v'$. Since the rotational velocity of the earth adds to the ether velocity, this can be done by waiting for the earth to rotate 180°. (It can also be done by moving the interferometer to a different latitude.) Show that

$$(\Delta t_A - \Delta t_{A'}) \approx (v^2 - v'^2)\frac{(D_1 - D_2)}{c^3}.$$

Such experiments were done and the result zero was always obtained. Thus Lorentz contraction alone cannot explain the null result. Time dilation as well as Lorentz contraction is necessary.

21. In the two simultaneous equations of Eq. 8-9 assume $x'$ and $t'$ are known and that $x$ and $t$ are unknowns. Solve the simultaneous equations for $x$ and $t$ in terms of $x'$ and $t'$.

22. Suppose in Figure 8-9 there is a third observer, Mr. Double Prime moving to the left with velocity $v'$. His $x$ and $t$ coordinates are $x''$ and $t''$. According to the Lorentz transformation the relation between $(x'', t'')$ and $(x', t')$ is

$$x'' = \frac{x' + v't'}{\sqrt{1 - v'^2/c^2}}$$

$$t'' = \frac{t' + v'x'/c}{\sqrt{1 - v'^2/c^2}}$$

Now use Eq. 8-9 to eliminate $x'$ and $t'$ from these two equations. After some algebraic manipulations you should obtain the result

$$x'' = \frac{x + v''t}{\sqrt{1 - v''^2/c^2}}$$

$$t'' = \frac{t + v''x/c}{\sqrt{1 - v''^2/c^2}} \quad \text{where} \quad v'' = \frac{v + v'}{1 + vv'/c^2}$$

The significance of this result is that the Lorentz transformation of a Lorentz transformation is also a Lorentz transformation with a velocity given by the Einstein addition of velocities formula.

23. Two flash bulbs $S_1$ and $S_2$ are mounted on the ends of a rod of length $l_0$ when at rest. The rod is moving to the right with velocity $v$. $S_1$ emits a flash of light before $S_2$ such that both flashes reach Mr. X simultaneously. $S_1$ was at $x = x_1$ and $S_2$ was at $x = x_2$ when their respective flashes were emitted. What is the distance $(x_2 - x_1)$ as calculated by Mr. X? This is the apparent length of the rod as viewed by eye or as photographed by a camera. Note that it is longer than $l_0$, not shorter. After correcting for the different times it takes light to come from the two ends, the calculated length is of course the Lorentz contracted length. A three-dimensional object when viewed by eye or camera will appear rotated when viewed at right angles to its direction of motion.

24. Derive a formula for the Doppler effect for sound when the source is moving and the detector stationary. In Figure 8-12 the primed system would be at rest with respect to the air.

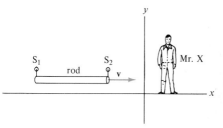

**Problem 23**

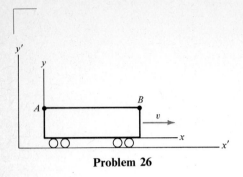

Problem 26

25. Repeat Problem 24 for a stationary source and moving detector. In Figure 8-12 the air would be moving to the right with velocity $v$.

26. A boxcar of length $l$ when at rest is moving to the right with velocity $v$ as measured in the primed system. In this system a distant star directly overhead emits a pulse of light that is detected simultaneously (in the primed system) at points $A$ and $B$. According to an observer in the boxcar, the pulse is received at point $A$ [before; after; simultaneously with] receiving the pulse at point $B$. Noting that $\Delta t' = \gamma \Delta t + \gamma(v/c^2)\Delta x$, we obtain $(t_B - t_A) = [vl/c^2; -vl/c^2; \gamma vl/c^2; -\gamma vl/c^2; 0]$ According to an observer moving with the boxcar, the star will appear shifted by an angle $\theta$ where $\theta = [0; \sin^{-1}(v/c); \tan^{-1}(v/c); \sin^{-1}(\gamma v/c); \tan^{-1}(\gamma v/c)]$. (You may assume that the light rays from the star are parallel.)

27. A rocket of rest length (that is, proper length) 200 m moves with respect to us at $v/c = \frac{3}{5}$. There are two clocks on the ship, at the nose and the tail, that have been synchronized with each other in their rest frame. We on the ground have a number of clocks, also synchronized with one another. Just as the nose of the ship reaches us, both our clocks and the clock in the nose of the ship read $t = 0$.
   (a) At this time $t = 0$ (to us) what does the clock in the tail of the ship read?
   (b) How long does it take (to us) for the tail of the ship to reach us?
   (c) At this time, when the tail of the ship is beside us, what does the clock in the tail read?

28. A stopwatch is located at $x = 0$. In another coordinate system (the primed system) it is moving to the right along the $x'$ axis with a velocity $v = 0.6c$. The stopwatch stops running at $t = 10$ s. (At $t' = t = 0$ it was at $x' = 0$.)
   (a) Where was the stopwatch in the primed system when it stopped running? (What was its $x'$ coordinate?)
   (b) At what time in the primed system did the watch stop?
   (c) If Mr. Prime is fixed at $x' = 0$, what is his velocity as measured by an observer moving along with the stopwatch?
   (d) Suppose that a second stopwatch located at $x = l$ was started at the same time $t = 0$ as the first stopwatch located at $x = 0$. What is the reading $t$ of the second stopwatch when the first one stops according to Mr. Prime who is at rest in the primed system? (That is, when $t'$ is the same as in part (b).)

29. Kinetic theory tells us that $\frac{1}{2}\overline{mv^2} = \frac{3}{2}kT$ for particles of mass $m$ and absolute temperature $T$ where $k = 1.38 \times 10^{-23}$ in S.I. units. An iron atom has $m = 9.3 \times 10^{-26}$ kg.
   (a) Calculate $\overline{\beta^2}$ for iron atoms at 300° absolute (room temperature).
   (b) What is $\overline{\gamma} = (1 - \overline{\beta^2})^{-1/2}$ for these atoms?
   (c) Because of time dilation, a sample of moving (or heated) iron atoms will emit a frequency $f' = (1/\overline{\gamma})f_0$ where $f_0$ is the frequency when at rest or at absolute zero. For iron atoms at 300° absolute, what is $(f' - f_0)/f_0$? What would be $\Delta f/f$, the fractional change in frequency of the Mossbauer effect, for a one degree change in temperature?

# Relativistic Dynamics

So far we have been discussing general properties of time and space. Now we want to treat material particles that have mass, momentum, and energy. We shall see that the laws of conservation of momentum and conservation of energy still hold, but that the classical definitions of momentum and energy must be modified. Of course, our new relativistic definitions of momentum and energy will become identical with the classical definitions as $v \to 0$. One of the new surprises will be that associated with each mass $m$ is an energy $E = mc^2$. Einstein proposed that locked up in 1 kg of any substance is $9 \times 10^{16}$ J. This is such a large amount of energy that it could run a 100 W light bulb for 30 million years.

Before dealing directly with mass, energy, and momentum, we must first learn how different observers view the same moving object; that is, how velocity transforms in relativity theory.

## 9-1 Einstein's Addition of Velocities

So far we have dealt with objects or particles that are at rest in one frame and thus moving with velocity $v$ in the other frame. Now we go to the case where the object already has velocity $u_x$ in one frame and $u'_x$ in the other. An example of such a situation is shown in Figure 9-1 where, according to

**Figure 9-1.** *Mr. Prime sees boxcar moving to the right with velocity $v$. Inside boxcar is a car with velocity $u_x$ with respect to boxcar.*

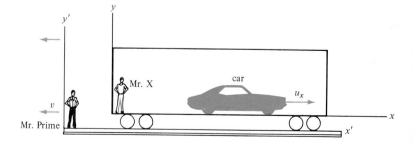

169

Mr. X, the car has velocity $u_x$, and, according to Mr. Prime, it is moving faster with velocity $u'_x$. Classically $u'_x = u_x + v$. The relativistic form is obtained by writing Eq. 8-9 in differential form:

$$dx' = \gamma\, dx + \gamma v\, dt$$

$$dt' = \gamma\, dt + \frac{\gamma v}{c^2}\, dx$$

Now divide the first equation by the second:

$$\frac{dx'}{dt'} = \frac{dx + v\, dt}{dt + (v/c^2)\, dx} = \frac{dx/dt + v}{1 + (v/c^2)(dx/dt)}$$

Replacing $dx/dt$ and $dx'/dt'$ with $u_x$ and $u'_x$ gives

$$u'_x = \frac{u_x + v}{1 + vu_x/c^2} \quad \text{(Einstein's addition of velocities)} \quad (9\text{-}1)$$

This is called the Einstein addition of velocities relation. We see that the resultant velocity is less than the sum of the two separate velocities; however, when both velocities are much smaller than the speed of light, the resultant velocity is very close to the sum of the two.

If the theory is to be self-consistent, Eq. 9-1 must forbid velocities greater than $c$. Consider a particle in the unprimed frame which is already traveling at the speed of light (it could be a particle of light, the photon, or a neutrino); that is, $u_x = c$. Then an observer in the primed system will see

$$u' = \frac{(c) + v}{1 + v(c)/c^2} = \frac{c + v}{(c + v)/c} = c$$

We see that light (or anything else) traveling with velocity $c$ must appear also to have velocity $c$ to all other observers, no matter how fast they are moving. As previously stated, the Lorentz transformation equations transform space and time in just such a way that a beam of light must travel with the same velocity $c$ as seen by all observers.

---

**Example 1.** Two supersonic jet planes are headed toward each other on a collision course as in Figure 9-2. If their ground speeds are exactly 2000 mph and 1000 mph, what is the velocity of the first plane as measured by the second plane?

ANSWER: In this case Mr. X (the unprimed system) is standing on the ground and Mr. Prime is the moving observer traveling at

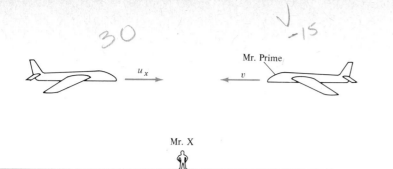

**Figure 9-2.** *Two jet planes traveling with speeds $u_x$ and $v$ with respect to the ground. The observer plane sees the jet on the left coming at him with a speed $u'_x$ that is less than $(u_x + v)$.*

$v = 2000$ mph. The first plane has $u_x = 1000$ mph according to Mr. X. Then Eq. 9-1 gives

$$u'_x = \frac{1000 + 2000}{1 + (2 \times 10^6)/c^2} \text{ mph} = \frac{3000}{1 + 4.5 \times 10^{-12}} \text{ mph}$$

$$= 2999.999999986 \text{ mph}$$

We see that classical physics is a very good approximation even for supersonic aircraft.

**Example 2.** The neutron is an unstable particle that decays into a proton, an electron, and an antineutrino: $N \rightarrow P + e^- + \bar{\nu}$. Suppose the decay electron has a velocity of $0.8c$ when its parent neutron is at rest. What will be the observed electron velocity if the decay occurs while the neutron is moving in the same direction at $0.9c$?

ANSWER: Our frame of reference is moving at $v = 0.9c$ to observe the electron traveling at $u_x = 0.8c$. Then Eq. 9-1 gives

$$u'_x = \frac{0.8c + 0.9c}{1 + 0.72c^2/c^2} = \frac{1.7}{1.72}c = 0.988c$$

**Example 3.** Suppose the car in Figure 9-1 is traveling to the left with a velocity whose magnitude is $u$. What will be the velocity of the car in the primed system?

ANSWER: In this case $u_x = -u$. Substitution into Eq. 9-1 gives

$$u'_x = \frac{v - u}{1 - uv/c^2}$$

The result of Example 3 should be used when $u_x$ and $v$ as observed in the primed system are of opposite sign.

## 9-2 Definition of Relativistic Momentum

According to classical physics, momentum is $\mathbf{p} \equiv m\mathbf{u}$ where $m$ is the mass of the particle and $\mathbf{u}$ is its velocity. The total $x$-component of momentum in a closed system is obtained by summing over all the particles:

$$(p_x)_{\text{total}} = \sum_j m_j u_{jx}$$

where $u_{jx}$ is the $x$-component of velocity of the $j$th particle. According to the classical law of conservation of momentum,

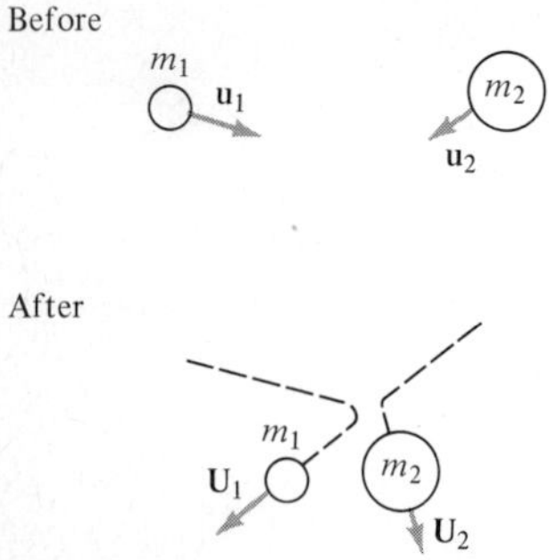

$$\sum_j m_j u_{jx} = \sum_j m_j U_{jx} \tag{9-2}$$

where the upper-case $U_j$ stands for velocity of the $j$th particle at some later time. The later time could be after a collision as shown in Figure 9-3. Now add $\sum_j m_j v$ to both sides of Eq. 9-2. Then

$$\sum m_j(u_{jx} + v) = \sum m_j(U_{jx} + v) \tag{9-3}$$

In classical mechanics an observer moving with velocity $v$ to the left sees $u'_{jx} = u_{jx} + v$ and $U'_{jx} = U_{jx} + v$. Substitution into Eq. 9-3 gives

$$\sum m_j u'_{jx} = \sum m_j U'_{jx}$$

which says that, if momentum is conserved in one frame of reference, it will be conserved in all others. But in relativity theory momentum, if defined as $m\mathbf{u}$, will be conserved in the primed frame of reference only if

$$\sum m_j \frac{u_{jx} + v}{1 + v u_{jx}/c^2} = \sum m_j \frac{U_{jx} + v}{1 + v U_{jx}/c^2}$$

In general it is not possible for this equation to be true when Eq. 9-2 (or 9-3) is true.

So Einstein then was faced with the problem of finding a new mathematical form for momentum that stays conserved when transforming to other frames of reference. He found that if momentum is defined as

**Figure 9-3.** *Elastic collision between $m_1$ and $m_2$.*

where
$$\mathbf{p} \equiv m\gamma(u)\mathbf{u} \quad \text{(relativistic momentum)} \quad (9\text{-}4)$$

$$\gamma(u) \equiv \left(1 - \frac{u^2}{c^2}\right)^{-1/2}$$

it will stay conserved for different observers if it is conserved in any one reference frame. In order to show that momentum as defined in Eq. 9-4 will be conserved as seen by moving observers, we must first see how it transforms from one coordinate system to another. This is shown in Appendix 9-1 with the result that

$$\begin{cases} p'_x = \gamma p_x + \gamma\beta \dfrac{E}{c} \\ p'_y = p_y \\ p'_z = p_z \\ \dfrac{E'}{c} = \gamma \dfrac{E}{c} + \gamma\beta p_x \end{cases} \quad (9\text{-}5)$$

where $E \equiv m\gamma(u)c^2$, $E' \equiv m\gamma(u')c^2$, and $\beta \equiv v/c$.

We see that the four quantities $(p_x, p_y, p_z, E/c)$ transform in exactly the same way as the Lorentz transformation of the four quantities $(x, y, z, ct)$. Einstein identified $\mathbf{p}$ with the particle momentum and $E$ with its energy. Justification for this identification is given in the next section. There we shall show that if relativistic momentum is conserved in the unprimed frame, it will also be conserved in the primed frame.

## 9-3 Conservation of Momentum and Energy

When the particle velocity $u$ is much less than the speed of light, Einstein's relativistic momentum $p_x = m\gamma(u)u_x \to mu_x$ because $\gamma(u) \to 1$ when $u \to 0$. So Einstein's new definition of momentum agrees with classical mechanics in the classical limit. But what about his definition of energy?

$$E \equiv m\gamma(u)c^2 \quad (9\text{-}6)$$

$$= m\left(1 - \frac{u^2}{c^2}\right)^{-1/2} c^2$$

$$\approx m\left(1 + \frac{u^2}{2c^2}\right)c^2 \quad \text{for } u/c \ll 1$$

Here we have used the binomial expansion to get $(1 - u^2/c^2)^{-1/2} \approx (1 + u^2/2c^2)$. So in the limit of small velocity, Einstein's energy becomes

$$E \approx mc^2 + \tfrac{1}{2}mu^2$$

We note that $\tfrac{1}{2}mu^2$ is the classical energy of a free particle of velocity $u$. Hence Einstein's new definition of energy agrees with classical physics if the fixed quantity $mc^2$ is added to the kinetic energy. In classical mechanics an additive constant to the energy is completely arbitrary. However, in Einstein's theory it is not. Einstein drew the physical conclusion that a particle when at rest still contains the amount of energy $E_0 = mc^2$. He called this the rest energy. Since 1905 there have been many verifications of such a daring prediction—one of them being the atomic bomb. Some of the verifications will be discussed in the next section.

Our final job for this section is to show that if the relativistic forms of $p_x$, $p_y$, $p_z$, and $E$ are conserved in the unprimed system, then they will be conserved in the primed system as well. Consider a system of $n$ interacting particles. The initial total momentum and energy is

$$(p_x)_{\text{total}} = \sum_j p_{jx} \quad \text{and} \quad e_{\text{total}} = \sum_j e_j$$

where lower-case letters stand for the initial values. We shall use upper-case letters for the final values after a given length of time. In order to get the momentum and energy in the primed system we use Eqs. 9-5:

$$p'_{jx} = \gamma p_{jx} + \gamma \beta \left(\frac{e_j}{c}\right)$$

$$\frac{e'_j}{c} = \gamma \frac{e_j}{c} + \gamma \beta p_{jx}$$

Now add together the $p'_{jx}$ equations for each of the $n$ particles:

$$\sum p'_{jx} = \gamma \sum p_{jx} + \gamma \beta \sum \frac{e_j}{c} \qquad (9\text{-}7)$$

Next we make use of the condition that momentum and energy be conserved in the unprimed system. Then

$$\sum p_{jx} = \sum P_{jx} \quad \text{and} \quad \sum e_j = \sum E_j$$

Substitution into Eq. 9-7 gives

$$\sum p'_{jx} = \gamma \left( \sum P_{jx} \right) + \gamma\beta \left( \sum \frac{E_j}{c} \right)$$

$$= \sum \left[ \gamma P_{jx} + \gamma\beta \frac{E_j}{c} \right]$$

$$= \sum [P'_{jx}]$$

This result says that in the primed system the total initial momentum equals the total final momentum; that is, momentum is conserved in the primed system. This completes our proof for momentum.

We can show the same thing for energy by adding together the $n$ energy equations:

$$\sum e'_j = \gamma \sum e_j + \gamma v \sum p_{jx}$$

Now use the fact that momentum and energy are conserved in the unprimed system:

$$\sum e'_j = \gamma \left( \sum E_j \right) + \gamma v \left( \sum P_{jx} \right)$$

$$= \sum [\gamma E_j + \gamma v P_{jx}] = \sum E'_j$$

This says that if momentum and energy (as defined by Einstein) are conserved in the unprimed system, then they must be conserved in the primed system.

We know from classical physics that Einstein's forms for momentum and energy are such that they are conserved in a frame of reference where all the particle velocities are much less than the speed of light. We have just shown that then the relativistic momentum and energy will be conserved even when viewed by an observer moving near the speed of light. But no matter how plausible the theory, the real proof lies in experimental verification of the predictions. Needless to say, conservation of relativistic momentum and energy has been very thoroughly verified by experiment. Some examples will be given in the next section.

## 9-4 Equivalence of Mass and Energy

Einstein's prediction (Eq. 9-6) that a mass $m$ when at rest should contain an enormous amount of energy $E_0 = mc^2$ was a very daring prediction having all kinds of possible practical applications including nuclear power and nuclear bombs. Einstein was proposing that if the rest mass of a particle or system of particles could be decreased by the amount $\Delta m$, there would be a release of energy $\Delta E = (\Delta m)c^2$.

---

**Example 4.** What is the energy contained in 1 g of sand? If it could be released, how would it compare with the 7000 cal of heat delivered by burning 1 g of coal? (1 cal = 4.18 J.)

ANSWER
$$E_0 = (10^{-3}\text{ kg})(3 \times 10^8\text{ m/s})^2 = 9 \times 10^{13}\text{ J}$$

The energy released by burning 1 g of coal is (7000 cal) × 4.18 J/cal) or $2.9 \times 10^4$ J. Thus the rest energy is $3.1 \times 10^9$ times as much as the chemical energy.

---

We see from Example 4 that even if only one thousandth of the rest energy is released, it is still millions of times more powerful than conventional energy sources.

---

**Example 5.** If 1 ton of TNT gives an energy release of $10^9$ cal when exploded, how much mass must be converted into energy in a 1-megaton (Mton) bomb?

ANSWER: One Mton of TNT releases $10^{15}$ cal or $4.18 \times 10^{15}$ J.

$$m = \frac{E}{c^2} = \frac{4.18 \times 10^{15}\text{ kg}}{9 \times 10^{16}} = 0.046\text{ kg} = 46\text{ g}$$

When a 1 Mton bomb is exploded, the nuclear explosive must decrease in mass by 46 gm. The total mass of nuclear explosive (fission and fusion) needed in such a bomb is about 1000 times as much. Hence 1-Mton H-bombs must weigh more than 50 kg.

---

The first experimental confirmation of the Einstein mass-energy relationship came from the comparison of energy release in radioactive decay with the mass difference between initial nucleus and final products. As an example of how $E_0 = mc^2$ can be checked in the laboratory, let us consider the simplest case of beta decay, the beta decay of the free neutron. The free neutron is observed to decay into a proton, an electron, and an antineutrino (of zero rest mass).

$$N \to P + e^- + \bar{\nu}$$

The decay products are observed to have a total of $1.25 \times 10^{-13}$ J of kinetic energy. The rest mass of the neutron is measured to be greater than that of the proton plus electron by $13.9 \times 10^{-31}$ kg. The energy corresponding to this amount of mass decrease should be $\Delta E = (13.9 \times 10^{-31})(3 \times 10^8)^2$ J, which is $1.25 \times 10^{-13}$ J. This amount of energy checks with the observed kinetic energy of the decay products, which is also $1.25 \times 10^{-13}$ J within the accuracy of measurement.

Another example of the vast energy locked up in rest mass is the annihilation of an electron and a positron (see Figure 9-4). The positron is a positive electron and is discussed in Chapter 31. When an electron and a positron are put into contact, they will annihilate each other, converting into two photons. The photon is a quantum of electromagnetic radiation. In this case rest energy of $2m_e c^2$ is converted into energy of electromagnetic radiation ($m_e$ is the electron rest mass).

Our third example is an elementary particle called the muon, which decays into an electron and two neutrinos:

$$\mu^- \rightarrow e^- + 2\nu$$

The muon has 208 times as much rest mass as the decay electron. The two neutrinos have zero rest mass. In this example about 99.5% of the muon rest mass is converted into kinetic energy of the electron and two neutrinos.

The reverse is also true—kinetic energy can be converted into rest mass. Usually when a particle of high kinetic energy collides with an atomic nucleus or with a single proton, new particles are produced where some of the kinetic energy is converted into the rest mass energy of the new particles. Figure 9-5 shows an example of this where a proton having 300 GeV ($3 \times 10^{11}$ eV) of kinetic energy collides with a proton that is at

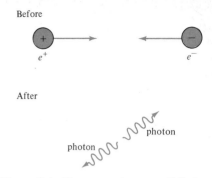

**Figure 9-4.** *Electron-positron annihilation into two photons.*

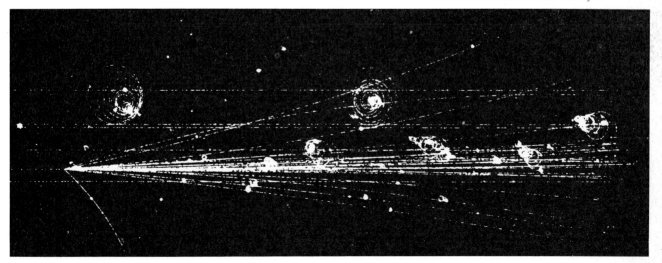

**Figure 9-5.** *300-GeV proton collides with a hydrogen nucleus in a liquid hydrogen bubble chamber. The tracks of 24 charged particles leave the point where the collision occurred. Photo was taken at the Fermi National Accelerator Laboratory.*

rest in a liquid hydrogen bubble chamber. Twenty-two new particles (mainly pions) are produced.

There is, however, a severe constraint on how much energy can be obtained by conversion of rest mass. In Chapter 31 we shall study a basic law of nature called the conservation of baryons. This law states that the total number of protons and neutrons in a sample of ordinary matter must stay the same. For this reason there is no way one can get the $9 \times 10^{13}$ J out of a gram of sand. However, in a large nucleus such as uranium, it is possible to redistribute the protons and neutrons so that the rest mass decreases by about 0.1%. In this process, called nuclear fission, a nucleus such as uranium will spontaneously split into two approximately half-sized nuclei plus a few neutrons. The total rest mass of the final products is about 0.1% less than the initial rest mass. In an inelastic collision of two particles, or a decay of a single particle, rest mass is clearly not conserved; however, the total energy $E = \Sigma m_j (1 - u_j^2/c^2)^{-1/2} c^2$ is conserved.

## 9-5 Kinetic Energy

The definition of kinetic energy in relativity theory is the same as in classical mechanics: kinetic energy of a particle is the energy due to the motion of the particle. The energy due to motion of a free particle can be obtained by subtracting the rest energy from Eq. 9-6:

$$K \equiv E - mc^2 = mc^2 \left[ \left(1 - \frac{u^2}{c^2}\right)^{-1/2} - 1 \right] \quad \text{(kinetic energy)}$$

As previously pointed out, if one makes use of the binomial expansion, namely, $(1 - \varepsilon)^n \to 1 - n\varepsilon$ as $\varepsilon \to 0$, then $K \to \frac{1}{2}mu^2$, which is the classical expression.

---

**Example 6.** A particle happens to have a kinetic energy equal to its rest energy. What is its velocity?

ANSWER:
$$K = mc^2 \left[ \left(1 - \frac{u^2}{c^2}\right)^{-1/2} - 1 \right] = mc^2$$

$$\left(1 - \frac{u^2}{c^2}\right)^{-1/2} = 2$$

$$\left(1 - \frac{u^2}{c^2}\right) = \frac{1}{4}$$

$$\frac{u}{c} = \sqrt{\frac{3}{4}} \quad \text{or} \quad u = 0.866c$$

At this point it is useful to summarize our relativistic expressions for momentum, energy, and velocity:

$$p = m\left(1 - \frac{u^2}{c^2}\right)^{-1/2} u \qquad (9\text{-}8)$$

$$E = m\left(1 - \frac{u^2}{c^2}\right)^{-1/2} c^2 \qquad (9\text{-}9)$$

$$E = K + mc^2$$

We can obtain velocity in terms of $p$ and $E$ by dividing Eq. 9-8 by Eq. 9-9:

$$\frac{p}{E} = \frac{u}{c^2}$$

or

$$u = \frac{pc^2}{E} \qquad (9\text{-}10)$$

Note that Eqs. 9-8 and 9-9 are consistent with the fact that a material particle cannot reach the speed of light. Putting $u = c$ into these equations would give infinite momentum or energy, which is impossible.

We can get a very useful relation between $E, p,$ and $m$ by squaring both sides of Eq. 9-9:

$$E^2\left(1 - \frac{u^2}{c^2}\right) = (mc^2)^2$$

$$E^2 - \frac{E^2 u^2}{c^2} = m^2 c^4$$

Now substitute the right-hand side of Eq. 9-10 for $u$:

$$E^2 - p^2 c^2 = m^2 c^4 \qquad (9\text{-}11)$$

---

**Example 7.** At the Fermi National Accelerator Laboratory near Chicago, Illinois, protons are accelerated until they have an energy 400 times their rest energy.
  (a) What is the velocity of these protons?
  (b) What is the ratio of $E$ to $pc$?

ANSWER: (a) Put $E = 400mc^2$ into Eq. 9-9 and use $\beta = u/c$:

$$400\, mc^2 = mc^2(1 - \beta^2)^{-1/2}$$

SEC. 9-5/KINETIC ENERGY

$$1 - \beta^2 = \frac{1}{400^2}$$

$$\beta = \sqrt{1 - \frac{1}{160,000}} \approx 1 - \frac{1}{320,000}$$

or

$$u = 0.999997c$$

(b) Inverting Eq. 9-10 gives

$$\frac{E}{pc} = \frac{c}{u} = \frac{1}{\beta} \approx 1 + \frac{1}{320,000}$$

## 9-6 Mass and Force

### Relativistic Mass

It was rather common in older textbooks to express relativistic momentum as $p = m(u)u$ where

$$m(u) \equiv \frac{m}{\sqrt{1 - u^2/c^2}} \quad \text{(relativistic mass)} \quad (9\text{-}12)$$

$m(u)$ is called the relativistic mass. In this book we shall reserve the symbol $m$ for rest mass. If ever we speak of relativistic mass we will use the notation $m(u)$. It is clear from Eq. 9-12 that relativistic mass increases with velocity in the very same way as energy $E$. For a free particle $m(u) = (1/c^2)E$. Relativistic mass is merely the relativistic energy times the proportionality constant $(1/c^2)$. Hence the relativistic mass of a closed system is conserved, whereas the total rest mass contained in the individual particles can change.

### Gravitational Mass

If some particles are sealed up in a box of perfectly reflecting walls and then the kinetic energy of the particles is increased, by how much will the weight of the box increase? If we so wish, the particles trapped inside the box could even be photons of zero rest mass. Then the gravitational mass of the box (defined as the weight divided by $g$) will be $E_{\text{total}}/c^2$. This result is predicted by the general theory of relativity and is confirmed experimentally. Note that the mass of a *system* of photons is not the sum of the separate photon rest masses. If photons are injected into a box, the gravitational mass will increase by the amount $\Delta m = \Delta E/c^2$ where $\Delta E$ is the total energy of the photons.

## Relativistic Force

It is useful to define force in such a way that Newton's third law of motion still holds for two interacting particles. Because of the law of conservation of momentum $d\mathbf{p}_1 = -d\mathbf{p}_2$. Then $d\mathbf{p}_1/dt = -d\mathbf{p}_2/dt$. In relativity theory the expression for force is thus

$$\mathbf{F} \equiv \frac{d\mathbf{p}}{dt}$$

We note that with this definition the magnitude and direction of the force will depend on the velocity of the moving observer, whereas in classical mechanics force is independent of the velocity of the observer. The change in direction and magnitude of force with the velocity of the observer leads to interesting effects, such as magnetic force when studying the electromagnetic interaction. Relativistic effects in the theory of electricity are discussed in Chapter 17.

## 9-7 General Relativity

To be precise, what we have been calling relativity theory should be called special relativity in order to distinguish it from general relativity. Einstein fully developed special relativity in 1905 and much of general relativity in 1911. The theory of general relativity is really a modern, relativistic theory of gravitation.

In Newton's theory of gravitation the force $F = Gm_1m_2/r^2$ is a force that acts instantaneously. If a force can act instantaneously, this means a signal, or energy, could be transmitted instantaneously from $m_1$ to $m_2$. This violates one of the basic tenets of relativity: that no energy, not even a signal, can travel faster than the speed of light. Thus Einstein tackled the problem of a relativistic theory of gravitation. He was determined that his new theory should satisfy both the principle of relativity and automatically give the result that gravitational mass is always equivalent to inertial mass. Einstein's determination led him to postulate what is called the principle of equivalence (see Section 5-5). This principle states that being in a gravitational field is equivalent to being in an accelerated reference system. For example, in a spacecraft blasting off and accelerating upward, a passenger has the impression that gravity has suddenly been increased. In a rocket taking off from the earth with acceleration $a = 2g$, the passengers and everything else in the rocket would weigh three times their normal weight. This "pseudogravitational" force is exactly proportional to the inertial mass. No physics experiment performed inside the rocket could tell the occupants whether the gravitational force of the earth had suddenly increased by a factor of three or whether the rocket was accelerating with respect to the earth.

In the general theory of relativity, Einstein mathematically incorporated the principle of equivalence using a form of mathematics well beyond the scope of this book. In this mathematical description any mass will "distort" the region of space around it so that all freely moving objects will follow paths curving toward the mass producing the distortion. Einstein's equations relate the amount of curvature to the strength (or mass) of the source.

Classically we would say that any object moving in a curved path is being accelerated and must be under the influence of some force. It is this acceleration, which in the general theory of relativity is a property of space, that explains the phenomenon of gravitation. Since it is the space itself that is "distorted," the effect on all inertial masses wll be the same and the principle of equivalence is automatically satisfied.

One of the consequences of this theory is that the wavelength of light is increased when leaving a mass. This effect is called the gravitational red shift. Such a shift is observed in the spectral lines of the sun and of heavy stars. Thus the ticks of an atomic clock on the surface of the sun are running slower than the very same atomic clock running here on earth. As we might expect, the general theory of relativity predicts that all clocks should be slowed down when in the presence of a gravitational field. In fact, if two identical clocks on the earth are 1 m apart in height, the lower clock should run slower by about one part in $10^{16}$. Frequency standards were first built in 1960 with this accuracy using the photons emitted from radioactive iron nuclei imbedded in a crystal. The phenomenon giving rise to such accurate frequencies is called the Mossbauer effect (see page 153).

So far it has been difficult to make experimental checks of general relativity. However, with these new frequency standards, it has finally been shown in laboratory experiments that gravity slows down time. The first such experiments were performed in a 70-ft tower at Harvard University in 1960.

Another effect predicted by general relativity is the bending toward the center of the sun of a beam of light that is passing near the edge of the sun. It takes general relativity to calculate the gravitational force between the sun and a photon moving at the speed of light. Only during eclipses of the sun is it possible to see stars whose apparent positions in the sky are near the edge of the sun. These positions are observed to shift by the amount predicted by Einstein.

Another prediction of general relativity, which seemed rather "far out" until the 1970s, was that under reasonably normal circumstances a star could collapse after it had used up its thermonuclear fuel and that the collapse could be so extreme that the final result would be a "black hole." By black hole we mean that no light or signals could get very far from the surface of the star. Such a star would suddenly and completely disappear and would never be seen again. The theory and experimental status of black holes is one of the main topics of Chapter 30. One more prediction of general relativity is that an accelerating mass (such as a collapsing star or a star–star collision) should emit gravitational waves just as an accelerating electrical charge emits electromagnetic radiation. Recently gravity wave detectors have been built that should be sensitive enough to detect a nearby supernova (see Chapter 30).

## Cosmology

General relativity has much to say in the field of astrophysics called cosmology. Cosmology deals with questions of the origin, size, and structure of the universe. Some of these questions are: Is the universe infinite or finite in size? Is it expanding in size? How and when were our solar system and galaxy formed? How many galaxies are there and how are they distributed? Where did they come from and what was the universe like before these galaxies were formed? In addition to relativity theory one must learn some nuclear physics in order to understand cosmology. Some of these questions are discussed further in Chapter 30 after the presentation of nuclear physics.

## Summary

The Einstein addition of velocities formula can be derived from the Lorentz transformation equations. The result is

$$u'_x = \frac{u_x \pm v}{1 \pm u_x v/c^2}$$

which corresponds to the classical expression $u'_x = u_x \pm v$.

If momentum is defined as $\mathbf{p} = m\gamma(u)\mathbf{u}$ and energy as $E = m\gamma(u)c^2$, then momentum and energy will appear conserved in all frames of reference if they are conserved in any one frame.

Rest mass has rest energy $E = mc^2$, and in those cases where rest mass can be decreased (such as electron-positron annihilation) the rest energy is converted to other forms such as kinetic energy.

For a free particle the kinetic energy is

$$K = E - mc^2 = mc^2[\gamma(u) - 1]$$

A useful relation between energy and momentum is

$$E^2 = p^2c^2 + (mc^2)^2$$

The relativistic mass of a body of velocity $u$ is defined as

$$m(u) = \frac{m}{\sqrt{1 - u^2/c^2}}$$

The relativistic mass of a closed system is conserved because energy is conserved. The gravitational force acts on relativistic mass. The relativistic theory of gravity, which is called general relativity, uses the principle of equivalence as a starting point.

## Appendix 9-1 Momentum–Energy Transformation

In order to see how the components of $\mathbf{p} = m\gamma(u)\mathbf{u}$ transform, we must first see how the components of $\gamma(u)\mathbf{u}$ transform from one coordinate system to another. We start with the differential form of the Lorentz transformation (Eq. 8-9) with both sides divided by $d\tau$ ($\tau$ is the proper time as measured by an observer attached to the moving particle):

$$\left| \begin{aligned} \frac{dx'}{d\tau} &= \frac{\gamma \, dx}{d\tau} + \gamma v \frac{dt}{d\tau} \\ \frac{dy'}{d\tau} &= \frac{dy}{d\tau} \\ \frac{dz'}{d\tau} &= \frac{dz}{d\tau} \\ \frac{dt'}{d\tau} &= \frac{\gamma \, dt}{d\tau} + \frac{\gamma v}{c^2} \frac{dx}{d\tau} \end{aligned} \right. \quad (9\text{-}13)$$

In Figure 9-1 the proper time $\tau$ would be the time measured by an observer inside the car that is moving with velocity $u_x$ (in the unprimed system). The relations between $t$ and $\tau$ and between $t'$ and $\tau$ are given by Eq. 8-6:

$$d\tau = \frac{1}{\gamma(u)} dt \quad \text{and} \quad d\tau = \frac{1}{\gamma(u')} dt'$$

Next we shall substitute these expressions for $d\tau$ into Eqs. 9-13 and also use the forms

$$\frac{dx'}{dt'} \equiv u'_x \qquad \frac{dx}{dt} \equiv u_x \qquad \cdots$$

Then we obtain

$$\gamma(u')\, u'_x = \gamma\gamma(u)\, u_x + \gamma v\, \gamma(u) \qquad \text{where } \gamma(u) \equiv \left(1 - \frac{u^2}{c^2}\right)^{-1/2}$$

$$\gamma(u')\, u'_y = \gamma(u)\, u_y$$

$$\gamma(u')\, u'_z = \gamma(u)\, u_z \qquad\qquad \gamma(u') \equiv \left(1 - \frac{u'^2}{c^2}\right)^{-1/2} \qquad (9\text{-}14)$$

$$\gamma(u') = \gamma\gamma(u) + \frac{\gamma v}{c^2} \gamma(u)\, u_x \qquad \gamma \equiv \left(1 - \frac{v^2}{c^2}\right)^{-1/2}$$

The preceding set of four equations shows how to transform all three components of velocity **u** from the unprimed system to the velocity **u'** as observed in a frame of reference moving with velocity $v$ to the left (see Figure 9-1).

In order to convert Eqs. 9-14 to relativistic momentum, merely multiply both sides by $m$. We shall use the notation $p_x \equiv m\gamma(u)\, u_x$, $p'_x \equiv m\gamma(u')\, u'_x$, etc. Then Eqs. 9-14 become

$$\begin{cases} p'_x = \gamma p_x + \gamma v m\, \gamma(u) \\ p'_y = p_y \\ p'_z = p_z \\ m\, \gamma(u') = \gamma m\, \gamma(u) + \dfrac{\gamma v}{c^2} p_x \end{cases}$$

Finally, we shall use the notation $E \equiv m\, \gamma(u)\, c^2$, $E' \equiv m\, \gamma(u')\, c^2$, and $\beta \equiv v/c$. Then the preceding equations become

$$\begin{cases} p'_x = \gamma p_x + \gamma\beta \dfrac{E}{c} \\ p'_y = p_y \\ p'_z = p_z \\ \dfrac{E'}{c} = \gamma\left(\dfrac{E}{c}\right) + \gamma\beta\, p_x \end{cases} \qquad (9\text{-}5)$$

## Exercises

1. Solve Eq. 9-1 for $u_x$ in terms of $u'_x$ and $v$.
2. Repeat Example 2 for the case where the decay electron is moving in the direction opposite to the neutron.
3. When coal or oil is burned, the resulting compounds have slightly less rest mass than the original. If a typical car is running continuously at 50 hp

(assume 250 hp for the fuel energy released), how much mass decrease will there be per year in the fuel products?

4. How much loss in rest mass does it take to supply the energy needs of the United States for 1 year? The United States uses about $7 \times 10^{12}$ kW-hr per year. If this energy were all supplied by nuclear fission, how many grams of fission products would be generated per year?

5. A proton of rest energy 938 MeV is given a kinetic energy of 47 MeV. By what percent is its relativistic mass increased?

6. Does the bending of light near the sun make the stars appear to move away from the sun or toward the sun?

7. How many micrograms does a 100-W light bulb radiate away in 1 year?

8. The velocity of a body is such that its relativistic mass increases by 10%.
   (a) By what fraction does its length decrease?
   (b) If its rest energy is $E_0$, what is its kinetic energy?

9. The rest energy of a proton is 938 MeV. Consider a proton traveling at one half the speed of light.
   (a) What is its kinetic energy in MeV according to classical mechanics?
   (b) What is its kinetic energy in MeV according to relativistic mechanics?

10. The energy flux from the sun at the earth is about 1 kW/m². How much of the sun's mass in grams reaches the earth per year?

11. To an observer at rest with respect to particle A, particle A appears to decay and emit particle B to the right with a velocity $v = 0.5c$. Suppose we observe this same event when particle A is moving with a velocity $v_A = 0.4c$ to the right. What velocity would we then measure for B? (We see A moving to the right when it decays.)

12. What mass of fissionable material is needed for a 20-kton nuclear bomb?

13. The kinetic energy of a pion is 35 MeV. By what factor is its half-life increased? The rest energy of a pion is 140 MeV.

14. A proton has total energy $E = 100\, m_p c^2$. What is its velocity?

15. For a particle of rest mass $m_0$ and velocity $v$, which is greater: $\frac{1}{2}m_0 v^2$, $p^2/2m_0$, or its kinetic energy?

16. If we define density as relativistic mass divided by the volume, by what factor is the density of a body increased when it is moving with velocity $v$?

17. A free proton has kinetic energy $K_0$.
    (a) What is its total relativistic energy $E$ in terms of $K_0$ and $m_p$, its rest mass?
    (b) What is its relativistic momentum $P$ in terms of $m_p$ and $K_0$?
    (c) What is its velocity in terms of $E$ and $P$?

18. Charge density is electric charge per unit volume. Electric charge is relativistically invariant. By what factor is the charge density of a body increased when it is moving with velocity $v$?

## Problems

19. Start with the Lorentz transformation equations for $x$ and $t$ in terms of $x'$ and $t'$. Derive a formula for $u_x$ in terms of $u'_x$ with $v$ as shown in Figure 9-1.

20. (a) Let $x_1 = x, x_2 = y, x_3 = z$, and $x_4 = ct$. Write the Lorentz transformation equations in terms of $x_1, x_2, x_3, x_4$.
    (b) Let $p_1 = p_x, p_2 = p_y, p_3 = p_z, p_4 = E/c$. Write Eqs. 9-5 in terms of $p_1, p_2, p_3, p_4$.

(c) Prove that $x_1'^2 - x_4'^2 = x_1'^2 - x_4'^2$.
(d) Prove that $x_1' p_1' - x_4' p_4' = x_1 p_1 - x_4 p_4$.

21. Suppose a quantity $A$ which has four components $(A_1, A_2, A_3, A_4)$ transforms the same way as the four components of $x$ or $p$ in Problem 20; that is, $A_1' = \gamma A_1 + \gamma \beta A_4$ and $A_4' = \gamma A_4 + \gamma \beta A_1$. The quantity $A$ is called a four-vector.
   (a) Prove that $A_1'^2 - A_4'^2 = A_1^2 - A_4^2$.
   (b) Prove that for two different four-vectors $A_1' B_1' - A_4' B_4' = A_1 B_1 - A_4 B_4$.
   (c) Prove that $(A_1' + B_1')^2 - (A_4' + B_4')^2 = (A_1 + B_1)^2 - (A_4 + B_4)^2$.

22. The rest energy of the $K$-meson is 495 MeV. Consider a 330-MeV $K$-meson beam (each $K$-meson has a kinetic energy of 330 MeV).
   (a) What is the total energy of each $K$-meson?
   (b) What is the rest mass in grams of the $K$-meson?
   (c) What is the velocity of these $K$-mesons?
   (d) What is the ratio of relativistic to rest mass for these $K$-mesons?
   (e) The half-life of the $K$-meson when at rest is $1.0 \times 10^{-8}$ s. What, then, is the observed half-life of the $K$-mesons in this beam?

23. Consider a beam of pions all of the same velocity. The average half-life of the pions in this beam is observed to be 67% longer than when at rest. The rest energy of a pion is 140 MeV.
   (a) What is the kinetic energy of each pion in this beam?
   (b) What is the velocity of each beam pion?
   (c) What is the ratio of the relativistic mass to rest mass?
   (d) What is $p/mc$ for each beam pion where $m$ is the rest mass?

24. (a) Consider a billiard ball collision with two balls of equal rest mass $m$. After the collision the balls have equal energy ($E_1 = E_2 = E$). What is the relativistic momentum $p$ of each ball after the collision in terms of $e_1$, the energy of the incoming ball?
   (b) Use energy and momentum conservation to derive the relation

$$\sin \theta = \sqrt{\frac{2mc^2}{e_1 + 3mc^2}}$$

25. Derive a formula for momentum $p$ in terms of kinetic energy $K$ and rest mass $m$.

26. Conventional high energy accelerators accelerate each beam particle of mass $m$ to a total energy $E_b$. These are then used to strike a stationary target, which in Figure (a) is of the same rest mass. However, the equivalent collision can be obtained by colliding two beams of lower energy $E'$ together. We wish to obtain a formula for $E'$ in terms of the equivalent $E_b$. That is, an observer moving to the right in Figure (a) sees both particles have the same energy $E'$. (Use the equation $E' = \gamma E + \gamma \beta pc$.)
   (a) Show that

$$\beta = \frac{E_b - mc^2}{p_b c} = \frac{p_b c}{E_b + mc^2}$$

   (b) Show that

$$E' = \sqrt{\frac{mc^2}{2}(E_b + mc^2)}$$

Before

After

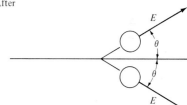

Problem 24

(a)

(b)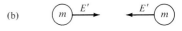

Problem 26

27. A proton-proton colliding beam accelerator with beams of 30 GeV each is in operation at CERN near Geneva, Switzerland. What would be the beam energy for an equivalent conventional accelerator? (Use the relation in Problem 26(b); $m_p c^2 = 0.938$ GeV.)

28. Electron-positron colliding beam machines are now under construction with beams up to $E' = 16$ GeV. What would be the beam energy for an equivalent conventional electron accelerator? ($m_e c^2 = 5.1 \times 10^{-4}$ GeV.)

29. Use the addition of velocities formula to show that

$$u'_x \approx v + \left(1 - \frac{v^2}{c^2}\right) u_x \qquad \text{when } u_x \ll c.$$

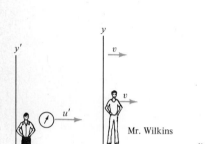

Problem 30

30. Mr. Prime who is at rest sees a clock moving to the right with velocity $u'$. An observer moving to the right with velocity $u'$ measures the time between ticks to be $\Delta \tau$. Mr. Wilkins, who is moving to the right with velocity $v$, sees the clock moving with velocity $u$.
    (a) What is $u$ in terms of $u'$ and $v$? ($u'$ and $v$ are positive numbers.)
    (b) In part (a) Mr. Wilkins measures $\Delta t$ for the time between ticks. What is $\Delta t / \Delta \tau$?

31. What is the limiting value for $u$ in Problem 37 for small $t$ ($a_0 t \ll c$), and for large $t$ ($a_0 t \gg c$)?

32. A muon has a rest mass $m_0 = 105$ MeV/$c^2$ and a lifetime $2 \times 10^{-6}$ s when at rest. At the Fermi National Accelerator a muon is created in a target at $t = 0$ with a kinetic energy of 10,395 MeV.
    (a) What is the total energy of the muon?
    (b) What is $\gamma$?
    (c) What are its velocity $v$ and its momentum $p$?
    (d) What is its lifetime in the lab frame?
    (e) How far does it travel before it decays?

33. A body of rest mass $m$ is moving with velocity $u$ and acceleration $a = du/dt$. The direction of $u$ is not changing.
    (a) What is the force $F$ in the unprimed system in terms of $m$, $u$, and $a$?
    (b) What is the $y$ component of force $F'_y = dp'_y/dt'$ according to Mr. Prime who is moving to the left with velocity $v$. Give answer in terms of unprimed quantities.

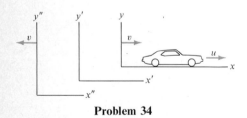

Problem 34

34. In the unprimed system a car is moving to the right with velocity $u$. With respect to the primed system, the unprimed system is moving to the right with velocity $v$. Also with respect to the primed system, the double primed system is moving to the left with velocity $v$.
    (a) With respect to the double primed system, what is the velocity of the unprimed system?
    (b) What would be the velocity of the car, $u''$, as measured by an observer in the double primed system?

35. (a) A photon has kinetic energy $E$. What is its relativistic or gravitational mass?
    (b) The above photon is part of a light beam that is aimed down an elevator shaft of height $h$. What is the fractional change in kinetic energy for such a photon? Give answer in terms of $g$, $h$, and $c$.

36. Show that for $\gamma \gg 1$ the momentum

$$p \approx \left(1 - \frac{1}{2\gamma^2}\right)\frac{E}{c}$$

37. Consider a constant force in the $x$ direction of strength $m_0 a_0$. This force is applied to a particle of rest mass $m_0$, which is initially at rest; $a_0$ is a constant. (It can be shown that $a_0$ is the acceleration as viewed by an observer who has the same instantaneous velocity as the particle.)
    (a) Starting from

    $$\frac{dp_x}{dt} = m_0 a_0 \quad \text{or} \quad \frac{d}{dt}\left[u\left(1 - \frac{u^2}{c^2}\right)^{-1/2}\right] = a_0$$

    show that

    $$u = \frac{a_0 t}{\sqrt{1 + a_0^2 t^2/c^2}}$$

    (b) Classically the particle would reach $u = c$ in a certain time $t_0$. What would be the true velocity at this time $t_0$?

# Rotational Motion

In the study of systems of interacting particles there is a great simplification if the rotational and translational motions are studied separately. In order to make this simplification we must first define two new quantities: angular momentum and torque. We shall show that for closed systems angular momentum is conserved just as is linear momentum and energy. Conservation of angular momentum is a conservation law of equal rank with conservation of momentum and energy. It permits the easy calculation of useful quantities without knowledge of the detailed forces and motions of the individual particles. In the last two sections of this chapter we shall study the special case of systems of particles where the particles all maintain the same relative positions. Such a system is called a rigid body or extended body. The study of rigid bodies is important because of their common occurrence in the everyday world.

## 10-1 Rotational Kinematics

Before discussing rotational dynamics (rotational forces and their effects), we must first work out the mathematics of rotational motion. We shall develop kinematic equations of angular quantities very similar to the kinematic equations of one dimensional motion developed in Chapter 2.

The angular analog to displacement $x$ is the angular displacement $\theta$. The angular analog to velocity $dx/dt$ is angular velocity $d\theta/dt$.

$$\omega \equiv \frac{d\theta}{dt} \quad \text{(angular velocity)} \quad (10\text{-}1)$$

It is common practice to use the Greek letter $\omega$ (omega) for $d\theta/dt$, which is the instantaneous angular velocity.

For motion in a circle there is a simple relation between $\omega$ and the velocity $v$ along the circumference. Consider a particle confined to a circle of radius $R$ as shown in Figure 10-1. According to the definition of radian measure, the distance traveled along the circumference of the circle is

$$s = R\theta$$

Now differentiate both sides with respect to $t$:

$$\frac{ds}{dt} = R\frac{d\theta}{dt}$$

or

$$v = R\omega \quad \text{(for circular motion)} \quad (10\text{-}2)$$

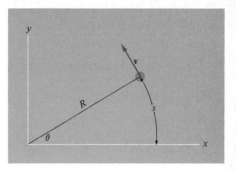

**Figure 10-1.** *Particle moving in circle travels distance $s = R\theta$.*

The quantities $v$ and $\omega$ may be changing with time, but $R$ is a constant.

In the case of uniform circular motion $\omega$ is also called the angular frequency. The velocity $v$ is the distance traveled in 1 second, which is the circumference $2\pi R$ times the number of revolutions per second $f$. If we put $v = 2\pi Rf$ into Eq. 10-2, we have

$$(2\pi Rf) = R\omega$$
$$\omega = 2\pi f \quad \text{(for uniform circular motion)} \quad (10\text{-}3)$$

The letter $f$ stands for frequency in revolutions per second and $\omega$ is radians per second.

### Angular Acceleration

Just as linear acceleration was defined as $d^2x/dt^2$, so is angular acceleration defined as

$$\alpha \equiv \frac{d^2\theta}{dt^2} \quad \text{(angular acceleration)} \quad (10\text{-}4)$$

One can obtain a relation between linear and angular acceleration by differentiating both sides of Eq. 10-2:

$$\frac{dv}{dt} = R\frac{d\omega}{dt}$$

$$a = R\alpha \quad \text{(for circular motion)} \quad (10\text{-}5)$$

where $a$ is the linear acceleration of the particle along the circumference of the circle. If the acceleration of the particle along the circumference is uniform, we have from Eq. 2-9:

$$s = s_0 + v_0 t + \tfrac{1}{2}at^2$$

Now replace $s$ with $(R\theta)$, $v_0$ with $(R\omega_0)$, and $a$ with $(R\alpha)$:

$$\theta = \theta_0 + \omega_0 t + \tfrac{1}{2}\alpha t^2 \quad \text{(for uniform angular acceleration)} \quad (10\text{-}6)$$

Similarly, we obtain the relation

$$2\alpha(\theta - \theta_0) = \omega^2 - \omega_0^2$$

from Eq. 2-10, when $\alpha$ and $R$ are constant.

## 10-2 The Vector Cross Product

The definitions of both angular momentum and torque make use of what in vector analysis is called the vector cross product. In Section 6-3 we have seen the definition of dot product of two vectors:

$$\mathbf{A} \cdot \mathbf{B} \equiv AB \cos \alpha$$

where the dot is "shorthand" for $\cos \alpha$. In the vector cross product the cross is "shorthand" for $\sin \alpha$:

$$\mathbf{A} \times \mathbf{B} \equiv \hat{\mathbf{n}} AB \sin \alpha \quad \text{(vector cross product)} \quad (10\text{-}7)$$

In the above equation, which defines vector cross product, the quantity $\hat{\mathbf{n}}$ is a unit vector that is perpendicular to the plane containing $\mathbf{A}$ and $\mathbf{B}$. But the plane containing $\mathbf{A}$ and $\mathbf{B}$ has two possible directions for its perpendicular. There is a convention for choosing one of the two; it is called the right-hand rule and is illustrated in Figure 10-2. Use the fingers of the right hand to "curl" or rotate the first vector into the second; then the thumb points in the direction of the vector cross product.

Note the following obvious applications of Eq. 10-7:

$$\mathbf{A} \times \mathbf{A} = 0$$
$$\mathbf{A} \times (\mathbf{B} + \mathbf{C}) = \mathbf{A} \times \mathbf{B} + \mathbf{A} \times \mathbf{C}$$
$$\mathbf{A} \times \mathbf{B} = -\mathbf{B} \times \mathbf{A}$$
$$\mathbf{i} \times \mathbf{i} = \mathbf{j} \times \mathbf{j} = \mathbf{k} \times \mathbf{k} = 0$$
$$\mathbf{i} \times \mathbf{j} = \mathbf{k}$$
$$\mathbf{j} \times \mathbf{k} = \mathbf{i}$$
$$\mathbf{k} \times \mathbf{i} = \mathbf{j}$$

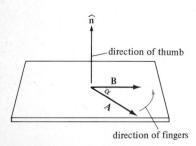

**Figure 10-2.** *Illustration of right-hand rule. Thumb of right hand points in direction of $\hat{\mathbf{n}}$, the normal to the plane containing $\mathbf{A}$ and $\mathbf{B}$. The fingers are curled in the plane of $\mathbf{A}$ and $\mathbf{B}$ as if to rotate $\mathbf{A}$ into $\mathbf{B}$. Then the thumb is in direction of $(\mathbf{A} \times \mathbf{B})$.*

where $\mathbf{i}$, $\mathbf{j}$, and $\mathbf{k}$ are the unit vectors along $x$, $y$, and $z$ respectively.

**Example 1.** What is $\mathbf{A} \times \mathbf{B}$ when $\mathbf{A} = \mathbf{i}A_x + \mathbf{j}A_y$ and $\mathbf{B} = \mathbf{i}B_x + \mathbf{j}B_y$? Also what is the sine of the angle between $\mathbf{A}$ and $\mathbf{B}$?

ANSWER: $\mathbf{A} \times \mathbf{B} = (\mathbf{i}A_x + \mathbf{j}A_y) \times (\mathbf{i}B_x + \mathbf{j}B_y)$
$= \mathbf{i} \times \mathbf{j} A_x B_y + \mathbf{j} \times \mathbf{i} A_y B_x$
$= \mathbf{k}(A_x B_y - A_y B_x)$

$$\sin \alpha = \frac{|\mathbf{A} \times \mathbf{B}|}{|\mathbf{A}||\mathbf{B}|} = \frac{A_x B_y - A_y B_x}{\sqrt{(A_x^2 + A_y^2)(B_x^2 + B_y^2)}}$$

Not only does the vector cross product appear in the definitions of angular momentum and torque but it is used in electricity to describe the force on a moving charge as well as to calculate the magnetic field produced by a current.

## 10-3 Angular Momentum

A single particle can have angular momentum even if it is traveling in a straight line. By definition the angular momentum $\mathbf{L}$ is

$$\mathbf{L} \equiv \mathbf{r} \times \mathbf{p} \quad \text{(angular momentum)} \quad (10\text{-}8)$$

where $\mathbf{p}$ is the particle's linear momentum and $\mathbf{r}$ is the position vector from the origin to the particle. For example, in Figure 10-3 the particle of mass $m$ has $L = rmv \sin \alpha$ for the magnitude of the angular momentum. According to the right-hand rule the direction of $\mathbf{L}$ is into the page or in the negative $z$ direction. The curved arrow in Figure 10-3(b) shows the orientation of the fingers of the right hand.

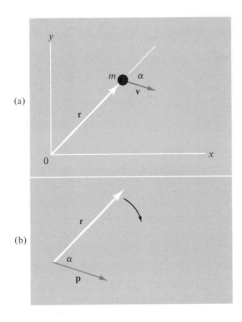

**Figure 10-3.** (a) Mass $m$ is moving in the xy plane with velocity $\mathbf{v}$. (b) The relative orientations of $\mathbf{r}$ and momentum $\mathbf{p}$. Curved arrow indicates position of fingers in application of right-hand rule.

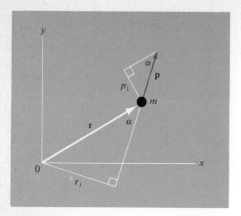

**Figure 10-4.** *Mass m is moving with momentum* **p** *in the xy plane.* $r_\perp$ *is called the moment arm.* $p_\perp$ *is the component of momentum perpendicular to* **r**.

Equation 10-8 is relativistically correct if one uses relativistic momentum for **p**. Note that the value of $L$ depends on the choice of origin or coordinate system. It can be seen using Figure 10-4 that

$$L = rp_\perp \quad \text{and} \quad L = r_\perp p$$

where $p_\perp = p \sin \alpha$ and $r_\perp = r \sin \alpha$.

The quantity $p_\perp$ is the component of $p$ that is perpendicular to $r$; $r_\perp$ is the perpendicular distance to the direction of motion of the particle; $r_\perp$ is also called the moment arm.

---

*__Example 2.__ The tidal forces cause a slowing down in the rotation or angular momentum of the earth. We want to show that the law of conservation of angular momentum requires that the moon slowly increase its distance from the earth; that is, we must show that the orbital angular momentum of the moon increases with increasing radius.

ANSWER: By definition the angular momentum of the moon is $L = Rmv$, where $R$ is the radius of its orbit and $m$ is the mass of the moon. In order to solve this problem we must express $v$ in terms of $R$. This can be done by equating the force on the moon to $ma$:

$$G \frac{M_e m}{R^2} = m \left( \frac{v^2}{R} \right)$$

or

$$v = \sqrt{\frac{GM_e}{R}}$$

Now we substitute this into the equation for angular momentum:

$$L = Rm \left( \sqrt{\frac{GM_e}{R}} \right) = m\sqrt{GM_e} R^{1/2}$$

We see that the angular momentum increases with increasing orbit radius. It is proportional to the square root of the radius. (At the same time the earth is losing rotational angular momentum and kinetic energy due to the tidal force of the moon on the earth.)

---

### Conservation of Angular Momentum for a Single Body

Before treating the general case of a closed system of $n$ interacting particles, we consider a single particle under the influence of a central force directed toward (or away from) the origin. An example of this situation would be a planet in orbit about the sun.

$$\mathbf{L} = \mathbf{r} \times \mathbf{p}$$

$$\frac{d\mathbf{L}}{dt} = \frac{d\mathbf{r}}{dt} \times \mathbf{p} + \mathbf{r} \times \frac{d\mathbf{p}}{dt}$$

$$= \mathbf{v} \times \mathbf{p} + \mathbf{r} \times \mathbf{F}$$

The term $\mathbf{v} \times \mathbf{p}$ is zero because these vectors are parallel. The term $\mathbf{r} \times \mathbf{F}$ is zero because $\mathbf{F}$ is a central force and these vectors are also parallel (or antiparallel). So

$$\frac{d\mathbf{L}}{dt} = 0 \quad \text{or} \quad \mathbf{L} = \text{a constant}$$

We have proved that a body under the influence of any kind of a central force will have a constant angular momentum. This result was used on page 85 when we derived Kepler's law of equal areas.

## 10-4 Rotational Dynamics

In this section we shall study the rotational equivalent to $\mathbf{F} = m\mathbf{a}$ and the conservation of angular momentum for general systems of particles.

### Torque

We shall now define a quantity which is the rotational analog of force. Torque $\mathbf{T}$ is angular force just as $\mathbf{L}$ is angular momentum. If a force $\mathbf{F}$ is acting on a particle, the corresponding torque is by definition

$$\mathbf{T} \equiv \mathbf{r} \times \mathbf{F} \quad \text{(torque)} \quad (10\text{-}9)$$

where $\mathbf{r}$ is the displacement vector from some reference point. In order to obtain the rotational equivalent to $\mathbf{F} = m\mathbf{a}$ we differentiate both sides of Eq. 10-8:

$$\frac{d\mathbf{L}}{dt} = \frac{d}{dt}(\mathbf{r} \times \mathbf{p})$$

$$= \frac{d\mathbf{r}}{dt} \times \mathbf{p} + \mathbf{r} \times \frac{d\mathbf{p}}{dt}$$

$$= \mathbf{v} \times \mathbf{p} + \mathbf{r} \times \mathbf{F}_{net}$$

The first term is zero because **v** and **p** are parallel. The second term is by definition the net torque. Hence,

$$\mathbf{T}_{net} = \frac{d\mathbf{L}}{dt} \tag{10-10}$$

We see that the net torque on a particle equals the time rate of change of angular momentum just as net force on a particle equals time rate of change of the momentum.

### Conservation of Angular Momentum

For a system of $n$ particles, we can sum Eq. 10-10 over all the particles:

$$\sum_{j=1}^{n} \mathbf{T}_j = \frac{d}{dt}\left(\sum_{j=1}^{n} \mathbf{L}_j\right) = \frac{d}{dt}\mathbf{L}_{total} \tag{10-11}$$

where $\mathbf{L}_{total}$ is the total angular momentum of the entire system. For a closed system there are no external torques and the left hand side is the sum of all the internal torques due to the interaction forces among the $n$ particles. According to Newton's third law each pair of interaction forces is equal and opposite and directed toward (or away from) each other. (Relativistic effects can produce noncentral forces, but the time-averaged effect is the same as for central forces.) Since $r_\perp$ is the same for each pair of interaction forces, their torques are equal and opposite. Because the left-hand side of Eq. 10-11 is the sum of all such pairs, it will be zero and Eq. 10-11 becomes

$$0 = \frac{d}{dt}\mathbf{L}_{total}$$

or

$$\mathbf{L}_{total} = \text{a constant} \quad \text{(conservation of angular momentum)} \tag{10-12}$$

We have just derived the conservation of angular momentum for a *closed system*. It is a direct consequence of Newton's laws of motion. There are many kinds of problems involving rotating systems where, even if the interaction forces are not known, the final velocities or angular momenta can be calculated using conservation of angular momentum.

* **Example 3.** A student holds two dumbbells with outstretched arms while standing on a turntable. He is given a push until he is rotating at a rate of $f_1 = 0.5$ revolution per second. Then the student pulls the dumbbells in toward his chest. See Figure 10-5. What is his new rate

of rotation? Assume the dumbbells are originally 60 cm from his axis of rotation and are pulled in to 10 cm from the axis of rotation. The mass of the dumbbells is such that the student and dumbbells have equal angular momentum when at the 60-cm distance.

ANSWER: The initial angular momentum of the dumbbells is

$$L_{d_1} = R_1 m v_1 = R_1 m(\omega_1 R_1) = m\omega_1 R_1^2$$

where $m$ is the mass of the two dumbbells. The initial angular momentum of the system is

$$L_1 = L_{s_1} + m\omega_1 R_1^2$$

where $L_{s_1}$ is the initial angular momentum of the student.

Since we are told that $L_{s_1} = L_{d_1}$, we have $L_{s_1} = m\omega_1 R_1^2$. When the dumbbells are at a distance $R_2$ the angular momentum of the system is

$$L_2 = L_{s_2} + m\omega_2 R_2^2$$

Applying conservation of angular momentum of the system gives

$$L_{s_2} + m\omega_2 R_2^2 = L_{s_1} + m\omega_1 R_1^2$$

Since the student's angular momentum is proportional to his rate of spin, we have

$$L_{s_2} = \frac{\omega_2}{\omega_1} L_{s_1}$$

Substitution in the preceding equation gives

$$\left(\frac{\omega_2}{\omega_1} L_{s_1}\right) + m\omega_2 R_2^2 = L_{s_1} + m\omega_1 R_1^2$$

Now use $L_{s_1} = m\omega_1 R_1^2$ and solve for $\omega_2$:

$$\omega_2 = \omega_1 \frac{2R_1^2}{R_1^2 + R_2^2}$$

$$f_2 = (0.5) \frac{2(0.6)^2}{0.6^2 + 0.1^2} \text{ s}^{-1} = 0.97 \text{ rev/s}$$

We see that the student's angular velocity has almost doubled.

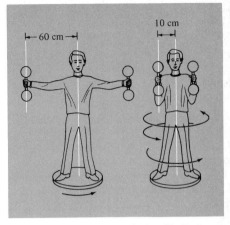

Figure 10-5. *Student pulls dumbbells in and spins faster.*

The same principle applies when a rotating figure skater pulls in his arms and legs.

> **Example 4.** A student stands on a turntable and holds a bicycle wheel above his head. He spins the wheel until he reaches an angular velocity of $\omega_1 = 5 \text{ s}^{-1}$. Now he steps off the turntable and then steps back on. When back on the turntable, he turns the wheel upside down. Now what is his angular velocity? See Figure 10-6.
>
> ANSWER: Since the initial angular momentum of the system is zero,
>
> $$0 = L_{s_1} + L_0$$
>
> or
>
> $$L_{s_1} = -L_0$$
>
> where $L_0$ is the angular momentum given to the wheel.
>
> When the student steps on the ground, $L_s$ drops to zero (his angular momentum is transferred to the earth). So now the angular momentum of the student plus wheel is $L_0$ when back on the turntable. Turning the wheel upside down changes the angular momentum of the wheel to $(-L_0)$. Since the angular momentum of the system must remain equal to $L_0$, we have
>
> $$L_0 = L_{s_2} + (-L_0)$$

**Figure 10-6.** (*a*) *Student spins wheel.* (*b*) *He returns to turntable after stepping on the ground.* (*c*) *He then turns the wheel upside down.*

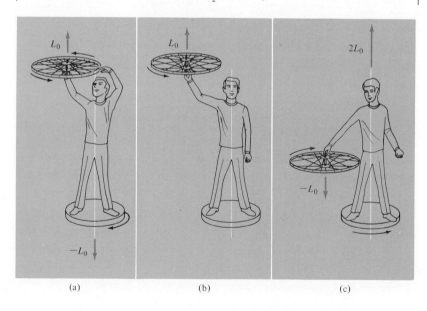

or
$$L_{s_2} = 2L_0$$

We conclude that the student spins twice as fast as originally and in the opposite direction; that is, $\omega_2 = 10 \text{ s}^{-1}$.

Equation 10-11 applies to a system under the influence of external torques, as well as to a closed system. If there are external torques the internal torques still cancel out and the summation

$$\sum_{j=1}^{n} \mathbf{T}_j \quad \text{becomes} \quad \mathbf{T}_{\text{ext}}$$

where $\mathbf{T}_{\text{ext}}$ is the vector sum of all the external torques acting on the system. In the next two examples, the system under discussion is a bicycle wheel where all the mass is assumed to be concentrated on the rim.

**Example 5.** A certain bicycle can go up a gentle incline with constant velocity when the force of the ground pushing on the rear wheel is $F_2 = 4$ N, as shown in Figure 10-7. With what force $F_1$ must the chain pull on the sprocket wheel if $R_2/R_1 = 6$?

ANSWER: Since the angular velocity of the wheel stays constant, $dL/dt = 0$ and

so
$$\mathbf{T}_{\text{net}} = (\mathbf{T}_2 + \mathbf{T}_1) = 0$$

$$|T_1| = |T_2|$$

Using Eq. 10-9:

$$R_1 F_1 = R_2 F_2$$

$$F_1 = \left(\frac{R_2}{R_1}\right) F_2 = (6)(4 \text{ N}) = 24 \text{ N}$$

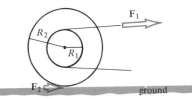

**Figure 10-7.** Bicycle wheel is pulled by chain with force $F_1$. The ground pushes the wheel with force $F_2$.

**Example 6.** Consider the bicycle wheel of Example 5 suspended above the ground. If a steady force of 20 N is applied to the chain, how long will it take to reach a rim velocity of 20 mph (8.94 m/s)? Assume $R_2 = 30$ cm and the entire mass of the wheel is 2 kg along the rim.

ANSWER: Using Eq. 10-10:

$$R_1 F_1 = \frac{\Delta L}{\Delta t}$$

$$\Delta t = \frac{\Delta L}{R_1 F_1}$$

$\Delta L$ equals the final angular momentum, which is $R_2 mv$:

$$\Delta L = R_2 mv = (0.3 \text{ m})(2 \text{ kg})(8.94 \text{ m/s}) = 5.36 \text{ kg m}^2/\text{s}$$

$$\Delta t = \frac{5.36 \text{ kg m}^2/\text{s}}{(0.05 \text{ m})(20 \text{ N})} = 5.36 \text{ s}$$

## 10-5 Center of Mass

In general the motion of a closed system of interacting particles is complicated. However, there is one point in the system that moves in a straight line with constant velocity. That point is the center of mass $R_{cm}$, which is defined as

$$\mathbf{R}_{cm} \equiv \frac{\sum m_j \mathbf{r}_j}{\sum m_j} \qquad \text{(position of center of mass)} \qquad (10\text{-}13)$$

Center of mass is merely the average position where mass is used as the weighting factor in forming the average. An observer at rest with respect to $\mathbf{R}_{cm}$ is said to be in the center-of-mass system. Next we differentiate both sides of Eq. 10-13 with respect to time:

$$\frac{d\mathbf{R}_{cm}}{dt} = \frac{\sum m_j \, d\mathbf{r}_j/dt}{\sum m_j}$$

The left-hand side is by definition the velocity of the center of mass, $v_{cm}$. So

$$\mathbf{v}_{cm} = \frac{\sum m_j \mathbf{v}_j}{M} = \frac{\sum \mathbf{p}_j}{M} = \frac{\mathbf{P}_{total}}{M} \qquad (10\text{-}14)$$

where $M$ is the total mass of the system.

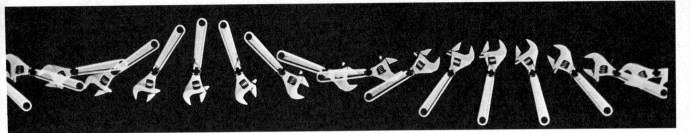

**Figure 10-8.** *A freely moving wrench. The net external force on this wrench is zero. Note that it rotates uniformly about its center of mass, which is marked with black tape.* [Courtesy Physical Science Study Committee]

Since $\mathbf{P}_{\text{total}}$ stays constant for a closed system, we have just proved that the velocity of the center of mass stays constant in magnitude and direction for a closed system. The moving wrench in Figure 10-8 is a closed system. Note that every point in the wrench follows a helical path except the center of mass, which moves uniformly in a straight line.

Another useful feature of the center of mass is the calculation of total kinetic energy. We shall now prove that the total kinetic energy of a system is the kinetic energy as measured in the center-of-mass system plus $\frac{1}{2}Mv_{cm}^2$.

$$K_{\text{total}} = \frac{1}{2} \sum m_j v_j^2$$

$$= \frac{1}{2} \sum m_j (\mathbf{v}_{cm} + \mathbf{v}_j') \cdot (\mathbf{v}_{cm} + \mathbf{v}_j')$$

where $\mathbf{v}_j'$ is the velocity of $m_j$ as measured in the center-of-mass system.

Performing the dot product gives

$$K_{\text{total}} = \frac{1}{2}\left(\sum m_j\right) v_{cm}^2 + \mathbf{v}_{cm} \cdot \sum (m_j \mathbf{v}_j') + \frac{1}{2} \sum m_j v_j'^2$$

The second term vanishes because $\sum m_j \mathbf{v}_j'$ equals $M$ times the velocity of the center of mass as measured in the center-of-mass system, which is zero. Hence

$$K_{\text{total}} = \tfrac{1}{2}Mv_{cm}^2 + K' \qquad (10\text{-}16)$$

where $K'$ is the total kinetic energy as measured in the center-of-mass system. This equation is especially useful when we consider rigid bodies in the next section. As measured in the center-of-mass system, a rigid body can only have kinetic energy of rotation. Then Eq. 10-16 becomes

$$K_{\text{total}} = \tfrac{1}{2}Mv_{cm}^2 + K'_{\text{rot}} \quad \text{(for rigid bodies)} \qquad (10\text{-}17)$$

where $K'_{\text{rot}}$ is the rotational kinetic energy as measured in the center-of-mass system.

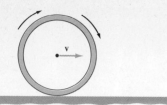

**Figure 10-9.** *Hoop rolling on a flat surface.*

**Example 7.** A hoop of mass $m$ is rolling along a flat surface as shown in Figure 10-9. The center of the hoop has velocity $v$. What is the hoop's kinetic energy?

ANSWER: Using Eq. 10-17 we have

$$K_{total} = \tfrac{1}{2}mv^2 + \tfrac{1}{2}mv'^2_{rim}$$

where $v'_{rim}$ is the rim velocity as viewed in the center-of-mass system. Since an observer moving along the center of the hoop will see the point of contact moving backward with velocity $v$, $v'_{rim} = v$.

$$K_{total} = \tfrac{1}{2}mv^2 + \tfrac{1}{2}m(v)^2 = mv^2$$

We note that the hoop has twice the energy of a nonrotating particle of mass $m$ moving along with the same velocity.

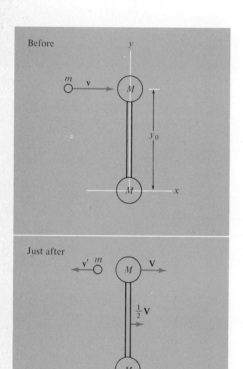

**Figure 10-10.** *Mass $m$ gives an impulse to one end of a rigid dumbbell. Just after the collision the velocities of the upper end, lower end, and center of mass are $V$, 0, and $\tfrac{1}{2}V$, respectively.*

\***Example 8.** A mass $m$ collides with one end of a rigid dumbbell as shown in Figure 10-10. If the dumbbell is initially at rest, what fraction of the kinetic energy it receives goes into rotation?

ANSWER: Just before the collision the total angular momentum measured from the origin is $(y_0 mv)$. Just after the collision it is $(y_0 MV - y_0 mv')$. According to the law of conservation of angular momentum these two are equal. We obtain

$$y_0 mv = y_0 MV - y_0 mv'$$

or

$$mv = MV - mv'$$

This says that all the linear momentum available goes into the top mass—there is none left over for the bottom mass. Hence after the collision the total kinetic energy of the dumbbell is

$$K_{total} = \tfrac{1}{2}MV^2$$

where $V$ is the velocity of the upper mass just after the collision. Using Eq. 10-17 we have

$$K'_{rot} = K_{total} - \tfrac{1}{2}M_{total}V^2_{cm}$$
$$= (\tfrac{1}{2}MV^2) - \tfrac{1}{2}(2M)(\tfrac{1}{2}V)^2$$
$$= \tfrac{1}{4}MV^2$$

which is one half of the kinetic energy given to the dumbbell. We shall see in Chapter 12 that when a dumbbell-shaped molecule is struck by other particles, it will receive on the average two thirds as much rotational kinetic energy as translational kinetic energy.

## 10-6 Rigid Bodies and Moment of Inertia

Up to now we have mainly dealt with particles or point masses. However, most masses in nature are extended rigid bodies that can rotate as well as translate. A rigid body can be divided into mass elements $\Delta m_j$. By rigid, we mean that the distance between any two mass elements remains the same in magnitude.

Let us now view a rigid body rotating with angular velocity $\omega$ about a fixed axis in the center-of-mass system. (See Figure 10-11.) If the mass element $\Delta m_j$ is a distance $r_j$ from the axis of rotation, its velocity will be $v_j = r_j \omega$. The magnitude of its angular momentum will be

$$L = \sum r_j \Delta m_j v_j = \sum r_j \Delta m_j (r_j \omega)$$

$$= \left( \sum r_j^2 \Delta m_j \right) \omega$$

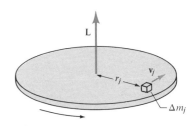

Figure 10-11. *A rotating disk showing a mass element* $\Delta m_j$.

We define the quantity in the parentheses as the moment of inertia $I$.

$$I \equiv \sum r_j^2 \Delta m_j \qquad (10\text{-}18)$$

or for a continuous mass distribution

$$I \equiv \int r^2\, dm \quad \text{(moment of inertia)} \qquad (10\text{-}18)$$

Then

$$L = I\omega \qquad (10\text{-}19)$$

Since torque is

$$T = \frac{dL}{dt}$$

we have

$$T = I\frac{d\omega}{dt} = I\alpha \qquad (10\text{-}20)$$

In the center-of-mass system the kinetic energy is

$$K = \frac{1}{2}\sum \Delta m_j v_j^2 = \frac{1}{2}\sum \Delta m_j (r_j \omega)^2$$

$$= \frac{1}{2}\left(\sum \Delta m_j r_j^2\right)\omega^2$$

**Table 10-1.** *Moments of inertia of some common bodies (about the axis shown)*

| Body | $I$ |
|---|---|
| Hoop or ring | $mR^2$ |
| Disk or cylinder | $\frac{1}{2}mR^2$ |
| Rod about center | $\frac{ml^2}{12}$ |
| Rod about end | $\frac{ml^2}{3}$ |
| Solid sphere | $\frac{2}{5}mR^2$ |
| Spherical shell | $\frac{2}{3}mR^2$ |
| Disk about edge | $\frac{3}{2}mR^2$ |

Hence
$$K = \tfrac{1}{2}I\omega^2$$
also
$$K = \frac{1}{2}\frac{(I\omega)^2}{I} = \frac{1}{2}\frac{L^2}{I} \qquad (10\text{-}21)$$

**Example 9.** What is the moment of inertia about the symmetry axis of a hoop and of a solid disk, both of mass $M$ and radius $R$?

ANSWER: The hoop has all its mass elements at $r = R$, hence
$$I_{\text{hoop}} = MR^2$$

For the disk in Figure 10-12, the area contained in a ring between $r$ and $(r + dr)$ is $dA = 2\pi r\, dr$. The total area is $\pi R^2$. Hence

$$\frac{dm}{M} = \frac{dA}{A} = \frac{2\pi r\, dr}{\pi R^2}$$

$$dm = M\frac{2r\, dr}{R^2}$$

$$I_{\text{disk}} = \int_0^R r^2\, dm = \int_0^R r^2 \left(\frac{2Mr\, dr}{R^2}\right) = \frac{2M}{R^2}\int_0^R r^3\, dr = \frac{2M}{R^2}\left[\frac{r^4}{4}\right]_0^R$$
$$= \tfrac{1}{2}MR^2$$

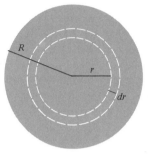

**Figure 10-12.** *Ring of thickness $dr$ inside solid disk of radius $R$.*

The moments of inertia of some common types of bodies are tabulated in Table 10-1.

**Example 10.** A hoop and then a disk of mass $m$ and radius $R$ are rolled down the same incline of angle $\theta$. (See Figure 10-13.) What are their accelerations?

ANSWER: When the hoop or disk reaches the bottom, $mgh$ of potential energy will be converted into kinetic energy of translation and rotation. Using Eq. 10-17 we have

$$mgh = \tfrac{1}{2}mv^2 + \tfrac{1}{2}I\omega^2 \qquad \text{where } \omega = v/R.$$

Substituting for $\omega$:

$$mgh = \tfrac{1}{2}mv^2 + \tfrac{1}{2}I\left(\frac{v}{R}\right)^2$$

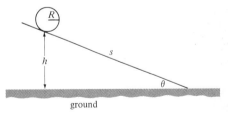

**Figure 10-13.** *Hoop or disk on an incline (Example 10).*

Solving for $v^2$:

$$v^2 = \frac{2mgh}{m + I/R^2}$$

For uniformly accelerated motion, $v^2 = 2as$; hence

$$2as = \frac{2mgh}{m + I/R^2}$$

$$a = \frac{m}{m + I/R^2} g \sin\theta$$

For the hoop,

$$\frac{I}{R^2} = m \quad \text{and} \quad a_{\text{hoop}} = \tfrac{1}{2}g \sin\theta$$

For the disk,

$$\frac{I}{R^2} = \tfrac{1}{2}m \quad \text{and} \quad a_{\text{disk}} = \tfrac{2}{3}g \sin\theta$$

Note that the answer is independent of the mass or radius. It only depends on the shape. We recall that a particle sliding down the incline will have $a = g \sin\theta$.

## 10-7 Statics

In this chapter we study the conditions for nonrotation as well as for rotation.

A whole field of mechanical engineering involves the study of rigid bodies under stress (force) while at rest. It is important to know what forces are necessary to keep a rigid body from moving or collapsing. A study of statics is necessary in order to design roofs and bridges that will not collapse when under maximum load. Since this subject is usually taught in a separate course to engineering students, we shall briefly treat the basic principles involved.

There are two conditions required to keep a body at rest.

CONDITION I: The vector sum of all the forces must be zero.

$$\sum \mathbf{F}_j = 0 \qquad (10\text{-}22)$$

CONDITION II: The vector sum of all the torques must be zero.

$$\sum \mathbf{T}_j = 0$$

The first condition is a consequence of Newton's first law of motion. The second is a consequence of the relation $\mathbf{T}_{net} = d\mathbf{L}/dt$. (If $\mathbf{L}$ is to stay equal to zero, $d\mathbf{L}/dt$ must be zero.)

Most problems in statics involve rigid bodies in a plane. If we call it the $xy$ plane, condition I gives two equations:

$$\sum (F_j)_x = 0 \quad \text{and} \quad \sum (F_j)_y = 0$$

Condition II gives $\Sigma T_z = 0$. $T_z$ is positive for counterclockwise rotation and is negative for clockwise rotation. We have three simultaneous equations, and thus we can solve problems having three unknown quantities. We shall now consider some examples.

> **Example 11.** Masses $m_1$ and $m_2$ are at the ends of a rod of length $l$ as in Figure 10-14. How far from $m_1$ must a fulcrum be placed so that the two masses are balanced—that is, the rod does not rotate?
>
> **Answer:** This problem can be solved by using condition II alone. In order to apply condition II we must choose an origin about which to calculate the torques. Calculations can be simplified by choosing the point of application of a torque (or torques) as the origin. If we choose the fulcrum as the origin, the torque due to $F$ is zero. The torque due to $m_1$ would cause a positive rotation and the torque due to $m_2$ would cause a negative rotation; hence
>
> $$T_1 + T_2 = (m_1 g x) + [-m_2 g(l - x)] = 0$$
>
> $$(m_1 g + m_2 g)x = m_2 g l$$
>
> $$x = \frac{m_2}{m_1 + m_2} l$$

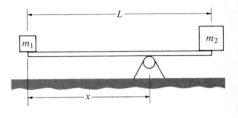

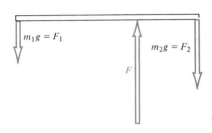

**Figure 10-14.** *The distance $x$ is chosen to balance $m_1$ and $m_2$.*

Example 11 illustrates the principle of the lever. Note that a small force $F_1$ can be used to balance (or slowly lift) a body exerting a larger force. The ratio

$$\frac{F_2}{F_1} = \frac{x}{l - x}$$

is called the mechanical advantage.

Another example involving mechanical advantage is the handcrank shown in Figure 10-15. Here condition II tells us that

$$|T_1| = |T_2|$$
$$R_1 F_1 = R_2 F_2$$

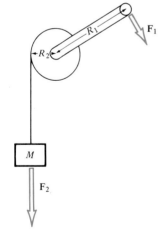

**Figure 10-15.** *Handcrank can hold mass $M$ with a force $F_1$ applied a distance $R_1$ from the axis.*

or

$$F_1 = \left(\frac{R_2}{R_1}\right) F_2$$

A heavy weight can be lifted by making the ratio $R_2/R_1$ sufficiently small. We note that this example of the handcrank, where a mass is being lifted at constant speed, is the same as the bicycle wheel in Example 3, which was also rotating at constant speed.

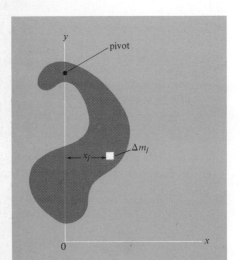

**Figure 10-16.** *Irregularly shaped body is hung from pivot.*

**Example 12.** A rigid body of arbitrary shape is hung from a frictionless pivot. Prove that its center of mass will be directly underneath the pivot point.

ANSWER: The torque about the pivot point exerted by the force of gravity acting on the mass element $\Delta m_j$ shown in Figure 10-16 is

$$T_z = -x_j \Delta m_j g$$

The sum of all the torques is

$$(T_z)_{\text{total}} = -\sum x_j \Delta m_j g$$

According to condition II this is zero. Thus

$$\sum x_j \Delta m_j = 0$$

$$\frac{\sum x_j \Delta m_j}{\sum \Delta m_j} = 0$$

The left-hand side is by definition the $x$ coordinate of the center of mass. We see that both the center of mass and the pivot point have the same $x$ coordinate.

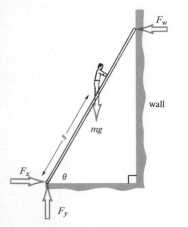

**Figure 10-17.** *Man on ladder of length l at a distance s from the bottom.*

**Example 13.** Suppose the coefficient of friction between the ladder and the ground in Figure 10-17 is $\mu = 0.4$. How far up the ladder can a man climb before it slips? Assume frictionless contact at the wall and that the man is much heavier than the ladder.

ANSWER: Here there are three unknowns: $F_x$, $F_y$, and $F_w$. Condition I gives

$$F_x + (-F_w) = 0$$
$$F_y + (-mg) = 0$$

For condition II choose the origin at the foot of the ladder, then

$$F_w l \sin\theta + (-mgs\cos\theta) = 0$$

or

$$F_w = \frac{mgs}{l}\cot\theta \qquad (10\text{-}23)$$

As the man reaches the point of slipping

$$F_x = \mu F_y$$

Now replace $F_x$ and $F_y$ with $F_w$ and $mg$ as obtained from condition I:

$$F_w = \mu mg$$

Equating to Eq. 10-23 gives

$$\frac{mgs}{l}\cot\theta = \mu mg$$

$$s = \mu l \tan\theta$$

For $\mu = 0.4$ and $\theta = 60°$ we have $s = 0.69l$. Under such conditions the man should not climb more than two thirds of the way up the ladder.

## 10-8 Flywheels

As we have seen, it is possible to store an amount of energy equal to $\frac{1}{2}I\omega^2$ in a rotating rigid body. For a rotating disk of radius $R$ this is $\frac{1}{4}mR^2\omega^2$.

Let us now consider an automobile with a flywheel in place of an engine. The flywheel could be energized by a small, highly efficient engine such as an electric motor while the car is parked. In this way a car could be "run" on a plentiful fuel such as coal rather than scarce gasoline. We shall briefly discuss how much energy can be stored in a flywheel, and how this compares with a conventional gasoline engine.

The limit on $\omega$ for a flywheel is determined by the breaking strength of the flywheel material. It is fairly easy to show for a rotating disk that

$$\tfrac{1}{2}I\omega^2_{max} = \frac{V}{4}S_{max}$$

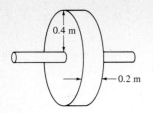

**Figure 10-18.** *Flywheel used in place of automobile engine.*

where $S_{max}$ is the tensile strength (force per unit area) and $V$ is the volume of the disk. Typical tensile strength for steel, fused quartz, and a few other strong materials is $S_{max} \approx 3 \times 10^9 \text{ N/m}^2$. Then a flywheel of the size shown in Figure 10-18, which has a volume of 0.1 m³, could store a kinetic energy

$$\tfrac{1}{2} I \omega^2 = \frac{(0.1 \text{ m}^3)}{4}(3 \times 10^9 \text{ N/m}^2) \approx 8 \times 10^7 \text{ J}$$

If made out of fused quartz or other material with a density $\sim 2 \times 10^3$ kg/m³, its mass would be $\sim 200$ kg, which is small compared to the total mass of $\sim 1000$ kg for a typical small car. Such a small car with a conventional engine would carry about 10 gal of gasoline with an energy content of $1.2 \times 10^8$ J/gal or $1.2 \times 10^9$ J. However, only about 20% of this can be converted into mechanical energy (see Chapter 13). So the conventional car has a power source of $24 \times 10^7$ J compared to the $8 \times 10^7$ J of the flywheel. The range of such a "motorless" car could be about one third the normal range or about 100 km.

So far most of this is theoretical speculation. Other shapes and types of materials might be even more efficient and economical. At the present time safety and cost considerations put such methods of propulsion in the experimental and developmental stage.

**Example 14.** How many turns per second does the flywheel just described make?

ANSWER: Since

$$\tfrac{1}{2} I \omega^2 = 8 \times 10^7 \text{ J}$$

$$\omega^2 = \frac{16 \times 10^7 \text{ J}}{(mR^2/2)} = \frac{16 \times 10^7 \text{ J}}{(200 \text{ kg})(0.4 \text{ m})^2/2} = 10^7 \text{ s}^{-2}$$

$$\omega = 3.16 \times 10^3 \text{ s}^{-1}$$

$$f = 503 \text{ rev/s}$$

**Example 15.** Suppose the 1000-kg car is traveling at 50 mph (22.4 m/s) and experiences a total frictional force of $F_f = 0.07mg = 686$ N. How far can it go with the initial stored energy of $8 \times 10^7$ J?

ANSWER

$$P = F_f \cdot v = (686) \cdot (22.4) \text{ W} = 1.54 \times 10^4 \text{ J/s}$$

If $K$ is the initial kinetic energy and $T$ the time of the trip, $P = K/T$ or

$$T = \frac{K}{P} = \frac{8 \times 10^7 \text{ J}}{1.54 \times 10^4 \text{ J/s}} = 5.2 \times 10^3 \text{ s} = 1.44 \text{ h}$$

The distance is $x = vT = 72$ mi $= 116$ km.

## Summary

Angular velocity is defined as $\omega = d\theta/dt$. Angular acceleration is defined as $\alpha = d\omega/dt = d^2\theta/dt^2$.

For circular motion $v = R\omega$ and the component of acceleration along the circumference is $a = R\alpha$. For uniform circular motion $\omega = 2\pi f$ and $\theta = \theta_0 + \omega_0 t + \frac{1}{2}\alpha t^2$.

The product $\mathbf{A} \times \mathbf{B}$ is a vector of magnitude $|A||B|\sin\alpha$ with a direction perpendicular to the plane determined by the right hand rule.

Angular momentum is defined as $\mathbf{L} = \mathbf{r} \times \mathbf{P}$ and torque as $\mathbf{T} = \mathbf{r} \times \mathbf{F}$. Because of Newton's laws they are related by $\mathbf{T}_{net} = d\mathbf{L}/dt$.

If the force on a body is a central force, $\mathbf{L}$ of the body stays constant. The law of conservation of angular momentum states that in a closed system the vector sum of angular momenta of all the particles stays constant. $\Sigma \mathbf{L}_j = $ constant.

The total momentum of a system $\mathbf{P}_{total} = M_{total}\mathbf{v}_{cm}$ where $\mathbf{v}_{cm} = d\mathbf{R}_{cm}/dt$ and $\mathbf{R}_{cm} = \Sigma m_j \mathbf{r}_j / \Sigma m_j$ is the position of the center of mass.

The moment of inertia of a rigid body is $I = \Sigma r_j^2 \Delta m_j = \int r^2 \, dm$, where $r$ is the distance from the axis of rotation. A rigid body of angular velocity $\omega$ has $L = I\omega$ about the axis of rotation.

If a rigid body is at rest (or rotating with constant $\omega$ about a fixed axis), it must obey two conditions:

I   $\sum \mathbf{F}_j = 0$

II  $\sum \mathbf{T}_j = 0$

The position, magnitude, and direction of an unknown force required to balance an extended body can be determined using these conditions.

## Exercises

1. The angular position of a particle is $\theta = a + bt + ct^2$. What are the angular velocity and angular acceleration at the time $t = t_0$?
2. If the particle in Exercise 1 is traveling in a circle of radius $R$, what is the

linear velocity and acceleration along the circumference? What is the acceleration component pointing toward the center?

3. If $\theta = \theta_0 + \omega_0 t + \frac{1}{2}\alpha_0 t^2$,
   (a) What is the average angular velocity $\bar{\omega}$ during the time $t$?
   (b) Express $\bar{\omega}$ in terms of $\theta$, $\theta_0$, and $t$.

4. A bicycle wheel of mass $m$ has all its mass on the rim at a distance $R$ from its center. It has an angular velocity $\omega$ about an axis normal to the plane of the wheel and passing through its center. What is its angular momentum in terms of $m$, $R$, and $\omega$?

5. Repeat Example 3 making the approximation that the student's angular momentum is much less than that of the dumbbells. Then compare the initial and final kinetic energy of the dumbbells.

6. Suppose in Example 4 that the student did not step off the turntable. What would be the direction and magnitude of his angular velocity after turning the wheel upside down?

7. Repeat Example 6 for a chain force of 10 N.

8. Repeat Example 10 for a solid sphere rolling down the incline.

9. A rigid body of moment of inertia $I$ has an angular acceleration $\alpha$ about its axis and an instantaneous angular velocity $\omega$. What will be the power supplied to the body?

10. In Example 10 what is the ratio of kinetic energy of rotation to kinetic energy of translation for the hoop? Repeat for the disk.

11. What force must be applied to the handle in order to lift mass $m$?

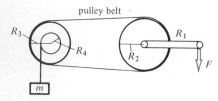

Exercise 11

12. Where must a 20-kg child sit in order to balance a 4-m seesaw if the father's mass is 70 kg and the mother's is 60 kg?

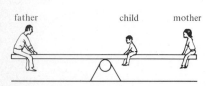

Exercise 12

13. An iron flywheel has density $\rho = 8 \times 10^3$ kg/m$^3$ and a fused quartz flywheel has $\rho = 2.4 \times 10^3$ kg/m$^3$. Both have the same breaking strength and both have the same mass. What is the ratio of maximum energy storage for the two flywheels?

14. The rim of a 0.8-m diameter bicycle wheel has a mass of 1.5 kg. What is the angular momentum of the bicycle wheel when the bicycle has a speed of 3 m/s? Ignore mass of the spokes.

15. A uniform meter stick has a mass of 100 g. A 50-g mass is attached to the 100-cm reading. What will be the reading of the center of mass?

16. Three masses under the influence of gravity are balanced on a meter stick as shown. What is $x$ in cm?

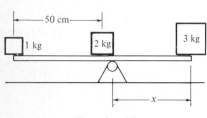

Exercise 16

17. A 10-kg ladder is resting at 45° against a frictionless wall. What is the force of the ladder against the wall?

18. Show that $\mathbf{A} \times \mathbf{B} = \mathbf{i}(A_y B_z - A_z B_y)$ when vectors $\mathbf{A}$ and $\mathbf{B}$ are in the $yz$ plane.

## Problems

19. Show that

$$\mathbf{A} \times \mathbf{B} = \begin{vmatrix} \mathbf{i} & \mathbf{j} & \mathbf{k} \\ A_x & A_y & A_z \\ B_x & B_y & B_z \end{vmatrix}$$

20. Show that $(\mathbf{A} \times \mathbf{B}) \cdot \mathbf{C} = \mathbf{A} \cdot (\mathbf{B} \times \mathbf{C})$.
21. Repeat Example 1 for $\mathbf{A} = \mathbf{i}A_x + \mathbf{j}A_y + \mathbf{k}A_z$ and $\mathbf{B} = \mathbf{i}B_x + \mathbf{j}B_y + \mathbf{k}B_z$.
22. Repeat Example 3 for a student who has twice the angular momentum of the dumbbells when they are at the 60-cm distance.
23. A neutron and a proton both of mass $m_0$ attract each other via gravity and are each in circular orbit about their center of mass.
    (a) If $R_0$ is the radius of the circular orbit and $v_0$ the velocity, what is the total angular momentum of the system about the center of mass in terms of $m_0$, $R_0$, and $v_0$?
    (b) What is the force on the neutron in terms of $G$, $m_0$, and $R_0$?
    (c) What is the force on the neutron in terms of $m_0$, $R_0$, and $v_0$?
    (d) If the total angular momentum is $L = \hbar$ where $\hbar$ is called Planck's constant, what is $R_0$ in terms of $\hbar$, $m_0$, and $G$?
24. Two small satellites of equal mass $m$ are in circular orbits at distances $R_1$ and $R_2$ from the center of the earth.
    (a) What is the angular momentum of satellite 1 in terms of $m$, $M_e$, $G$, and $R_1$?
    (b) If satellite 2 has twice the angular momentum of satellite 1, what is the ratio of $R_2$ to $R_1$?
    (c) What is the ratio of kinetic energies of satellite 2 to satellite 1?
25. A solid disk of mass $m$ is rolling along a surface. Its center has velocity $v$. What is the kinetic energy of the disk?
26. A solid sphere of mass $m$ is rolling along a surface. Its center has velocity $v$. What is the kinetic energy of the sphere?
27. Repeat Example 8 for the case where $m$ strikes the dumbbell at a distance $y$ from the origin ($y$ is less than $y_0$).
28. What is the ratio of the angular momentum of the earth's rotation to the orbital angular momentum of the moon? Give a numerical value.
29. (a) What is the total angular momentum of the earth-moon system?
    (b) When the earth stops rotating what will be the maximum distance of the moon from the earth? (See Example 2.)
    (c) What then will be the time for the moon to make one revolution?
30. The period of revolution of the sun is 27 days. When its nuclear fuel runs out it will undergo gravitational collapse. Angular momentum will be conserved. What is the smallest radius it can reach before breaking apart? (It will break apart when the centripetal acceleration exceeds the gravitational acceleration at the surface.)
31. A rod of mass $m$ and length $l$ is rotating about an axis which is a distance $x$ from the end. What is the moment of inertia?
32. The moment of inertia of a body about its center of mass is known to be $I$. What is the moment of inertia about the axis $y'$, which is a distance $R$ from the center of mass? (The result $I' = I + mR^2$ is called the parallel axis theorem.) (Hint: $I' = \Sigma \Delta m_j(\mathbf{R} + \mathbf{r}_j)^2$.)
33. Consider the collision of two identical, uniform cylindrical pucks of radius $r_0$ and mass $m$ on a frictionless air table as shown in the figure. Puck A has a linear velocity $v_A$ and an angular velocity $\omega_A$ about its center of mass. The linear velocity $v_A$ is directed toward the center of puck B, which is initially at rest. At the instant of collision the pucks stick together to form a rigid body.
    (a) What is the linear velocity of the system (of two pucks) after the collision?

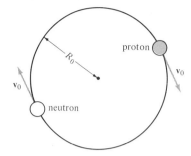

Problem 23

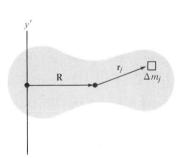

Problem 32

Problem 33

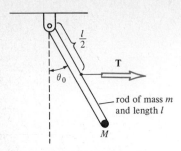

Problem 34

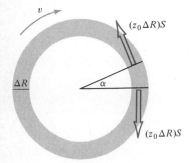

Problem 39

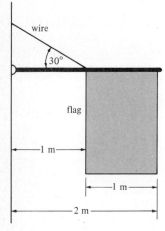

Problem 41

(b) What is the angular momentum of the system about its center of mass (located half way between the two pucks) before the collision?
(c) Calculate the angular velocity of the system about its center of mass (the point of contact of the pucks) after the collision.
(d) What is the total mechanical energy of the system before the collision?
(e) How much energy was lost in the collision?

34. A pendulum consists of a mass point of mass $M$ at the end of a uniform rod of mass $m$ and length $l$. The pendulum is suspended by a frictionless support attached to the ceiling. The pendulum is held at an angle $\theta_0$ by a horizontal string attached half way along the rod. The string has tension $T$. At $t = 0$ the string is cut and the pendulum swings freely.
   (a) Draw a free body diagram of the pendulum before the string is cut. Clearly label all forces acting on the pendulum. Assume the force of the support to be $F$ not necessarily directed along the pendulum bar.
   (b) Solve for $T$ in terms of $M$, $m$, $g$, and $\theta_0$.
   (c) At $t = 0$ the string is cut. Find the net torque as a function of $\theta$ acting on the pendulum about the fixed support. What is the moment of inertia of the pendulum about this axis?
   (d) Assuming $\theta$ to be small, write the equation of motion for the pendulum. What is the frequency of oscillation?
   (e) Is the angular momentum of the pendulum about the fixed support constant in time? If it is not, why does it change?

35. Derive the formula $I = ml^2/3$ for a rod rotating about its end.
36. Derive the formula $I = ml^2/12$ for a rod rotating about its center.
37. Repeat Example 13 for the case where the mass of the ladder is one quarter that of the man. Use $\theta = 60°$.
38. The angle of a ladder resting against a wall is slowly decreased. If the coefficient of friction is $\mu = 0.25$, at what angle will the ladder start slipping?
39. Consider a rotating hoop of width $z_0$ and thickness $\Delta R$ in the radial direction. The two stress forces on a piece of the hoop are shown in the figure and have maximum magnitude equal to the tensile strength $S_{max}$ times the area $(z_0 \Delta R)$. The vector sum of these two stress forces must equal the centripetal force. Show that $\frac{1}{2} I \omega_{max}^2 = vS_{max}/2$.
40. (a) A 40-cm diameter flywheel of 25-kg mass can store 10 kW h of energy. What is its angular velocity?
    (b) What is the centripetal acceleration of a point on the rim?
    (c) A car of total mass $10^3$ kg contains the described flywheel. Assuming that all the energy goes into lifting the car up a mountain road, what is the maximum height it can reach?
41. A flagpole of mass $m$ is pivoted to the wall of a building and supported by a wire as shown. The flag has a mass $M$ and the dimensions are as shown in the diagram. Treat the flag as rigid. What is the tension in the wire?

# Oscillatory Motion

So far we have studied both linear and rotational motion usually under constant acceleration. We have also studied one- and two-dimensional motion resulting from an inverse square law force (gravitation). In this chapter we shall study motion where the body under consideration moves back and forth sinusoidally with time. (This means a sine or cosine function of the time.) First we shall show that a body that obeys Hooke's law must oscillate sinusoidally. We shall study such oscillatory motion in detail and give some common examples. Then we will show that any body in stable equilibrium must oscillate sinusoidally when given a small displacement. Sinusoidal motion is the most common form of motion in the everyday world and is thus an important topic in physics.

## 11-1 The Harmonic Force

If the force on a body is proportional to its displacement from the origin and always directed toward the origin, such force is defined as a harmonic force. If we choose the direction of displacement as the $x$ axis, the equation

$$F = -kx$$

expresses a harmonic force where $x$ is the displacement from the equilibrium position. As discussed on page 110, the force exerted by a stretched (or compressed) spring has this property as long as the spring is not stretched beyond its elastic limit.

The observation that a spring exerts a harmonic force if it is not stretched too far is called Hooke's law.

$$F = -k(x - x_1) \quad \text{(Hooke's law)} \quad \text{(11-1)}$$

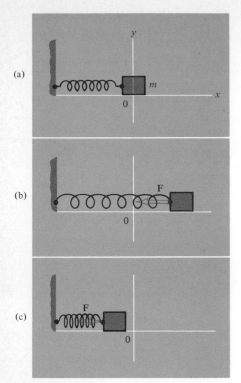

**Figure 11-1.** *Mass m is sliding on a frictionless surface. The spring (a) unstretched, (b) stretched, (c) compressed.*

where $x_1$ is the equilibrium position. In Figure 11-1 the origin is chosen as the equilibrium position ($x_1 = 0$). Then if the spring is stretched a distance $x_0$ and the mass $m$ released at $t = 0$, we shall show that its position as a function of time will be

$$x = x_0 \cos \omega t \quad \text{where } \omega = \sqrt{k/m} \quad (11\text{-}2)$$

and $k = -F/x$ is the spring constant. Such sinusoidal motion is commonly called simple harmonic motion (SHM). We start with $F_{\text{net}} = ma$ where $F_{\text{net}}$ is the spring force $(-kx)$:

$$-kx = ma$$

$$-kx = m\left(\frac{d^2x}{dt^2}\right)$$

$$\frac{d^2x}{dt^2} = -\frac{k}{m}x \quad (11\text{-}3)$$

This is called a second-order differential equation. A common procedure for solving differential equations is to "guess" the answer and check to see if the "guess" is indeed a solution. So for our "guess" we shall try

$$x = x_0 \cos \omega t$$

Then

$$\frac{dx}{dt} = -x_0 \omega \sin \omega t \quad \text{(velocity in SHM)} \quad (11\text{-}4)$$

and

$$\frac{d^2x}{dt^2} = -x_0 \omega^2 \cos \omega t \quad \text{(acceleration in SHM)} \quad (11\text{-}5)$$

Now substitute this into the left-hand side of Eq. 11-3 and $(x_0 \cos \omega t)$ for $x$ in the right-hand side:

$$(-x_0 \omega^2 \cos \omega t) = -\frac{k}{m}(x_0 \cos \omega t)$$

$$\omega^2 = \frac{k}{m}$$

We see that $x = x_0 \cos \omega t$ is indeed a solution but only if $\omega = \sqrt{k/m}$. The function $x = x_0 \sin \omega t$ is also a legitimate mathematical solution; however, it does not satisfy the initial condition that $x = x_0$ when $t = 0$. The most general solution is $x = x_0 \cos(\omega t + \phi)$ where $\phi$ is an arbitrary phase angle. The constants $x_0$ and $\phi$ are determined by the initial conditions.

The velocity as a function of time is given in Eq. 11-4 and the accelera-

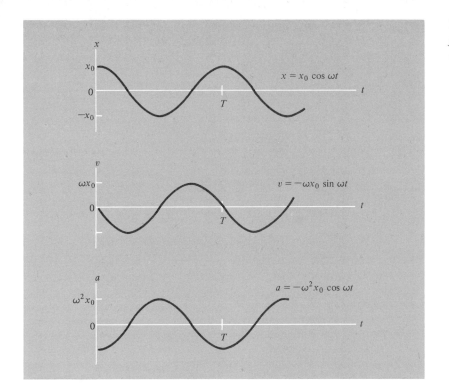

**Figure 11-2.** *Plots of x, v, and a as a function of time for simple harmonic motion. The period T has the value $2\pi/\omega$.*

tion in Eq. 11-5. These functions are plotted one above the other in Figure 11-2. We see from Eq. 11-4 that the maximum velocity is

$$v_{\max} = \omega x_0 \qquad (11\text{-}6)$$

and it occurs whenever $x = 0$.

We see from Eq. 11-3 that the acceleration is always $-\omega^2$ times the displacement:

$$\frac{d^2x}{dt^2} = -\omega^2 x$$

This is a very useful relationship: if the equation of motion of a body can be put into the form $d^2x/dt^2 = -Cx$, where $C$ is some constant, then $x = x_0 \cos \omega t$ and $\omega = \sqrt{C}$.

If

$$\left. \begin{array}{c} \dfrac{d^2u}{dt^2} = -Cu \\[1em] \text{then} \qquad \omega = \sqrt{C} \quad \text{and} \quad u = u_0 \cos \omega t \end{array} \right\} \qquad (11\text{-}7)$$

where $u$ is any displacement. This can be demonstrated by differentiating $u = u_0 \cos \omega t$. Then $d^2u/dt^2 = -\omega^2 u$, which is also equal to $-Cu$; that is,

$$-\omega^2 u = -Cu$$
$$\omega = \sqrt{C}$$

## 11-2 The Period of Oscillation

We can easily show that $\omega = 2\pi/T$ where $T$ is the period of oscillation. The function $\cos \omega t$ or $\sin \omega t$ completely repeats itself when the angle $\omega T = 2\pi$ or $T = 2\pi/\omega$. This special value of $t$ is by definition called the period $T$.

$$T = \frac{2\pi}{\omega} \quad \text{(period of oscillation)} \qquad (11\text{-}8)$$

The number of oscillations in a time $t$ is

$$n = \frac{t}{T}$$

If we divide both sides by $t$ we have the number of oscillations per unit time:

$$\frac{n}{t} = \frac{1}{T}$$

The left-hand side is by definition the frequency of oscillation and usually denoted by the letter $f$:

$$f = \frac{1}{T} \quad \text{(frequency of oscillation)} \qquad (11\text{-}9)$$

Comparing Eqs. 11-8 and 11-2 gives

$$T = 2\pi \sqrt{\frac{m}{k}} \qquad (11\text{-}10)$$

for the period of oscillation of a mass $m$ attached to the end of a spring that has a spring constant equal to $k$.

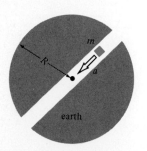

**Figure 11-3.** *Mass m is dropped into hole through the center of the earth.*

**Example 1.** Assume a hole could be dug through the center of the earth and out the other side as shown in Figure 11-3. If a mass $m$ is dropped in the hole, how long would it take to reach the other side? Assume uniform density for the earth and ignore air resistance. (The hole could be evacuated.) What is its velocity when passing the center of the earth?

ANSWER: In Chapter 5 we learned that the gravitational force inside a solid sphere is proportional to the distance from the center. The acceleration due to gravity inside the earth is, according to Eq. 5-10,

$$a = -\frac{g}{R}r \quad \text{or} \quad \frac{d^2r}{dt^2} = -\left(\frac{g}{R}\right)r$$

This is the same form as Eq. 11-7 with $r$ as the displacement $u$ and $g/R$ as the constant $C$. Since $\omega$ is always the square root of the proportionality constant between acceleration and displacement, we have

$$\omega = \sqrt{C} = \sqrt{\frac{g}{R}}$$

$$T = \frac{2\pi}{\omega}$$

$$T = 2\pi\sqrt{\frac{R}{g}} = 2\pi\sqrt{\frac{6.37 \times 10^6 \text{ m}}{9.8 \text{ m/s}^2}} = 5.06 \times 10^3 \text{ s} = 84 \text{ min}$$

This is the time for mass $m$ to reach the other side of the earth and return to its original position. The time to reach the other side is one half this or 42 min. To get the velocity while passing the center of the earth we use Eq. 11-6:

$$v_{max} = \omega R = \sqrt{gR} = 7.9 \times 10^3 \text{ m/s}$$

We note an interesting coincidence that the period and $v_{max}$ are the same as for a low-flying earth satellite (see Eq. 3-11).

---

\* **Example 2.** Two masses $m_1$ and $m_2$ are connected to the opposite ends of a spring. If the spring is stretched and both masses are released simultaneously, what will be the period of oscillation? Assume a spring constant $k$.

ANSWER: Let $x_1$ be the displacement of $m_1$ from equilibrium and $x_2$ the displacement of $m_2$ from equilibrium. We note that the center of mass must stay fixed, hence

**Figure 11-4.** Masses $m_1$ and $m_2$ are at ends of stretched spring.

$$m_1 x_1 = -m_2 x_2 \quad \text{or} \quad x_1 = -\frac{m_2}{m_1} x_2$$

Now apply $F_{\text{net}} = ma$ to $m_2$. The net force on $m_2$ is the force $F$ shown in Figure 11-4, which is $F = -k(x_2 - x_1)$ where $(x_2 - x_1)$ is the net stretching of the spring.

$$-k(x_2 - x_1) = m_2 \frac{d^2 x_2}{dt^2}$$

Now substitute $-(m_2/m_1)x_2$ for $x_1$ on the left-hand side:

$$-k\left[x_2 - \left(-\frac{m_2}{m_1} x_2\right)\right] = m_2 \frac{d^2 x_2}{dt^2}$$

or

$$\frac{d^2 x_2}{dt^2} = -\frac{k(m_1 + m_2)}{m_1 m_2} x_2$$

$$\frac{d^2 x_2}{dt^2} = -\left(\frac{k}{\mu}\right) x_2$$

where $\mu \equiv m_1 m_2 / (m_1 + m_2)$ is defined as the reduced mass. The preceding equation is the same as Eq. 11-7 with $x_2$ corresponding to $u$ and $(k/\mu)$ corresponding to $C$. Hence

$$\omega = \sqrt{\frac{k}{\mu}}$$

or

$$T = 2\pi \sqrt{\frac{\mu}{k}} \qquad (11\text{-}11)$$

We note that in all our examples of simple harmonic motion, the period $T$ is independent of the amplitude of oscillation $x_0$ (or $u_0$) as long as Hooke's law is obeyed. Galileo observed this property of simple harmonic motion and used it to construct a clock based on the motion of a pendulum.

## 11-3 The Pendulum

In Section 4-8 we have already seen that the period of oscillation of a simple pendulum should be $T = 2\pi \sqrt{l/g}$ when undergoing small oscillations. In this section we shall obtain the same result by a more general

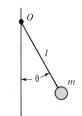

(a)

**Figure 11-5.** (a) A physical pendulum pivoted at O. (b) A simple pendulum.

(b)

method and also expressions for the position and velocity of a pendulum bob as a function of time.

For the sake of generality we shall consider any arbitrary rigid body pivoted at a point $O$ as shown in Figure 11-5a. Its center of mass is at $O'$ at a distance $l$ from the pivot. In order to calculate the period of oscillation all we need know is the moment of inertia $I$ about point $O$. The torque on the body is $T = -mgl \sin \theta$. Using the relation $T = I\alpha$ (Eq. 10-20), we have

$$(-mgl \sin \theta) = I \frac{d^2\theta}{dt^2}$$

$$\frac{d^2\theta}{dt^2} = -\frac{mgl}{I} \sin \theta$$

For small oscillations $\sin \theta \approx \theta$ and then

$$\frac{d^2\theta}{dt^2} = -\left(\frac{mgl}{I}\right)\theta$$

This is of the same form as Eq. 11-7 with $u$ corresponding to $\theta$ and $C = mgl/I$; hence

$$\theta = \theta_0 \cos \omega t$$

and

$$\omega = \sqrt{\frac{mgl}{I}} \quad \text{or} \quad T = 2\pi \sqrt{\frac{I}{mlg}} \qquad (11\text{-}12)$$

In the case of a simple pendulum all the mass is at a distance $l$ and $I = ml^2$. Putting this into Eq. 11-12 gives

$$T = 2\pi \sqrt{\frac{(ml^2)}{mlg}}$$

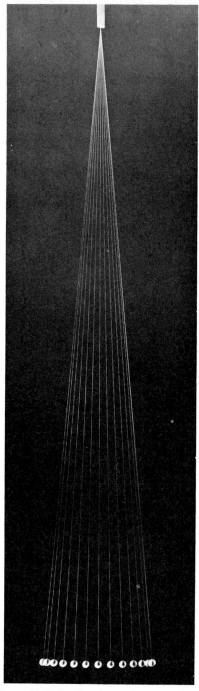

**Figure 11-6.** Strobe photo of a simple pendulum. [Courtesy Educational Development Center]

SEC. 11-3/THE PENDULUM

and

$$T = 2\pi \sqrt{\frac{l}{g}} \quad \text{(period of simple pendulum)} \quad (11\text{-}13)$$

Note that the period of oscillation is not only independent of amplitude, but also of mass. However, if $\theta_0$ is a large angle, the approximation $\sin\theta \approx \theta$ breaks down. But even for $\theta_0$ as large as 20°, Eq. 11-13 is correct to within 1%. A strobe photograph of a simple pendulum is shown in Figure 11-6.

**Example 3.** Assume the pendulum of a grandfather clock is a simple pendulum that swings 10 cm from side to side each second. The full period is then $T = 2$ s.
  (a) What is the length of the pendulum?
  (b) What is the peak velocity of the bob?

ANSWER: (a) Solving Eq. 11-13 for $l$ gives

$$l = g\left(\frac{T}{2\pi}\right)^2 = 9.8 \,\frac{\text{m}}{\text{s}^2}\left(\frac{2\text{ s}}{2\pi}\right)^2 = 0.99 \text{ m}$$

(b) According to Eq. 11-6,

$$v_{max} = \omega x_0 = \frac{2\pi}{T}x_0 \quad \text{where } x_0 = \frac{10 \text{ cm}}{2} = 5 \text{ cm}$$

$$v_{max} = \frac{2\pi}{2\text{ s}}(5 \text{ cm}) = 15.7 \text{ cm/s}$$

**Example 4.** What is the period of oscillation of a rod of mass $m$ swinging from one end? See Figure 11-7.

ANSWER: We see from Table 10-1 that the moment of inertia of a rod of length $l$ about one end is $I = \frac{1}{3}ml^2$. Putting this into Eq. 11-12 gives

$$T = 2\pi \sqrt{\frac{(\frac{1}{3}ml^2)}{mlg}} = 2\pi \sqrt{\frac{l}{3g}}$$

which is $1/\sqrt{3}$ times that of a simple pendulum of the same length.

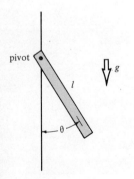

**Figure 11-7.** *Rod of mass m pivoted at end.*

## 11-4 Energy of Simple Harmonic Motion

The potential energy of a mass attached to the end of a spring was discussed in Chapters 6 and 7 (see pages 110 and 116). If the spring is displaced a distance $x$, the potential energy is $U = \frac{1}{2}kx^2$. If the mass is pulled an initial distance $x_0$ and then released, the initial energy of the system is $\frac{1}{2}kx_0^2$. Note that the energy of simple harmonic motion goes as the square of the amplitude $x_0$. Assuming no frictional or resistive forces, the sum of the kinetic energy and potential energy must equal $\frac{1}{2}kx_0^2$. At any instant

$$\tfrac{1}{2}mv^2 + \tfrac{1}{2}kx^2 = \tfrac{1}{2}kx_0^2$$

or

$$v^2 = \frac{k}{m}(x_0^2 - x^2)$$

Since $k/m = \omega^2$, we have

$$v = \omega\sqrt{x_0^2 - x^2} \qquad (11\text{-}14)$$

The time average of the potential energy is

$$\overline{U} = \tfrac{1}{2}k\overline{x^2}$$

We replace $x$ with $(x_0 \cos \omega t)$ to obtain

$$\overline{U} = \tfrac{1}{2}kx_0^2 \,\overline{\cos^2 \omega t}$$

The time average of the kinetic energy is $\overline{K} = \tfrac{1}{2}m\overline{v^2}$. Now replace $m$ with $k/\omega^2$ and $v$ with $(-\omega x_0 \sin \omega t)$:

$$\overline{K} = \frac{1}{2}\left(\frac{k}{\omega^2}\right)\overline{(-\omega x_0 \sin \omega t)^2} = \tfrac{1}{2}kx_0^2 \,\overline{\sin^2 \omega t}.$$

It should now be apparent that $\overline{U} = \overline{K}$. This is because a plot of $\sin^2 \omega t$ looks the same as $\cos^2 \omega t$ only shifted by $T/4$. The average height of both curves must be the same (it is equal to $\tfrac{1}{2}$) and then

$$\overline{K} = \overline{U} = \tfrac{1}{4}kx_0^2$$

The average height of $\sin^2 \omega t$ is $\tfrac{1}{2}$ because

$$\sin^2 \omega t + \cos^2 \omega t = 1$$

and the average of each term is the same.

When we study specific heat in Chapter 13, we will make use of the fact that vibrating molecules or atoms in a solid have the same amount of potential energy as vibrational kinetic energy.

## 11-5 Small Oscillations

Most rigid bodies in the everyday world are at rest and thus in stable equilibrium. For a body in stable equilibrium the net force on the body is zero and if the body is displaced a distance $x$ away from the position of stable equilibrium the force $F$ will be of opposite sign to $x$. In general the force will be of the form

$$F = a_1 x + a_2 x^2 + a_3 x^3 + \cdots$$

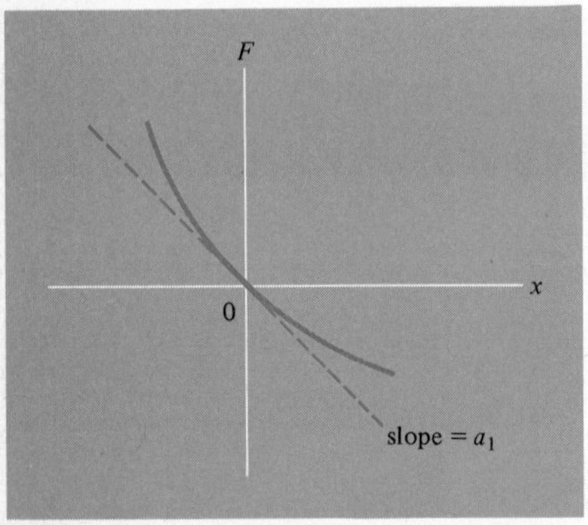

where $a_1$ is negative. This generalized function is plotted in Figure 11-8. The slope at the origin is

$$\left(\frac{dF}{dx}\right)_0 = a_1$$

For small oscillations about the origin any such curve can be considered as the straight line:

$$F = a_1 x$$

or

$$\frac{d^2 x}{dt^2} = -\left(-\frac{a_1}{m}\right) x \quad \text{where } a_1 = \left(\frac{dF}{dx}\right)_0$$

**Figure 11-8.** *Force on a body in stable equilibrium as a function of displacement $x$ from equilibrium.*

This is in the form of Eq. 11-7 with $C = (-a_1/m)$; hence

$$\omega = \sqrt{-\frac{a_1}{m}}$$

or

$$T = 2\pi \sqrt{-\frac{m}{(dF/dx)_0}} \quad \text{(period of small oscillation)} \quad \text{(11-15)}$$

Thus we can calculate the period of small oscillation of any body about its equilibrium position if the slope of the force function is known.

We can also obtain the period of oscillation if the potential energy is known as a function of the displacement. Since $F = -dU/dx$, the equivalent spring constant is

$$k = -\left(\frac{dF}{dx}\right)_0 = -\left[\frac{d(-dU/dx)}{dx}\right]_0 = \left(\frac{d^2 U}{dx^2}\right)_0 \quad \text{(11-16)}$$

This second derivative must be evaluated at the point of equilibrium where $U$ is a minimum.

**Example 5.** An ice cube is given a small displacement from the bottom of a spherical bowl of 10-cm radius. Assuming no friction, what will be its period of oscillation after being released?

ANSWER: According to Eq. 6-13 the $x$ component of force is $F_x = -dU/dx$ where $U = mgy$. We choose the axes such that the motion is in the $xy$ plane, as shown in Figure 11-9. The equation of the circle in this figure is

$$x^2 + (R - y)^2 = R^2$$

Solving for $y$:
$$y = R - \sqrt{R^2 - x^2}$$

Then
$$\frac{dU}{dx} = mg \frac{d}{dx}(R - \sqrt{R^2 - x^2}) = \frac{mgx}{\sqrt{R^2 - x^2}}$$

In the limit of $x \ll R$, $dU/dx \approx mgx/R$ and then

$$F_x = -\frac{mg}{R} x \quad \text{or} \quad \frac{dF_x}{dx} = -\frac{mg}{R}$$

Substitution of this into Eq. 11-15 gives

$$T = 2\pi \sqrt{-\frac{m}{(-mg/R)}} = 2\pi \sqrt{\frac{R}{g}} = 2\pi \sqrt{\frac{0.1 \text{ m}}{9.8 \text{ m/s}^2}} = 0.635 \text{ s}$$

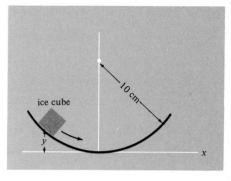

**Figure 11-9.** *Frictionless ice cube is released from position shown.*

\***Example 6.** Suppose the potential energy in a diatomic molecule can be expressed as the sum of an attractive term $b/r^3$ plus a shorter-range repulsive term $a/r^5$; that is, $U = (a/r^5) - (b/r^3)$. See Figure 11-10. What is the equilibrium position $r_0$, and what is the spring constant $k$ between the two atoms? What is the vibrational frequency of this diatomic molecule if each atom has mass $m$?

ANSWER: The equilibrium position can be obtained by solving for $r$ in the equation $dU/dr = 0$.

$$\frac{dU}{dr} = -\frac{5a}{r^6} + \frac{3b}{r^4}$$

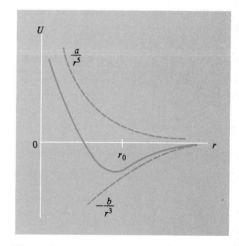

**Figure 11-10.** *Potential energy between two atoms consisting of attractive part $-b/r^3$ and repulsive part $a/r^5$.*

Setting this equal to zero gives

$$-\frac{5a}{r_0^6} + \frac{3b}{r_0^4} = 0$$

$$r_0 = \sqrt{\frac{5a}{3b}}$$

From Eq. 11-16 the spring constant is

$$k = \left(\frac{d^2U}{dr^2}\right)_{r_0}$$

$$\frac{d^2U}{dr^2} = \frac{30a}{r^7} - \frac{12b}{r^5}$$

$$\left(\frac{d^2U}{dr^2}\right)_{r_0} = \frac{1}{r_0^5}\left(\frac{30a}{r_0^2} - 12b\right) = \left(\frac{3b}{5a}\right)^{5/2}\left(\frac{30a}{5a/3b} - 12b\right)$$

$$k = 6b\left(\frac{3b}{5a}\right)^{5/2}$$

Using $\mu = m/2$ in Eq. 11-11 gives

$$f = \frac{1}{2\pi}\sqrt{\frac{k}{m/2}} = \frac{1}{2\pi}\sqrt{\frac{12b}{m}\left(\frac{3b}{5a}\right)^{5/2}}$$

For typical atomic values of $a$ and $b$ and $m$, this has a value $\sim 10^{14}$ Hz. The two atoms vibrating back and forth at this frequency emit electromagnetic radiation of this frequency, which is about one fifth the frequency of red light and is called infrared radiation.

Any solid is made of atoms that are bound to each other by an attractive force. However, if the atoms are pushed closer together they are repelled. Details of these interatomic forces are presented in Chapter 27. We see from Example 6 that such a basic interatomic force behaves as a harmonic force for small displacements. Hence most solids execute simple harmonic motion when displaced slightly from their equilibrium positions. Such motion goes as $\cos \omega t$ where $\omega$ is determined by the mass and force constant of the solid body under consideration. Of course such sinusoidal motion does not continue forever once it has been initiated. This is because, in addition to the elastic forces discussed previously, there are various kinds of resistive forces that convert energy of the oscillation into heat. In some solids such as a piano wire or a vibrating crystal, the resistive forces are small and the energy of oscillation is damped slowly over hundreds of oscillations. On the other hand a stretched string will lose half its vibrational energy in just a few oscillations.

Not only do small objects oscillate sinusoidally (many of them at frequencies detectable by the human ear) but also large objects such as bridges and skyscrapers. The upper part of a tall skyscraper will sway back and forth sinusoidally when hit by a gust of wind. On a typical windy day the Eiffel Tower sways more than a meter with a period of a few seconds. Understanding of small oscillations is understanding of the source of most sounds. Equation 11-15 predicts that the characteristic frequency emitted by a vibrating body should not depend on the amplitude of the vibration. This is certainly the case with most sounds that we hear around us.

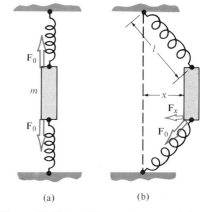

Figure 11-11. (a) Mass m is suspended between two stretched springs. (b) The mass is displaced a distance x to the side.

**Example 7.** A mass $m = 10$ g is suspended between two stretched springs, each exerting $F_0 = 15$ N as shown in Figure 11-11. The mass is given a small lateral displacement $x$ as shown in (b). If the length of the stretched spring is $l = 10$ cm, what will be the period of oscillation when released? Assume any additional stretching of the spring is negligible.

ANSWER: As seen from Figure 11-11(b), $F_x = -F_0 x/l$. Since there are two springs, the net force on $m$ is

$$F = \frac{2F_0}{l} x$$

$$\frac{dF}{dx} = -\frac{2F_0}{l}$$

If this is substituted into Eq. 11-15, one obtains

$$T = 2\pi \sqrt{-\frac{m}{(-2F_0/l)}} = 2\pi \sqrt{\frac{ml}{2F}} = 0.0363 \text{ s}$$

and

$$f = \frac{1}{T} = 27.6 \text{ Hz}$$

Example 7 gives us some idea of why a stretched spring (or wire or string) will vibrate when plucked. We see that the frequency of vibration is approximately $\sqrt{F_0/ml}$ where $F_0$ is the tension, $m$ is the effective mass, and $l$ is the length. We also see that for reasonable values of tension, the frequency of vibration is within the range of the human ear ($\sim$20 to 15,000 Hz). Example 7 also bears correspondence to a typical loudspeaker cone, which carries a coil of mass $\sim$10 g at the center. Then $2F_0 = 30$ N is the total tension with which the coil is suspended. If such a speaker cone is displaced and then released, it will have a natural frequency of oscillation on the order of 30 Hz.

## 11-6 Intensity of Sound

### Traveling Waves

If a portion of a continuous medium such as the end of a string (or a layer of air) is set into simple harmonic motion, the motion will be transmitted to the neighboring portion of the medium and then to the neighbor of the neighbor, and so on, with the result that the sinusoidal disturbance travels out along the medium away from the original source of motion. The resulting motion is called a traveling wave. A traveling wave moving along a string is shown in Figure 20-10. It is quite appropriate to study traveling waves on a string at this time. Those wishing to do so should at this point study Sections 20-6 and 20-7 and then return to this section. Whether waves on a string are studied now, or later in Chapter 20, is merely a matter of taste.

### Sound Waves

A vibrating flat plate will transmit a traveling wave in the air that moves away from the source with a wave velocity $u$. This is called a sound wave. Suppose the mass in Figure 11-11 is a thin flat plate of area $A$ and that it is oscillating back and forth in simple harmonic motion with amplitude $x_0$ and frequency $\omega/2\pi$. It will transfer energy to the layer of air shown in Figure 11-12, which has mass $\Delta m$. The maximum kinetic energy of this layer of air is

$$\tfrac{1}{2}\Delta m v_0^2 = \tfrac{1}{2}\Delta m \omega^2 x_0^2$$

$$\Delta E = \tfrac{1}{2}(\rho A\, \Delta x)\omega^2 x_0^2 \tag{11-17}$$

where $\rho$ is the density of air.

Since in simple harmonic motion the average potential energy equals the average kinetic energy, Eq. 11-17 describes the energy content of the layer of air of area $A$ and thickness $\Delta x$. If the vibration is started at $t = 0$, the oscillation in the air moves to the right in Figure 11-12 with a speed $u = \Delta x/\Delta t$ where $\Delta x$ is the distance the disturbance moves in a time $\Delta t$. The rate of energy transfer to each successive layer of thickness $\Delta x$ is obtained by dividing Eq. 11-17 by $\Delta t$:

$$\frac{\Delta E}{\Delta t} = \frac{1}{2}\rho A \frac{\Delta x}{\Delta t} \omega^2 x_0^2$$

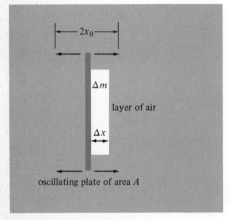

**Figure 11-12.** *Oscillating plate causes layer of air to oscillate with same amplitude $x_0$.*

Thus the power $P$ radiated in the positive $x$ direction by the vibrating plate is

$$P = \tfrac{1}{2}\rho A \omega^2 x_0^2 u$$

The intensity $I$ of any kind of a traveling wave is defined as the power per unit area. If we divide both sides of the preceding equation by $A$ we obtain

$$I = \tfrac{1}{2}\rho \omega^2 x_0^2 u \quad \text{(intensity of a sound wave)} \quad (11\text{-}18)$$

The S.I. units would be watts per square meter. Physically it is the energy flow in joules per second through a cross-sectional area of 1 m². Note that the intensity is proportional to the square of the amplitude. This is true of all kinds of traveling waves such as water waves, wave on a string, electromagnetic waves, and so on.

The smallest sound intensity detectable by the human ear is $I_0 \approx 10^{-12}$ W/m². This is called the threshold of hearing. As we shall see in the following example, this is not too much larger than the best that could ever be done consistent with the laws of physics. Biologically the human ear is almost as efficient as is physically possible.

---

*Example 8.* According to the kinetic theory of heat presented in the next chapter, any particle or body such as an ear drum will have a kinetic energy $\sim 6 \times 10^{-21}$ J due to molecular collisions when at room temperature. What is the kinetic energy given to the ear drum by a sound wave at the threshold of hearing? Compare with the intrinsic thermal energy of $6 \times 10^{-21}$ J. The speed of sound is $u = 330$ m/s and the density of air is $\rho = 1.3$ kg/m³. Use $m_e = 0.1$ g for the mass of an eardrum.

ANSWER: If we assume the eardrum oscillates freely with the air, it will have the same velocity of SHM as the air. Thus the average kinetic energy given to the eardrum is

$$K = \tfrac{1}{2} m_e \overline{v_{\text{air}}^2}$$

Since $v_{\text{air}} = \omega x_0 \sin \omega t$, we have

$$\overline{v_{\text{air}}^2} = \tfrac{1}{2}\omega^2 x_0^2$$

and

$$K = \tfrac{1}{2} m_e (\tfrac{1}{2}\omega^2 x_0^2)$$

Solving Eq. 11-18 for $\omega^2 x_0^2$ and substituting in the preceding equation gives

$$K = \tfrac{1}{4} m_e \left( \frac{2I}{\rho u} \right)$$

$$K = \tfrac{1}{4}(10^{-4} \text{ kg}) \left( \frac{2 \times 10^{-12} \text{ W/m}^2}{1.3 \text{ kg/m}^3 \times 330 \text{ m/s}} \right) = 1.16 \times 10^{-19} \text{ J}$$

We note that this is within a factor of 20 of the intrinsic thermal limit. This is even more remarkable when we consider that many common sounds are $\sim 10^{10}$ times $I_0$.

### The Decibel Scale

A jet plane taking off a few meters away can cause a sound intensity $\sim 10^{15}$ times $I_0$. A nearby subway train gives a sound intensity of $10^{10} I_0$. Constant exposure to such a sound can endanger one's hearing. Electronically amplified rock concerts typically give sound intensities up to $10^{12} I_0$. This is just at the pain threshold and can reduce one's hearing ability. Many municipalities are concerned with sound pollution and have ordinances forbidding outdoor sounds greater than $10^{10} I_0$.

The exponents in the preceding paragraph, when multiplied by 10, give what is defined as the decibel measure of sound. If a sound intensity is $\beta$ in decibels, then

$$\beta = 10 \log \frac{I}{I_0} \quad \text{(intensity } I \text{ in decibels)} \quad (11\text{-}19)$$

The abbreviation for decibel is db. Thus the threshold of hearing is 0 db and a typical rock concert is 120 db. Some towns and cities forbid street noises exceeding 100 db. Certainly workers exposed to over 100 db should use ear plugs. It is foolish for a rock musician, a chain saw operator, or a riveter to go without ear plugs.

## Summary

If $F = -kx$, $F$ is a harmonic force and the resulting motion is a sine wave of the form

$$x = x_0 \cos \omega t \quad \text{where } \omega = 2\pi/T = \sqrt{k/m}$$

Then the average potential energy, $\overline{U} = \tfrac{1}{2} \overline{kx^2}$, is equal to the average kinetic energy, $\overline{K} = \tfrac{1}{2} \overline{mv^2}$.

If any quantity $u(t)$ has its second time derivative proportional to $-u$, then $u = u_0 \cos \omega t$; that is, if $d^2u/dt^2 = -Cu$, then $u = u_0 \cos \omega t$ and $\omega = \sqrt{C}$. (Problems such as Exercise 3 can immediately be solved using $2\pi/T = \sqrt{C}$.)

The frequency $f = 1/T = \omega/2\pi$.

For a simple pendulum, $\omega = \sqrt{g/l}$ or $T = 2\pi\sqrt{l/g}$.

Any body in stable equilibrium when given a small displacement will oscillate in simple harmonic motion with a period

$$T = 2\pi\sqrt{\frac{m}{k_{\text{eff}}}} \quad \text{where } k_{\text{eff}} = -(dF/dx)_0 = (d^2U/dx^2)_0$$

This is useful if either the function $F(x)$ or $U(x)$ is known. The point of stable equilibrium is the $x$ value where the function $U(x)$ has a minimum.

The intensity of a sound wave is proportional to the square of the amplitude and is defined as the time rate of energy flow per unit area. The sound intensity is $I = \frac{1}{2}\rho\omega^2 x_0^2 u$ where $\rho$ is the air density and $u$ the wave velocity. The intensity in decibels is $\beta = 10\log(I/I_0)$.

## Exercises

1. A particle is executing simple harmonic motion. The displacement $x$ is shown as a function of time in the figure. What are the amplitude, period, maximum velocity, and maximum acceleration?
2. A mass $m$ is in simple harmonic motion according to the equation $d^2x/dt^2 = -Kx$
   (a) What is its period of oscillation?
   (b) If the amplitude is $x_0$, what is the instantaneous velocity in terms of $x$, $x_0$, and $K$?
3. The $x$ coordinate of a particle obeys the equation

$$A^2 \frac{d^2x}{dt^2} + x = 0$$

What is the period of oscillation?

4. A superball is dropped from a height of 1 m and bounces up and down with perfectly elastic collisions.
   (a) What is the period of oscillation?
   (b) Is this simple harmonic motion?
5. The pendulum of a desktop "grandfather" clock has a period $T = 1$ s. What is the length of the pendulum?
6. The $x$ coordinate of a particle obeys the equation $d^2x/dt^2 = -\omega^2 x$ and the $y$ coordinate obeys $d^2y/dt^2 = -\omega^2 y$. Its maximum $x$ value is $R_0$, and its maximum $y$ value is $R_0$. At $t = 0$, $x = R_0$ and $y = R_0$.
   (a) What are $x$, $y$, and $r$ as functions of $t$?
   (b) What is $v$ as a function of $t$?
   (c) What is the path of the particle?
7. A pendulum bob swings a total distance of 4 cm from end to end and reaches a speed of 10 cm/s at the midpoint. Assuming small oscillation, what is the period of oscillation in seconds?

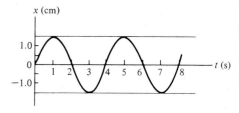

Exercise 1

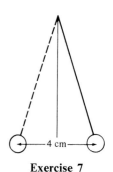

Exercise 7

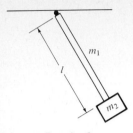

Exercise 8

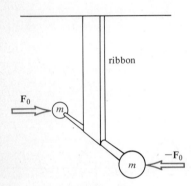

Problem 19

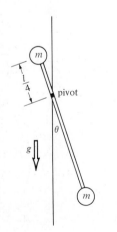

Problem 20

8. A pendulum consists of a rod of mass $m_1$ with a mass $m_2$ attached to the end. For small oscillations, what is the period of oscillation in terms of $m_1$, $m_2$, $l$, and $g$?
9. In the hydrogen molecule the force on each hydrogen atom is $F = -A(r - r_0)$ where $A = 0.057$ N/m and $r_0 = 7.4 \times 10^{-11}$ m is the equilibrium distance between the centers of the two hydrogen atoms. What is the frequency of oscillation?
10. If the position of a particle is given by $x = x_0 \sin \omega t$, what is its acceleration in terms of $x_0$, $\omega$, and $t$?
11. The amplitude of a sound wave is doubled. What is its increase in decibels?
12. Sound intensity from a loudspeaker is proportional to the square of the applied voltage. If the voltage is increased by a factor of 10, by how many decibels is the sound increased?

## Problems

13. A mass $m$ attached to the end of a spring is pulled a distance $x_0$ and released at a time $t_1$. The force constant is $k$.
    (a) What is $x$ as a function of $k$, $m$, $x_0$, $t_1$, and $t$?
    (b) What is $v$ as a function of the above quantities?
14. In Problem 13 the velocity is positive and maximum at $t = 0$.
    (a) What is $x$ as a function of $k$, $m$, $x_0$, and $t$?
    (b) What is $v$ as a function of $k$, $m$, $x_0$, and $t$?
15. Repeat Problem 14 for the case where the velocity is a maximum in magnitude, but is negative in sign at $t = 0$.
16. Repeat Exercise 6 for the case where $x = 0$ and $y = R_0$ at $t = 0$ and $v_x$ is positive.
17. Repeat Exercise 6 for the case where $x = R_0/\sqrt{2}$, $y = R_0/\sqrt{2}$, and $v_x$ and $v_y$ are both negative at $t = 0$.
18. A mass $m$ is attached to a spring of unknown spring constant. It is oscillating back and forth between $x = -x_0$ and $+x_0$. At a time $t_1$, its displacement is $x = x_1$ and its acceleration is $a = a_1$.
    (a) What is the period of oscillation in terms of $a_1$ and $x_1$.
    (b) What is the force constant $k$ in terms of $a_1$, $x_1$, and $m$?
    (c) What is the maximum velocity of $m$ in terms of $x_0$, $x_1$, and $a_1$?
19. A massless rod of length $l$ has a mass $m$ at either end. The rod is fastened to a fairly stiff metal ribbon as shown.
    (a) Ignoring the mass of the ribbon, what is the moment of inertia of the system about a vertical axis through the center of the rod? (Give answers in terms of one or more of the quantities $m$, $l$, $F_0$, and $\theta_0$.)
    (b) If a force $F_0$ is applied to each mass normal to the rod, the metal ribbon twists by an angle $\theta_0$. The ribbon will exert a reverse torque $T = -k\theta$. Express $k$ in terms of the above quantities.
    (c) If the rod is released, how many oscillations will there be per second?
20. Consider two masses of $m$ each connected by a massless rod of length $l$. The rod is pivoted at a distance $\frac{1}{4}l$ from the top and oscillates with small amplitude about this point.
    (a) What is the moment of inertia $I$ about the point of oscillation?
    (b) What is the torque $T$ about the pivot point when at a small angle $\theta$ from the vertical?

(c) What is $d^2\theta/dt^2$ in terms of $I$ and $T$?
(d) What is the period of oscillation in terms of $l$ and $g$?
(e) What would be the period of oscillation if the top mass were removed?

21. Repeat Example 2 by applying $F_{\text{net}} = ma$ to $m_1$. Find $d^2x_1/dt^2$ as a function of $x_1$ and find the period of oscillation of $m_1$.

22. Three masses of $m$ each are connected by two identical springs as shown. Each spring has a force constant $k_0$. Let $\Delta x_1$, $\Delta x_2$, and $\Delta x_3$ be their displacements from their equilibrium positions. The two outer masses are pulled out the same amount and released; that is, $\Delta x_1 = -\Delta x_3$ and $\Delta x_2 = 0$. What is the period of oscillation in terms of $m$ and $k_0$?

23. In Problem 22 the end masses are held fixed and the central mass is displaced to the right. Then all three are released simultaneously. Because the center of mass must stay fixed, we have $\Delta x_1 + \Delta x_2 + \Delta x_3 = 0$. Use $(F_2)_{\text{net}} = ma_2$ to find $d^2x_2/dt^2$ as a function of $x_2$. Show that the period of oscillation is $T = 2\pi\sqrt{m/3k_0}$.

24. At $t = 0$ the bob of a "circular" pendulum has velocity $v_0 = iv_0$ when $x = 0$ and $y = y_0$ as shown. Assume that $x \ll l$ and $y \ll l$ for all values of $t$. (The bob is moving in an ellipse.)
(a) What is $x$ as a function of $t$ in terms of $l$, $y_0$, $g$, and $v_0$?
(b) What is $y$ as a function of $t$?
(c) What is the speed $v = \sqrt{v_x^2 + v_y^2}$ as a function of $t$?
(d) For what value of $v_0$ is $v$ independent of $t$?

25. Suppose the potential energy between two ions of mass $m$ each is of the form

$$U = \frac{a}{r^5} - \frac{b}{r}$$

(a) Find the equilibrium position in terms of $a$ and $b$.
(b) Show that the frequency of oscillation is

$$\omega = \sqrt{\frac{8b}{m}\left(\frac{b}{5a}\right)^{1/4}}$$

26. The potential energy due to the force between two atoms is $U(r) = B\{1 - \exp[-A(r - r_0)^2]\}$. If the mass of each atom is $m$ and the atoms are oscillating about the equilibrium position with small amplitude, what is the frequency of oscillation in terms of $m$, $A$, and $B$?

27. In Example 7 suppose the 10-g mass is suspended by six stretched springs of 10-m length, as shown, where $F_0 = 5$ N for each spring. If $m$ is pulled slightly out of the page and then released, what will be its frequency of oscillation?

28. The amplitude of a sound wave is 1000 times that of the minimum amplitude which can be heard. What is its intensity in decibels?

29. The sound intensity emitted by a loudspeaker is proportional to the square of the audio signal voltage. What is the difference in decibels between two audio signal voltages $V_1$ and $V_2$?

30. Suppose the eardrum of a rabbit has 10 times the area of a human eardrum and that it has the same thickness. Also assume that the eardrum kinetic energy at the threshold of sound is the same as for a human. What then is the intensity in decibels of the threshold of hearing for a rabbit?

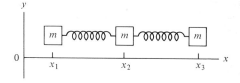

Problem 22

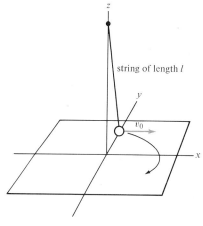

Problem 24

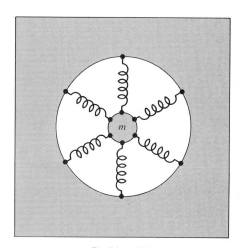

Problem 27

# Kinetic Theory

So far we have studied simple mechanical systems usually involving just one or two bodies or particles. However, a typical volume of gas contains an extremely large number of particles. It would be more than overwhelming to follow the motions of each gas particle or molecule; however, there are macroscopic quantities of practical value that can be calculated. Some such quantities are density, pressure, temperature, heat, entropy, internal energy, and mechanical energy. These quantities will be defined in this and the next chapter. Applying Newtonian mechanics to the particles of such a large system, we shall derive useful relations among the macroscopic quantities. The study of relationships among the macroscopic quantities is called thermodynamics. In this and the next chapter we shall show how the "laws" of thermodynamics can be derived from Newtonian mechanics. The microscopic or particle approach to thermodynamics is called kinetic theory. In Chapters 13 and 14 the laws of thermodynamics will be applied to practical systems, such as engines and cooling and heating systems. We shall see that according to the laws of thermodynamics most of the present-day engines and heating (and cooling) systems are considerably wasteful of energy.

## 12-1 Pressure and Hydrostatics

Any gas or liquid under pressure transmits a force to any surface containing the fluid. We shall restrict our discussion to fluids at rest (hydrostatics). The transmitted force will be perpendicular to the confining surface. The pressure on the surface is defined as

$$P \equiv \frac{\Delta F}{\Delta A} \quad \text{(pressure)} \tag{12-1}$$

where $\Delta F$ is the force on the surface area $\Delta A$.

One can also speak of the pressure inside the fluid. This could be measured by inserting in the fluid a small cube with thin walls filled with the fluid as shown in Figure 12-1. Since the fluid is at rest, the force $\Delta F$ of the fluid on each wall will be the same. The pressure in the region of the cube is $\Delta F/\Delta A$ where $\Delta A$ is the area of a face of the cube. We see that the internal pressure is the same in all directions and, ignoring gravity, must be the same throughout the entire volume no matter what its shape. We have just deduced what is called **Pascal's principle: If a pressure $P_0$ is applied to a portion of the surface of a fluid, it will be transmitted equally to all portions of the containing surface.**

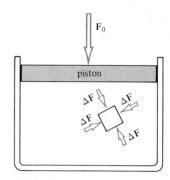

Figure 12-1. *Forces on a cube immersed in a fluid under pressure.*

**Example 1.** An automobile rack is lifted by a hydraulic jack that consists of two pistons connected by a pipe as shown in Figure 12-2. The large piston is 1 m in diameter and the small piston is 10 cm in diameter. If the weight of the car is $F_2$, how much smaller a force is needed on the small piston to lift the car?

ANSWER: Both pistons are walls of the same container and according to Pascal's principle experience the same pressure. Let $P_1 = F_1/A_1$ be the pressure on the small piston and $P_2 = F_2/A_2$ be the pressure on the large piston. Then

$$\frac{F_1}{A_1} = \frac{F_2}{A_2}$$

$$F_1 = F_2 \frac{A_1}{A_2} = 0.01 F_2$$

We see that the force on the pump piston only need be 1% of the weight of the car.

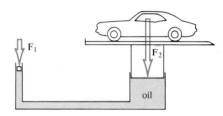

Figure 12-2. *Hydraulic jack used in lifting automobile.*

In the presence of gravity Pascal's principle applied to an incompressible liquid becomes

$$P = P_0 + \rho g h \quad (12\text{-}2)$$

where $P_0$ is the applied external pressure at the top, $\rho$ is the density, and $h$ is the distance from the top. Consider, for example, the column of liquid in Figure 12-3. In addition to the transmitted force $P_0 A$ at the bottom, there is the weight of the column of liquid, which is

$$mg = (\rho A h) g$$

The total force is

$$F = P_0 A + \rho g h A$$

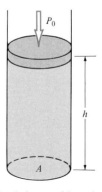

Figure 12-3. *Column of liquid of height h. External pressure $P_0$ applied at top.*

Dividing both sides by $A$ gives

$$P = P_0 + \rho g h$$

which is the total pressure at the bottom. Because each volume element cannot move sideways, this relation is independent of the shape of the container.

### The Barometer

The height of the earth's atmosphere is a few hundred kilometers. Since $P = \rho g h$ there should be a pressure $P_0$ at the surface of the earth equal to the height of the atmosphere times $g$ times the density of the air averaged over the height of the atmosphere. The numerical result is

$$P_0 = 1.01 \times 10^5 \text{ N/m}^2 \quad \text{(atmospheric pressure)}$$

This value of pressure is also called 1 atmosphere (atm):

$$1 \text{ atm} = 1.01 \times 10^5 \text{ N/m}^2$$

Suppose we take a tube filled with mercury ($\rho = 13.6 \times 10^3$ kg/m³) and invert it over a beaker of mercury as shown in Figure 12-4. The pressures at points $A$ and $B$ must be the same because these two points are at the same height. According to Eq. 12-2 $P_A = \rho g h$, where $h$ is the height of the mercury column. The pressure on the mercury–air surface must be $P_B = P_{atm}$, the atmospheric pressure. Thus

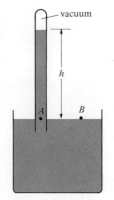

Figure 12-4. *A mercury barometer.*

$$\rho g h = P_{atm}$$

$$h = \left(\frac{1}{\rho g}\right) P_{atm} \qquad (12\text{-}3)$$

$$= \frac{1.01 \times 10^5}{13.6 \times 10^3 \times 9.8} \text{ m} = 0.76 \text{ m}$$

The height of the mercury column is proportional to the atmospheric pressure. A device, called a barometer, is used to measure the value of atmospheric pressure. The set-up in Figure 12-4 makes an excellent barometer.

According to Eq. 12-3 the height of a water barometer would be 13.6 times greater because water is 13.6 times less dense than mercury. Now replace the mercury in Figure 12-4 with water. The height of water in the tube would be 10.3 m or 34 ft. If the tube had originally contained air and a vacuum pump were attached to the top of the tube, as shown in Figure 12-5, the water could be pumped up only to a height of 10.3 m. Many wells are more than 10 m deep. How then can the water be pumped out? The trick is to use a submersible pump in the water at the bottom of the well.

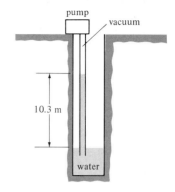

**Figure 12-5.** *The wrong way to get water out of a well, but it could be used as a barometer.*

**Example 2.** A house is connected to a city water main that is 100 m above the house in altitude (Figure 12-6). If the city water pressure is 4 atm, what will be the water pressure at the house?

ANSWER: Using Eq. 12-2 we have $P = P_0 + \rho g h$, where $P_0 = 4$ atm

$\rho g h = (10^3 \text{ kg/m}^3)(9.8 \text{ m/s}^2)(100 \text{ m}) = 9.8 \times 10^5 \text{ N/m}^2 = 9.8$ atm

The total static pressure is then

$$(4 + 9.8) \text{ atm} = 13.8 \text{ atm}$$

This is an uncomfortably high pressure for household use. A pressure reduction valve would be used at the entrance to the house.

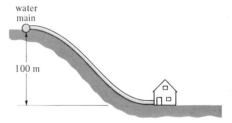

**Figure 12-6.** *Water pressure at house is higher than at water main when house is below water main.*

**Example 3.** What is the pressure at the center of the earth and at the center of the sun? Assume they are spheres of constant density where $R_e = 6.36 \times 10^6$ m, $R_s = 6.95 \times 10^8$ m, $\rho_e = 5.52 \times 10^3$ kg/m³, and $\rho_s = 1.42 \times 10^3$ kg/m³. The acceleration due to gravity at the surface of the sun is 274 m/s².

ANSWER: According to Eq. 12-2 the pressure at a depth $h$ is $P = \rho \bar{g} h$ where $\bar{g}$ is the average acceleration due to gravity. For a depth $h = R$, $P = \rho \bar{g} R$. We saw in Chapter 5 that $g$ increases linearly from the center of a sphere of uniform density; hence $\bar{g} = \tfrac{1}{2} g$ and $P = \tfrac{1}{2} \rho g R$.

For the earth

$$P = \tfrac{1}{2}(5.52 \times 10^3)(9.8)(6.36 \times 10^6) = 1.72 \times 10^{11} \text{ N/m}^2.$$

For the sun

$$P = \tfrac{1}{2}(1.42 \times 10^3)(274)(6.95 \times 10^8) = 1.35 \times 10^{14} \text{ N/m}^2.$$

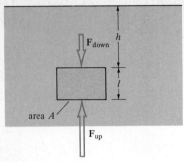

**Figure 12-7.** Block of volume $lA$ submerged in liquid of density $\rho$. $F_{up}$ and $F_{down}$ are forces of the liquid pushing on the block.

### Archimedes' Principle

Suppose a block of height $l$ and base area $A$, as shown in Figure 12-7, is submerged a distance $h$ in a liquid of density $\rho$. The force on the bottom surface will be

$$F_{up} = PA = \rho g(h + l)A$$

The force on the top surface will be

$$F_{down} = (\rho g h)A$$

The resultant force of the liquid on the block will be

$$F_{up} - F_{down} = \rho g l A = (m_{liq})g$$

where $m_{liq} = \rho l A$ is the mass of the liquid displaced by the block. Thus the block experiences an upward force equal to the weight of the water displaced. **Archimedes' principle** states: **A body immersed in a fluid is buoyed up by a force equal to the weight of the fluid displaced.**

$$F_b = (m_{liq})g \tag{12-4}$$

where $F_b$ is the buoyant force.

This principle, when applied to the special case of a floating body, states that a floating body must displace its own weight in water.

A popular thought question is: What happens to the water level in a glass of ice water as the ice melts? Does the melted ice raise or lower the water level? The answer is that the level remains the same, assuming that the ice had originally been floating in the water. Because an ice cube displaces its own weight in water, it will exactly fill up its own space in the water when it becomes water itself. A related type of problem concerns a boat on the Erie canal. There are a few places where the canal flows across a road on an aqueduct. Does the load on the bridge increase when the boat sails across the bridge? The answer is that the load on the bridge is the same with or without the boat as long as the boat is somewhere in the canal.

**Example 4.** Consider a hot-air balloon 10 m in diameter. If the air inside the balloon has a density 75% that of the outside air, how many passengers could it safely carry? The density of the outside air is 1.3 kg/m³.

ANSWER: Archimedes' principle applies to a body immersed in any fluid whether it be a liquid or a gas. The mass of the gas displaced is

$$M_{gas} = \rho_{air}(\tfrac{4}{3}\pi R^3) = (1.3)\tfrac{4}{3}\pi(5)^3 \text{ kg} = 680 \text{ kg}$$

According to Eq. 12-4 the bouyant force is

$$F_b = M_{gas}g = 680(9.8) \text{ N} = 6664 \text{ N}$$

Such a balloon would be in equilibrium if it carried a load of 680 kg including the mass of the air inside. Since the mass of the air inside is 510 kg, 170 kg is available for the external load. Such a balloon could carry two adults plus a lightweight passenger compartment.

## 12-2 The Ideal Gas Law

An ideal gas satisfies two conditions: (1) the volume of the gas molecules is much less than the volume occupied by the gas, and (2) the range of the force between two molecules is much less than the average distance between molecules. In this section we shall prove that for such a gas

$$PV = NkT \quad \text{(ideal gas law)} \quad (12\text{-}5)$$

where $P$ is the gas pressure, $N$ is the number of gas molecules in a volume $V$, $T$ is the absolute temperature and $k$ is a universal physical constant the same for all ideal gases. The ideal gas law is very useful. It permits the calculation of pressure, volume, density, or temperature of a sample of gas independent of the kind of gas. Temperature will be defined in the next section.

**Example 5.** An air bubble triples in volume as it rises from the bottom of a lake to the surface. How deep is the lake?

ANSWER: Let $P_1$ be the pressure and $V_1$ the volume at the bottom of the lake. Then $P_2 = P_0$, the atmospheric pressure, and $V_2 = 3V_1$. Assuming the temperature of the bubble does not change as it rises, Eq. 12-4 gives

$$P_1V_1 = P_2V_2$$
$$= (P_0)(3V_1)$$

after substitutions for $P_2$ and $V_2$. We see that $P_1 = 3P_0$ at the bottom of the lake. The pressure difference between top and bottom is $2P_0$. According to Eq. 12-2 this must equal $\rho g h$. The answer is twice the height of a water barometer or 20.6 m.

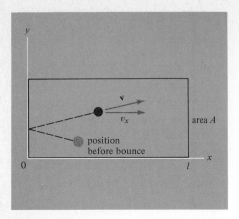

**Figure 12-8.** *Particle in box of volume lA just after bouncing off the left-hand side.*

In our derivation of the ideal gas law we shall treat the molecules as $N$ small, hard spheres confined to a box of volume $V$. By hard spheres we mean the collisions will be elastic. Let us first consider one such particle as it bounces off the left-hand side of the box in Figure 12-8. The average force the particle exerts on the wall in a time $\Delta t$ is

$$\bar{F} = \frac{\Delta p_x}{\Delta t}$$

The change in momentum due to the collision is

$$\Delta p_x = mv_x - (-mv_x) = 2mv_x$$

Since the time between collisions with this wall is

$$\Delta t = \frac{2l}{v_x}$$

the average force on the wall is

$$\bar{F} = \frac{(2mv_x)}{(2l/v_x)} = \frac{mv_x^2}{l} \quad \text{per particle}$$

For all $N$ particles in the box, the total force on the wall is

$$F = N\frac{\overline{mv_x^2}}{l}$$

where $\overline{v_x^2}$ is $v_x^2$ averaged over all the particles. It is called the mean squared velocity in the $x$ direction. Dividing both sides by the area $A$ of the wall gives the pressure:

$$P = \frac{Nm\overline{v_x^2}}{Al}$$

Now replace ($Al$) with the volume $V$:

$$P = \frac{Nm\overline{v_x^2}}{V}$$

or

$$PV = Nm\overline{v_x^2} \tag{12-6}$$

We see already that the product $PV$ for a given sample of gas stays constant as long as the kinetic energy of the particles stays constant. This result is called **Boyle's law.** Example 5 was solved using Boyle's law. We can express Eq. 12-6 in terms of $v$ rather than $v_x$ by noting that

$$\overline{v^2} = \overline{v_x^2} + \overline{v_y^2} + \overline{v_z^2}$$

Because particles bounce against all six sides the same way, we have

$$\overline{v_x^2} = \overline{v_y^2} = \overline{v_z^2}$$

Then

$$\overline{v^2} = 3\overline{v_x^2} \quad \text{or} \quad \overline{v_x^2} = \frac{\overline{v^2}}{3}$$

Now substitute $\overline{v^2}/3$ for $\overline{v_x^2}$ in Eq. 12-6:

$$PV = Nm\left(\frac{\overline{v^2}}{3}\right) \tag{12-7}$$

In the next section we shall see that the right hand side is by definition equal to $NkT$, where $T$ is the absolute temperature. Then $PV = NkT$, which is the ideal gas law. Starting with Newtonian mechanics at the microscopic level, we have derived an important relationship between the macroscopic quantities $P$, $V$, and $T$.

We shall define absolute temperature as being directly proportional to the average kinetic energy of the particles in the box:

## 12-3 Temperature

$$T \equiv \left(\frac{2}{3k}\right)\frac{m\overline{v^2}}{2} \quad \text{(definition of temperature)} \tag{12-8}$$

$$= \left(\frac{2}{3k}\right)\overline{K}$$

where $\overline{K}$ is the average kinetic energy per particle. The factor $(2/3k)$ is the proportionality constant. The size of $k$ is determined by the unit of temperature. The unit of temperature is established by assigning 100 degrees between the freezing and boiling temperatures of water at 1 atm of pressure. So $k$, which is called the **Boltzmann constant,** depends on measurements of the properties of water. The experimentally obtained value is

$$k = 1.38 \times 10^{-23} \text{ J/degree} \quad \text{(Boltzmann constant)}$$

If we eliminate $\overline{v^2}$ between Eqs. 12-7 and 12-8 we obtain

$$PV = NkT \quad \text{(ideal gas law)}$$

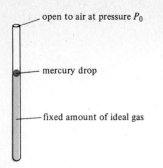

**Figure 12-9.** *Constant-pressure gas thermometer.*

### Thermometers

A primary measurement of temperature would use the defining Eq. 12-8, which requires measurement of translational kinetic energy of gas molecules. It is very difficult to "see" a gas molecule, much less measure its kinetic energy. Instead, we can use Eq. 12-5 to determine the temperature of a sample of ideal gas. It is easy to measure the product $PV$ of such a sample. As an example, we show a simple constant pressure gas thermometer in Figure 12-9. The volume of the gas in the tube will be

$$V = \left(\frac{Nk}{P_0}\right)T \quad \text{or} \quad V \propto T \tag{12-9}$$

Since the height of the mercury drop is proportional to $V$ it is also proportional to $T$. It is important to use an ideal gas in a gas thermometer. If in Figure 12-9 the fixed amount of ideal gas were replaced with a fixed amount of mercury, we would have the familiar mercury thermometer. Even though mercury is far from an ideal gas, its increase in volume is almost proportional to temperature when near room temperature. Thermometers that do not use ideal gases are ultimately calibrated in terms of accurate gas thermometers.

It is also possible to construct a constant-volume thermometer using an ideal gas. According to Eq. 12-5 $P = (Nk/V_0)T$. We see that if a pressure $P$ is applied such that the volume is kept at a value $V_0$, this pressure will be proportional to $T$. At normal outdoor temperatures and above, most gases behave as ideal gases. At very low temperatures, where air and even hydrogen gas liquefy, helium still behaves as an ideal gas. However,

**Figure 12-10.** *Comparison of Fahrenheit, Celsius, and Kelvin temperature scales.*

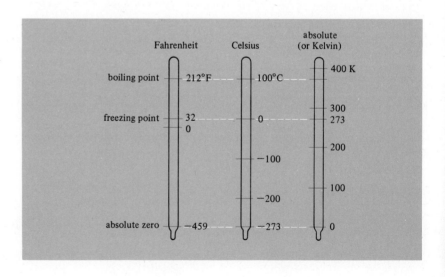

at 4° above absolute zero, helium liquefies and no longer behaves as an ideal gas.

**Example 6.** At the freezing temperature of water the density of air (and of nitrogen) at sea level is observed to be 1.255 kg/m³. The mass of a nitrogen molecule is $4.68 \times 10^{-26}$ kg. What is the absolute (Kelvin) temperature at 0° Celsius (where water freezes)?

ANSWER: Solving Eq. 12-5 for $N$ we have $N = PV/kT$. Now multiply both sides by $m/V$:

$$\frac{Nm}{V} = \frac{mP}{kT}$$

Since $Nm$ is the total mass of the gas contained in a volume $V$, the left hand side is the gas density $\rho$:

$$\rho = \frac{mP}{kT} \quad \text{(density of ideal gas)} \quad (12\text{-}10)$$

Solving for $T$:

$$T = \frac{mP}{k\rho} = \frac{(4.68 \times 10^{-26})(1.01 \times 10^5)}{(1.38 \times 10^{-23})(1.255)} = 273 \text{ deg}$$

We see that according to our definition of temperature water freezes at 273 degrees absolute. The absolute scale is also called the Kelvin scale. The symbol K is used and is the SI unit for temperature. For example, the temperature of freezing water or melting ice would be written 273 K. In the Celsius temperature scale, which is now used worldwide (although only unofficially in the United States), the temperature of freezing water is defined as zero degrees (written as 0°C). In both the Kelvin scale and the Celsius scale there are 100 degrees between the freezing and boiling of water, hence the degree size is the same in both scales. Comparison of the Celsius and Kelvin temperature scales is shown in Figure 12-10. Also shown is the Fahrenheit scale (still used in the United States). One Fahrenheit degree equals 5/9 of a Celsius degree.

**Example 7.** Estimate the temperature at the center of the sun by assuming it is a sphere of constant density consisting of an ideal gas of hydrogen atoms. $M = 2.00 \times 10^{30}$ kg and $R = 6.96 \times 10^8$ m. $m_H = 1.67 \times 10^{-27}$ kg is the mass of a hydrogen atom.

ANSWER: Solving Eq. 12-10 for $T$ gives

$$T = \frac{m_H P}{k\rho} \quad \text{where } \rho = \frac{M}{\frac{4}{3}\pi R^3} = 1.41 \times 10^3 \text{ kg/m}^3$$

For the pressure at the center we use the result from Example 3: $P = 1.35 \times 10^{14}$ N/m². Then

$$T = \frac{(1.67 \times 10^{-27})(1.35 \times 10^{14})}{(1.38 \times 10^{-23})(1.41 \times 10^3)} \text{ K} = 1.16 \times 10^7 \text{ K}$$

As will be discussed in Chapter 30, such high temperatures are sufficient to keep a slow but steady thermonuclear reaction going.

According to Eq. 12-8 the temperature $T = 0$ K occurs when all molecular motion stops. This is called absolute zero and is at $-273\,°$C.

We conclude this section on temperature with two examples that may help to reinforce this new concept. For our first example consider a box of volume $V_1$ with $N_1$ particles of average velocity $\overline{v_1^2}$. Now we double the number of particles in the box keeping $V_1$ and $\overline{v_1^2}$ the same and we ask what is the new temperature? According to the defining equation $T \propto \frac{1}{2}m\overline{v^2}$ per particle, there is no change in temperature. For our second example we allow the gas to expand freely into an evacuated chamber of equal volume by suddenly removing the partition in Figure 12-11. Then the final volume is twice the original volume. Now what is the final temperature? Again $\overline{v^2}$ remains the same. So if the average kinetic energy per particle remains the same, so does the temperature. From this example we see that the free expansion of an ideal gas results in no temperature change.

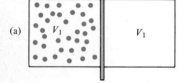

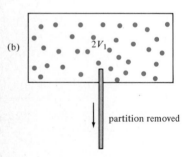

**Figure 12-11.** *Free expansion of a gas. In (b) the partition has been removed.*

## 12-4 Equipartition of Energy

We take it for granted that if two or more bodies at differing temperatures are put into contact and isolated from everything else, they will, after a sufficient time, all reach the same temperature. The bodies may be solids or liquids as well as gases. This is so common an everyday experience that it was not considered worthy of stating as a basic "law" of physics until after the first and second laws of thermodynamics had been formulated. Rather than renumbering the first and second law, the law that isolated bodies in contact reach thermal equilibrium was designated as the zeroth law of thermodynamics.

## Zeroth Law of Thermodynamics

If bodies 1 and 2 are in thermal equilibrium, and 2 and 3 are in thermal equilibrium, then 1 and 3 will be in the same thermal equilibrium as if they were in contact. If the bodies in Figure 12-12 are containers of three different kinds of ideal gases, we are claiming that the average translational kinetic energy per molecule will be the same for all three. Ignoring everyday experience, it is not at all obvious that $\frac{1}{2}m_1\overline{v_1^2} = \frac{1}{2}m_2\overline{v_2^2}$ when two different gases are in contact (or mixed in the same container). As part of our program of deriving thermodynamics from Newtonian mechanics, we should prove that $m_1\overline{v_1^2} = m_2\overline{v_2^2}$ when particles of masses $m_1$ and $m_2$ are mixed together in the same box. In the next paragraph we shall attempt to justify this result.

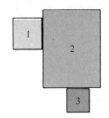

Figure 12-12. *Three bodies in thermal equilibrium.*

Consider a particle of mass $m_2$ in a box containing $N_1$ particles of mass $m_1$. The relative velocity between a particular particle of mass $m_1$ and $m_2$ is $\mathbf{v}_{rel} = (\mathbf{v}_1 - \mathbf{v}_2)$. This has the same magnitude and direction whether viewed in the lab system or any other frame of reference. The direction of motion of the center of mass of the two particles is in the direction of the vector $(\mathbf{p}_1 + \mathbf{p}_2)$. Since this direction is independent of the direction of $\mathbf{v}_{rel}$, we have

$$\overline{(\mathbf{p}_1 + \mathbf{p}_2) \cdot \mathbf{v}_{rel}} = 0$$

$$\overline{(\mathbf{p}_1 + \mathbf{p}_2) \cdot (\mathbf{v}_1 - \mathbf{v}_2)} = 0$$

$$\overline{\mathbf{p}_1 \cdot \mathbf{v}_1} - \overline{\mathbf{p}_1 \cdot \mathbf{v}_2} + \overline{\mathbf{p}_2 \cdot \mathbf{v}_1} - \overline{\mathbf{p}_2 \cdot \mathbf{v}_2} = 0$$

The average $\overline{\mathbf{p}_1 \cdot \mathbf{v}_2}$ is zero because for a given $\mathbf{p}_1$ each component of $\mathbf{v}_2$ is as often negative as positive. Similarly, $\overline{\mathbf{p}_2 \cdot \mathbf{v}_1} = 0$. Then

$$\overline{\mathbf{p}_1 \cdot \mathbf{v}_1} - \overline{\mathbf{p}_2 \cdot \mathbf{v}_2} = 0$$

$$\overline{p_1 v_1} = \overline{p_2 v_2}$$

$$m_1 \overline{v_1^2} = m_2 \overline{v_2^2}$$

which completes the "proof."

## Equipartition of Energy

The preceding derivation shows that all particles in thermal equilibrium have the same average translational kinetic energy per particle independent of the particle mass. But what about rotational and vibrational kinetic energy? Certainly two dumbbell-shaped molecules would start

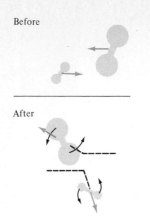

**Figure 12-13.** *Two molecules set into rotation by a collision.*

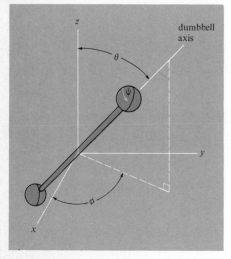

**Figure 12-14.** *Three angles—$\theta$, $\phi$, and $\psi$—are required to specify the orientation of a dumbbell.*

## 12-5 Kinetic Theory of Heat

rotating after a collision as seen in Figure 12-13. In Example 8 of Chapter 10 we found that when a dumbbell-shaped molecule was hit by a single particle, the dumbbell molecule acquired equal amounts of translational and rotational kinetic energy in the collision. A generalization of the derivation in the previous paragraph gives the result that the average kinetic energy for each degree of freedom will be the same for all particles. This result is called the equipartition of energy. The general derivation is too complex to present here. The number of degrees of freedom available to a body equals the number of independent coordinates needed to specify its position in space.

The average translational kinetic energy is obtained from Eq. 12-8:

$$\tfrac{1}{2}m\overline{v^2} = \tfrac{3}{2}kT$$

This corresponds to three degrees of freedom since the three coordinates $(x, y, z)$ are needed to specify the position of the center of a molecule. Thus the average energy per degree of freedom is $\tfrac{1}{2}kT$ per particle.

$$\text{av. energy per degree of freedom} = \tfrac{1}{2}kT \quad \begin{array}{c}\text{(law of equipartition}\\ \text{of energy)}\end{array} \quad (12\text{-}11)$$

There are also three degrees of freedom to describe the angular orientation of a rigid body about its center. For example, the orientation of the dumbbell in Figure 12-14 requires polar angle $\theta$ and azimuth angle $\phi$ to specify the orientation of the dumbbell axis. Then the angle $\psi$ is needed to specify the angular position of the dumbbell with respect to its axis. So we would expect small, hard dumbbells colliding in a box to have $\tfrac{3}{2}kT$ of rotational kinetic energy per particle on the average in addition to the $\tfrac{3}{2}kT$ of translational kinetic energy. The total kinetic energy per particle in a box containing $N$ molecules would be $3NkT$.

Actually the early models of molecules hypothesized that the molecules would be smooth; that is, there would be no surface friction. Then there would be no way to induce rotation about the dumbbell axis and there would be only two effective rotational degrees of freedom. Thus the total kinetic energy per particle for a diatomic molecule was expected to be $\tfrac{5}{2}kT$ on the average. Some violations of the equipartition of energy have been observed. These will be discussed in Section 13-3 of the next chapter.

On pages 132–133 internal energy is defined as the sum of kinetic and potential energy of the individual particles exclusive of any macroscopic energy (mass motion). Internal energy is the additional energy of the individual particles that is not tallied up when the system is viewed macroscopically. For a box containing $N$ rotation-free particles, the total internal energy would just be the translational kinetic energy or $U = \frac{3}{2}NkT$. The symbol $U$ is commonly used for internal energy of a body or system of particles. Unfortunately the same symbol is also commonly used for potential energy, which can lead to some confusion. If the box contains particles, such as $H_2O$ molecules, that are free to rotate in all three directions

$$U = K_{\text{translation}} + K_{\text{rotation}}$$
$$U = \tfrac{3}{2}NkT + \tfrac{3}{2}NkT = 3NkT \tag{12-12}$$

Diatomic molecules (smooth dumbbells) are expected to have

$$U = \tfrac{3}{2}NkT + \tfrac{2}{2}NkT = \tfrac{5}{2}NkT$$

So far we have been giving ideal gas examples. Clearly more complicated systems have well-defined internal energies. If the internal energy of a system is increased, its temperature increases. The internal energy of a system or sample of gas depends on its mass, temperature, and pressure (or volume). It is some definite function of these variables.

## Heat Energy

If one rubs the surface of a water container, one is doing work against a dissipative or frictional force. The temperature (or internal energy) of the water will increase. The rubbing produces heat energy that is transferred to the water. In addition to the joule, there is a unit of heat energy called the calorie (cal), which is defined as that amount of energy required to raise the temperature of 1 g of water by 1°C (or 1 K). Specifically,

1 cal ≡ heat to raise temperature of 1 g
             water from 14.5 to 15.5°C     (definition of calorie)

In the SI system we have 1 kcal = heat to raise 1 kg of water by 1°C (or 1 K). A third unit of heat is the food calorie which is sometimes written with a capital C:

1 food calorie = 1 Calorie = 1 kcal

This is an obsolete unit which can cause considerable confusion and its use should be avoided. The burning of 1 g of body fat will release about 10 kcal of heat energy. Put another way, a 1000 Calorie (= 1000 kcal) meal could produce 100 g of body fat if all the food were converted and stored as body fat. We shall see in the next paragraph that 1000 kcal is enough energy to raise a human body a height of 7 km.

## Mechanical Equivalent of Heat

To avoid confusion and mixed units it is better for us to use the joule as a unit of heat energy rather than the calorie. The conversion between the two units is called the mechanical equivalent of heat. It is easily measured by doing a fixed amount of work ($F_f \cdot \Delta s$) on a sample of water. A standard laboratory experiment for physics students consists of a sample of water in a thin-walled copper vessel. A known frictional force is applied to the surface. The temperature increase caused by the known amount of work done by the frictional force is measured. The result is that 4.185 J of work is converted into 1 cal of heat.

$$1 \text{ cal} = 4.185 \text{ J} \qquad \text{(mechanical equivalent of heat)}$$

Heat is a transfer of "hidden" particle energy from one body to another. The internal energy (particle energy) of a first body may be increased by doing mechanical work on the body, or it may be increased by putting it into contact with a hotter second body. In the latter case heat energy flows into the first body from the second. The mechanism is equipartition of energy via particle collisions. The particles of the first body pick up energy via collisions with the more energetic particles of the second body.

---

**Example 8.** Assume an average person must dissipate energy at an average rate of 120 W in order to stay alive.
(a) How many kilocalories per day must this person consume to stay alive?
(b) If the person's mass is 60 kg, how high could he climb on the energy equivalent of one day's minimum kilocalorie intake? (Assume 10% conversion efficiency to gravitational potential energy.)

ANSWER: (a) The energy used in one day is (120 W)(86400 s/day) = $1.04 \times 10^7$ J/day
The calorie equivalent is

$$\frac{1.04 \times 10^7 \text{ J/day}}{4.185 \text{ J/cal}} = 2.48 \times 10^6 \text{ cal/day} = 2480 \text{ kcal/day}$$

(b) We equate $mgh$ to 10% of the result of part (a):

$$mgh = 1.04 \times 10^6 \text{ J}$$

$$h = \frac{1.04 \times 10^6 \text{ J}}{(60 \text{ kg})(9.8 \text{ m/s}^2)} = 1.76 \text{ km}$$

A mountain climber should be able to climb such heights per day on "double-sized" meals without losing weight. If the climber did not increase his diet, the extra 2480 kcal would have to come from his body fat. At 10 kcal per gram, he would lost about 250 g per day of body fat.

## Summary

The pressure on a surface $\Delta A$ is the force $\Delta F$ pushing out divided by $\Delta A$: $P \equiv \Delta F/\Delta A$

For a fluid of constant density in the presence of gravity, $P = P_0 + \rho g h$, where $P_0$ is the pressure at the top and $h$ is the depth or distance from the top.

The pressure at the bottom of the earth's atmosphere is $P_{atm} = 1.01 \times 10^5$ N/m².

If a body is immersed in a fluid, the fluid will exert a net upward force on the body equal to the weight of the fluid displaced.

An ideal gas consisting of $N$ small, hard spheres of mass $m$ contained in a volume $V$ will exert a pressure $P$, where

$$PV = Nm\frac{\overline{v^2}}{3}$$

In kinetic theory the definition of absolute temperature $T$ of an ideal gas is

$$T \equiv \left(\frac{2}{3k}\right)\frac{m\overline{v^2}}{2}$$

where $k \equiv 1.38 \times 10^{-23}$ J/K is the Boltzmann constant. If $\overline{v^2}$ is eliminated in the two preceding equations,

$$PV = NkT \quad \text{(the ideal gas law)}$$

Temperature can be measured by the height of a column of ideal gas at constant pressure or by the pressure of a constant volume of ideal gas.

It can be proved that temperature as defined above will be the same for two bodies left in contact long enough to reach equilibrium. This is called the zeroth law of thermodynamics. A related result is equipartition of energy which says that all degrees of freedom have the same energy content when equilibrium is reached.

The "hidden" energy of the individual particles is called internal energy. Heat energy is produced when work is done by frictional forces; 1 cal of heat raises the temperature of 1 g of water by 1°C (or 1 K).

## Exercises

1. A phonograph needle has a radius $R = 10^{-2}$ mm. It is rated at 2 grams. (The force it exerts is $mg$ where $m = 2$ g.) Assuming the area of contact is $\pi R^2$, what is the pressure exerted on the record?

2. An air bubble doubles in volume as it rises from the bottom of a lake. What is the depth?

3. If the earth's atmosphere were of uniform density ($\rho = 1.25$ kg/m³), what would be its thickness?

4. Suppose the pump in Figure 12-5 could only pump down to a pressure of 0.1 atm. What would be the height of the water column?

5. The density of ice is 0.9 g/cm³. What fraction of the volume of an ice cube is above the water line? Is the answer independent of the shape of the piece of ice?

6. How far below the surface of a lake must one dive in order to find the pressure 50% greater than on the surface?

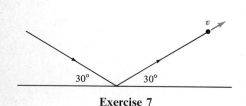

Exercise 7

7. A particle of mass $m$ and velocity $v$ strikes a surface at an angle of 30° as shown. It is reflected also at 30° with no change in velocity. What is its change in momentum?

8. A balloon has a total mass of 50 kg and a volume of 110 m³. It is tied to the earth with a rope.
   (a) If the rope stands vertical, what is the tension in it?
   (b) If there is a wind causing the rope to assume a direction at an angle of 30° to the vertical, what is the tension in the rope? (Assume the bouyant force is the same.)

9. A body when weighed under water weighs 200 N. Its normal weight is 300 N. What is its density and what is its volume?

10. A block of lead of density 11.5 g/cm³ floats on a pool of mercury of density 13.6 g/cm³.
    (a) What fraction of the block is submerged?
    (b) If the block has a mass of 2 kg, what force must be applied to keep it totally submerged?

11. A piece of wood whose density is 0.8 g/cm³ floats in a liquid of density 1.2 g/cm³. The total volume of the wood is 36 cm³.
    (a) What is the mass of the wood?
    (b) What is the mass of the displaced liquid?
    (c) What volume of wood appears above the surface of the liquid?

12. A person wishes to reduce weight by dieting. If he reduces his intake so that his body fat must supply 1000 kcal per day, how much weight does he lose per week?

13. An overweight person spends 1 h per day in reducing exercises. His body expends 400 W in additional effort during the exercise. Assuming no increase in diet, how much weight is lost per week?

14. A student takes a shower for 10 min using hot water at the rate of 15 liters per minute. The hot water temperature is 60°C and the cold water is 20°C. What is the total heat energy in joules used in this shower? If this energy costs 10¢ per kW h, what is the cost of the shower? What is the rate of heat energy used in watts?

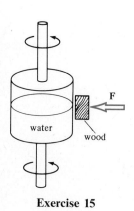

Exercise 15

15. A thin-walled copper drum contains 100 cm³ of water. It is of 1-cm radius and is being rotated at 2 rev/s. An external force of 10 N is pushing a wooden

block against the copper surface. Assuming all the heat produced flows into the water, what will be the temperature increase after 2 min?

16. What is the temperature increase in water flowing over a 50-m high waterfall?
17. A stream drops 200 m in altitude over a distance of 2 km. Assuming no heat flow into or out of the stream, what would be the temperature increase?
18. (a) Small hard disks are colliding inside a box. Assume the surface friction between the disks is nonzero. What will be the ratio of average rotational kinetic energy to translational energy?
    (b) The disks are floating on a thin film of air on an air table and are undergoing collisions. Now what is the ratio of average rotational to translational kinetic energy?
19. In a box of volume $V$ we have $N$ particles where the average kinetic energy per particle is $\varepsilon$. Give all answers in terms of $V$, $N$, $\varepsilon$, and $k$.
    (a) What is the total kinetic energy in the box?
    (b) What is the temperature in the box?
    (c) What is the pressure in the box?
    (d) If the volume is doubled by connecting to an empty box of the same volume, what happens to the temperature and pressure?
20. An ideal gas is contained in a vessel of fixed volume of 0°C and 1 atm pressure.
    (a) If the average velocity per molecule is doubled, what will be the new temperature?
    (b) If the average velocity per molecule is doubled, what will be the new pressure?
    (c) If the volume of the vessel is 1 liter, how many molecules does it contain?
21. An ideal gas in a box has $N$ molecules. Now double the number of molecules in the same box, while keeping the total kinetic or heat energy of the gas the same as before (the total energy content of the new amount of gas is the same as that of the original amount of gas).
    (a) What will be the ratio of the new pressure to the original pressure?
    (b) What will be the ratio of the new temperature to the original temperature?
22. Estimate the pressure and temperature at the center of Jupiter. Its mass and radius are $1.9 \times 10^{27}$ kg and $7.2 \times 10^4$ km, respectively.

## Problems

23. We wish to design a building intended for the surface of the moon to hold 0.5 atm of pressure. The roof is a cylindrical surface of 2-m radius. What must be the force $F$ per meter of length required to hold down the roof? If the wall and roof are of material of tensile strength $2 \times 10^9$ N/m², what must be the minimum permissible thickness $d$? (Tensile strength is the maximum force per unit area that can be applied to a rod of the material without incurring deformation or breakage.)
24. A steel gas cylinder is of 0.5-m diameter. If it is to hold a pressure of 150 atm, what must be the wall thickness $d$? The tensile strength of steel is $1 \times 10^9$ N/m². *Hint:* See Problem 23.
25. A sphere of radius $R$ is to hold a pressure $P$. If the tensile strength of the wall material is $S$ in force per unit area, what is the minimum wall thickness $d$ in terms of $R$, $P$, and $S$? (*Hint:* See Problem 24.)

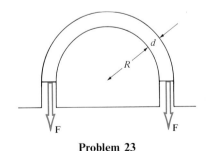

Problem 23

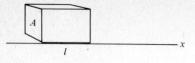

**Problem 26**

26. Consider a single photon in a box of volume $V = Al$. Assume the photon is moving parallel to the $x$ axis and the walls are perfectly reflecting. The photon has energy $E$ and momentum $p = E/c$.
   (a) What is the momentum transfer to the wall when the photon is reflected off of it?
   (b) What is the average pressure on the wall due to continued reflections of the photon?
   (c) Assume a total of $N$ photons of energy $E$ each and that $N/3$ are moving parallel to each axis. What is the product $PV$ in terms of $N$ and $E$?

27. Assume the density of the earth's atmosphere is proportional to the pressure; that is, $\rho/\rho_0 = P/P_0$ where $\rho_0$ and $P_0$ are the density and pressure at the surface. Then Eq. 12-2 gives $dP = -\rho g dh$ for the pressure change in moving a distance $dh$ away from the earth.
   (a) Prove that the pressure is of the form $P(h) = P_0 \exp(-h/h_0)$, where $P_0 = \rho_0 g h_0$.
   (b) At what height is the air pressure $\tfrac{1}{2}P_0$?
   (c) What would be the pressure at the top of Mt. Everest (8.8 km)?

28. Suppose all the particles in the box in Figure 12-7 are extremely relativistic. Then momentum $p \approx E/c$ where $E$ is the particle energy. Using $\overline{p_x v_x} = \tfrac{1}{3}\overline{pv}$, prove that $PV = E_t/3$ where $E_t$ is the total energy in the box. How does this compare with the nonrelativistic result where $E_t$ is the total kinetic energy in the box?

29. Prove that the exact relativistic result for Eq. 12-7 is

$$PV = \frac{N}{3}\overline{pv}$$

30. A 1-Mton bomb is exploded in an underground cavity of 200-m diameter.
   (a) If the energy released by 1 Mton is $4 \times 10^{15}$ J, what is the pressure in the cavity?
   (b) The cavity will rupture if the pressure excedes that of the surrounding rock. Assuming the rock density is $3 \times 10^3$ kg/m$^3$, how deep should the cavity be?

31. Derive a numerical relationship between the radius $R$ of the cavity in problem 30 and the underground depth $y$ of the cavity.

32. Consider a star made up of $N$ hydrogen atoms of mass $m_H$. Assume it is of uniform density of radius $R$.
   (a) What is the pressure at the center in terms of $m_H$, $N$, $R$, and $G$?
   (b) What is the temperature at the center in terms of $m_H$, $N$, $R$, and $G$?

33. How does the atmospheric density decrease with height $h$? Assume the atmosphere is all at the same temperature. Derive the formula

$$\rho = \rho_0 \exp\left(-\frac{mgh}{kT}\right)$$

# 13

# Thermodynamics

In this chapter we shall deal further with the macroscopic quantities of pressure, volume, temperature, heat, and energy. We shall discuss practical applications such as gasoline engines and their efficiency.

## 13-1 First Law of Thermodynamics

The first law of thermodynamics is just another version of the law of conservation of energy. Now that we have defined both internal energy and the energy of heat flow, we can classify the energy of a body into two parts: a macroscopic part and a microscopic part. The macroscopic energy is the energy of mass motion that we call mechanical energy, and the microscopic is the "hidden" particle energy that we call internal energy. When two bodies or systems at different temperatures are put into contact, heat energy $\Delta Q$ flows from the hotter to the colder. According to the law of conservation of energy, the heat that has flowed into a system must equal the increase in internal energy of the system plus the work done by the system on its external environment:

$$\begin{pmatrix}\text{heat into}\\ \text{system}\end{pmatrix} = \begin{pmatrix}\text{increase in}\\ \text{internal energy}\end{pmatrix} + \begin{pmatrix}\text{work done}\\ \text{by system}\end{pmatrix}$$

or

$$\Delta Q = \Delta U + \Delta W \quad \text{(1st law of thermodynamics)} \quad (13\text{-}1)$$

This law works backward as well as forward: if work is done on the system, heat can flow out of the system—then both $\Delta W$ and $\Delta Q$ would be negative. It is important to note here an inconsistency in notation: $\Delta Q$ and $\Delta U$ are changes of the system, whereas $\Delta W$ is not the work done on the system, but work done *by* the system. It might have made things easier to remember had $\Delta W$ been defined as work done on the system, but it is important in this case to adhere to common usage and convention.

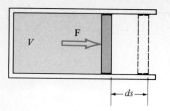

**Figure 13-1.** *Gas is pushing on a piston with force F. Gas does work F ds in pushing the piston a distance ds.*

Eq. 13-1 often appears in the form

$$dU = dQ - dW \tag{13-2}$$

If the system under study is a cylinder of gas pushing on a piston with force $F$, as shown in Figure 13-1,

$$dW = F\,ds = \frac{F}{A}(A\,ds) = P\,dV$$

is the work done by the gas. Then

$$dU = dQ - P\,dV \tag{13-3}$$

## 13-2 Avogadro's Hypothesis

In thermodynamics one very often works with moles and Avogadro's number. We devote this section to defining these quantities.

### The Mole

The mole is a standardized amount of gas or other substance especially used by chemists. **A mole of gas or other chemical element or compound is defined as having a mass in grams equal to its molecular weight.**

$$1\text{ mole} \equiv \text{molecular weight in grams} \quad \text{(definition of mole)}$$

The molecular weight of a compound is the sum of its constituent atomic weights. A table of atomic weights is given on page 600. The carbon isotope carbon-12 (symbol, $^{12}C$) is defined as having atomic weight of 12. Then hydrogen has an atomic weight of 1.008. This means that the mass ratio

$$\frac{M(^1H)}{M(^{12}C)} = \frac{1.008}{12}$$

So 1 mole of $^{12}C$ has a mass of 12 g, and 1 mole of hydrogen gas ($^1H_2$) has a mass of $(2 \times 1.008)\text{ g} = 2.016$ g.

### Avogadro's Number

In later chapters we shall learn of techniques by which the masses of elementary particles such as protons and electrons are measured. The result for the hydrogen atom is $m_H = 1.673 \times 10^{-24}$ g. Let $N_0$ be the number of atoms in a mole of atomic hydrogen ($M = 1.008$ g). Then

$$N_0 = \frac{M \text{ g/mole}}{m_H \text{ g/atom}} = \frac{1.008 \text{ g/mole}}{1.673 \times 10^{-24} \text{ g/atom}} = 6.02 \times 10^{23} \text{ atoms/mole}$$

This number is called Avogadro's number and is the number of molecules of any chemical compound in a mole of that compound.

---

**Example 1.** What volume is occupied by a mole of ideal gas at atmospheric pressure and $T = 273$ K ($0°C$)?

ANSWER: Put $N = N_0$ into Eq. 12-5 and solve for $V$:

$$V = \frac{N_0 kT}{P_0} = \frac{(6.02 \times 10^{23})(1.38 \times 10^{-23})(273)}{(1.01 \times 10^5)} \text{ m}^3$$

$$= 2.24 \times 10^{-2} \text{ m}^3 = 22.4 \text{ liters}$$

One liter is $10^3$ cm$^3$ or $10^{-3}$ m$^3$.

---

In 1811 Avogadro proposed that any two gases at the same temperature, pressure, and volume contain the same number of particles. This proposal was (and still is) known as Avogadro's hypothesis. Using Eq. 12-5 we have

$$N_1 = \frac{P_1 V_1}{kT_1} \quad \text{and} \quad N_2 = \frac{P_2 V_2}{kT_2}$$

Now if $P_1 = P_2$, $V_1 = V_2$, and $T_1 = T_2$, we have $N_1 = N_2$. We see that Avogadro's hypothesis is a direct consequence of the ideal gas law.

The ideal gas law applied to 1 mole of gas becomes

$$PV = N_0 kT \quad \text{(for 1 mole of ideal gas)}$$

The product $N_0 k$ is commonly called the gas constant $R$:

$$R \equiv N_0 k = (6.02 \times 10^{23})(1.38 \times 10^{-23})$$
$$= 8.31 \text{ J/(mole K)} = 1.99 \text{ cal/(mole K)}$$

## 13-3 Specific Heat

Specific heat is defined as $dQ/dT$ per gram or per mole of material as the case may be. The molar specific heat of a gas is the amount of heat required to raise the temperature of 1 mole of gas by 1 degree.

## Specific Heat at Constant Volume

It is common to use the symbol $C_v$ for the specific heat of one mole of a gas kept at a constant volume. Putting $dV = 0$ into Eq. 13-3 gives $dQ = dU$; hence

$$C_v = \frac{dU}{dT}$$

We saw in Eq. 12-12 that $U = \frac{3}{2}N_0 kT$ for a mole of a monatomic ideal gas. Then $dU/dT = \frac{3}{2}N_0 k$ or

$$C_v = \tfrac{3}{2}R = 3 \text{ cal/(mole K)}$$

Since diatomic molecules are of a dumbbell shape, we might expect three additional rotational degrees of freedom. Then $U$ would equal $3N_0 kT$ and $C_v$ would be $3R$ or twice as large as for a monatomic gas. However, measurements of $C_v$ at room temperature give $C_v \approx \tfrac{5}{2}R$ for 1 mole of a diatomic gas. This would make sense if one of the rotational degrees of freedom were missing. We shall see when we study quantum mechanics that if the atoms in the diatomic molecule are to be in their ground states, they cannot have angular momentum about any axis. This would correspond to the classical model of smooth dumbbells. Polyatomic molecules, on the other hand, do have the predicted three rotational degrees of freedom with $C_v = 3R$.

Another difficulty with classical mechanics is that it predicts that specific heat should be independent of $T$. However, for all but monatomic gases,

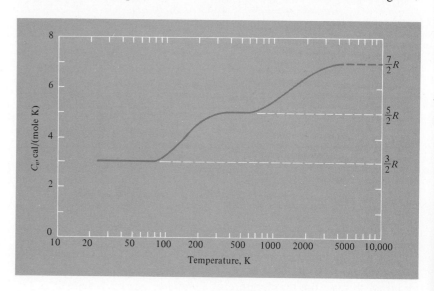

**Figure 13-2.** *Molar specific heat of hydrogen gas at constant volume as a function of T.*

$C_v$ increases with temperature. The specific heat of $H_2$ as a function of temperature is shown in Figure 13-2. At temperatures below 100 K, $C_v \approx \frac{3}{2}R$, which indicates that there are no rotational degrees of freedom at such low temperatures. Again, quantum mechanics must be invoked in order to explain why collisions between low temperature $H_2$ molecules do not excite any rotational motion. As we shall see in Chapter 26 on quantum mechanics, if a molecule is to have angular momentum, it must be at least $L_{min} = h/2\pi \approx 10^{-34}$ kg m² s⁻¹, where $h$ is Planck's constant. According to Eq. 10-21 the corresponding kinetic energy of rotation is $(K_{rot})_{min} = \frac{1}{2} L_{min}^2 / I$, where $I$ is the moment of inertia of the molecule. If $\frac{1}{2}kT$ is less than this, the energy of collisions is not sufficient to excite rotational motion in a typical collision.

---

**Example 2.** At what temperature is $\frac{1}{2}kT$ equal to the minimum rotational energy permitted for a hydrogen molecule?

ANSWER: Then

$$\frac{1}{2}kT = \frac{1}{2}\frac{L_{min}^2}{I}$$

$$T = \frac{L_{min}^2}{kI} \tag{13-4}$$

where $I = 2m\overline{R^2}$ is the moment of inertia. In the $H_2$ molecule $m = 1.67 \times 10^{-27}$ kg and $R \approx 5 \times 10^{-11}$ m. Then $I = 2(1.67 \times 10^{-27})(5 \times 10^{-11})^2 = 8.3 \times 10^{-48}$ kg m². Evaluating Eq. 13-4:

$$T = \frac{(10^{-34})^2}{(1.38 \times 10^{-23})(8.3 \times 10^{-48})} = 87 \text{ K}$$

---

We see from Example 2 why $C_v$ for $H_2$ starts increasing at about 100 K.

At temperatures above 2000 K, $C_v$ for $H_2$ has another increase: this time from $\sim \frac{5}{2}R$ to $\sim \frac{7}{2}R$. This experimental result suggests that two more degrees of freedom have become available. As explained in Section 11-5 on small oscillations, we should expect the two hydrogen atoms to oscillate back and forth at a frequency determined by the curvature of the potential energy curve. As explained in Example 4 of Chapter 11, these molecular vibrational frequencies are typically in the infrared or $\sim 10^{14}$ Hz. In chapter 26 we shall see that vibrational motion is also quantized with $(E_{vib})_{min} = hf$. For $f = 10^{14}$ Hz this is $(E_{vib})_{min} \approx 6 \times 10^{-20}$ J. If the average kinetic energy per molecule is greater than this, we would expect the vibrational motion to be excited. This situation would occur when

$$kT \approx 6 \times 10^{-20} \text{ J} \quad \text{or} \quad T \approx 4 \times 10^3 \text{ K}$$

So at temperatures above ~4000 K, the law of equipartition of energy tells us that the average $K_{vib}$ per molecule will be $\tfrac{1}{2}kT$. In addition to vibrational kinetic energy there is vibrational potential energy. We saw on page 223 that the average vibrational potential energy equals the average vibrational kinetic energy. Then the average internal energy per molecule is

$$U = \bar{K}_{trans} + \bar{K}_{rot} + \bar{K}_{vib} + \bar{U}_{vib}$$
$$U = \tfrac{3}{2}kT + \tfrac{2}{2}kT + \tfrac{1}{2}kT + \tfrac{1}{2}kT \quad \text{(per molecule)}$$
$$U = \tfrac{7}{2}N_0 kT = \tfrac{7}{2}RT \quad \text{(for 1 mole)}$$

and

$$C_v = \tfrac{7}{2}R$$

Our predictions for specific heats of different gases are summarized in the first column of Table 13-1. Table 13-2 shows that they agree remarkably well with the experimental measurements.

---

*__Example 3.__ What should be the molar specific heat of a solid crystal?

ANSWER: In this case each atom is "frozen" into a crystal lattice and can have vibrational motion in all three directions, which contributes $\tfrac{3}{2}kT$ of kinetic energy per atom. Since there is an equal amount of vibrational potential energy, the average internal energy per atom is $U = 3kT$. Per mole, $U = 3N_0kT$ and $C_v = 3R = 6$ cal/(mole K) independent of the atomic weight. This is called the law of Dulong and Petit (see Table 13-2).

---

From measurements of $C_v$ as a function of $T$, the temperature behavior of rotational and vibrational degrees of freedom was known well before the discovery of quantum mechanics. Of course such "strange" behavior was not understood until after the development of quantum mechanics.

## Specific Heat at Constant Pressure

If a mole of gas is held at constant pressure as in Figure 13-3 and heat is allowed to flow into the gas, there will be an increase in volume and an

**Figure 13-3.** *Gas in a cylinder is kept under constant pressure. The piston is free to move whenever the gas is heated or cooled.*

**Table 13-1.** *Specific heats per mole of different kinds of ideal gases (theoretical)*

| Type of gas | $C_v$ | $C_p$ | $C_p/C_v = \gamma$ |
|---|---|---|---|
| Monatomic | $\frac{3}{2}R$ | $\frac{5}{2}R$ | $\frac{5}{3}$ |
| Diatomic with rotation | $\frac{5}{2}R$ | $\frac{7}{2}R$ | $\frac{7}{5}$ |
| Diatomic with rotation and vibration | $\frac{7}{2}R$ | $\frac{9}{2}R$ | $\frac{9}{7}$ |
| Polyatomic with rotation and no vibration | $\frac{6}{2}R$ | $\frac{8}{2}R$ | $\frac{4}{3}$ |

**Table 13-2.** *Specific heats per mole at $20°C$ and $1.0$ atm, cal/(mole K)*

| Substance | $C_v$ | $C_p$ | $C_p/C_v$ |
|---|---|---|---|
| Monatomic gas | | | |
| He | 2.98 | 4.97 | 1.67 |
| Ar | 2.98 | 4.97 | 1.67 |
| Diatomic gas | | | |
| $H_2$ | 4.88 | 6.87 | 1.41 |
| $N_2$ | 4.96 | 6.95 | 1.40 |
| Polyatomic gas | | | |
| $CO_2$ | 6.80 | 8.83 | 1.30 |
| $NH_3$ | 6.65 | 8.80 | 1.31 |
| Solid | | | |
| Al | | 5.82 | |
| Cu | | 5.85 | |
| Ag | | 6.09 | |

additional amount of heat equal to $P\,\Delta V$ will be converted into mechanical work. According to Eq. 13-3

$$dQ = dU + P\,dV$$

Since $U$ depends only on the temperature, we have $dU = C_v\,dT$ and

$$dQ = C_v\,dT + P\,dV \tag{13-5}$$

For an ideal gas

$$V = \frac{RT}{P}$$

and

$$dV = \frac{R}{P}\,dT$$

Substitute this into Eq. 13-5:

$$dQ = C_v\,dT + P\left(\frac{R}{P}\,dT\right)$$

Dividing both sides by $dT$ gives

$$\frac{dQ}{dT} = C_v + R$$

which, by definition, is $C_p$, the specific heat at constant pressure. Hence,

$$C_p - C_v = R \quad \text{(for an ideal gas)} \tag{13-6}$$

This prediction agrees well with the measured values, some of which are displayed in Table 13-2.

## 13-4 Isothermal Expansion

The basis of most engines is a cylinder of gas with a moving wall or piston as in Figure 13-3. The gas could be a mixture of a hydrocarbon vapor and air. When ignited, this gas would be under pressure and would push the piston out. The piston could be connected by appropriate mechanical linkage to an engine crankshaft converting the mechanical energy $P\,dV$ to rotational energy.

We wish to calculate the mechanical work delivered to the outside when the piston moves out so that the volume expands from $V_1$ to $V_2$. We shall consider two common conditions: (1) isothermal expansion (the temperature of the gas is kept constant) and (2) adiabatic expansion (the gas is thermally insulated from the environment).

In order to have an isothermal expansion as shown in Figure 13-4, the cylinder walls must be kept at a constant temperature and the piston must move slowly in order to give the gas time to remain in thermal equilibrium with the walls. If the gas were allowed to expand quickly, it would tend to cool as it expanded. This is because some of the internal energy of the gas would be converted into the mechanical energy, $W = \int P\,dV$. Clearly, for an isothermal expansion, heat must flow from the constant temperature bath into the expanding gas in order to maintain its temperature. The heat flow must equal the mechanical work done by the gas. This result also follows from the first law of thermodynamics, $dQ = dU + P\,dV$. Since $dU = 0$ for an isothermal expansion, we have

$$dQ = P\,dV = dW$$

$$\Delta Q = \Delta W = \int_{V_1}^{V_2} P\,dV$$

For an ideal gas we substitute $P = NkT/V$ for the integrand:

$$\Delta Q = \Delta W = \int \left(\frac{NkT}{V}\right) dV = NkT \int \frac{dV}{V}$$

$$\Delta Q = \Delta W = NkT \ln \frac{V_2}{V_1} \qquad \text{(isothermal expansion of ideal gas)} \qquad (13\text{-}7)$$

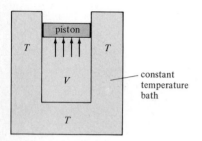

**Figure 13-4.** *Isothermal expansion. Gas in the cylinder is in continuous thermal equilibrium with the bath.*

## 13-5 Adiabatic Expansion

Normally, if the piston is allowed to expand quickly when the gas is under pressure, there will not be enough time for the gas to stay in thermal equilibrium with the cylinder walls; however, it will stay in thermal equilibrium with itself unless the expansion is unusually fast. So, for most expansions in engines, there is not enough time for heat to flow from the cylinder walls into the gas and we can put $dQ = 0$ into the equation $dQ = dU + P\,dV$:

$$dU + P\,dV = 0$$

For $dU$ substitute $C_v\,dT$:

$$C_v\,dT + P\,dV = 0 \quad \text{(per mole)} \quad (13\text{-}8)$$

We can get an expression for $dT$ by differentiating $RT = PV$:

$$R\,dT = P\,dV + V\,dP$$

Now solve this for $dT$ and substitute into Eq. 13-8:

$$C_v\left(\frac{P\,dV}{R} + \frac{V\,dP}{R}\right) + P\,dV = 0$$

$$\left(\frac{C_v + R}{R}\right) P\,dV + \frac{C_v V}{R}\,dP = 0$$

Now we make use of Eq. 13-6 to replace $(C_v + R)$ with $C_p$:

$$C_p P\,dV + C_v V\,dP = 0$$

$$\gamma\,\frac{dV}{V} + \frac{dP}{P} = 0 \quad \text{where } \gamma \equiv C_p/C_v$$

Integrating:

$$\gamma \int \frac{dV}{V} + \int \frac{dP}{P} = 0$$

$$\gamma \ln V + \ln P = \ln K$$

where $\ln K$ is a constant of integration.

$$\ln(PV^\gamma) = \ln K$$
$$PV^\gamma = K$$

We have shown that, for an ideal gas, the product $P$ times $V^\gamma$ must stay constant during an adiabatic expansion, thus

$$P_1 V_1^\gamma = P_2 V_2^\gamma \quad \text{(adiabatic ideal gas)} \quad (13\text{-}9)$$

It will prove useful to plot pressure versus volume for a fixed quantity of gas. Figure 13-5 shows both an isothermal and an adiabatic expansion from $V_1$ to $V_2$. For the adiabatic $P \propto 1/V^\gamma$ where $\gamma$ is always greater than 1; hence $P$ drops off faster than $1/V$ and thus the adiabatic curve drops below the isothermal.

In the case of an isothermal expansion we saw that heat was converted into mechanical work. In fact the amount of work done in expanding 1 mole of ideal gas from $V_1$ to $V_2$ is, according to Eq. 13-7:

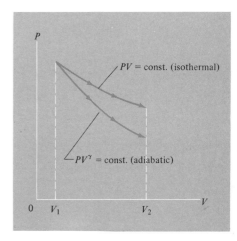

**Figure 13-5.** *Comparison of isothermal and adiabatic expansions both starting from same volume and pressure.*

$$\Delta W = RT \ln \frac{V_2}{V_1} \quad \text{(isothermal expansion of 1 mole)} \quad (13\text{-}10)$$

In an adiabatic expansion some of the internal energy of the gas is converted into mechanical work. The work done by 1 mole of ideal gas in expanding from $V_1$ to $V_2$ is

$$\Delta W = \int_{V_1}^{V_2} P\,dV \quad (13\text{-}11)$$

which is the shaded area under the curve in Figure 13-6. Since $PV^\gamma = P_1 V_1^\gamma$,

$$P = (P_1 V_1^\gamma) V^{-\gamma}$$

Now substitute this for $P$ in Eq. 13-11:

$$\Delta W = \int_{V_1}^{V_2} [(P_1 V_1^\gamma) V^{-\gamma}]\,dV = P_1 V_1^\gamma \left[ \frac{1}{-\gamma + 1} V^{-\gamma+1} \right]_{V_1}^{V_2}$$

$$\Delta W = \frac{P_1 V_1^\gamma}{\gamma - 1} \left[ \frac{1}{V_1^\gamma} - \frac{1}{V_2^\gamma} \right]$$

$$\Delta W = \frac{P_1 V_1}{\gamma - 1} \left[ 1 - \left(\frac{V_1}{V_2}\right)^{\gamma - 1} \right] \quad \begin{array}{l}\text{(for adiabatic expansion}\\\text{of ideal gas)}\end{array} \quad (13\text{-}12)$$

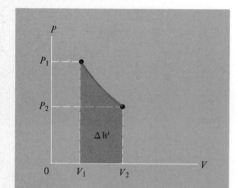

**Figure 13-6.** *Shaded area is work done by gas in adiabatically expanding from $V_1$ to $V_2$.*

**Example 4.** A gasoline engine has a compression ratio of 8 to 1, that is, $V_2/V_1 = 8$. What is the ratio of exhaust temperature to combustion temperature?

ANSWER: We assume an ideal gas adiabatic expansion. Then

$$P_2 V_2^\gamma = P_1 V_1^\gamma$$

$$\frac{P_2}{P_1} = \left(\frac{V_1}{V_2}\right)^\gamma$$

According to the ideal gas law

$$\frac{P_2}{P_1} = \frac{V_1 T_2}{V_2 T_1}$$

Equating this to $(V_1/V_2)^\gamma$ gives

$$\frac{T_2}{T_1} = \left(\frac{V_1}{V_2}\right)^{\gamma-1} \qquad (13\text{-}13)$$

The gas is mainly air, which is diatomic, and according to Table 13-1 has $\gamma = 1.4$. Hence

$$\frac{T_2}{T_1} = \left(\frac{1}{8}\right)^{0.4} = 0.435$$

**Example 5.** A one-cylinder motorcycle engine has a 6 to 1 compression ratio and maximum displacement (volume) of 200 cm³. When it runs at 3000 rpm (50 Hz), how much power is being delivered while the gas is expanding? Assume ideal gas adiabatic expansion and $P_1 = 20$ atm $\approx 2 \times 10^6$ N/m².

ANSWER: Using Eq. 13-12 we have

$$\Delta W = \frac{(2 \times 10^6 \text{ N/m}^2)(200 \times 10^{-6} \text{ m}^3)}{1.4 - 1}\left[1 - \left(\frac{1}{6}\right)^{0.4}\right] = 511 \text{ J}$$

This is during the time $\Delta t = 10^{-2}$ s. So the power

$$P = \frac{\Delta W}{\Delta t} = \frac{511}{0.01} = 5.1 \times 10^4 \text{ W} = 68 \text{ hp}$$

during expansion. If no work was done on the gas while being compressed, this would be an average power of 34 hp. However, as we shall see in the next section, some of this power must be used in compressing a new, cold gasoline–air mixture. Thus, the final average horsepower may be $\sim 10$ hp.

We see that the simple theory we have developed can be applied to engine specifications to predict engine performance. We know which parameters to change in order to increase the power output and, most importantly, we know the reasons why. This kind of basic understanding is necessary for applied scientists who are trying to design better engines.

## Gas Compression

Isothermal and adiabatic expansions are said to be reversible. This means that if one takes a movie of the expansion and then runs the film

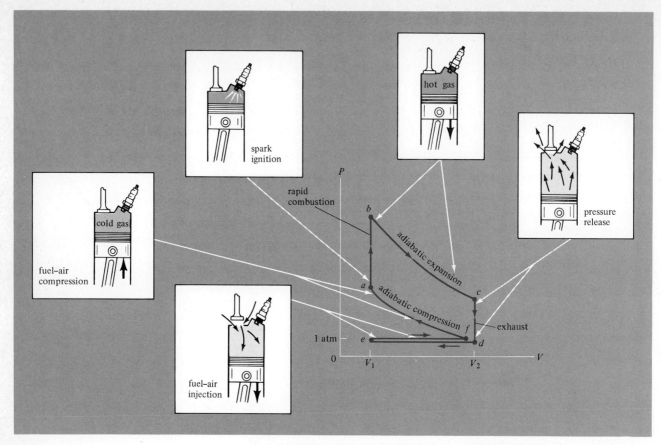

**Figure 13-7.** *PV diagram of the Otto cycle (four-stroke gasoline engine). Inserts show positions of piston and intake and exhaust valves.*

backwards, what one sees on the movie screen is a physical possibility. Clearly, if the piston is pushed in, work is being done on the gas, and in the case of an adiabatic compression this additional energy shows up as a temperature increase. Equations 13-7 to 13-13 hold whether the process is an expansion or a compression.

## 13-6 The Gasoline Engine

In this section we shall study the operation and efficiency of the common four-stroke gasoline engine used in automobiles. It is called four-stroke because the piston moves down twice and up twice during each complete cycle. Different stages of one complete cycle are shown in Figure 13-7. The stages are as follows:

$a$         Compressed fuel–air mixture is ignited by the spark plug.
$a \to b$   Sudden pressure increase after ignition.

$c$     End of adiabatic expansion; then the exhaust valve is opened.
$c \to d$   Heated, pressurized gas escapes quickly via open exhaust valve.
$e$     Piston has pushed out the remaining used gas; then exhaust valve closes and intake valve opens.
$f$     Fresh gas–air mixture has been sucked in; then the intake valve closes.
$f \to a$   The fresh mixture is compressed adiabatically.

A gasoline–air mixture releases 7.4 kcal of heat energy per gram of fuel that is burned. It is of great importance to know what fraction of this fuel energy can be converted into useful mechanical energy. This fraction is called the efficiency $\varepsilon$ of the engine.

$$\varepsilon \equiv \frac{\Delta W}{\Delta Q_{ab}}$$

where $\Delta W$ is the net mechanical work done by the engine in one cycle and $\Delta Q_{ab}$ is the heat of combustion of the fuel used in one cycle. Using Eq. 13-12 the net work done is

$$\Delta W = \frac{P_b V_1}{\gamma - 1}\left[1 - \left(\frac{V_1}{V_2}\right)^{\gamma-1}\right] - \frac{P_a V_1}{\gamma - 1}\left[1 - \left(\frac{V_1}{V_2}\right)^{\gamma-1}\right]$$

$$= \frac{(P_b - P_a)V_1}{\gamma - 1}\left[1 - \left(\frac{V_1}{V_2}\right)^{\gamma-1}\right]$$

For each mole of gas $(P_b - P_a)V_1 = R(T_b - T_a)$, so

$$\Delta W = \frac{R(T_b - T_a)}{(C_p - C_v)/C_v}\left[1 - \left(\frac{V_1}{V_2}\right)^{\gamma-1}\right]$$

$$\Delta W = C_v(T_b - T_a)\left[1 - \left(\frac{V_1}{V_2}\right)^{\gamma-1}\right] \quad (13\text{-}14)$$

where we have used the relation $R = (C_p - C_v)$ from Eq. 13-6. The heat required to raise the mole of gas from $T_a$ to $T_b$ is

$$\Delta Q_{ab} = C_v(T_b - T_a)$$

The efficiency is obtained by dividing Eq. 13-14 by this:

$$\varepsilon = 1 - \left(\frac{V_1}{V_2}\right)^{\gamma-1} \quad (13\text{-}15)$$

**Example 6.** What is the theoretical efficiency of a gasoline engine with an 8 to 1 compression ratio?

ANSWER: Putting $V_1/V_2 = \frac{1}{8}$ and $\gamma = 1.4$ into Eq. 13-15 gives

$$\varepsilon = 1 - (\tfrac{1}{8})^{0.4} = 0.56$$

It is important to note in Example 6 that 56% is the *theoretical* upper limit. In actual practice gasoline engines have about half this efficiency or less. This is due to several causes. Not all the fuel is completely burned. The cylinder walls are cooled; hence, some heat flows into the cooling system. In addition there are friction and turbulence. A gasoline or fuel oil heater can be close to 100% efficient in converting fuel energy to heat energy for heating a building; however, if gasoline is used in an internal combustion engine it is only ~25% efficient in converting fuel energy to mechanical energy. Most of the energy is being used to heat the outdoors.

## Summary

The first law of thermodynamics is a special case of the law of conservation of energy which takes into account the internal energy of a system. The first law states $\Delta Q = \Delta U + \Delta W$ where $\Delta Q$ is the heat flow into a system, $\Delta U$ is the increase of its internal energy, and $\Delta W$ is the work done by the system. For a gas pushing a piston, $dW = P\,dV$.

One mole of a compound is the molecular weight in grams. One mole of any compound contains $N_0 = 6.02 \times 10^{23}$ molecules (Avogadro's number).

Specific heat per mole is $dQ/dT$ where $dQ$ is the heat flow into the substance. $C_v$ is the specific heat per mole at constant volume and $C_p$ at constant pressure. For an ideal gas $(C_p - C_v) = N_0 k \equiv R = 1.99$ cal/K. $C_v = \tfrac{3}{2}R$ for a monatomic gas and $\tfrac{5}{2}R$ for a diatomic gas. Because of quantum mechanics, diatomic molecules cannot rotate or vibrate at very low temperatures; hence, $C_v$ is a function of temperature.

If an ideal gas is allowed to expand at constant temperature, the heat flow into the gas is

$$\Delta Q = NkT \ln \frac{V_2}{V_1} \qquad \text{(isothermal expansion)}$$

If the same gas is allowed to expand with no heat flow in or out, then

$$P_1 V_1^\gamma = P_2 V_2^\gamma \qquad \text{(adiabatic expansion)}$$

where $\gamma \equiv C_p/C_v$.

## Exercises

1. A 20-cm diameter air-filled balloon is 10 m under water. It is lowered a small distance such that the diameter is decreased to 19.8 cm. Using the definition on page 253, what is $\Delta W$ for the balloon? Is it positive or negative?
2. If the internal energy of the balloon in the above exercise increased 10 J while it was being lowered, what was the heat flow $\Delta Q$?
3. How many molecules are in one gram of (a) hydrogen gas, (b) water, (c) glucose ($C_6H_{12}O_6$)?
4. (a) A good vacuum pump can evacuate a 10-liter jar to $10^{-12}$ atm. How many molecules would be in the evacuated jar at room temperature?
   (b) If 1 liter of a certain gas has a mass of 0.0894 g at 0°C and a pressure of 1 atm, what is the gas?
5. The best vacuum man can achieve on earth is about $10^{-14}$ cm of mercury. At $T = 300$ K, how many molecules still remain in 1 cm³ of this "vacuum"? The vacuum of interstellar space contains about one proton per cm³.
6. The atomic weight of oxygen is 16. Consider 8 g of $O_2$ contained in an 8-liter vessel. The pressure is 1 atm.
   (a) How many moles of oxygen gas ($O_2$) are in the vessel?
   (b) How many molecules of $O_2$ are in the vessel?
   (c) What is the temperature and what is the total kinetic energy of the molecules?
7. What is the specific heat per gram of helium, hydrogen, and nitrogen when these gases are kept at constant volume?
8. Compare the minimum possible rotational energy permitted by quantum mechanics to the minimum possible vibrational energy for the $H_2$ molecule.
9. Consider a polyatomic molecule that has two independent modes of vibration. What would be $C_v$ for a mole of such a gas?
10. What is the specific heat per gram of copper?
11. Repeat Example 4 for a compression ratio of 6 to 1.
12. What is the theoretical gain in efficiency in going from a 6 to 1 to an 8 to 1 compression ratio for a gasoline engine?
13. Consider 1 mole of an ideal monatomic gas and 1 mole of an ideal diatomic gas adiabatically compressed, separately, by the same volume ratio. If both were initially at the same temperature, how will the temperatures compare after the compression?
14. Prove that $(T_1^\gamma P_1^{1-\gamma}) = (T_2^\gamma P_2^{1-\gamma})$ for an adiabatic expansion.

## Problems

15. (a) If 1 mole of oxygen gas is mixed with 2 moles of hydrogen, what is the specific heat at constant volume of this mixture of 18 g of gas?
    (b) If the mixture in (a) is ignited to form 18 g of water vapor, what is the specific heat at constant volume?
16. Suppose the equation of state is $P(V - V_0) = RT$ for 1 mole of a nonideal gas where $V_0$ is the volume of the $N_0$ molecules. What is $(C_p - C_v)$ for this gas?
17. In Problem 16, what is $\Delta Q$ when the gas expands isothermally from $V_1$ to $V_2$?
18. For the gas in Problems 16 and 17, show that the relation between $P$ and $V$ for an adiabatic expansion $P(V - V_0)^\gamma$ is a constant.

19. One mole of $N_2$ gas at $V_1 = 22.4$ liters and atmospheric pressure is expanded adiabatically to a volume $V_2 = 2V_1$. Then it is compressed isothermally back to its original volume.
   (a) What are $P_2$ and $T_2$?
   (b) What is $\Delta W_{12}$, the work done during the adiabatic expansion?
   (c) What is $\Delta W_{23}$, the work done during the isothermal compression?
   (d) What is the net work transferred to the outside?
   (e) What is $T_3$, the final temperature?
   (f) What is $C_v(T_1 - T_3)$?

20. One mole of $N_2$ gas at atmospheric pressure and $V_1 = 22.4$ liters is compressed adiabatically to $V_2 = \frac{1}{2}V_1$ and then expanded isothermally to its original volume.
   (a) What are $P_2$ and $T_2$?
   (b) What is the net work transferred to the outside?
   (c) What is $T_3$, the final temperature?
   (d) What is $C_v(T_3 - T_1)$?

21. Let us consider a gas where the particles are small rough spheres rather than small smooth spheres. (They can take up rotational motion.) They are of mass $m$.
   (a) How many degrees of freedom per particle are there?
   (b) At equilibrium what would be the ratio of average kinetic energy of rotation to translation?
   (c) What would be $C_v$ (specific heat per mole at constant volume)? Give answer in terms of the gas constant $R$.
   (d) What would be $C_p$ in terms of $R$?
   (e) The mean square velocity is $v_0^2$. What is the temperature in terms of $v_0$ and any other necessary constants?

22. One mole of an ideal monatomic gas goes through the reversible cycle illustrated in the figure. The gas is initially at $P_0$, $V_0$, and $T_0$ at point $a$. The volume of the gas at $b$ is $V = 32V_0$.

$a \to b$ constant $T$ process
$b \to c$ constant $P$ process
$c \to a$ adiabatic compression

(a) Complete the following table in terms of $P_0$, $V_0$, $R$, and $T_0$.

|  | $\Delta U$ | $\Delta Q$ | $\Delta W$ |
|---|---|---|---|
| $a \to b$ |  |  |  |
| $b \to c$ |  |  |  |
| $c \to a$ |  |  |  |
| $a \to b \to c \to a$ |  |  |  |

(b) Complete the following table using, among other things, the information from the table in (a).

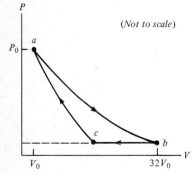

Problem 22

|   | P | V | T |
|---|---|---|---|
| a | $P_0$ | $V_0$ | $T_0$ |
| b |   | $32V_0$ |   |
| c |   |   |   |

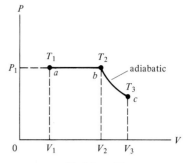

Problem 23

23. One mole of an ideal monatomic gas expands at constant pressure $P_1$ from $a$ to $b$.
   (a) What is $T_2$ in terms of $T_1$, $V_1$, and $V_2$?
   (b) How much work is done by the gas in expanding from $a$ to $b$? Give answer in terms of $P_1$, $V_1$, and $V_2$.
   (c) How much heat is supplied to the gas in going from $a$ to $b$? Give answer in terms of $R$, $T_1$, and $T_2$.
   (d) What is $V_3$ in terms of $V_2$, $T_2$, and $T_3$?

24. A heat engine has the cycle $a \rightarrow b \rightarrow c \rightarrow a$. The path $c \rightarrow a$ is along an isothermal. At point $c$ the pressure is $P_0$ times [$\frac{1}{2}$; $1/\sqrt{2}$; $\ln 2$; $1 - \ln 2$]. At point $b$ the temperature $T' = T_0$ times [$\ln 2$; $\sqrt{2}$; $2$; $2\sqrt{2}$]. The heat taken in by the engine in going from $a$ to $b$ is [$C_p(T' - T_0)$; $C_v(T' - T_0)$; $\int_a^b P\,dV$; none of these]. The mechanical energy delivered in one cycle is $P_0 V_0$ times [$1$; $\ln 2$; $1/\sqrt{2}$; $\frac{1}{2}$; $1 - \ln 2$; none of these]. (Assume an ideal gas.)

25. What is the average power output of the gasoline engine of Example 5 over the entire Otto cycle? Give answer in horsepower.

26. What is $T_1$ in Example 5? (Refer to Figure 13-7.)

27. Consider the reversible cycle of an *ideal monatomic gas* plotted on the $PV$ diagram. $V_0 = 100$ liters; $P_0 = 1$ A; $R = 0.082$ liter atm (mole K)$^{-1}$. The steps consist of:

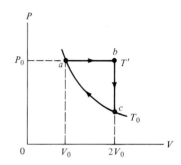

Problem 24

1 An isobaric (constant pressure) expansion ($a \rightarrow b$) at $P = P_0$.
2 An isothermal (constant temperature) expansion ($b \rightarrow c$) at $T = 600$ K.
3 An isochoric (constant volume) cooling ($c \rightarrow d$) at $V = 2V_0$.
4 An isothermal (constant temperature) compression ($d \rightarrow a$) at $T = 400$ K.

(a) Find $W_{ab}$, $W_{bc}$, $W_{cd}$.
(b) Given that $W_{da} = -(200/3)\ln 3$ liter-atm.
   (i) Find $Q_{ab}$, $Q_{bc}$, $Q_{cd}$, and $Q_{da}$ (taken as positive quantities).
   (ii) Indicate whether the heat is absorbed or rejected by the system for each $Q$.
(c) Write out an expression for the efficiency in terms of the symbols $Q_{ab}$, $Q_{bc}$, $Q_{cd}$, and $Q_{da}$. Give a numerical value.

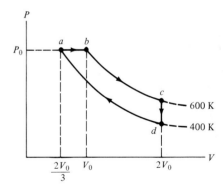

Problem 27

# Second Law of Thermodynamics

The second law of thermodynamics is a basic law of nature that pervades much of the everyday world and has deep practical and philosophical implications. We shall see in Section 14-5 that it can be derived from classical mechanics (or from quantum mechanics) by taking a microscopic rather than a macroscopic approach. Sir C. P. Snow in his book *The Two Cultures* gives an example of the cultural split between scientists and nonscientists that involves the second law of thermodynamics. He observes that both scientists and nonscientists can engage in a discussion of Shakespeare, but if the conversation should turn to relevant aspects of the second law of thermodynamics only those who are scientifically literate could follow the discussion.

In order to design optimal systems of fuel utilization and energy production, it is necessary to understand the severe restrictions imposed by the second law of thermodynamics. These restrictions are closely related to the Carnot engine, which is discussed next.

## 14-1 The Carnot Engine

In this section we shall learn about an engine that has a higher theoretical efficiency than an internal combustion engine; in fact, it is the most efficient that a heat engine could ever be. It is called the Carnot engine, or Carnot cycle. Its $PV$ diagram is shown in Figure 14-1. The Carnot engine uses a cylinder and piston; however, there are no valves and the same sample of gas or working substance is reused each cycle. The energy source (which could be gasoline or fuel oil) is used to keep a heat reservoir at a temperature $T_1$. There is a second reservoir at a lower temperature $T_2$. As an example the Carnot engine could be on the shore of a lake, which would serve as the low temperature reservoir at about 290 K, and it could use boiling water for the high temperature reservoir. Water from each

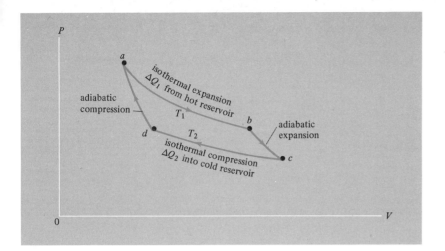

**Figure 14-1.** *The Carnot cycle. Heat $\Delta Q_1$ is extracted from $T_1$ reservoir. Heat $\Delta Q_2$ flows into the $T_2$ reservoir. Enclosed area is work done.*

reservoir is alternately circulated around the gas cylinder. Referring to Figure 14-1, we see that heat energy $\Delta Q_1$ is removed from the high temperature reservoir during an isothermal expansion and a smaller amount of heat $\Delta Q_2$ is transferred into the low temperature reservoir during the isothermal compression. This is illustrated schematically in Figure 14-2. Note that we have defined $\Delta Q_2$ to be positive when heat is delivered to the cold reservoir.

According to the first law of thermodynamics the heat loss over one cycle $(\Delta Q_1 - \Delta Q_2)$ must show up as mechanical energy $\Delta W$:

$$\Delta Q_1 - \Delta Q_2 = \Delta W$$

The efficiency is the fraction of heat extracted from the hot reservoir that goes into mechanical energy:

$$\varepsilon = \frac{\Delta W}{\Delta Q_1} = \frac{\Delta Q_1 - \Delta Q_2}{\Delta Q_1} = 1 - \frac{\Delta Q_2}{\Delta Q_1}$$

**Figure 14-2.** *Schematic representation of a Carnot engine. Quantities of heat and work are proportional to the width of the arrows.*

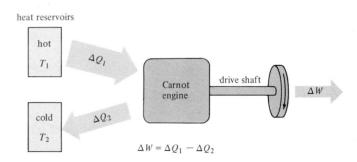

Using Eq. 13-7 we obtain, for an ideal gas,

$$\Delta Q_1 = NkT_1 \ln \frac{V_b}{V_a} \quad \text{(heat from hot reservoir into engine)}$$

$$\Delta Q_2 = NkT_2 \ln \frac{V_c}{V_d} \quad \text{(heat from engine into cold reservoir)}$$

Then

$$\varepsilon = 1 - \frac{T_2}{T_1} \frac{\ln(V_c/V_d)}{\ln(V_b/V_a)} \quad (14\text{-}1)$$

We can obtain these volume ratios by writing the equations of state for each of the four parts to the cycle:

$$P_a V_a = P_b V_b \quad \text{isothermal expansion}$$
$$P_b V_b^\gamma = P_c V_c^\gamma \quad \text{adiabatic expansion}$$
$$P_c V_c = P_d V_d \quad \text{isothermal contraction}$$
$$P_d V_d^\gamma = P_a V_a^\gamma \quad \text{adiabatic contraction}$$

Now we multiply all four equations together to obtain

$$P_a P_b P_c P_d V_a V_b^\gamma V_c V_d^\gamma = P_a P_b P_c P_d V_b V_c^\gamma V_d V_a^\gamma$$
$$V_b^{\gamma-1} V_d^{\gamma-1} = V_c^{\gamma-1} V_a^{\gamma-1}$$

then

$$\frac{V_b}{V_a} = \frac{V_c}{V_d}$$

Substitution into Eq. 14-1 gives

$$\varepsilon = 1 - \frac{T_2}{T_1} = \frac{T_1 - T_2}{T_1} = \frac{\Delta W}{\Delta Q_1} \quad \text{(efficiency of Carnot engine)} \quad (14\text{-}2)$$

If a Carnot engine operates between boiling and freezing water, the efficiency would be

$$\varepsilon = \frac{373 \text{ K} - 273 \text{ K}}{373 \text{ K}} = 0.27$$

In order to compare with the internal combustion engine of the previous chapter, we note that gasoline can heat a hot reservoir up to a temperature of $\sim 2700$ K. The outdoor air ($T_2 \sim 300$ K) could be used for the cold temperature reservoir. Then

$$\varepsilon = \frac{2700 - 300}{2700} = 0.89 \qquad (14\text{-}3)$$

compared to the theoretical maximum of 0.56 given by Eq. 13-15 for the internal combustion engine. We see that the Carnot cycle is 59% more efficient than the Otto cycle in this case. Such theoretical efficiencies are usually not achieved in practice because of losses due to friction, heat leakage, and irreversibility. Adiabatic and isothermal expansions and compressions are reversible only in the limit of very slow change.

## 14-2 Thermal Pollution

We shall see in Section 14-4 that the Carnot engine is the most efficient of all possible heat engines. Since most electric power plants use boiling water as the high temperature source, one might think that the efficiency could not exceed 27% as indicated in the above paragraph. However, if water is heated under pressure it will boil at considerably higher temperatures. Fossil fuel power plants use pressure to superheat steam to temperatures of 500 K or higher. Then the efficiency can be greater than 40%. Power plants that use nuclear fuel run at lower pressures and temperatures for safety reasons; hence, they are typically ~30% efficient as compared to the ~40% efficiency of fossil fuel plants.

In either case most of the energy supplied by the fuel is returned as heat to the low temperature reservoir. Such energy is completely wasted and ends up heating part of the local environment, such as a nearby body of water or the outside air (if cooling towers are used). Usually heating of the local environment is undesirable and is called thermal pollution.

We note that electric heating of homes is wasteful of fuel. If the fuel were used directly in the home for heating, one could achieve close to 100% efficiency, but the electric power plant is delivering electrical energy to the home at ~30% efficiency because of Eq. 14-2. So the same fuel ends up producing approximately one third as much heat in a home using electric heating. However, whether or not electric heating should be used in the home is not a question that physics can answer by itself. It is a question for society and governments. There is a complex of factors, such as pollution of the environment and depletion of limited natural resources, that must be considered. For example, coal may be more plentiful than fuel oil or natural gas, but using coal directly in the home might cause unacceptable air pollution.

## 14-3 Refrigerators and Heat Pumps

### Refrigerators

Since all isothermal and adiabatic expansions are reversible, we can run a Carnot engine backward. For example, at point $a$ in the cycle (see Figure

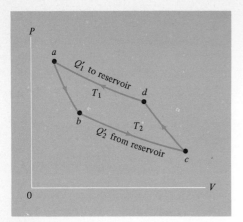

**Figure 14-3.** *The reverse Carnot cycle (refrigerator). Heat $Q'_2$ is extracted from $T_2$. Heat $Q'_1$ flows into $T_1$ reservoir.*

14-3), instead of letting the piston expand isothermally, we let it expand adiabatically to point $b$. Then from $b$ to $c$ we have an isothermal expansion. This is followed by an adiabatic and an isothermal compression until point $a$ is reached, which completes the cycle. Since everything here is reversible, Eq. 14-2 still holds:

$$\frac{\Delta W}{\Delta Q_1} = \frac{\Delta W}{\Delta W + \Delta Q_2} = \frac{T_1 - T_2}{T_1}$$

except that the quantities $\Delta Q$ and $\Delta W$ are now negative. Let us now use the notation $W' \equiv$ work done on the engine, $Q'_1 \equiv$ heat delivered to the hot reservoir, and $Q'_2 \equiv$ heat extracted from the cold reservoir. Then $W' = -\Delta W$, $Q'_1 = -\Delta Q_1$, and $Q'_2 = -\Delta Q_2$. Now substitute these quantities in the above equation:

$$\frac{(-W')}{(-W') + (-Q'_2)} = \frac{T_1 - T_2}{T_1}$$

Then

$$\frac{Q'_2}{W'} = \frac{T_2}{T_1 - T_2} \quad \text{(refrigerator coefficient of performance)} \quad (14\text{-}4)$$

The ratio $Q'_2/W'$ is an important quantity in refrigeration engineering. It is the heat extracted from a cold sample divided by the mechanical work used to extract that amount of heat. We are pleased to see that this ratio is usually greater than 1.

For a typical home refrigerator, the cold reservoir (including ice cube tray and freezer compartment) is $T_2 \approx 250$ K. The hot reservoir is the room air in the region of the heat exchanger which is $T_1 \approx 310$ K. Then Eq. 14-3 yields

$$\frac{Q'_2}{W'} = \frac{250}{310 - 250} = 4.17$$

We see that for every joule of electrical energy used to run the compressor, 4.17 J of heat is removed from the freezer compartment, provided an efficient Carnot cycle is used.

### Air Conditioners

In the case of air conditioners, the heat exchanger is put outside and the entire room is cooled. Then the ratio $Q'_2/W'$ is called the EER or energy efficiency ratio by engineers. Unfortunately, these engineers use a mixture of British and metric units for the EER. The convention is

$$\text{EER} = \frac{dQ'_2/dt \text{ (in Btu/h)}}{dW'/dt \text{ (in W)}} \qquad \text{where } 1 \text{ Btu/h} = 0.293 \text{ W}$$

The highest EER one can find for home air conditioners is 12 (Btu/h)/W.

$$12 \frac{\text{Btu/h}}{\text{W}} = 12 \frac{(0.293 \text{ W})}{\text{W}} = 3.5$$

According to Eq. 14-3 the best this could ever be is $T_2/(T_1 - T_2)$. Assuming a cooling range of $T_1 - T_2 \approx 20$ K,

$$\frac{T_2}{T_1 - T_2} \approx \frac{300 \text{ K}}{20 \text{ K}} \approx 15$$

which is significantly larger than the actual 3.5. This is due in part to the fact that $T_2$ inside the air conditioner is considerably lower than the desired final room temperature.

## Heat Pumps

Heat pump is just another name for a refrigerator which as we have just seen is a Carnot engine run backward. A refrigerator pumps heat out of a cooling compartment and into the surrounding air. A house could be heated by putting a refrigerator outdoors and extracting heat $Q'_2$ from the outdoors and delivering heat $Q'_1$ to the indoors. The heat delivery ratio is

$$\frac{Q'_1}{W'} = \frac{1}{\varepsilon} = \frac{T_1}{T_1 - T_2} \qquad (14\text{-}4a)$$

For example, assume an outdoor temperature of 250 K and indoor of 300 K, then

$$\frac{Q'_1}{W'} = \frac{300}{300 - 250} = 6$$

Then 5 J of heat would be removed from the cold outdoor air by using 1 J of mechanical energy to drive the compressor resulting in a delivery of 6 J of heat to the indoor air. In actual practice, home heat pumps are less than half as efficient. We see that a heat pump is an air conditioner connected backward to a house—it is a heat engine run backward and connected backward. By "connected backward," we mean that the refrigerator is connected to cool the outdoors rather than the indoors. A typical home installation is shown in Figure 14-4.

**Figure 14-4.** *A working heat pump connected to a home in Ithaca, N.Y.*

The heating of a building is usually accomplished by burning fuel oil or coal at the site (with about 70% conversion of chemical energy to useful heat), or at an electric power plant (with about 30% of the chemical energy resulting as heat at the building). In this sense a home furnace is more than twice as efficient as electric heating. However, we see that by using idealized heat pumps one could achieve much more heating from the chemical energy. Because oil or coal burns at a high temperature, about 85% of the chemical energy could be converted into mechanical energy (see Eq. 14-3). Then this mechanical energy $W'$ could be used to drive an idealized heat pump to deliver $Q'_1$ to the building in question.

If $T_1$ is room temperature (300 K) and $T_2$ is an outdoor temperature of 273 K, then Eq. 14-4a gives

$$\frac{Q'_1}{W'} = \frac{300 \text{ K}}{300 \text{ K} - 273 \text{ K}} = 11$$

So an original 1 J of chemical energy could deliver 0.85 J times 11, or 9.4 J of heat compared with the present 0.7 J. The ratio is 0.075. By this kind of figuring a typical furnace has a $7\frac{1}{2}$% efficiency and electric heating has an efficiency of ~3%. The American Physical Society has proposed measuring efficiencies of energy systems in this way, which compares the energy or heat delivered to the theoretical upper limit that could be delivered assuming idealized Carnot engines or heat pumps. This new way of measuring efficiency is called the second-law efficiency. The old way is called the first-law efficiency.

## 14-4 The Second Law of Thermodynamics

We have seen that heat engines can be built that will convert some $\Delta Q$ into $\Delta W$. Why can't we convert the heat stored in the oceans into $\Delta W$? Even with a 1% conversion efficiency we could get ~$10^{24}$ J, whereas only ~$10^{18}$ J are needed to supply 1 year of electric power for the United States. Solar radiation would resupply whatever small amount of heat would be removed from the oceans. It turns out that there is a fundamental reason why the enormous amount of heat energy stored in the oceans cannot be used. As we shall now see, the second law of thermodynamics prevents the direct conversion of heat into mechanical energy.

We start by listing four versions of the second law of thermodynamics that are mathematically equivalent:

1 Perpetual motion machines of the second kind are forbidden.
2 When two bodies of different temperatures are put into thermal contact, heat will flow from the hot to the cold temperature.

3 No cyclic heat engine operating between an upper temperature $T_1$ and a lower temperature $T_2$ can have efficiency greater than $(T_1 - T_2)/T_1$.
4 In a closed system the entropy cannot decrease.

We shall defer the discussion of entropy until Section 14-5. Perpetual motion machines of the first and second kinds are symbolized in Figure 14-5. A perpetual motion machine of the first kind would be a machine that runs by itself (is closed off from the environment) and delivers energy forever to the environment. According to the law of conservation of energy, this would require that an infinite source of energy be contained in a finite box. Clearly perpetual motion machines of the first kind directly violate the law of conservation of energy.

However, perpetual motion machines of the second kind do not violate the law of conservation of energy and are thus intellectually more intriguing. Such a machine would convert heat energy into mechanical energy. The source of the heat energy would continually cool down as it was delivering mechanical energy to the environment. If such devices could be designed, they could be put in the oceans, which contain $\sim 10^{26}$ J of heat energy, and transform this energy into mechanical energy. This is much more energy than has been used by mankind up until now. However the second law of thermodynamics states that it is impossible to transform directly the disordered thermal motion of molecules into the ordered motion of a machine or electric generator.

Actually it is possible to extract some energy from the ocean by making use of the fact that the surface temperature is higher than that of deeper water. Heat engines are being designed to cycle between upper and lower layers of water. Then one would have a heat engine operating between two temperatures $T_1$ and $T_2$ with a maximum efficiency $\varepsilon = (T_1 - T_2)/T_1$. The upper limit of efficiency would be $\sim \frac{1}{30}$ because $T_1 - T_2$ is $\sim 10$ K or less.

Now that we have discussed version 1 of the second law of thermodynamics, we would like to demonstrate that the other three versions are logically equivalent. If version 2 were violated, heat could flow from a cold to a hot reservoir. If this heat were used to run a heat engine we would have perpetual motion of the second kind, thus violating version 1.

Next we shall show that if there were a cyclic heat engine more efficient than the Carnot engine, heat could flow from cold to hot; that is, a violation of version 3 implies a violation of version 2. Suppose there was a super engine with efficiency $\varepsilon_s > \varepsilon$ where $\varepsilon = (T_1 - T_2)/T_1$ is the Carnot efficiency. If we connect the mechanical output or driveshaft of the super engine to the driveshaft of a Carnot engine and then use the super engine to drive the Carnot engine as a refrigerator between the same two heat reservoirs, the net effect will be a transfer of heat from the cold reservoir to the hot reservoir as shown in Figure 14-6.

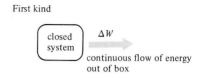
First kind

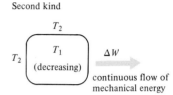
Second kind

**Figure 14-5.** *Schematic representation of perpetual motion machines of first and second kinds.*

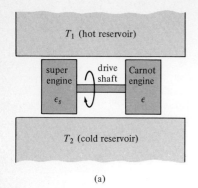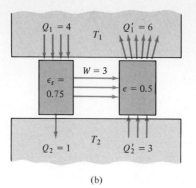

**Figure 14-6.** (a) Super engine having $\varepsilon_s = 0.75$ drives a Carnot engine of $\varepsilon = 0.5$ backward. (b) Here 4 units of heat power the super engine that powers the Carnot refrigerator with 3 units of mechanical energy. Net result is transfer of 2 units of heat from $T_2$ to $T_1$.

> **Example 1.** If heat $Q_1$ is extracted from the hot reservoir by the super engine, what is the net heat transferred by the system of two engines from the cold to the hot reservoir?
>
> ANSWER: The super engine will deliver mechanical energy $W = \varepsilon_s Q_1$ to drive the Carnot refrigerator which, in turn, delivers $Q'_1 = W/\varepsilon$ to the hot reservoir. The net heat delivered to the hot reservoir is
>
> $$Q'_1 - Q_1 = \frac{W}{\varepsilon} - Q_1 = \frac{\varepsilon_s Q_1}{\varepsilon} - Q_1 = \left(\frac{\varepsilon_s - \varepsilon}{\varepsilon}\right) Q_1$$
>
> Since no net work is done, this must be the same as the net heat extracted from the cold reservoir.

We see from Example 1 that if $\varepsilon_s - \varepsilon > 0$, version 2 will be violated.

The equivalence of version 4 to versions 1, 2, and 3 will be discussed in Section 14-5. A variation of the preceding reasoning leads to the conclusion that any two reversible engines must have the same efficiency. Merely connect the two engines as in Figure 14-6 by driving the lower efficiency engine backward with the higher efficiency engine running forward. The net effect would be a transfer of heat from cold to hot.

### Thermodynamic Temperature

Our original definition of temperature is based on particle energies (see page 241). There is an equivalent macroscopic definition. We have just proved that a Carnot engine using *any* working substance has efficiency

$$\frac{W}{Q_1} = \frac{T_1 - T_2}{T_1}$$

Using $W = Q_1 - Q_2$ from the first law of thermodynamics, we have

$$\frac{Q_1 - Q_2}{Q_1} = \frac{T_1 - T_2}{T_1}$$

or

$$\frac{T_1}{T_2} = \frac{Q_1}{Q_2} \qquad (14\text{-}5)$$

So the temperature ratio of any two heat reservoirs may be measured by measuring the heat transfers during one cycle of a Carnot engine. Actually Eq. 14-5 defines what is known as the thermodynamic temperature scale. Since we have derived Eq. 14-5 in terms of our original microscopic definition of temperature, we have proved that the two definitions of temperature are equivalent.

## 14-5 Entropy

Entropy is a measure of the amount of disorder in a system of particles. The greater the state of disorder in the positions and velocities of the particles in a system, the higher the probability $p$ that the system be in that particular state. The entropy $S$ of a system is by definition

$$S \equiv k \ln p \qquad \text{(definition of entropy)} \qquad (14\text{-}6)$$

where $k$ is the Boltzmann constant. According to the definition of probability a system will be in a state of higher probability more often than in a state of lower probability. So a system which is initially in a state of low probability will "seek" states of higher probability. Since $S$ increases with $p$, we have

$$\Delta S \geq 0 \qquad (14\text{-}7)$$

This is all there is to the derivation of version 4 of the second law of thermodynamics. We must still show that the other versions are equivalent. This is done at the end of this section.

From the defining equation we have

$$\Delta S = S_2 - S_1 = k \ln p_2 - k \ln p_1$$

$$\Delta S = k \ln \left(\frac{p_2}{p_1}\right) \qquad (14\text{-}8)$$

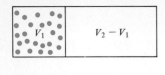

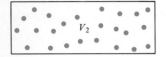

**Figure 14-7.** *Free expansion from volume $V_1$ to volume $V_2$ by removal of partition.*

We see that only probability ratios or relative probabilities are needed in order to calculate entropy change.

We shall now apply our formula for entropy change to the free expansion of a gas from an initial volume $V_1$ to a final volume $V_2$ as shown in Figure 14-7. The relative probability for a particle to be in $V_1$ compared to $V_2$ is

$$\left(\frac{p_1}{p_2}\right)_{\text{one particle}} = \frac{V_1}{V_2}$$

For $N$ particles in $V_1$ compared to $V_2$, we have the joint probability

$$\frac{p_1}{p_2} = \left(\frac{V_1}{V_2}\right)^N$$

Putting this into Eq. 14-8 gives

$$\Delta S = Nk \ln\left(\frac{V_2}{V_1}\right) \tag{14-9}$$

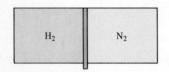

**Figure 14-8.** *Box with two gases.*

**Example 2.** A 2-liter box has a center barrier dividing it into two equal parts as shown in Figure 14-8. One side is filled with hydrogen gas and the other with nitrogen. Both gases are at room temperature and atmospheric pressure. The barrier is removed and the gases mix. What is the entropy increase due to the mixing?

ANSWER: The entropy increase for each gas is given by Eq. 14-9:

$$\Delta S = Nk \ln\left(\frac{V_2}{V_1}\right) = Nk \ln 2$$

The total entropy increase is twice this or

$$\Delta S = 2Nk \ln 2$$

At normal pressure and temperature 1 mole of gas occupies 22.4 liters (see page 255). Hence 1 liter contains $(1/22.4) N_0$ molecules. So

$$\Delta S = 2\left(\frac{N_0}{22.4}\right) k \ln 2 = 0.062 R = 0.124 \text{ cal/K}$$

We can simplify Eq. 14-9 by multiplying and dividing by $T$:

280 CH. 14/SECOND LAW OF THERMODYNAMICS

$$\Delta S = \frac{NkT \ln(V_2/V_1)}{T}$$

We note that the numerator is the same as $\Delta Q$ in Eq. 13-7. This is the heat that must be put into the initial system in order to arrive at the final system by a reversible process (an isothermal expansion). Making the substitution gives

$$\Delta S = \frac{\Delta Q}{T}$$

or

$$dS = \frac{dQ}{T} \tag{14-10}$$

where $dQ$ is heat put into the system along a reversible path. We have derived Eq. 14-10 for the special case of free expansion of an ideal gas. Using more complicated mathematics including mathematical statistics, it is possible to give a general proof of Eq. 14-10. This statistical approach to thermodynamics is called statistical mechanics.

Now that we have a macroscopic formula for entropy change, we can prove that heat must flow from hot to cold and not from cold to hot. In Figure 14-9 we consider two identical bodies initially at temperatures $T_1$ and $T_2$. These bodies are put into thermal contact. A short time later the temperatures will be $T_1 - dT_1$ and $T_2 + dT_2$ owing to the flow of heat $dQ_1 = -mc\, dT_1$ and $dQ_2 = +mc\, dT_2$, where $c$ is the specific heat per unit mass. Since $dQ_1 = -dQ_2$, we have $dT_1 = -dT_2 = dT$. The entropy change for each body is according to Eq. 14-10

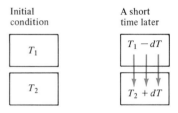

**Figure 14-9.** *Two identical bodies are put into thermal contact, and heat flows from one to the other.*

$$dS_1 = -\frac{mc\, dT}{T_1} \quad \text{and} \quad dS_2 = \frac{mc\, dT}{T_2}$$

The net entropy change is

$$dS = mc\, dT \left(\frac{1}{T_2} - \frac{1}{T_1}\right) \tag{14-11}$$

The temperature change is

$$dT = \frac{T_1 T_2}{mc}\left(\frac{dS}{T_1 - T_2}\right)$$

Since we have already proved that $dS$ must be positive, we have now shown that $dT$ must have the same sign as $(T_1 - T_2)$; that is, if $T_1 > T_2$, then heat flows from $T_1$ into $T_2$.

**Example 3.** Suppose the heat flow $dQ_1$ in Figure 14-9 was instead used to run a Carnot engine between $T_1$ and $T_2$. How much mechanical energy could have been obtained?

ANSWER: According to Eq. 14-2

$$dW = dQ_1 \left( \frac{T_1 - T_2}{T_1} \right) = T_2 \, dQ_1 \left( \frac{1}{T_2} - \frac{1}{T_1} \right)$$

According to Eq. 14-11 this is

$$dW = T_2 \, dS$$

We see that, by allowing the entropy increase in Figure 14-9, an amount of available mechanical energy equal to $T_2$ times the entropy increase was lost.

It can be shown that the result in Example 3 is just a special case of a more general theorem: In a closed system containing bodies of differing temperatures, whenever there is an entropy increase $dS$, there is a corresponding loss of available mechanical energy equal to $dS$ times the temperature of the coldest body. So we have another physical interpretation of entropy increase: loss of available energy per unit temperature. All this results from our starting definition of entropy increase as increase in probability (or increase in disorder of the constituent particles).

**Example 4.** Consider 1 kg of iron at 100°C put into contact with another kilogram of iron at 0°C. What is the entropy change of the system after the equilibrium temperature of 50°C has been reached?

ANSWER: Let $T_1$ be the initial temperature of the cold piece of iron and let $c$ be the specific heat per kilogram. Then

$$dS_1 = mc \frac{dT}{T}$$

$$\Delta S_1 = mc \int_{T_1}^{T_f} \frac{dT}{T} = mc \ln \frac{T_f}{T_1} \quad \text{where } T_f \text{ is the final temperature}$$

If $T_2$ is the initial temperature of the hot piece, its entropy change is

$$\Delta S_2 = mc \ln \frac{T_f}{T_2}$$

The net entropy change is

$$\Delta s = mc\left(\ln\frac{T_f}{T_1} + \ln\frac{T_f}{T_2}\right) = mc\ln\left(\frac{T_f^2}{T_1 T_2}\right)$$

$$= mc\ln\frac{323^2}{(273)(373)} = 0.024\,mc$$

We can use the law of Dulong and Petit to approximate the specific heat of iron (see page 258). One mole of iron would have a specific heat of 6 cal/K. Then 1 kg has $mc = 107$ cal/K and $\Delta S = 2.57$ cal/K.

**Example 5.** A motor delivers 1 J of mechanical energy to a Carnot refrigerator that removes heat from ice trays at 0°C and delivers it to the kitchen at 27°C. (a) What is the entropy change of the ice trays? (b) What is the entropy change of the total system?

ANSWER: (a) Using Eq. 14-4 we have

$$Q'_2 = \frac{T_2}{T_1 - T_2}W' = \frac{273}{27}(1\,\text{J}) = 10.1\,\text{J}$$

is the heat extracted from the ice trays. The entropy change is

$$\Delta S_2 = \frac{\Delta Q_2}{T_2}$$

where $\Delta Q_2 = -Q'_2 = -10.1$ J is the heat delivered to the ice trays.

$$\Delta S_2 = -\frac{10.1}{273}\,\text{J/K} = -3.7 \times 10^{-2}\,\text{J/K}$$

is the entropy change of the ice tray. Note that the entropy has been decreased rather than increased.
(b) We can use Eq. 14-5 to obtain the entropy change for the total system. According to this equation

$$\frac{\Delta Q_1}{T_1} + \frac{\Delta Q_2}{T_2} = 0$$

Hence the entropy change of the ice trays plus room is zero.

Example 5 shows that it is possible to decrease the entropy of a body without violating the second law. This is because the second law applies only to a closed system; when all parts of the system are tallied up, the net entropy change will either be zero or an increase. Certainly human activity on earth produces local entropy decreases. Refrigerators and heat pumps can move heat from a cold body to a hot body. A person can sort out bad peanuts from good by hand or by machine. Life itself consists of processes involving local entropy decreases. Whenever we observe a local increase in order as opposed to disorder, there is a local entropy decrease. But the total system, which must include our ultimate energy source, the sun, is increasing in net entropy.

## 14-6 Time Reversal

The second law of thermodynamics seems to imply a definite direction for the flow of time. If the direction of time were reversed, the total entropy of a closed system would decrease, heat would flow from cold to hot, and so on. Let us now consider in more detail the free expansion of a gas under time reversal. We shall study the situation shown in Figure 14-10 of two boxes, each 1 cm³, with a partition between them. If the pressure in box 1 is 1 atm, the number of particles in the box will be $6.02 \times 10^{23}$ divided by the number of cubic centimeters in 22.4 liters, or $2.7 \times 10^{19}$ particles/cm³. At the start box 2 is empty. We now remove the partition, and within a short time we shall find half of the particles in box 2. The gas has expanded into the vacuum. No matter how long we wait, the reverse of this process will never occur. Actually the number of particles in box 2 will fluctuate slightly. Mathematical statistics tell us that about 70% of the time the number of particles in a given volume will lie between $N - \sqrt{N}$ and $N + \sqrt{N}$, where $N$ is the average number. In this case we have

**Figure 14-10.** *An irreversible process. The gas is originally in box 1. When the partition is removed, it expands into empty box 2.*

$$1.35 \times 10^{19} \pm \sqrt{1.35 \times 10^{19}} = (1.35 \pm 0.00000000037) \times 10^{19}$$

We see that the fluctuations are so small that they are incapable of detection and that it is virtually impossible to have a fluctuation so large that no particles are left in box 2.

Suppose, however, that after removing the partition, and after half the particles have escaped from box 1 to box 2, time was suddenly stopped and made to run backwards. Physically time can never run backwards, but we can observe what it would be like by taking a movie of the experiment and running the film backwards in the projector. Box 2 would then spontaneously empty and produce a vacuum. We are now faced with a paradox. We know that in nature a box would never spontaneously empty and produce a vacuum when open to the outside air; yet, in the movie film run backwards, none of Newton's laws was violated. In fact, the movie film told us of a specific configuration of particle positions and velocities in

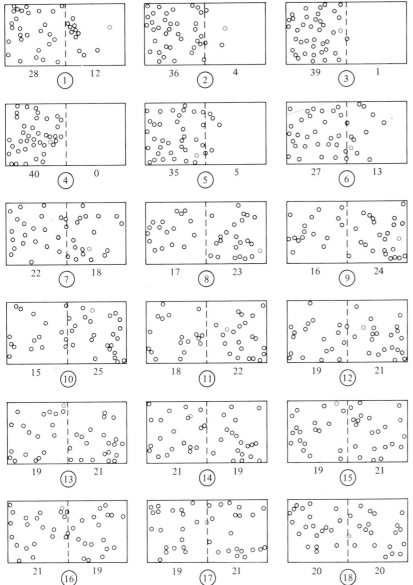

**Figure 14-11.** *Computer plots of 40 hard spheres colliding in a box. Plot 1 is a highly improbable state such that three plots later all particles are in left half of box. Then it would take ~$10^{12}$ more plots for this to happen again. There are equal time intervals between each plot. One of the spheres is in color so its movement can be followed. [Adapted from Berkeley Physics Course, Vol. 5, by F. Reif. Copyright © 1965 by McGraw-Hill, Inc. Used with permission of McGraw-Hill Book Company.]*

box 2 that would require the particles there to move and collide in such a way that they all would leave the box. During this process there would be no violation of any law of physics. The paradox is resolved by noting that for this one special configuration of particles in box 2 there are virtually an infinite number of configurations where the particles will stay almost equally in both boxes. Thus in practice this configuration, which allows all

SEC. 14-6/TIME REVERSAL  285

the particles to leave box 2, although permitted, never occurs. Hence the process of a gas expanding into a vacuum is irreversible, even though in principle it is possible to have a situation where the vacuum is "spontaneously produced."

Suppose somebody played a trick and carefully prepared an initial state such that after many collisions the particles did all empty out of one side of the box. Such an improbable initial state is shown in Figure 14-11 using a system of 40 hard spheres. However, if the computer calculations were continued even further, we would have to look over $\sim 10^{12}$ computer pictures before finding the 40 particles again all on the left side. At the rate of one computer plot per second this would take $\sim 10^5$ years.

The probability that $N$ particles all appear on the left side is $(\frac{1}{2})^N$. With $10^{19}$ particles, even if one started with the unnatural prepared state so that the right hand side of the box emptied itself, it would be quickly refilled and then "never" empty itself a second time. In this sense the second law of thermodynamics works whether time is flowing forward or backward. Put another way, take a movie of a free expansion of a large number of particles starting just as the partition is being removed. Now run the film backward and one will see half of the box emptying itself. This would be interpreted as a rare fluctuation or momentary decrease in entropy, but if the film could keep running backward (a computer could continue calculating collisions in "past time") entropy would increase and the second law of thermodynamics would be obeyed even though time was indeed running backward.

So far all the truly fundamental laws of physics that we have encountered are time reversible. Time reversibility (or time reversal invariance) means that if one reverses the directions of motion of all the particles (including their rotations), the same equations or laws of physics still hold. This very fundamental symmetry principle of nature has recently been checked by special experiments designed to show up possible violations. In 1964 a violation was found in the weak interactions. If the violation is confined to the weak interaction, it will not affect strong and electromagnetic interactions that determine nuclear and atomic physics. In addition, violations were found of two other very basic symmetry principles (conservation of parity and antiparticle symmetry) that were checked for the same reasons as time reversibility. The recent overthrow of these three symmetry principles is discussed in the last chapter.

## Summary

A Carnot engine operates between two heat reservoirs at temperatures $T_1$ and $T_2$. When run forward, in one cycle it extracts heat $Q_1$ from the $T_1$ reservoir and delivers $Q_2$ to the $T_2$ reservoir. The work done is $W = Q_1 - Q_2$.

The efficiency $\varepsilon \equiv W/Q_1 = 1 - T_2/T_1$. Also $T_2/T_1 = Q_2/Q_1$, which can be used as a method for measuring temperatures (the thermodynamic temperature scale).

When a Carnot engine is run backward, heat $Q_2'$ is extracted from the colder reservoir and $Q_1'$ is delivered to the warmer one. Then

$$\frac{Q_2'}{W'} = \frac{T_2}{T_1 - T_2}$$

The second law of thermodynamics can be derived by applying mathematical statistics to classical mechanics with the result that heat cannot spontaneously flow from a cold to a hot body. An equivalent statement is that no heat engine can be more efficient than a Carnot engine. Another equivalent statement of the second law is that the net entropy of a closed system cannot decrease. Entropy is defined as

$$S \equiv k \ln p$$

where $p$ is the probability that the system be in that particular state. An equivalent statement is that $dS = dQ/T$ where $dQ$ is the heat put into the system along a reversible path. Another equivalent statement is that whenever the entropy of a closed system increases, available mechanical energy $\Delta W$ is lost where $\Delta W = T' \Delta S$ and $T'$ is the lowest temperature body in the system.

## Exercises

1. Reverse the directions of all the arrows in Figure 14.2. Then which of the following are negative: $\Delta W, \Delta Q_1, \Delta Q_2$? Does the relation $\Delta W = \Delta Q_1 - \Delta Q_2$ still hold?
2. In Figure 14-1 let $\Delta Q_{ab}$ and $\Delta Q_{cd}$ be the heat flowing into the working substance in going from $a$ to $b$ and from $c$ to $d$, respectively. Express $\Delta Q_1$ and $\Delta Q_2$ in terms of $\Delta Q_{ab}$ and $\Delta Q_{cd}$. What is $\Delta Q_{bc}$?
3. A Carnot engine extracts energy from the ocean by operating on a 5° temperature difference between the surface water and cooler water underneath. If $10^6$ cal of heat energy is transferred to the surface by this engine every second, what is the maximum power output in watts?
4. In Figure 14-3 express $\Delta Q_{ad}$ and $\Delta Q_{bc}$ in terms of $Q_1'$ and $Q_2'$.
5. A Carnot refrigerator is designed to cool helium gas to 4 K. How many joules of mechanical energy are needed to extract 1 J from the helium when it is at 4 K? (The hot reservoir is at room temperature.)
6. Repeat Exercise 5 for a sample of helium at 0.1 K rather than 4 K.
7. A Carnot cycle refrigerator extracts 140 J of heat from a body that is being cooled. This heat is delivered to a heat exchanger at 27°C. The body being cooled is at an average temperature of 7°C while the 140 J is being extracted. How many joules of work are required?
8. In Figure 14-6 suppose that $\varepsilon_s = 0.55$ instead of 0.75. If $Q_1 = 4$, what is the net transfer of heat from $T_2$ to $T_1$?

9. Suppose in Example 2 that there are 0.5 liter of $H_2$ and 1.5 liters of $N_2$ instead of 1 liter of each. What then would be the entropy increase due to mixing?
10. In Figure 14-11 what is the probability that all 40 particles be in the left half of the box?
11. Suppose Figure 14-11 contained a total of five particles. What is the probability that all five be in the left half of the box?
12. Assume the Carnot engine in Figure 14-1 is using an ideal gas. Show that the efficiency is $\varepsilon = 1 - (V_b/V_c)^{\gamma-1}$.
13. Show that for a Carnot engine $\Delta W = \Delta Q_2[(T_1/T_2) - 1]$.
14. If 100 W of electrical power is used in a house to run a heat pump with an EER of 12, how many watts of heat are being delivered to the house?

## Problems

15. Show that the efficiency of the Otto engine in Figure 13-7 is $\varepsilon = 1 - (T_a/T_b)$.
16. The fuel used in a 100-MW power station generates $10^8$ W of mechanical power. Its overall efficiency is 0.4.
    (a) What is the rate of generation of waste heat?
    (b) If this waste heat is removed by cooling water, what must be the rate of flow if the water temperature is raised by 5°C?
17. A household refrigerator in order to make a tray of ice cubes extracts 50 kcal from the freezing compartment at 260 K. If the temperature of the room is 300 K, what is the minimum mechanical energy required to freeze the tray? (Assume an ideal Carnot refrigerator.) If this refrigerator can extract heat at the rate of 3 kcal/min, what is the electric power requirement in watts?
18. One mole of air at 1 atm and 300 K is adiabatically compressed to a pressure of 2 atm. What are its final volume and temperature? What is its change of entropy?
19. Two Carnot engines are run in series as shown. Engine 1 takes $Q_1$ from reservoir $T_1$ and puts $Q_2$ into $T_2$, which is then used as the heat input for engine 2. Engine 2 exhausts heat $Q_3$ into $T_3$. Find the overall efficiency; that is, the total work output divided by $Q_1$, the heat used to run the two engines.
20. It is desired to cool 1 mole of helium gas from room temperature (300 K) to 100 K by an ideal Carnot refrigerator. Assuming a constant specific heat of $\tfrac{5}{2}R$, how much work in joules must be done?
21. Repeat Problem 20 for a final temperature of 10 K rather than 100 K.
22. Estimate how many joules of heat must be removed from the air of a room 10 m × 5 m × 3 m in order to reduce its temperature by 20 K. If an air conditioner with an EER of 6 Btu h⁻¹ W⁻¹ must do this job in 30 min, how many watts of electricity does it use? The room is initially at 35°C.
23. Suppose an idealized heat pump is run by electricity rather than by a heat engine. If indoor temperature is 300 K and outdoor temperature is 273 K, what would be the second-law efficiency of such a heating system?
24. A car encounters a net opposing force of 500 N when traveling at 80 km/h. If its gas mileage is 40 mi/gal, what is the second-law efficiency of the system?
25. Suppose in Example 4 that 2 kg at 100°C is put into contact with 1 kg at 0°C. What is the final temperature and what is the net entropy change of the system?
26. What is the entropy decrease of the gas in Problem 20?

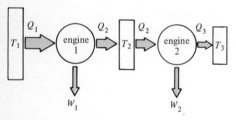

Problem 19

27. The following experiment is done to measure the specific heat of a metal alloy sample.

   The alloy sample has a mass of 200 g and is immersed in boiling water for a long time. It is then quickly removed from the boiling water and placed in an insulated calorimeter, which initially contained 300 g of cool water at a temperature of 20°C (the temperature of the room). The calorimeter temperature is observed to increase to 30°C, at which point it stops rising.
   (a) What is the specific heat of the alloy sample? Neglect the heat capacity of the calorimeter. (Specific heat of water = 1 cal g$^{-1}$ K$^{-1}$.)
   (b) Assuming that the specific heats of the alloy and the water are constant over the temperature intervals, calculate
      (i) The entropy change of the alloy, $\Delta S_A$
      (ii) The entropy change of the water, $\Delta S_W$
      (iii) The total change in entropy.

# Electrostatic Force

In the next six chapters we study what is perhaps the most important subject in all of physics—electromagnetic interaction. Not only does it explain all electrical phenomena, but it provides the force that holds matter together at the atomic and molecular level. Even the chapters on radiation and optics that follow the six on electricity are indirectly a study of the electromagnetic interaction, because light itself consists of electromagnetic radiation. The chapters that follow optics apply quantum mechanics to the electromagnetic interaction in order to explain atoms, molecules, and solids. So, in a sense, the remainder of this book is a study of electricity and its applications.

## 15-1 Electric Charge

So far the only basic interaction we have studied is the gravitational interaction. If we calculated the gravitational force of attraction between an electron and proton at a distance equal to the radius of the hydrogen atom, we would find

$$F = G\frac{m_\mathrm{p} m_\mathrm{e}}{R_\mathrm{H}^2} = 3.61 \times 10^{-47} \text{ N}$$

However, there is another attractive force between the electron and proton which is $8.19 \times 10^{-8}$ N, or $2.27 \times 10^{39}$ times larger! This *much* stronger force, which also follows an inverse square law, is called the electrostatic force (it is also called the electric force).

We know that all ordinary matter is made up of electrons, protons, and neutrons. So, if the forces between electrons and protons as well as between electrons and electrons are so many orders of magnitude larger than the gravitational force, why do we observe that the gravitational force between large bodies is stronger than the electrostatic force? It is because the electrostatic force between two electrons (or between two protons) is repulsive and of exactly the same strength as the attractive force between an electron and proton at the same distance. Since large bodies have the same number of electrons and protons, their very large attractive and repulsive electrostatic forces cancel, and all that is left over is the very weak gravitational force.

The source of gravitational force is what we called gravitational mass (or gravitational charge) on page 90. In the same way the source of the electrostatic force is called electric charge. (It is often called charge without the adjective "electric.") The mass or charge of a particle is just a mathematical attribute that tells how strongly that particle is affected by a gravitational force or by an electrostatic force, respectively. The two kinds of force are independent; hence, there is no fixed relationship between the mass and the charge of a body. Unlike mass, electric charge can be either positive or negative. Two charges of opposite sign experience an attractive force, whereas two charges of the same sign experience a repulsive force (see Figure 15-1).

The repulsive force between charges of the same sign can easily be demonstrated by rubbing two balloons with a wool cloth. A few of the outer electrons from the wool atoms will be captured by the atoms of the balloon; hence both balloons will be negatively charged. If one balloon is brought near the other, it will push the other one away without ever touching it (a case of force at a distance).

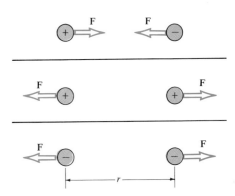

**Figure 15-1.** *Dependence of the direction of the electrostatic force on the signs of the two charges.*

### Quantization of Charge

Experiments show that no charged particle can have less charge than that of the proton or electron. This elementary unit of charge, equal to $1.60 \times 10^{-19}$ coulomb, is represented by the symbol $e$. Some elementary particles such as the neutron, photon, and neutrino have zero charge. Charged bodies can only have integral multiples of the charge $e$.

### Conservation of Charge

One of the most basic laws of physics is the law of conservation of charge first proposed by Benjamin Franklin in 1747. The law says that, in a closed system, the net amount of charge (the amount of positive charge minus the amount of negative charge) will stay constant. Even in the extreme case of annihilation of charged particles the law holds. If an electron is annihilated by a positron, both negative and positive charges are destroyed; however, the net charge of zero was the same before annihilation as it is after annihilation. The law of conservation of charge has been well tested by many sensitive experiments.

## 15-2 Coulomb's Law

As with Newton's universal law of gravitation, the force between two charged particles is proportional to the product of the two charges $q_1$ and $q_2$ and inversely proportional to the square of the distance $r$ between them:

$$F = k_0 \frac{q_1 q_2}{r^2} \quad \text{(Coulomb's law)} \tag{15-1}$$

where $k_0$ is a proportionality constant to be determined by experiment. It is called the Coulomb constant. The dependence of $F$ on $q$ and $r$ has been tested to extreme accuracy.

In the cgs system Eq. 15-1 is used to define the unit of charge by setting $k_0$ equal to 1 and adjusting the unit of charge so that two such charges 1 cm apart will experience a force of 1 dyne. This unit of charge is called the **statcoulomb.** Thus in the cgs system

$$F = \frac{q_1 q_2}{r^2} \quad \text{when } q \text{ is in units of statcoulombs}$$

In the mks or SI system, charge is defined via magnetic force between two identical currents. As we shall see in Chapter 17, this gives a much larger unit of charge, which is related to the statcoulomb via the speed of light. The mks unit of charge is called the coulomb, and we use the abbreviation C for it. As we shall prove in Chapter 17, the relation between coulomb and statcoulomb is

$$1 \text{ C} = 2.998 \times 10^9 \text{ statcoul}$$

We shall see that the conversion constant of $2.998 \times 10^9$ is exactly 10 times the speed of light.

Now we can evaluate $k_0$ in mks by solving Eq. 15-1 for $k_0$:

$$k_0 = \frac{Fr^2}{q_1 q_2}$$

Putting $q_1 = q_2 = 1$ statcoul, $r = 1$ cm, and $F = 1$ dyne:

$$k_0 = \frac{(1 \text{ dyne})(1 \text{ cm})^2}{(1 \text{ statcoul})^2} = \frac{(10^{-5} \text{ N})(10^{-2} \text{ m})^2}{[1 \text{ C}/(2.998 \times 10^9)]^2}$$

$$k_0 = 8.988 \times 10^9 \text{ N m}^2/\text{C}^2 \approx 9 \times 10^9 \text{ N m}^2/\text{C}^2$$

Not only is $9 \times 10^9$ of sufficient accuracy, but it is easy to remember. In the mks system it is a common convention to write the constant $k_0$ in the form $1/4\pi\varepsilon_0$. Then

$$F = \frac{1}{4\pi\varepsilon_0} \frac{q_1 q_2}{r^2}$$

where

$$\varepsilon_0 \equiv \frac{1}{4\pi k_0} = 8.854 \times 10^{-12} \text{ C}^2/(\text{N m}^2) \qquad (15\text{-}2)$$

This is called the permittivity of free space.

In this book we shall usually write equations of electricity using $k_0$ rather than $\varepsilon_0$. Not only will it simplify some of the calculations, but our equations will have the same form both in mks and cgs. To convert from mks to cgs, merely put $k_0 = 1$. For a person who goes deeper into physics, it is necessary to learn both the cgs (also called gaussian) and the mks versions of electromagnetic theory. In our presentation we will learn cgs at the same time as mks with no extra work.

---

\* **Example 1.** Two carbon spheres both have a small excess of electrons. What must be the ratio of electrons to protons in order to just cancel out the attractive gravitational force?

ANSWER

$$F_E = F_G$$

$$k_0 \frac{q_1 q_2}{r^2} = G \frac{m_1 m_2}{r^2}$$

where $q_1$ and $q_2$ are the charges and $m_1$ and $m_2$ are the masses of the two spheres.

$$\left(\frac{q_1}{m_1}\right)\left(\frac{q_2}{m_2}\right) = \frac{G}{k_0}$$

If we make the electron-to-proton ratio the same in both, we have

$$\frac{q_1}{m_1} = \sqrt{\frac{G}{k_0}}$$

Now

$$q_1 = (N_e - N_p)e$$

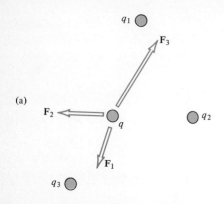

(a)

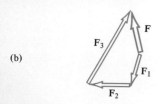

(b)

**Figure 15-2.** (a) *Forces on q due to $q_1$, $q_2$, and $q_3$.* (b) *Vector sum gives net force* **F**.

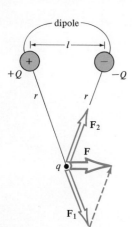

**Figure 15-3.** *Forces on charge q due to dipole $p = Ql$.*

where $N_e$ and $N_p$ are the numbers of electrons and protons.

$$m_1 = N_p m_p + N_n m_n + N_e m_e$$

where $m_p$, $m_n$, and $m_e$ are the proton, neutron, and electron masses. Using $m_p \approx m_n \gg m_e$ and $N_p = N_n$, we have $m_1 \approx 2N_p m_p$. Then

$$\frac{q_1}{m_1} = \frac{(N_e - N_p)e}{2N_p m_p} = \sqrt{\frac{G}{k_0}}$$

$$\frac{N_e - N_p}{N_p} = \frac{2m_p}{e}\sqrt{\frac{G}{k_0}} = 1.8 \times 10^{-18}$$

We see that the effects of gravity can be canceled out if there is only one extra electron for every $\sim 5 \times 10^{17}$ protons.

### Principle of Superposition

So far we have discussed the force on a charged body due to one other charged body. But suppose the charged body in question is in the presence of several charged bodies. What then is the electrostatic force on the first body? We are tempted to treat this problem the same way we did with gravitational force; namely, to add vectorially the separate two-body forces to get the resultant force. In Figure 15-2 we have $\mathbf{F} = \mathbf{F}_1 + \mathbf{F}_2 + \mathbf{F}_3$ as the force on $q$. This may seem obvious, but we cannot derive it from something more basic. The principle of superposition for the electrostatic force must be tested by experiment. Fortunately, it does pass the test.

We shall encounter problems where the source of the electrostatic force is an extended body uniformly charged, such as a charged wire or a charged rectangular plate. Then $\mathbf{F} = \int d\mathbf{F}$ where $d\mathbf{F}$ is the force from each element of charge. We shall deal with linear charge density $\lambda$ in coulombs per meter of length, surface charge density $\sigma$ in coulombs per square meter of surface area, and volume charge density $\rho$ in coulombs per cubic meter of volume.

**Example 2.** An electric dipole consists of two charges $+Q$ and $-Q$ separated by a distance $l$. Dipole strength $p$ is defined as $p \equiv Ql$. What is the force exerted on a charge $q$ positioned as shown in Figure 15-3?

ANSWER: Since the force triangle in Figure 15-3 is similar to the triangle formed by the three charges, we have

$$\frac{F}{F_1} = \frac{l}{r}$$

$$F = \frac{l}{r} F_1 = \frac{l}{r}\left(k_0 \frac{Qq}{r^2}\right) = qk_0 \frac{Ql}{r^3}$$

$$= qk_0 \frac{p}{r^3}$$

We see in this case that the force exerted by a dipole on a charge $q$ varies as the inverse cube of the distance. The general case where $q$ is at any angle from the dipole is given on page 320.

## 15-3 Electric Field

On page 91 we introduced the concept of gravitational field. The gravitational field at any point in space can be obtained by putting a mass $m$ at that point and measuring the net gravitational force $F_G$ on that mass. The gravitational field is by definition equal to $F_G/m$. We can do the same sort of thing for electric force by defining electric field as the electric force on a test charge divided by the charge. In order to measure the electric field $E$ at any point $P$, place a test charge $q$ at that point and measure the net electric force **F** on it. Make sure the presence of $q$ does not alter the positions of the other charges. Then

$$\mathbf{E} \equiv \frac{\mathbf{F}}{q} \quad \text{(definition of electric field)} \quad (15\text{-}3)$$

The direction of **E** is the same as the direction of force on a positive test charge. The units of $E$ are newtons per coulomb which, as we shall see in the next chapter, are the same as volts per meter.

As an example consider the electric field at point $P$ in Figure 15-4. Point $P$ is on the perpendicular bisector of the line joining the two charges $Q$ and $-Q$. If we substitute the result of Example 2 into Eq. 15-3, we obtain

$$E = \frac{k_0 q(p/r^3)}{q} = k_0 \frac{p}{r^3}$$

The direction of $E$ at point $P$ is to the right.

$$= \frac{1}{4\pi\varepsilon_0} \frac{p}{r^3}$$

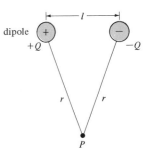

Figure 15-4. *Same as Figure 15-3 except that no charge is at point P.*

The electric field at a distance $r$ from a single point charge $Q$ is

$$\mathbf{E} = \frac{1}{q}\mathbf{F} = \frac{1}{q}\left(k_0 \frac{Qq}{r^2}\hat{\mathbf{r}}\right)$$

$$\mathbf{E} = k_0 \frac{Q}{r^2}\hat{\mathbf{r}} \quad \text{(electric field of a point charge } Q\text{)} \quad (15\text{-}4)$$

where $\hat{\mathbf{r}}$ is the unit vector from $Q$ to the point $P$.

The electric field due to $n$ point charges is the following vector summation:

$$\mathbf{E} = k_0 \sum_{j=1}^{n} \frac{Q_j}{r_j^2}\hat{\mathbf{r}}_j$$

**Example 3.** A charged ring of radius $R$ has a total charge $Q$. What is the electric field along the axis at a distance $x_0$ from the center?

ANSWER: In Figure 15-5,

$$dE_x = dE(\cos\alpha)$$

is produced by $dl$ of the ring, where $\cos\alpha = x_0/r$. If $\lambda = Q/2\pi R$ is the linear charge density, then

$$dE = k_0 \frac{\lambda\,dl}{r^2}$$

and

$$dE_x = k_0 \frac{\lambda\,dl}{r^2}\frac{x_0}{r}$$

$$E = E_x = \frac{k_0 \lambda x_0}{r^3}\int dl = \frac{k_0 \lambda x_0}{r^3}(2\pi R)$$

$$= \frac{k_0 x_0 Q}{(x_0^2 + R^2)^{3/2}}$$

At the center of the ring $x_0 = 0$ and $E = 0$. For $x_0 \gg R$, $E \to k_0 Q/x_0^2$, which is the same as for a point charge $Q$ at the same distance.

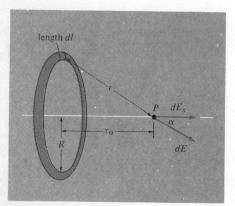

**Figure 15-5.** Field produced by uniformly charged ring of total charge $Q$.

One advantage of using electric field is that we do not have to be concerned with the details of the source of the field. For example, the field in region I in Figures 15-6(a) and (b) is $E = k_0 Q/r^2$ in both views. There are many possibilities for the source of such a field. It could be a point

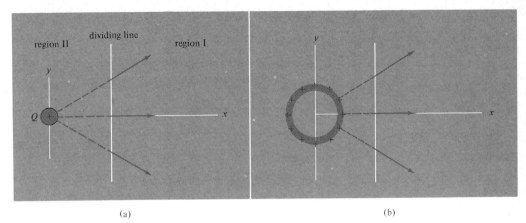

(a)            (b)

**Figure 15-6.** *In both* (a) *and* (b) *the field everywhere in region I is the same. There is no way to determine the charge distribution in region II by field measurements in region I alone.*

charge as in Figure 15-6(a) or a uniformly charged shell as in (b). From detailed measurements in the region of interest (region I) there is no way to determine just what charge distribution is the source of the field. For all we know the source could be moving. Then the use of electric field will allow us to account for relativistic effects such as the fact that signals cannot travel faster than the speed of light. We shall see that electric field is a real physical entity with local energy and momentum of its own. Using the field formalism, all forces become local, thus avoiding the concept of force at a distance.

The direction of **E** as one travels through space can be represented by continuous lines such as are shown in Figure 15-7. Such lines, which indicate the direction of **E** at any point on the line, are called lines of force or lines of flux. Not only are such lines useful in visualizing the field directions, but they can be drawn in such a way as to display the magni-

## 15-4 Lines of Flux

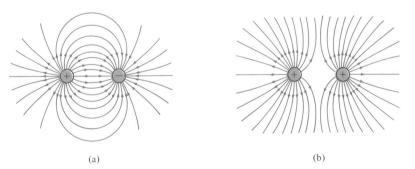

(a)            (b)

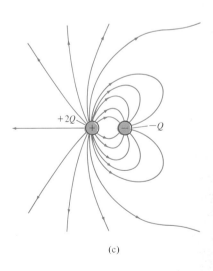

(c)

**Figure 15-7.** *Lines of flux diagrams for* (a) *two charges of opposite sign (a dipole),* (b) *two charges of the same sign, and* (c) *two charges whose values are* $-Q$ *and* $+2Q$.

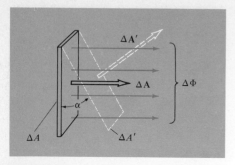

**Figure 15-8.** *Four lines of flux cutting through areas $\Delta A$ and $\Delta A'$, which are at an angle $\alpha$ to each other. The surface $\Delta A$ is perpendicular to the lines of flux.*

tude of **E** in each region of space. The trick is to draw the lines so that the area density of lines is equal numerically to the magnitude of the electric field. In Figure 15-8 we choose an area $\Delta A$ whose surface is perpendicular to **E**. The *vector* direction $\Delta \mathbf{A}$ is by convention perpendicular to the surface, and hence parallel to **E**. The magnitude of the vector $\Delta \mathbf{A}$ is the area $\Delta A$.

Let $\Delta \Phi$ be the number of lines cutting through $\Delta A$. Then

$$E = \frac{\Delta \Phi}{\Delta A}$$

or

$$\Delta \Phi = E\, \Delta A$$

Now consider the area $\Delta A'$ in Figure 15-8, which is tilted at an angle $\alpha$ to $\Delta A$ and which encloses the same lines $\Delta \Phi$. Then

$$\Delta A' = \frac{\Delta A}{\cos \alpha}$$

We note that the vector dot product

$$\mathbf{E} \cdot \Delta \mathbf{A}' = E(\Delta A') \cos \alpha = E \left( \frac{\Delta A}{\cos \alpha} \right) \cos \alpha = E\, \Delta A = \Delta \Phi$$

So, in general, the number of flux lines is

$$d\Phi = \mathbf{E} \cdot d\mathbf{A} \quad \text{(electric flux or number of lines of force)} \quad \text{(15-5)}$$

The total flux cutting through an extended surface $S$ is the quantity $\mathbf{E} \cdot d\mathbf{A}$ summed over the surface:

$$\Phi = \sum_{\text{surface}} \mathbf{E} \cdot \Delta \mathbf{A}$$

This can be expressed as a surface integral.

$$\Phi = \int_S \mathbf{E} \cdot d\mathbf{A}$$

A surface $S$ is shown in Figure 15-9 along with three of the many possible surface elements. The quantity $\Phi$, which is the number of lines cutting through $S$, is also called the flux through $S$. Flux is merely another name for the number of lines of flux (or lines of force).

In order to maintain the quantitative relation in Eq. 15-5 we might expect to have new lines appearing (or old ones disappearing) as we move

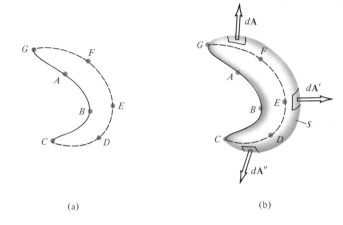

**Figure 15-9.** *Boundary curve containing points A to G encloses surface S. Three vector surface elements are shown. (a) Boundary curve alone. (b) Surface S. To emphasize the fact that S need not be a flat surface, we have chosen a surface which bulges out from the page.*

away from the source of the field. However, we shall now show that for a single point charge the number of lines stays the same for all values of $r$. We draw an imaginary sphere of radius $r_1$ about the charge $Q$ as shown in Figure 15-10. Since the total area is $4\pi r_1^2$, the number of lines cutting through this spherical shell is $E$ times the total area or

$$\Phi = E \cdot (4\pi r_1^2) = \left(k_0 \frac{Q}{r_1^2}\right)(4\pi r_1^2)$$
$$= 4\pi k_0 Q$$

Note that this result is independent of $r_1$ and thus holds for all values of $r$. So, the total number of lines leaving a point charge $Q$ is $4\pi k_0 Q$, and these lines are continuous all the way to infinity.

Now we shall show that the number of lines is still $\Phi = 4\pi k_0 Q$ even if the closed surface is not a sphere. We have already shown that $\mathbf{E} \cdot d\mathbf{A} = \mathbf{E} \cdot d\mathbf{A}'$, where $d\mathbf{A}$ and $d\mathbf{A}'$ enclose the same lines; consequently

$$\Phi = \int_{\text{sphere}} \mathbf{E} \cdot d\mathbf{A} = \int_{S'} \mathbf{E} \cdot d\mathbf{A}'$$

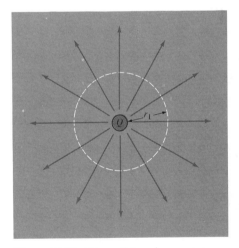

**Figure 15-10.** *Lines of flux from point charge cut through imaginary sphere of radius $r_1$.*

where $S'$ is a surface of any shape that completely encloses $Q$. We let the symbol $\oint \mathbf{E} \cdot d\mathbf{A}$ stand for the integral of $\mathbf{E}$ over a completely closed surface of any shape. Then

$$\oint \mathbf{E} \cdot d\mathbf{A} = 4\pi k_0 Q \qquad (15\text{-}6)$$

if the surface encloses a single point charge. Such a completely closed surface is called a gaussian surface.

## 15-5 Gauss' Law

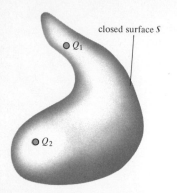

**Figure 15-11.** *Two point charges in volume enclosed by surface S.*

To derive Gauss' law we let the closed surface enclose two point charges $Q_1$ and $Q_2$ as shown in Figure 15-11. The total number of lines leaving the surface is

$$\Phi_{total} = \oint \mathbf{E} \cdot d\mathbf{A} = \oint (\mathbf{E}_1 + \mathbf{E}_2) \cdot d\mathbf{A}$$

$$= \oint \mathbf{E}_1 \cdot d\mathbf{A} + \oint \mathbf{E}_2 \cdot d\mathbf{A}$$

where $\mathbf{E}_1$ is the field due to $Q_1$ by itself and $\mathbf{E}_2$ is due to $Q_2$ by itself. According to Eq. 15-6, $\oint \mathbf{E}_1 \cdot d\mathbf{A} = 4\pi k_0 Q_1$ and $\oint \mathbf{E}_2 \cdot d\mathbf{A} = 4\pi k_0 Q_2$. Hence

$$\Phi_{total} = (4\pi k_0 Q_1) + (4\pi k_0 Q_2)$$
$$= 4\pi k_0 (Q_1 + Q_2)$$

We have shown that the total number of lines leaving the closed surface is equal to $(4\pi k_0)$ times the total charge enclosed by the surface in the case of two point charges. The same derivation can be extended to $n$ point charges sitting inside a closed surface. Then

$$\oint \mathbf{E} \cdot d\mathbf{A} = 4\pi k_0 Q_{in} \quad \text{(Gauss' law)} \quad (15\text{-}7)$$

where $Q_{in}$ is the net charge inside the closed surface. In general the total number of lines leaving a charged body is $4\pi k_0$ times the net charge of the body. If $Q$ is negative, the lines enter the body. Lines can only start or stop on charge; otherwise they are continuous. If $4\pi k_0 Q_{in}$ is a small number, one can draw micro-lines; i.e., $10^6$ micro-lines is equivalent to one full line.

Gauss' law works whether or not there are charges outside the closed surface. For example, consider the closed surface in Figure 15-12, which has $Q_{in} = 0$. In this figure there must be some external charges producing the lines of flux that pass through the surface. We can break up the total flux into pieces:

$$\Phi_{total} = \Phi_{ab} + \Phi_{bc} + \Phi_{cd} + \Phi_{da}$$

In Figure 15-12 we see that three lines leave the "area" between $a$ and $b$, hence $\Phi_{ab} = +3$. A total of five lines enter between $b$ and $c$, hence $\Phi_{bc} = -5$. Six lines leave between $c$ and $d$, hence $\Phi_{cd} = +6$. $\Phi_{da} = -4$ because four lines enter between $d$ and $a$. Adding these four parts of the flux together gives

$$\Phi_{total} = (+3) + (-5) + (+6) + (-4) = 0$$

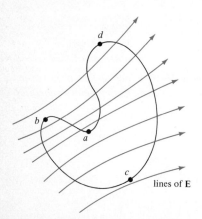

**Figure 15-12.** *Two-dimensional representation of a closed surface with external source of* **E**.

which checks with Eq. 15-7. Clearly, any line that enters must leave, so the net number of lines leaving is zero. (A line entering is a negative flux line, and a line leaving is a positive flux line.)

Since the left-hand side of Eq. 15-7 is the net number of lines leaving a closed surface, we can write

$$\Phi = 4\pi k_0 Q_{in}$$

or

(number of lines leaving closed surface) = $4\pi k_0$(net enclosed charge)

This is an alternate version of Gauss' law. We can also express Gauss' law in terms of $\varepsilon_0$ by replacing $k_0$ with $1/4\pi\varepsilon_0$,

$$\oint \mathbf{E} \cdot d\mathbf{A} = \frac{1}{\varepsilon_0} Q_{in} \qquad (15\text{-}8)$$

Equation 15-7 or 15-8 is one of the four basic Maxwell equations that encompass all of electromagnetic theory. Although it is mathematically equivalent to Coulomb's law, Gauss' law is in a form that often is easier to use for calculating electric fields or charge distributions. We shall see this in the next chapter when we use Gauss' law to calculate the field:

1 Outside a charged spherical shell.
2 Inside and outside a uniformly charged solid sphere.
3 Outside a charged wire.
4 Inside a uniformly charged solid cylinder.
5 Outside a charged sheet.
6 Between two charged sheets.
7 Inside a uniformly charged solid slab.

If we were to use Coulomb's law to calculate the field outside a uniformly charged solid sphere, we would have a quite complicated triple vector integral. This is the way Newton proved that the mass of the earth behaves as if it is all concentrated at the center. If he had known Gauss' law, he could have done the proof in two lines rather than taking several pages of messy calculation.

Most solid bodies can be classified into two categories: conductors and nonconductors (also called insulators). Excess charge can be placed on the surface or interior of a nonconductor and it will stay there. Conductors, however, contain a large number of free electrons that are not attached to any particular atom. For this reason an electric field could exist inside a conductor only for a short time because the free electrons would move under the influence of the field and accumulate on the surface until an equal and opposite field was built up.

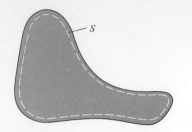

**Figure 15-13.** *Dashed line indicates closed surface just inside outside surface.*

Before closing this section we shall use Gauss' law to prove that if a charge is applied to a conductor, it must all appear on the surface of the conductor even if the charge is applied to the inside. In Figure 15-13 we have a conductor of arbitrary shape (it can even be hollow). Just below the surface we take a closed surface $S$ indicated by the dashed line in the figure. Now apply Gauss' law to this closed surface:

$$\oint_S \mathbf{E} \cdot d\mathbf{A} = 4\pi k_0 Q_{\text{in}}$$

At any point on the surface $S$ inside the conductor the field must be zero because, if it were not, conduction electrons would move. (They do not move because we have waited until all charges and conduction electrons have redistributed themselves.) Since the charges in the conductor are no longer moving, there is zero electric force on a charge inside the conductor; thus $E = 0$ everywhere over the surface $S$. Then

$$\oint \mathbf{E} \cdot d\mathbf{A} = 0$$

Thus the left-hand side of Eq. 15-7 is zero:

$$0 = 4\pi k_0 Q_{\text{in}}$$

So $Q_{\text{in}} = 0$. The net charge everywhere inside the closed surface must be zero. Since we can take all possible closed surfaces and always get $Q_{\text{in}} = 0$, we have proved that the net charge over any small region inside a conductor must be zero.

---

**Example 4.** Typically the earth has a small electric field that, when measured just above the surface, is $\sim 100$ N/C.
  (a) What would be the electric field just below the surface?
  (b) What surface charge would cause $E = 100$ N/C just above the surface? How many excess electrons per square centimeter would be needed?

ANSWER: (a) Since the earth is a conductor rather than an insulator, just below the surface is inside the conductor, and no steady field can exist inside a conductor.
  (b) Apply Gauss' law to a sphere of radius slightly greater than $R_e$, which encloses the earth. Since $E$ is constant over the sphere, the integral becomes $E$ times the total area of the earth, $A_e$:

$$\oint \mathbf{E} \cdot d\mathbf{A} = E \cdot A_e$$

Then Gauss' law becomes

$$E \cdot A_e = 4\pi k_0 Q_e$$

where $Q_e$ is the total surface charge.
The surface charge density is

$$\sigma = \frac{Q_e}{A_e} = \frac{E}{4\pi k_0} = \frac{100}{4\pi(9 \times 10^9)} \text{ C/m}^2 = 8.84 \times 10^{-14} \text{ C/cm}^2$$

Since $e = 1.6 \times 10^{-19}$ C is the electronic unit of charge, we can substitute $e/(1.6 \times 10^{-19})$ for C in the preceding equation

$$\sigma = 8.84 \times 10^{-14} \frac{e/(1.6 \times 10^{-19})}{\text{cm}^2} = 5.52 \times 10^5 \frac{e}{\text{cm}^2}$$

---

**Example 5.** If dry air is exposed to an electric field greater than $1 \times 10^6$ N/C, ions will be rapidly formed resulting in thin sparks (the air then becomes a conductor). What is the maximum charge that can be stored on a 1-cm radius sphere? On a sphere of 1 m radius?

ANSWER: Take a sphere of radius slightly greater than that of the charged sphere as a gaussian surface. Since $E$ is constant, the surface integral is $E \cdot (4\pi R^2)$. According to Gauss' law this must equal $4\pi k_0 Q$:

$$E \cdot (4\pi R^2) = 4\pi k_0 Q$$

$$Q = \frac{E \cdot R^2}{k_0}$$

For $R = 10^{-2}$ m:

$$Q = \frac{(1 \times 10^6)(10^{-2})^2}{9 \times 10^9} \text{ C} = 1.1 \times 10^{-8} \text{ C}$$

For $R = 1$ m, $Q$ is $10^4$ times larger or

$$Q = 1.1 \times 10^{-4} \text{ C}$$

We see that one coulomb is so large an amount of charge that it cannot be "held" on a sphere in air.

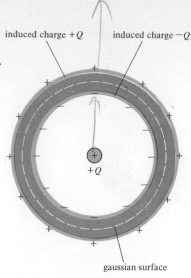

**Figure 15-14.** *A charge $+Q$ is placed inside a hollow cavity within a solid spherical conductor. Induced charges will appear on the inner and outer surfaces as shown.*

### Electrical Induction

If an electrically neutral conductor has an empty cavity, the net charge everywhere inside the conductor must still be zero. However, a fixed charge could have been placed inside the cavity. Then according to Gauss' law there must be an equal and opposite amount of induced charge on the cavity wall as shown in Figure 15-14. This is because a gaussian surface, as indicated by the dashed line, would enclose zero net charge which is consistent with zero flux inside the conductor.

Note that since the conductor is electrically neutral, there must be a charge on the outer surface that is equal and opposite to the induced charge on the cavity wall. This is an example of electrical induction. Whenever a neutral body is brought into a region of electric field, induced charges will appear on the surfaces that cancel out the field inside the body, if the body is a conductor (even a poor conductor). For perfect insulators there will still be induced charges, but they will not completely cancel out the field inside the body. Such an insulator is called a dielectric and is discussed in Section 16-6.

An electrically neutral conductor can also be given a net charge by making use of electrical induction. This is illustrated in Figure 15-15 where a previously charged glass rod is brought near two uncharged spherical conductors. Then, as shown in Figure 15-15(a), conduction electrons from the far sphere are attracted to the near sphere by the positive charges on the rod. If the spheres are now separated, they each retain the net charges shown in Figure 15-15(b). In view (c) the rod has been withdrawn with the result that previously uncharged conductors have become charged. This process could be repeated using the original charged insulator to charge up as many conductors as desired with no loss of charge of the original rod.

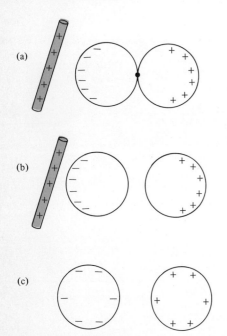

**Figure 15-15.** *Charging by induction. (a) A charged rod is brought near two uncharged conductors. (b) The conductors are moved apart. (c) The charged rod is taken away, leaving the two spheres with equal and opposite charge.*

## Summary

All elementary particles have an intrinsic charge of zero, $+e$, $-e$, or integral multiples of $\pm e$ where $e = 1.60 \times 10^{-19}$ coulomb. Coulomb's law states that the force between any two charged particles is

$$\mathbf{F} = k_0 \frac{q_1 q_2}{r^2} \hat{\mathbf{r}} \quad \text{where } k_0 = 9.00 \times 10^9 \text{ N m}^2/\text{C}^2$$

Electric field is electric force per unit charge or $\mathbf{E} = \mathbf{F}/q$. The electric field produced by a point charge $Q$ is $\mathbf{E} = k_0(Q/r^2)\hat{\mathbf{r}}$. Electric field produced by a volume element $dV$ of charge density $\rho$ is $d\mathbf{E} = k_0(\hat{\mathbf{r}}/r^2)\rho \, dV$. The net field produced by an extended body can be obtained by integrating this over the volume.

Electric flux, which is the same as the number of lines of force, is $d\Phi = \mathbf{E} \cdot d\mathbf{A}$. The net flux leaving a body is $\Phi = \oint \mathbf{E} \cdot d\mathbf{A}$ where the integral is over an entire surface that completely encloses the body. Gauss' law states that the value of such a surface integral is $4\pi k_0$ times the net charge enclosed by the surface:

$$\oint \mathbf{E} \cdot d\mathbf{A} = 4\pi k_0 Q_{\text{in}}$$

One application of Gauss' law is that there can be no net charge inside a conductor.

## Exercises

1. What is the ratio of the electric force to gravitational force between two electrons?
2. Repeat Example 1 for two spheres of frozen hydrogen.
3. What is the electric field at point $P$ in terms of $q$, $l$, and $r$?

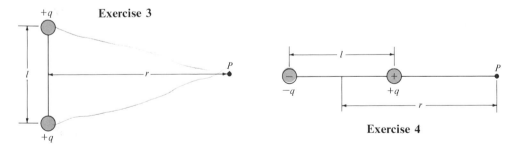

Exercise 3

Exercise 4

4. What is $E$ at point $P$ in terms of $q$, $r$, and $l$?
5. A positive charge is applied to a metal sphere. Will the mass of the sphere increase, decrease, or remain the same?
6. A charge of $-4 \times 10^{-5}$ C is placed 10 cm from a charge of $+5 \times 10^{-5}$ C. What is the electrostatic force? How many lines of force go to infinity (assuming no other charges anywhere)?

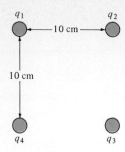

Exercise 8

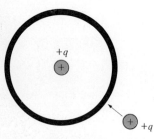

Exercise 12

7. A charge of $-1 \times 10^{-6}$ C sits at the center of a hollow metal sphere that contains a positive surface charge of $+1.5 \times 10^{-6}$ C on its outer surface.
   (a) Draw a lines of force diagram to represent the resultant electric field.
   (b) What is the net flux radiating out from the sphere?
   (c) How much excess charge is on the conductor?
8. A 10-cm by 10-cm square has charges of $10^{-8}$ C magnitude on all four corners. Give the magnitude and direction of $E$ in the center of the square if the signs of $q_1$, $q_2$, $q_3$, and $q_4$ are
   (a) $+, +, +, +$.
   (b) $+, -, +, -$.
   (c) $+, +, -, -$.
9. What is the ratio of gravitational to electrostatic force for two protons?
10. Suppose the earth had an excess surface charge of one electron per square centimeter.
    (a) What would be the electric field just below the surface?
    (b) What would be the electric field just above the surface?
11. Draw a lines of force diagram for Figure 15-14.
12. A charge $+q$ is placed at the center of a hollow conducting sphere. An amount of charge $+q$ is deposited on the outer surface of the sphere. After the charges come into equilibrium, what will be the net charge on
    (a) the inner surface of the sphere?
    (b) the outer surface of the sphere?
13. Repeat Exercise 4 for the case where both charges $q$ are positive.
14. How large a sphere is needed to hold 1 C of charge in air?

## Problems

15. Let us explore the hypothesis that the attractive force between two unlike charges is very slightly stronger than the repulsive force between like charges. Assume the excess attraction is $4.04 \times 10^{-37}$; that is,

$$F = \begin{cases} -(1 + 4.04 \times 10^{-37})k_0 \dfrac{Q_1 Q_2}{r^2} & \text{when } Q_1 \text{ and } Q_2 \text{ of opposite sign} \\ +k_0 \dfrac{Q_1 Q_2}{r^2} & \text{when } Q_1 \text{ and } Q_2 \text{ of same sign} \end{cases}$$

What then would be the net force between two hydrogen atoms 1 m apart? Compare this with the value of the gravitational force. Why couldn't this hypothesis explain gravitation? (Give an example where the hypothesis gives the wrong answer for the gravitational force.)

16. Assume that the hydrogen atom consists of an electron of charge $-e$ in circular orbit about a proton of charge $+e$. The orbit radius is $0.53 \times 10^{-10}$ m.
    (a) What is the ratio of the speed of light to the electron velocity?
    (b) How many revolutions per second does the electron make?

17. Two charged rings of radius $R$ are a distance $R$ apart as shown. If the charge of each ring is $Q$, what is $E$ along the $x$ axis as a function of $x$, $Q$, and $R$?

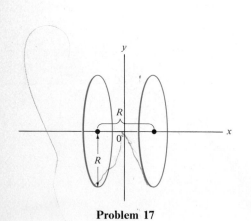

Problem 17

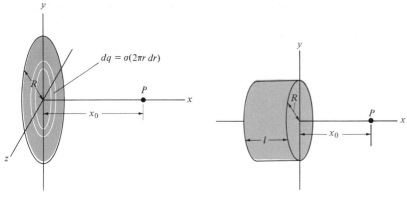

Problem 18  Problem 20

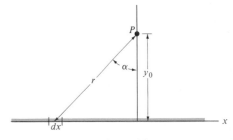

Problem 21

18. What is the electric field at point P produced by a uniformly charged disk of radius R and surface charge density $\sigma = Q/\pi R^2$. (*Hint:* A ring of thickness $dr$ and radius $r$ has a charge $dq = 2\pi r\sigma\, dr$.)
19. In Problem 18 let R approach infinity to get the result for the field produced by an infinite plane of surface charge $\sigma$.
20. What is the electric field at point P produced by a charged cylindrical surface of radius R and length $l$? Its total charge is Q.
21. What is the electric field at a distance $y_0$ from an infinitely long charged wire of linear charge density $\lambda$? Note that the contribution from $dx$ is

$$dE_y = k_0 \frac{\lambda\, dx}{r^2} \cos\alpha = \frac{k_0 \lambda}{y_0} \cos\alpha\, d\alpha$$

22. Consider two concentric hollow spheres of thickness $d$, each of which is a conductor. The inner radii of the two shells are $R_1$ and $R_3$, respectively. A point charge $q_1$ is located in the center. A thin spherical shell of radius $R_2$ holding a total charge $q_2$ is located in between the two previously described shells.
    (a) What is the charge on the surface of radius $R_1$?
    (b) What is the charge on the surface of radius $R_3$?
    (c) What is the charge on the surface of radius $(R_1 + d)$?
    (d) What is the charge on the surface of radius $(R_3 + d)$?
23. Charges of magnitude $q$ are placed at the corners of a square as shown. Point P is at a distance $x$ from the center. What is **E** at point P when $x \gg l$?
24. Two dipoles of strength $p$ are a distance $x_0$ apart and pointing in opposite directions. Show that in the limit of $x$ much greater than $x_0$, $E = 3k_0 p x_0/x^4$. (This kind of charge distribution is called an electric quadrupole. Note that the field drops off as the inverse fourth power of the distance.)
25. In Problem 17 the two rings are $\sqrt{2}R$ apart. Evaluate $\partial E/\partial x$ and $\partial^2 E/\partial x^2$ at $x = y = 0$.
26. In Example 3, at what distance $x_0$ is $E$ a maximum?

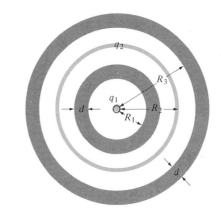

Problem 22

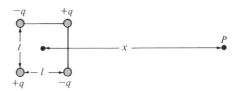

Problem 23

Problem 24

# 16

# Electrostatics

In electricity one is constantly dealing with charged surfaces in devices such as electric capacitors, antennas, transmission lines, wave guides, solid state devices, and so on. In order to calculate the amount of charge and voltage involved, one must first be able to calculate the electric field produced by common types of charge distributions. In this chapter we shall calculate the electric field produced by spherical, cylindrical, and flat charge distributions. Then we shall define electric potential or voltage, and show how to determine the voltage once the charges are known. The chapter ends with a presentation of capacitance and dielectrics.

## 16-1 Spherical Charge Distributions

Our first charge distribution will be a charged spherical surface of total charge $Q$, as shown in Figure 16-1. We shall calculate **E** both inside and outside this spherical shell. Because of the symmetry, the lines of **E** must point radially out from the center. (A given line of **E** as it leaves the surface cannot bend to the side because there is no preference of left over right.) We take the dashed-line sphere of radius $r$ for the surface of integration (also called a gaussian surface).

At any point on this sphere we have $\mathbf{E} \cdot d\mathbf{A} = E\, dA$ and

$$\oint \mathbf{E} \cdot d\mathbf{A} = E \oint dA = E(4\pi r^2)$$

Using Gauss' law we equate this to $4\pi k_0 Q_{\text{in}}$:

$$E(4\pi r^2) = 4\pi k_0 Q$$

$$E = k_0 \frac{Q}{r^2} \quad \text{for } r > R \tag{16-1}$$

We note that this is the same as if all the charge were concentrated at $r = 0$ rather than being at $r = R$.

For the field inside we have

$$\oint \mathbf{E}_{in} \cdot d\mathbf{A} = 0$$
$$E_{in}(4\pi r^2) = 0$$
$$E_{in} = 0 \quad \text{for } r < R.$$

This is the same result as we had on page 92 for the gravitational field inside a hollow sphere.

## Uniformly Charged Sphere

Since a uniformly charged solid sphere can be broken up into a series of concentric spherical shells, Eq. 16-1 remains valid for the field outside. We note that if we put $r = R$ into Eq. 16-1, we obtain

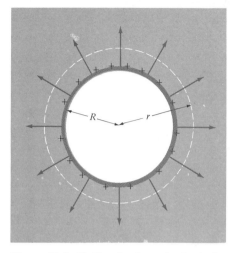

**Figure 16-1.** *Uniformly charged spherical surface of radius R. Dashed line is imaginary sphere of radius r. Lines of flux in color.*

$$E = k_0 \frac{Q}{R^2} \quad \text{(at the surface of a charged sphere)} \quad (16\text{-}2)$$

where $Q$ is the total charge of the solid sphere. This result can be used to prove that a spherical mass produces a gravitational force as if all its mass were concentrated at its center. Since gravity is also an inverse square law, Gauss' law for gravity would give the same result as Eq. 16-2, except that $k_0$ is replaced by $G$ and $Q$ is replaced by $M$, the mass of the sphere. Then

$$\text{(gravitational field)} = G \frac{M}{R^2}$$

Expressing gravitational field as force per unit gravitational charge, we have

$$\frac{F}{m} = G \frac{M}{R^2}$$

or

$$F = G \frac{Mm}{R^2}$$

where $m$ is a small mass sitting on the surface of $M$. This is a solution to the problem that had given Newton so much trouble. If $m$ is the mass of an apple, the earth attracts it as if all the mass of the earth were concentrated at its center. Enough for gravity. Now we go back to electrostatics.

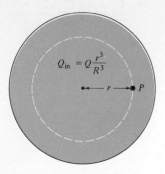

**Figure 16-2.** *Imaginary spherical surface is drawn through point P and encloses charge $Q_{in}$.*

Our next mission is to calculate $E$ at a point $P$ inside a uniformly charged sphere. We take an imaginary sphere as the gaussian surface through point $P$, which is inside the sphere as shown in Figure 16-2. It encloses a volume $\frac{4}{3}\pi r^3$, which is $(r/R)^3$ times the total volume. Hence the charge inside the imaginary sphere is $Q_{in} = Q(r/R)^3$ and Gauss' law gives

$$\oint \mathbf{E} \cdot d\mathbf{A} = 4\pi k_0 \left( Q \frac{r^3}{R^3} \right)$$

$$E(4\pi r^2) = 4\pi k_0 Q \frac{r^3}{R^3}$$

$$E = k_0 \frac{Q}{R^3} r \qquad (16\text{-}3)$$

is the field inside a uniformly charged sphere of radius $R$ and total charge $Q$. The field as a function of $r$ is plotted in Figure 16-3.

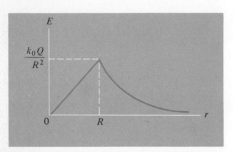

**Figure 16-3.** *Plot of E as a function of the distance from the center of a uniformly charged sphere.*

*Example 1.* We shall learn in Chapter 25 that a fairly reasonable model of the hydrogen atom represents the electron as a rigid, uniformly charged sphere of radius $R \sim 10^{-10}$ m with total charge $Q = -e = -1.6 \times 10^{-19}$ C and mass $m_e = 9.1 \times 10^{-31}$ kg. Normally the proton of charge $+e$ sits at the center of the electron cloud. Suppose the proton is displaced from the center of the electron charge cloud by a small distance $x_0$ as shown in Figure 16-4. If the proton and electron are released, they will oscillate with amplitude $x_0$ about their equilibrium position. What will be the frequency of oscillation?

ANSWER: According to Eq. 16-3 the restoring force on the proton will be

$$F = eE = e\left[k_0 \frac{(-e)}{R^3} x\right]$$

$$F = -k_0 \frac{e^2}{R^3} x$$

According to Newton's third law of motion this same force acts on the electron, and we have

$$m_e \frac{d^2 x}{dt^2} = -\frac{k_0 e^2}{R^3} x$$

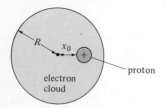

**Figure 16-4.** *Electron represented as a uniformly charged sphere is pulled a distance $x_0$ from the proton.*

[Strictly speaking, we should use reduced mass $\mu = M_p m_e/(M_p + m_e)$ rather than $m_e$; however, $\mu \approx m_e$ (see page 220).] Dividing both sides by $m_e$ gives

$$\frac{d^2x}{dt^2} = -\left(\frac{k_0 e^2}{m_e R^3}\right)x$$

According to Eq. 11-7

$$\omega = \sqrt{\frac{k_0 e^2}{m_e R^3}} \qquad (16\text{-}3a)$$

$$f = \frac{1}{2\pi}\sqrt{\frac{(9 \times 10^9)(1.6 \times 10^{-19})^2}{(9.1 \times 10^{-31})(10^{-10})^3}} \text{ s}^{-1} = 2.5 \times 10^{15} \text{ Hz}$$

This result is quite close to the frequency of electromagnetic radiation emitted by a hydrogen atom in its first excited state and confirms our assertion that this model of the hydrogen atom is a reasonable one.

---

*Example 2.** What is the induced dipole moment of an atom when placed in an electric field $E_0$? Again assume that the outer electron is a uniformly charged cloud of radius $R$.

ANSWER: When a neutral atom is placed in an electric field $E_0$ (Figure 16-5), the outer electron cloud experiences a force $F = -eE_0$ and is pulled so that its center is displaced a distance $x_0$ from the atomic core (which has charge $Q = +e$). Then the atom has an induced dipole moment $p = ex_0$. If the outer electron is represented as a uniformly charged sphere of radius $R$, we can express the dipole moment in terms of $R$, $e$, and $E_0$.

According to Eq. 16-3 the field produced by the electron cloud at the position of the core is

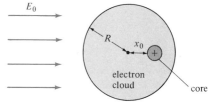

**Figure 16-5.** *In an atom the electron cloud and atomic core are pulled a distance $x_0$ apart by external field $E_0$.*

$$E_{\text{cloud}} = -\frac{k_0 e}{R^3}x_0$$

The net field on the core is

$$E_{\text{net}} = E_0 + E_{\text{cloud}} = E_0 - \frac{k_0 e}{R^3}x_0$$

Since the core is at equilibrium, the net force or net field acting on it must be zero, hence

$$0 = E_0 - \frac{k_0 e}{R^3} x_0$$

$$x_0 = \frac{R^3}{ek_0} E_0$$

The dipole moment is

$$p = ex_0 = \frac{R^3}{k_0} E_0 \qquad (16\text{-}4)$$

## 16-2 Linear Charge Distributions

We shall first consider the field at a distance $r$ from a uniformly charged wire or rod of length much greater than $r$. Let $\lambda$ be the charge per unit length along the rod. For the gaussian surface we take the cylinder of length $L$ shown in Figure 16-6. It encloses a charge $Q_{in} = \lambda L$. Then Gauss' law (Eq. 15-7) gives

$$\oint \mathbf{E} \cdot d\mathbf{A} = 4\pi k_0 (\lambda L)$$

Using the same symmetry arguments as before, the lines of $E$ can only point out radially. Hence the vectors $\mathbf{E}$ and $d\mathbf{A}$ are at right angles on the two end surfaces of the imaginary cylinder and they are parallel over the curved surface of the cylinder. Since $\mathbf{E} \cdot d\mathbf{A}$ is zero on the two end surfaces, we have

$$\oint \mathbf{E} \cdot d\mathbf{A} = E(2\pi r L)$$

Equating this to $4\pi k_0 \lambda L$ gives

$$2\pi r L E = 4\pi k_0 \lambda L$$

$$E = \frac{2k_0 \lambda}{r} \quad \text{(due to line charge)} \qquad (16\text{-}5)$$

In order to evaluate the field inside a uniformly charged rod, we again

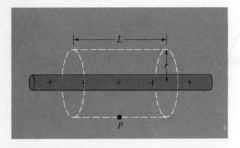

**Figure 16-6.** *Portion of long charged rod. Gaussian surface is imaginary cylinder of length L and radius r.*

take a cylinder of length $L$ for the surface of integration (the gaussian surface), but this time it has a radius $r < R$. If $\rho$ is the charge per unit volume in the rod, the charge enclosed by the dashed-line cylinder in Figure 16-7 is $Q_{in} = \rho \pi r^2 L$. Then Gauss' law gives

$$\oint \mathbf{E} \cdot d\mathbf{A} = 4\pi k_0 (\rho \pi r^2 L)$$

$$E(2\pi r L) = 4\pi k_0 \rho \pi r^2 L$$

$$E = 2k_0 \rho \pi r \quad \text{(inside rod)} \quad (16\text{-}6)$$

This can also be expressed in terms of $\lambda$ by using $\lambda = \rho \pi R^2$, then

$$E = \frac{2k_0 \lambda}{R^2} r \quad (16\text{-}6)$$

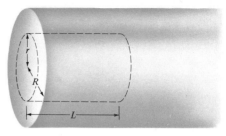

**Figure 16-7.** *Portion of uniformly charged rod. Surface of integration is cylinder of length $L$ and base radius $r$.*

---

**Example 3.** A coaxial cable consists of an inner wire surrounded by a hollow cylindrical conductor. Suppose they have linear charge densities $\lambda$ and $-\lambda$ as shown in Figure 16-8. (a) What is $E$ in region I? (b) What is $E$ in region II?

ANSWER: For part (a) we could take for the surface of integration a cylinder that encloses both conductors. Since the net charge enclosed is zero, $E_I = 0$.

For part (b) where $r < R$, the field due to the outer conductor alone is zero for the same reason that the field inside a hollow sphere is zero. The field due to the inner conductor is given by Eq. 16-5:

$$E_{II} = \frac{2k_0 \lambda}{r} \quad (16\text{-}7)$$

---

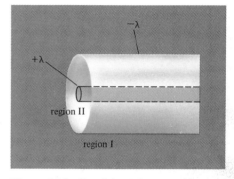

**Figure 16-8.** *Coaxial conductors with equal and opposite charge.*

Now we wish to calculate the field produced by a uniformly charged infinite plane where $\sigma$ is the charge per unit area. (Actually it could be a thin finite sheet of metal, provided the distance from the sheet is much less than the distance to the edge of the sheet.) For the surface of integration we take a pillbox or cylinder with flat ends of area $A_0$ each at a distance $a$ from the sheet as shown in Figure 16-9. The enclosed charge is $Q_{in} = \sigma A_0$. Since there is no preference of left over right, the lines of flux must flow out equally from both sides. The surface integral from each end is $EA_0$.

## 16-3 Flat Charge Distributions

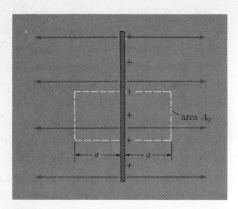

**Figure 16-9.** *Edge view of infinite charged sheet having σ coulombs per meter squared. Dashed line is edge view of cylinder of length 2a and base area $A_0$.*

Now there are two ends, so we have
$$\oint \mathbf{E} \cdot d\mathbf{A} = 2EA_0$$

Gauss' law gives
$$\oint \mathbf{E} \cdot d\mathbf{A} = 4\pi k_0 \sigma A_0$$
$$2EA_0 = 4\pi k_0 \sigma A_0$$
$$E = 2\pi k_0 \sigma \quad \text{(charged sheet)} \quad (16\text{-}8)$$

Many practical devices consist of two parallel plates with equal and opposite charges, as shown in Figure 16-10. For plate $a$ alone $E_a = 2\pi k_0 \sigma$ pointing toward plate $a$. For plate $b$ alone $E_b = 2\pi k_0 \sigma$ pointing out from plate $b$. In region I:

$$E_\text{I} = E_{a_\text{I}} + E_{b_\text{I}}$$
$$= 2\pi k_0 \sigma + (-2\pi k_0 \sigma) = 0$$

In region II:
$$E_\text{II} = E_{a_\text{II}} + E_{b_\text{II}}$$
$$= (-2\pi k_0 \sigma) + (-2\pi k_0 \sigma)$$
$$E_\text{II} = -4\pi k_0 \sigma \quad (16\text{-}9)$$

In region III:
$$E_\text{III} = E_{a_\text{III}} + E_{b_\text{III}}$$
$$= (-2\pi k_0 \sigma) + (2\pi k_0 \sigma) = 0$$

We see that there is no field on the outside and that the field anywhere between the two plates is $4\pi k_0 \sigma$.

### Surface of a Conductor

We showed in the last chapter that a single charged conductor holds all its charge on its outer surface and has $E = 0$ everywhere inside. Furthermore, the lines of $E$ must be perpendicular to the surface as they leave it. This is because there can be no component of $\mathbf{E}$ along the surface (if there were, conduction electrons would flow). In Figure 16-11 we take a small pillbox about $\Delta A$ of the surface area. Then the integral over this pillbox is

$$\oint \mathbf{E} \cdot d\mathbf{A} = E\,\Delta A$$

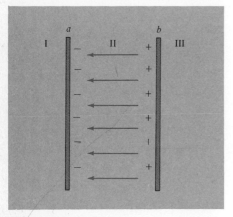

**Figure 16-10.** *Field between two sheets of equal and opposite charge.*

The enclosed charge is $Q_{in} = \sigma \Delta A$. Gauss' law gives

$$E \Delta A = 4\pi k_0(\sigma \Delta A)$$
$$E = 4\pi k_0 \sigma \quad \text{(at the surface of a conductor)}$$

## Uniformly Charged Slab

We wish to evaluate $E$ at a distance $x_0$ from the center of a uniformly charged solid slab which is infinite in two of its three dimensions. For the gaussian surface we take the rectangular box shown in Figure 16-12 of length $x_0$ in the $x$ direction and surface area $A_0$ in the $yz$ plane. The box has volume $x_0 A_0$ and encloses a charge $Q_{in} = \rho x_0 A_0$. The only contribution of $\oint \mathbf{E} \cdot d\mathbf{A}$ is the right end surface, since by symmetry $E = 0$ at $x = 0$. Hence

$$\oint \mathbf{E} \cdot d\mathbf{A} = EA_0$$

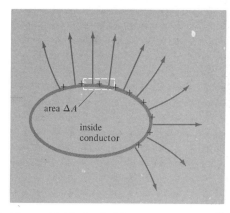

**Figure 16-11.** Dashed line represents small pillbox of base area $\Delta A$ enclosing charge $\sigma \Delta A$.

Equating this to $4\pi k_0 Q_{in}$ gives

$$EA_0 = 4\pi k_0(\rho x_0 A_0)$$
$$E = 4\pi k_0 \rho x_0 \quad (16\text{-}10)$$

---

**Example 4.** Suppose an electron is placed at $x = x_0$ inside a uniformly charged slab and then released. Describe its motion. Ignore any frictional forces.

ANSWER: Using Eq. 16-10 the force on the electron at any position $x$ will be

$$F = (-e)E = -4\pi k_0 e\rho x$$

$$\frac{d^2 x}{dt^2} = -\left(\frac{4\pi k_0 e\rho}{m_e}\right) x$$

This is an example of the harmonic force law. According to Eq. 11-7 the solution is $x = x_0 \cos \omega t$ where

$$\omega = \sqrt{\frac{4\pi k_0 e\rho}{m_e}}$$

---

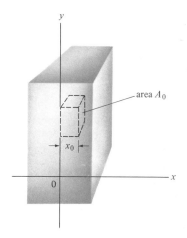

**Figure 16-12.** Uniformly charged slab centered on $yz$ plane. Dashed rectangular box has left edge at $x = 0$ along $y$ axis.

SEC. 16-3/FLAT CHARGE DISTRIBUTIONS

We note that the frequency of oscillation in Example 4 is independent of the thickness of the slab. One example of a layer or "slab" of positive charge is a plasma layer such as the ionosphere above the earth's atmosphere. Then the density of positive charge is $\rho = \mathfrak{N}_+ e$ where $\mathfrak{N}_+$ equals the number of ions per unit volume. Then

$$f = \frac{1}{2\pi}\sqrt{\frac{4\pi k_0 \mathfrak{N}_+ e^2}{m_e}}$$

is the frequency of oscillation of a single electron inside the charge layer. It can be shown that the preceding equation still holds when the number of electrons equals the number of ions. If the electrons in a plasma are disturbed, they will oscillate with this frequency (called the plasma frequency).

Radio waves with a lower frequency than this are reflected by the ionosphere. Then the ionosphere behaves as if it is a conductor. Such reflection makes long distance radio communication possible all over the surface of the earth. However, in order to communicate with space vehicles one must use frequencies higher than the plasma frequency.

**Table 16-1.** *Electric fields produced by various types of charged bodies (Solid bodies are assumed to be uniformly charged throughout the volume.)*

| Location | Type of body | $E$ (using $k_0$) | $E$ (using $\epsilon_0$) |
|---|---|---|---|
| outside | solid or hollow sphere | $k_0 \dfrac{Q}{r^2}$ | $\dfrac{1}{4\pi\epsilon_0}\dfrac{Q}{r^2}$ |
| inside | hollow sphere | 0 | 0 |
| inside | solid sphere | $k_0 \dfrac{Q}{R^3}r$ | $\dfrac{1}{4\pi\epsilon_0}\dfrac{Q}{R^3}r$ |
| outside | wire or rod | $2k_0 \dfrac{\lambda}{r}$ | $\dfrac{1}{2\pi\epsilon_0}\dfrac{\lambda}{r}$ |
| inside | solid rod | $2k_0 \dfrac{\lambda}{R^2}r$ | $\dfrac{1}{2\pi\epsilon_0}\dfrac{\lambda}{R^2}r$ |
| either side | single sheet | $2\pi k_0 \sigma$ | $\dfrac{1}{2\epsilon_0}\sigma$ |
| between two sheets | two sheets with $\sigma$ and $-\sigma$ | $4\pi k_0 \sigma$ | $\dfrac{1}{\epsilon_0}\sigma$ |
| distance $x$ from center inside | slab | $4\pi k_0 \rho x$ | $\dfrac{1}{\epsilon_0}\rho x$ |
| just above surface | conductor | $4\pi k_0 \sigma$ | $\dfrac{1}{\epsilon_0}\sigma$ |

The fields we have determined for various kinds of charged bodies are summarized in Table 16-1.

## 16-4 Electric Potential

Before studying this section it is advisable to reread pages 106–110 and Section 7-3.

According to Eq. 6-12 the electrical potential energy difference between point $a$ and point $b$ for a body of charge $q$ is

$$U_b - U_a = -\int_a^b \mathbf{F} \cdot d\mathbf{s} = -q \int_a^b \mathbf{E} \cdot d\mathbf{s} \qquad \text{(electric potential energy)} \qquad (16\text{-}11)$$

where $\mathbf{F}$ is the electrostatic force on $q$. As with gravitational potential energy we generally define $U = 0$ when the body is at infinity. Then

$$U(r) = -q \int_\infty^r \mathbf{E} \cdot d\mathbf{s}$$

If we bring a charge $q$ from infinity to a distance $r$ from a point charge $Q$, the potential energy is the work done against the electric force or

$$U = -q \int_\infty^r k_0 \frac{Q}{r^2} dr$$

$$= -qQk_0 \left[ -\frac{1}{r} \right]_\infty^r$$

$$U = k_0 \frac{qQ}{r} \qquad \text{(potential energy of point charges } Q \text{ and } q) \qquad (16\text{-}12)$$

Electric potential $V$ is defined as the electric potential energy per unit charge:

$$V \equiv \frac{U}{q} \qquad \text{(definition of electric potential)}$$

The unit of electric potential is the joule per coulomb, which is given a special name: the volt. Dividing both sides of Eq. 16-12 by $q$ gives the potential of a point charge $Q$:

$$V = \frac{k_0 Q}{r} \qquad \text{(potential due to point charge)} \qquad (16\text{-}13)$$

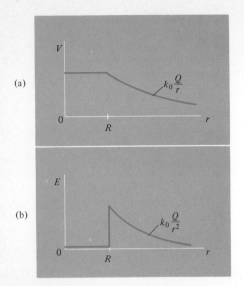

(a)

(b)

**Figure 16-13.** (a) *Potential of hollow sphere of radius R.* (b) *Corresponding electric field.*

**Physically electric potential is the work required to bring a unit charge from infinity to a distance $r$ from the point charge $Q$.** Electric potential is potential energy per unit charge just as electric field is force per unit charge.

**The potential difference between two points is the work required to move a unit charge from one point to the other.** The general expression for potential difference is obtained by dividing Eq. 16-11 by $q$:

$$V_b - V_a = -\int_a^b \mathbf{E} \cdot d\mathbf{s} \qquad \text{(potential difference)} \qquad (16\text{-}14)$$

Consider, for example, the potential difference between the surface and the center of a charged spherical shell. Since $E = 0$ over the region of integration, $V_b - V_a = 0$; that is, the potential is the same at the center as at the surface. $V$ as a function of $r$ is plotted in Figure 16-13a. Note that the corresponding electric field (Figure 16-13b) is at each point the negative slope of curve $a$. This is because Eq. 16-14 tells us that $dV = -E\,dr$ or $E = -dV/dr$. More generally,

$$dV = -\mathbf{E} \cdot d\mathbf{s}$$

**Figure 16-14.** (a) *Professor subjects himself to an electric potential of $\sim 10^5$ volts by touching electrode of a Van de Graaf generator.* (b) *Experiment is repeated by a student with longer hair. Hairs follow the electric lines of force.*

(a)

(b)

For $ds$ in the $x$ direction we have

$$dV = -E_x\, dx$$

Thus

$$E_x = -\frac{\partial V}{\partial x}, \quad E_y = -\frac{\partial V}{\partial y}, \quad E_z = -\frac{\partial V}{\partial z} \quad (16\text{-}15)$$

We see that the unit volt per meter can be used for electric field as well as newton per coulomb and that **E** points in the direction of decreasing $V$.

The maximum electric field possible in air at atmospheric pressure is measured to be about $10^6$ V/m. At field strengths above this there is electrical breakdown (a chain reaction where each ion forms new ions resulting in sparks or corona discharge). The effects of a corona discharge are shown in Figure 16-14a. The professor is touching the electrode of a Van de Graaf generator, which can achieve a potential of $\sim 10^5$ V (see Example 5). Thin sparks discharge from the ends of his hair. The charged hairs repel each other, displaying the lines of force pattern of his charged head. As seen in Figure 16-14b, the effect is more dramatic with longer hair.

**Example 5.** What are the maximum charge and voltage that a 30-cm diameter sphere can hold in air?

ANSWER: Since the field produced by a sphere is the same as that for a point charge, we can use Eq. 16-13 for the potential at the surface.

$$V = \frac{k_0 Q}{R} = \left(k_0 \frac{Q}{R^2}\right) R$$

Since the quantity in parentheses is the field $E$, we have

$$V = (E)R$$

Since the maximum value of $E$ is $10^6$ V/m, we have

$$V_{max} = (10^6 \text{ V/m})(0.15 \text{ m}) = 1.5 \times 10^5 \text{ V}$$

Solving $V = k_0 Q/R$ for $Q$ gives

$$Q_{max} = \frac{V_{max} R}{k_0} = \frac{(1.5 \times 10^5)(0.15)}{9 \times 10^9} \text{ C} = 2.5 \times 10^{-6} \text{ C}$$

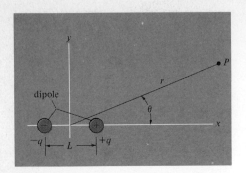

**Figure 16-15.** *Electric dipole pointing along x axis.*

**\* Example 6.** An electric dipole $p = qL$ is pointing along the x axis as shown in Figure 16-15. What are $V$, $E_x$, and $E_y$ when $r \gg L$?

ANSWER: When point $P$ is far compared to $L$, its distance from $+q$ is $(r - \tfrac{1}{2}L \cos \theta)$ and its distance from $-q$ is $(r + \tfrac{1}{2}L \cos \theta)$. Then the total potential at point $P$ is the sum of the separate potentials:

$$V = k_0 \frac{q}{r - \tfrac{1}{2}L \cos \theta} + k_0 \frac{(-q)}{r + \tfrac{1}{2}L \cos \theta} = k_0 \frac{qL \cos \theta}{r^2 - (L^2/4) \cos^2 \theta}$$

For $r \gg L$ this is

$$V \approx k_0 \frac{p \cos \theta}{r^2} = k_0 p \frac{x}{r^3}$$

$$E_y = -\frac{\partial V}{\partial y} = -k_0 p x \frac{\partial (r^{-3})}{\partial y} = -k_0 p x (-3) r^{-4} \frac{\partial r}{\partial y}$$

$$= 3 k_0 p \frac{x}{r} \frac{1}{r^3} \frac{\partial}{\partial y}(x^2 + y^2 + z^2)^{1/2} = 3 \frac{k_0 p \cos \theta}{r^3} \left(\frac{y}{r}\right)$$

$$= \frac{3 k_0 p \cos \theta \sin \theta}{r^3}$$

$$E_x = -\frac{\partial V}{\partial x} = -\frac{k_0 p}{r^3} - k_0 p x \frac{\partial (r^{-3})}{\partial x} = \frac{k_0 p}{r^3} \left(-1 + \frac{3x}{r} \frac{\partial r}{\partial x}\right)$$

$$-\frac{k_0 p}{r^3}(-1 + 3\cos^2 \theta)$$

$$= \frac{k_0 p}{r^3}(3 \cos^2 \theta - 1)$$

We see that for a fixed angle $\theta$ the field drops off as the inverse cube of the distance. A plot of the field lines is given in Figure 15-7(a).

We next consider the field and potential difference of two oppositely charged parallel plates of area $A$ a distance $x_0$ apart as shown in Figure 16-16. If they have total charge $+Q$ and $-Q$, their charge densities are $\sigma = Q/A$ and $-Q/A$.

According to Eq. 16-14 the potential difference is

$$\Delta V = -E x_0$$

Since lines of $E$ go from positive to negative, the minus sign tells us that

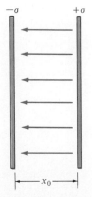

**Figure 16-16.** *Two parallel plates with equal and opposite charge density σ.*

the higher potential is at the positive plate. According to Eq. 16-9 $E = -4\pi k_0 \sigma$, so

$$\Delta V = 4\pi k_0 \sigma x_0$$

If the plates are of area $A$ and have charges $Q$ and $-Q$, respectively, we have $\sigma = Q/A$ and

$$\Delta V = \frac{4\pi k_0 x_0}{A} Q \qquad (16\text{-}16)$$

Our next example is a coaxial cable with a charge $\lambda$ per unit length on the central conductor and $-\lambda$ on the outer conductor. If the radius of the central conductor is $a$ and the outer conductor is $b$, what is the potential difference between the two conductors? See Figure 16-17. According to Eq. 16-14 the magnitude of the potential difference is

$$\Delta V = \int_a^b E \, dr$$

Using Eq. 16-7 for $E$, we have

$$\Delta V = \int_a^b \left(\frac{2k_0 \lambda}{r}\right) dr = 2k_0 \lambda [\ln r]_a^b$$

$$\Delta V = 2k_0 \lambda \ln \frac{b}{a} \qquad (16\text{-}17)$$

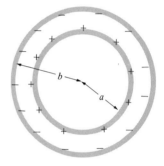

**Figure 16-17.** *Coaxial conductors with equal and opposite charge.*

If there are several charged bodies at distances $r_1, r_2, \ldots, r_n$ from point $P$, the electric potential at point $P$ is just the sum of the separate potentials. This is a consequence of the principle of superposition:

$$V = -\int \mathbf{E} \cdot d\mathbf{s} = -\int (\mathbf{E}_1 + \mathbf{E}_2 + \cdots + \mathbf{E}_n) \cdot d\mathbf{s}$$

$$= \left(-\int \mathbf{E}_1 \cdot d\mathbf{s}\right) + \left(-\int \mathbf{E}_2 \cdot d\mathbf{s}\right) + \cdots + \left(-\int \mathbf{E}_n \cdot d\mathbf{s}\right)$$

$$= V_1 + V_2 + \cdots + V_n$$

The potential difference between two points is unique because the electric force is a conservative force. As we saw in Chapter 6, the integral $\int \mathbf{F} \cdot d\mathbf{s}$ around a closed path is zero. Hence

$$\oint \mathbf{E} \cdot d\mathbf{s} = 0 \quad \text{(for electrostatics)} \qquad (16\text{-}18)$$

## The Electron Volt

A useful unit of energy in physics is the energy given to an electron (or any other particle of the same charge) when accelerated by a potential difference of 1 V. The electric field acting on the particle increases the kinetic energy of the particle by

$$\Delta K = -\Delta U = e\,\Delta V = (1.60 \times 10^{-19}\text{ C})(1\text{ V}) = 1.60 \times 10^{-19}\text{ J}$$

This amount of energy is defined as 1 electron volt:

$$1\text{ eV} = 1.60 \times 10^{-19}\text{ J} \quad \text{(electron volt)}$$

The abbreviation eV is used for electron volt. Related units are the MeV and the GeV;

$$1\text{ MeV} = 10^6\text{ eV} = 1.6 \times 10^{-13}\text{ J}$$
$$1\text{ GeV} = 10^9\text{ eV} = 1.6 \times 10^{-10}\text{ J}$$

---

**Example 7.** In the Bohr model of the hydrogen atom the electron is in a circular orbit of radius $R = 0.53 \times 10^{-10}$ m with the proton at the center. (a) What is the electron velocity? (b) What is the electric potential energy in electron volts? (c) What is the total mechanical energy in electron volts?

ANSWER: To get the velocity we write $F = ma$ for the electron, where $F = k_0 e^2/R^2$ is the electrostatic force and $a = v^2/R$ is the acceleration. Then

$$k_0 \frac{e^2}{R^2} = m \frac{v^2}{R} \tag{16-19}$$

$$v = \sqrt{\frac{k_0 e^2}{mR}} = \sqrt{\frac{(9 \times 10^9)(1.6 \times 10^{-19})^2}{(9.11 \times 10^{-31})(0.53 \times 10^{-10})}}\text{ m/s}$$

$$= 2.18 \times 10^6 \text{ m/s} = \frac{c}{137}$$

To get the potential energy we use Eq. 16-12 with $q_p = e$ and $q_e = -e$:

$$U = -k_0 \frac{e^2}{R} = -(9 \times 10^9)\frac{(1.6 \times 10^{-19})^2}{0.53 \times 10^{-10}}\text{ J} = -27.2\text{ eV}$$

We can get the kinetic energy by multiplying both sides of Eq. 16-19 by $\frac{1}{2}R$:

$$\tfrac{1}{2}mv^2 = \tfrac{1}{2}k_0\frac{e^2}{R} = -\tfrac{1}{2}U$$

We see that the kinetic energy is one half the magnitude of the potential energy. The total mechanical energy is

$$E = K + U = (-\tfrac{1}{2}U) + U = \tfrac{1}{2}U = -13.6 \text{ eV}$$

The magnitude of this is the amount of energy that must be given to the electron in order to move it out to infinity. This is called the ionization energy.

We close this section by noting that the surface of any conductor is an equipotential surface. If this were not so, two points on the surface separated by a distance $\Delta s$ could have a potential difference $\Delta V$. Then $E_s$ along the surface would equal $-\Delta V/\Delta s$ and conduction electrons would move in the direction opposite to $E_s$. The electric lines of force are always perpendicular to the equipotential contours and they "flow" downhill.

We could go on calculating potentials, charge distributions, and fields of various geometries of conductors and nonconductors. However, we are mainly interested in developing the basic concepts and laws, so we will skip this part of electrostatics.

## 16-5 Capacitance

The capacitor is a common device in almost all electronic equipment. A capacitor has two terminals and it has the property that it can hold or store electric charge when a potential difference is applied across the two terminals. The capacitance $C$ is defined as the stored charge $Q$ divided by the potential difference $\Delta V$:

$$C \equiv \frac{Q}{\Delta V} \quad \text{(capacitance)}$$

The units are coulomb per volt. This unit is given a special name called the **farad**. The farad is too large a unit to achieve by a normal-sized capacitor; hence, the unit $\mu F$ (microfarad) is commonly used.

Capacitors usually consist of two conducting surfaces separated by a

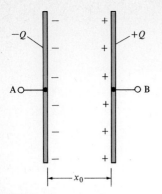

**Figure 16-18.** *Parallel plate capacitor. A potential difference V is applied across the terminals A and B.*

thin insulating layer. The charges on the two surfaces will be equal and opposite. The capacitor in Figure 16-18 has its two surfaces as flat sheets a distance $x_0$ apart. According to Eq. 16-16 the potential difference between the two sheets is

$$\Delta V = \frac{4\pi k_0 x_0}{A} Q$$

Then

$$\frac{Q}{\Delta V} = \frac{A}{4\pi k_0 x_0}$$

which by definition is the capacitance $C$:

$$C = \frac{A}{4\pi k_0 x_0} \qquad (16\text{-}20)$$

for a parallel plate capacitor with no material between the plates.

As a second example we shall calculate the capacitance per unit length of a coaxial cable. The capacitance per unit length is $\lambda$ divided by $\Delta V$:

$$C_l = \frac{\lambda}{\Delta V}$$

The denominator is given by Eq. 16-17:

$$\Delta V = 2k_0 \lambda \ln \frac{b}{a}$$

Then

$$C_l = \frac{1}{2k_0 \ln(b/a)} \qquad (16\text{-}21)$$

---

**Example 8.** What is the capacitance in picofarads (1 pF = $10^{-12}$ F) of a 1-m long hi fi cable where the center conductor is 1 mm in diameter and the shield is 5 mm in diameter?

ANSWER: Using Eq. 16-21 we have

$$C_l = \frac{1}{2k_0 \ln(b/a)} \quad \text{where} \quad \frac{b}{a} = 5$$

$$C_l = \frac{1}{2(9 \times 10^9) \ln 5} \text{ F/m} = 34.5 \times 10^{-12} \text{ F/m} = 34.5 \text{ pF/m}$$

---

324 CH. 16/ELECTROSTATICS

For a 1-m long cable this is 34.5 pF if there is no material between the conductor and shield. However, hi fi cables usually have polyethylene between the center conductor and shield. As we shall see in the next section, one must then multiply by the dielectric constant of polyethylene, which is 2.3.

## Energy Storage

Suppose an uncharged capacitor is charged by gradually increasing the potential difference from 0 to $V_0$. The charge will increase from 0 to $Q_0$ where $Q_0 = CV_0$. The work done in transferring $dq$ from the negative terminal to the positive terminal is

$$dU = V\,dq$$

The total work or stored energy is

$$U = \int_0^{Q_0} V\,dq$$

$$= \int \left(\frac{q}{C}\right) dq$$

$$U = \frac{1}{2}\frac{Q_0^2}{C} \qquad \text{(energy stored in a capacitor)} \qquad (16\text{-}22)$$

**Example 9.** How much work must be done to charge up a spherical shell of total charge $Q$ and radius $R$?

ANSWER: If the surface already holds a charge $q$, the work done in bringing an additional $dq$ from infinity is

$$dU = V\,dq = \left(k_0 \frac{q}{R}\right) dq$$

The total work done in assembling the total charge $Q$ is

$$U = \int \frac{k_0}{R} q\,dq = \frac{k_0}{R}\left[\frac{q^2}{2}\right]_0^Q = \frac{k_0 Q^2}{2R} \qquad (16\text{-}23)$$

**Example 10.** Assume the radius of the electron is the same as the radius of the proton, which is $10^{-15}$ m. Assume that the electron's charge ($q_e = -1.6 \times 10^{-19}$ C) resides on the surface of the electron. (a) What is the potential energy of such a charge distribution? (b) What is the relativistic mass equivalent of this energy?

ANSWER: Equation 16-23 gives

$$U = \frac{k_0 q_e^2}{2R} = \frac{(9 \times 10^9)(1.6 \times 10^{-19})^2}{2 \times 10^{-15}} \text{ J}$$
$$= 1.15 \times 10^{-13} \text{ J} = 0.72 \text{ MeV}$$

The equivalent mass is

$$m = \frac{U}{c^2} = \frac{1.15 \times 10^{-13}}{(3 \times 10^8)^2} \text{ kg} = 1.28 \times 10^{-30} \text{ kg}$$

We note that this is already greater than the measured mass of $0.91 \times 10^{-30}$ kg. Recent experiments indicate that the electron radius is at least 10 times smaller than $10^{-15}$ m; hence, its rest mass should be more than ten times what is measured. This serious discrepancy is a current problem in physics. There still are some serious basic questions which present-day physics cannot answer and this is one of them.

It is of interest to express Eq. 16-22 in terms of the electric field rather than the charge. This is easy to do for the parallel plate capacitor in Figure 16-16. Then

$$E = 4\pi k_0 \frac{Q_0}{A} \quad \text{or} \quad Q_0 = \frac{EA}{4\pi k_0}$$

Now we substitute this into Eq. 16-22 to obtain

$$U = \frac{1}{2C} \left(\frac{EA}{4\pi k_0}\right)^2$$

Use Eq. 16-20 to eliminate $C$:

$$U = \frac{1}{2(A/4\pi k_0 x_0)} \left(\frac{EA}{4\pi k_0}\right)^2 = \frac{E^2}{8\pi k_0} A x_0$$

Now we divide both sides by $\mathcal{V} = Ax_0$, which is the volume occupied by the field $E$:

$$\frac{U}{\mathcal{V}} = \frac{E^2}{8\pi k_0} \quad \text{(energy density of electric field)} \quad (16\text{-}24)$$

Suppose we were to assert that the energy used to move the charge into its final position was stored in the electric field and that the electric field contained $E^2/8\pi k_0$ J/m³. Clearly, such a procedure for calculating the stored energy works in the case of a parallel plate capacitor. A more complicated, general derivation proves that the total energy required to build up *any* charge distribution is exactly equal to the integral of $E^2/8\pi k_0$ over all space, where $E$ is the field produced by that charge distribution. So Eq. 16-24 is completely general and it is safe to adopt the physical interpretation of energy per unit volume stored in the electric field at the rate of $E^2/8\pi k_0$. Thus far the use of Eq. 16-24 seems to be just an alternate way to calculate the energy required to assemble a group of charges; however, the interpretation of energy stored in the field takes on even more meaning when we study the energy radiated by an accelerating charge via an electric and magnetic field moving away with the speed of light. As we shall see in Chapter 20, the radiated energy agrees with Eq. 16-24.

## 16-6 Dielectrics

In previous sections we have considered the fields of charges placed on conductors that are situated in otherwise empty space. It is observed that when material is inserted between the plates of a capacitor, the capacitance increases. If we call the new capacitance $C'$, the ratio of $C'$ to $C$ is defined as the dielectric constant $\kappa$ of the material:

$$\kappa \equiv \frac{C'}{C} \quad \text{(definition of dielectric constant)} \quad (16\text{-}25)$$

where $C$ is the capacitance with no material between the plates.

The material (a nonconductor) placed between capacitor plates is called the dielectric. Figure 16-19 shows why the capacitance goes up when a dielectric is inserted between the plates. When a nonconductor is put in an external electric field, induced charges appear at either end, as shown in Figure 16-20. The explanation of what causes the induced charges will be given in the next paragraph where we shall "derive" the dielectric constant in terms of basic atomic constants. First we shall express $\kappa$ in terms of the induced charge $q'$. The capacitance in Figure 16-19 is

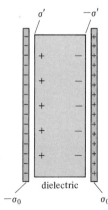

**Figure 16-19.** Induced charge $q' = \sigma' A$ appears on dielectric slab when inserted between capacitor plates. $q'$ reduces $E$ and the potential difference between plates.

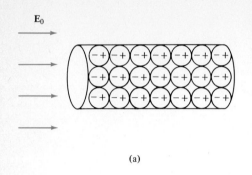
(a)

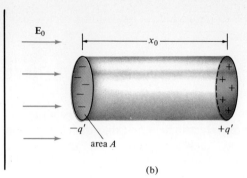
(b)

**Figure 16-20.** *When a dielectric cylinder is placed in an external field* $E_0$, *the atomic dipole moments line up as in* (a). *The net charge distribution is shown in* (b).

where

$$C' = \frac{q_0}{V} = \frac{q_0}{Ex_0}$$

$$E = 4\pi k_0 \sigma = 4\pi k_0 \left(\frac{q_0}{A} - \frac{q'}{A}\right)$$

So

$$C' = \frac{q_0}{4\pi k_0(q_0/A - q'/A)x_0} = \frac{1}{1 - q'/q_0} \frac{A}{4\pi k_0 x_0} = \frac{1}{1 - q'/q_0} C$$

Dividing by $C$ gives

$$\kappa = \frac{1}{1 - q'/q_0} \qquad (16\text{-}26)$$

Now we shall explain why an induced charge $q'$ must appear at the two ends of a nonconductor when it is placed in an external electric field. This is because the individual molecules or atoms must have dipole moments. Some molecules have permanent dipole moments and are called polar molecules. But even those molecules or atoms with zero dipole moment must have an induced dipole moment as shown in Example 2. If each particle has an average dipole moment $\bar{p}$ pointing in the direction of $E$ and there are $N$ particles in the cylinder of Figure 16-20, its total dipole moment is

$$p_{\text{total}} = N\bar{p}$$

Since the particles or atoms are still electrically neutral, the average charge over any volume inside the cylinder is zero; however, effective charge $q'$ appears at the ends giving

$$p_{\text{total}} = q'x_0$$

for the total dipole moment of the cylinder. Equating these two expressions for $p_{\text{total}}$ gives

$$q'x_0 = N\bar{p}$$
$$q'x_0 = (\mathfrak{N}Ax_0)\bar{p}$$

where $\mathfrak{N}$ is the number of particles per unit volume. Then

$$q' = \mathfrak{N}A\bar{p}$$

Substituting into Eq. 16-26 gives

$$\kappa = \frac{1}{1 - \mathfrak{N}(A/q_0)\bar{p}} \qquad (16\text{-}27)$$

For most materials the ratio $\bar{p}/q_0$ is a constant, and then $\kappa$ is independent of the applied voltage. For polar molecules $\bar{p}$ decreases with temperature. All atoms and nonpolar molecules have zero dipole moment; however, when in the presence of an electric field, they will have an induced dipole moment

$$p \approx \frac{R^3}{k_0} E$$

as shown in Eq. 16-4. Using $E = 4\pi k_0[q_0 - q')/A]$, we have

$$p \approx 4\pi R^3 \left(\frac{q_0 - q'}{A}\right)$$

Substitution in Eq. 16-27 gives

$$\kappa \approx \frac{1}{1 - 4\pi \mathfrak{N}R^3(1 - q'/q_0)}$$

By using Eq. 16-26 we can replace the expression in parentheses with $1/\kappa$:

$$\kappa \approx \frac{1}{1 - 4\pi \mathfrak{N}R^3(1/\kappa)}$$

Solving for $\kappa$ gives

$$\kappa \approx 1 + 4\pi \mathfrak{N}R^3$$

This derivation of dielectric constant in terms of the atomic radius $R$ and the number of atoms per unit volume $\mathfrak{N}$ is not exact because of the approximation that the outer electrons can be represented by a uniformly charged sphere. Also the atoms must not be too close. However, it does work reasonably well for atomic gases, and it gives us a physical understanding of why materials made out of nonpolar molecules or atoms have a dielectric constant greater than 1.

## Summary

A simple application of Gauss' law shows that the field outside a spherical charge (or mass) distribution is the same as if all the charge (or mass) were concentrated at the center, $E = k_0 Q_{\text{total}}/r^2$. Since the field inside a uniformly charged cloud increases proportional to the displacement from the center, a point charge inside the cloud will oscillate in SHM. Similarly, a spherical atomic cloud, when in an external electric field $E_0$, will be displaced proportional to $E_0$; that is, there will be an induced dipole moment $p = (R^3/k_0)E_0$ where $R$ is the radius of the sphere.

The field produced by a line charge goes as $1/r$, $E = 2k_0\lambda/r$, whereas the field produced by a charged infinite plane is constant everywhere in space, $E = 2\pi k_0 \sigma$.

The electric potential energy of a charge $q$ is given by $U = -q\int_{\infty}^{r} \mathbf{E} \cdot d\mathbf{s}$ when $U$ is defined equal to zero at infinity. Electric potential at a given point in space is the potential energy per unit charge $V = U/q$ and is thus the work required to bring a unit charge from infinity to the given point. The electric potential of a point charge is $V = k_0 Q/r$.

The potential difference between two capacitor plates of equal and opposite charge $Q$ is $\Delta V = 4\pi k_0 x_0 (Q/A)$. Since capacitance is $C = Q/\Delta V$, a parallel plate capacitor has $C = A/(4\pi k_0 x_0)$.

An electron picks up a kinetic energy of one electron volt (1 eV) when it is accelerated by a potential difference of 1 V. 1 eV = $1.60 \times 10^{-19}$ J.

The energy stored in a capacitor is $U = \tfrac{1}{2}Q^2/C$, which is also equal to the integral of $E^2/(8\pi k_0)$ over all space. The energy density of an electric field is $dU/dV = E^2/(8\pi k_0)$. The dielectric constant $\kappa = C'/C$ is the ratio of capacity with and without the dielectric. It is due to induced charges $q'$, which reduce the electric field when the dielectric is inserted; that is, $E' = E/\kappa$ is the reduced electric field.

## Exercises

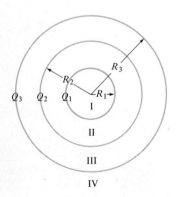

Exercise 1

1. Consider three concentric thin shells with charges $Q_1$, $Q_2$, and $Q_3$, as shown. What are $E$ and $V$ in regions I, II, III, and IV?
2. Suppose $P = 10^{-3}$ atm and $T = 300$ K in the ionosphere and that $10^{-4}$ of the atoms are ionized. What would be the plasma frequency?
3. How large a sphere is needed to hold half a million volts in air? How much charge would it hold?
4. Consider two infinite slabs of charge density $\rho$ each. They are of thickness $x_0$ and a distance $d$ apart. What is $E$ in regions I, II, III, IV, and V?
5. Repeat Exercise 4 for the case where the slabs are of equal but opposite charge.
6. Use the results of Example 6 to show that the electric potential at a distance $r$ from a dipole $p$ is $V = k_0 \mathbf{p} \cdot \hat{\mathbf{r}}/r^2$ where $\hat{\mathbf{r}}$ is the unit vector in the direction of $\mathbf{r}$.
7. A parallel plate capacitor has a plate separation of 0.1 mm. What must be the plate area to achieve a capacitance of 1 F?
8. The capacitance of a single conductor is defined as $C = Q/V$ where $V$ is the electric potential of the conductor with respect to infinity. What then is the

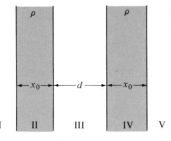

Exercise 4

Exercise 9

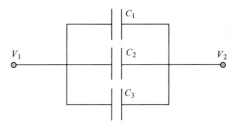

Exercise 10

capacitance of a sphere of radius $R$? What would be the numerical value in picofarads when $R = 1$ cm?

9. The capacitance of capacitors in series is defined as $C = Q/(V_2 - V_1)$. What is $C$ in terms of $C_1$, $C_2$, and $C_3$?

10. The capacitance of capacitors in parallel is defined as

$$C = \frac{Q_1 + Q_2 + Q_3}{V_2 - V_1}$$

What is $C$ in terms of $C_1$, $C_2$, and $C_3$?

11. An electric dipole consists of two charges $q$ and $-q$ of $10^{-7}$ C each, 2 cm apart.
    (a) What is the total flux or net number of lines of force leaving the dashed spherical surface of radius 2 cm?
    (b) What is the electric potential at the center of the sphere?
    (c) What is the electric field at the center of the sphere?

12. Suppose that in the helium nucleus there are two protons separated by $1.5 \times 10^{-15}$ m.
    (a) What is the electrostatic force between them?
    (b) How much work must be done in order to bring two protons this close together?

13. An electron is $5.3 \times 10^{-11}$ m from a proton. How much velocity must it have in order to escape to infinity?

14. Consider an electron of charge $-e$ and a neutron of zero charge separated by a distance $R$. Let $m$ be the electron mass and $M$ the neutron mass.
    (a) What will be the force between them in terms of the distance $R$ and any other universal physical constants?
    (b) Suppose the electron is in a circular orbit about the neutron. What is the force between them in terms of $m$, $R$, and $v$ (the electron's circular velocity)?
    (c) What is the kinetic energy of the electron in terms of $G$, $m$, $M$, and $R$?
    (d) What is the electron potential energy?

15. Two parallel plates are separated by 2 cm. The field between the plates is 20,000 N/C. What is the potential difference between the plates?

16. Consider the case of two parallel infinite planes separated by 8 cm. Both have $10^{-6}$ C of positive charge per square meter. What is the electric field between the planes?

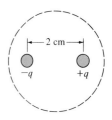

Exercise 11

17. Suppose a 1-cm diameter carbon sphere has one extra electron per million protons.
    (a) If the density is 1.7 g/cm³, what is the charge?
    (b) What is the electric field and potential at the surface of the sphere?
18. An electron is in a circular orbit around a proton. What is the ratio of the potential energy to the kinetic energy of this electron? Is this ratio positive or negative? What is the ratio of the binding energy to the kinetic energy?
19. Suppose the earth had an excess surface charge of 1 electron/cm². What would be the earth's potential?
20. Estimate the dielectric constant of helium gas at 1 atm and 300 K. Use $R = 10^{-10}$ m.
21. Estimate the dielectric constant of liquid helium. Use $R = 10^{-10}$ m and $\rho = 150$ kg/m³.
22. What is the total capacity of the system of four identical capacitors?

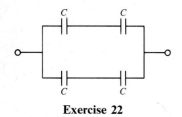

Exercise 22

## Problems

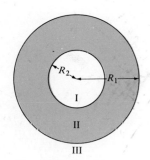

Problem 23

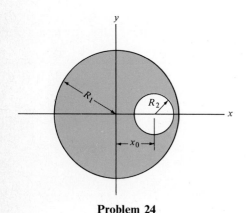

Problem 24

23. Consider a solid uniformly charged sphere of radius $R_1$ with a hole of radius $R_2$ at the center. If the total charge is $Q$, what is $E$ in regions I, II, and III?
24. Consider a uniformly charged sphere of radius $R_1$ with a spherical hole of radius $R_2$ displaced a distance $x_0$ from the center. What is $E$ on the $x$ axis as a function of $x$? Let $\rho$ be the charge density. (*Hint:* The superposition of two spheres of radii $R_1$ and $R_2$ and of equal and opposite charge densities has the same solution.)
25. Repeat Problem 24 for $E$ along the $y$ axis. Specify direction as well as magnitude.
26. What is the electric potential as a function of $r$ and $\theta$ at any point inside the sphere for Problem 24?
27. What is $V(r)$ in all three regions of Problem 23?
28. Consider a single-wire power line of 1-cm diameter bare wire. Near each point of suspension the electric potential is zero at a distance of 20 cm. What is the highest voltage this line can carry before breakdown in air? What is the maximum charge per meter that this wire could hold? Assume zero potential at all points 20 cm from the wire.
29. Two metal plates of 100 cm² area are separated by 2 cm. The charge on the left-hand plate is $-2 \times 10^{-9}$ C, and the charge on the right-hand plate is $-4 \times 10^{-9}$ C.
    (a) What is the field immediately to the left of the left plate?
    (b) What is the field between the plates?
    (c) What is the field immediately to the right of the right plate?
    (d) What is the potential difference between the plates?
30. Consider a solid uniformly charged cylinder of radius $R_1$ with a cylindrical hole of radius $R_2$. If the charge density is $\rho$, what is $E$ both inside and out?
31. Consider two oppositely charged, infinitely long wires separated by $x_0$ as shown. If the line charges are $\lambda$ and $-\lambda$, show that

$$V(r, \alpha) = 2k_0 \frac{\lambda x_0}{r^2} \cos \alpha \quad \text{when } r \gg x_0$$

32. For Problem 31, derive exact expressions for $E$ along the $x$ axis and along the $y$ axis.
33. A high energy accelerator shoots a beam of protons of 1 MeV kinetic energy into hydrogen gas.
    (a) What is the total kinetic energy of one beam proton and one target proton in their center-of-mass system?
    (b) What will be the electric potential energy in joules between a beam proton and a hydrogen nucleus when they are at their closest possible distance of approach?
    (c) What is the closest distance of approach?
34. Consider the three charged planes as shown. The potential of plane A is zero. $E_3 = 0$.
    (a) What is $V_B$?
    (b) What is $V_C$?
    (c) What are the charge densities on each of the three planes?
35. Assume that a proton as it approaches a certain atomic nucleus will "see" a potential energy curve $U(r)$ as shown. Suppose a proton is trapped in the nucleus of radius $R$ with a total energy $E_0 = 2$ MeV as shown.
    (a) What will be its kinetic energy in the nucleus?
    (b) What will be its potential energy in the nucleus?
    (c) According to classical physics how much additional energy does it need in order to get out of the nucleus?
    (d) If external protons are shot at the nucleus, classically how much initial kinetic energy would they need in order to penetrate the nucleus? Assume the protons start from infinity.
36. The gap of a parallel plate capacitor is filled with polyethylene ($\kappa = 2.3$). If the gap is 1 mm and the capacitor voltage is 1000 V, what is the induced charge per square meter on the surface of the polyethylene?
37. A slab of dielectric constant $\kappa$ and thickness $x_1$ is placed in the gap of a parallel plate capacitor. What is the capacitance $C'$ if the gap $x_0$ is greater than $x_1$?
38. Two conducting spheres are connected by a conducting wire of length $L$ where $L \gg R_1 > R_2$. A charge is placed on the conducting system raising it to a potential $V_0$.
    (a) What is the ratio of $E_1$ (at the surface of $R_1$) to $E_2$ (at the surface of $R_2$)? (This explains why $E$ is stronger at sharp corners or edges of a conductor.)
    (b) What is the ratio of $\sigma_1$ to $\sigma_2$?

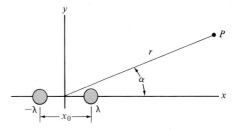

Problem 31

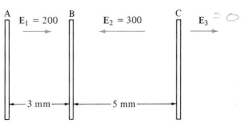

Problem 34

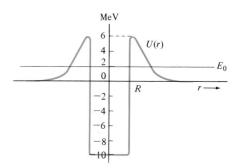

Problem 35

Problem 38

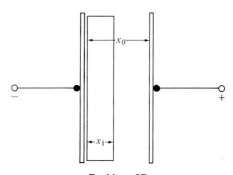

Problem 37

39. In Problem 38 what are $Q_1$ and $Q_2$ in terms of $V_0$, $R_1$, and $R_2$?

40. Consider a solid sphere of uniform charge density of total charge $Q$ and radius $R$. What is the electrical potential energy of the charge distribution? Use $dU = V\, dq$ and build up the sphere by adding concentric shells of $dq = (4\pi r^2\, dr)\rho$.
41. In Problem 40 show that the electric potential energy is $U = \frac{3}{5}k_0 Q^2/R$ by integrating $E^2/8\pi k_0$ over all space.
42. What is the electric potential at the center of the sphere in Problem 40?

# Current and Magnetic Force

In the study of electricity, the subjects of magnetism and magnetic field are of importance comparable to electrostatics and electric field. In Chapter 18 we shall see that the force on a magnet is due to the magnetic force on moving atomic electrons within the magnet. In this chapter we shall show that magnetic force is simply a consequence of special relativity. We shall demonstrate that Coulomb's law plus special relativity requires that, in addition to the electrostatic force, there must be a "new" force on a charge $q$ proportional to its velocity $v$. This new force is called the magnetic force, and this magnetic force divided by $(qv)$ will be defined as the magnetic field. Practical applications of magnetic effects will be discussed in the following two chapters.

## 17-1 Electric Current

Magnetic effects occur when there are moving charges or currents. Before we can study magnetism we must define electric current. The current passing through a given area is defined as the charge per unit time passing through that area:

$$I \equiv \frac{Q}{t} \quad \text{(definition of electric current)} \quad (17\text{-}1)$$

The unit is coulombs per second, which is given the special name ampere (abbreviated A).

A related quantity is the electric current density, which is defined as charge density $\rho$ times its velocity $v$:

$$\mathbf{j} \equiv \rho \mathbf{v} \quad \text{(definition of current density)} \quad (17\text{-}2)$$

This is the current per unit area and has the unit $C\ m^{-2}\ s^{-1}$ or $A/m^2$.

If we multiply by an area $A$ whose surface is perpendicular to $\mathbf{j}$, we obtain the current $I$:

$$I = \mathbf{j} \cdot \mathbf{A}$$

where the vector direction of $\mathbf{A}$ is perpendicular to the surface. If $\mathbf{j}$ varies over $A$, then

$$I = \int \mathbf{j} \cdot d\mathbf{A}$$

In a metal wire the positive charges (atomic nuclei) cannot move; they are bound in a crystalline structure. However, the outer electrons or conduction electrons are no longer bound to their particular atoms, but are quite free to move along the wire. (This is contrary to all ideas of classical physics and can only be explained using quantum mechanics as is done in Chapter 28.) In the absence of an external electric field, conduction electrons move randomly in all directions with zero net velocity. If $\mathcal{N}$ equals the number of conduction electrons per unit volume, then $\rho = \mathcal{N}e$ is the magnitude of the charge density and $j = (\mathcal{N}e)\bar{v}_d$, where $\bar{v}_d$ is the net drift velocity of the conduction electrons. The magnitude of the current is obtained by multiplying by the area $A$:

$$I = \mathcal{N}e\bar{v}_d A \tag{17-3}$$

But what is the direction of the current? According to the convention established by Benjamin Franklin, a current flowing into a capacitor plate would supply positive charge to the plate. We now know, however, that a capacitor plate is made positive by conduction electrons flowing away from the plate. Hence the conduction electrons always flow in the direction opposite to that of the current. If the charge of the electron had been chosen as positive rather than negative, this difficulty would not have arisen. In this book (and most other books) whenever current is indicated by an arrow it means the direction that positive charges would be flowing. If said current is actually due to moving electrons, they will be moving against the arrow.

**Example 1.** A current of 1 A flows through a copper wire of 1 mm² cross section. What is the average drift velocity of the conduction electrons?

ANSWER: Solving Eq. 17-3 for $\bar{v}_d$ gives

$$\bar{v}_d = \frac{I}{\mathcal{N}eA}$$

Assuming one conduction electron per atom, $\mathcal{N} = DN_0/M$ where $D = 8.9$ g/cm$^3$ is the density, $N_0 = 6.02 \times 10^{23}$ atoms/mole, and $M = 63.6$ g/mole. Then

$$\mathcal{N} = \frac{(8.9 \text{ g/cm}^3)(6.02 \times 10^{23} \text{ atoms/mole})}{63.6 \text{ g/mole}} = 8.42 \times 10^{22} \frac{\text{atoms}}{\text{cm}^3}$$

$$= 8.42 \times 10^{28} \text{ m}^{-3}$$

$$\bar{v}_d = \frac{1 \text{ A}}{(8.42 \times 10^{28} \text{ m}^{-3})(1.6 \times 10^{-19} \text{ C})(10^{-6} \text{ m}^2)} = 7.4 \times 10^{-5} \text{ m/s}$$

$$= 0.074 \text{ mm/s}$$

We see that the typical drift velocity of conduction electrons is ~0.1 mm/s. Currents can also flow in gases and liquids. Neon signs and fluorescent lights are examples of currents in gases. In these the current is due to moving positive ions as well as the moving electrons. However, the lighter electrons are much faster and are the major contributor to the current. When an electron collides with a gas ion or atom, the kinetic energy of the collision can be absorbed by the atom and then radiated away in the form of electromagnetic radiation, some of which can be seen by the eye as light.

## 17-2 Ohm's Law

If a potential difference $V$ is applied across a conductor, some current $I$ will flow. Early in the nineteenth century Georg Ohm discovered that the magnitude of the current in metals was proportional to the applied voltage as long as the temperature was kept constant. Ohm defined the resistance of a conductor as the voltage divided by the current.

$$R \equiv \frac{V}{I} \quad \text{(definition of resistance)}$$

This equation is the definition of resistance. The following statement is Ohm's law. The ratio

$$R = \frac{V}{I} \text{ is independent of } I \text{ for a metal when at constant } T$$
$$\text{(Ohm's law)}$$

In the mks system $V$ is in volts and $I$ in amperes. The unit of resistance is volts per ampere. This unit was given the special name ohm (by someone other than Georg Ohm). The abbreviation is $\Omega$.

Just how basic is Ohm's law? Is it some new fundamental law of nature, or is it merely the consequence of the structure of matter and basic laws of interaction? Fortunately, the latter is the case. The resistance of different substances under various conditions is explained quite well by the quantum theory of solids (Chapter 28). In the following paragraph Ohm's law is derived using two results from the theory of metals.

### Derivation of Ohm's Law

The quantum theory of metals tells us that, because of the wave nature of electrons, the outer atomic electrons are not bound to individual atoms in the metallic crystal. The quantum theory also tells us that these conduction electrons can travel many atomic diameters before having an atomic collision. Let $L$ be the average distance between collisions, which is called mean free path. The average time between collisions will be $\Delta t = L/u$, where $u$ is the average speed of the conduction electrons ($u$ is in all random directions; hence it does not give rise to a net current). If a voltage or potential difference is applied across a piece of metal, there will be a force of magnitude $eE$ on each conduction electron, and during the time $\Delta t$ each conduction electron will achieve a drift velocity $v_d = \Delta u$ given by

$$m \frac{\Delta u}{\Delta t} = eE \quad \text{(Newton's second law)}$$

$$\Delta u = v_d = \frac{eE \, \Delta t}{m}$$

Replacing $\Delta t$ by the average time $L/u$ gives

$$\overline{v_d} = \frac{eL}{mu} E$$

The drift velocity is in the same direction (direction of $-\mathbf{E}$) for all the electrons and thus gives rise to a net current. Each electron loses its drift velocity after each collision. The mean free path $L$ is so small that $v_d$ is always much less than $u$. As we saw in Example 1 the drift velocity is usually less than 1 mm/s.

We can obtain the current in a wire of cross-sectional area $A$ by substituting the above expression for $\overline{v_d}$ into Eq. 17-3:

$$I = \frac{\mathcal{N}e^2 LA}{mu} E \qquad (17\text{-}4)$$

We can obtain a formula for the resistance of the wire of length $x_0$ shown in Figure 17-1 by noting that the voltage across it will be $Ex_0$. If we substitute $V/x_0$ for $E$ in Eq. 17-4, we obtain

$$I = \left(\frac{\mathcal{N}e^2 L}{mu}\frac{A}{x_0}\right)V$$

and

$$R = \frac{mux_0}{\mathcal{N}e^2 LA}$$

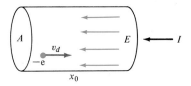

**Figure 17-1.** *Wire of length $x_0$ showing one of its conduction electrons with drift velocity $v_d$.*

is the resistance. We see that the resistance is proportional to the length $x_0$ of a rod and inversely proportional to the area $A$. We note that our expression for $R$ is a constant as long as $u$ remains constant. The average electron velocity $u$ will stay constant if the temperature is held constant. The proportionality constant is defined as the resistivity $\rho$; that is,

$$R = \rho \frac{x_0}{A}$$

---

**Example 2.** Electrical conductivity $\sigma$ is defined by the relation $j = \sigma E$. What is $\sigma$ in terms of $\mathcal{N}$, $e$, $m$, and $u$?

ANSWER: Take a cylinder of length $x_0$ and cross-sectional area $A$. Multiply both sides of the above equation by $A$:

$$jA = \sigma AE$$

The left-hand side is by definition the current $I$, so

$$\sigma = \frac{I}{AE}$$

Now substitute Eq. 17-4 for $I$:

$$\sigma = \frac{\left(\dfrac{\mathcal{N}e^2 LAE}{mu}\right)}{AE} = \frac{\mathcal{N}e^2 L}{mu} = \frac{1}{\rho} \qquad (17\text{-}5)$$

---

### Heat Loss

Each time a conduction electron has an atomic collision, it loses the extra energy it picked up from the electric field. This energy becomes

SEC. 17-2/OHM'S LAW 339

disordered atomic motion or heat. Since there is no net gain in kinetic energy, the energy loss to collisions for conduction electrons of charge $dq$ moving across a potential difference $V$ is

$$dE_{\text{heat}} = V\, dq$$

Now divide both sides by $dt$:

$$\frac{dE_{\text{heat}}}{dt} = V\frac{dq}{dt} = VI$$

$$P = VI \quad \text{(electric power loss)} \quad (17\text{-}6)$$

Equation 17-6 can be expressed as $P = I^2 R$ by replacing $V$ with $IR$ or as $P = V^2/R$ by replacing $I$ with $V/R$. The quantity $P$ is electrical power being converted into heat. Note that power, which has units of watts, also has the units of volts times amperes. A 100-W light bulb draws a current $I = P/V = 100\text{ W}/120\text{ V} = 0.83\text{ A}$. The electrical energy goes into heat and light. In a home nearly all the light is absorbed and converted to heat by indoor objects. Thus electric light bulbs are just about as efficient as electric heaters in heating a home.

---

**Example 3.** A 60-W light bulb operates at 120 V. What is its resistance?

ANSWER: The current is $I = 60\text{ W}/120\text{ V} = 0.5\text{ A}$. The resistance can be determined using the definition of resistance:

$$R = \frac{V}{I} = \frac{120\text{ V}}{0.5\text{ A}} = 240 \text{ ohms}$$

This result could be obtained more directly by using the relation $P = V^2/R$.

---

### Electromotive Force (EMF)

In order to maintain a steady current in a resistor or a wire, we need a steady source of electric energy. Two common sources are electric batteries and electric generators. Such a source of electric energy is called a seat of electromotive force or simply, an **emf**. In such a device energy of some form is converted into electric energy. In a battery it is chemical energy, and in a generator it is mechanical energy. In a solar cell, light energy is converted into electric energy. We shall use the symbol $\mathcal{E}$ for emf:

$$\mathcal{E} = \frac{\Delta W}{\Delta q} \quad \text{(definition of emf)}$$

where $\Delta W$ is the electrical energy given to a charge $\Delta q$ when it passes through the seat of emf. We shall use the symbol labeled $\mathcal{E}$ in Figure 17-2 to designate a battery. A charge $\Delta q$ in passing from the negative to positive terminal will gain energy equal to $(\Delta q)\mathcal{E}$.

## 17-3 Direct Current Circuits

Although it is rightfully a topic for applied physics or engineering, we shall say a few words about circuit theory. We use the abbreviations dc for direct current and ac for alternating current. A brief introduction to ac circuit theory is given in Section 19-6. So many everyday applications of simple circuits exist that it is useful to have some discussion of circuits in a general physics textbook. (For further discussion see the Appendix to Chapter 28.) In our everyday life it is useful to know how to hook up speakers and record players to a stereo system, how to connect burglar alarms, Christmas tree lights, battery chargers, automobile tape players, and so on.

Most circuits involve series and/or parallel combinations of resistances. The total resistance $R_t$ of a combination of resistances is obtained by dividing the voltage applied across the combination by the total current flowing through the combination. In each of the resistance combinations shown in Figure 17-2 the total resistance is

$$R_t = \frac{V}{I}$$

In a series circuit the same current flows through all the resistors. In a

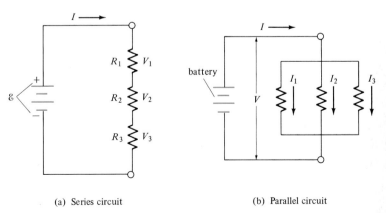

**Figure 17-2.** Resistances $R_1$, $R_2$, and $R_3$ (a) in series combination and (b) in parallel combination.

(a) Series circuit

(b) Parallel circuit

parallel circuit the total current is the sum of separate currents through each of the resistors. In the series combination the total potential difference $V$ is

$$V = V_1 + V_2 + V_3$$

Dividing both sides by $I$

$$\frac{V}{I} = \frac{V_1}{I} + \frac{V_2}{I} + \frac{V_3}{I}$$

or

$$R_t = R_1 + R_2 + R_3 \quad \text{(series combination)} \quad (17\text{-}7)$$

We see that in a series circuit the total resistance is the sum of the individual resistances.

In the parallel combination (Figure 17-2b)

$$I = I_1 + I_2 + I_3$$

Dividing both sides by $V$

$$\frac{I}{V} = \frac{I_1}{V} + \frac{I_2}{V} + \frac{I_3}{V}$$

or

$$\frac{1}{R_t} = \frac{1}{R_1} + \frac{1}{R_2} + \frac{1}{R_3} \quad \text{(parallel combination)} \quad (17\text{-}8)$$

In a parallel circuit, the reciprocal of the total resistance is obtained by adding the reciprocals of the individual resistances.

Many complicated circuits can be solved by breaking them down into combinations of simple series and parallel circuits. This procedure is illustrated in the following example.

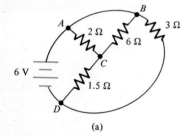

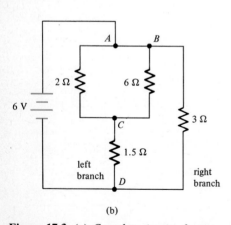

**Figure 17-3.** (a) *Complex circuit of resistances.* (*The symbol* $\Omega$ *stands for ohm.*) (b) *The same circuit redrawn to make series and parallel combinations more apparent.*

**Example 4.** In the circuit shown in Figure 17-3: (a) what is the total current supplied by the battery? (b) How much current flows through the 6-ohm resistor?

ANSWER: In order to find $I$ we must first calculate $R_t$. This is facilitated by redrawing exactly the same circuit as shown in Figure 17-3(b), so that the series and parallel combinations are made more apparent. We start with the parallel combination of the 2- and 6-ohm resistors. Let $R$ be the resistance of this parallel combination.

Then

$$\frac{1}{R} = \frac{1}{2} + \frac{1}{6} = \frac{2}{3}$$

$$R = 1.5 \text{ ohms}$$

This is in series with a 1.5-ohm resistor; hence the total resistance $R'$ of the left-hand branch

$$R' = R + 1.5 = 3 \text{ ohms}$$

Finally this left-hand branch ($R'$) is in parallel with the 3-ohm resistor. Thus

$$\frac{1}{R_t} = \frac{1}{R'} + \frac{1}{3} = \frac{2}{3}$$

$$R_t = 1.5 \text{ ohms}$$

and

$$I = \frac{V}{R_t} = \frac{6}{1.5} \text{A} = 4 \text{ A}$$

is the total current supplied by the battery.

To find the current through the 6-ohm resistor we must first determine the current $I'$ through the left-hand branch, which is

$$I' = \frac{6 \text{ V}}{R'} = 2 \text{ A}$$

This current splits up in such a way that the voltage across the 6-ohm resistor will be the same as across the 2-ohm resistor. Thus 75% of $I'$ goes through the 2-ohm resistor and 25% of $I'$, or 0.5 A, goes through the 6-ohm resistor.

## The Short Circuit

In Figure 17-4 the bottom of the voltage source is connected to ground (the potential of the earth). In circuit theory, the potential of the earth is defined as zero potential. Note that if we connect just one point of a circuit to ground, no current will flow through this connection. In Figure 17-4 what is the potential at point $A$? Note that it is possible to trace a zigzag path of wires from point $A$ to ground without passing through any resistor. In these diagrams, the connecting wires are assumed to have zero resist-

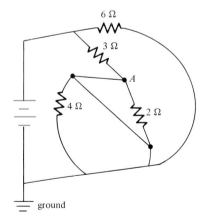

**Figure 17-4.** *Illustration of a short circuit. Point A is short-circuited to ground.*

ance. According to Ohm's law, the potential drop, or potential difference, across a wire will be

$$V = IR_{wire}$$
$$= I \times 0$$
$$= 0 \text{ V}$$

Thus the potential at point $A$ is also zero. In fact, any two points connected by a resistanceless wire will be at the same potential. Since the potential difference across the 2- and 4-ohm resistors is zero, no current will flow through them. These two resistors are then said to have been short-circuited. If you were to short circuit a 120-V home outlet, the current would be

$$I = \frac{V}{R} = \frac{120 \text{ V}}{0 \text{ ohm}} = \infty$$

In actual practice the current would not be infinite since the wires do have a small amount of resistance; however, the current would be large enough for sparks to fly and fuses to blow.

### Kirchhoff's Laws

In more complex circuits the resistors may be connected in such a way that they cannot be resolved into series and parallel combinations. Figure 17-5(a) shows such a situation. Also more complex circuits may have more than one battery or voltage source as in Figure 17-5(b). There is a general procedure for dealing with complex circuits that invokes what is called Kirchhoff's laws. They are as follows:

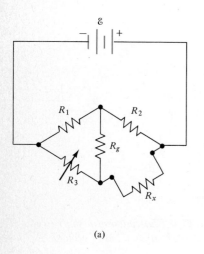

(a)

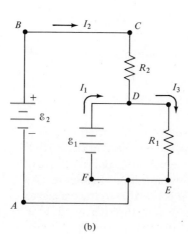

(b)

**Figure 17-5.** (a) *A Wheatstone bridge used to measure $R_x$. (b) A voltage regulator. $R_1$ gets its voltage from $\mathcal{E}_1$ and its current from $\mathcal{E}_2$.*

**The Loop Theorem.** The algebraic sum of voltage gains around any closed loop is zero. (A voltage drop is a negative voltage gain.)

**The Junction Theorem.** The algebraic sum of currents flowing into any junction is zero.

The loop theorem is a consequence of the conservation of energy, and the junction theorem of the conservation of charge.

In each branch a separate current and current direction is assumed. In using the loop theorem, a voltage drop occurs when a resistor is traversed in the same direction as the assumed current direction and a voltage gain in traversing an emf from $-$ to $+$. If the final solution for a current is negative, it means the actual current flows opposite to the assumed direction.

Let us now solve for the currents in Figure 17-5(b) as an example. Application of the loop theorem to the loop $ABCDEA$ gives

$$\mathcal{E}_2 - I_2 R_2 - I_3 R_1 = 0$$

Application to the loop $EFDE$ gives

$$\mathcal{E}_1 - I_3 R_1 = 0$$

Subtracting these two equations gives

$$\mathcal{E}_2 - \mathcal{E}_1 - I_2 R_2 = 0$$

$$I_2 = \frac{\mathcal{E}_2 - \mathcal{E}_1}{R_2}$$

$I_1$ can be obtained by applying the junction theorem to junction $D$:

$$I_1 + I_2 - I_3 = 0.$$

$$I_1 = I_3 - I_2 = \frac{\mathcal{E}_1}{R_1} - \frac{\mathcal{E}_2 - \mathcal{E}_1}{R_2} = \mathcal{E}_1 \left( \frac{1}{R_1} + \frac{1}{R_2} \right) - \frac{\mathcal{E}_2}{R_2}$$

We see that if $\mathcal{E}_1(1/R_1 + 1/R_2) = \mathcal{E}_2/R_2$, then $I_1 = 0$ and no current is drawn from $\mathcal{E}_1$. This arrangement where $I_1 \approx 0$ has an important practical application. Suppose it is necessary to have a precise voltage $\mathcal{E}_1$ across $R_1$ even though $R_1$ is drawing a large current $I_3$. Then the arrangement in Figure 17-5(b) acts as a voltage regulator. Small voltage variations in the larger power supply $\mathcal{E}_2$ will not show up in the voltage $\mathcal{E}_1$ applied to $R_1$. $\mathcal{E}_1$ could be a low-current standard cell, even though $R_1$ draws a high current. (It draws its current mainly from $\mathcal{E}_2$ but has its voltage "clamped" by $\mathcal{E}_1$.) If the battery $\mathcal{E}_1$ were used by itself, it would have a short lifetime.

Figure 17-5(a) has a practical application as an accurate method for measuring an unknown resistor $R_x$. If $R_g$ represents the resistance of a galvanometer (a sensitive current meter), we have what is called a Wheatstone bridge. $R_x$ is determined by adjusting $R_3$ such that zero current passes through $R_g$. Then $R_2/R_1 = R_x/R_3$, or $R_x = R_3(R_2/R_1)$.

## 17-4 Magnetic Force—Experimental

Everyday experience with magnetic force involves permanent magnets, electromagnets, solenoids, relays, electric motors, deflection of electron beams in TV picture tubes, and so on. All such phenomena can be reduced to a fundamental force between moving charges (or, since moving charge is current, a force between currents). We shall see in Chapter 18 that, from an atomic point of view, a magnet contains permanent circulating currents and that the force on a magnet can be explained in terms of the fundamental force between moving charges. We shall now discuss the fundamental force on a single charge $q$.

Whether or not it is moving, there will always be a force $\mathbf{F}_E = q\mathbf{E}$ on a charge $q$. However, if it is moving with velocity $v$, there will often be an *additional* force $\mathbf{F}_{\text{mag}}$ which is measured to be proportional to the product $qv$. Some introductory physics textbooks introduce this additional force (called the magnetic force) as a result of experimental measurement and imply that it is a new fundamental force of nature. Such a presentation is misleading. In the next section we shall show that an "additional" force proportional to $qv$ is a necessary consequence of relativity (if such force did not exist, relativity would be violated). There is nothing new or fundamental about it—it is merely a relativistic consequence of Coulomb's law. So if $\mathbf{F}$ is the net electromagnetic force on a moving charge $q$,

$$\mathbf{F} = q\mathbf{E} + \mathbf{F}_{\text{mag}}$$

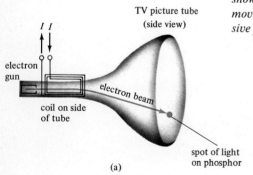
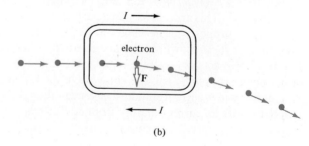

**Figure 17-6.** (a) *TV picture tube showing vertical deflection coil on side of tube. (There is an identical coil on the other side as well.)* (b) *Enlarged view of (a) showing a single turn of deflection coil. There is an attractive force between the moving electron and the current in the bottom part of coil, and there is a repulsive force between the electron and the upper part of the coil.*

**Figure 17-7.** *When placed a few centimeters apart, two uncharged wires or rods connected to a storage battery will be visibly attracted to or repelled by each other, depending on whether the currents are parallel or antiparallel. This demonstration should be done quickly before the wires get too hot. (This is an example of a short circuit.)*

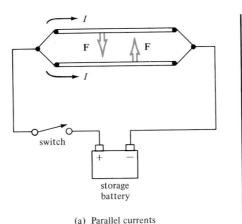

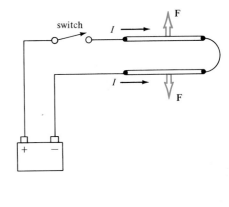

(a) Parallel currents

(b) Antiparallel currents

Since $\mathbf{F}_{mag}$ is proportional to $qv$, a vector quantity $\mathcal{B}$ can be defined such that

$$\mathbf{F}_{mag} = q\mathbf{v} \times \mathcal{B}$$

In the next section we shall see that this is the defining equation for magnetic field $\mathcal{B}$, and we shall learn how to calculate $\mathcal{B}$ in terms of the currents that produce it.

It is interesting to note that this "new" force can exist even in cases where the net electrostatic force is zero (when the net charge on the wires is zero). Figures 17-6, 17-7, and 17-8 show such situations. Figure 17-6 illustrates the observation that moving electrons in the vacuum of a TV picture tube are attracted by the conduction electrons that are moving parallel to the electron beam and are repelled by the conduction electrons that are moving antiparallel. A simple demonstration to illustrate the experimental observation of an attractive force between parallel currents is shown in Figure 17-7.

If the lower current in Figure 17-7(a) is replaced by a single moving charge $q$, we observe an attractive force on $q$ as illustrated in Figure 17-8. This force is measured to be proportional to $I$ and inversely proportional to the distance $y$:

$$F_{mag} = (10^{-7}) \left(\frac{2I}{y}\right)(qv) \qquad (17\text{-}9)$$

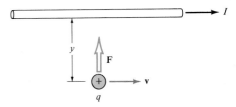

**Figure 17-8.** *A single point charge $q$ moving with velocity $\mathbf{v}$ parallel to a current $I$ will be attracted to it.*

The proportionality constant is $10^{-7}$ when all the quantities are in SI units. In actual practice this equation (or its equivalent in the force between two currents) is used to define the mks unit of current. So the size of the coulomb is determined by adjusting $I$ until the measured force agrees in magnitude with Eq. 17-9. The value of $k_0$ is determined by taking two of these coulombs and placing them 1 m apart and using Coulomb's law to measure the force: $F = k_0(Q^2/R^2)$. When $k_0$ is so determined by measurement, it turns out that it has the magnitude $k_0 = c^2/10^7$, where $c = 2.998 \times 10^8$. This is the same value as the speed of light. In the next section, not only shall we derive Eq. 17-9, but we shall prove that $k_0$ must equal $c^2/10^7$ in magnitude.

## 17-5 Derivation of Magnetic Force

Let us examine the situation in Figure 17-8 more closely. The current $I$ consists of conduction electrons moving to the left with drift velocity $v_d$. In order for the wire to be electrically neutral there must be a string of stationary positive ions of equal and opposite charge density as shown in Figure 17-9. We let $\lambda^-$ be the linear charge density of the conduction electrons and $\lambda^+$ that of the positive ions where $\lambda^+ = -\lambda^-$. Our goal is to calculate the net force on $q$, if indeed there is any force on it. So far all we know is how to calculate electric forces on stationary charges.

However, we can "make" $q$ be stationary by moving along with it. The moving observer sees both the electrons and the positive ions moving to the left when $q$ is at rest. At this point we invoke a famous result of relativity theory called the Lorentz contraction. The positive ions will appear more closely spaced by the factor $\sqrt{1 - v^2/c^2}$. We ask any readers who have skipped the chapters on relativity to accept this on faith. (It is not necessary to have studied Chapters 8 and 9 in order to follow the remaining discussion.) Since the electrons are moving faster than the positive ions, their spacing will be contracted even more than that of the positive ions. So, for the moving observer $\lambda^-$ will be greater in magnitude than $\lambda^+$, and the net charge density of the wire will be negative rather than zero. It is shown in Appendix 17-1 that the net charge density of the wire as seen by the moving observer is

$$\lambda' = -\frac{1}{\sqrt{1 - v^2/c^2}} \left(\frac{v}{c^2}\right) I \qquad (17\text{-}10)$$

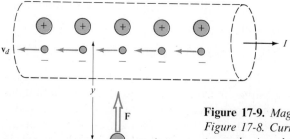

Figure 17-9. *Magnified view of wire in Figure 17-8. Current to the right is due to conduction electrons moving to the left.*

where $I = \lambda^- v_d$ is the original current in the wire (see Eq. 17-25). So according to the moving observer the negatively charged wire attracts $q$ with a force

$$F' = qE' = q\left(\frac{2k_0\lambda'}{y}\right) \quad \text{(see Eq. 16-5)}$$

Using the value given in Eq. 17-10 for $\lambda'$ we obtain

$$F' = \frac{qv}{\sqrt{1 - v^2/c^2}}\left(\frac{2k_0}{c^2}\frac{I}{y}\right) \quad (17\text{-}11)$$

for the magnitude of force seen by the moving observer. We have now shown that according to relativity theory there must be a force on $q$. And if the moving observer measures a force, so must any other nonaccelerating observer. (According to relativity theory the stationary observer will measure $F = \sqrt{1 - v^2/c^2}\, F'$.) So the original stationary observer sees a force

$$F = \frac{k_0}{c^2}\left(\frac{2I}{y}\right)qv \quad (17\text{-}12)$$

To recapitulate, an exact calculation using Coulomb's law (Eq. 16-5) and results from relativity theory yields expression 17-12 for the force shown in Figure 17-8. This force is seen to be proportional to $q$, $v$, and $I$ and is inversely proportional to $y$. It agrees, as it should, with the experimental observations (Eq. 17-9). Comparing Eq. 17-9 with Eq. 17-12 yields $k_0/c^2 = 10^{-7}$ N/A² or

$$k_0 = 10^{-7}c^2 \text{ N/A}^2 = 8.988 \times 10^9 \text{ N m}^2/\text{C}^2$$

This, of course, is what is measured within the accuracy of the measurements.

## 17-6 Magnetic Field

Just as we have defined electric field as electrostatic force per unit charge $q$, so can we define a magnetic field $\mathcal{B}$ as magnetic force per unit of $(qv)$; that is,

$$\mathcal{B} \equiv \frac{F_{\text{mag}}}{qv}$$

The units of $\mathcal{B}$ are N/(A m) which is given the special name **tesla**. The SI abbreviation is T for tesla.

Dividing Eq. 17-12 by $qv$ gives the result

$$\mathcal{B} = \frac{k_0}{c^2} \frac{2I}{y} \qquad (17\text{-}13)$$

for the magnitude of magnetic field at a distance $y$ from a single wire. With $\mathcal{B}$ written in this form some of the common equations involving $\mathcal{B}$ would contain a factor of $4\pi$. This factor can be eliminated by writing

$$\mathcal{B} = \frac{\mu_0}{4\pi} \frac{2I}{y} \qquad \text{where} \qquad \frac{\mu_0}{4\pi} \equiv \frac{k_0}{c^2} = 10^{-7} \text{ N/A}^2 \qquad (17\text{-}14)$$

and since $k_0 \equiv 1/4\pi\epsilon_0$, we also have

$$\mu_0\epsilon_0 = \frac{1}{c^2} \qquad (17\text{-}15)$$

Whether to use $k_0/c^2$ or $\mu_0/4\pi$ is a matter of taste. We shall usually use $k_0/c^2$ in place of $\mu_0/4\pi$. Then we have only one proportionality constant, $k_0$, to deal with rather than both $k_0$ and $\mu_0$. Also we immediately see just where the speed of light enters into the equations of electricity.

So far we have only discussed the force when $q$ is moving parallel to $I$. If $q$ is moving in other directions relative to $I$, relativity theory gives directions for the magnetic force as shown in Figure 17-10.

This dependence of $\mathbf{F}_{\text{mag}}$ on the direction of $\mathbf{v}$ can be summarized by using the vector equation

$$\mathbf{F}_{\text{mag}} = q\mathbf{v} \times \mathcal{B} \qquad (17\text{-}16)$$

along with the following right hand rule to define the direction of $\mathcal{B}$.

**Figure 17-10.** *Magnetic force* **F** *for four different directions of* **v**.

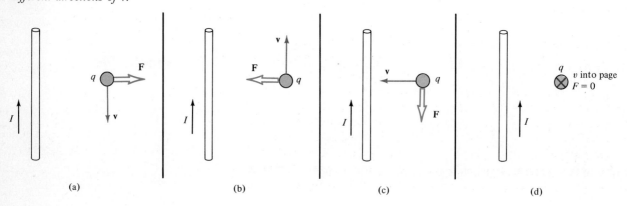

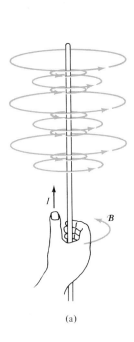

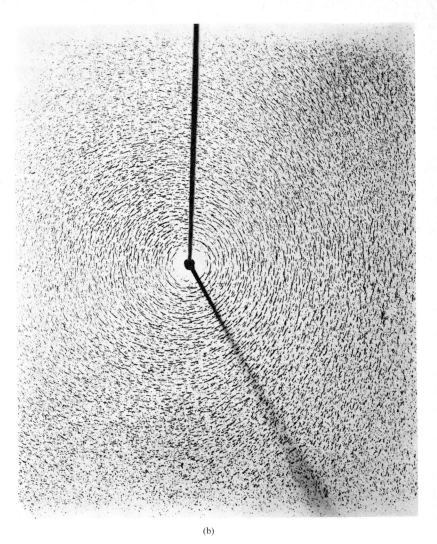

**Figure 17-11.** (a) *Right-hand rule and lines of ℬ produced by a long straight current.* (b) *Photograph of iron filings sprinkled about a long straight wire. After the current is turned on, each iron filing behaves as a miniature magnet and lines up in the direction of ℬ.* [*Courtesy Educational Development Center.*]

### Right-Hand Rule

The right-hand rule is illustrated in Figure 17-11. If the thumb of the right hand is pointed along $I$, the fingers curl in the direction of ℬ (the lines of ℬ will circle around the current). So in Figure 17-8, ℬ would be pointing into the page at the position of $q$ and the cross product $\mathbf{v} \times$ ℬ would be pointing up, thus giving the correct direction for $\mathbf{F}$.

In the next chapter we shall develop two formulas (Ampere's law and the Biot–Savart law) that allow us to calculate magnetic field for currents other than a single wire. Note that $\mathbf{F}_{mag}$ is always perpendicular to $\mathbf{v}$. Then according to the work-energy theorem (Eq. 6-10) the magnetic force can neither increase nor decrease the kinetic energy of a moving charge. A charge moving in a uniform magnetic field will go around in a circle. In such a case the acceleration is $v^2/R$ and the force is

$$m\frac{v^2}{R} = qv\mathcal{B}$$

$$R = \frac{mv}{q\mathcal{B}} \quad \text{(radius of the circle)}$$

The track left by an electron moving in a uniform magnetic field is shown in Figure 31-3.

**Example 5.** What is the period of revolution of a charged particle moving with velocity $v$ in a uniform magnetic field $\mathcal{B}$?

ANSWER: Let $T$ be the time for one revolution. Then

$$T = \frac{2\pi R}{v}$$

Putting $R = mv/(q\mathcal{B})$ into this equation gives

$$T = \frac{2\pi}{v}\left(\frac{mv}{q\mathcal{B}}\right) = \frac{2\pi m}{q\mathcal{B}}$$

We note that the period is independent of both $R$ and $v$.

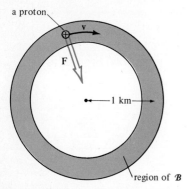

**Figure 17-12.** Top view of main ring of 500-GeV proton synchrotron. Protons circulate clockwise between magnet polefaces (not shown). $\mathcal{B}$ is pointing out of the page and ($\mathbf{v} \times \mathcal{B}$) is in the direction indicated by vector **F**.

**Example 6.** The world's highest energy accelerator at Batavia, Illinois, consists of a donut-shaped vacuum chamber 2 km in diameter inserted between magnet polefaces giving a maximum field of $\mathcal{B} = 1.8$ T (tesla). When viewed from above as in Figure 17-12, the beam of protons circulates clockwise with a speed very close to the speed of light.
(a) What is the direction of $\mathcal{B}$?
(b) What is the energy of the protons when $\mathcal{B}$ has reached 1.8 T?

ANSWER: (a) The force is a centripetal force pointing toward the center. If $\mathcal{B}$ is pointing out of the paper, ($\mathbf{v} \times \mathcal{B}$) will always point toward the center.
(b) The centripetal force is $F_c = m_r(v^2/R)$ where $m_r$ is the relativistic mass. Since this force is supplied by the magnetic field, it has the magnitude ($ev\mathcal{B}$). So

$$m_r \frac{v^2}{R} = ev\mathcal{B}$$

Now replace $v$ with $c$, the speed of light:

$$m_r c^2 = ec\mathcal{B}R$$

Since the left-hand side is the total relativistic energy, we have

$$E = (1.6 \times 10^{-19})(3 \times 10^8)(1.8)(10^3) \text{ J} = 8.64 \times 10^{-8} \text{ J}$$
$$= 540 \text{ GeV}$$

Note that the magnetic force does not change the velocity or energy of the particles. Separate, short regions of electric field are used to accelerate the protons each time they pass around the ring.

## 17-7 Magnetic Field Units

When a charge $q$ is moving it still experiences the electrostatic force $q\mathbf{E}$ as well as the magnetic force, hence the complete electric force on a moving charge is

$$\mathbf{F} = q\mathbf{E} + q\mathbf{v} \times \mathcal{B} \tag{17-17}$$

This complete electric force is called the electromagnetic force. We see immediately from Eq. 17-17 that $\mathcal{B}$ and $E$ do not have the same units. (This is why we use different type faces for $\mathcal{B}$ and $E$.) We see that the units of $[v\mathcal{B}]$ are the same as the units of $E$. Thus $\mathcal{B}$ has units of $E$ divided by m/s or $[\mathcal{B}] = \text{V s m}^{-2}$. The unit of $\mathcal{B}$ is given a special name called the tesla. In older books the equivalent unit weber per square meter is used instead of tesla. For historical reasons the term magnetic induction is used for $\mathcal{B}$. However, it is common usage to refer to $\mathcal{B}$ as magnetic field.

In the cgs system of units (also called the gaussian system) magnetic field is defined via the equation

$$\mathbf{F}_{\text{mag}} = \frac{q\mathbf{v}}{c} \times \mathbf{B}_{\text{cgs}} \quad \text{(cgs system)} \tag{17-18}$$

In the cgs system the unit of $B$ is called "gauss."

$$1 \text{ tesla} = 10^4 \text{ gauss}$$

In the cgs system $B$ and $E$ have the same units. As we shall see in the next section, there are some advantages to a system where $B$ and $E$ have the same units. Comparison of Eqs. 17-17 and 17-18 shows that

$$\mathcal{B}_{mks} \rightleftarrows \frac{\mathbf{B}_{cgs}}{c}$$

Hence any electricity equation in this book can be written in cgs form by replacing $\mathcal{B}$ with $\mathbf{B}/c$.

We are effectively using a dual presentation: all the electricity equations in this book are effectively in both sets of units at the same time. A cgs-based course merely reads $k_0$ as 1 and $\mathcal{B}$ as $\mathbf{B}/c$. Basically, the equations are written in the mks system assuming one uses $\mathcal{B}$ as magnetic field in teslas and $k_0 = 9.00 \times 10^9$. Then all other units will be the practical units of volts, amperes, ohms, and so on. Moreover, writing all the mks equations in terms of $k_0$ and $c$ (rather than $\mu_0$) makes many of the pedagogical and physical advantages associated with the cgs or gaussian system apparent for those who are using mks throughout.

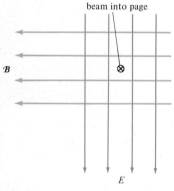

**Figure 17-13.** *Electric and magnetic fields are perpendicular to each other.*

**Example 7.** A beam of protons of velocity $v = 0.1c$ travels through a region of crossed magnetic and electric field, as shown in Figure 17-13. The protons are moving perpendicularly into the page. The electrostatic force on the protons is $3 \times 10^{-13}$ N.

(a) What must be the ratio $E/B$ so that the net force on the protons would be zero? Give answer in both mks and cgs units.

(b) What would be the magnitude of $B$ in gauss?

(c) Suppose $E/B$ was as above, what would be the direction and magnitude of the net force on a particle of charge $+e$ and velocity $v = 0.2c$? (This arrangement of crossed fields is used as a device to select only those particles of a given velocity. Only those particles will be undeflected.)

ANSWER:

(a) 
$$\mathbf{F}_{net} = 0 = e\mathbf{E} + e\mathbf{v} \times \mathcal{B}$$
$$\mathbf{E} = -\mathbf{v} \times \mathcal{B}$$

Since $v$ and $\mathcal{B}$ are perpendicular, $E = v\mathcal{B}$ or $E/\mathcal{B} = v$ for mks. For cgs replace $\mathcal{B}$ with $(B/c)$ to get $E/B = v/c$.

(b) $$E = \frac{F}{e} = \frac{3 \times 10^{-13} \text{ N}}{1.6 \times 10^{-19} \text{ C}} = 1.875 \times 10^6 \text{ N/C}$$

$$\mathcal{B} = \frac{E}{v} = \frac{1.875 \times 10^6}{3 \times 10^7} \text{T} = 0.0625 \text{ T} = 625 \text{ gauss}.$$

(c) Let $v_1 = 0.1c$.

$$\begin{aligned} F_{\text{net}} &= e\mathbf{E} + 2\, e\mathbf{v}_1 \times \mathcal{B} \\ &= e\mathbf{E} + e\mathbf{v}_1 \times \mathcal{B} + e\mathbf{v}_1 \times \mathcal{B} \\ &= (0) + (-e\mathbf{E}) \\ F_{\text{net}} &= 3 \times 10^{-13} \text{ N pointing up}. \end{aligned}$$

## 17-8 Relativistic Transformation of $\mathcal{B}$ and E

In this section we shall see that $\mathcal{B}$ and $\mathbf{E}$ transform back and forth into each other when viewed by moving observers. Physically they should be thought of as one and the same thing, which is called the electromagnetic field. The electromagnetic field has six components: $(E_x, E_y, E_z, B_x, B_y, B_z)$, all of which get "mixed up" when viewed by a moving observer. From this point of view the cgs system of units is more physical since all components of the same physical entity should at least have the same units.

We shall consider only two common examples of the field transformations. For the first example consider the case where $\mathcal{B} = 0$ everywhere in the lab system. There may be stationary charges and an electric field $\mathbf{E}$. We shall show that a moving observer sees $\mathcal{B}' = (\mathbf{v}'/c^2) \times \mathbf{E}'$, where $\mathbf{v}'$ is the velocity of the moving charges in the primed system. We first consider a charged wire at rest in the lab system. Then $\mathcal{B} = 0$ everywhere and $E = 2k_0\lambda/y$ at point $P$ in Figure 17-14. Now consider Mr. Prime moving by with velocity $\mathbf{v}$ as shown. He sees a current $I' = \lambda' v$ moving by to the left where $\lambda'$ is the charge density of the wire in his frame of reference. According to Eq. 17-13 he sees a magnetic field

$$\mathcal{B}' = \frac{k_0}{c^2} \frac{2(\lambda' v)}{y}$$

According to the equation $E' = 2k_0\lambda'/y$ he sees

$$\mathcal{B}' = \frac{v}{c^2} E'$$

The vector form is

$$\mathcal{B}' = \frac{\mathbf{v}'}{c^2} \times \mathbf{E}'$$

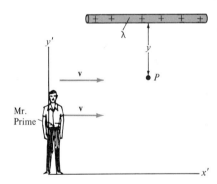

Figure 17-14. *Charged wire at rest in lab system (unprimed)*.

where **v′** is the velocity of the source as viewed in the primed system. This result must be independent of the nature of the source if the transformed local field at point $P$ is to be unique. It will be true in general that if any system of charges is moving all with the same velocity $v$, then there will be a magnetic field

$$\mathcal{B} = \frac{\mathbf{v}}{c^2} \times \mathbf{E} \quad \text{(when all charges have the same velocity } v\text{)} \tag{17-19}$$

We have dropped all the primes because we are now talking about observing the moving charges in an unprimed system. Equation 17-19 will be of use to us in the beginning of the next chapter when we develop the Biot–Savart law. Note that in the cgs system, when there is moving charge, $B/E = v/c$, and that when the moving charge is near the speed of light the magnetic and electric fields are almost equal.

**Figure 17-15.** *Rod with surface charge is moving to the right with velocity $v_0$. Charge per unit length is $Q_l$ when rod is at rest.*

**Example 8.** A long charged rod of radius $R$ holds a surface charge $Q_l$ for every meter of its length. The rod is then given a velocity $v_0$ to the right as shown in Figure 17-15. What are $E$ and $\mathcal{B}$ just outside the rod?

ANSWER: When the rod is moving, its linear charge density will be $\lambda = \gamma Q_l$ where $\gamma = 1/\sqrt{1 - v_0^2/c^2}$. According to Eq. 16-5

$$E = k_0 \frac{2\lambda}{R} = 2k_0 \frac{\gamma Q_l}{R}$$

According to Eq. 17-18

$$\mathcal{B} = \frac{v_0}{c^2} E = \frac{2k_0}{c^2} \gamma \frac{v_0 Q_l}{R}$$

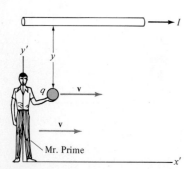

**Figure 17-16.** *Uncharged stationary wire carries current I. It is observed by Mr. Prime, who is moving with the velocity v.*

For our second example of field transformations we shall consider the reverse situation where in the lab system $\mathbf{E} = 0$ everywhere. Our source will be an uncharged wire carrying a current $I$ as shown in Figure 17-16. We saw in Section 17-5 that Mr. Prime would observe a net negative charge on the wire and according to Eq. 17-11, if he held a charge $q$ in his hand, the force on it would be

$$F' = qv\left(\frac{2k_0\gamma I}{c^2 y}\right) \quad \text{where} \quad \gamma = \frac{1}{\sqrt{1-v^2/c^2}}$$

Dividing both sides by $q$ gives

$$E' = v\left(\frac{2k_0}{c^2}\frac{\gamma I}{y}\right)$$

According to Eq. 17-13 Mr. Prime also observes a magnetic field

$$\mathcal{B}' = 2\frac{k_0}{c^2}\frac{I'}{y}$$

pointing into the page. Now use Eq. 17-26 (Appendix 17-1) to replace $I'$ with $\gamma I$. Then

$$\mathcal{B}' = 2\frac{k_0 \gamma I}{c^2 y}$$

and we see that

$$E' = -v' \times \mathcal{B}'$$

where $v'$ is the velocity of the current source observed in the primed system. Again, because local field transformations must be unique, we have the general relation for any shape of current source that

$$E = -v \times \mathcal{B} \quad \text{(when current sources are moving with velocity } v\text{)} \quad (17\text{-}20)$$

is the electric field produced by a source of current moving by with velocity **v**. As before, we have dropped all the primes because we are now talking about moving current sources as observed in an unprimed system. Such an electric field is a form of what is called electromagnetic induction (to be discussed in more detail in Chapter 19). This result will lead us directly into Faraday's law of induced electric field produced by a changing magnetic field.

**Example 9.** A jet plane is flying due north at a latitude where the vertical component of the earth's magnetic field is 0.6 gauss pointing down. Its speed is 278 m/s.
  (a) What is the external electric field as measured by the pilot?
  (b) Will the wings be electrically charged?

ANSWER: (a) Since the pilot is moving with respect to the source of the earth's magnetic field, Eq. 17-20 applies. Then $E = v\mathcal{B} = (278 \text{ m/s})(0.6 \times 10^{-4} \text{ T}) = 0.167 \text{ V/m}$.
(b) According to Eq. 17-20, **E** is in the direction of $-\mathbf{v} \times \mathcal{B}$, which is to the east. Thus positive charges would tend to accumulate on the east wing and negative on the west wing. According to an observer at rest on the ground there would be a potential difference between the wings. If the wingspan is 20 m, the potential difference from one end to the other is

$$V = Ex_0 = (0.167 \text{ V/m})(20 \text{ m}) = 3.34 \text{ V}.$$

## Summary

Electric current $I = Q/t = \int \mathbf{j} \cdot d\mathbf{A}$ where $\mathbf{j} = \rho \mathbf{v}$ is the current density or amperes per square meter. Then $I = \rho v A = \mathcal{N} e v A$ where $\mathcal{N}$ is the number of conduction electrons per meter cubed.

Ohm's law states that $I \propto V$ for a metal kept at constant temperature. $R = V/I$ is the definition of resistance. Electric power dissipation $P = VI$.

The magnetic force between a moving charge $q$ and a parallel current $I$ at a distance $y$ is measured to be

$$F_{\text{mag}} = qv \times \left(\frac{\mu_0}{4\pi}\right) \frac{2I}{y}$$

where the constant $\mu_0/4\pi$ is set equal to $10^{-7}$ N/A².

When relativity theory is applied to Coulomb's law, one can derive the result

$$F_{\text{mag}} = qv \left(\frac{k_0}{c^2}\right) \frac{2I}{y}$$

Hence $k_0/c^2 = 10^{-7}$ N/A² $= (\mu_0/4\pi)$ or $\mu_0 \epsilon_0 = 1/c^2$. The magnitude of magnetic field $\mathcal{B}$ is $F_{\text{mag}}/qv$, which gives

$$\mathcal{B} = \frac{k_0}{c^2} \frac{2I}{y}$$

for a straight line current $I$. The direction of $\mathcal{B}$ is given by fingers of the right hand when the thumb is pointed along $I$. The direction of $\mathbf{F}_{\text{mag}}$ is obtained from the relation $\mathbf{F}_{\text{mag}} = q\mathbf{v} \times \mathcal{B}$. This force is always perpendicular to the velocity and if a particle is moving in a region of uniform $\mathcal{B}$, it will move in a circle of radius $R = mv/q\mathcal{B}$ when **v** is perpendicular to $\mathcal{B}$.

When expressing equations in the cgs or gaussian system one sets $k_0 = 1$ and replaces $\mathcal{B}$ with $(\mathbf{B}/c)$.

In a system where there are only currents and all currents are moving by with velocity $\mathbf{v}$, $\mathbf{E} = -\mathbf{v} \times \mathcal{B}$.

In a system where all charges have the same velocity $\mathbf{v}$ and there are no currents, $\mathcal{B} = (\mathbf{v}/c^2) \times \mathbf{E}$.

## Appendix 17-1 Transformation of Current and Charge

We wish to show that the neutrally charged wire in Figure 17-8 will have a charge

$$\lambda' = \frac{-1}{\sqrt{1 - v^2/c^2}} \frac{v}{c^2} I$$

when viewed by an observer moving along with $q$. Let the primed system be that of the moving observer. As shown in Figure 17-17 he sees the positive ions moving to the left with velocity $v$ and the conduction electrons with a larger velocity $v'_-$. The charge density of the wire is

$$\lambda' = \lambda'_+ + \lambda'_- \qquad (17\text{-}21)$$

Because of the Lorentz contraction,

$$\lambda'_+ = \frac{1}{\sqrt{1 - v^2/c^2}} \lambda_+ \qquad (17\text{-}22)$$

and

$$\lambda'_- = \frac{1}{\sqrt{1 - v'^2_-/c^2}} (\lambda_-)_0 \qquad (17\text{-}23)$$

**Figure 17-17.** Same as Figure 17-8 except as seen by an observer moving along with the charge $q$.

$$v'_- = \frac{v + v_d}{1 + v v_d / c^2}$$

where $(\lambda_-)_0$ is the charge density of the conduction electrons at rest. Because of the Lorentz contraction

$$\lambda_- = \frac{1}{\sqrt{1 - v_d^2/c^2}}(\lambda_-)_0 \quad \text{or} \quad (\lambda_-)_0 = \sqrt{1 - v_d^2/c^2}\,\lambda_-$$

We use this for $(\lambda_-)_0$ in Eq. 17-23 to obtain

$$\lambda'_- = \frac{\sqrt{1 - v_d^2/c^2}}{\sqrt{1 - v'^2_-/c^2}}\lambda_-$$

Let $\beta = v/c$, $\beta_d \equiv v_d/c$, and $\beta' \equiv v'_-/c$. Then

$$\lambda'_- = \frac{\sqrt{1 - \beta_d^2}}{\sqrt{1 - \beta'^2}}\lambda_-$$

Now use the Einstein addition of velocities formula (Eq. 9-1) to eliminate $\beta'$:

$$\beta' = \frac{\beta_d + \beta}{1 + \beta_d\beta}$$

$$\lambda'_- = \frac{\sqrt{1 - \beta_d^2}}{\sqrt{1 - \left(\frac{\beta_d + \beta}{1 + \beta_d\beta}\right)^2}}\lambda_- = \frac{(1 + \beta_d\beta)\sqrt{1 - \beta_d^2}}{\sqrt{(1 - \beta^2)(1 - \beta_d^2)}}\lambda_-$$

$$\lambda'_- = \frac{(1 + \beta_d\beta)}{\sqrt{1 - \beta^2}}\lambda_- \tag{17-24}$$

Substitute this and Eq. 17-22 into Eq. 17-21:

$$\lambda' = \frac{1}{\sqrt{1 - \beta^2}}\lambda_+ + \frac{1 + \beta_d\beta}{\sqrt{1 - \beta^2}}\lambda_-$$

Replace $\lambda_+$ with $(-\lambda_-)$:

$$\lambda' = \frac{\beta_d\beta}{\sqrt{1 - \beta^2}}\lambda_- = \frac{1}{\sqrt{1 - \beta^2}}\frac{v}{c^2}\lambda_- v_d$$

Using $I = -\lambda_- v_d$ gives the result

$$\lambda' = -\frac{1}{\sqrt{1 - v^2/c^2}}\frac{vI}{c^2} \tag{17-25}$$

## The Transformed Current

The current $I'$ is $I' = \lambda'_+ v + \lambda'_- v'_-$. Now substitute Eqs. 17-22 and 17-24 into this:

$$I' = \left(\frac{1}{\sqrt{1-\beta^2}}\lambda_+\right)v + \left(\frac{1+\beta_d\beta}{\sqrt{1-\beta^2}}\lambda_-\right)v'_-.$$

Replace $\lambda_+$ with $(-\lambda_-)$ and $v'_-$ with $(v + v_d)/(1 + \beta_d\beta)$.

$$I' = -\frac{\lambda_- v}{\sqrt{1-\beta^2}} + \left(\frac{1+\beta_d\beta}{\sqrt{1-\beta^2}}\lambda_-\right)\left(\frac{v + v_d}{1+\beta_d\beta}\right) = \frac{\lambda_- v_d}{\sqrt{1-\beta^2}}$$

Thus

$$I' = \gamma I \quad \text{where } \gamma = 1/\sqrt{1-\beta^2}. \tag{17-26}$$

## Exercises

1. Conduction electrons in a copper wire of 2 mm² cross section have an average drift velocity of 0.1 mm/s. What is the current?
2. The conductivity of copper is $5.9 \times 10^7$ (ohm m)$^{-1}$. The average electron speed is $u = 1.3 \times 10^6$ m/s. What is the collision mean free path?
3. Resistivity $\rho$ is defined such that the resistance of a cylinder of length $x_0$ and cross-sectional area $A$ is $R = \rho(x_0/A)$. Express resistivity in terms of $\mathcal{N}$, $e$, $L$, $m$, and $u$.
4. A home is wired for 100 A at 120 V.
   (a) What is the maximum power it can draw?
   (b) If it runs at maximum power for 1 month, how many kilowatt-hours does it use and what would be the electric bill at 10¢/kW h?
5. A 12-V storage battery has an internal resistance of 0.05 ohm. By mistake a jumper cable of 0.1 ohm resistance is connected across the two terminals.
   (a) What current is drawn from the battery?
   (b) How much power is dissipated in the jumper cable? Will it produce heat at a rate faster than a 100-W light bulb?
6. If each of the two rods in Figure 17-6 has a resistance of 0.2 ohm and the remaining resistance of the circuit is 0.1 ohm, what will be the force per meter on either rod. Assume a 1-cm separation and a 12-V battery.
7. Two long, straight parallel wires are 16 cm apart and carry currents of 4 A each. Determine $\mathcal{B}$ at a point midway between them.
   (a) When the currents are in the same direction.
   (b) When the currents are in opposite directions.
8. An electron moving horizontally from east to west enters a magnetic field and is deflected downward. What is the direction of the magnetic field?

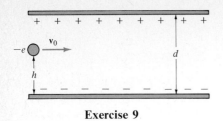

**Exercise 9**

9. An electron of mass $m$ enters between two parallel charged plates with velocity $v_0$; $v_0 \ll c$.
   (a) What will be the direction of the acceleration? What will be its magnitude in terms of $e$, $E$, and $m$?
   (b) How long will it take until the electron hits one of the plates? Give answer in terms of $e$, $E$, $m$, $k$, $d$, and $v_0$.
   (c) Suppose the lines of $E$ were magnetic rather than electric field pointing in the same direction as $E$ now points. Then what would be the magnitude and direction of force on the electron? Give answer in terms of $e$, $v_0$, and $\mathcal{B}$.

10. Derive a formula for the period of revolution of an electron having velocity $\mathbf{v}$ in a uniform magnetic field $\mathcal{B}$. $\mathbf{v}$ is perpendicular to $\mathcal{B}$.

## Problems

11. Assume that the hydrogen atom consists of an electron in circular orbit of radius $R = 5.3 \times 10^{-11}$ m about a proton. What is the current circulating about the proton?

12. A uniformly charged solid sphere of radius $R$ and total charge $Q$ is rotating with period $T$. $\rho = Q/(\tfrac{4}{3}\pi R^3)$.
    (a) What is $j$ at the equator?
    (b) What is the total current circulating around the axis?

13. Repeat Problem 12 for a rotating spherical shell that has uniform surface charge. In this case $j$ at the equator will be infinite, but the total current will be finite. $\sigma = Q/4\pi R^2$.

14. What must be the value of $r$ in order to maximize the power delivered to $r$?

15. A black box contains resistors connected to three terminals as shown. The resistance between terminals 1 and 2 is measured to be $R_{12}$. Between terminals 1 and 3 it is $R_{13}$ and between 2 and 3 it is $R_{23}$. What are $R_1$, $R_2$, and $R_3$?

16. Suppose the resistors inside the box are as shown? What now is $R_A$ in terms of $R_{12}$, $R_{13}$, and $R_{23}$? What are $R_B$ and $R_C$?

17. An ohmmeter is a device to measure unknown resistances. It contains a meter movement that reads full scale when a current of 60 mA flows through

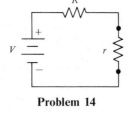

**Problem 14**

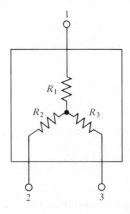

Problem 15

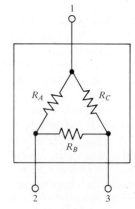

Problem 16

**Problem 17.** *Ohmmeter.*

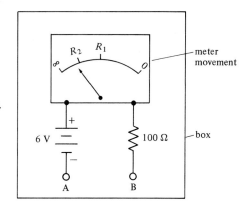

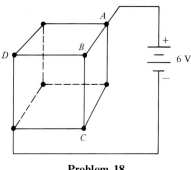

Problem 18

it. Also inside the box are a small 6-V battery and a 100-ohm resistor connected as shown. The scale of the meter movement is calibrated from infinity to zero. (When terminals A and B are shorted together, $I = 60$ mA and a full-scale deflection is obtained giving a reading of 0 ohm.)
  (a) What is the scale reading $R_1$ for a half-scale deflection of the meter movement?
  (b) What is the scale reading $R_2$ for a $\frac{1}{4}$-scale deflection?
18. Each of the edges of a cube is a 10-ohm resistor. What is the current drawn from the 6-V battery? (*Hint:* The current entering at $A$ splits three equal ways. The current entering $B$ splits two equal ways.)
19. The Cornell electron synchrotron has a radius of 100 m. The electrons circulate counterclockwise as viewed from above with almost the speed of light. They have an energy of 12 GeV.
  (a) What is the direction of $\mathcal{B}$?
  (b) What is the magnitude of $\mathcal{B}$ assuming it is uniform around the ring?
20. A beam of electrons of velocity $10^6$ m/s is to be deflected 90° by a magnet as shown.
  (a) What must be the direction of $\mathcal{B}$ to give this downward deflection?
  (b) What is the radius of curvature of the electron path when between the magnet poles? (Assume $\mathcal{B}$ is constant in region between poles and zero outside.)
  (c) What is the force in newtons on the electrons when in the magnetic field?
  (d) What is the value of $\mathcal{B}$?

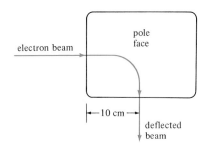

Problem 20

21. What is the relativistic mass of an electron moving in a circle of $R = 10$ cm and $\mathcal{B} = 1$ tesla? What is the ratio of this mass to the rest mass?
22. In the Bevatron at Berkeley the protons reach seven times their rest mass and have a velocity $v = 0.99c$. What must be the diameter of the Bevatron magnet if it has a magnetic field of 1.8 tesla?
23. Two infinite sheets of surface charge density $+\sigma$ and $-\sigma$ are separated by a distance $y_0 = 2$ cm. The surface charge is $\sigma = 10^{-5}$ C/m².
  (a) What is $E$ at point $P$ when the sheets are at rest?
  (b) What is $E$ at point $P$ when the sheets are moving to the right with velocity $v = 0.6c$?
  (c) What is $\mathcal{B}$ at point $P$ when the sheets are moving to the right at $0.6c$? Give direction and magnitude.

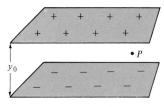

Problem 23

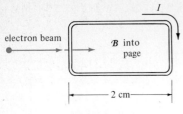

Problem 24

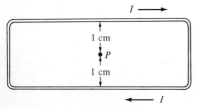

Problem 25

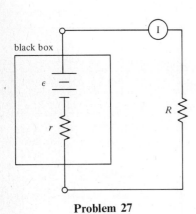

Problem 27

24. Assume that the vertical deflection coils of a TV picture tube produce a uniform magnetic field of 50 gauss pointing into the page over the area of the coil. $\mathcal{B} = 0$ outside the coil. The electrons in the beam have kinetic energy of 20,000 eV.
   (a) Will the beam be deflected up or down?
   (b) By what angle will the beam be deflected?

25. A long, rectangular coil carries a circulating current $I = 10$ A as shown. Assume the coil is so long that end effects can be ignored.
   (a) What is $\mathcal{B}$ at point $P$? Give direction and magnitude. ($P$ is 1 cm from each wire.)
   (b) If the coil is moving to the right with velocity $v$ the current will be increased by the factor $\gamma = (1 - v^2/c^2)^{-1/2}$. If $v = 0.6c$, what now is $\mathcal{B}$ at point $P$?
   (c) What is $E$ at point $P$ in part (b). Give direction and magnitude.

26. A 3-V battery has an internal resistance of 0.2 ohm. Three 1.5-ohm light bulbs are connected in parallel across the battery terminals. What voltage appears across the terminals?

27. A black box contains a resistor $r$ and a battery $\varepsilon$ in series as shown. When a resistor $R = 20\,\Omega$ is connected across the terminals, the ammeter reads 250 mA. When $R = 80\,\Omega$ the reading is 100 mA. What is the battery voltage $\varepsilon$, and what is the resistance $r$?

28. In Figure 17-5(a), $\varepsilon = 6$ V, $R_1 = R_2 = R_3 = 3\,\Omega$, $R_x = 3.01\,\Omega$, and $R_g = 0.1\,\Omega$. What is the current through $R_g$?

29. A voltmeter has an internal resistance $r$.
   (a) What will be its reading if it is connected across $R_2$ as shown?
   (b) What is the voltage across $R_2$ before the voltmeter is connected?

30. When the voltmeter of Problem 29 is connected across $R_2$, the voltage across $R_2$ drops by a factor of two. What is $r$ in terms of $R_1$ and $R_2$?

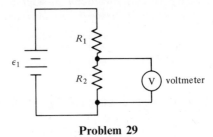

Problem 29

# Magnetic Fields

In the previous chapter we found that the magnetic field produced by a current $I$ flowing in an infinitely long wire is

$$\mathcal{B} = \frac{2k_0 I}{c^2 r} \quad \text{(at a distance } r \text{ from the wire)}$$

Now we wish to find the magnetic field produced by other common current distributions such as solenoids, coils, solid rods, and current sheets. For this task we need a magnetic equivalent to Gauss' law. There is a general equation for $\mathcal{B}$ produced by any current, and it is called Ampere's law. Instead of an integral of **E** over a closed surface, in Ampere's law we take an integral of $\mathcal{B}$ around a closed path. This is called a path integral and is written as

$$\oint \mathcal{B} \cdot d\mathbf{s}$$

Let us first evaluate the path integral around a single long wire where we already know the answer. In Figure 18-1 for the closed path we take the circle of radius $r$. Because $\mathcal{B}$ and $d\mathbf{s}$ are always parallel, we have

$$\oint_{\text{circle}} \mathcal{B} \cdot d\mathbf{s} = \oint \mathcal{B}\, ds = \mathcal{B} \oint ds = \left(\frac{2k_0 I}{c^2 r}\right)(2\pi r)$$

$$= \frac{4\pi k_0}{c^2} I$$

This result is independent of $r$ and holds for any circle having the wire at its center.

## 18-1 Ampere's Law

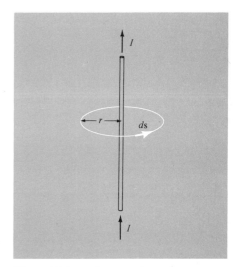

**Figure 18-1.** *Path integral is taken around imaginary circle of radius r. One element d**s** of the path is shown.*

365

We can show that the preceding relation holds for any shape of path that completely encloses the wire; that is,

$$\oint_{\text{circle}} \mathcal{B} \cdot d\mathbf{s} = \oint_{\text{any path}} \mathcal{B} \cdot d\mathbf{s}$$

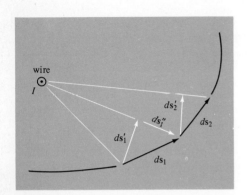

**Figure 18-2.** A current $I$ is pointing out of the page. $d\mathbf{s}_1$ and $d\mathbf{s}_2$ are elements of an arbitrary path; $d\mathbf{s}_1'$ and $d\mathbf{s}_2'$ are at right angles to radius vector from the wire.

According to Figure 18-2

$$\mathcal{B}_1 \cdot d\mathbf{s}_1 = \mathcal{B}_1 \cdot (d\mathbf{s}_1' + d\mathbf{s}_1'')$$

The term $\mathcal{B}_1 \cdot d\mathbf{s}_1'' = 0$ because the vectors are at right angles. Hence

$$\mathcal{B}_1 \cdot d\mathbf{s}_1 = \mathcal{B}_1 \cdot d\mathbf{s}_1'$$

Then

$$\mathcal{B}_1 \cdot d\mathbf{s}_1 + \mathcal{B}_2 \cdot d\mathbf{s}_2 = \mathcal{B}_1 \cdot d\mathbf{s}_1' + \mathcal{B}_2 \cdot d\mathbf{s}_2'$$

Note that $d\mathbf{s}_1$ and $d\mathbf{s}_2$ are elements of an arbitrary path and $d\mathbf{s}_1'$ and $d\mathbf{s}_2'$ are elements of corresponding circles that subtend the same angle from the wire. The left-hand side of the above equation is part of the path integral around the arbitrary path, and the right-hand side is part of corresponding circular path integrals. Since $\mathcal{B} \cdot d\mathbf{s}$ is the same for all circular paths subtending the same angle, we have

$$\oint_{\text{any path}} \mathcal{B} \cdot d\mathbf{s} = \oint_{\text{circle}} \mathcal{B} \cdot d\mathbf{s} = 4\pi \frac{k_0}{c^2} I$$

If our arbitrary path encloses $n$ different currents $I_1$ to $I_n$, we have

$$\oint \mathcal{B}_1 \cdot d\mathbf{s}_1 + \cdots + \oint \mathcal{B}_k \cdot d\mathbf{s} = \frac{4\pi k_0}{c^2}(I_1 + \cdots + I_n)$$

where $\mathcal{B}_n$ is the field produced by $I_n$ alone. Thus

$$\oint (\mathcal{B}_1 + \cdots + \mathcal{B}_n) \cdot d\mathbf{s} = \frac{4\pi k_0}{c^2} I_{\text{in}}$$

or

$$\oint_{\mathcal{C}} \mathcal{B} \cdot d\mathbf{s} = \frac{4\pi k_0}{c^2} I_{\text{in}} \quad \text{(Ampere's law)} \qquad (18\text{-}1)$$

where $I_{\text{in}}$ is the net current enclosed by the closed path $\mathcal{C}$ and $\mathcal{B} = \mathcal{B}_1 + \cdots + \mathcal{B}_n$. Since $I = \int \mathbf{j} \cdot d\mathbf{A}$, this can also be written as

$$\oint_{\mathcal{C}} \mathcal{B} \cdot d\mathbf{s} = \frac{4\pi k_0}{c^2} \int_S \mathbf{j} \cdot d\mathbf{A} \qquad \text{(Ampere's law, alternate form)} \qquad (18\text{-}2)$$

where **j** is the current density and $S$ is any surface whose boundary is the curve $\mathcal{C}$. Ampere's law can also be expressed in terms of $\mu_0$ by replacing $k_0/c^2$ with $\mu_0/4\pi$. Then

$$\oint \mathcal{B} \cdot d\mathbf{s} = \mu_0 I_{\text{in}}$$

We note that the use of $\mu_0$ has eliminated the $4\pi$.

Though our derivation of Ampere's law used only straight-line currents, there is a more general derivation that includes all steady currents moving in curves as well as straight lines. In the next section we shall apply Ampere's law to several common current distributions.

## Magnetic Flux

Just as

$$\Phi_E = \int_S \mathbf{E} \cdot d\mathbf{A}$$

is the electric flux or the number of lines of $E$ passing through a surface $S$, so also we define

$$\Phi_B = \int_S \mathcal{B} \cdot d\mathbf{A}$$

as the magnetic flux or the number of lines of $\mathcal{B}$ passing through $S$. Then magnetic field strength is quantitatively the number of lines per unit area. As seen in Figure 17-9 the lines of $\mathcal{B}$ are continuous around a single current. Then $\oint \mathcal{B} \cdot d\mathbf{A}$ over any closed surface must be zero (just as much flux enters it as leaves it). If there are $n$ separate currents

$$\oint \mathcal{B} \cdot d\mathbf{A} = \oint \mathcal{B}_1 \cdot d\mathbf{A} + \cdots + \oint \mathcal{B}_n \cdot d\mathbf{A}$$

where $\mathcal{B}_n$ is the field produced by the $n$th current all by itself. Since each term on the right-hand side is separately zero, we have

$$\oint \mathcal{B} \cdot d\mathbf{A} = 0 \qquad (18\text{-}3)$$

As was the case with Ampere's law, there exists a more general proof that covers currents following curved paths.

## 18-2 Current Distributions

### Solid Rod

Suppose we have a solid rod of radius $R$ carrying a uniform current density $\mathbf{j}$. The total current will be $I = j\pi R^2$. We wish to find $\mathcal{B}$ both inside and outside the rod. If we take the dashed-line path in Figure 18-3, we have

$$\oint \mathcal{B} \cdot d\mathbf{s} = \frac{4\pi k_0}{c^2} I$$

$$\mathcal{B}(2\pi r) = \frac{4\pi k_0}{c^2} I$$

$$\mathcal{B} = \frac{2k_0}{c^2} \frac{I}{r} \qquad \text{for } r > R.$$

We see that the field outside the rod is the same as if all the current had been concentrated at the center. Now if we take a second path of integration where $r < R$, we have

$$\oint \mathcal{B} \cdot d\mathbf{s} = \frac{4\pi k_0}{c^2} \int \mathbf{j} \cdot d\mathbf{A}$$

$$\mathcal{B}(2\pi r) = \frac{4\pi k_0}{c^2} j(\pi r^2)$$

$$\mathcal{B} = \frac{2\pi k_0}{c^2} jr = \frac{2k_0}{c^2} \frac{I}{R^2} r \qquad \text{for } r < R \qquad (18\text{-}4)$$

We find that the field inside the rod increases linearly from the center.

We note that the formulas for $\mathcal{B}$ for inside and outside a solid rod have the same form as the corresponding formulas for $E$. We could have foreseen this result by using Eq. 17-19, which says

$$\mathcal{B} = \frac{\mathbf{v} \times \mathbf{E}}{c^2}$$

is the magnetic field produced by a moving charged body where $\mathbf{E}$ is the electric field of that same body. The preceding result for $\mathcal{B}$ inside a solid rod could have been obtained by starting with a moving charged rod of uniform charge density $\rho$. According to Table 16-1 the electric field is $E = 2\pi k_0 \rho r$. Then the magnetic field must be

$$\mathcal{B} = \frac{v}{c^2}(2\pi k_0 \rho r) = \frac{2\pi k_0}{c^2}(\rho v) r$$

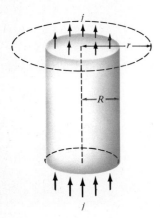

**Figure 18-3.** *Solid rod of radius $R$ has uniform current density $j$ flowing through it. Path integral is taken along dashed line circle of radius $r$.*

**Table 18-1.** *Electric and magnetic fields of charged bodies moving along their lengths with velocity v*

| Position | Body | $E$ | $\mathcal{B}$ (using $k_0$) | | $\mathcal{B}$ (using $\mu_0$) |
|---|---|---|---|---|---|
| outside | rod or wire | $\dfrac{2k_0\lambda}{r}$ | $\dfrac{v}{c^2} \times \left(\dfrac{2k_0\lambda}{r}\right) =$ | $\dfrac{2k_0}{c^2}\dfrac{I}{r}$ | $\dfrac{\mu_0}{2\pi}\dfrac{I}{r}$ |
| inside | rod | $\dfrac{2k_0\lambda}{R^2}r$ | $\dfrac{v}{c^2} \times \left(\dfrac{2k_0\lambda}{R^2}r\right) =$ | $\dfrac{2k_0}{c^2}\dfrac{I}{R^2}r$ | $\dfrac{\mu_0}{2\pi}\dfrac{I}{R^2}r$ |
| either side | single sheet with $\mathcal{J} = \sigma v$ | $2\pi k_0 \sigma$ | $\dfrac{v}{c^2} \times (2\pi k_0 \sigma) =$ | $\dfrac{2\pi k_0}{c^2}\mathcal{J}$ | $\dfrac{\mu_0}{2}\mathcal{J}$ |
| between | 2 sheets with $\mathcal{J}$ and $-\mathcal{J}$ | $4\pi k_0 \sigma$ | $\dfrac{v}{c^2} \times (4\pi k_0 \sigma) =$ | $\dfrac{4\pi k_0}{c^2}\mathcal{J}$ | $\mu_0 \mathcal{J}$ |
| $x$ from center | slab | $4\pi k_0 \rho x$ | $\dfrac{v}{c^2} \times (4\pi k_0 \rho x) =$ | $\dfrac{4\pi k_0}{c^2}jx$ | $\mu_0 j x$ |
| $r$ from charge | point charge $q$ ($v \ll c$) | $\dfrac{k_0 q \hat{\mathbf{r}}}{r^2}$ | $\dfrac{\mathbf{v}}{c^2} \times \left(\dfrac{k_0 q \hat{\mathbf{r}}}{r^2}\right) =$ | $\dfrac{k_0}{c^2}\dfrac{q\mathbf{v} \times \hat{\mathbf{r}}}{r^2}$ | $\dfrac{\mu_0}{4\pi}\dfrac{q\mathbf{v} \times \hat{\mathbf{r}}}{r^2}$ |

Eq. 18-4 is obtained by replacing ($\rho v$) with $j$.

In Table 18-1 we have summarized $E$ and $\mathcal{B}$ produced by various charged bodies that are moving along their length. In each case the formula for $\mathcal{B}$ can be obtained by replacing $\rho$, $\lambda$, or $\sigma$ in the formula for $E$ with $j$, $I$, or $\mathcal{J}$, respectively, and multiplying by $1/c^2$. In the last column, $k_0$ has been replaced with its equivalent $\mu_0 c^2/4\pi$.

## Current Sheets

A sheet of current can be obtained by moving a charged sheet along its length with velocity $v$. Then we have what is called a surface current $\mathcal{J}$ where $\mathcal{J} \equiv \sigma v$ amperes per meter. In Figure 18-4 the current is flowing up in the $y$ direction. For every meter along the sheet in the $x$ direction there are $\mathcal{J}$ amperes flowing up.

As shown in Table 18-1, we can use the equation

$$\mathcal{B} = \frac{\mathbf{v}}{c^2} \times \mathbf{E}$$

to obtain the result

$$\mathcal{B} = \frac{2\pi k_0}{c^2}\mathcal{J} \quad \text{(current sheet)} \tag{18-5}$$

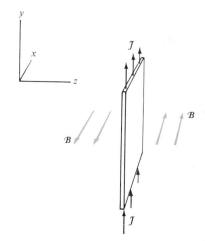

**Figure 18-4.** *Sheet with surface current $\mathcal{J}$ flowing up. $\mathcal{B}$ is into page on right side and out of page on left side.*

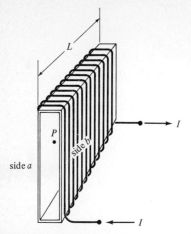

**Figure 18-5.** *Rectangular coil of length L with N turns each carrying current I.*

**Example 1.** What is the magnetic field inside the long rectangular coil shown in Figure 18-5? The coil is of length $L$ and has $N$ turns of wire. The current $I$ flows through each turn.

ANSWER: This coil can be considered as four sheets each carrying a surface current

$$\mathcal{J} = I\frac{N}{L} \text{ A/m}$$

A point $P$ deep inside sees the effects of sides $a$ and $b$ as two "infinite" current sheets with $\mathcal{J}$ and $-\mathcal{J}$ (see fourth entry in Table 18-1). The field due to sides $a$ and $b$ is then

$$\mathcal{B} = \frac{4\pi k_0}{c^2}\mathcal{J} = \frac{4\pi k_0}{c^2}\frac{IN}{L}$$

The effects of the top and bottom sheets are small because the distance of point $P$ from these sheets is much greater than their width. (It turns out that the effect of the top and bottom sheets exactly corrects for the finite height of sheets $a$ and $b$ so that the above result is exact.) This rectangular coil is a special case of the solenoid discussed below.

## The Solenoid

The equation $\mathcal{B} = \mathbf{v} \times \mathbf{E}/c^2$ is of no use when the current is flowing in a circle rather than a straight line. A common example of a current flowing in a circle is the solenoid where the current flows around the surface of a cylinder as shown in Figure 18-6. Let $n_l$ be the number of turns per unit length along the solenoid. Then the surface current is $\mathcal{J} = n_l I$ A/m. Now we apply Ampere's law to the rectangular closed path $ABCD$:

$$\oint_{ABCD} \mathcal{B}\cdot d\mathbf{s} = 4\pi\frac{k_0}{c^2}I_{\text{in}}$$

$$\mathcal{B}_{\text{in}}\int_{AB} ds + \int_{BC}\mathcal{B}\cdot d\mathbf{s} + \mathcal{B}_{\text{out}}\int_{CD} ds + \int_{DA}\mathcal{B}\cdot d\mathbf{s} = 4\pi\frac{k_0}{c^2}(\mathcal{J}x_0)$$

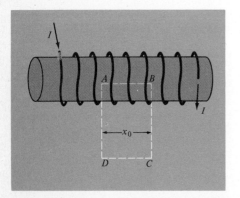

**Figure 18-6.** *A solenoid. Rectangular dashed line in path of integration for Ampere's law.*

The first integral is just $\mathcal{B}$ times $x_0$ because $\mathcal{B}$ and $d\mathbf{s}$ are parallel. The second and fourth integrals are zero because $\mathcal{B}$ and $d\mathbf{s}$ are at right angles. The third integral is zero because $\mathcal{B}_{\text{out}}$ is close to zero, as can be seen by the low density of flux lines in Figure 18-7(b). (For an infinitely long solenoid $\mathcal{B}_{\text{out}} = 0$.) Hence

$$\mathcal{B}_{\text{in}}(x_0) + 0 + 0 + 0 = \frac{4\pi k_0}{c^2} \mathcal{J} x_0$$

$$\mathcal{B}_{\text{in}} = \frac{4\pi k_0}{c^2} \mathcal{J} \quad \text{(field inside a long solenoid)} \quad \text{(18-6)}$$

Using $\mathcal{J} = n_l I$ we have

$$\mathcal{B} = \frac{4\pi k_0}{c^2} n_l I = \frac{4\pi k_0}{c^2} \frac{N}{L} I \quad \text{(18-7)}$$

where $N$ is the total number of turns and $L$ the length of the solenoid. Note that this result is independent of the position inside the solenoid because the side $AB$ need not be on the axis. We also note that Eq. 18-7 is the same as the answer to the rectangular solenoid in Example 1. The field inside a solenoid is uniform and independent of the coil shape as long as it is very long.

**Figure 18-7.** (a) *Lines of $\mathcal{B}$ from magnetized rod.* (b) *Lines of $\mathcal{B}$ from solenoid of identical shape.* (c) *Magnetic lines of force can be "seen" by sprinkling iron filings on a sheet of paper placed over a bar magnet. The filings tend to line up along the lines of force.*

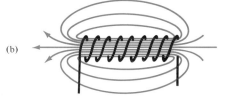

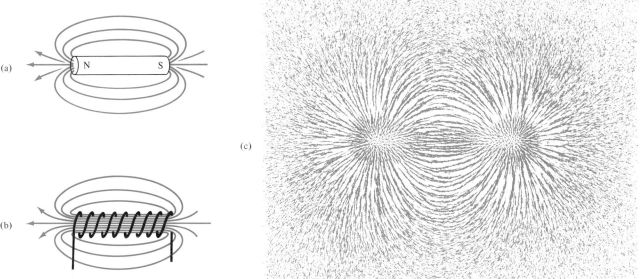

## 18-3 The Biot–Savart Law

There is another equation called the Biot–Savart law that permits the calculation of $\mathcal{B}$ from a current distribution. Clearly, the Biot–Savart law and Ampere's law must be mathematically equivalent. In some circumstances the calculation of $\mathcal{B}$ is easier using the Biot–Savart law.

Actually the last entry in Table 18-1 is in effect a "derivation" of the Biot–Savart law. If a charge $q$ is moving with velocity $v$ where $v \ll c$, the electric field will be $\mathbf{E} = k_0 q \hat{\mathbf{r}}/r^2$ and, according to Eq. 17-1, the magnetic field will be

$$\mathcal{B} = \frac{\mathbf{v}}{c^2} \times \left(\frac{k_0 q \hat{\mathbf{r}}}{r^2}\right) = \frac{k_0}{c^2} \frac{q\mathbf{v} \times \hat{\mathbf{r}}}{r^2}$$

If the charge $q$ is the amount of moving charge $dq$ in a length $dl$ of a wire, we have

$$d\mathcal{B} = \frac{k_0}{c^2} dq \frac{\frac{d\mathbf{l}}{dt} \times \hat{\mathbf{r}}}{r^2} = \frac{k_0}{c^2} \frac{dq}{dt} \frac{d\mathbf{l} \times \hat{\mathbf{r}}}{r^2}$$

$$d\mathcal{B} = \frac{k_0}{c^2} I \frac{d\mathbf{l} \times \hat{\mathbf{r}}}{r^2} \qquad \text{(the Biot–Savart law)} \qquad (18\text{-}8)$$

where $d\mathbf{l}$ is the vector length of the current element. Because of the condition $v \ll c$, our "derivation" of Eq. 18-8 is not rigorous. There is, however, a general derivation that gives Eq. 18-8 as an exact expression, keeping in mind that the current must complete a closed path and that the total field at any point is obtained by integrating Eq. 18-8 over the entire closed path. There is no way a current element $I\,d\mathbf{l}$ can contribute a particular $d\mathcal{B}$ because there is no way to isolate a particular current element from all the other current elements.

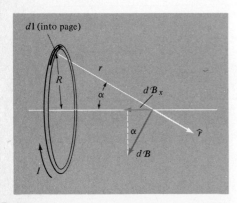

**Figure 18-8.** $d\mathbf{B}$ is magnetic field produced by the current element $I\,d\mathbf{l}$ from a ring of current of radius $R$.

**Example 2.** What is the magnetic field along the axis of a single ring of current?

ANSWER: We see in Figure 18-8 that $\hat{\mathbf{r}}$ and $d\mathbf{l}$ are perpendicular. So according to Eq. 18-8

$$d\mathcal{B} = \frac{k_0}{c^2} I \frac{dl}{r^2}$$

The component along the axis is

$$d\mathcal{B}_x = \frac{k_0}{c^2} I \frac{dl}{r^2} \sin \alpha = \frac{k_0}{c^2} I \frac{R}{r^3} dl$$

Integration around the ring gives

$$\mathcal{B}_x = \frac{k_0}{c^2} I \frac{R}{r^3} (2\pi R) = \frac{k_0}{c^2} \frac{2I}{r^3} \pi R^2$$

$$\mathcal{B} = \frac{k_0}{c^2} \frac{2IA}{r^3} \qquad (18\text{-}9)$$

where $A$ is the area enclosed by $I$.

We see that, just as with an electric dipole (Example 2, Chapter 15), the field decreases as the inverse cube of the distance. At large distances the lines of $E$ from an electric dipole and the lines of $\mathcal{B}$ from a current loop have the same pattern. The current loop behaves as a magnetic dipole. This is discussed in Section 18-4.

**Example 3.** The earth's magnetic field is essentially that of a magnetic dipole. Suppose the earth's magnetic field is produced by a current circling around underneath the equator at an average distance of 5000 km from the center. What would be the value of this current if the earth's magnetic field at either magnetic pole is $\sim 1$ gauss?

ANSWER: We can use Eq. 18-9 to solve for $I$:

$$\mathcal{B} = \frac{k_0}{c^2} \frac{2(IA)}{r^3}$$

where $r = 8100$ km is the distance from current to magnetic pole.

$$I = \frac{\mathcal{B} r^3}{2A(k_0/c^2)}$$

The area $A = \pi(5 \times 10^6)^2$ m$^2$ and $k_0/c^2 = 10^{-7}$ N/A$^2$. Then

$$I = \frac{(10^{-4})(8.1 \times 10^6)^3 \text{ A}}{2\pi(5 \times 10^6)^2(10^{-7})} = 3.38 \times 10^9 \text{ A}$$

So somehow there must be a current of over a billion amperes flowing inside the earth. Geophysicists have possible explanations for the cause of this current.

**Example 4.** Repeat the calculation in Example 3 using the cgs system of units.

ANSWER: In order to obtain the correct expression in the cgs system we replace $\mathcal{B}$ with $B/c$ and put $k_0 = 1$:

$$\frac{B}{c} = \frac{1}{c^2}\frac{2IA}{r^3}$$

or

$$I = \frac{Bcr^3}{2A}$$

where $B = 1$ gauss, $c = 3 \times 10^{10}$ cm/s, $R = 8.1 \times 10^8$ cm, and $A = \pi(5 \times 10^8 \text{ cm})^2$.

$$I = \frac{(1)(3 \times 10^{10})(8.1 \times 10^8)^3}{2\pi(5 \times 10^8)^2} = 1.01 \times 10^{19} \frac{\text{statcoul}}{\text{s}}$$

$$= 1.10 \times 10^{19} \left(\frac{1}{3 \times 10^9}\frac{\text{C}}{\text{s}}\right)$$

$$= 3.38 \times 10^9 \text{ A}$$

While on the subject of current elements, we can get an expression for the force $d\mathbf{F}$ on a current element $I\,d\mathbf{l}$ sitting in a magnetic field $\mathcal{B}$. Using Eq. 17-16 we have

$$d\mathbf{F} = dq\,\mathbf{v} \times \mathcal{B}$$

$$= dq\left(\frac{d\mathbf{l}}{dt}\right) \times \mathcal{B} = \frac{dq}{dt}d\mathbf{l} \times \mathcal{B}$$

$$d\mathbf{F} = I\,d\mathbf{l} \times \mathcal{B} \qquad \text{(force on a current element)} \qquad (18\text{-}10)$$

The following example shows how the size of the ampere is established using Eq. 18-10. If two equal currents are 1 m apart and are adjusted in magnitude until the force per unit length is $2 \times 10^{-7}$ N/m, then the amount of current in each wire is 1 A.

**Example 5.** What is the force per unit length between two equal currents a distance $r$ apart? Evaluate for $r = 1$ m and $I = 1$ A.

ANSWER: Using Eq. 18-10 we have

$$F = I\mathbf{l} \times \mathcal{B}$$

We see in Figure 18-9 that $\mathcal{B}$ is pointed into the page and thus perpendicular to $\mathbf{l}$. The vector cross product ($\mathbf{l} \times \mathcal{B}$) is pointing to the right. Using Eq. 17-10 for $\mathcal{B}$:

$$\frac{F}{l} = I\left(\frac{2k_0}{c^2}\frac{I}{r}\right) = \frac{2k_0}{c^2}\frac{I^2}{r}$$

For $r = 1$ m and $I = 1$ A:

$$\frac{F}{l} = 2\left(10^{-7}\,\frac{\text{N}}{\text{A}^2}\right)\frac{(1\text{ A})^2}{1\text{ m}} = 2 \times 10^{-7}\text{ N/m}$$

In fact, this is how the standard ampere is defined. The constant $k_0/c^2$, which is the same as $\mu_0/4\pi$, is set exactly equal to $10^{-7}$ so that the force is exactly $2 \times 10^{-7}$ N/m when both wires have 1 A of current.

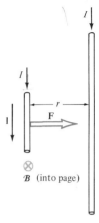

**Figure 18-9.** *Attractive force between two equal currents I.*

**Example 6.** A rectangular coil as shown in Figure 18-10 is in a uniform magnetic field. What is the torque on the coil?

ANSWER: Examination of the vector cross products $\mathbf{l} \times \mathcal{B}$ for each of the four sides shows that the magnetic forces on the two sides of length $l_1$ exert a net torque

$$T = F(l_2 \sin \alpha)$$

The magnetic force on the length $l_1$ is

$$F = Il_1\mathcal{B}$$

Hence

$$\begin{aligned}T &= (Il_1\mathcal{B})(l_2 \sin \alpha) = Il_1 l_2 \mathcal{B} \sin \alpha \\ &= (IA)\mathcal{B} \sin \alpha\end{aligned} \quad (18\text{-}11)$$

(The forces on the two sides of length $l_2$ are in line and cancel each other out.)

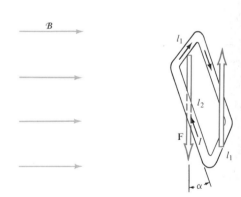

**Figure 18-10.** *Rectangular coil of area ($l_1 l_2$) in uniform magnetic field. Forces on sides of length $l_1$ are shown.*

SEC. 18-3/THE BIOT-SAVART LAW

## 18-4 Magnetism

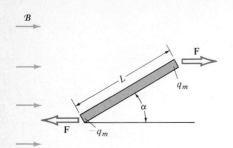

**Figure 18-11.** Bar magnet of length L at angle α to external magnetic field ℬ. Magnetic pole strength is $q_m$. Forces on magnetic poles are shown.

A bar magnet when placed in a uniform magnetic field tends to line up with the field as shown in Figure 18-11. A magnetic compass is a small bar magnet that lines up with the magnetic field of the earth. The effect can be described by assigning a magnetic pole strength (or magnetic charge) $q_m$ to one end and $-q_m$ to the other. Magnetic charge, if it exists, has the property that the force on it is $\mathbf{F} = q_m \mathcal{B}$ in analogy with electric charge. Although magnetic charge does not really exist (see following discussion), it is commonly used as a convenient mathematical way to describe magnets.

The torque on the magnet in Figure 18-11 is

$$T = FL \sin \alpha$$
$$T = q_m L \mathcal{B} \sin \alpha \qquad (18\text{-}12)$$

The product $(q_m L)$ is by definition the magnetic moment $\mu$. The torque on a magnetic moment is then

$$T = \mu \mathcal{B} \sin \alpha$$

In vector notation it is $\mathbf{T} = \boldsymbol{\mu} \times \boldsymbol{\mathcal{B}}$.

We note that the current loop in Eq. 18-11 behaves in the same way and, as we saw in Example 1, produces the same kind of a field. We can find the effective magnetic moment of a current loop by equating the torques in Equations 18-11 and 18-12:

$$\mu \mathcal{B} \sin \alpha = (IA) \mathcal{B} \sin \alpha$$
$$\mu = IA \qquad \text{(magnetic moment of current loop)} \qquad (18\text{-}13)$$

Before the discovery of the electron and atomic electron currents, the hypothesis of magnetic charge explained why magnets behave the way they do in magnetic fields. But then, if magnetic charge exists as does electric charge, it should be possible to isolate some net magnetic charge of either positive or negative polarity. One might expect to isolate a magnetic pole by breaking the end off a bar magnet. Whenever such an attempt is made, however, another equal and opposite pole appears at the other end as shown in Figure 18-12. So far isolated magnetic poles have never been found in nature. The attempts to isolate magnetic charge have been so extensive that by now most physicists believe that they do not exist. All physicists agree that the properties of a bar magnet are explained by what is called amperian currents.

The claim is that any bar magnet is actually a solenoid with a net surface current $\mathcal{J}'$ circling around the outside as shown in Figure 18-13. We shall now calculate what $\mathcal{J}'$ must be for a bar magnet of pole strength $q_m$. We saw in Example 6 that the torque on a single turn of a solenoid is

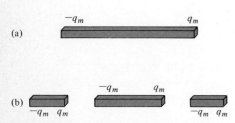

**Figure 18-12.** Bar magnet in view (a) is broken into three pieces in view (b).

**Figure 18-13.** (a) *Bar magnet showing net amperian current circulating around outside surface.* (b) *End view showing individual atomic currents. Net current is shown in color.*

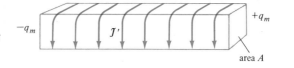

(a)    (b)

$T_1 = IA\mathcal{B} \sin \alpha$. For a solenoid of $N$ turns the torque is $T = NIA\mathcal{B} \sin \alpha$. Since the surface current is $\mathcal{J}' = NI/L$ we have $T = \mathcal{J}'LA\mathcal{B} \sin \alpha$. Equating this to the torque in Eq. 18-12 gives

$$q_m L \mathcal{B} \sin \alpha = \mathcal{J}' LA \mathcal{B} \sin \alpha$$
$$q_m = \mathcal{J}'A \qquad (18\text{-}14)$$

We see that a bar magnet of pole strength $q_m$ can be described as a piece of metal with a permanent surface current $\mathcal{J}' = q_m/A$ circulating around the outer surface. Eq. 18-14 also tells us that a solenoid with a surface current $\mathcal{J}'$ behaves as a bar magnet of pole strength $q_m = \mathcal{J}'A$.

Let us now see if the atomic currents in an iron atom can account for the field $\mathcal{B} \approx 2$ tesla that is observed inside a fully magnetized piece of iron. If each atom by virtue of its electron currents has a magnetic moment $\mu_a$ and there are $\mathcal{N}$ atoms per unit volume, the total magnetic moment of a bar magnet of cross sectional area $A$ and length $L$ is $\mu = \mathcal{N}AL\mu_a$, assuming all the atomic magnetic moments are lined up. $\mu_a$ for the iron atom can be calculated from the electron currents as is done in the next paragraph, which predicts a value of $1.86 \times 10^{-23}$ A m². Since the total magnetic moment by definition is the pole strength $q_m$ times the length $L$, we have

$$(q_m L) = \mathcal{N}AL\mu_a$$
$$q_m = \mathcal{N}A\mu_a$$

We use Eq. 18-14 to substitute $(\mathcal{J}'A)$ for $q_m$:

$$(\mathcal{J}'A) = \mathcal{N}A\mu_a$$
$$\mathcal{J}' = \mathcal{N}\mu_a$$

For iron $\mathcal{N} = 8.51 \times 10^{22}$ atoms/cm³. Using the predicted value $\mu_a = 1.86 \times 10^{-23}$ A m² we have

$$\mathcal{J}' = (8.51 \times 10^{28} \text{ m}^{-3})(1.86 \times 10^{-23} \text{ A m}^2) = 1.58 \times 10^6 \text{ A/m}$$

This is a much larger current than can be achieved in a "man-made" solenoid. Insertion of an iron core into a solenoid will boost the field by a corresponding amount.

According to Eq. 18-6 the field inside the solenoid (bar magnet) is

$$\mathcal{B} = \frac{4\pi k_0}{c^2} \mathcal{J}' = \frac{4\pi(9 \times 10^9)}{(3 \times 10^8)^2}(1.58 \times 10^6)\ \mathrm{T} = 1.99\ \mathrm{T}$$

This result agrees with the measurements of fully magnetized iron.

### Magnetic Moment of the Electron

If we invoke some quantum mechanics, there is no need to measure $\mu_a$ for the iron atom: it can be calculated from first principles. We shall now briefly outline how this goes. Suppose an electron is in a circular orbit of radius $r$. The electron current is its charge $e$ times its orbital frequency: $I = e(v/2\pi r)$. According to Eq. 18-13 its magnetic moment is

$$\mu_e = I(\pi r^2) = \left(\frac{ev}{2\pi r}\right)(\pi r^2) = \frac{evr}{2}$$

$$= \frac{e}{2m}(mvr)$$

Note that the quantity $mvr$ is the angular momentum $L$. Then

$$\mu_e = \frac{e}{2m} L \qquad (18\text{-}16)$$

As we shall learn in quantum mechanics, the orbital angular momentum of an electron is quantized in units of $L = h/2\pi$, where $h = 6.63 \times 10^{-34}$ J s is Planck's constant. Then the minimum magnetic moment is

$$\mu_e = \frac{e}{2m}\frac{h}{2\pi} = 9.3 \times 10^{-24}\ \mathrm{A\ m^2}$$

According to Eq. 18-16 this result is independent of the distance $r$ from the center of rotation so a rotating electron might be expected to have the same magnetic moment. This indeed happens to be the case in quantum mechanics. All electrons have intrinsic magnetic moment $\mu_e = 9.3 \times 10^{-24}$ A m². In the iron atom all the magnetic moments of the 26 orbital electrons cancel out except for two of them. The magnetic moment of iron is $\mu_{\mathrm{Fe}} = 2\mu_e = 1.86 \times 10^{-23}$ A m², which agrees with the measured value.

It remains to be explained why iron atoms prefer to line up so that their magnetic moments are all pointed in the same direction. A full explanation involves quantum mechanics and solid state physics. The modern theory of solid state physics predicts that iron, cobalt, and nickel (the ferromagnetic metals) are made up of macroscopic domains (on the order of a few thousandths of an inch) where the atoms are completely lined up. In an unmagnetized sample the domains are randomly oriented. In the process of magnetization the domains line up by movement of the domain boundaries, the domains favorably oriented with respect to the field growing at the expense of the others.

## 18-5 Maxwell's Equations with Steady Currents

Up to now we have accumulated a total of four equations for $\mathcal{B}$ and $\mathbf{E}$ in integral form. As we have stressed, they can be derived from Coulomb's law and relativity theory. These four equations taken together are called Maxwell's equations. We shall list them in the following order: Eq. 15-7 (Gauss' law), Eq. 16-7 (potential difference is independent of the path), Eq. 18-3 (lines of $\mathcal{B}$ are continuous), Eq. 18-1 (Ampere's law). For the sake of completeness they are listed in both cgs and mks forms. The cgs form is obtained by putting $k_0 = 1$ and replacing $\mathcal{B}$ with $(\mathbf{B}/c)$. The mks form is also listed using $\varepsilon_0$ and $\mu_0$ by replacing $k_0$ with $1/4\pi\varepsilon_0$ and $(k_0/c^2)$ with $(\mu_0/4\pi)$. Equations I and III are integrals taken over completely closed surfaces. The left-hand sides are respectively the net electric and magnetic fluxes leaving the closed surface. Equations II and IV are path integrals around any completely closed path.

So far we have only dealt with steady currents. Starting with the next chapter we shall consider the more general situation where the current can change with time. Then we shall find that there will be an additional term on the right-hand side of equations II and IV.

**Table 18-2.** *Maxwell's equations with steady currents*

| | mks using $k_0$ | mks using $\varepsilon_0$ and $\mu_0$ | cgs |
|---|---|---|---|
| I | $\oint \mathbf{E} \cdot d\mathbf{A} = 4\pi k_0 Q_{in}$ | $\oint \mathbf{E} \cdot d\mathbf{A} = \dfrac{1}{\varepsilon_0} Q_{in}$ | $\oint \mathbf{E} \cdot d\mathbf{A} = 4\pi Q_{in}$ |
| II | $\oint \mathbf{E} \cdot d\mathbf{s} = 0$ | $\oint \mathbf{E} \cdot d\mathbf{s} = 0$ | $\oint \mathbf{E} \cdot d\mathbf{s} = 0$ |
| III | $\oint \mathcal{B} \cdot d\mathbf{A} = 0$ | $\oint \mathcal{B} \cdot d\mathbf{A} = 0$ | $\oint \mathbf{B} \cdot d\mathbf{A} = 0$ |
| IV | $\oint \mathcal{B} \cdot d\mathbf{s} = \dfrac{4\pi k_0}{c^2} I_{in}$ | $\oint \mathcal{B} \cdot d\mathbf{s} = \mu_0 I_{in}$ | $\oint \mathbf{B} \cdot d\mathbf{s} = \dfrac{4\pi}{c} I_{in}$ |

## Summary

Magnetic field produced by steady currents can be determined using Ampere's law, which gives for the magnetic field integrated around any closed path:

$$\oint \mathcal{B} \cdot d\mathbf{s} = 4\pi \frac{k_0}{c^2} I_{in}$$

$I_{in}$ is the net current enclosed by the path and is also equal to $\int \mathbf{j} \cdot d\mathbf{A}$ integrated over any surface bounded by the path. A second basic relation involving magnetic field says that the net flux over any closed surface is zero:

$$\oint \mathcal{B} \cdot d\mathbf{A} = 0$$

Another relation which is in some cases more useful is the Biot–Savart law:

$$d\mathcal{B} = \frac{k_0}{c^2} I \frac{d\mathbf{l} \times \hat{\mathbf{r}}}{r^2}$$

These relations give

$$\mathcal{B} = 4\pi \frac{k_0}{c^2} I n_l$$

for the field anywhere inside a solenoid having $n_l$ turns per unit length. The field produced by a single straight wire is

$$\mathcal{B} = \frac{2k_0}{c^2} \frac{I}{r}$$

and the field produced by a current sheet having $\mathcal{J}$ amperes per meter is

$$\mathcal{B} = 2\pi \frac{k_0}{c^2} \mathcal{J}$$

The magnetic field produced by a permanent bar magnet is explained in terms of atomic current loops whose net effect is a surface current $\mathcal{J}'$ circling around the outside so that the bar magnet is in actuality a permanent solenoid. In terms of this surface current, the effective (but fictitious) magnetic pole strength is $q_m = \mathcal{J}'A$. The magnetic moment of a single current loop of area $A$ is $\mu = IA$. The magnetic moment of a single electron is

$$\mu_e = \frac{e}{2m}\frac{h}{2\pi}$$

where $h$ is Planck's constant.

## Exercises

1. Consider two parallel infinite sheets each with surface current $\mathcal{J}$. What is the direction and magnitude of $\mathcal{B}$ in regions I, II, and III?
2. Consider two adjacent slabs with current densities $j$ flowing in opposite directions. Each slab has a thickness $x_0$. What is the direction and magnitude of $\mathcal{B}$ in regions I, II, III, and IV?
3. Consider two concentric long solenoids of radii $R_1$ and $R_2$ with surface currents $\mathcal{J}_1$ and $\mathcal{J}_2$. What is the direction and magnitude of $\mathcal{B}$ in the regions where $r < R_1$, $R_1 < r < R_2$, and $r > R_2$?
4. A solenoid 1 m long and 8 cm in diameter is wound with 500 turns of wire.
   (a) What is $\mathcal{B}$ inside the solenoid when it carries a current of 5 A?
   (b) What is the total number of flux lines generated by this current?
5. An electron is aimed down the axis of a solenoid. Describe the motion of the electron.
6. A coaxial cable consists of an inner and an outer cylinder of radii $R_1$ and $R_2$ respectively. A current $I$ flows along the surface of each cylinder but in opposite directions (parallel and antiparallel to the axis). What is $\mathcal{B}$ at a distance $r$ from the common axis
   (a) when $r$ is between the two cylinders?
   (b) when $r$ is outside the outer cylinder?

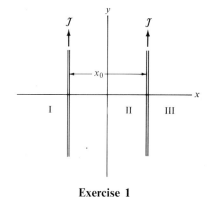

Exercise 1

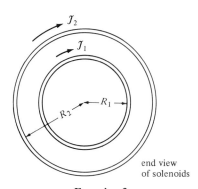

Exercise 2

Exercise 3

7. A vertical coaxial cable has a current $I_1 = 3$ A flowing down the center conductor and $I_2 = 2$ A flowing up the outer cylindrical conductor. The outer cylinder is 3 cm diameter. The center conductor is 4 mm diameter.
   (a) What is $\mathcal{B}$ at a distance of 5 cm from the center?
   (b) What is $\mathcal{B}$ inside the cable at a distance of 0.5 cm from the center?
8. Suppose the current in Example 3 is flowing across an area equal to one fourth the cross-sectional area of the earth.
   (a) What is the current density $j$?
   (b) What is the current $I$ per square centimeter of cross-sectional area?
   (c) If the current is due to moving electrons having drift velocity of $10^{-4}$ m/s, how many moving electrons would there be per cubic centimeter?
9. What is the force in dynes per centimeter between two parallel currents of 1 A each a distance of 1 cm apart?
10. What is $\mathcal{B}$ at the end of a solenoid? (*Hint:* Consider a second identical solenoid butted up against the end of the first giving the effect of one long continuous solenoid.)

## Problems

11. Consider a hollow conducting rod. The inner radius is $R_1$ and the outer is $R_2$. A current $I$ is flowing uniformly throughout the conductor (into the page). What is $\mathcal{B}$ in the regions where $r < R_1$, $R_1 < r < R_2$, and $r > R_2$.
12. This is the same as Problem 11 except that the cylindrical hole is off center by a distance $x_0$. What is $\mathcal{B}$ at every point along the $x$ axis? Express answer in terms of $I$, the total current flowing in the conductor. *Hint:* The solution to this problem is the same as to the problem of two solid cylinders of equal and opposite current density of radii $R_1$ and $R_2$ and displaced by $x_0$.
13. Repeat Problem 12, but this time give $\mathcal{B}$ at every point along the $y$ axis.
14. Derive a formula for $\mathcal{B}$ inside a toroidal solenoid. The inner radius is $R_1$ and the outer is $R_2$. There are a total of $N$ turns. (It is like winding $N$ turns around a donut.)
15. Consider a solenoid of finite length $L$ with surface current $\mathcal{J}$. Then Eq. 18-9 gives

$$d\mathcal{B} = \frac{k_0}{c^2} \frac{2A\mathcal{J}dx}{r^3}$$

Integrate over $x$ to get $\mathcal{B}$ at point $P$. Answer is

$$\mathcal{B} = \frac{2\pi k_0}{c^2} \mathcal{J}(\cos\theta_2 - \cos\theta_1)$$

where $\theta_1$ and $\theta_2$ are the angles subtended by the two ends.
16. Repeat Problem 15 for the case where point $P$ is inside the solenoid.
17. Consider two parallel rings of radius $R$ and each carrying current $I$. They are separated by a distance also equal to $R$. Use Eq. 18-9 to get an expression for $\mathcal{B}$ everywhere along the $x$ axis.

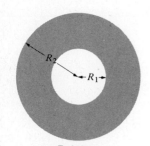

End view
Problem 11

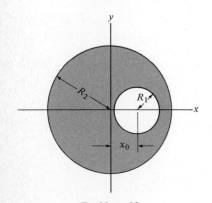

Problem 12

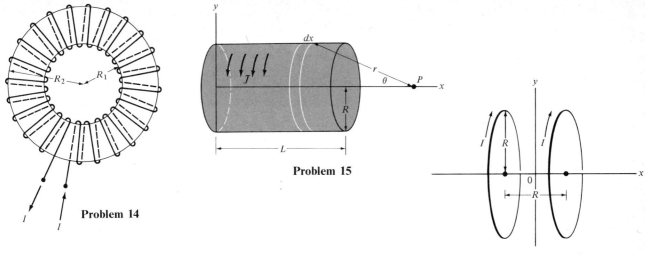

Problem 14

Problem 15

Problem 17

18. In Problem 17 what is $\mathcal{B}$ at $x = 0$? What are the first three derivatives of $\mathcal{B}$ with respect to $x$ at $x = 0$? Note that such an arrangement gives a quite uniform field in the central region between the coils. These are called Helmholtz coils.

19. Suppose the coil shown in Figure 18-10 has a moment of inertia $I_0$ and a current $I$ in the opposite direction to that shown in the figure. If the coil is rotated a small angle $\alpha$ and released, what will be the period of oscillation in terms of $I_0$, $I$, $A$, and $\mathcal{B}$?

20. Assuming the existence of isolated magnetic charge, Gauss' law for magnetic charge would be of the form

$$\oint_S \mathcal{B} \cdot d\mathbf{A} = KQ_m$$

where $Q_m$ is the net magnetic charge enclosed by the surface $S$ and $K$ is a proportionality constant. What is $K$? (See Exercise 4b.)

21. A certain metallic alloy has $8 \times 10^{22}$ atoms/cm$^3$ and an average of one aligned electron per two atoms. What is the magnetic field inside this magnetized substance?

22. Consider a ring of mass $M$ with inner and outer radii $R_1$ and $R_2$. It is uniformly charged with total charge $Q$. It is rotating about its axis with angular velocity $\omega$.
    (a) What is its angular momentum $L$ in terms of $M$, $R_1$, $R_2$, and $\omega$?
    (b) What is its magnetic moment $\mu$ in terms of $Q$, $R_1$, $R_2$, and $\omega$?
    (c) What is the ratio $\mu/L$? Is this the same as for an electron in circular orbit?

23. Repeat Problem 22 for a disk of radius $R$.

24. Consider three infinite slabs with current density $j$ as shown. What are the magnitudes and directions of $\mathcal{B}$ in regions I, II, III, and IV?

25. Repeat Exercise 2 for the case where the two slabs are separated by a distance $x_1$.

26. What is the total magnetic flux leaving the charge $q_m$ in Figure 18-11?

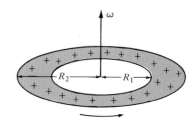

Problem 22

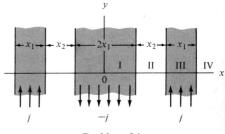

Problem 24

# Electromagnetic Induction

The central topic in this chapter is Faraday's law, which has the simple form emf $= -d\Phi/dt$ where emf is work per unit charge done in taking a charge around a closed loop and $\Phi$ is the magnetic flux through that loop. We shall show that Faraday's law applies to three different physical situations:

1. A wire loop moving through a region of magnetic field—sometimes called magnetic emf.
2. A stationary wire loop (or an imaginary loop) where the source of magnetic field is moving by—sometimes called electric emf.
3. A stationary wire loop (or an imaginary loop) where the source of magnetic field is also not moving, but the current that is the source of the magnetic field is changing—also electric emf.

In situation 1 the emf is equal to $(1/q)\oint \mathbf{F}_m \cdot d\mathbf{s}$, which is the integral of magnetic force on charge $q$ around a closed path. In situations 2 and 3 the emf is equal to $\oint \mathbf{E} \cdot d\mathbf{s}$ around the closed path. In situation 3 we shall see that a changing magnetic field produces an accompanying electric field.

## 19-1 Motors and Generators

Figure 19-1 illustrates the basic principle of the electric generator. A simple rectangular loop is placed in a region of uniform magnetic field. If the loop is rotated about the axis shown, a voltage difference will appear across the ends $P_1$ and $P_2$. As we shall see, this voltage will alternate in polarity at the frequency of rotation of the loop. We shall now calculate the output voltage in terms of the size of the loop, $\mathcal{B}$, and the angular velocity $\omega$.

Consider the force $F$ on a small charge $q$ inside the lower arm $l_1$. Since

**Figure 19-1.** (a) *Rectangular coil of area* $A = l_1 l_2$ *is rotating counterclockwise in uniform magnetic field. Charge q in lower arm experiences force* $\mathbf{F} = q\mathbf{v} \times \mathcal{B}$. (b) *Side view of* (a).

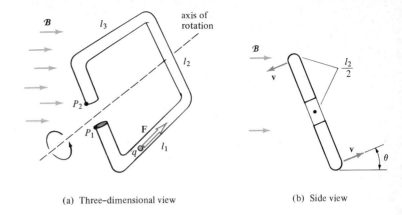

(a) Three-dimensional view     (b) Side view

$E = 0$ everywhere the only force it can experience is the magnetic force $\mathbf{F}_1 = q\mathbf{v} \times \mathcal{B}$ or

$$F_1 = q\left(\frac{\omega l_2}{2}\right)\mathcal{B} \sin\theta$$

where we have used $(\omega l_2/2)$ for $v$. The work done on $q$ as it drifts along $l_1$ is

$$W_1 = \mathbf{F}_1 \cdot \mathbf{l}_1 = \left(q\frac{\omega l_2}{2}\mathcal{B} \sin\theta\right)l_1$$

No work is done on $q$ as it moves along $l_2$ since $\mathbf{F}$ and $\mathbf{l}_2$ are perpendicular. When $q$ moves along $l_3$ the work $W_3 = W_1$. Hence the total work done on $q$ by magnetic forces in moving from $P_1$ to $P_2$ is

$$W = q l_1 l_2 \mathcal{B} \omega \sin\theta$$

**The work per unit charge in going around the loop from $P_1$ to $P_2$ is defined as the electromotive force (emf).**

$$\text{emf} = \frac{W}{q} = \mathcal{B} l_1 l_2 \omega \sin\theta$$

Using $l_1 l_2 = A$, the area of the loop, and $\theta = \omega t$ gives

$$\text{emf} = \mathcal{B} A \omega \sin \omega t \qquad (19\text{-}1)$$

Clearly, electromotive force is not a force. It has units of joules per coulomb or volts. It is the energy per unit charge given to a conduction electron each time it completes the circuit (assuming "terminals" $P_1$ and $P_2$ are connected to an external circuit). Emf is a voltage source in the same sense that an electric battery is a voltage source. Every conduction electron gains energy per unit charge equal to the emf each time it passes through the rotating coil of the generator. This additional energy can be transferred to atoms in an external circuit by collisions resulting in ohmic heating.

We can now relate the emf to the magnetic flux passing through the loop (or single-turn coil). The flux is

$$\Phi_B = \int \mathcal{B} \cdot d\mathbf{A} = \mathcal{B}l_1 l_2 \cos\theta = \mathcal{B}A \cos\omega t$$

The time derivative of the flux passing through the rotating coil is

$$\frac{d\Phi_B}{dt} = -\mathcal{B}A\omega \sin\omega t$$

Using Eq. 19-1 we have

$$\text{emf} = -\frac{d\Phi_B}{dt}$$

which is our first version of Faraday's law. So far we have shown that Faraday's law applies to a rectangular loop rotating in a stationary and uniform magnetic field.

If the rotating coil has $n$ turns connected in series, the emf is $n$ times as much. If a capacitor is connected across terminals $P_1$ and $P_2$, positive charges would quickly accumulate on $P_2$ under the situation shown in Figure 19-1. When the charge reaches the point where $\int_{P_1}^{P_2} \mathbf{E} \cdot d\mathbf{s}$ along the loop is equal and opposite to the magnetic emf, the charges will no longer flow. Thus the voltage which appears across $P_1$ and $P_2$ is equal to the emf.

---

**Example 1.** A 200-turn coil of 5 cm by 6 cm is rotating at 60 rev/s in a field of 5000 gauss (0.5 T). Plot the output voltage as a function of time.

ANSWER: The voltage per turn is given by Eq. 19-1. Putting $\mathcal{B} = 0.5$ T, $A = 30 \times 10^{-4}$ m², and $\omega = 120\pi$ s$^{-1}$ into this equation gives

$$\text{emf/turn} = (.5)(30 \times 10^{-4})(120\pi) \sin\omega t \text{ V}$$
$$= 0.565 \sin\omega t \text{ V}$$

Multiplying by the 200 turns gives

$$\text{emf} = 113 \sin \omega t \text{ V}$$

This is plotted in Figure 19-2. For comparison the household voltage used in the United States is $V = 170 \sin (120\pi t)$ in volts. The root-mean-square average of this is $\sqrt{\overline{V^2}} = 120$ V. Any current or voltage that varies sinusoidally with time is called an ac current or ac voltage.

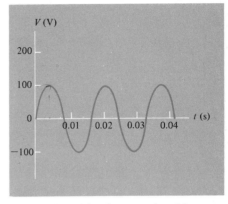

**Figure 19-2.** *AC voltage produced by generator described in Example 1.*

### Electric Motors

If the rectangular loop in Figure 19-1 is turned by hand (or by other mechanical means), we have an electric generator. If, instead, a current $I$ is sent through the loop from the outside, there will be a torque given by Eq. 18-11:

$$T = IA\mathcal{B} \sin \theta$$

If the current enters at $P_1$ and leaves at $P_2$ the torque will tend to rotate the coil clockwise.

---

**Example 2.** Using the same coil as in Example 1, assume a sinusoidal current is causing the coil to rotate at the same speed. The peak value of this current is 2 A. What is the peak instantaneous power delivered by this motor?

ANSWER: In linear motion power $P = Fv$. So for rotation about a circle of radius $R$, $P = (RF)(v/R) = T\omega$. For $n$ turns the torque will be $T = nIA\mathcal{B} \sin \theta$. Thus

$$P = (nIA\mathcal{B} \sin \theta)\omega$$

The peak power is

$$P_0 = nIA\mathcal{B}\omega$$

Using $N = 200$, $I = 2$ A, $A = 30 \times 10^{-4}$ m², $\mathcal{B} = 0.5$ T, and $\omega = 120\pi$ s$^{-1}$ gives

$$P_0 = 226 \text{ W} = 0.30 \text{ hp}$$

---

In Examples 1 and 2 the coil terminals $P_1$ and $P_2$ would have to be connected to slip rings to avoid twisting of wires.

## 19-2 Faraday's Law

So far we have shown that Faraday's law works for the limited case of a rectangular coil rotating in a uniform magnetic field. In Appendix 19-1 we show that the emf induced in a coil of arbitrary shape moving with linear velocity $v$ through a region of nonuniform magnetic field is also equal to $-d\Phi_B/dt$.

Appendix 19-1 shows that for a coil moving through a region of magnetic field the work done per unit charge on a charge that moves once around the circuit is

$$\text{emf} = -\frac{d\Phi_B}{dt} \qquad (19\text{-}2)$$

For a stationary coil there can be no magnetic force because $v = 0$. However, if the source of the magnetic field is moving, there will be an electric field in the region of the coil that obeys the relation

$$\oint \mathbf{E} \cdot d\mathbf{s} = -\frac{d\Phi_B}{dt} \qquad \text{(Faraday's law)} \qquad (19\text{-}3)$$

Equation 19-3 can be derived directly from Eq. 19-2 by imposing the principle of relativity on the system of coil and source of magnetic field. Clearly, the measured emf can depend only on the relative velocity between the coil and the source of magnetic field. An observer at rest with respect to the coil must see the same force on a charge $q$ sitting on the coil as does an observer moving along with the source of the magnetic field. By definition this force divided by $q$ is the electric field according to the observer at rest with respect to the coil. This result is mathematically consistent with Eq. 17-20, which says that $\mathbf{E} = -\mathbf{v} \times \mathcal{B}$ where $v$ is the velocity of the moving source.

To summarize, if the coil moves through a region of magnetic field, there is no emf due to electric field around the closed path, but the magnetic force produces an emf equal to $-d\Phi_B/dt$. This situation is shown in Figure 19-3(a).

However, if, according to an observer, the coil is fixed and the source of magnetic field is moving, there will be an electric field and it can be calculated using Eq. 19-3. This latter situation is shown in Figure 19-3(b). According to the principle of relativity the two meter readings must be the same. In experiment (b) the changing magnetic field in the region of coil B has produced a force on conduction electrons in the wire that results in the meter deflection. This force per unit charge is by definition electric field.

But what about experiment (c) in Figure 19-3 where neither coil moves? The magnetic field at coil B can be made to decrease at the same rate as in experiment (b) by increasing the resistance $R$ at an appropriate rate. The magnetic field and its time derivatives in the region around coil B is the same in the two cases. In experiment (b) we concluded that changing magnetic field has produced a force on the conduction electrons. Since the

**Figure 19-3.** (a) *Coil B is moved through the magnetic field produced by coil A. The meter detects an induced emf.* (b) *The relative velocity between coils A and B is the same. Meter gives same reading.* (c) *The resistance R is increased at a rate such that $\mathcal{B}$ decreases at coil B at the same rate as in (a) and (b).*

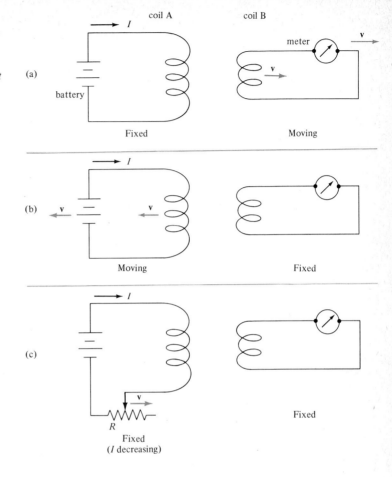

changing magnetic field is exactly the same in experiment (c), it must give rise to the same force on charges in coil B. Thus the meter readings must be the same in experiments (b) and (c). If this were not true, identical conditions in the region of coil B would give rise to nonunique results. Our original purpose in constructing $E$ and $\mathcal{B}$ fields was to be able to calculate forces on charged particles. If such a mathematical construct is to have any usefulness or meaning, specific values of **E**, **$\mathcal{B}$**, and their derivatives must give rise to specific forces on charged particles. Just what the "distant" source of $\mathcal{B}$ is doing cannot affect the local situation. So we are forced to conclude that the equation

$$\oint \mathbf{E} \cdot d\mathbf{s} = -\frac{d\Phi_B}{dt}$$

applies to experiment (c) as well as to experiment (b).

We have just deduced that Eq. 19-3 (Faraday's law) applies also to a system of stationary circuits. A changing current in one circuit can induce an emf and electric field in a second circuit. In fact the second circuit is unnecessary. The electric field will be there whether or not the circuit is there. The equation

$$\oint \mathbf{E} \cdot d\mathbf{s} = -\frac{d\Phi_B}{dt}$$

is completely general and holds for any imaginary fixed closed path in space. It can also be written as

$$\oint_\mathcal{C} \mathbf{E} \cdot d\mathbf{s} = -\frac{d}{dt}\left(\int_S \mathcal{B} \cdot d\mathbf{A}\right)$$

$$\oint_\mathcal{C} \mathbf{E} \cdot d\mathbf{s} = -\int_S \frac{\partial \mathcal{B}}{\partial t} \cdot d\mathbf{A} \quad (19\text{-}4)$$

where $S$ is any surface enclosed by curve $\mathcal{C}$. We have converted to $\partial \mathcal{B}/\partial t$ because the boundaries of $d\mathbf{A}$ do not change with time.

## 19-3 Lenz's Law

In order for the minus sign in Eq. 19-4 to have meaning, it is important to establish the correct direction for $d\mathbf{A}$ in the surface integral. This is done using a right-hand circulation rule, illustrated in Figure 19-4. Curl the fingers of the right hand around the direction of the path integral (direction of $d\mathbf{s}$). Then the thumb indicates the positive direction for the enclosed surface $S$. We see from Eq. 19-4 and Figure 19-4 that, if the flux in the direction of $d\mathbf{A}$ is increasing, the induced emf will be negative. It would cause a current to flow opposite to the circulating arrow $d\mathbf{s}$ in Figure 19-4. Such a current would produce flux of its own opposite to the direction of increase of flux. **This observation that induced current produces flux which resists the original change in flux is called Lenz's law.**

A graphic example of Lenz's law is the behavior of a closed loop of superconducting wire. No matter how much any external magnetic fields are varied, the net flux through a superconducting loop stays constant. (If the net flux changed, there would be a nonzero emf or an infinite current, which is impossible.) If a superconducting ring is brought near a magnet, there will be a finite current induced in the ring that produces flux of just the right amount to oppose the flux from the magnet (see Figure 19-5). Also each element of the current loop will have a force $I d\mathbf{l} \times \mathcal{B}$ on it repelling it from the magnet. This force can exceed the weight of the ring. A ring made of a good conductor, when placed above the pole of a magnet, will actually float for a short time. Another application of Lenz's law is to release a bar magnet over a superconducting bowl. The bar

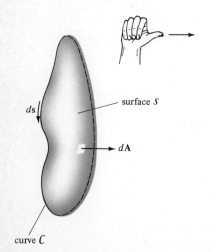

**Figure 19-4.** *Method to determine positive direction of surface $S$ using right hand.*

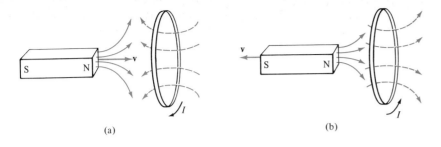

**Figure 19-5.** (a) *The bar magnet is moving to the right, increasing the flux in the closed loop of wire. The induced I produces the dashed lines of $\mathcal{B}$ which oppose the increase in flux due to the bar magnet.* (b) *A bar magnet originally at rest is moved to the left, which decreases its flux through the loop. There will be an induced current I that produces the dashed lines of $\mathcal{B}$, which opposes the change; that is, the dashed lines try to maintain the original amount of flux through the coil. In* (a) *the net force on the coil is to the right and in* (b) *it is to the left.*

magnet will then float. There are plans to build high speed trains that use superconducting coils to float the trains over a special bed or "track."

It is not necessary for the conductor to be superconducting. Because of Lenz's law, any conductor that is pushed into a magnetic field will experience an opposing force. The induced currents in such situations are called eddy currents.

## 19-4 Inductance

### Transformers

If two separate coils are wound around the same core, a changing current in one will produce an emf in the other that can be calculated using Faraday's law. Such a device is called a transformer. In most transformers the primary and secondary coils are wound almost on top of each other, so that they both enclose the same lines of $\mathcal{B}$ (see Figure 19-6). Thus both coils have the same value of $d\Phi_B/dt$. Let $n_1$ be the number of turns of the primary coil and $n_2$ be the number of turns in the secondary coil. According to Eq. 19-3 the emf or induced voltage $V_2$ in coil 2 is

$$V_2 = -n_2 \frac{d\Phi_B}{dt}$$

Similarly, the emf for coil 1 is

$$V_1 = -n_1 \frac{d\Phi_B}{dt}$$

The ratio of the two voltages is then

$$\frac{V_2}{V_1} = \frac{n_2}{n_1}$$

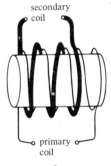

**Figure 19-6.** *A transformer.*

When an ac voltage $V_{ac}$ is first applied to the primary, the current increases until $n_1 d\Phi_B/dt$ reaches a value almost equal in magnitude to $V_{ac}$; hence $V_{ac} \approx V_1$.

We see that if an ac voltage is applied to the primary, the induced secondary voltage can be made larger or smaller by choosing the appropriate turns ratio. This convenient method for converting small voltages into large ones and vice versa is one of the advantages of using ac as compared to dc. This advantage is very important when it comes to electrical power production and transmission. The most economical generators produce a rather low ac voltage. As we shall see in Example 3, it is necessary to use a high voltage in order to reduce power loss over long transmission lines. A transformer can boost the low voltage to a high voltage with very little power loss. Then, at the receiving end, another transformer must be used to transform the high voltage down to a safer and more convenient low voltage.

In order to see why a high voltage is needed, we shall consider the specific example of a generator that delivers 10 MW via a transmission line having a resistance of 10 ohms.

**Example 3.** Calculate the transmission loss for a generator that delivers 10 MW via a transmission line having a resistance of 10 ohms. Consider the two cases, where the voltage at the generator is (a) $1.4 \times 10^4$ V and (b) $10^5$ V.

ANSWER: $P = IV$, so the current transmitted is given by $I = P/V$. The power loss in the transmission line is $P_{loss} = I^2 R = (P/V)^2 R = RP^2/V^2 = (10)(10^7)^2/V^2$, or, $P_{loss} = 10^{15}/V^2$. Thus the transmission losses are reduced as the square of the output voltage. This gives
(a) $P_{loss} = 10^{15}/(1.4 \times 10^4)^2 = 5$ MW which is half the original power.
(b) $P_{loss} = 10^{15}/(10^5)^2 = 10^5$ W which is a 1% loss.
It is clear that more than ~20 kV should be used to transmit electrical power in this case.

### Self-inductance

When the current through a coil or a solenoid is changing, the flux through each turn is changing and, according to Faraday's law, there must be an induced emf in each turn equal to

$$\frac{\text{emf}}{\text{turn}} = -\frac{d\Phi}{dt}$$

where $d\Phi/dt$ is the rate of change of flux through the turn. This is called a back emf. If the flux is the same through all $N$ turns, the total back emf is

$$\text{emf} = -N\frac{d\Phi}{dt}$$

The quantity $N\Phi$ is the total flux enclosed by the circuit and is called the flux linkage. It must be proportional to the current through the coil:

$$N\Phi = LI \qquad (19\text{-}5)$$

or

$$L \equiv \frac{N\Phi}{I} \quad \text{(definition of inductance)}$$

The proportionality constant $L$ is by definition the inductance. Differentiation of Eq. 19-5 leads to an equivalent definition of $L$:

$$N\frac{d\Phi}{dt} = L\frac{dI}{dt}$$

Hence

$$\text{emf} = -L\frac{dI}{dt} \quad \text{(back emf)} \qquad (19\text{-}6)$$

We see that $L$ has units of volt-seconds per ampere or ohm-seconds. In the mks system this unit has the special name **henry.** The SI abbreviation is H.

As an example we calculate the inductance of a long solenoid of length $x_0$ having $N$ turns (see Figure 19-7). Using Eq. 19-5,

$$L = N\frac{\Phi_1}{I}$$

The flux through any one turn is

$$\Phi_1 = \mathcal{B}A \quad \text{where } \mathcal{B} = \frac{4\pi k_0}{c^2}\frac{NI}{x_0} \text{ is given by Eq. 18-7.}$$

So

$$\Phi_1 = \frac{4\pi k_0 NA}{c^2 x_0} I.$$

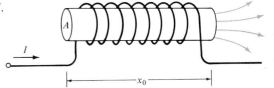

**Figure 19-7.** *Solenoid of length $x_0$ produces flux $\Phi_1$, due to current $I$.*

$L$ is obtained by multiplying this by $N/I$:

$$L = \frac{4\pi k_0 N^2 A}{c^2 x_0} \quad \text{(inductance of a solenoid)} \quad (19\text{-}7)$$

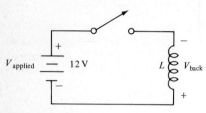

**Figure 19-8.** *Switch is closed at $t = 0$.*

**Example 4.** A superconducting solenoid of 10-cm length, 1000 turns, and core area of 2.0 cm² is connected to a 12-V battery (Figure 19-8). What is the current 0.01 s after the switch is closed?

ANSWER: Use Eq. 19-7 to obtain the inductance $L$:

$$L = 4\pi \left(\frac{k_0}{c^2}\right) \frac{N^2 A}{x_0} = 4\pi (10^{-7}) \frac{(1000)^2 (2.0 \times 10^{-4})}{0.1} \text{ H}$$
$$= 2.51 \times 10^{-3} \text{ H}$$

According to Ohm's law the net emf is $(V_{\text{applied}} + V_{\text{back}}) = IR$. Since $R = 0$ for a superconductor, we have

$$V_{\text{back}} = -V_{\text{applied}}$$

Now replace $V_{\text{back}}$ with $-L \, dI/dt$:

$$\left(-L \frac{dI}{dt}\right) = -V_{\text{applied}}$$

$$L \frac{dI}{dt} = V_{\text{applied}}$$

$$dI = \frac{12 \text{ V}}{L} dt$$

$$I = \frac{12}{2.51 \times 10^{-3}} t = 4780 t$$

For $t = 0.01$ s, $I = 47.8$ A. The current will continue to increase linearly with time until the superconducting limit is reached, at which time there is a sudden transition from zero resistance to normal resistance.

## 19-5 Magnetic Field Energy

### The LC Oscillator

Not only are capacitors used for storing charge, but they are commonly used in combination with inductors to generate ac voltages and currents.

We shall analyze the simplest case of a capacitor and inductor in parallel as shown in Figure 19-9. We assume zero resistance in the circuit. Suppose at $t = 0$ a charged capacitor is connected across an inductor as shown. The voltage across the capacitor is

$$V_{ab} = \frac{q}{C}$$

According to Eq. 19-6 the voltage across the inductor is

$$V_{cd} = -L\frac{dI}{dt}$$

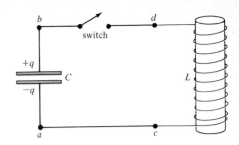

Figure 19-9. *Switch is closed at $t = 0$, thus connecting charged capacitor C across inductor L.*

At any time $t$ after the switch is closed these two voltages must be the same:

$$\frac{q}{C} = -L\frac{dI}{dt}$$

Now substitute $dq/dt$ for $I$:

$$\frac{q}{C} = -L\frac{d^2q}{dt^2}$$

$$\frac{d^2q}{dt^2} = -\frac{1}{LC}q$$

This differential equation has the same form as the equation for simple harmonic motion. The solution is given in Eq. 11-7:

$$q = q_0 \cos \omega t \qquad \text{where } \omega = \frac{1}{\sqrt{LC}}$$

The voltage across the capacitor will be

$$V = \frac{q}{C} = \frac{q_0}{C}\cos \omega t = V_0 \cos \omega t$$

which oscillates with a frequency

$$f = \frac{1}{2\pi\sqrt{LC}}$$

The ac current is obtained by differentiation of the equation $q = q_0 \cos \omega t$:

$$I = \frac{dq}{dt} = -q_0\omega \sin \omega t$$

The minus sign indicates that the current or charge $q$ initially flows out of the capacitor plate, which started out with charge $q_0$. An equal amount of current must flow into the lower capacitor plate since there is no place to store charge in an inductor. We see that an ac current acts as if it is flowing through a capacitor and that it is 90° out of phase with the voltage across the capacitor. (See Figure 19-11, page 399.)

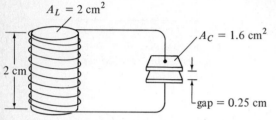

**Figure 19-10.** *Coil and capacitor are as drawn (to scale). Coil has 10 turns and capacitor gap is 2.5 mm.*

---

**Example 5.** What is the natural resonant frequency of the coil and capacitor drawn to scale in Figure 19-10?

ANSWER: Using Eq. 19-7 for $L$:

$$L = \frac{4\pi k_0 N^2 A_L}{c^2 x_L} = 1.26 \times 10^{-6} \text{ H}$$

Using Eq. 16-20 for $C$:

$$C = \frac{A_C}{4\pi k_0 x_0} = 5.67 \times 10^{-13} \text{ F}$$

$$\sqrt{LC} = \sqrt{(1.26 \times 10^{-6})(5.67 \times 10^{-13})} = 8.43 \times 10^{-10} \text{ s}$$

$$f = \frac{1}{2\pi\sqrt{LC}} = 1.88 \times 10^8 \text{ s}^{-1} = 188 \text{ MHz}$$

This resonant frequency corresponds to Channel 9 of the VHF television band. Circuit components similar to these are used inside TV tuners.

---

### Energy of the Magnetic Field

In Examples 4 and 5 the initial energy of the system was the energy stored in the capacitor; according to Eq. 16-22 this energy is

$$V = \frac{1}{2}\frac{q_0^2}{C} = \frac{1}{2}CV_0^2.$$

But as $V$ decreases, the energy stored in the electric field of the capacitor decreases. According to the law of conservation of energy this initial energy cannot disappear—it must be stored somewhere else. We shall now show that it is stored in the magnetic field associated with the inductor.

As a charge $dq$ flows through an inductor it picks up an energy $V\,dq$ where $V = -L\,dI/dt$. Thus the energy lost by the charge (delivered to the inductor) is

$$dU = \left(L\frac{dI}{dt}\right)dq = L\frac{dI\,dq}{dt} = L\,dI\left(\frac{dq}{dt}\right) = LI\,dI$$

If we start with zero current and build up to a current $I_0$, the energy stored in the inductor is

$$U = \int_0^{I_0} LI\,dI = \frac{1}{2}LI_0^2 \qquad (19\text{-}8)$$

---

**Example 6.** What is the instantaneous energy stored in the inductor and the capacitor of Figure 19-7 at a time $t$?

ANSWER: The energy stored in the capacitor is

$$U_c = \tfrac{1}{2}CV^2 = \tfrac{1}{2}C(V_0\cos\omega t)^2 = \tfrac{1}{2}CV_0^2\cos^2\omega t$$

According to Eq. 19-8 the energy stored in the inductor is

$$U_L = \tfrac{1}{2}LI^2 = \tfrac{1}{2}L(-q_0\omega\sin\omega t)^2 = \tfrac{1}{2}Lq_0^2\omega^2\sin^2\omega t$$

Now substitute $(CV_0)$ for $q_0$:

$$U_L = \tfrac{1}{2}L(CV_0)^2\omega^2\sin^2\omega t$$

Replace $\omega^2$ with $1/LC$:

$$U_L = \tfrac{1}{2}CV_0^2\sin^2\omega t$$

The sum

$$U_C + U_L = \tfrac{1}{2}CV_0^2\cos^2\omega t + \tfrac{1}{2}CV_0^2\sin^2\omega t = \tfrac{1}{2}CV_0^2$$

equals the initial energy put into the system.

It is of interest to reexpress Eq. 19-8 in terms of the magnetic field associated with the inductor. This is easy to do for a simple, long solenoid. We replace $L$ with its value given by Eq. 19-7:

$$U = \tfrac{1}{2}\left(\frac{4\pi k_0 N^2 A}{c^2 x_0}\right) I^2$$

Now use the relation $\mathcal{B} = 4\pi k_0 NI/(c^2 x_0)$ (Eq. 18-7) to eliminate $I$:

$$U = \tfrac{1}{2}\left(\frac{4\pi k_0 N^2 A}{c^2 x_0}\right)\left(\frac{\mathcal{B}}{\frac{4\pi k_0 N}{c^2 x_0}}\right)^2$$

$$= \frac{c^2 A x_0 \mathcal{B}^2}{8\pi k_0}$$

If we divide by $A x_0$, which is the volume $\mathcal{V}$ of the solenoid, we obtain

$$\frac{U}{\mathcal{V}} = \frac{c^2 \mathcal{B}^2}{8\pi k_0} \qquad \text{(energy density of magnetic field)} \qquad (19\text{-}9)$$

Although this derivation was for a solenoid, there is a general proof showing that, for any shape coil, the integral of $c^2 \mathcal{B}^2/8\pi k_0$ over all space equals $\tfrac{1}{2}LI^2$, where $L$ is the inductance of the coil.

So just as we interpreted $E^2/8\pi k_0$ as the energy per unit volume stored in the electric field, we can interpret $c^2\mathcal{B}^2/(8\pi k_0)$ as the energy stored in the magnetic field. In general there may be both electric and magnetic fields in space, so the total energy density of the electromagnetic field is

$$\frac{dU}{d\mathcal{V}} = \frac{1}{8\pi k_0}(E^2 + c^2 \mathcal{B}^2) \qquad (19\text{-}10)$$

We shall see in the next chapter that the electromagnetic wave which is radiated by a changing current has $E = c\mathcal{B}$. Then the radiated energy contained in the electric field equals the energy contained in the magnetic field.

## 19-6 Alternating Current Circuits

We saw in the analysis of Figure 19-9 that there is a definite relationship between ac voltage and current for a capacitor or inductor. Our goal in this section is to be able to calculate the currents when an ac voltage is applied across a circuit consisting of capacitors, inductors, and resistors.

## Capacitive Reactance

Let us first calculate the ac current when an ac voltage, $V = V_0 \sin \omega t$, is applied across a capacitor. The instantaneous voltage is

$$V = \frac{q}{C}$$

Then

$$\frac{dV}{dt} = \frac{1}{C}\frac{dq}{dt}$$

Now substitute ($V_0 \sin \omega t$) for $V$ and $I$ for $dq/dt$:

$$\omega V_0 \cos \omega t = \frac{1}{C} I$$

$$I = \omega C V_0 \cos \omega t = \omega C V_0 \sin(\omega t + 90°)$$

or

$$I = I_0 \sin(\omega t + 90°)$$

where $I_0 = \omega C V_0$ is the peak value of the current. We note that the current flowing into a capacitor leads the voltage across it by 90°; that is, the peak current arrives a quarter period earlier (Figure 19-11). According to the above equation the relation between peak voltage and peak current is

$$V_0 = \frac{1}{\omega C} I_0$$

The proportionality constant $1/(\omega C)$ is called the capacitive reactance $X_C$:

$$X_C \equiv \frac{1}{\omega C} \quad \text{(capacitive reactance)} \quad (19\text{-}11)$$

Then

$$V_0 = I_0 X_C \quad (19\text{-}12)$$

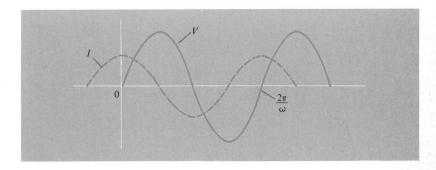

**Figure 19-11.** Plot of current and voltage across a capacitor. Note that current leads voltage by 90°.

Note that Eq. 19-12 is of the same form as Ohm's law where $V = IR$. The reactance $X_C$ in ac circuits plays the same role as does resistance in dc circuits. It is merely the proportionality constant between the peak ac current and the peak ac voltage.

> **Example 7.** What is the current flowing into a 1 $\mu$F capacitor for an applied ac voltage of peak value 100 V and frequency 60 Hz? Repeat for frequencies of 1 kHz and 1 MHz.
>
> ANSWER: Since $\omega = 2\pi f$ the three values of $\omega$ are 377, 6283, and $6.283 \times 10^6$ s$^{-1}$. The three values of reactance for $C = 10^{-6}$ F are $X_C = 1/\omega C = 2653, 159,$ and 0.159 ohms. The three values of peak current are $I_0 = V_0/X_C = 27.4$ mA, 629 mA, and 629 A.

### Inductive Reactance

If an ac generator is connected to an inductance, the circuit will be as shown in Figure 19-12 where $V_{\text{applied}} = V_0 \sin \omega t$. We saw in Example 4 that $V_{\text{applied}} = L\, dI/dt$. Thus

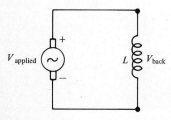

**Figure 19-12.** An ac generator is connected to an inductance.

$$\frac{dI}{dt} = \frac{1}{L} V_0 \sin \omega t$$

$$dI = \frac{1}{L} V_0 \sin \omega t\, dt$$

$$I = \frac{V_0}{L} \int \sin \omega t\, dt$$

$$= -\frac{V_0}{\omega L} \cos \omega t$$

$$= \frac{V_0}{\omega L} \sin(\omega t - 90°)$$

The constant of integration is zero because there is no dc current flowing. In this case the current lags the applied voltage by 90° (or the voltage leads the current by 90°). Since $V_0 = I_0 \omega L$, we have for the inductive reactance

$$X_L \equiv \omega L \qquad \text{(inductive reactance)} \qquad (19\text{-}13)$$

## Series Circuit

Suppose an ac generator is connected across a circuit consisting of a resistor $R$, inductor $L$, and capacitor $C$ all connected in series as shown in Figure 19-13. Let the current be $I = I_0 \sin \omega t$. Then the instantaneous voltages are

$$V_R = I_0 R \sin \omega t$$
$$V_L = X_L I_0 \sin(\omega t + 90°) \quad \text{(voltage across } L \text{ leads current by } 90°\text{)}$$
$$V_C = X_C I_0 \sin(\omega t - 90°) \quad \text{(voltage across } C \text{ lags current by } 90°\text{)}$$

The total instantaneous voltage is

$$V = V_R + V_C + V_L$$
$$V = I_0 R \sin \omega t - X_C I_0 \cos \omega t + X_L I_0 \cos \omega t$$

$$\frac{V}{I_0} = R \sin \omega t + (X_L - X_C) \cos \omega t \tag{19-14}$$

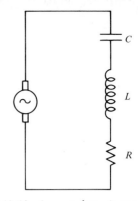

**Figure 19-13.** *An ac voltage is connected across an LCR series circuit.*

The sine and cosine can be added with the help of Figure 19-14, which defines a phase angle $\phi$:

$$\phi \equiv \tan^{-1} \frac{X_L - X_C}{R}$$

The hypotenuse of the triangle in Figure 19-14 is

$$Z = \sqrt{R^2 + (X_L - X_C)^2}$$

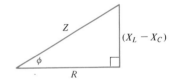

**Figure 19-14.** *Relation between phase angle $\phi$, resistance $R$, reactances $X_L$ and $X_C$, and impedance $Z$.*

Now divide both sides of Eq. 19-14 by $Z$:

$$\frac{1}{Z}\frac{V}{I_0} = \frac{R}{Z} \sin \omega t + \frac{X_L - X_C}{Z} \cos \omega t$$

$$= \cos \phi \sin \omega t + \sin \phi \cos \omega t = \sin(\omega t + \phi)$$

$$V = ZI_0 \sin(\omega t + \phi) \tag{19-15}$$

We see that $V_0 = ZI_0$ and that the voltage source leads the current by the angle $\phi$. The proportionality constant $Z$ between $V_0$ and $I_0$ is called the impedance and is the analog of $R$ in Ohm's law.

$$\frac{V_0}{I_0} = Z = \sqrt{R^2 + \left(\omega L - \frac{1}{\omega C}\right)^2} \quad \begin{array}{l}\text{(impedance of}\\ \text{series circuit)}\end{array} \tag{19-16}$$

## Resonance

Note that the formula for impedance of a series ac circuit has a minimum when

$$\left(\omega L - \frac{1}{\omega C}\right) = 0$$

or

$$\omega = \frac{1}{\sqrt{LC}}$$

This special value of $\omega$ is called the resonant frequency $\omega_0$:

$$\omega_0 = \frac{1}{\sqrt{LC}}$$

or

$$f_0 = \frac{1}{2\pi\sqrt{LC}}$$

**Example 8.** The TV tuner coil and capacitor of Example 4 are connected to a TV antenna as shown in Figure 19-15(a). Assume the ac voltage picked up by the antenna for any TV channel has an amplitude of 100 $\mu$V. The inductance and resistance of the coil are 1.26 $\mu$H (microhenries) and 20 ohms, respectively. The capacitance is $C = 0.567$ pF (picofarad).

(a) What is the ac current and voltage across the capacitor at the resonant frequency? ($f_0 = 188$ MHz.)

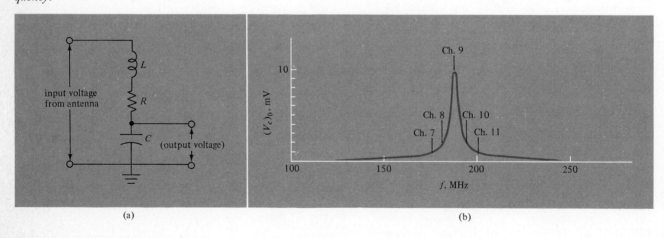

**Figure 19-15.** (a) *Series resonant circuit for Example 8.* (b) *The output voltage across the capacitor as a function of frequency.*

(b) Plot the output voltage as a function of frequency for fixed input voltage of 100 μV amplitude.

(c) If the resonant frequency is at TV Channel 9 and Channel 10 is 6 MHz higher, by what factor is the signal from Channel 10 suppressed?

ANSWER: (a) At the resonant frequency $Z = \sqrt{R^2 + 0} = R = 20$ ohms.

$$I_0 = \frac{V_0}{Z} = \frac{10^{-4} \text{ V}}{20\,\Omega} = 5\,\mu\text{A}$$

The peak voltage across the capacitor is $(V_C)_0 = I_0 X_C = I_0(1/\omega C) = 7.46$ mV. We note that this simple circuit gives a voltage gain of a factor of 74.6.

(b) See Figure 19-15(b).

(c) At any frequency $f$ the output voltage is

$$(V_C)_0 = I_0 X_C = \frac{V_0}{\sqrt{R^2 + \left(\omega L - \frac{1}{\omega C}\right)^2}} \frac{1}{\omega C}$$

$$= \frac{V_0}{\sqrt{(2\pi fRC)^2 + \left(\frac{f^2}{f_0^2} - 1\right)^2}}$$

For $f = 194$ MHz, $(V_C)_0 = 1.54$ mV, which is a factor 4.84 times smaller than the peak value of 7.46 mV, which is obtained at the resonant frequency. Thus the rejection factor for adjacent channels is 4.84.

## Power

The instantaneous power dissipated in Figure 19-13 is $P(t) = V(t)I(t)$. According to Eq. 19-15, $V(t) = V_0 \sin(\omega t + \phi)$, when $I = I_0 \sin \omega t$. Thus

$$P(t) = V_0 I_0 \sin(\omega t + \phi) \sin \omega t$$
$$= V_0 I_0 (\sin \omega t \cos \phi - \cos \omega t \sin \phi) \sin \omega t$$

Since $\cos \omega t \sin \omega t = (\sin 2\omega t)/2$, the average power dissipated is

$$\bar{P} = V_0 I_0 \left(\overline{\sin^2 \omega t} \cos \phi - \frac{\overline{\sin 2\omega t}}{2} \sin \phi\right)$$

This is

$$P = \frac{V_0 I_0}{2} \cos \phi \qquad (19\text{-}17)$$

because $\overline{\sin^2 \omega t} = \frac{1}{2}$ and $\overline{\sin 2\omega t} = 0$. The average $\overline{\sin^2 \omega t} = \frac{1}{2}$ because $\sin^2 \omega t + \cos^2 \omega t = 1$ and because $\overline{\sin^2 \omega t} = \overline{\cos^2 \omega t}$. (This is because sine and cosine are the same curves only shifted by 90°.) The average $\overline{\sin 2\omega t}$ is zero because it is positive as often as it is negative. The factor $\cos \phi$ is called the power factor and, as shown in Figure 19-14, it is $\cos \phi = R/Z$. Then

$$P = \frac{(Z I_0) I_0}{2} \frac{R}{Z} = \tfrac{1}{2} I_0^2 R$$

As we shall next see, all the power is dissipated in the resistor and none in the capacitor or inductor. We shall now show that $\tfrac{1}{2} I_0^2$ is $I_{\text{rms}}^2$ where $I_{\text{rms}}$ stands for the root mean square current. The definition of root mean square is

$$I_{\text{rms}} \equiv \sqrt{\overline{I^2}}$$

The mean squared current is

$$\overline{I^2} = I_0^2 \overline{\sin^2 \omega t} = \tfrac{1}{2} I_0^2$$

Then

$$I_{\text{rms}} = \frac{1}{\sqrt{2}} I_0$$

We have just shown that the average power dissipated is $I_{\text{rms}}^2 R$ for the entire circuit. But we already know that for a resistor alone the power dissipation will be $\overline{I^2} R$. All the power loss is accounted for in the resistor alone—there is none left over for the capacitor or inductor (see Example 9).

It is a convention that ac ammeters read $I_{\text{rms}}$ and ac voltmeters read $V_{\text{rms}}$. Thus a household reading of 120 volts ac means that the peak voltage is $\sqrt{2}$ times 120 volts or 170 volts. In United States homes the voltage swings back and forth between $+170$ V and $-170$ V 120 times a second ($f = 60$ Hz). It is convenient to have ac meters reading rms because then the power loss in a resistive element is just the product of the ac voltmeter reading times the ac ammeter reading.

**Example 9.** A 10-μF capacitor is plugged into an ac outlet in a United States home.
 (a) What is the current drawn as read by an ac ammeter?
 (b) What is the average power dissipated?
 (c) What is the maximum instantaneous power?

ANSWER:

$$I_{rms} = \frac{V_{rms}}{X_C} \quad \text{where} \quad X_C = \frac{1}{2\pi f C} = \frac{1}{2\pi(60)(10^{-5})}\Omega = 265\ \Omega$$

$$I_{rms} = \frac{120\ V}{265\ \Omega} = 0.452\ A$$

$$\bar{P} = V_{rms}I_{rms}\cos\phi$$

where $\phi = -90°$ is the phase angle (the voltage lags the current by 90°). Since $\cos\phi = 0$, we have $\bar{P} = 0$. The instantaneous power is

$$P(t) = V_0 I_0 \sin(\omega t - 90°)\sin\omega t = -\tfrac{1}{2}V_0 I_0 \sin 2\omega t$$

which has a maximum value of $\tfrac{1}{2}V_0 I_0 = 54.2$ W. Note that the instantaneous power is just as often negative as positive so that its average is zero. When negative, it is delivering stored energy back into the circuit.

## 19-7 RC and RL Circuits

Whenever a voltage is suddenly applied to a circuit that contains a resistor and capacitor (or resistor and inductor), there will be a current that varies exponentially with time. Such a nonrepetitive current is called a transient. We shall now consider three examples.

### RC Circuits

Our first example is similar to that in Figure 19-9 where a charged capacitor is connected across an inductor. In this case at $t = 0$ the switch in Figure 19-16 is closed and the capacitor $C$ discharges through the resistor $R$. At any instant of time the voltage across the capacitor equals the $IR$ drop in the resistor:

$$\frac{q}{C} = IR$$

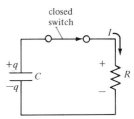

**Figure 19-16.** At $t = 0$ the charged capacitor is connected across resistor $R$ by closing the switch.

The arrow direction in Figure 19-16 is such that $I = -dq/dt$, so

$$\frac{q}{C} = -R\frac{dq}{dt}$$

$$\frac{dq}{q} = -\frac{dt}{RC}$$

Integrating both sides gives

$$\ln q = -\frac{t}{RC} + \text{constant}$$

or

$$q = q_0 \exp\left(\frac{-t}{RC}\right)$$

Note that the capacitor does not discharge instantaneously, but in a time $\tau = RC$ the charge drops to $1/e$ of its initial value.

In our next example we shall connect a battery of emf $\mathcal{E}$ across a capacitor and resistor that are in series.

At a time $t = 0$, a voltage $\mathcal{E}$ is applied by closing the switch in Figure 19-17. We shall show that at $t = 0$ the full voltage $\mathcal{E}$ appears across the resistor, but it decreases exponentially with time; namely, we shall show that

$$V_R = \mathcal{E} \exp\left(-\frac{t}{RC}\right)$$

In Figure 19-17(b) the potential of point $B$ with respect to $A$ is the emf $\mathcal{E}$, which is also the sum of the voltage across the resistor plus the voltage across the capacitor:

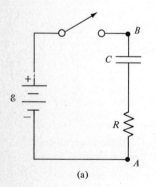

(a)

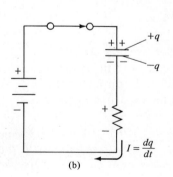

(b)

**Figure 19-17.** (a) *Before $t = 0$.* (b) *After $t = 0$. In (b) the same amount of charge that is stored in the capacitor has flowed through the resistor.*

$$\mathcal{E} = IR + \frac{1}{C}q \qquad (19\text{-}18)$$

Now differentiate both sides with respect to time:

$$0 = R\frac{dI}{dt} + \frac{1}{C}\frac{dq}{dt}$$

$$= R\frac{dI}{dt} + \frac{1}{C}I$$

$$\frac{dI}{I} = -\frac{1}{RC}dt$$

Integrating both sides gives

$$\ln I = -\frac{t}{RC} + \text{constant}$$

Taking the antilog of both sides gives

$$I = I_0 \exp\left(-\frac{t}{RC}\right) \qquad (19\text{-}19)$$

The constant of integration $I_0$ can be found by substitution into Eq. 19-18:

$$\mathcal{E} = R(I_0 e^{-t/RC}) + \frac{1}{C}q$$

At $t = 0$, $q = 0$ and this equation becomes

$$\mathcal{E} = RI_0 \exp(-0) + 0$$
$$= RI_0$$
$$I_0 = \frac{\mathcal{E}}{R}$$

The voltage drop across the resistor is

$$V_R = RI$$
$$= R(I_0 e^{-t/RC})$$

Replacing $I_0$ with $(\mathcal{E}/R)$ gives

$$V_R = \mathcal{E} \exp\left(-\frac{t}{RC}\right) \qquad (19\text{-}20)$$

The product $RC$ has the units of time and is called the time constant. For example, suppose $R = 1\ \text{M}\Omega$ and $C = 10\ \mu\text{F}$. Then

$$RC = (10^6\ \Omega)(10 \times 10^{-6}\ \text{F}) = 10\ \text{s}$$

In this case 10 s after the switch is closed the voltage across $R$ will drop to $1/e$ of its initial value.

### LR Circuits

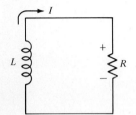

**Figure 19-18.** *At $t = 0$ the current is $I_0$.*

Our last example (Figure 19-18) is an inductor $L$ connected across a resistor $R$. If at $t = 0$ the current is $I_0$, then by using the same mathematical approach as in the preceding section we can show that

$$I = I_0 \exp\left(-\frac{Rt}{L}\right)$$

In this case the voltage across the inductor equals the voltage drop across the resistor:

$$-L\frac{dI}{dt} = IR$$

$$\frac{dI}{I} = -\frac{R}{L}dt$$

or

$$I = I_0 \exp\left(-\frac{Rt}{L}\right)$$

The initial current $I_0$ can be established using the circuit in Figure 19-19 where $R_L$ is the internal resistance of the inductor $L$. Assuming the switch has been closed for a time much greater than $L/R_L$, the current $I_0$ through $L$ will be $\mathcal{E}/R_L$. Immediately after the switch is opened $I_0$ must flow through $R_1$ as well as $R_L$. (Otherwise $\Delta I/\Delta t$ would approach infinity as $\Delta t$ approaches zero.) Hence just after the switch is opened the voltage drop across $R_1$ is

$$V_1 = I_0 R_1 = \mathcal{E}\frac{R_1}{R_L}$$

The voltage drop across the switch contacts as the switch is opened will be

$$(V_1 + \mathcal{E}) = \mathcal{E}\left[\frac{R_1}{R_L} + 1\right]$$

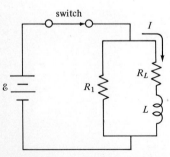

**Figure 19-19.** *When switch is opened, a very large voltage will appear across the switch contacts if $R_1 \gg R_L$.*

We note that if $R_1$ is considerably greater than $R_L$, the voltage appearing across the switch will be considerably greater than the battery voltage. Sparks would fly and the switch could be damaged. So in practical circuits it is necessary to put a resistor $R_1$ across an inductor before disconnecting it from a voltage source.

We see that inductors insist on maintaining the same current in the face of a sudden change, while capacitors behave as short circuits during a sudden change.

## Summary

Faraday's law states that there will be an emf around a closed path equal to the rate of change of magnetic flux enclosed by it:

$$\oint \mathbf{E} \cdot d\mathbf{s} = -\frac{d\Phi_B}{dt}$$

The changing $\Phi_B$ can be produced by moving coils (or magnets) or by changing currents that are fixed in position. This same expression for emf also applies to the case where a closed wire loop is moving through a region of magnetic field. A famous example of this latter case is a coil of area $A$ rotating with angular frequency $\omega$ in a region of uniform $\mathcal{B}$; then

$$\text{emf} = \mathcal{B}A\omega \sin \omega t$$

This method of producing an emf or voltage is the basis of the electric generator.

The minus sign in Faraday's law means that the induced voltage is trying to produce a current whose flux lines would oppose the original change in $\Phi_B$. This is called Lenz's law.

In a transformer consisting of a secondary and a primary coil both wound around the same core, the ratio

$$\frac{V_{sec}}{V_{pri}} = \frac{N_{sec}}{N_{pri}}$$

is the turns ratio. If $L$ is the self-inductance of a coil, the back emf is

$$V_{back} = -L\frac{dI}{dt}.$$

For a long solenoid of length $x_0$,

$$L = 4\pi \frac{k_0}{c^2} \frac{N^2 A}{x_0}$$

When an inductor $L$ is connected across a charged capacitor $C$, the charge will oscillate back and forth with a frequency $f = 1/(2\pi \sqrt{LC})$. The energy stored in an inductor carrying a current $I$ is

$$U = \tfrac{1}{2} L I^2$$

This is mathematically equivalent to integrating

$$\frac{dU}{d\mathcal{V}} = \frac{c^2 \mathcal{B}^2}{8\pi k_0}$$

over all space. The general energy density of any electromagnetic field at any point in space is

$$\frac{dU}{d\mathcal{V}} = \frac{1}{8\pi k_0}(E^2 + c^2 \mathcal{B}^2)$$

For an ac voltage source $V = V_0 \sin \omega t$, the peak current flowing into a capacitor is

$$I = \frac{V_0}{X_C} \sin(\omega t + 90°) \quad \text{where } X_C = \frac{1}{\omega C}$$

The current flowing into an inductor is

$$I = \frac{V_0}{X_L} \sin(\omega t - 90°) \quad \text{where } X_L = \omega L \text{ is the inductive reactance}$$

In a series circuit with $L$, $C$, and $R$, where the current is $I = I_0 \sin \omega t$, the applied voltage is $V = Z I_0 \sin(\omega t + \phi)$ where the impedance

$$Z = \sqrt{R^2 + \left(\omega L - \frac{1}{\omega C}\right)^2} \quad \text{and} \quad \tan \phi = \frac{\omega L - 1/(\omega C)}{R}$$

The average power is $V_{rms} I_{rms} \cos \phi = I_{rms}^2 R$ where $I_{rms} = \sqrt{\overline{I^2}} = I_0/\sqrt{2}$.

If an emf $\mathcal{E}$ is connected across $R$ and $C$ in series the current will be $I = (\mathcal{E}/R) \exp(-t/RC)$.

# Appendix 19-1  Loop of Arbitrary Shape

We wish to show that the moving wire loop (of arbitrary shape) in Figure 19-20 will experience an emf equal to $-d\Phi_B/dt$. $\mathcal{B}$ can be an arbitrary function of position. The work done against the magnetic force in moving a charge $q$ the distance $d\mathbf{s}$ shown in Figure 19-20 is

$$dW = \mathbf{F}_{\text{mag}} \cdot d\mathbf{s}$$
$$= (q\mathbf{v} \times \mathcal{B}) \cdot d\mathbf{s}$$
$$= q\left(\frac{\Delta \mathbf{x}}{\Delta t} \times \mathcal{B}\right) \cdot d\mathbf{s}$$

where $\Delta \mathbf{x}$ is a vector in the $x$ direction of length $\Delta x$.

$$dW = -q\frac{(\mathcal{B} \times \Delta \mathbf{x}) \cdot d\mathbf{s}}{\Delta t}$$
$$= -q\frac{\mathcal{B} \cdot \Delta \mathbf{x} \times d\mathbf{s}}{\Delta t}$$

(We have used the vector relation $\mathbf{A} \times \mathbf{B} \cdot \mathbf{C} = \mathbf{A} \cdot \mathbf{B} \times \mathbf{C}$.) According to Figure 19-20 we can replace $(\Delta \mathbf{x} \times d\mathbf{s})$ with the element of area $d\mathbf{A}$:

$$dW = -q\frac{\mathcal{B} \cdot (d\mathbf{A})}{\Delta t}$$

The total work done in moving $q$ from $a$ to $b$ along path 1 is

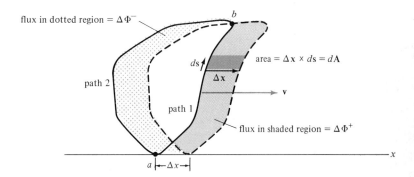

**Figure 19-20.** *Closed loop consisting of path 1 and path 2 is moving with velocity v along x axis. Dashed line is position of loop after a time $\Delta t$. Then flux increases by $\Delta \Phi^+$ and decreases by $\Delta \Phi^-$.*

$$W_{ab1} = \int_a^b dW = -q\frac{\int_a^b \mathcal{B} \cdot d\mathbf{A}}{\Delta t} = -q\frac{\Delta \Phi^+}{\Delta t}$$
path 1

Likewise

$$W_{ba2} = \int_b^a dW = q\frac{\Delta \Phi^-}{\Delta t}$$
path 2

The emf is the work done in moving around the entire path divided by $q$:

$$\text{emf} = \frac{W_{ab1} + W_{ba2}}{q} = -\frac{\Delta \Phi^+}{\Delta t} + \frac{\Delta \Phi^-}{\Delta t} = -\frac{\Delta \Phi^+ - \Delta \Phi^-}{\Delta t} = -\frac{\Delta \Phi_B}{\Delta t}$$

$$= -\frac{d\Phi_B}{dt}$$

This is Faraday's law for a coil moving through a region of magnetic field.

## Exercises

1. Consider a circular loop of radius $R$ in a region of uniform magnetic field $\mathcal{B}$. The loop is initially in the $xy$ plane as shown in the figure. If the loop were rotated through 180° about the $z$ axis in $\tfrac{1}{2}$ sec, what would be the average induced emf?
2. If the loop shown in the figure for Exercise 1 were moved along the $x$ axis with a constant velocity $v$, without rotation, through the region of uniform magnetic field, what would be the induced emf?
3. If the loop in the figure for Exercise 1 were held stationary and the external field $\mathcal{B}$ were decreased, what would be the direction of the induced current as seen from above?
4. Consider an ideal circuit consisting of an emf source $\varepsilon_0$ and an inductance $L$, as shown. Assume that the total resistance around the circuit is zero. What would be the current 1 s after the switch is closed, if $L = 0.1$ henry and $\varepsilon_0 = 1.5$ V?
5. A bent wire of radius $R$ is rotated with a frequency $f$ in a uniform field $\mathcal{B}$. What are the amplitudes of the induced voltage and current when the internal resistance of the meter M is $R_m$ and the remainder of the circuit has negligible resistance?

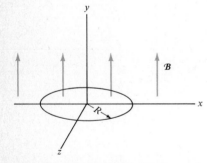

**Exercise 1.** *Circular loop of radius R in xy plane. $\mathcal{B}$ is uniform everywhere and parallel to y axis.*

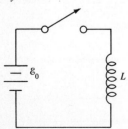

**Exercise 4.** *At $t = 0$ switch is closed.*

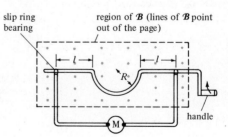

**Exercise 5.** *Bent wire of radius R is rotated uniformly by hand crank. Lines of $\mathcal{B}$ are pointing out of page. Electrical contact is made to meter M via slip rings.*

6. A conducting rod of length 1 m, weight $mg = 1$ N, and resistance $R = 10$ ohms falls while making electrical contact that completes a circuit loop with vertical rails having a negligible resistance. As shown in the figure, the rod falls through a region of space containing a magnetic field $\mathcal{B} = 2$ T that is perpendicular to the plane of the rails (directed into the page). Neglecting friction, what will be the final velocity of the rod? What will be the direction of current flow?

7. A square loop with 1-m edges and a resistance of 0.5 ohm is fixed in a region of space containing a uniform magnetic field $\mathcal{B}$, which makes an angle of $45°$ with the plane of the loop and increases uniformly at the rate of 0.1 T/s. What is the power dissipated in the loop?

8. At the time $t = 0$ the switch is thrown from $a$ to $b$ in the figure. Give a formula for the charge on the capacitor in terms of $\varepsilon$, $C$, $L$, and $t$.

9. An inductance $L$ has an internal resistance $R$. At what frequency $f$ of ac voltage will the current lag the voltage by $45°$?

10. A 300-turn coil of 100-cm² area is rotating in a magnetic field of 0.5 T at 1800 rpm. What is the peak value of the emf generated?

11. A 1000-turn coil whose area of 100 cm² is perpendicular to the earth's magnetic field is flipped $90°$ in 1 s. The average emf coming from the coil during that second was measured to be 0.6 mV. What was the strength of the earth's magnetic field?

12. Consider a transformer that has 10 turns in the primary and 25 turns in the secondary windings. A 60-Hz ac voltage is applied to the primary, which produces a maximum of four lines of $\mathcal{B}$ passing through both coils. What are the peak voltages across the primary and the secondary?

13. A generalization of the law of equipartition of energy predicts that the average energy stored in the magnetic field in interstellar space should be equal to the average kinetic energy of the particles which consist mainly of hydrogen atoms of thermal velocity of about $10^3$ m/s. The particle density is about 1 cm$^{-3}$. Make an estimate of the average magnetic field.

14. In Example 1 the output voltage is $V = 113 \sin \omega t$ in volts. What is $V_{\text{rms}}$?

$$V_{\text{rms}} \equiv \sqrt{\overline{V^2}}; \quad \overline{V^2} \equiv \frac{1}{T}\int_0^T V^2\, dt$$

15. The coil in Figure 19-1 is rotated by just $180°$ from $\theta = 0°$ to $\theta = 180°$ in $\frac{1}{2}$ sec. What is the average voltage between $P_1$ and $P_2$ during this time? Give answer in terms of $l_1$, $l_2$, and $\mathcal{B}$.

16. Repeat Exercise 15 for $\theta$ from $90°$ to $270°$.

17. In Figure 19-1 suppose the coil is fixed in the position shown, and that $\mathcal{B}$ is decreasing. Which will be at the higher potential, point $P_1$ or $P_2$?

18. Use conservation of energy and assume no power losses. A 12-V transformer delivers 10 A to a motor.
    (a) What is the horsepower of the motor under these conditions?
    (b) What current is drawn from the 120-V power line?

19. A 100-MW ac generator supplies power via a 5-ohm transmission line to a city. What ac voltage should be used in order to suffer only a 2% transmission loss?

20. In Example 5, what is $L$ in henrys and what is $C$ in farads?

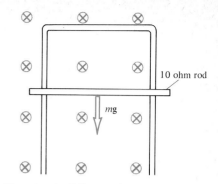

**Exercise 6.** *Falling rod makes contact with side rails. Lines of $\mathcal{B}$ are pointing into page.*

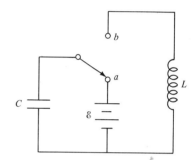

**Exercise 8.** *At $t = 0$ capacitor is connected across inductance.*

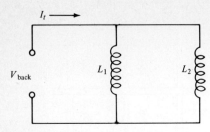

**Exercise 21.** *Two inductances are connected in parallel.*

21. Two solenoids of inductance $L_1$ and $L_2$ are connected in parallel as shown in the figure. Prove that

$$\frac{1}{L_t} = \frac{1}{L_1} + \frac{1}{L_2} \quad \text{where} \quad V_{back} = L_t \frac{dI_t}{dt}$$

22. Express $Z = \sqrt{R^2 + (\omega L - 1/\omega C)^2}$ in terms of the resonant frequency $\omega_0$, $R$, $L$, and $\omega$.
23. In Exercise 22, let $Q \equiv \omega_0 L/R$. Express $Z$ in terms of $\omega_0$, $R$, $Q$, and $\omega$.
24. A 1-henry inductor with $R = 0$ ohm is plugged into a household ac outlet. What is the current and what is the power dissipated?

## Problems

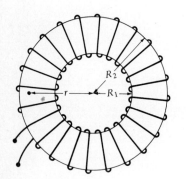

**Problem 25.** *Top view of core of toroidal coil.*

25. A toroidal coil has a total of $N$ turns with inner radius $R_1$ and outer radius $R_2$. (It is like a coil being wound on a donut.) What is $\mathcal{B}$ inside the toroid as a function of $r$? What is the inductance of this toroid (shown in the accompanying figure)? (Assume a square cross section.)
26. In Problem 25 integrate $c^2 \mathcal{B}^2/(8\pi k_0)$ over all space. Compare with $\frac{1}{2}LI^2$.
27. If a 10-ohm resistor is connected from $P_1$ to $P_2$ in Figure 19-1, how many coulombs pass through the resistor when the coil is flipped from 0° to 180°? Assume $l_1 = l_2 = 0.1$ m and $\mathcal{B} = 1.5$ T. *Hint:* Starting with $I = (1/R)V$, show that $dq = (1/R)\, d\Phi$.
28. In Example 1 suppose one end of the coil is connected to one side of a split copper ring and the other end to the other side as shown in the figure. This split commutator ring is in contact with two copper "brushes" as shown. Plot the potential difference between terminals A and B as a function of time. What is

$$\bar{V} = \frac{1}{T} \int_0^T V\, dt$$

(This is the basic principle of the dc generator.)

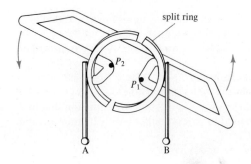

**Problem 28.** *Side view of coil of Figure 19-1. Split ring rotates with coil. Hence terminals (a) and (b) are alternately connected to $P_1$ and $P_2$.*

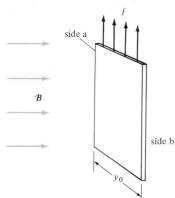

**Problem 29.** *Current-carrying strip in uniform magnetic field.*

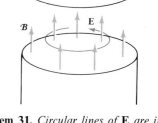

**Problem 31.** *Circular lines of E are induced when cyclotron magnet is turned on.*

29. A conducting strip carrying a current density **j** is in a region of uniform $\mathcal{B}$ as shown in the figure. If the current is due to conduction electrons there will be a transverse magnetic force on each electron.
    (a) Which side is more positive, side a or side b?
    (b) The conducting strip could be a *p*-type semiconductor where the current carriers are positive charges. Then which side would be more positive? (This phenomenon is called the Hall effect and permits one to determine the sign of the current carrier.)
30. If there are $\mathcal{N}$ conduction electrons per unit volume in Problem 29, what will be the potential difference between sides a and b in terms of $j$, $\mathcal{B}$, $\mathcal{N}$, $y_0$ and $e$? (*Hint:* The net transverse force on a conduction electron must be zero.)
31. A cyclotron magnet consists of circular polefaces of radius 50 cm as shown in the accompanying figure. Because of the large inductance, when the magnet is turned on the current increases linearly for 2 s until the peak field of 2 T is obtained. During this time there is an induced $E$ field.
    (a) What is $E$ in terms of $\partial \mathcal{B}/\partial t$ and $r$?
    (b) What is $E$ at $r = 40$ cm?
32. Repeat Problem 31(a) for the case where $r$ is greater than the radius $R$ of the magnet.
33. When the terminals of a low voltage battery are touched to the tongue, one can literally taste the electricity. For example, 1.5 V spaced 6 cm apart gives a strong taste. How fast must one shake one's head between the pole faces of the cyclotron magnet of Problem 31 in order to experience the same taste? ($\mathcal{B} = 2$ T.)
34. A square copper coil, which has a resistance $R = 0.5$ ohms, is dropped into a region of magnetic field $\mathcal{B} = 1.6$ T as shown in the figure. The mass per unit length of the wire is 2 g/cm. What will be its terminal velocity? (Magnetic force offsets gravitational force.) What will be the direction of current flow?
35. In the figure an inductor and a resistor are in series. At $t = 0$ the switch is thrown putting a voltage $V$ across the two. Prove that the current as a function of time is

$$I = \frac{V}{R}\left[1 - \exp\left(-\frac{Rt}{L}\right)\right]$$

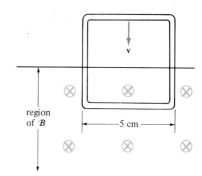

**Problem 34.** *Square coil is falling into region of uniform $\mathcal{B}$. Lines of $\mathcal{B}$ are into page.*

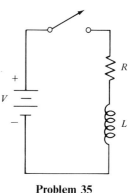

Problem 35

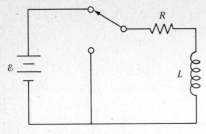

**Problem 36**

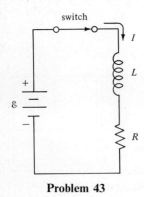

**Problem 43**

36. (a) In the figure, what is the energy stored in L with the switch in position 1 and the current in steady state flow?
    (b) What is the potential across the resistor R as a function of time after the switch is thrown from position 1 to 2?
    (c) What is the total energy dissipated in the resistor as heat (from $t = 0$ to $t = \infty$) after the switch is thrown from position 1 to 2?

37. Design a series circuit similar to that in Example 8 such that $f_0 = 1$ MHz and that $V_C$ drops by a factor of two when $f = f_0 + \Delta f$ where $\Delta f = 5$ kHz. (This is the kind of response curve an AM radio receiver would have.)

38. A 1-henry inductor with $R = 1$ ohm is plugged into a household ac outlet. What is the current and what is the power dissipated?

39. Repeat Example 9 for the case where a 100-ohm resistor is in series with the capacitor.
    (a) What is the ac current?
    (b) What is the average power dissipated?
    (c) What is the maximum instantaneous power flowing into the capacitor?

40. A 1 µF capacitor is defective and has an internal resistance of 100 MΩ. At $t = 0$ it is charged up to 100 V. At what time will the voltage have decayed down to 10 V?

41. In Figure 19-16 at what time will the voltage across the resistor be $\frac{1}{2}\varepsilon$? Give answer in terms of $R$ and $C$.

42. In Figure 19-16 express the voltage across the capacitor in terms of $\varepsilon$, $R$, $C$, and $t$.

43. The switch in the accompanying figure is closed at time $t = 0$. Derive a formula for the voltage drop across R in terms of $\varepsilon$, $L$, $R$, and $t$. [The sum of the voltages around the loop is $\varepsilon - L(dI/dt) - IR = 0$.]

# Electromagnetic Radiation and Waves

This and the next three chapters are on the subject of waves and optics. In this chapter we start the discussion with traveling electromagnetic waves and then traveling waves on a string.

In Table 18-2 we presented a set of four equations that are the basis of all of electrical phenomena dealing with steady currents and charges at rest or moving with constant velocity. These four equations taken together are called Maxwell's equations for electrostatics and magnetostatics.

Now we are interested in the more general situation where currents may be changing or charges accelerating. We must modify the four equations in Table 18-2 to take into account changing magnetic and electric fields. Actually this job was half done in the last chapter. We saw in Chapter 19 that, if the magnetic field is changing, it produces an electric field described by Faraday's law:

$$\oint \mathbf{E} \cdot d\mathbf{s} = -\frac{d}{dt}\Phi_B \quad \text{where} \quad \Phi_B = \int \mathcal{B} \cdot d\mathbf{A}$$

is the enclosed magnetic flux. In the next section we shall see that the equation for $\oint \mathcal{B} \cdot d\mathbf{s}$ (Ampere's law) must have a similar term $(1/c^2)(d/dt)\Phi_E$ on the right-hand side if the electric field is changing. By adding these two terms to Equations II and IV (of Table 18-2), respectively, the Maxwell equations are put into their most general form and then are the basis of all electrical phenomena. We shall see later in this chapter that the Maxwell equations predict a radiated electric and magnetic field moving out with $v = c$ whenever a current is turned on or off, or is varied in any manner with time.

The example in Figure 20-1 illustrates why an extra term $(1/c^2)(d/dt)\Phi_E$ must be added to Ampere's law. Here we have a capacitor consisting of

## 20-1 The Displacement Current

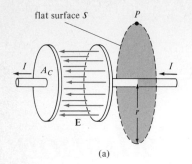
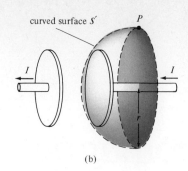

**Figure 20-1.** *Current $I$ flows into a circular capacitor.* (a) *The electric field between the capacitor plates. The flat surface in (a) enclosed by the dashed-line circle includes the current, while the curved surface in (b) does not.*

two circular plates. The capacitor is being charged by the current $I$, which sends charge into the right plate and out of the left plate. The magnetic field at point $P$ can be obtained by drawing a circle of radius $r$ through point $P$ and applying Ampere's law. The flat surface enclosed by the circle in view (a) includes the current $I$, so Ampere's law (Eq. 18-2) gives

$$\oint_{\text{circle}} \mathcal{B} \cdot d\mathbf{s} = 4\pi \frac{k_0}{c^2} \int_S \mathbf{j} \cdot d\mathbf{A}$$

In Figure 20-1(a), $\int \mathbf{j} \cdot d\mathbf{A} = I$. Hence

$$\mathcal{B} \cdot 2\pi r = 4\pi \frac{k_0}{c^2} I$$

$$\mathcal{B} = \frac{2k_0}{c^2} \frac{I}{r} \quad \text{(at point } P\text{)}$$

However, Ampere's law must apply to all possible surfaces enclosed by the same circle, and thus it should work for the curved surface $S'$ of view (b). But here $\int_{S'} \mathbf{j} \cdot d\mathbf{A} = 0$ (the surface has no current flowing through it) and Ampere's law would then give $\oint \mathcal{B} \cdot d\mathbf{s} = 0$, which contradicts our previous result that we know to be correct. Maxwell in 1860, using similar examples, realized that Ampere's law as written in Chapter 18 is mathematically inconsistent for situations involving changing electric fields. He found the inconsistencies could be removed by adding the term $(1/c^2)\int(\partial \mathbf{E}/\partial t) \cdot d\mathbf{A}$ to the right-hand side; that is, the corrected Ampere's law is

$$\oint \mathcal{B} \cdot d\mathbf{s} = 4\pi \frac{k_0}{c^2} \int \mathbf{j} \cdot d\mathbf{A} + \frac{1}{c^2} \int \frac{\partial \mathbf{E}}{\partial t} \cdot d\mathbf{A}$$

Let us now show that this equation gives the same result for $\mathcal{B}$ at point $P$ whether $S$ or $S'$ is used as the surface of integration. In the part of the surface of $S'$ between the capacitor plates

$$E = 4\pi k_0 \frac{Q}{A_C}$$

Thus

$$\frac{\partial E}{\partial t} = \frac{4\pi k_0}{A_C}\frac{\partial Q}{\partial t} = \frac{4\pi k_0}{A_C} I \qquad (20\text{-}1)$$

over the capacitor area $A_C$. The integral of this over the surface $S'$ is

$$\int_{S'} \frac{\partial \mathbf{E}}{\partial t} \cdot d\mathbf{A} = 4\pi k_0 I$$

and then Maxwell's added term becomes

$$\frac{1}{c^2}\int_{S'} \frac{\partial \mathbf{E}}{\partial t} \cdot d\mathbf{A} = \frac{1}{c^2}(4\pi k_0 I)$$

which we know gives the correct answer when equated to $\oint \mathcal{B} \cdot d\mathbf{s}$. Thus Ampere's law as corrected by Maxwell is

$$\oint \mathcal{B} \cdot d\mathbf{s} = \frac{4\pi k_0}{c^2} \int \mathbf{j} \cdot d\mathbf{A} + \frac{1}{c^2}\int \frac{\partial \mathbf{E}}{\partial t} \cdot d\mathbf{A} \qquad (20\text{-}2)$$

The first term on the right-hand side is a real current flowing through any surface enclosed by the closed path. The second term can be thought of as a pseudocurrent. Maxwell called it the displacement current.

---

**Example 1.** What is $\mathcal{B}$ as a function of $r$ between the capacitor plates of Figure 20-1?

ANSWER: Applying Eq. 20-2 to the dashed line path in Figure 20-2, we have

$$\mathcal{B} \cdot (2\pi r) = 0 + \frac{1}{c^2}\left(\frac{\partial E}{\partial t}\right)(\pi r^2)$$

$$\mathcal{B} = \frac{r}{2c^2}\frac{\partial E}{\partial t}$$

We use Eq. 20-1 for $\partial E/\partial t$:

$$\mathcal{B} = \frac{r}{2c^2}\left(\frac{4\pi k_0 I_c}{\pi R^2}\right) = \frac{2k_0 I_0}{c^2 R^2} r$$

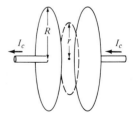

**Figure 20-2.** *A circular parallel plate capacitor with current $I_c$ "flowing through." The lines of $\mathcal{B}$ will curl around the axis.*

> Note that this is the same result as given in Eq. 18-4 for the magnetic field inside a solid rod of radius $R$ which carries a current $I_c$. The solution for $\mathcal{B}$ is the same as if the gap between the capacitor plates were filled with a conductor.

## 20-2 Maxwell's Equations in General Form

Now that we have both Faraday's law and the displacement current term to Ampere's law, we can list the four Maxwell equations in their most general form, as is done in Table 20-1. With the equations of Table 20-1, $\mathbf{E}$ and $\mathcal{B}$ can be determined as a function of position and time if the positions and velocities of the charges producing the field are known. Then $\mathbf{E}$ and $\mathcal{B}$ are known at the position of each particle and the force on each particle ($\mathbf{F} = q\mathbf{E} + q\mathbf{v} \times \mathcal{B}$) is known. This allows one to calculate the future positions and velocities of the interacting charged particles. So, in principle, we have fulfilled the goal of electromagnetic theory, which is to be able to determine the future positions and velocities of any system of interacting charged particles. Maxwell discovered, in addition, that there can be radiated energy and momentum contained in the fields.

We recall that these equations were obtained in Chapters 16 to 19 as a result of Coulomb's law and the requirement of consistency with special relativity. The four equations taken together can properly be thought of as the relativistic form of Coulomb's law. Actually it is possible using a higher form of mathematics to express all four Maxwell equations in one tensor equation.

These equations hold inside matter as well as empty space if atomic

**Table 20-1.** *Maxwell's equations*

|   |   | mks with $k_0$ | mks with $\epsilon_0$ and $\mu_0$ | cgs or gaussian system |
|---|---|---|---|---|
| I | Gauss's law | $\oint \mathbf{E} \cdot d\mathbf{A} = 4\pi k_0 \int \rho \, dV$ | $\oint \mathbf{E} \cdot d\mathbf{A} = \dfrac{1}{\epsilon_0} \int \rho \, dV$ | $\oint \mathbf{E} \cdot d\mathbf{A} = 4\pi \int \rho \, dV$ |
| II | Faraday's law | $\oint \mathbf{E} \cdot d\mathbf{s} = -\int \dfrac{\partial \mathcal{B}}{\partial t} \cdot d\mathbf{A}$ | $\oint \mathbf{E} \cdot d\mathbf{s} = -\int \dfrac{\partial \mathcal{B}}{\partial t} \cdot d\mathbf{A}$ | $\oint \mathbf{E} \cdot d\mathbf{s} = -\dfrac{1}{c} \int \dfrac{\partial \mathcal{B}}{\partial t} \cdot d\mathbf{A}$ |
| III | no magnetic poles | $\oint \mathcal{B} \cdot d\mathbf{A} = 0$ | $\oint \mathcal{B} \cdot d\mathbf{A} = 0$ | $\oint \mathcal{B} \cdot d\mathbf{A} = 0$ |
| IV | Ampere's law (corrected) | $\oint \mathcal{B} \cdot d\mathbf{s} = \dfrac{4\pi k_0}{c^2} \int \mathbf{j} \cdot d\mathbf{A} + \dfrac{1}{c^2} \int \dfrac{\partial \mathbf{E}}{\partial t} \cdot d\mathbf{A}$ | $\oint \mathcal{B} \cdot d\mathbf{s} = \mu_0 \int \mathbf{j} \cdot d\mathbf{A} + \mu_0 \epsilon_0 \int \dfrac{\partial \mathbf{E}}{\partial t} \cdot d\mathbf{A}$ | $\oint \mathcal{B} \cdot d\mathbf{s} = \dfrac{4\pi}{c} \int \mathbf{j} \cdot d\mathbf{A} + \dfrac{1}{c} \int \dfrac{\partial \mathbf{E}}{\partial t} \cdot d\mathbf{A}$ |

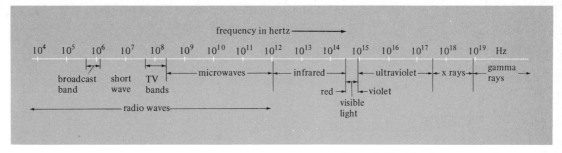

Figure 20-3. *The electromagnetic spectrum.*

charges and currents (and magnetic moments) are explicitly included in $\rho$ and $j$. These equations hold over all possible surfaces and paths and give a unique solution for **E** and $\mathcal{B}$ for a given distribution of charge and current ($\rho$ and **j**). One can make use of this uniqueness property of the solution by guessing at the form of the solution, substituting into Maxwell's equations, and showing that the guess satisfies them. If a certain mathematical expression satisfies Maxwell's equations for all possible surfaces and paths, then it must be *the* solution. In the next section we shall apply this uniqueness theorem to the case of a metal sheet which carries a sinusoidal current.

Not only did Maxwell explain all of electricity in four simple equations but he derived mathematical consequences from them that no one had previously identified with "electricity." In 1864 he showed that an accelerating charge must radiate an electric and magnetic field that moves away from the source at the velocity $u = 1/\sqrt{\mu_0 \varepsilon_0} = c$ and, further, that the radiated electric and magnetic fields are at right angles to each other and also at right angles to the direction of wave propagation. He also showed that in this radiated wave $E = c\mathcal{B}$ ($E = B$ in the cgs system) and that actual energy and momentum are radiated away or given up by the accelerating charge and contained in the field. If the charge were oscillating, the waves would have the same frequency of oscillation. He proposed that light consists merely of such electromagnetic waves of the appropriate frequency (4 to $7 \times 10^{14}$ Hz) and that there should be electromagnetic waves of lower and higher frequencies all the way from zero on up. (See Figure 20-3.)

We see that not only did Maxwell explain the big mystery of just what light is but he proposed that the oscillating charge in resonant circuits should radiate electromagnetic waves and that such waves should be detectable. He predicted the possibility of radio communication well before any such phenomenon had ever been seen! For such a great synthesis of diverse physical phenomena, many think this work of Maxwell's was the greatest achievement in classical physics. Certainly Maxwell did for the electromagnetic interaction what Newton did for the gravitational interaction, and moreover Maxwell has the advantage over Newton in that the vast bulk of physical phenomena are dominated by the

## 20-3 Electromagnetic Radiation

electromagnetic interaction rather than the gravitational. Although he was not aware of it, he developed the complete relativistic theory of the electromagnetic interaction. In effect he invented field theory, which is necessary in order to avoid problems with action at a distance.

Visual inspection of Maxwell's equations already suggests to us the possibility that electric and magnetic fields can maintain themselves even after a source is turned off. Of course a steady source of charge or current merely gives steady fields (Coulomb's law for $\mathbf{E}$ and Ampere's law for $\mathcal{B}$). However, a changing current (accelerated charge) produces a changing magnetic field; that is, $\partial \mathcal{B}/\partial t$ is nonzero. Then according to Eq. II an electric field must be produced even if $\rho = 0$ everywhere in space. For the electric field that is being so produced, $\partial \mathbf{E}/\partial t$ is not zero, so it must produce a contribution to $\mathcal{B}$ according to Eq. IV *even after the current source j is turned off*. Of course, this new contribution to $\mathcal{B}$ has $\partial \mathcal{B}/\partial t \neq 0$, and it produces a new contribution to $\mathbf{E}$, and so on. It is like a dog chasing its tail. If there is no way to dissipate the energy in the fields, the process keeps going on forever, although, as we shall soon see, the disturbance moves out in space with a velocity determined by the proportionality constants in Maxwell's equations. The dog keeps chasing his tail forever, but his outer boundary moves out with velocity $u = c$.

### Radiation from a Current Sheet

Consider an infinite sheet (the $yz$ plane) that carries a surface current $\mathcal{J}$ flowing in the negative $y$ direction. $\mathcal{J}$ is the surface current per unit distance along $z$ (Figure 20-4). Even if the current is changing with time, we can calculate the magnetic field near the sheet by taking a rectangular path integral around the current as shown in Figure 20-5. Let $a$ be the width of the rectangle and $b$ be the height. One half of $a$ is the distance at which we are measuring $\mathcal{B}$. As $a$ approaches zero the area of the rectangle approaches zero, and we can then ignore the term $\int (\partial \mathbf{E}/\partial t) \cdot d\mathbf{A}$ in Eq. 20-2. Since $\mathcal{J}$ is pointing into the page we should take the path integral in the clockwise direction. Then Eq. 20-2 gives

$$\oint \mathcal{B} \cdot d\mathbf{s} = \frac{4\pi k_0}{c^2} I_{\text{in}} = \frac{4\pi k_0}{c^2} \mathcal{J} b$$

$$2\mathcal{B} \cdot b = \frac{4\pi k_0}{c^2} \mathcal{J} b$$

$$\mathcal{B} = \frac{2\pi k_0}{c^2} \mathcal{J} \quad \text{(field near the source)} \tag{20-3}$$

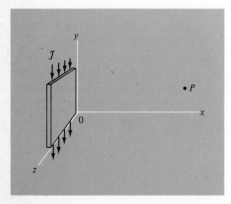

**Figure 20-4.** *A rectangular portion of an infinite sheet of surface current $\mathcal{J}$.*

This is the same result as in Eq. 18-5 for the case where $\mathcal{J}$ is constant over the entire sheet. However, in the present case $\mathcal{J}$ may be changing with time and our result holds only very close to the source.

The field at point $P$, which is not close to the source, can be determined by taking two mutually perpendicular rectangular path integrals around point $P$. (One of them is shown in Figure 20-5.) Since the mathematics is a bit lengthy, this derivation is presented in Appendix 20-1. The result obtained there is

$$\frac{\partial E_y}{\partial x} = -\frac{\partial \mathcal{B}_z}{\partial t} \qquad (20\text{-}4)$$

and

$$\frac{\partial^2 \mathcal{B}_z}{\partial x^2} = \frac{1}{c^2}\frac{\partial^2 \mathcal{B}_z}{\partial t^2} \quad \text{(the wave equation)} \qquad (20\text{-}5)$$

This is a famous differential equation called the wave equation. We shall study it in more detail in Section 20-6. Its solution is a traveling wave moving out from the source with wave velocity $u = c$. Equation 20-4 gives the additional result that there is an accompanying electric field $E = c\mathcal{B}$ and that $\mathbf{E}$ and $\mathcal{B}$ are mutually perpendicular. These solutions are worked out in more detail in the next two sections.

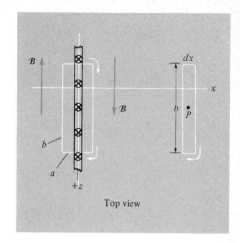

**Figure 20-5.** *Top view of Figure 20-4. The current is flowing into the page. Path integral is taken in clockwise direction around the current. A second path integral is taken around point P.*

## 20-4 Radiation from a Sinusoidal Current Sheet

Suppose the surface current in Figure 20-4 is

$$\mathcal{J} = \mathcal{J}_0 \cos \omega t$$

where $\mathcal{J}_0$ is in the negative $y$ direction. Such sinusoidal currents are easy to generate using electronic circuits (see page 395). We wish to determine $\mathcal{B}(x, t)$ for all values of $x$ and $t$. We have already obtained the solution for small $x$ in Eq. 20-3:

$$\mathcal{B}_z(x, t) = \frac{2\pi k_0}{c^2} \mathcal{J}_0 \cos \omega t \qquad \text{(for small x)}$$

Our procedure for large $x$ is to "guess" a solution that is consistent with the small-$x$ solution and show that it also obeys Eq. 20-5 and thus is the unique solution for the current source under consideration. Our "guess," which agrees with the small-$x$ solution, is

$$\mathcal{B}_z(x, t) = \frac{2\pi k_0}{c^2} \mathcal{J}_0 \cos \omega \left(t - \frac{x}{c}\right) \qquad (20\text{-}6)$$

The left-hand side of Eq. 20-5 becomes

$$\frac{\partial^2 \mathcal{B}_z}{\partial x^2} = -\left(\frac{\omega}{c}\right)^2 \frac{2\pi k_0}{c^2} \mathcal{J}_0 \cos\left(\omega t - \frac{\omega}{c}x\right) = -\frac{\omega^2}{c^2}\mathcal{B}_z$$

The right-hand side becomes

$$\frac{1}{c^2}\frac{\partial^2 \mathcal{B}_z}{\partial t^2} = \frac{1}{c^2}(-\omega^2)\frac{2\pi k_0}{c^2}\mathcal{J}_0 \cos\left(\omega t - \frac{\omega}{c}x\right) = -\frac{\omega^2}{c^2}\mathcal{B}_z$$

We see that the left-hand side equals the right-hand side when Eq. 20-6 is used for $\mathcal{B}$. Also Eq. 20-6 gives $\mathcal{B}_z = (2\pi k_0/c^2)\mathcal{J}_0 \cos \omega t$ when $x$ approaches zero; that is, it obeys what is called the boundary condition and is thus the unique solution to our problem.

As will be shown in Section 20-6, the form $\cos \omega(t - x/c)$ is a sine wave traveling along the $x$ axis with wave velocity $u = c$ and with wavelength $\lambda = 2\pi c/\omega$.

### The Radiated Electric Field

Now that we know $\mathcal{B}$, we can obtain $E$ by substituting our solution for $\mathcal{B}$ into Eq. 20-4.

$$\frac{\partial E_y}{\partial x} = -\frac{\partial}{\partial t}\left[\mathcal{B}_0 \cos \omega\left(t - \frac{x}{c}\right)\right]$$

$$= \omega \mathcal{B}_0 \sin \omega\left(t - \frac{x}{c}\right)$$

$$E_y = \omega \mathcal{B}_0 \int \sin \omega\left(t - \frac{x}{c}\right) dx$$

$$= c\mathcal{B}_0 \cos \omega\left(t - \frac{x}{c}\right) + \text{constant}$$

The constant of integration is zero because there is no net charge to give us a steady electric field. Hence

$$E_y = c\mathcal{B}_z = \frac{2\pi k_0}{c}\mathcal{J}_0 \cos \omega\left(t - \frac{x}{c}\right) \qquad \text{(radiation fields)} \qquad \textbf{(20-7)}$$

In this derivation $\mathcal{J}_0$ was in the negative $y$ direction. Since the amplitudes of $E_y$ and $\mathcal{B}_z$ both turned out to be positive, $E_y$ and $\mathcal{J}$ are in opposite direction near the source. A helpful way of remembering this is to ask

**Figure 20-6.** *A plane electromagnetic wave traveling to the right with velocity $u = c$. Wave is produced by sinusoidal surface current $\mathcal{J}$ in yz plane.*

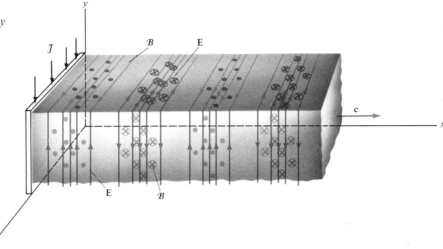

what happens at the top and bottom "edges" of the current sheet. Positive charges would pile up at the bottom edge and negative at the top. The lines of $E$ would go from bottom to top, which is in the opposite direction to the current flow.

We have now shown that $E = c\mathcal{B}$ (or $E = B$ in the cgs system). We have also shown that the electric and magnetic fields are at right angles to each other. (We could take $\oint \mathbf{E} \cdot d\mathbf{s}$ around the rectangle in Figure 20-5 to show that $E_z = 0$.) The solution we have obtained for $\mathbf{E}$ and $\mathcal{B}$ satisfies all possible paths of integration of Maxwell's equations and is the unique solution. The electric and magnetic lines of flux for the sinusoidal current sheet are represented in Figure 20-6.

---

\* **Example 2.** A 3-W square flashlight gives a square beam of 10 cm by 10 cm as in Figure 20-7. The beam strikes a polished metal surface and is reflected back. The 3-W reflected beam is produced by a surface current

$$\mathcal{J} = \frac{I_0}{z_0} \cos \omega t$$

(a) What is $I_0$ in amperes? (The total current in the plate.)
(b) What is $E_0$ in V m$^{-1}$?
(c) What is $B_0$ in gauss?

ANSWER: The $10^{-2}$ m$^2$ beam travels $3 \times 10^8$ m in 1 s. This volume of $3 \times 10^6$ m$^3$ contains 3 J of energy. Hence

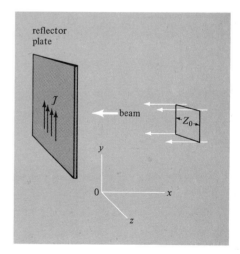

**Figure 20-7.** *Square beam coming from the right strikes the mirror and is reflected back. Induced current $\mathcal{J}$ produces reflected beam.*

$$\frac{dU}{dV} = \frac{3 \text{ J}}{3 \times 10^6 \text{ m}^3} = 10^{-6} \text{ J/m}^3$$

As seen from Eq. 19-10, the energy is split equally between the electric and magnetic fields. Hence

$$\frac{\overline{E^2}}{8\pi k_0} = \frac{1}{2} \times 10^{-6} \text{ J/m}^3$$

Since $\overline{E^2} = \tfrac{1}{2} E_0^2$, we have

$$E_0^2 = 8\pi k_0 \times 10^{-6} \text{ J/m}^3$$
$$E_0 = \sqrt{8\pi (9 \times 10^9) \times 10^{-6}} \text{ V/m} = 475 \text{ V/m}$$
$$\mathcal{B}_0 = \frac{E_0}{c} = \frac{475}{3 \times 10^8} \text{ T} = 1.58 \times 10^{-6} \text{ T}$$

or

$$B = 1.58 \times 10^{-2} \text{ gauss}$$

Solving Eq. 20-7 for $\mathcal{J}_0$ gives

$$\mathcal{J}_0 = \frac{cE_0}{2\pi k_0} = \frac{(3 \times 10^8)(475)}{2\pi(9 \times 10^9)} \text{ A/m} = 2.52 \text{ A/m}$$

$$I_0 = z_0 \mathcal{J}_0 = (0.1 \text{ m})(2.52 \text{ A/m}) = 0.252 \text{ A}$$

---

\* **Example 3.** An ac generator is connected across a large metal sheet causing a surface current $\mathcal{J}_0 \cos \omega t$ to oscillate up and down the sheet. Since the radiated electric field $E_{\text{rad}}$ is continuous from one side of the sheet to the other, there will be a force $eE_{\text{rad}}$ on each conduction electron in the sheet and this force will contribute to an energy transfer from the generator to the electrons. What is the power per square meter drawn from the generator? Ignore any resistive losses.

ANSWER: Take an area $y_0$ by $z_0$ of the sheet. Then the electric power per unit area is

$$\frac{P}{A} = \frac{VI}{y_0 z_0} = \frac{(E_{\text{rad}} y_0)(\mathcal{J} z_0)}{y_0 z_0} = E_{\text{rad}} \mathcal{J}$$

Using $\mathcal{B}_{rad} = (2\pi k_0/c^2)\mathcal{J}$ gives

$$\frac{P}{A} = \frac{c^2}{2\pi k_0} E_{rad}\mathcal{B}_{rad}$$

This is energy radiated away into the electric and magnetic fields. (See Section 21-1.) The energy radiated to the left or to the right is one-half of this. Hence

$$\text{power per unit area} = \frac{c^2}{4\pi k_0} E\mathcal{B}$$

for a traveling electromagnetic wave.

* **Example 4.** What is the effective resistance of a 1-m × 1-m piece of the sheet in Example 3?

ANSWER:

$$R = \frac{V}{I} = \frac{Ey_0}{\mathcal{J}z_0}$$

For $y_0 = z_0 = 1$ m,

$$R = \frac{E}{\mathcal{J}} = \frac{2\pi k_0}{c} = 2\pi \frac{9 \times 10^9}{3 \times 10^8} \Omega = 188.5 \, \Omega$$

In a real metal sheet the current on each of the two surfaces is $\frac{1}{2}\mathcal{J}$. The effective resistance per square meter for each surface is then twice the preceding value or

$$4\pi \frac{k_0}{c} = \frac{1}{\varepsilon_0 c} = \sqrt{\frac{\mu_0}{\varepsilon_0}} = 377 \, \Omega$$

This is called the impedence of free space.

## 20-5 Nonsinusoidal Current Sources—Fourier Analysis

Instead of a sinusoidal source of current, let us take the current source as any function of time. It will radiate out $E$ and $\mathcal{B}$ fields having the same function of time as the source. As an example we shall take the case where the surface current is a sawtooth function of period $\tau$. Then $\omega = 2\pi/\tau$.

Figure 20-8 shows that a sawtooth function can be expanded as an infinite sum of sine waves;

$$F(t) = \sum_{n=1}^{\infty} \left(\frac{1}{n} \sin n\omega t\right)$$

This is called the Fourier expansion of the periodic function $F(t)$. In general any periodic function of frequency $1/\tau$ can be expanded as a sum of sine and cosine waves of frequencies $n(1/\tau)$, where $n$ goes from 1 to $\infty$. It is amazing that perfectly straight lines with infinitely sharp corners can be constructed out of pure sine waves which are continuously curving and have no straight sections. Of course, an infinite number of sine waves is required to achieve perfectly straight lines and sharp corners.

In order to generate a sawtooth electromagnetic wave the surface current should be

$$\mathcal{J} = \mathcal{J}_0 \sum_{n=1}^{\infty} \left(\frac{1}{n} \sin n\omega t\right) \quad \text{where } \omega = \frac{2\pi}{\tau}$$

Each sine wave term in the preceding expansion has the solution given by Eqs. 20-6 and 20-7. Since Maxwell's equations are linear in $E$, $\mathcal{B}$, and $j$, the resultant solution will be the sum of the separate solutions. This is known as the principle of superposition. The resultant solution is then

$$E = c\mathcal{B} = \frac{2\pi k_0}{c} \mathcal{J}_0 \sum \left[\frac{1}{n} \sin n\omega \left(t - \frac{x}{c}\right)\right]$$

Since each term in the sum travels along the $x$ axis with the same velocity $u = c$, the component sine waves travel along together and will always add up in the same way as they do in Figure 20-8. Thus the field at any point in space will have the same sawtoothed time dependence as the source, only retarded in time by the amount $x/c$.

The most general periodic function for the source can be expressed as

$$\mathcal{J} = \sum_{n=1}^{\infty} A_n \sin(n\omega t + \phi_n)$$

The solution would have the same time dependence:

$$E = c\mathcal{B} = \frac{2\pi k_0}{c} \sum A_n \sin\left[n\omega\left(t - \frac{x}{c}\right) + \phi_n\right]$$

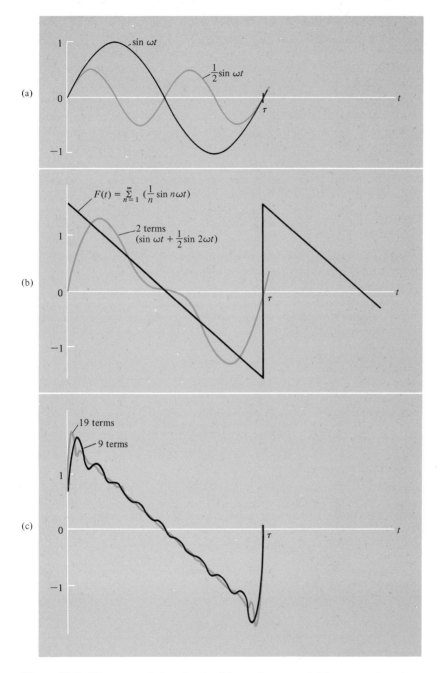

**Figure 20-8.** *The sawtooth function in* (b) *can be expanded in terms of an infinite number of sine waves.* (a) *The first two sine waves.* (b) *The two sine waves of* (a) *added together.* (c) *The sum of the first nine (and also the first 19) sine waves.*

As a final example we show the fields radiated by a source that is abruptly turned on and off (rectangular pulses or square waves). This is shown in Figure 20-9. Even if the source is turned on only for one single pulse, the same approach may be used. As is shown in Appendix 21-2, a single pulse may be expanded mathematically as an infinite sum of pure sine waves (called a Fourier integral). The principle of superposition still holds, and we still have the characteristic solution that $E = c\mathcal{B}$, that **E** and $\mathcal{B}$ are perpendicular to each other and to the direction of propagation, and that they travel away from the source with $u = c$. Also at any fixed position $x$, the fields have the same time dependence as the source, only retarded in time by $x/c$.

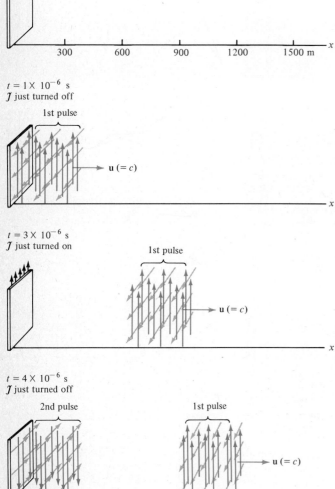

**Figure 20-9.** *Electromagnetic pulses of $1 \times 10^{-6}$ s duration being radiated by turning on and off an infinite plane of current. A square section of the infinite plane and electromagnetic wave is shown. Lines of E are in full color and lines of $\mathcal{B}$ in half color.*

430 CH. 20/ELECTROMAGNETIC RADIATION AND WAVES

Since it is easier to visualize a traveling wave along a string as opposed to a traveling electromagnetic wave, we shall start the discussion with waves on a string. The concepts and the mathematics are the same for all kinds of traveling waves including the electromagnetic wave of Eq. 20-7.

If the end of a long stretched string is moved up and down sinusoidally, a sine wave is observed to travel along the string as shown in Figure 20-10. We shall prove that the velocity of the wave along the string will depend on the tension $T$ in the string and the mass per unit length $\mu$. Since we use the symbol $v$ for the up and down velocity of a point on the string, we shall use the symbol $u$ for the wave velocity, that is, the speed with which the crest of a wave moves along the string.

First we must convince ourselves that the following function represents a sine wave traveling in the positive $x$ direction with wave velocity $u$:

$$y(x, t) = y_0 \cos \frac{2\pi}{\lambda}(x - ut) \quad \text{(traveling wave)} \quad (20\text{-}8)$$

For any fixed value of $t$ this is of the form

$$\cos\left(\frac{2\pi x}{\lambda} + \phi\right) \quad \text{where } \phi \text{ is some fixed angle.}$$

Note that if $x$ increases by $\lambda$, the angle $(2\pi x/\lambda + \phi)$ increases by $2\pi$; hence $\lambda$ is the wavelength (the wave repeats itself after a distance $\lambda$). We can show that $u$ in Eq. 20-8 is the wave velocity by following the crest of a wave. When the angle $[2\pi/\lambda)(x - ut)]$ in Eq. 20-8 is zero, $y(x, t)$ is maximum; that is, we have the crest of the wave. Take the crest where the angle $[(2\pi/\lambda)(x - ut)]$ is zero (this is crest 1 in Figure 20-10):

$$\frac{2\pi}{\lambda}(x - ut) = 0$$

Then

$$x = ut \quad \text{or} \quad u = \frac{x}{t}$$

where $x$ is the position of this crest at a time $t$. By definition $x/t$ is the velocity of the crest, which we have just shown is equal to $u$.

The velocity of the crest is by definition the wave velocity.

Now that we have shown that Eq. 20-8 represents a sine wave moving with wave velocity of $u$, we shall rewrite it in terms of $\omega$ by noting that

$$\lambda f = u \quad \text{(wave velocity)} \quad (20\text{-}9)$$

The wavelength times the frequency equals the wave velocity. This is

## 20-6 Traveling Waves

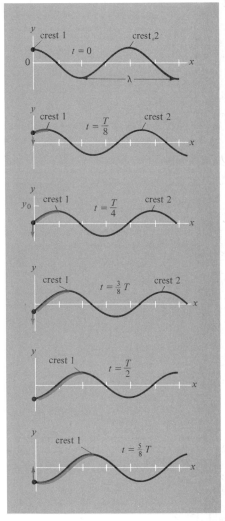

**Figure 20-10.** *Traveling wave on a string. Point at $x = 0$ is moving according to $y = y_0 \cos \omega t$. Successive positions are shown at $\frac{1}{8}$-period intervals. The crest of the wave moves $\frac{9}{8}\lambda$ to the right.*

because the length of one wave ($\lambda$) times the number of waves in 1 s ($f$) is the distance the wave moves in 1 s ($u$). Rewriting Eq. 20-9:

$$\frac{\lambda}{2\pi}(2\pi f) = u$$

or

$$\frac{\lambda}{2\pi}\omega = u$$

If we replace $u$ in Eq. 20-8 with ($\lambda/2\pi)\omega$ we obtain

$$y(x,t) = y_0 \cos\left(\frac{2\pi}{\lambda}x - \omega t\right) = y_0 \cos\left(\omega t - \frac{2\pi}{\lambda}x\right)$$

The notation $k = 2\pi/\lambda$ is often used. The quantity $k$ is called the wavenumber. Then

$$y(x,t) = y_0 \cos(\omega t - kx)$$

where

$$k \equiv \frac{2\pi}{\lambda} \quad \text{(wavenumber)} \tag{20-10}$$

and

$$\frac{\omega}{k} = u \quad \text{(wave velocity)} \tag{20-11}$$

Note that for any fixed value of $x$ this is of the form $y_0 \cos(\omega t - \phi)$, where $\phi$ is a fixed phase angle. This says that each point on the string is oscillating up and down with simple harmonic motion.

> **Example 5.** A traveling wave is of the form $y = \cos(Ax + Bt)$. What is its wave velocity?
>
> ANSWER: Comparing with Eq. 20-8 we have $A = 2\pi/\lambda$ and $B = -2\pi u/\lambda$. Then $B/A = -u$ or $u = -B/A$ is the wave velocity. If $A$ and $B$ are positive, the wave will travel in the negative $x$ direction.

Now we are ready to derive the formula for the wave velocity in terms of $T$ and $\mu$ of the string. We take a small length of string $\Delta x$ which has small angles $\alpha_1$ and $\alpha_2$ between its ends and the $x$ axis (see Figure 20-11). By small we mean $\sin \alpha \approx \alpha \approx \partial y/\partial x$. The net vertical force on the string is $F_{\text{net}} = (T\alpha_2 - T\alpha_1)$. This must equal the mass ($\mu \Delta x$) times the vertical acceleration $\partial^2 y/\partial t^2$. (The notation $\partial y/\partial t$ means to treat the independent

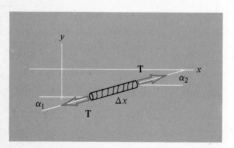

**Figure 20-11.** *Forces acting on element $\Delta x$ of the string.*

variable $x$ as a constant when taking the time derivative.) So

$$F_{\text{net}} = (T\alpha_2 - T\alpha_1) = (\mu \Delta x)\frac{\partial^2 y}{\partial t^2}$$

$$T\frac{\Delta \alpha}{\Delta x} = \mu \frac{\partial^2 y}{\partial t^2}$$

$$\frac{\partial \alpha}{\partial x} = \frac{\mu}{T}\frac{\partial^2 y}{\partial t^2}$$

Now substitute $\alpha = \partial y/\partial x$ into the left-hand side:

$$\frac{\partial^2 y}{\partial x^2} = \frac{\mu}{T}\frac{\partial^2 y}{\partial t^2} \quad \text{(wave equation for a string)} \quad (20\text{-}12)$$

This equation is known as the wave equation for a string. We can get the wave velocity by substituting the appropriate derivatives of Eq. 20-8 into Eq. 20-12:

$$\frac{\partial^2 y}{\partial x^2} = -y_0\left(\frac{2\pi}{\lambda}\right)^2 \cos\frac{2\pi}{\lambda}(x - ut) \quad (20\text{-}13)$$

$$\frac{\partial^2 y}{\partial t^2} = -y_0\left(\frac{2\pi}{\lambda}u\right)^2 \cos\frac{2\pi}{\lambda}(x - ut) \quad (20\text{-}14)$$

Substitute Eq. 20-13 into the left-hand side of Eq. 20-12, and 20-14 into the right-hand side. Then

$$\left(\frac{2\pi}{\lambda}\right)^2 = \frac{\mu}{T}\left(\frac{2\pi u}{\lambda}\right)^2$$

$$u = \sqrt{\frac{T}{\mu}} \quad \text{(wave velocity on a string)} \quad (20\text{-}15)$$

Not only have we derived a formula for the wave velocity but we have shown that Eq. 20-8 (which represents a traveling wave) is a solution of Eq. 20-12; namely, we have proved that a sine wave can be transmitted along a string with a wave velocity that is independent of the amplitude and of the frequency.

---

**Example 6.** A 30-cm guitar string has a mass of 100 g and is a half wavelength long. What value of tension must be put on the string in order to tune it to middle C (262 Hz)?

> ANSWER: Solving Eq. 20-15 for $T$ gives
>
> $$T = \mu u^2$$
>
> where $\mu$ is 0.1 kg per 0.3 m = 0.333 kg/m. We can find $u$ from the relation $u = \lambda f$:
>
> $$u = (0.6 \text{ m})(262 \text{ Hz}) = 157.2 \text{ m/s}.$$
>
> Then
>
> $$T = (0.333 \text{ kg/m})(157.2 \text{ m/s})^2 = 8.24 \times 10^3 \text{ N}$$
>
> This is a rather large force to be supported by such a light string. For this reason strings of musical instruments are usually made of strong metallic alloys.

If we substitute $1/u^2$ for $\mu/T$ in Eq. 20-12 we obtain

$$\frac{\partial^2 y}{\partial x^2} = \frac{1}{u^2} \frac{\partial^2 y}{\partial t^2} \quad \text{(the wave equation)} \quad (20\text{-}16)$$

Note that this is the same form as Eq. 20-5 with $\mathcal{B}_z$ in place of $y$ and $c$ in place of $u$. Eq. 20-16 is known as the wave equation and it applies to all kinds of traveling waves such as sound waves, electromagnetic waves, water waves, torsional waves, waves along rods, strings, and springs. We have already shown that

$$y = y_0 \cos \frac{2\pi}{\lambda}(x - ut) = y_0 \cos \omega \left(t - \frac{x}{u}\right) \quad (20\text{-}17)$$

is a solution of the wave equation. This is true for any frequency $\omega$. We see that

$$E_y = \frac{2\pi k_0}{c} \mathcal{J}_0 \cos \omega \left(t - \frac{x}{c}\right)$$

which is the field radiated by a sinusoidal surface current of amplitude $\mathcal{J}_0$, is a traveling wave moving along the $x$ axis with wave velocity $u = c$ and wavelength $\lambda = 2\pi c/\omega$.

The wave theory presented in this and the next section is quite general and applies to all kinds of waves.

If one starts to vibrate the end of a long taut string by hand, one is doing work which shows up as kinetic and potential energy of points on the string farther and farther away as time progresses. The energy put into the string is literally transmitted away at the wave velocity and could be received and utilized by someone at the other end of the string. We can calculate the rate of energy transmission by calculating the force with which the string must be pushed up and down. We shall apply the relation $P = \mathbf{F} \cdot \mathbf{v}$ (Eq. 6-3) to Figure 20-12 in order to calculate the power supplied to the string:

## 20-7 Energy Transmission By Waves

$$P = \mathbf{T} \cdot \mathbf{v} = T\left(\frac{\partial y}{\partial t}\right) \sin \alpha$$

using $v = \partial y/\partial t$. Since $\alpha$ is always small, $\sin \alpha \approx -\partial y/\partial x$. Then

$$P = -T\left(\frac{\partial y}{\partial t}\right)\left(\frac{\partial y}{\partial x}\right)$$

Since $y = y_0 \cos \omega(t - x/u)$, we have

$$\frac{\partial y}{\partial t} = -y_0 \omega \sin \omega \left(t - \frac{x}{u}\right)$$

$$\frac{\partial y}{\partial x} = y_0 \frac{\omega}{u} \sin \omega \left(t - \frac{x}{u}\right)$$

$$P = T y_0^2 \frac{\omega^2}{u} \sin^2 \omega t$$

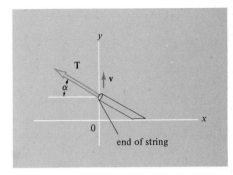

**Figure 20-12.** *End of string is being pulled up in order to transmit a traveling wave.*

is the instantaneous power supplied at time $t$ and $x = 0$. The average power supplied would be one half this because the average of $\sin^2 \omega t$ is $\tfrac{1}{2}$:

$$\bar{P} = \frac{T\omega^2}{2u} y_0^2 \qquad \text{(av. power transmitted by a string)} \qquad (20\text{-}18)$$

The average power transmitted is defined as the wave intensity (see Section 11-6). We see that the wave intensity is proportional to the square of the wave amplitude. For waves in three-dimensional space, such as sound waves and electromagnetic waves, intensity is the average power transmitted across a square meter of wave front. For all kinds of waves the intensity is proportional to the square of the amplitude. This checks with our previous conclusion that the energy carried by an electromagnetic wave is proportional to $E^2$. The proportionality constant connecting the source current $\mathscr{J}$ and the radiated power is worked out in the next chapter.

> **Example 7.** A long metal wire has $\mu = 4$ g/cm and tension $T = 5 \times 10^3$ N. A 10-W sinusoidal vibrator is fastened at one end and tuned to middle C (262 Hz). The other end is fastened to an energy absorber (no reflected wave). (a) What is the wave velocity? (b) What is the wavelength? (c) What is the maximum transverse displacement?
>
> ANSWER: The wave velocity is
>
> $$u = \sqrt{\frac{T}{\mu}} = \sqrt{\frac{5 \times 10^3 \text{ N}}{0.4 \text{ kg/m}}} = 112 \text{ m/s}$$
>
> The wavelength is
>
> $$\lambda = \frac{u}{f} = \frac{112 \text{ m/s}}{262 \text{ s}^{-1}} = 0.427 \text{ m}$$
>
> Solving Eq. 20-18 for $y_0$:
>
> $$y_0 = \frac{1}{\omega}\sqrt{\frac{2u\overline{P}}{T}} = \frac{1}{2\pi(262)}\sqrt{\frac{2(112)(10)}{5 \times 10^3}} \text{ m}$$
> $$= 4.06 \times 10^{-4} \text{ m} = 0.406 \text{ mm}$$

## Summary

An additional term is needed on the right-hand side of Ampere's law when dealing with changing currents. The corrected form is

$$\oint \mathcal{B} \cdot d\mathbf{s} = \frac{4\pi k_0}{c^2} I_{\text{in}} + \frac{1}{c^2}\frac{d\Phi_E}{dt}$$

Since both $I_{\text{in}}$ and $\Phi_E$ can be written as integrals over the enclosed area, we have

IV
$$\oint \mathcal{B} \cdot d\mathbf{s} = \frac{4\pi k_0}{c^2}\int \mathbf{j} \cdot d\mathbf{A} + \frac{1}{c^2}\int \frac{\partial \mathbf{E}}{\partial t} \cdot d\mathbf{A}$$

This plus the following three equations make up the four Maxwell equations:

I
$$\oint \mathbf{E} \cdot d\mathbf{A} = 4\pi k_0 \int \rho \, dV \quad \text{(Gauss's law)}$$

II $$\oint \mathbf{E} \cdot d\mathbf{s} = -\int \frac{\partial \mathcal{B}}{\partial t} \cdot d\mathbf{A} \quad \text{(Faraday's law)}$$

III $$\oint \mathcal{B} \cdot d\mathbf{A} = 0 \quad \text{(magnetic charge does not exist)}$$

These four equations completely determine the fields produced by a system of moving charges (including acceleration). When the charges accelerate, these equations show that there will be a radiated electromagnetic field moving away from the source with a velocity $c = 3 \times 10^8$ m/s where $E = c\mathcal{B}$ and $\mathbf{E}$ and $\mathcal{B}$ are at right angles to each other and to the direction of wave propagation.

The specific example of a surface charge accelerating up and down an infinite sheet is worked out. The result is that both $E$ and $\mathcal{B}$ satisfy the wave equation:

$$\frac{\partial^2 y}{\partial x^2} = \frac{1}{c^2} \frac{\partial^2 y}{\partial t^2} \quad \text{(Where } y = \text{either } E_y \text{ or } \mathcal{B}_z\text{)}$$

If the surface current is sinudoidal ($\mathcal{J} = \mathcal{J}_0 \cos \omega t$) with $\mathcal{J}_0$ pointing in the negative $y$ direction, the result is

$$E_y = c\mathcal{B}_z = \frac{2\pi k_0}{c} \mathcal{J}_0 \cos \omega \left(t - \frac{x}{c}\right)$$

which is a traveling wave moving out from the sheet with wave velocity $u = c$.

If the current in the current sheet is periodic, but not sinusoidal, it can be Fourier-analyzed as a sum of sine waves, and the radiated field at any fixed point will be

$$E = c\mathcal{B} = \frac{2\pi k_0}{c} \mathcal{J}(t')$$

where $t' = (t - x/c)$ is the retarded time.

The relation between frequency $f$, wavelength $\lambda$, and wave velocity $u$ is $\lambda f = u$; this can also be written $\omega/k = u$.

A displacement $y$ (on a string, for example) which is the following function of $x$ and $t$ is a solution of the wave equation:

$$y = y_0 \cos(\omega t - kx) \quad \text{(traveling wave)}$$

$$\frac{\partial^2 y}{\partial x^2} = \frac{1}{u^2} \frac{\partial^2 y}{\partial t^2} \quad \text{(the wave equation)}$$

The wave velocity of a traveling wave on a string is $u = \sqrt{T/\mu}$, where $T$ is the tension and $\mu$ is the mass per unit length. The average power transmitted along a string is

$$\bar{P} = \frac{T\omega^2}{2u}y_0^2$$

## Appendix 20-1  Derivation of Wave Equations

In order to determine $\mathcal{B}$ at point $P$ in Figure 20-5 we take a rectangular path integral around the point as shown in Figure 20-5. If we take the path integral in the clockwise direction, the area of the rectangle, $d\mathbf{A}$, will be pointing into the page in the negative $y$ direction; hence the dot product $\mathbf{E} \cdot d\mathbf{A}$ would be $-E_y dA = -E_y(b\,dx)$. Then Eq. 20-2 which is

$$\oint \mathcal{B} \cdot d\mathbf{s} = 0 + \frac{1}{c^2} \int \frac{\partial \mathbf{E}}{\partial t} \cdot d\mathbf{A}$$

becomes

$$(\mathcal{B}_z + d\mathcal{B}_z)b - \mathcal{B}_z b = -\frac{1}{c^2}\frac{\partial E_y}{\partial t}(b\,dx)$$

where $\mathcal{B} = \mathcal{B}_z$ at the left side and $\mathcal{B} = (\mathcal{B}_z + d\mathcal{B}_z)$ at the right side of the rectangle. (The top and bottom sides don't contribute to $\oint \mathcal{B} \cdot d\mathbf{s}$.) Thus

$$d\mathcal{B}_z = -\frac{1}{c^2}\frac{\partial E_y}{\partial t}dx$$

$$\left(\frac{d\mathcal{B}_z}{dx}\right)_{t\text{ const.}} = -\frac{1}{c^2}\frac{\partial E_y}{\partial t}$$

$$\frac{\partial \mathcal{B}_z}{\partial x} = -\frac{1}{c^2}\frac{\partial E_y}{\partial t} \tag{20-19}$$

This is a partial derivative because Figure 20-5 is a snapshot in time ($t$ is held constant when taking the derivative).

We can get a second relation between $\mathcal{B}$ and $E$ by using a second Maxwell equation. Now we use Eq. II in Table 20-1 and take a counterclockwise path integral around point P in the $xy$ plane as shown in Figure 20-13.

$$\oint \mathbf{E} \cdot d\mathbf{s} = -\int \frac{\partial \mathcal{B}}{\partial t} \cdot d\mathbf{A}$$

$$(E_y + dE_y) \cdot h - E_y \cdot h = -\frac{\partial \mathcal{B}_z}{\partial t} \cdot (h\, dx)$$

$$dE_y = -\frac{\partial \mathcal{B}_z}{\partial t} \cdot dx$$

$$\left(\frac{dE_y}{dx}\right)_{t\text{ const}} = -\frac{\partial \mathcal{B}_z}{\partial t}$$

$$\frac{\partial E_y}{\partial x} = -\frac{\partial \mathcal{B}_z}{\partial t} \qquad (20\text{-}4)$$

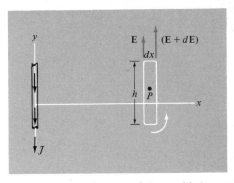

**Figure 20-13.** *Side view of Figure 20-4.*

Our goal is to evaluate $\mathcal{B}$ at point $P$. We now have two simultaneous equations with two unknowns ($\mathcal{B}_z$ and $E_y$). We can eliminate $E_y$ by taking the $x$ derivative of the first equation and the $t$ derivative of the second equation. Taking the $x$ derivative of Eq. 20-19 gives

$$\frac{\partial}{\partial x}\left(\frac{\partial \mathcal{B}_z}{\partial x}\right) = \frac{\partial}{\partial x}\left(-\frac{1}{c^2}\frac{\partial E_y}{\partial t}\right)$$

$$\frac{\partial^2 \mathcal{B}_z}{\partial x^2} = -\frac{1}{c^2}\frac{\partial^2 E_y}{\partial x\, \partial t} \qquad (20\text{-}20)$$

Now take the $t$ derivative of Eq. 20-4:

$$\frac{\partial}{\partial t}\left(\frac{\partial E_y}{\partial x}\right) = \frac{\partial}{\partial t}\left(-\frac{\partial \mathcal{B}_z}{\partial t}\right)$$

$$\frac{\partial^2 E_y}{\partial x\, \partial t} = -\frac{\partial^2 \mathcal{B}_z}{\partial t^2}$$

Substitute this into the right-hand side of Eq. 20-20:

$$\frac{\partial^2 \mathcal{B}_z}{\partial x^2} = -\frac{1}{c^2}\left[-\frac{\partial^2 \mathcal{B}_z}{\partial t^2}\right]$$

$$\frac{\partial^2 \mathcal{B}_z}{\partial x^2} = \frac{1}{c^2}\frac{\partial^2 \mathcal{B}_z}{\partial t^2} \qquad (20\text{-}5)$$

## Exercises

1. Rewrite Maxwell's Equations in Table 20-1 in terms of magnetic flux $\Phi_B$, electric flux $\Phi_E$, enclosed current $I_{in}$, and enclosed charge $Q_{in}$.
2. Use Figure 20-3 to classify electromagnetic radiation of the following wavelengths: (a) 1 m. (b) 1 cm. (c) 1 μm. (d) 0.5 μm. (e) 1 Å = $10^{-10}$ m.
3. Suppose **E** in Figure 20-5 is $E = E_0 \cos \omega t$ pointing out of the page. Then what is $\int (\partial \mathbf{E}/\partial t) \cdot d\mathbf{A}$ for the rectangular path around the current source? Give answer in terms of $a$, $b$, $E_0$, and $\omega$.
4. What is the value of $\int \mathbf{E} \cdot d\mathbf{s}$ around the rectangle in Figure 20-5?
5. Plot the function

$$y = \sum_{n \text{ odd}} \frac{1}{n} \sin nx = \sum_{j=0}^{\infty} \frac{1}{2j+1} \sin(2j+1)x$$

6. Does Eq. 20-7 hold for negative $x$; that is, for the traveling wave radiated to the left of the plate in Figure 20-4? If not, make whatever changes are necessary to describe $E_y$ and $\mathcal{B}_z$ for negative $x$.
7. In Eq. 20-7 what is the direction of $\mathbf{E} \times \mathcal{B}$?
8. Consider the traveling wave $y = \sin(Ax - Bt)$. What is the wave velocity $u$?
9. A vibrating string has a mass of 10 g/m and a frequency of vibration of 30 Hz. What must be the tension in order for the wavelength to be 20 cm?
10. A wave traveling along a string is of the form $y = \sin(at - bx)$. Express $a$ and $b$ in terms of $\lambda$ and $f$.
11. Repeat Example 5 for an octave above middle C; that is, $f = 2 \times 262$ Hz.
12. The tempered scale in music has twelve equally spaced notes per octave; that is, the ratio of any two successive notes is the same. What is this ratio?
13. A large sheet of current is oscillating sinusoidally at a frequency of 100 MHz. The peak value of the radiated electric field is 5 V/cm.
    (a) What is the peak value of the current in amperes per meter?
    (b) What is the peak value of $\mathcal{B}$ in tesla?
14. In an electromagnetic wave what is the distance in wavelengths between two successive intensity maxima?
15. What are the wavelength limits of visible light in terms of angstroms? 1 Å (angstrom) is defined as $10^{-10}$ m. Use Figure 20-3.
16. The lowest and highest frequencies detectable by the human ear are about 20 and 15,000 Hz, respectively. What are the corresponding wavelengths in air? Velocity of sound in air is 330 m/s.

## Problems

17. In Figure 20-1 the current $I_C$ has been flowing for a time $\Delta t$.
    (a) What is $E$ between the capacitor plates at a distance $r$ from the axis?
    (b) What is $\mathcal{B}$ between the capacitor plates at a distance $r$ from the axis? Give answer in terms of $R$, $r$, $E$, and $\Delta t$.
18. A horseshoe electromagnet is turned on at $t = 0$. Assume the flux increases linearly with time: that is, $\Phi = K_1 t$. What are the magnitude and direction of **E** between the pole faces? Give answer in terms of $K_1$, $t$, $r$, and $R$.
19. Suppose the coil in Problem 18 has $N$ turns and the current is $I = K_2 t$. What is the inductance in terms of $K_1$, $K_2$, and $N$?

20. Substitute $B_z = B_0 \cos(kx - \omega t)$ into Eq. 20-5. Solve for the ratio $\omega/k$.
21. Substitute $B_z = f(x - ut)$ into Eq. 20-5. What must be the value of $u$ in order for this to be a solution?
22. Substitute $B_z = B_1 \cos \omega_1(t - x/c) + B_2 \cos \omega_2(t - x/u)$ into Eq. 20-5. For what value of $u$ will it be a solution?
23. Two parallel sheets are separated by a distance $\lambda/2$ where $\lambda = 2\pi c/\omega$. Each sheet has a surface current $\mathcal{J} = \mathcal{J}_0 \cos \omega t$.
    (a) What is $\mathcal{B}$ at a distance $x$ from the second sheet?
    (b) What is $\mathcal{B}$ halfway between the two sheets?
24. Suppose the metal plate in Figure 20-7 is replaced by a vertical grid of wires spaced 2 cm apart. Wire diameter = 1 mm.
    (a) If the radiation striking the grid is the same as in Example 2, what is the induced current per wire? Assume $I_{wire} = \mathcal{J}$ times the wire diameter.
    (b) If the wires are copper (resistivity $\rho = 1.72 \times 10^{-8}$ ohm m) of 1 mm diameter, what is the power dissipated per wire?
    (c) What fraction of the energy in the light beam is lost in the wire grid?
25. A sheet of germanium is 1 mm thick and has an internal electric field $E_{in} = 400 \cos \omega t$ in volts per meter. The internal electric field is produced by connecting the sheet to an ac generator at a frequency $\omega/2\pi = 10^9$ Hz. If the resistivity is $\rho = 0.5$ ohm m, what is the radiated electric field? What is the ratio of radiated field to internal field?
26. Suppose $\mathcal{J} = \mathcal{J}_1 \sin \omega t + 2\mathcal{J}_1 \cos \omega t$. We want to express $\mathcal{J}$ in the form $\mathcal{J} = \mathcal{J}_0 \sin(\omega t + \phi)$.
    (a) What is $\mathcal{J}_0$ in terms of $\mathcal{J}_1$?
    (b) What is the angle $\phi$?
    Hint: $\mathcal{J}_0 \sin(\omega t + \phi) = (\mathcal{J}_0 \cos \phi) \sin \omega t + (\mathcal{J}_0 \sin \phi) \cos \omega t$.
27. Suppose $\mathcal{J} = A_1 \sin \omega t + A_2 \cos \omega t$. Express $\mathcal{J}$ in the form $\mathcal{J} = \mathcal{J}_0 \sin(\omega t + \phi)$.
    (a) What is $\mathcal{J}_0$ in terms of $A_1$ and $A_2$?
    (b) What is $\phi$ in terms of $A_1$ and $A_2$?
28. A thin slab of thickness $x_0$, width $z_0$, and length $y_0$ has resistivity $\rho$. An ac generator giving a voltage $V_0 \cos \omega t$ is connected across the slab. Show that the radiated electric field at point $P$ is

$$E_{rad} = \frac{2\pi k_0}{c} \frac{x_0}{\rho y_0} V_0 \cos(\omega t - kx) \quad \text{when } x \ll y_0 \text{ and } x_0 \ll \lambda$$

29. In Problem 28
    (a) What is the average power dissipated in the slab per unit area? Use $P = V^2/R$.
    (b) What is the average power radiated per unit area? Use

$$\frac{dU}{dV} = \frac{\overline{E^2}}{8\pi k_0} + \frac{c^2}{8\pi k_0}\mathcal{B}^2$$

30. What is the wave velocity of a solution to the following equation?

$$A \frac{\partial^2 y}{\partial t^2} = B \frac{\partial^2 y}{\partial x^2}$$

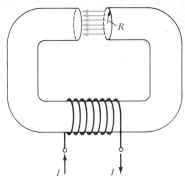

Problem 18

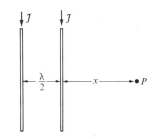

Problem 23

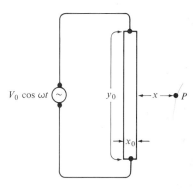

Problem 28

# Interaction of Radiation with Matter

In this chapter we shall study how an electromagnetic wave interacts with a slab of material. We shall deal with four cases: a poor conductor, a good conductor, a nonconductor, and a plasma. We shall find that the poor conductor absorbs energy and momentum from the radiated wave, thus permitting an evaluation of the energy and momentum content of an electromagnetic wave. A good conductor reflects the wave with 100% efficiency. A nonconductor such as a gas allows the wave to pass through without any attenuation; however, the wave appears to travel slower than the speed of light in vacuum. On the other hand, a wave entering a plasma appears to travel faster than the speed of light. These apparent paradoxes are resolved by taking a microscopic or atomic approach. Finally, we shall study the radiation from a single oscillating point charge.

## 21-1 Radiated Energy

We saw in Example 3 of Chapter 20 that an oscillating current suffers a power loss per unit area of $c^2 E\mathcal{B}/(4\pi k_0)$, where $E$ and $\mathcal{B}$ are the fields of the radiated electromagnetic wave. This energy must go somewhere. We suspect it is carried away by the radiated electric and magnetic fields. If the traveling electromagnetic wave is transporting this much energy away from the source, we should be able to catch this energy by intercepting the plane wave with an absorbing slab. We will use an electric conductor of finite resistance so that we can calculate the total ohmic heat generated in the slab. According to the law of conservation of energy, the total heat generated will be the energy contained in the traveling electromagnetic wave that enters the slab. We already have a clue from Chapter 16 that the field itself contains energy. It was pointed out that the total electrostatic energy of a system is $E^2/(8\pi k_0)$ integrated over all space. Similarly, in Chapter 18 there was the claim that the magnetic energy per unit volume is $c^2\mathcal{B}^2/(8\pi k_0)$. If this energy is actually in the field itself, we would expect the ohmic heating in an absorbing slab to be equal to the sum of these two terms.

Figure 21-1 shows an electromagnetic wave impinging on a rectangular piece of an "infinite" slab. If the induced current density is $j$, the current $I$ in the rectangular piece is $j(z_0 \Delta x)$. The voltage difference from top to bottom is $V = Ey_0$. Hence the power loss is

$$\frac{dU}{dt} = IV = (jz_0 \Delta x)(Ey_0)$$

$$= jE(y_0 z_0 \Delta x) \quad \text{where } (y_0 z_0 \Delta x) \text{ is the volume}$$

We have just shown that the power loss per unit volume inside a conducting medium is $jE$. The sinusoidal plane electromagnetic wave hitting the conducting slab not only generates heat at the rate of $jE$ W/m³, but the induced current $j$ must radiate its own electromagnetic wave according to Eq. 20-7. Let $\Delta E$ be the field radiated by the induced current $j$. This is shown in Figure 21-1 for a thin slab of thickness $\Delta x$. We shall use the notation $E_{in}$ for the original field of the incoming plane wave. The equivalent surface current is $\mathcal{J} = j \Delta x$; hence, according to Eq. 20-7

$$\Delta E = -\frac{2\pi k_0}{c} j \Delta x \qquad (21\text{-}1)$$

We use the minus sign to indicate that $\Delta \mathbf{E}$ is in the opposite direction to $\mathbf{j}$ as it leaves the slab. Let $\Delta S$ stand for the power loss per unit area. For this thin slab of thickness $\Delta x$, the power loss per unit area is

$$\Delta S = \frac{1}{y_0 z_0} \frac{dU}{dt} = jE \Delta x$$

Now solve Eq. 21-1 for $\Delta x$ and substitute into the above equation:

$$\Delta S = \frac{-c}{2\pi k_0} E \Delta E \qquad (21\text{-}2)$$

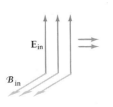 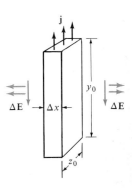

**Figure 21-1.** *Incoming wave $\mathbf{E}_{in}$ traveling to the right. When the wave hits the slab, the induced current in the slab radiates its own field $\Delta \mathbf{E}$ to the right and to the left.*

Now we take a stack of such thin slabs. Each $\Delta E$ radiated to the right is always in the opposite direction to the net field that produced it and thus reduces the net field by an amount $\Delta E$. Each $\Delta E$ radiated to the left is a weak reflected wave. If the conductivity $\sigma$ is small enough so that $E$ is hardly reduced after traveling one wavelength into the conductor, these reflected waves will all cancel each other out because for any one reflected wave there will be a partner produced further along, which, by the time it gets back to its mate, will be $\lambda/2$ out of phase with it. (The crest of one wave lines up with the trough of its partner and the sum of the two amplitudes is zero everywhere.) The entire system of reflected waves cancel themselves out. We now have the condition that the conductor not be too good. Graphite or an ionized gas will do. A good conductor such as silver will be discussed later. If the stack of slabs is infinitely thick, the entire field $E_{\text{in}}$ will be absorbed and the total power per unit area is easily obtained by integrating Eq. 21-2.

$$S = -\frac{c}{2\pi k_0} \int_{E_{\text{in}}}^{0} E\, dE = \frac{c}{4\pi k_0} E_{\text{in}}^2$$

$$= \frac{c^2}{4\pi k_0} E_{\text{in}} \mathcal{B}_{\text{in}}$$

The power radiated per unit area is given a special name called the Poynting vector and the special symbol $\mathbf{S}$. Since the direction of energy flow is given by the vector cross product $\mathbf{E} \times \mathcal{B}$, the vector $\mathbf{S}$ is

$$\mathbf{S} = \frac{c^2}{4\pi k_0} \mathbf{E} \times \mathcal{B} \quad \text{(Poynting vector)} \quad (21\text{-}3)$$

**Example 1.** Consider the same 3-W square flashlight of Example 2 in Chapter 20, which gives a square beam of 10 cm by 10 cm. (a) What is the magnitude of $\bar{S}$? (b) Use Eq. 21-3 and $\mathcal{B} = E/c$ to obtain $E_0$.

ANSWER: Since the magnitude of $\bar{S}$ is the power per unit area,

$$\bar{S} = \frac{3 \text{ W}}{(0.1 \text{ m})^2} = 300 \text{ W/m}^2$$

In Eq. 21-3 we replace $\mathcal{B}$ with $E/c$:

$$\bar{S} = \frac{c^2}{4\pi k_0} E\left(\frac{E}{c}\right) = \frac{c}{4\pi k_0} \overline{E^2}$$

Since $\overline{E^2} = \frac{1}{2}E_0^2$, we have

$$\overline{S} = \frac{c}{8\pi k_0}E_0^2$$

$$E_0 = \sqrt{\frac{8\pi k_0}{c}\overline{S}} = 475 \text{ V/m}$$

This is the same result as that of Example 2 in Chapter 20.

Finally, we wish to see if our result for $S$ is consistent with our earlier expressions for the field energy per unit volume. Consider a plane wave passing through an area $A$. From the definition of $S$ the energy flow in a time $dt$ will be

$$dU = S \cdot A\, dt$$

where $dU$ is the energy contained in a volume $dV = A\, dx$

$$dU = S \cdot A \frac{dx}{c}$$

$$= \frac{S}{c} dV$$

or

$$\frac{dU}{dV} = \frac{S}{c}$$

Using Eq. 21-3:

$$\frac{dU}{dV} = \frac{1}{c}\left(\frac{c^2}{4\pi k_0}E\mathcal{B}\right)$$

Now replace $E$ with $(c\mathcal{B})$:

$$\frac{dU}{dV} = \frac{c^2\mathcal{B}^2}{8\pi k_0} + \frac{(c\mathcal{B})^2}{8\pi k_0}$$

Now replace $(c\mathcal{B})$ with $E$:

$$\frac{dU}{dV} = \frac{E^2}{8\pi k_0} + \frac{c^2\mathcal{B}^2}{8\pi k_0}$$

The two terms on the right are indeed the electric and magnetic field energies per unit volume. We see by the preceding demonstration that the amount of heating in our absorbing slab is numerically equal to our earlier expressions for energy in the electric and magnetic fields. Our condition that the reflected wave cancel itself out is supported by the fact that graphite appears black (it does not reflect the electromagnetic radiation striking it). Our condition that the transmitted wave be absorbed is supported by the fact that a thick enough layer of graphite is opaque.

## 21-2 Momentum of the Radiation Field

We shall now show that not only does the plane wave in Figure 21-1 deliver energy to the slab of thickness $\Delta x$, but it also delivers momentum to it. Let us consider the rectangular piece of an "infinite" slab with area $y_0 z_0$ as shown in Figure 21-2. Since $jE\,dt$ is the ohmic heating per unit volume in a time $dt$, the energy delivered to this piece of the slab in a time $dt$ is $jE\,dt$ times the volume $y_0 z_0 \Delta x$:

$$dU = (jE\,dt)(y_0 z_0 \Delta x)$$

Now replace $E$ with $c\mathcal{B}$:

$$dU = cjz_0 \Delta x\, y_0 \mathcal{B}\, dt$$

The current flowing in this piece of slab is

$$I = j(z_0 \Delta x)$$

hence

$$dU = cI y_0 \mathcal{B}\, dt$$

Since we have a current element of length $y_0$ perpendicular to the incoming magnetic field, there will be a magnetic force $\mathbf{F}_m = I\mathbf{y}_0 \times \mathcal{B}$ on it in the direction of $\mathbf{E} \times \mathcal{B}$ which is the direction of the incoming wave. Replacing $(Iy_0 \mathcal{B})$ with $F_m$ gives

$$dU = cF_m\, dt$$

The momentum given to the piece of slab is $dp = F_m\, dt$ or

$$dU = c\, dp$$

$$dp = \frac{1}{c} dU \qquad (21\text{-}4)$$

As before, we could integrate over $x$ for a thick slab and would obtain the

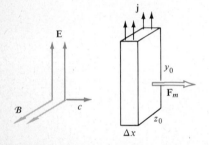

**Figure 21-2.** *When the incoming wave strikes the slab, there is an induced current $I = jz_0 \Delta x$. Magnetic force on this current is $\mathbf{F}_m = I\mathbf{y}_0 \times \mathcal{B}$.*

result $p = U/c$. The momentum given to the slab will be $1/c$ times the energy dissipated in the slab.

Not only is there energy contained in the radiation field, but there is also momentum. Any volume element $dV$ of the radiation field will have energy content

$$dU = \frac{E^2}{4\pi k_0} dV$$

and it will have $1/c$ times as much momentum; that is, the vector momentum contained in the volume $dV$ is

$$d\mathbf{p} = \frac{1}{c}\left(\frac{\mathbf{S}}{c} dV\right) \quad \left(\text{using the relation } dU = \frac{S}{c} dV\right)$$

We see that the radiation field that is emitted by a source of accelerating charge has a physical reality of its own. It has at every point in space both energy and momentum, which can be measured. You can measure the energy by putting your hand in a beam of light. Most of the light is absorbed by the hand and converted into heat. Because of the smallness of $1/c$, it is difficult to measure the momentum of a beam of light (see Figure 21-3).

**Figure 21-3.** A radiometer. Light bouncing back from the silvered side of each vane transfers twice as much momentum as light being absorbed on the black-coated side. Hence the vanes in the figure should rotate clockwise as seen from above. However, in actual practice, they rotate the other way! This is because of a stronger physical effect—the residual gas near the black surface is heated to a higher pressure. If the bulb is evacuated further to produce a better vacuum, the direction of rotation will be reversed.

**Example 2.** Suppose 100 W from a bright lamp is concentrated onto the reflecting vane of a radiometer. With what force will the light push the vane?

ANSWER: According to Eq. 21-4 the momentum in the beam is

$$dp = \frac{1}{c} dU$$

Because the beam reverses its direction, the momentum transferred to the vane is twice this:

$$dp_{\text{vane}} = \frac{2}{c} dU$$

The force is

$$F = \frac{dp_{\text{vane}}}{dt} = \frac{2}{c}\frac{dU}{dt} = \frac{2}{3 \times 10^8}(100) \text{ N} = 6.67 \times 10^{-7} \text{ N}$$

Even with a small amount of friction, it is difficult for such a small force to move the vane.

This physical reality of the field will become even more meaningful when we study quantum theory and photons in Chapter 24. Then we will see that the radiation field is made up of physical particles called photons, which are just as real as electrons or protons. Each photon has an energy $U = hf$ and a momentum $p = hf/c$, where $f$ is the frequency of the wave and $h$ is a very small number called Planck's constant.

## 21-3 Reflection by a Good Conductor

As was mentioned previously, if the conductivity $\sigma$ is too large, not all of the electromagnetic wave will be absorbed because some of it will be reflected. This is easily seen by taking the limiting case of $\sigma = \infty$ (a superconductor). We recall that the electric field inside a superconductor must always be zero (otherwise there would be an infinite current). Hence the induced surface current will adjust itself to give a radiation field $\Delta E = -E_{in}$. Then the net field inside the slab will be $E = E_{in} + \Delta E = 0$. The net field to the left of the slab will not be zero everywhere because we have the situation of two equal strength sine waves traveling in opposite directions. In such a situation there will be a standing wave with nodes and antinodes to the left of the slab in Figure 21-4. These standing waves are discussed on page 473.

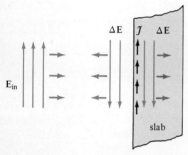

**Figure 21-4.** *Incoming wave striking a superconducting slab. The induced surface current $\mathcal{J}$ adjusts itself to give a radiated $\Delta E$ exactly equal in magnitude to $E_{in}$.*

**Example 3.** In Figure 21-4, what will be the magnitude of the induced surface current and what will be the magnitudes and directions of $\Delta \mathcal{B}$, the magnetic field produced by the surface current?

ANSWER: According to Eq. 20-7:

$$\Delta E = c\,\Delta \mathcal{B} = \frac{2\pi k_0}{c}\mathcal{J}$$

$$\mathcal{J} = \frac{c}{2\pi k_0}\Delta E = \frac{c}{2\pi k_0}E_{in}$$

This is the induced surface current in terms of the incoming electric field. The magnetic field produced by $\mathcal{J}$ is obtained by solving the first equation for $\Delta \mathcal{B}$:

$$\Delta \mathcal{B} = \frac{2\pi k_0}{c^2}\mathcal{J} = \frac{E_{in}}{c}$$

The direction of $\Delta \mathcal{B}$ near $\mathcal{J}$ is given by the right-hand rule and is illustrated in Figure 20-5. To the right of $\mathcal{J}$ it is pointing into the page. To the left of $\mathcal{J}$ it is pointing out of the page. Note that in both cases the vector product $\Delta \mathbf{E} \times \Delta \mathbf{\mathcal{B}}$, which is along the direction of propagation, is pointing away from $\mathcal{J}$. One final observation: the magnitude of $\mathcal{J}$ is the same as that of the original source which produced $E_{in}$.

We have just completed our discussion of the interaction of radiation with matter that contains free electrons. Such interaction results in electron currents with the effects discussed above. Now we shall turn to materials that have no free electrons. These materials are called nonconductors or insulators.

## 21-4 Interaction of Radiation with a Nonconductor

The atoms of nonconductors have outer electrons that are loosely bound and will experience some displacement when under the influence of an external field. We shall now try to estimate the magnitude of such a displacement.

We will take an atomic model close to that of modern quantum theory where an outer electron is represented as a spherical cloud of radius $R$. We make the approximation that the charge density is uniform over the sphere. In Eq. 16-3 we saw that if such a sphere is displaced by a distance $y$ from the atomic center, there will be a restoring force proportional to $y$ and the electron cloud will oscillate in simple harmonic motion about the atomic center (see Eq. 16-3a). The atomic center consists of the atomic nucleus shielded by the inner electron clouds that are tightly bound to it. According to Eq. 16-3a, the force on the outer electron cloud is

$$F_{atomic} = -m\omega_0^2 y \qquad (21\text{-}5)$$

where $\omega_0/(2\pi)$ is the frequency of oscillation.

If an incoming wave of field strength $E_{in}$ strikes the electron cloud, the net force on the atomic electron will be

$$F_{net} = F_{atomic} + (-e)E_{in}$$

$$m\frac{d^2y}{dt^2} = -m\omega_0^2 y - eE_{in}$$

We have used $y$ for the displacement rather than $x$ because we will be considering $\mathbf{E}_{in}$ pointing in the $y$ direction only. Our usual incoming wave is $E_{in} = E_0 \cos \omega(t - x/c)$, where $x$ is the distance from the source. Then the preceding equation becomes

$$\frac{d^2y}{dt^2} = -\omega_0^2 y - \frac{eE_0}{m} \cos \omega\left(t - \frac{x}{c}\right)$$

where $\omega_0$ is the natural frequency of oscillation of the atomic electron. The solution to the preceding differential equation is

$$y = -\frac{eE_0}{m(\omega_0^2 - \omega^2)} \cos \omega \left(t - \frac{x}{c}\right) \qquad (21\text{-}6)$$

This solution can be checked by substitution into the differential equation.

So far we have seen how a single atom interacts with radiation. Now we shall put many such atoms together into a slab of a solid or gas.

## 21-5 The Origin of the Index of Refraction

Consider now the incoming plane wave striking a slab of thickness $\Delta x$ of the above atoms. Let us try to guess what will happen. As seen in Eq. 21-6 the incoming electric field $E_{\text{in}}$ will make the atomic electrons oscillate sinusoidally. Any oscillating electron must radiate an electromagnetic wave of its own. So, as in the case of the conducting slab, there will be a reflected and a transmitted wave, but in this case there will be no ohmic energy loss to the slab. All the energy remains in the form of electromagnetic waves; thus the slab will be transparent rather than opaque. Furthermore, if we draw upon a famous result in optics, the electromagnetic wave (or light wave) will travel with a wave velocity $u < c$ while inside the slab. The ratio $c/u = n$ is defined as the index of refraction. Most solids have an index of refraction of approximately 1.5; that is, the speed of light is slowed down by $\sim 33\%$. The index of refraction of a few common materials is shown in Table 21-1.

How can it be that electromagnetic waves can travel with $u < c$? Equations 20-5 and 20-7 say that the $E$ and $\mathcal{B}$ fields must obey the wave equation with a wave velocity equal to $c$. From a particle point of view this is certainly true—the radiated field from an accelerating charged particle moves out with $u = c$. The paradox is resolved by noting that the field

Table 21-1. *Some indices of refraction (for $\lambda = 5.9 \times 10^{-7}$ m—sodium yellow light)*

| Medium | Index of Refraction | Medium | Index of Refraction |
|---|---|---|---|
| air | 1.0003 | glass, zinc crown | 1.52 |
| carbon disulfide | 1.63 | polyethylene | 1.52 |
| diamond | 2.42 | quartz, fused | 1.46 |
| glass, heaviest flint | 1.89 | sapphire | 1.77 |
| glass, light barium flint | 1.58 | sodium chloride | 1.53 |
| | | water | 1.33 |

**Figure 21-5.** *Incoming wave $E_{in}$ strikes slab and induces a current density j, which contributes to the outgoing wave E'.*

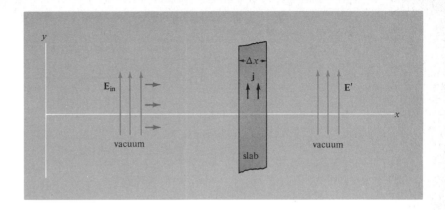

inside the slab is not the radiated field of a single particle (or of the original current sheet that generated the incoming wave) but the sum of the incident field plus the radiated fields of all the atomic electrons. Each separate field must move with $u = c$, but perhaps the resultant field can behave as if it is propagating more slowly. We will show, using Eq. 21-6, that the radiated field of each atomic electron lags 90° in phase behind the incident field which moves the atomic electrons. Then one might expect the phase of the resultant outgoing wave to lag behind the phase of the incoming wave taken by itself. The resultant wave front appears to travel slower than $u = c$, even though each individual wave by itself is traveling with $u = c$.

We shall make a quantitative derivation of the index of refraction of the slab shown in Figure 21-5. Our derivation can be broken into five parts:

1. Specify the electric field of the incident wave.
2. Calculate the velocity of the atomic electrons in the slab due to the incoming electric field.
3. From this velocity (or electron current density) calculate the secondary radiation produced by these electrons.
4. Add the incident and secondary waves to get the resultant outgoing wave.
5. Relate the phase of the outgoing wave to the index of refraction.

## 1 The Incident Wave

According to Eq. 20-7 the incident wave is

$$E_{in} = E_0 \cos \omega \left(t - \frac{x}{c}\right)$$

## 2 Velocity of the Electrons

Differentiation of Eq. 21-6 with respect to time yields

$$v_y = \frac{dy}{dt} = \frac{eE_0\omega}{m(\omega_0^2 - \omega^2)} \sin \omega \left(t - \frac{x}{c}\right)$$

for the velocity of the loosely bound outer electrons, where $\omega_0$ is their natural frequency of oscillation. The current density in the slab is $j = \mathcal{N}(-e)v_y$, where $\mathcal{N}$ is the number of oscillating atomic electrons per unit volume. Using the preceding result we have

$$j = -\frac{\mathcal{N}e^2\omega E_0}{m(\omega_0^2 - \omega^2)} \sin \omega \left(t - \frac{x}{c}\right)$$

## 3 Radiation Produced by the Atomic Electrons

According to Eq. 20-7, the radiation field at the slab produced by the electrons in the slab is

$$\Delta E = -\frac{2\pi}{c} k_0 j \, \Delta x$$

Minus sign indicates $j$ and its radiated field in opposite directions.

$$\Delta E = -\frac{2\pi k_0}{c} \left[ -\frac{\mathcal{N}e^2\omega E_0}{m(\omega_0^2 - \omega^2)} \sin \omega \left(t - \frac{x}{c}\right) \right] \Delta x$$

$$\Delta E = \Delta E_0 \cos \left(\omega t - kx - \frac{\pi}{2}\right) \tag{21-7}$$

where

$$\Delta E_0 = \frac{2\pi k_0 \mathcal{N} e^2 \omega}{cm(\omega_0^2 - \omega^2)} E_0 \, \Delta x \tag{21-8}$$

## 4 The Resultant Outgoing Wave

The resultant outgoing electric field is the sum of the original field plus the additional field produced by the atomic electrons:

$$E' = E_{\text{in}} + \Delta E$$

Using Eq. 21-7 for $\Delta E$ gives

$$E' = E_0 \cos\theta + \Delta E_0 \cos\left(\theta - \frac{\pi}{2}\right) \quad \text{where } \theta \equiv \omega\left(t - \frac{x}{c}\right).$$

Although $\theta$ is increasing uniformly with time, these two sine waves are "locked together" with a fixed phase difference of $\pi/2$ radians. The two sine waves can easily be added together using the method of phasors described in Appendix 21-1. At a given time $t$ the first sine wave is the $x$ component of the vector $\mathbf{E}_0$ in Figure 21-6. The second sine wave is the $x$ component of the vector labeled $\Delta\mathbf{E}_0$, which is at an angle $-\pi/2$ to the first vector. We see from Figure 21-6a that the resultant vector labeled $E_0'$ lags the incoming sine wave by an angle

$$\phi = \frac{\Delta E_0}{E_0}$$

where we have used the small angle approximation, assuming $\Delta E_0/E_0 \ll 1$.

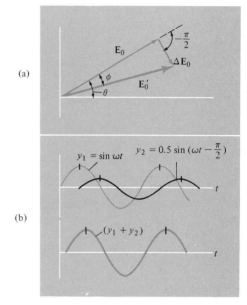

Figure 21-6. (a) Phasor diagram for adding two sine waves $E_0 \cos\theta$ and $\Delta E_0 \cos(\theta - \pi/2)$. (b) Direct addition of two sine waves that are 90° out of phase.

## 5 Relation Between Phase Lag and Index of Refraction

The incoming wave would take a time $t = \Delta x/c$ to go through the slab, whereas a wave of velocity $u = c/n$ would take a longer time $t' = n(\Delta x/c)$. The resultant wavefront would be slowed down by the time difference $\Delta t = (n-1)\Delta x/c$, when passing through the slab. This is a phase lag of

$$\phi = \omega \Delta t = \omega\left[(n-1)\frac{\Delta x}{c}\right]$$

Since $\phi = \Delta E_0/E_0$, we have

$$\omega(n-1)\frac{\Delta x}{c} = \frac{\Delta E_0}{E_0}$$

Now substitute Eq. 21-8 for $\Delta E_0$ and solve for $n$:

$$n = 1 + \frac{2\pi k_0 \mathcal{N} e^2}{m(\omega_0^2 - \omega^2)} \quad \text{(index of refraction)} \quad (21\text{-}9)$$

This is the index of refraction for a slab of thickness $\Delta x$. Note that $\Delta x$ has canceled out. One might expect this since a thick slab can be thought of as a series of thin slabs just repeating the same process over again. One warning: we have been using the approximation that the modification to the incoming field is small; that is, $(n - 1) \ll 1$. For large $n$, one should use the resultant field in place of $E_{\text{in}}$ inside the slab. Since this makes the calculation more complicated, we will not do it here.

### Dispersion

Our result does give the correct dependence on $\omega$, the frequency of the incoming light (see Figure 21-7). We saw in Example 1 of Chapter 16 that for typical atoms $\omega_0 > \omega$, when $\omega$ is in the region of the visible spectrum. This gives an index of refraction greater than 1 or a wave velocity less than $c$. Also it gives what is called normal dispersion: as the frequency of light goes from red to violet, the index of refraction increases and so does the amount of bending of light in a prism. Hence a prism can convert a beam of white light into a spectrum of colors.

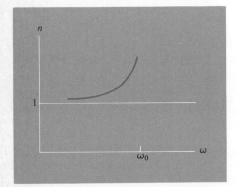

**Figure 21-7.** *A plot of Eq. 21-9: the normal dispersion curve.*

**Example 4.** Use Eq. 21-9 and Table 21-1 to obtain the index of refraction of air for ultraviolet light having $\lambda = 3.0 \times 10^{-7}$ m. Assume $\omega_0 = 7.5 \times 10^{15}$ s$^{-1}$.

ANSWER: Eq. 21-9 can be written in the form

$$\frac{n_2 - 1}{n_1 - 1} = \frac{\omega_0^2 - \omega_1^2}{\omega_0^2 - \omega_2^2}$$

From Table 21-1, $n_1 = 1.00030$ when $\omega_1 = 2\pi c/\lambda_1 = 3.19 \times 10^{15}$ s$^{-1}$.

$$\omega_2 = \frac{2\pi c}{\lambda_2} = 6.28 \times 10^{-15} \text{ s}^{-1} \text{ for } \lambda_2 = 3.0 \times 10^{-7} \text{ m}$$

$$\frac{n_2 - 1}{1.00030 - 1} = \frac{(7.5 \times 10^{15})^2 - (3.19 \times 10^{15})^2}{(7.5 \times 10^{15})^2 - (6.28 \times 10^{15})^2} = 2.74$$

$$n_2 = 1.00082$$

Starting with Maxwell's equations and simple atomic models of conductors and nonconductors we have been able to calculate many of the common properties of light and electromagnetic radiation observed in nature.

## 21-6 Electromagnetic Radiation in an Ionized Medium

In a plasma or ionized gas the electrons are no longer bound and $\omega_0 = 0$. Setting $\omega_0 = 0$ in Eq. 21-9 gives the result

$$n = 1 - \frac{2\pi k_0 \mathcal{N} e^2}{m\omega^2}$$

In order for this result to be correct, the electron mean free path must be greater than $y_{max} = eE_0/(m\omega^2)$, which is the case for radio waves passing through the ionosphere and outer space. A careful reader at this point should be horrified to see $n < 1$, which means the wave velocity is greater than $c$. But it is well known that no signals or particles can travel faster than the speed of light. Actually any individual particle of light (a photon) always travels with $u = c$. What we are discussing here are the resultant fields obtained by adding together the effects of many photons being radiated by many electrons. We must wait until Chapter 24 for a proper discussion of photons.

### Modulation and Group Velocity

A reader determined to overthrow Einstein might say why not increase and decrease the amplitude of an electromagnetic wave passing through an ionized gas. These modulations of the electromagnetic wave could contain coded information (Morse code for example) and these signals would be transmitted faster than the speed of light. Our answer to this is that the modulated sine wave is not a pure sine wave; it can be Fourier analysed into a group of pure sine waves, all with somewhat different frequencies. Each pure sine wave is traveling with $u > c$; however, it is shown in Appendix 21-2 that in such a case the modulation pattern travels with what is called the group velocity $v_g$, which may be quite different from the velocities of the component pure sine waves and in this case $v_g < c$. In this appendix we derive the formula

$$v_g = \frac{d\omega}{dk} \qquad \text{where } k = \frac{2\pi}{\lambda}$$

*Example 5. What is the group velocity in normal nonconductors corresponding to Eq. 21-9? Is it always less than $c$?

ANSWER: We must express $k$ as a function of $\omega$ and then differentiate:

$$\frac{\omega}{k} = u = \frac{c}{n}$$

$$k = \frac{\omega}{c} n$$

Equation 21-9 is of the form

$$n = 1 + \frac{A}{\omega_0^2 - \omega^2} \qquad \text{where } A \equiv \frac{2\pi k_0 \mathcal{N} e^2}{m}$$

$$k = \frac{\omega}{c}\left(1 + \frac{A}{\omega_0^2 - \omega^2}\right)$$

$$\frac{dk}{d\omega} = \frac{1}{c} + \frac{A}{c} \frac{\omega_0^2 + \omega^2}{(\omega_0^2 - \omega^2)^2}$$

Using

$$\frac{dk}{d\omega} = \frac{1}{v_g}$$

gives

$$\frac{1}{v_g} = \frac{1}{c}\left[1 + A\frac{\omega_0^2 + \omega^2}{(\omega_0^2 - \omega^2)^2}\right] \qquad (21\text{-}10)$$

Hence $v_g < c$.

Eq. 21-10 tells us that $v_g$ is always less than $c$ whether $\omega_0 > \omega$ or $\omega_0 < \omega$. This also applies to the case when $\omega_0 = 0$; that is, radio waves in the ionosphere. We have just shown that $v_g$ is less than the speed of light even when the wave velocity is greater than the speed of light.

*Example 6. A certain pulsar is observed to emit a pulse of radio waves every 3.6 seconds. It can be detected only by radio telescopes. When the radio receiver tuning is changed from 150 MHz to 240 MHz, the pulse arrives 1.3 seconds earlier. Assuming $\mathcal{N} = 0.03$ electron/cm³ in interstellar space, what is the distance of this pulsar?

ANSWER: Let $D$ be the distance to the pulsar. The time difference is

$$\Delta t = \frac{D}{v_{g_2}} - \frac{D}{v_{g_1}}$$

$1/v_g$ is obtained by setting $\omega_0 = 0$ in Eq. 21-10. Then

$$\frac{1}{v_g} = \frac{1}{c}\left[1 + \frac{A}{\omega^2}\right]$$

$$\Delta t = \frac{DA}{c}\left(\frac{1}{\omega_2^2} - \frac{1}{\omega_1^2}\right)$$

$$\frac{D}{c} = \frac{m\,\Delta t\,4\pi^2}{2\pi k_0 \mathcal{N} e^2}\left(\frac{1}{f_2^2} - \frac{1}{f_1^2}\right)^{-1}$$

$$= \frac{2\pi(9.1 \times 10^{-31}\text{ kg})(1.3\text{ s})}{(9 \times 10^9\text{ N m}^2/\text{C}^2)(.03 \times 10^6\text{ m}^{-3})(1.6 \times 10^{-19}\text{ C})^2}$$

$$\times \left(\frac{1}{(150 \times 10^6)^2\text{ s}^{-2}} - \frac{1}{(240 \times 10^6)^2\text{ s}^{-2}}\right)^{-1}$$

$$= 3.97 \times 10^{10}\text{ s} = 1260\text{ yr}$$

Thus it takes light 1260 yr to travel from this pulsar to the earth. It is at a distance of 1260 light years.

Example 6 illustrates a new, independent method to determine astronomical distances, but it only works for pulsars (rotating neutron stars). Fortunately, one of these can be seen visually as well (see Figure 30-8). This provides astronomers with an opportunity to check out and calibrate this new method against the old.

## 21-7 Radiation by Point Charges

So far we have derived the radiation field for a large number of charges oscillating together in a current sheet. Assume there are $\mathcal{N}$ of them per unit volume. If each charge $q$ is oscillating according to $y = y_0 \sin \omega t$, the current density in the sheet is $j = \mathcal{N} q \omega y_0 \cos \omega t$ and the surface current for a sheet of thickness $\Delta x$ is

$$\mathcal{J} = j\,\Delta x = (\mathcal{N} q \omega y_0 \Delta x) \cos \omega t$$

Then according to Eq. 20-7 the radiated field is

$$E_y = \frac{2\pi k_0}{c} \mathcal{N} q\,\Delta x\,\omega y_0 \cos(\omega t - kx) \qquad (21\text{-}11)$$

But suppose, instead of an entire sheet of charges, we had just a single charge $q$ oscillating up and down with $y = y_0 \sin \omega t$. It can be shown using

Maxwell's equations that the radiated field at a distance $r$ from $q$ is

$$E = k_0 \frac{q\omega^2 y_0}{c^2 r} \sin \omega \left(t - \frac{r}{c}\right) \sin \theta \qquad (21\text{-}12)$$

where $\theta$ is the angle between the acceleration direction and the vector $\mathbf{r}$ as shown in Figure 21-8. Noting that $(-\omega^2 y_0 \sin \omega t)$ is the acceleration, we have

$$E = -k_0 \frac{q}{c^2 r} a(t - r/c) \sin \theta \quad \text{(radiation by point charge)} \qquad (21\text{-}13)$$

where $a(t - r/c)$ is the value of the acceleration at the time $(t - r/c)$. The direction of $\mathbf{E}$ is perpendicular to $\mathbf{r}$. Not only does Eq. 21-13 hold for an oscillating charge, but for any moving charge as long as $v \ll c$.

The direction of $\mathcal{B}$ is perpendicular to both $\mathbf{E}$ and $\mathbf{r}$. As before $\mathcal{B} = E/c$ in magnitude. Note that in order to determine the field at a distance $r$ from the point charge, one must use the value of the acceleration at an earlier time $(t - r/c)$. This is designated by $a(t - r/c)$. If this were not the case, we could tell what the charge is doing by measuring the field at point $P$ at the same time, and this would violate the condition that signals cannot travel faster than $v = c$. We can only tell what the charge was doing at an earlier time $(t - r/c)$.

**Figure 21-8.** *Direction of radiation field* $\mathbf{E}$ *produced by point charge $q$ with acceleration* $\mathbf{a}$ *is as shown.*

---

*__Example 7.__ What is the power radiated by a charge $q$ whose acceleration is $a$?

ANSWER: The radiated power per unit area is given by the Poynting vector (Eq. 21-3). To get the total power we integrate this over the surface of a sphere.

$$S = \frac{c^2}{4\pi k_0} E\mathcal{B} = \frac{c^2}{4\pi k_0} E\left(\frac{E}{c}\right) = \frac{c}{4\pi k_0} E^2$$

$$= \frac{c}{4\pi k_0}\left(-\frac{k_0 q a}{c^2 r} \sin \theta\right)^2$$

$$= k_0 \frac{q^2 a^2}{4\pi c^3 r^2} \sin^2 \theta$$

$$\frac{dU}{dt} = P = \oint \mathbf{S} \cdot d\mathbf{A}$$

is the total power radiated. We use $dA = 2\pi r^2 \sin \theta \, d\theta$ for the element of area:

$$P = \oint \left(k_0 \frac{q^2 a^2}{4\pi c^3 r^2} \sin^2 \theta\right)(2\pi r^2 \sin \theta \, d\theta)$$

$$= k_0 \frac{q^2 a^2}{2c^3} \int_0^\pi \sin^3 \theta \, d\theta$$

$$P = \frac{2}{3} k_0 \frac{q^2}{c^3} a^2 \qquad (21\text{-}14)$$

*Example 8.* (a) What is the radiation field of an oscillating electric dipole of dipole strength $p = p_0 \sin \omega t$? (b) At what average rate does this dipole radiate energy?

ANSWER: We can represent the dipole by a negative charge $-q$ at the origin and a positive charge $q$ moving along the $z$ axis, where $z = z_0 \sin \omega t$. Then $p_0 = qz_0$ and the acceleration is

$$a = -\omega^2 z_0 \sin \omega t = -\frac{\omega^2 p_0}{q} \sin \omega t$$

The radiated field is obtained by substituting this into Eq. 21-13:

$$E = -k_0 \frac{q}{c^2 r} \left[-\frac{\omega^2 p_0}{q} \sin \omega \left(t - \frac{r}{c}\right)\right] \sin \theta \qquad (21\text{-}15)$$

$$= k_0 \frac{\omega^2 p_0 \sin \theta}{c^2 r} \sin \omega \left(t - \frac{r}{c}\right)$$

Note that the radiated dipole field decreases only as $1/r$ compared to the $1/r^3$ decrease for a static dipole field.

For part (b) we use Eq. 21-14:

$$\bar{P} = \frac{2}{3} k_0 \frac{q^2}{c^3} \overline{a^2}$$

where $\overline{a^2}$ is the average of $(\omega^4 p_0^2/q^2) \sin^2 \omega t$. Since the average of $\sin^2 \omega t$ is $\frac{1}{2}$, we have

$$\bar{P} = \frac{1}{3} k_0 \frac{p_0^2 \omega^4}{c^3} \qquad (21\text{-}16)$$

Note that for fixed amplitude of oscillation, the radiated power increases with the fourth power of the frequency.

**Example 9.** Electromagnetic radiation from a 500-kHz radio transmitter has a field strength $E_0 = 10$ mV/cm when it encounters a free electron. (a) What is the amplitude of oscillation of the electron? (b) What is the peak velocity of the electron? (c) What is the average power radiated by the electron?

ANSWER: (a) Putting $\omega_0 = 0$ into Eq. 21-6 gives the amplitude

$$y_0 = \frac{eE_0}{m\omega^2} = \frac{(1.6 \times 10^{-19} \text{ C})(1.0 \text{ V/m})}{(9.11 \times 10^{-31} \text{ kg})(2\pi \times 5 \times 10^5 \text{ s}^{-1})^2}$$

$$= 1.78 \times 10^{-2} \text{ m}$$

(b) In simple harmonic motion

$$v_0 = \omega y_0 = (2\pi \times 5 \times 10^{-5} \text{ s}^{-1})(1.78 \times 10^{-2} \text{ m}) = 5.58 \times 10^4 \text{ m/s}$$

(c) We can use Eq. 21-16 where the dipole strength is

$$p_0 = ey_0 = (1.6 \times 10^{-19} \text{ C})(1.78 \times 10^{-2} \text{ m}) = 2.85 \times 10^{-21} \text{ C m}$$

$$\bar{P} = \frac{1}{3}k_0 \frac{p_0^2 \omega^4}{c^3} = 8.79 \times 10^{-32} \text{ W}$$

## Atomic Transitions

When an atomic spectrum is viewed with an ideal spectrometer of infinite resolving power, the radiation is found to have certain discrete values or "lines." Each of these lines has a natural line width $\Delta f$.

We can use Eq. 21-16 to estimate the time duration of an excited atomic state which is related to the natural line width in the atomic spectrum. We shall use the approximation that the emitting electron is bound to the atomic center by a spring with a force constant $k = m\omega_0^2$. The frequency of the emitted radiation is then $f_0 = \omega_0/(2\pi)$ and the dipole strength is $p_0 = ex_0$. It is shown on page 223 that the total mechanical energy of the oscillator is $E = \frac{1}{2}kx_0^2 = \frac{1}{2}m\omega_0^2 x_0^2$, where $x_0$ is the amplitude of the oscillation. According to Eq. 21-16, the rate of energy loss is

$$\frac{dE}{dt} = \frac{1}{3}k_0(ex_0)^2 \frac{\omega_0^4}{c^3}$$

Then

$$\frac{dE}{E} = \frac{2}{3}k_0 \frac{e^2}{c^3} \frac{\omega_0^2}{m} dt$$

Integrating both sides gives

$$E = E_0 e^{-t/\tau} \quad \text{where } \tau = \frac{3mc^3}{2k_0 e^2 \omega_0^2}$$

The mean life $\tau$ of the radiation for yellow light ($f = 6 \times 10^{14}$ Hz) is then

$$\tau = \frac{3}{2} \frac{(9.11 \times 10^{-31})(3 \times 10^8)^3}{(9 \times 10^9)(1.6 \times 10^{-19})^2(2\pi)^2(6 \times 10^{14})^2} \, \text{s} = 1.13 \times 10^{-8} \, \text{s}$$

In spite of the fact that we have used a classical physics approximation, this result turns out to be close to the correct quantum mechanical result. The width of such a yellow spectral line can be obtained from Eq. 21-16 in Appendix 21-2.

$$\Delta \omega \approx \frac{1}{\tau} \quad \text{or} \quad \Delta f \approx \frac{1}{2\pi \tau}$$

The fractional line width is

$$\frac{\Delta f}{f_0} \approx \frac{1}{2\pi f_0 \tau} \approx 2.35 \times 10^{-8}$$

In the next example the Bohr model of the hydrogen atom is used in a similar type of calculation to estimate how long it would take the electron to radiate an amount of energy comparable to its binding energy.

**Example 10.** In the Bohr model of the hydrogen atom the electron is in a circular orbit of radius $R = 0.53 \times 10^{-10}$ m. At what rate is the electron radiating energy in electron volts per second? At this rate how long would it take to radiate an energy equal to 7 eV, which is about one half of its binding energy?

ANSWER: The electron acceleration is obtained using Coulomb's law:

$$ma = k_0 \frac{e^2}{R^2}$$

$$a = k_0 \frac{e^2}{mR^2}$$

The rate of energy loss is obtained by substituting this into Eq. 21-14:

$$\frac{dU}{dt} = \frac{2}{3}k_0\frac{e^2}{c^3}\left(k_0\frac{e^2}{mR^2}\right)^2 = \frac{2}{3}k_0^3\frac{e^6}{c^3m^2R^4}$$

$$= \frac{2}{3}(9\times 10^9)^3\frac{(1.6\times 10^{-19})^6}{(3\times 10^{-8})^3(9.11\times 10^{-31})^2(0.53\times 10^{-10})^4}\text{ J/s}$$

$$= 4.61\times 10^{-8}\text{ J/s}$$

$$= 2.88\times 10^{11}\text{ eV/s}$$

$$\Delta t = \frac{\Delta U}{dU/dt} = \frac{7}{2.88\times 10^{11}}\text{ s} = 2.43\times 10^{-11}\text{ s}$$

We see that according to classical physics an electron in a circular orbit about a proton will quickly radiate what to it is a large amount of energy. A more detailed calculation shows that it will spiral in and collide with the proton in $\sim 10^{-11}$ s (see Problem 21). The modern theory of quantum mechanics is free from this trouble because the electron cloud is stationary; it has zero acceleration.

## Summary

The energy density of an electromagnetic wave is

$$\frac{E^2 + (c\mathcal{B})^2}{8\pi k_0}$$

The energy flow per unit area per unit time is

$$\mathbf{S} = \frac{c^2}{4\pi k_0}\mathbf{E}\times\mathcal{B}$$

This is called the Poynting vector and is the power loss per unit area when a wave front hits an absorbing surface.

A portion of a traveling electromagnetic wave that contains energy $\Delta U$ also contains momentum $\Delta U/c$.

A plane electromagnetic wave striking the surface of a good conductor induces a surface current in the conductor that has the net effect of reflecting the incoming wave back on itself.

When an electromagnetic wave of frequency $\omega$ enters a nonconductor, the outer atomic electrons oscillate at the same frequency, but emit secondary electromagnetic waves that lag the primary wave by 90°. The overall effect is a wave velocity $u$ less than $c$. The index of refraction is defined as $n \equiv c/u$, and in such a medium (if not too dense) it has the value

$$n = 1 + \frac{2\pi k_0 \mathcal{N} e^2}{m(\omega_0^2 - \omega^2)}$$

where $\omega_0$ is the natural atomic vibrational frequency. The increase of $n$ with $\omega$ is called normal dispersion. A wave packet travels at an effective velocity

$$v_g = \frac{d\omega}{dk}$$

which is called the group velocity. Since $n$ is a function of $\omega$, $v_g$ is different from $u$ when in such a dispersive medium. It turns out to be less than $c$ even when $u$ is greater than $c$.

A single point charge $q$ with acceleration $a(t)$ radiates an electromagnetic wave of amplitude

$$E = -k_0 \frac{q}{c^2 r} a(t - r/c) \sin \theta$$

## Appendix 21-1  Method of Phasors

In this appendix we shall develop a method for the addition of two or more sine waves that are locked in phase with each other. In order to maintain a constant phase difference as a function of time, they must of course have the same frequency. So we wish to find the sum

$$S(t) = A_1 \cos(\omega t + \phi_1) + A_2 \cos(\omega t + \phi_2)$$

The result will also be a sine wave of the same frequency:

$$S(t) = S_0 \cos(\omega t + \phi_s)$$

So the problem is to find $S_0$ and $\phi_s$ in terms of $A_1$, $A_2$, $\phi_1$, and $\phi_2$.

This is easily done by noting that the $x$ component of the vector in Figure 21-9 is $A_1 \cos(\omega t + \phi_1)$, the first of our two sine waves. In this figure we have plotted a vector of length $A_1$ at an angle $\theta = (\omega t + \phi_1)$ to the $x$ axis. As $t$ increases, this vector will rotate counterclockwise, but its $x$ component is always the quantity we want. Now plot a second vector of

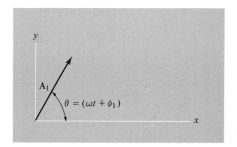

**Figure 21-9.** *The x component of rotating vector* $\mathbf{A}_1$ *is one of the two sine waves to be added.*

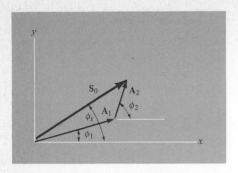

**Figure 21-10.** *The vector sum* $\mathbf{S} = \mathbf{A}_1 + \mathbf{A}_2$ *yields* $S_0$ *and* $\phi_s$.

length $A_2$ and at an angle $\theta_2 = \omega t + \phi_2$. It will always be at an angle $\phi_2 - \phi_1$ to the first vector. We are after the sum of the two $x$ components. Our problem is solved by noting that the sum of the $x$ components is the $x$ component of the vector sum. So the prescription will be to form the vector sum $(\mathbf{A}_1 + \mathbf{A}_2)$. The length of this vector will be the amplitude $S_0$ of the resultant sine wave. The procedure to use is shown in Figure 21-10. Plot vector $\mathbf{A}_1$ at an angle $\phi_1$ and vector $\mathbf{A}_2$ at angle $\phi_2$ to the $x$ axis. The length of their vector sum is $S_0$ and the angle between $\mathbf{S}$ and the $x$ axis is $\phi_s$. Such vectors are called phasors. If there are $n$ sine waves to be added, just take the vector sum of the $n$ phasors.

## Appendix 21-2  Wave Packets and Group Velocity

### Wave Packets

In the previous appendix we were adding sine waves of exactly the same frequency. In this appendix we shall add sine waves of different frequencies, but close to each other. In this case, as time goes on, the sine waves would get increasingly out of phase with each other. The simplest case would be the sum of two equal sine waves of frequencies $\omega_1$ and $\omega_2$, as shown in Figure 21-11. We will use the notation $\bar{\omega} \equiv (\omega_1 + \omega_2)/2$ and $\Delta\omega = (\omega_2 - \omega_1)/2$. The sum is

$$S(t) = \cos(\bar{\omega} + \Delta\omega)t + \cos(\bar{\omega} - \Delta\omega)t$$

This can be put in a more useful form by using the relation for the sum of two cosines from trigonometry:

$$\cos A + \cos B = 2\cos\left(\frac{A-B}{2}\right)\cos\left(\frac{A+B}{2}\right)$$

Hence

$$S(t) = 2\cos(\Delta\omega t)\cos\bar{\omega}t$$
$$= A(t)\cos\bar{\omega}t \quad \text{where } A(t) = 2\cos(\Delta\omega)t$$

is the modulation function or "envelope" (the white curve in Figure 21-11). In this case the modulation function happens to be a sine wave of lower frequency.

It is possible to obtain modulation functions of other shapes by adding

**Figure 21-11.** (a) *Two sine waves of slightly different frequency are in phase at the origin and successively get out of and in phase moving away from the origin.* (b) *The sum of the two sine waves.*

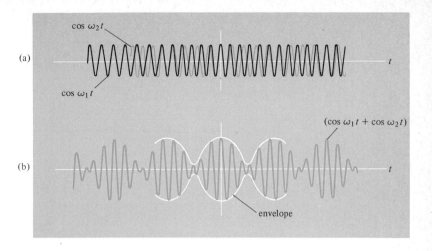

more sine waves of slightly different frequencies. An important example is that of a single pulse of oscillation. Such a single pulse would look like the central hump in Figure 21-11b and is called a wave packet. We shall now show how a wave packet can be built up out of a group of neighboring sine waves. Starting with the result in Figure 21-11b, we could "kill off" the neighboring bunches of oscillation by adding a third sine wave of frequency $\bar{\omega}$ with amplitude equal to the height of the bunches [note that $A(t)$ changes sign for alternate bunches]. This new sine wave will add to the central bunch, and will be 180° out of phase with its nearest neighbors as shown in Figure 21-12. The plot $G(\omega)$ shows the relative strengths of the three sine waves that are being added together. In order to kill off the next set of neighboring bunches, we could add the two sine waves $\cos(\bar{\omega} - \Delta\omega/2)t$ and $\cos(\bar{\omega} + \Delta\omega/2)t$ with appropriate amplitude as shown

**Figure 21-12.** (a) *The sum of three sine waves.* (b) *Plot of the relative amplitudes.*

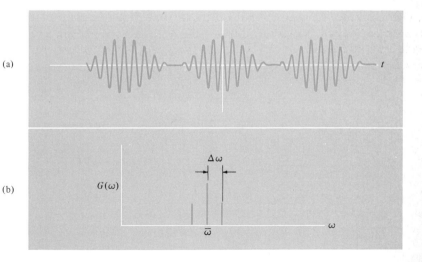

in Figure 21-13. These next two sine waves will hardly affect the central bunch, but their sum will be 180° out of phase with the next set of neighboring bunches.

In actual fact an infinite number of neighboring sine waves must be added together to form a single wave packet with no neighboring bunches. This situation is shown in Figure 21-14, where the function $G(\omega)$ displays the relative amplitudes of the component sine waves. The shape of $G(\omega)$ is called a gaussian function:

$$G(\omega) = \exp\left[-\frac{(\omega - \bar{\omega})^2}{2(\Delta\omega)^2}\right]$$

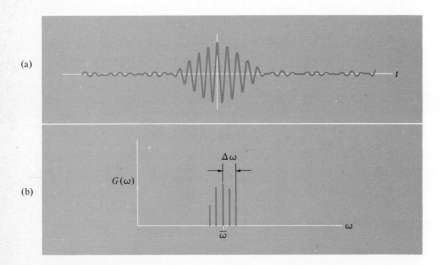

**Figure 21-13.** (a) *The sum of five sine waves.* (b) *Plot of the relative amplitudes.*

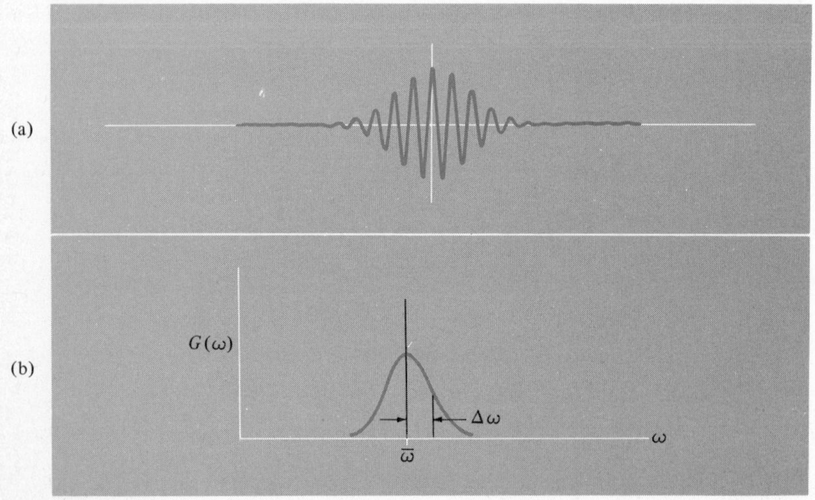

**Figure 21-14.** (a) *The sum of an infinite number of sine waves.* (b) *Plot of the relative amplitudes. $G(\omega)$ is a gaussian function with mean equal to $\bar{\omega}$ and standard deviation $\Delta\omega$.*

where $\Delta\omega$ is called the standard deviation. It is the root mean squared deviation of $\omega$ from $\bar{\omega}$. We shall call it the frequency spread. In order to add the "infinite" number of pure sine waves, we must calculate the integral

$$\int G(\omega) \cos \omega t \, d\omega$$

This can be done using a table of integrals with the result:

$$\int \exp\left[-\frac{(\omega - \bar{\omega})^2}{2(\Delta\omega)^2}\right] \cos \omega t \, d\omega = \exp\left[-\frac{t^2}{2(1/\Delta\omega)^2}\right] \cos \bar{\omega} t$$

Note that the right-hand side is a sine wave, $\cos \bar{\omega} t$, modulated by a gaussian envelope function, $\exp\{-t^2/[2(1/\Delta\omega)^2]\}$. The standard deviation of this gaussian is $\Delta t = 1/\Delta\omega$, which is called the "width" of the wave packet. We have

$$\Delta t = \frac{1}{\Delta\omega} \quad \text{(width of wave packet)} \quad (21\text{-}16)$$

We see that the frequency spread of the component sine waves is just the reciprocal of the "width" of the wave packet. The function $G(\omega)$ is called the Fourier transform of the wave packet.

## Group Velocity

We can also add sine waves of different frequencies that are traveling waves. Then we can get the interesting result that the velocity of the envelope can be quite different from the wave velocity of the component sine waves. The velocity of a wave packet as a whole can be quite different from the velocities of its component pure sine waves. The velocity of the wave packet or of the envelope function is called the group velocity. It is easy to see why the envelope can travel at a different speed by taking the two sine waves of Figure 21-11 as traveling waves that have close to the same wavelength as well as close to the same frequency. The sum of two such traveling waves is

$$y(x, t) = \cos[(\bar{\omega} + \Delta\omega)t - (\bar{k} + \Delta k)x] + \cos[(\bar{\omega} - \Delta\omega)t - (\bar{k} - \Delta k)x]$$

where $\bar{k} = 2\pi/\bar{\lambda}$ is the average wave number. Two such waves are plotted as a function of $x$ for four successively increasing values of $t$ in Figure 21-15. In this particular case the envelope function moves to the right twice as fast as the separate component sine waves. We can use the trigonometric relation for $\cos A + \cos B$ to obtain

$$y(x, t) = 2 \cos[(\Delta\omega)t - (\Delta k)x] \cos(\bar{\omega}t - \bar{k}x).$$

Now the envelope function is $A(x, t) = 2 \cos[(\Delta\omega)t - (\Delta k)x]$. This keeps its maximum value when

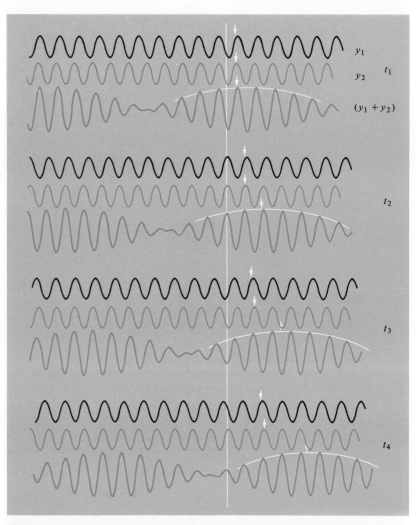

**Figure 21-15.** *Two sine waves, $y_1$ and $y_2$, are traveling to the right with slightly different velocities. In this case the envelope of their sum $(y_1 + y_2)$ travels to the right twice as fast. Four successive views are shown from top to bottom. Position of wave crests as a function of time are shown as white arrows.*

or

$$[(\Delta\omega)t - (\Delta k)x] = 0$$

$$\frac{x}{t} = \frac{\Delta\omega}{\Delta k}$$

is the velocity of the crest of the envelope. It is what we have defined as the group velocity.

If there is a group of neighboring pure sine waves where $\omega$ depends on $k$—that is, $\omega = \omega(k)$—the group velocity is

$$v_g = \frac{d\omega}{dk} \quad \text{(group velocity)} \quad (21\text{-}17)$$

The most common example is transmission of light in a nonconductor, which is worked out in Example 5. Another very important application is quantum mechanics where particles are represented as wave packets. The particle velocity is given by the group velocity and not by the velocity of the component waves, which is usually very different.

## Exercises

1. Express the magnitude of the Poynting vector in terms of $\mathcal{B}$, $k_0$, and $c$. Also express it in terms of $\varepsilon_0$, $\mu_0$, $E$, and $\mathcal{B}$.
2. A 1-ton spaceship is using a light beam for propulsion. If the beam power is 10 kW, what will be the increase in velocity after 1 day of operation? (Ignore the effect of any gravitational force.)
3. A possible method of spaceship propulsion is using radiation pressure from the sun for solar sailing. Assume the sail is made of aluminized Mylar with a density of 2 g/cm³. If the flux at the earth from the sun is 1.35 kW/m², how thick must the sail be in order for the radiation force to cancel out the gravitational force of the sun on the sail?
4. Sunlight striking the earth has $S = 1.35$ kW/m². What is the root mean squared value of $E$ and $\mathcal{B}$ in the sunlight?
5. The maximum magnetic field at a distance of 1 km from an oscillating dipole is $10^{-15}$ tesla.
   (a) What is the maximum value of the electric field?
   (b) What is the maximum value of the Poynting vector?
   (c) What is the average power output from the dipole?
6. Consider a free electron in the presence of an oscillating electric field of amplitude 0.1 V/m. What will be the amplitude of oscillation of the electron
   (a) for $f = 1$ kHz? (b) for $f = 100$ MHz?
7. In Exercise 6, calculate the corresponding electron velocities.

8. What is the angle $\phi$ in degrees in Figure 21-6(b)?
9. What is the index of refraction in air for infrared radiation having $\lambda = 2\ \mu m$? (See Example 4.)
10. Suppose light of two frequencies $f_1$ and $f_2$ strikes an air molecule causing it to oscillate with a dipole moment $p = p_0(\sin \omega_1 t + \sin \omega_2 t)$. If $f_2$ is blue light and $f_1$ is red light, what is the ratio of energy radiated in the blue to energy radiated in the red? Give a numerical answer. (Such a sky would appear blue.)

**Problems**

11. A 1-ton spaceship has a 100-m by 100-m aluminized Mylar sail that can be pointed in any direction. The spaceship is initially in a circular orbit of $10^5$ km radius. The sun's energy flux is 1.35 kW/m².
    (a) Make an approximate estimate of the energy increase that could be obtained using radiation pressure in one orbit of the spaceship.
    (b) Approximately how long would it take to get to the moon by solar sailing?
12. Assume that the weakly conducting slab in Figure 21-1 has resistivity $\rho$. How thick must it be so that $E_{\text{in}}$ is reduced by a factor of 2? Give answer in terms of $\rho$, $k_0$, and $c$. (By weakly conducting, we mean that the slab thickness is many wavelengths long.)
13. A 1-kW arc lamp has all its energy focussed in a beam of electromagnetic radiation of circular area 10-cm diameter.
    (a) If this beam strikes a mirror, with what force does the beam push the mirror?
    (b) What will be the induced surface current $\mathcal{J}$ in amperes per meter?
    (c) What is the energy density of the beam in joules per cubic meter?
14. A free particle of mass $m$ is oscillating in the $y$ direction under the influence of an external force, $F_0 \cos \omega t$.
    (a) What is $y$ as a function of $t$?
    (b) What is the maximum displacement as a function of $F_0$, $\omega$, and $m$?
15. The net force on a particle of mass $m$ is

$$F_{\text{net}} = -ky + F_{\text{ext}} \qquad \text{where } F_{\text{ext}} = F_0 \sin \omega t$$

   (a) What is $y$ as a function of $t$?
   (b) What is $y$ as a function of $t$ if $F_{\text{ext}} = F_0 \cos \omega t$?
   (c) What is the natural frequency of oscillation, $\omega_0$, in the absence of $F_{\text{ext}}$?
   (d) If the driving frequency $\omega$ is greater than $\omega_0$, is the external force in phase or out of phase with the displacement?
16. In Example 6 what would be the shift in arrival time when the receiver tuning is changed from 150 MHz to 160 MHz?
17. In Example 6 suppose the distance to the pulsar is known to be 1500 light years. What, then, must be the average density of electrons in interstellar space to give the observed shift in arrival times?

18. Suppose a conduction electron experiences a field $E = E_0 \cos \omega t$ where $E_0 = 100$ V/m and $f = 100$ Hz. What will be its amplitude of oscillation?

19. An incoming electromagnetic wave $E_{in}$ strikes a thin, poorly conducting sheet that radiates a field $\Delta E = -0.01 E_{in}$. Then the net transmitted wave will have amplitude $(1 - 0.01)E_{in}$ and the reflected wave will have amplitude $0.01 E_{in}$). Consider two such sheets spaced $\lambda/4$ apart. What will be the net transmitted wave and what will be the net reflected wave? Note that when the reflected wave from sheet 2 reaches sheet 1, it will be 180° out of phase with the reflected wave leaving sheet 1.

20. What is the velocity of the electron in Problem 18?

21. Consider the classical model of the hydrogen atom with the electron in a circular orbit of radius $R = 0.53 \times 10^{-10}$ m. The electron kinetic plus potential energy is

$$E = K + U = -k_0 \frac{e^2}{2R} = -13.6 \text{ eV}$$

(a) How much energy would it radiate in one turn? Give a numerical answer in electron volts.
(b) Show that $dE/dt = -\frac{2}{3}(k_0^3 e^6)/(m^2 c^3 R^4)$.
(c) Show that $dE/dR = (k_0 e^2)/(2R^2)$.
(d) Show that $dR/dt = -\frac{4}{3}(k_0^2 e^4)/(m^2 c^3 R^2)$.
(e) When the electron reaches $R = 10^{-15}$ m, it will crash into the proton. How long will this take: that is, what is the "lifetime" of a classical hydrogen atom? Use $t = \int_{R_1}^{R_2} (dt/dR)\, dR$.

22. In Example 6 what is $(c - v_g)$ for 150 MHz?

23. For electromagnetic radiation traveling through an ionized medium show that

$$u v_g = c^2 \left[ 1 - \left( \frac{2\pi k_0 \mathcal{N} e^2}{m \omega^2} \right)^2 \right]^{-1} \approx c^2$$

24. Use the method of phasors to show that
$\sin \omega t + \sin(\omega t + \phi) + \sin(\omega t + 2\phi) = (1 + 2 \cos \phi) \sin(\omega t + \phi)$

25. A quadrupole radiator can be represented as two dipoles $p$ a distance $z_0$ apart, where $z_0 \ll \lambda$. The two dipoles are 180° out of phase. If the upper dipole varies with time as $p = p_0 \cos \omega t$, the field radiated by it alone is

$$E_1 = k_0 \frac{p_0 \omega^2}{c^2 r} \sin \theta \cos \omega \left( t - \frac{r_1}{c} \right)$$

For the lower dipole

$$E_2 = k_0 \frac{p_0 \omega^2}{c^2 r} \sin \theta \cos \left[ \omega \left( t - \frac{r_2}{c} \right) - \pi \right]$$

What is the resultant radiated field from the two dipoles? Express the answer in terms of $\omega$, $c$, $r$, $\theta$, and the quadrupole moment $Q_0 \equiv p_0 z_0$. (Hint: $k(r_2 - r_1) \approx k z_0 \cos \theta$.)

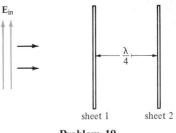

**Problem 19**

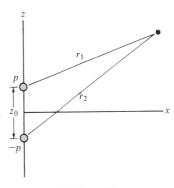

**Problem 25**

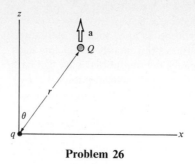

Problem 26

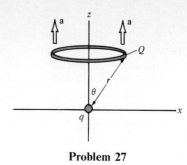

Problem 27

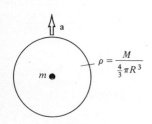

Problem 28

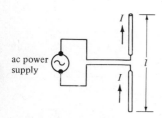

Problem 29

26. A charge $Q$ has constant acceleration $a$ pointing along the $z$ axis. In the unprimed frame it is at the position shown at $t = 0$. What is the magnitude and direction of the force on $q$ due to the radiation field of $Q$?

27. Repeat Problem 26 for the case where $Q$ is a ring of charge of radius $R = r \sin \theta$.

28. Repeat Problem 27 for a "solid" sphere of total charge $Q$ and radius $R$.

29. Repeat Problem 28 for a small mass $m$ and a sphere of total mass $M$. Assume there is a gravitational force proportional to $a/r$ for the same relativistic reason that there is an electric force. Show that the net force on $m$ is

$$F = ma\left[\frac{4\pi}{3} G\rho \frac{R^2}{c^2}\right]$$

(*Note:* In Chapter 30 we shall see that the factor $\frac{4}{3}\pi G\bar{\rho}(R^2/c^2)$ might be close to 1 for our universe of radius $R$ and average density $\bar{\rho}$. In that case we have derived $F = ma$! This explanation of inertial mass is called the strong version of Mach's principle.)

30. An ac power supply is connected to a dipole antenna of length $l$ where $l \ll \lambda$. The charge supplied by the ac generator (or oscillator) is $q = q_0 \cos \omega t$. (The peak current is then $I_0 = \omega q_0$.) Assume this charge runs back and forth from one end to the other of the dipole. Then the dipole strength is $p_0 = q_0 l$.

(a) Show that the radiated field is

$$E = -\frac{k_0 l \omega I_0}{c^2 r} \cos \omega \left(t - \frac{r}{c}\right) \sin \theta$$

(b) Using the relation $P = I^2 R$, show that the effective resistance seen by the power supply is $R = \frac{2}{3}(k_0/c)(kl)^2 = (kl)^2 0.20$ ohms. (For $l = \lambda/2$ the effective distance covered by $q_0$ is $\bar{l} = 0.61\lambda/2$. This gives the result $R = 72$ ohms for a half-wave dipole antenna.)

# Wave Interference

In the last chapter we studied the effects of a single source of electromagnetic waves whether it was an oscillating surface current or an oscillating point charge. In this chapter we shall study the effects of two or more sources of waves oscillating in phase with each other. The resultant wave amplitude is the sum of the separate amplitudes. The resultant wave exhibits what is called wave interference. The wave interference phenomena we shall study apply to all kinds of waves, not just electromagnetic waves. The mathematics will be the same because mechanical waves (such as the waves on a stretched string of Section 20-6) obey the same wave equation as do electromagnetic waves.

## 22-1 Standing Waves

When a traveling wave is totally reflected, the sum of the initial traveling wave plus the reflected wave gives what is called a standing wave. We saw on page 448 that an electromagnetic wave striking a perfectly reflecting surface induces a surface current $\mathcal{J}$, which radiates a field $E'$ that is always the negative of $\mathbf{E}_{in}$ just in front of the reflecting surface. In Figure 22-1 the reflected electric field is $\mathbf{E}'$ and the reflected magnetic field is $\mathcal{B}'$. Note that $(\mathbf{E}' \times \mathcal{B}')$, which is the direction of propagation, points in the negative $x$ direction. If the incoming wave is $E_{in} = E_0 \cos(\omega t - kx)$, then the radiated wave must be $E'_{left} = -E_0 \cos(\omega t + kx)$ to the left of the reflector and $E'_{rt} = -E_0 \cos(\omega t - kx)$ to the right of the reflector. This meets the condition that everywhere to the right of the reflector $E = E_{in} + E'_{rt} = 0$. To the left of the reflector the net electric field will be

$$E = E_{in} + E'_{left} = E_0[\cos(\omega t - kx) - \cos(\omega t + kx)]$$

Now we apply the trigonometric relation

$$\cos A - \cos B = 2 \sin \frac{B + A}{2} \sin \frac{B - A}{2}$$

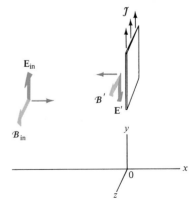

**Figure 22-1.** *Incoming wave* $\mathbf{E}_{in}$ *and* $\mathcal{B}_{in}$ *induces surface current* $\mathcal{J}$, *which radiates* $\mathbf{E}'$, *and* $\mathcal{B}'$, *propagating to the left.*

to the preceding equation; then

$$E = 2E_0 \sin \omega t \sin kx \qquad (22\text{-}1)$$
$$= A(t) \sin kx \qquad \text{where } A(t) \equiv 2E_0 \sin \omega t.$$

This is plotted for successive values of $t$ in Figure 22-2. The standing wave is a fixed function in $x$ (it is $\sin kx$) which increases and decreases sinusoidally with time.

Note that whenever $kx_n = n\pi$ where $n$ is a positive or negative integer we have a node $E = 0$ for all $t$. The nodes occur at

$$x_n = \frac{n\pi}{k} = \frac{n\pi}{(2\pi/\lambda)} = n\frac{\lambda}{2}$$

where $n$ is any integer. Any two successive nodes are a half wavelength apart.

---

**Example 1.** The microwave oscillator in Figure 22-3 emits plane electromagnetic waves to the right that are reflected back to the left. $P_1$ and $P_2$ are the positions of two successive intensity minima and are 5 cm apart. What is the frequency of the microwave oscillator?

ANSWER: Since any two successive nodes are a half wavelength apart, $\lambda = 10$ cm. The frequency is

$$f = \frac{c}{\lambda} = \frac{3.00 \times 10^8 \text{ m/s}}{0.1 \text{ m}} = 3.00 \times 10^9 \text{ Hz} = 3000 \text{ MHz}$$

---

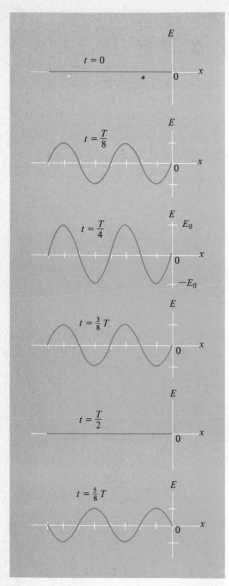

**Figure 22-2.** *Time sequence of views of Eq. 22-1. Physically it is a sine wave traveling to the right and being reflected at $x = 0$.*

### Standing Waves on a String

If the end of a stretched string is held rigid, it behaves as a perfect reflector. Just as with electromagnetic waves, a traveling wave of opposite sign is reflected. Then the displacement $y$ will always be zero at the end of the string. The function

$$y = y_0[\cos(\omega t - kx) - \cos(\omega t + kx)] = 2y_0 \sin \omega t \sin kx$$

is always zero at $x = 0$. Figure 22-4 shows standing waves on a string of length $L$. Since nodes are at both ends of the string we have $L = \tfrac{1}{2}\lambda$ in view (a), $L = 2(\lambda/2)$ in view (b), $L = 3(\lambda/2)$ in view (c), and $L = 4(\lambda/2)$

in view (d). Such pure standing waves are called resonances and obey the relation

$$n\frac{\lambda_n}{2} = L \quad \text{or} \quad \lambda_n = \frac{2L}{n} \quad \text{(standing waves)} \quad (22\text{-}2)$$

Standing waves on a plucked string are the basis of stringed musical instruments. Similar standing waves in columns of air are the basis of wind instruments.

**Figure 22-4.** *First four resonances of a vibrating string.* (a) *The string is one-half wavelength long.* (d) *The string is two wavelengths long.* [*Courtesy Educational Development Center*]

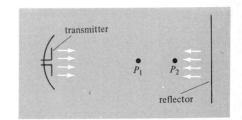

**Figure 22-3.** *Microwave oscillator emits electromagnetic waves to the right which are reflected back on themselves. Nodes are observed at $P_1$ and $P_2$.*

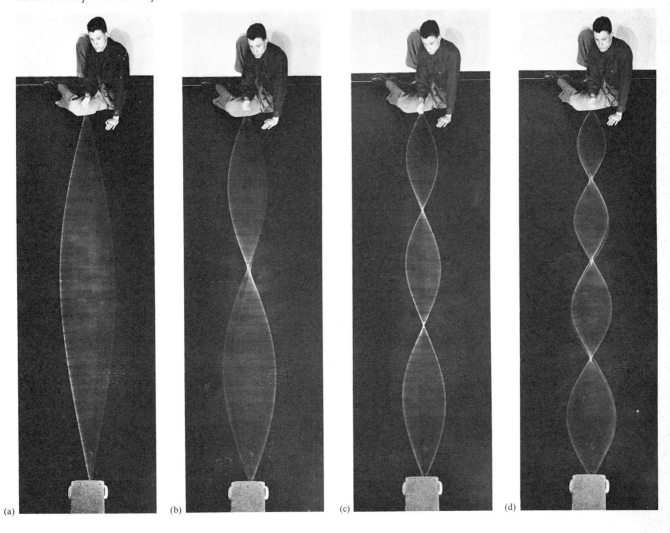

(a)  (b)  (c)  (d)

**Example 2.** A 12-cm vibrating string has nodes 4 cm apart. The wave velocity is $u = 30$ m/s. (a) What is the frequency of vibration? (b) List all possible lower resonant frequencies. (c) What is the tension on the string if its mass is 24 g?

ANSWER: (a) $\lambda/2 = 4$ cm is the distance between nodes, so $\lambda = 0.08$ m

$$f = \frac{u}{\lambda} = \frac{(30 \text{ m/s})}{(0.08 \text{ m})} = 375 \text{ Hz}$$

(b) The possible wavelengths are $\lambda_n = 2L/n$. Thus

$$\lambda_1 = \frac{2(12 \text{ cm})}{1} = 24 \text{ cm} \qquad \lambda_2 = \frac{2(12 \text{ cm})}{2} = 12 \text{ cm}$$

The corresponding frequencies are $f_1 = u/\lambda_1 = 125$ Hz, $f_2 = u/\lambda_2 = 250$ Hz. $f_3$ is the same as in part (a).

(c) According to Eq. 20-15, the tension is

$$T = \mu u^2 = (0.2 \text{ kg/m})(30 \text{ m/s})^2 = 180 \text{ N}$$

## 22-2 Interference with Two Point Sources

Consider two electric dipoles $S_1$ and $S_2$ oscillating up and down together in the $z$ direction, as shown in Figure 22-5. Let each dipole strength be $p = p_0 \cos \omega t$. According to Eq. 21-15, the field at point $P$ will be

$$E' = E_1 + E_2 = E_0 \cos(kr_1 - \omega t) + E_0 \cos(kr_2 - \omega t)$$

where

$$E_0 = \frac{k_0 \omega^2 p_0}{c^2 r}$$

Since these two sources are oscillating in phase, we can use the method of phasors to obtain the sum of the two cosines (see Appendix 21-1). This is illustrated in Figure 22-6. The vectors $E_1$ and $E_2$ both have magnitude $E_0$. The angle $\phi$ between them is the difference in phase between $E_2$ and $E_1$; namely,

$$\phi = (kr_2 - \omega t) - (kr_1 - \omega t) = k(r_2 - r_1)$$

Their vector sum $E'$ is the amplitude of the resultant field. Using the law

**Figure 22-5.** Two point sources $S_1$ and $S_2$ are a distance $d$ apart. An enlarged view of the source region is shown at (b). The path difference $(r_2 - r_1) \approx d \sin \theta$.

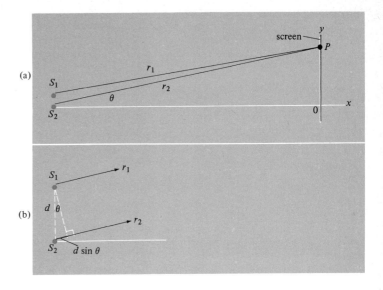

of cosines on the isosceles triangle in Figure 22-6 gives

$$E'^2 = E_0^2 + E_0^2 + 2E_0^2 \cos \phi$$
$$= 2E_0^2(1 + \cos \phi)$$

Since the wave intensities are proportional to the square of the amplitudes, we have

$$I = 2I_0[1 + \cos k(r_2 - r_1)]$$

From the right triangle in Figure 22-5(b) we have that the path difference $(r_2 - r_1) = d \sin \theta$, if the distance to the screen is great enough. This condition, that the path difference equal $d \sin \theta$, is called the Fraunhofer approximation. Then

$$I = 2I_0[1 + \cos(kd \sin \theta)] \quad \text{(two-source interference)} \quad (22\text{-}3)$$

Equation 22-3 is plotted in Figure 22-7. There is an intensity maximum whenever the argument

$$(kd \sin \theta) = n2\pi$$

or

$$\sin \theta = \frac{n\lambda}{d} \quad \text{(condition for a maximum)} \quad (22\text{-}4)$$

Then the path difference, which according to Figure 22-5 equals $d \sin \theta$, is

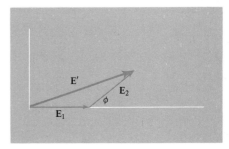

**Figure 22-6.** Phasor diagram for field from two dipoles where phase difference is $\phi$.

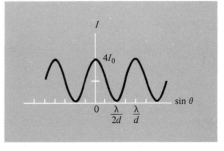

**Figure 22-7.** Two-source interference pattern. Intensity is plotted versus $\sin \theta$.

SEC. 22-2/INTERFERENCE WITH TWO POINT SOURCES

**Figure 22-8.** *Two-source interference pattern generated by two vibrators tapping the surface of water in phase. [From Educational Development Center]*

$n\lambda$. Certainly we expect to have a maximum at the screen when the crest of one wave lines up with the crest of another. This can only occur when the path difference is an integral number of wavelengths. When the path difference is one half wavelength or $(n + \frac{1}{2})\lambda$, the crest of one wave lines up with the trough of the other and we have an intensity minimum. The locus of such minima is called a line of nodes. This is illustrated in Figure 22-8 using water waves.

**Example 3.** Two stereo speakers are 2 m apart and playing the same 1000-Hz musical note. A listener is 4 m away (see Figure 22-9). How far can he be from the center line before reaching the first node? Speed of sound is 330 m/s.

ANSWER: The first node occurs when the path difference is $\frac{1}{2}\lambda$ or when

$$(r_2 - r_1) = \frac{\lambda}{2}$$

If $\theta$ is less than 30°, $d \sin \theta$ is a good approximation for the path difference. Then $\sin \theta = \lambda/2d$. We can use $\lambda = u/f$ to get the wavelength:

$$\lambda = \frac{330 \text{ m/s}}{1000 \text{ Hz}} = 0.33 \text{ m}$$

$$\sin \theta = \frac{0.33 \text{ m}}{2(2 \text{ m})} = 0.0825$$

$$y = D \tan \theta = 0.33 \text{ m}$$

So if the listener moves about a foot to either side, he or she will encounter a node at 1000 Hz. (Also frequencies of 3000 Hz, 5000 Hz, 7000 Hz will have nodes at the same position.) In an actual room the nodes are partly filled in because of large amounts of wall reflection. However, the effect is quite noticeable if one plays a pure sine wave into a stereo system and then walks across the room.

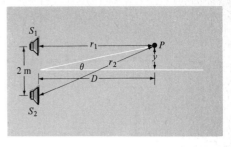

Figure 22-9. *Stereo speakers for Example 3.*

## 22-3 Interference from Multiple Sources

Suppose an observer is off at an angle $\theta$ from the normal to the line joining $N$ equally spaced sources, as shown in Figure 22-10. The observer sees a phase difference $\phi = k(r_2 - r_1) = kd \sin \theta$ between any two successive sources. The corresponding $N$ phasors are shown in Figure 22-11. Using the right triangle in Figure 22-11(a), we obtain

$$\frac{E'}{2} = R \sin N\frac{\phi}{2}$$

From the right triangle in Figure 22-11b we have

$$\frac{E_1}{2} = R \sin \frac{\phi}{2}$$

Dividing the previous equation by this equation gives

$$\frac{E'}{E_1} = \frac{\sin N(\phi/2)}{\sin(\phi/2)}$$

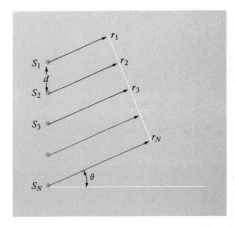

Figure 22-10. *N sources in phase each separated by a distance d.*

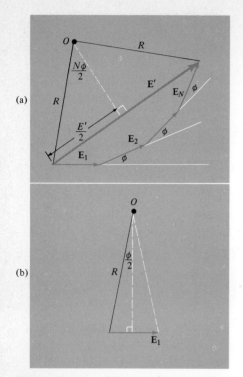

Squaring both sides leads to

$$\begin{cases} I = I_0 \dfrac{\sin^2 N(\phi/2)}{\sin^2(\phi/2)} \\ \phi = kd \sin\theta \end{cases} \quad (22\text{-}5)$$

where $I_0$ is the intensity from one source alone and $\phi = kd \sin\theta$. This intensity pattern is plotted in Figure 22-12. Note that as $\phi \to 0$, $\sin N(\phi/2) \to N(\phi/2)$ and $\sin \phi/2 \to \phi/2$, so that Eq. 22-5 becomes

$$I \to I_0 \frac{[N(\phi/2)]^2}{(\phi/2)^2} = N^2 I_0$$

Then the intensity is $N^2$ times the intensity of one source.

**Figure 22-11.** (a) *Phasor diagram for the N sources in Figure 22-10. Ends of phasors must lie on circle of radius R.* (b) *First phasor alone with center of the circle.*

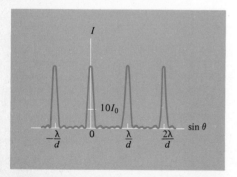

**Figure 22-12.** *Interference pattern for six sources in a row. Intensity I is plotted versus $\sin\theta$ from Eq. 22-5.*

**Example 4.** Small town radio stations in the broadcast band are normally not permitted to broadcast at night because the hundreds of such stations in the United States would interfere with each other. (The ionosphere at night is at high altitude so long distance reception via reflection is possible.) However, station WTKO in Ithaca, N.Y., has approval to broadcast at night because it uses an antenna system that beams a relatively strong signal to Ithaca, but a very weak signal to the rest of the country. It uses four dipole radiators in a row pointed toward Ithaca with each successive radiator or antenna 90° out of phase with its neighbor. The spacing is $d = \frac{1}{4}\lambda$. What, then, is the intensity as a function of the angle $\theta$ shown in Figure 22-13?

ANSWER: In this case $\phi = (kd \sin\theta - \pi/2)$ rather than $kd \sin\theta$, so

$$\phi = \left(k\frac{\lambda}{4}\sin\theta - \frac{\pi}{2}\right) = \frac{\pi}{2}(\sin\theta - 1)$$

The intensity pattern is given by substituting $N = 4$ and $\phi = (\pi/2)(\sin\theta - 1)$ into Eq. 22-5:

$$I = I_0 \frac{\sin^2[\pi(\sin\theta - 1)]}{\sin^2[(\pi/4)(\sin\theta - 1)]}$$

This equation is plotted as $I$ versus $\theta$ in Figure 22-13(b).

**Figure 22-13.** (a) *Orientation of four antenna towers of station WTKO in Ithaca, N.Y.* (b) *Plot of radiated intensity I versus angle θ.*

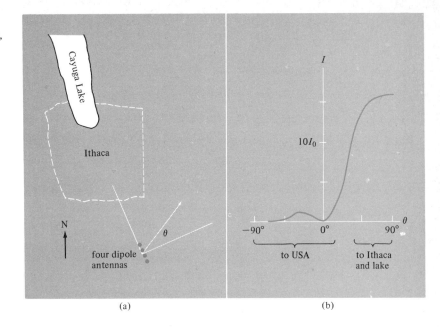

## 22-4 The Diffraction Grating

A source of $N$ slits can be prepared by scratching parallel lines on a flat piece of glass. The line of glass between each scratch will behave as a separate slit. If parallel monochromatic light from a single source strikes the slits as shown in Figure 22-14, we have the case of $N$ sources oscillating in phase. For such a case the intensity pattern at the screen is given by Eq. 22-5:

$$I = I_0 \frac{\sin^2 N(\phi/2)}{\sin^2(\phi/2)} \qquad \text{where } \phi = kd \sin \theta.$$

**Figure 22-14.** (a) *Enlarged view of slits of a diffraction grating.* (b) *Corresponding intensity pattern on screen.*

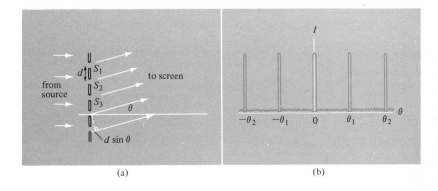

This has the value $I = N^2 I_0$ whenever the denominator is zero or when

$$\phi_n = 2\pi n \quad (22\text{-}6)$$
$$kd \sin \theta_n = 2\pi n$$

$$\sin \theta_n = n\frac{\lambda}{d} \quad (22\text{-}7)$$

At other values of $\theta$, $I$ is approximately $I_0$, which is a factor $N^2$ times smaller. For a typical grating $N$ is several thousand. In Figure 22-14 the parallel rays leaving the grating can be focussed by a lens to give a sharp line image on the screen. The condition in Eq. 22-7 is easily seen from Figure 22-14. In order for all the parallel rays to be in phase, each pair of adjacent rays must have a path difference equal to $n\lambda$. Since each path difference is $d \sin \theta$, we have

$$d \sin \theta = n\lambda$$

$$\sin \theta = n\frac{\lambda}{d}$$

A spectral line of wavelength $\lambda$ will appear at the angle given by $\sin \theta = \lambda/d$. A second-order image of the line will be at $\sin \theta = 2\lambda/d$; a third-order at $\sin \theta = 3\lambda/d$; and so on.

---

**Example 5.** It is possible to purchase replica diffraction gratings for about 10¢ each. They have 13,400 lines per inch. Suppose a sodium source is viewed through such a grating. Sodium atoms when ionized emit nearly all their light at $\lambda = 5893$ Å where Å stands for angstrom (1 Å $= 10^{-10}$ m). At what angle will this yellow light appear?

ANSWER: The spacing between rulings on the grating is

$$d = \frac{2.54 \text{ cm}}{13,400} = 1.90 \times 10^{-6} \text{ m}$$

Using Eq. 22-7 we have

$$\sin \theta_1 = \frac{\lambda}{d} = \frac{5893 \times 10^{-10}}{1.90 \times 10^{-6}} = 0.31 \qquad \theta_1 = 18.1$$

$$\sin \theta_2 = \frac{2\lambda}{d} = 0.62 \qquad \theta_2 = 38.3°$$

$$\sin \theta_3 = \frac{3\lambda}{d} = 0.93 \qquad \theta_3 = 68.5°$$

If the source is a line source, three lines will appear on either side of the direct image at $\pm 18.1°$, $\pm 38.3°$, and $\pm 68.5°$.

## Resolution

In an ideal spectroscope, light of an exact wavelength $\lambda_0$ should appear as a spectral line of zero width. However, a diffraction grating introduces an artificial width $\Delta \lambda_0$, which we shall now calculate.

The intensity pattern of the source $\lambda_0$ is given by Eq. 22-5 and is plotted as the heavy curve in Figure 22-15. In this figure $\Delta \theta_0$ is the difference in angle from the intensity maximum to the first zero. We call it the artificial width. The change $\Delta \theta_0$ corresponds to a change in phase $\Delta \phi_0$ such that the angle $N\phi/2$ in the numerator of Eq. 22-5 changes by $180°$. Hence

$$\left(\frac{N\Delta\phi_0}{2}\right) = \pi$$

$$\Delta\phi_0 = \frac{2\pi}{N}$$

Differentiation of the equation $\phi_0 = kd \sin \theta_0$ gives $\Delta\phi_0 = kd \cos \theta_0 \Delta\theta_0$. Solve this for $\Delta\theta_0$ to get

$$\Delta\theta_0 = \frac{2\pi}{Nkd \cos \theta_0}$$

$$\Delta\theta_0 = \frac{\lambda_0}{Nd \cos \theta_0} \quad (22\text{-}8)$$

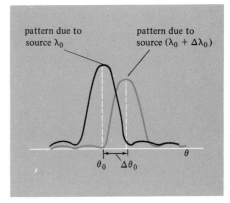

**Figure 22-15.** *Diffraction grating intensity patterns of two sources $\Delta\lambda_0$ apart in wavelength.*

The angle scale of the spectrometer can also be labeled in wavelength. We wish to know the $\Delta\lambda_0$ corresponding to this artificial line width. This can be obtained from the relation $\sin \theta = \lambda/d$ giving $\Delta\lambda = d \cos \theta \, \Delta\theta$. If we use Eq. 22-8 for $\Delta\theta$, we obtain

$$\Delta\lambda_0 = \frac{\lambda_0}{N} \quad (22\text{-}9)$$

for the artificial line width.

Now suppose the source consists of two exact wavelengths $\lambda_0$ and $\lambda_0 + \Delta\lambda$. These two lines will not quite appear separated on the screen if $\Delta\lambda$ equals the $\Delta\lambda_0$ of Eq. 22-9. This situation is shown in Figure 22-15, where the solid curve is the intensity pattern corresponding to $\lambda_0$ and the colored curve is the pattern for $\lambda_0 + \Delta\lambda_0$. Hence, if two lines are separated by the artificial line width $\Delta\lambda = \lambda/N$, the maximum of one will lie on the edge of the other. They would be blurred together into one line. However, if $\Delta\lambda$ were greater than $\lambda/N$ the two lines could be resolved. The ratio of $\lambda$ to this critical $\Delta\lambda$ is called the resolution and for a diffraction grating is merely the total number of scratches $N$.

> **Example 6.** The famous sodium D line at 5893 Å is really two separate lines at 5890 and 5896 Å. Can the 10¢ diffraction grating of Example 5 resolve these two lines?
>
> ANSWER: The required resolution is
>
> $$\frac{\lambda}{\Delta\lambda} = \frac{5893}{6} = 982$$
>
> The resolution of the grating is the total number of rulings, which is the 13,400 lines per inch times the width of the grating, which happens to be 1 inch. We see that the grating's resolution is about 13 times the minimum necessary resolution.

## 22-5 Huygens' Principle

Actually it is not necessary to have separate sources in order to observe interference effects. Two (or more) sources can effectively be obtained by using a single source and a barrier with holes in it. Figure 22-16 shows plane waves of water (from a single source) striking a barrier with two holes. The waves leaving each hole behave as if they originate from two sources located at the holes, oscillating in phase.

A famous electromagnetic equivalent of the water waves in Figure 22-16 is the double slit experiment with light shown in Figure 22-17. A single source of light shines on two fine slits. The pattern observed on the screen is given by Eq. 22-3 and is the same as if the sources were at the slits. This experiment was first done by Thomas Young in 1803.

In the eighteenth century, Christian Huygens after experimenting with water waves proposed without proof that **when a wave front enters a hole or holes, each part of the wave front behaves as if it were a new source of radiation.** At first glance this might seem strange because there is no current source in the holes. In fact, there should be induced currents everywhere in the barrier except the holes. We shall now show that current sources that just fill up the holes radiate a field mathematically equal to the incoming field striking the barrier with holes in it. Let $E_{\text{barrier}}$ be the radiated field due to the induced currents in the barrier. Then the net field to the right of the barrier is

$$E_{\text{net}} = E_{\text{in}} + E_{\text{barrier}} \tag{22-10}$$

where $E_{in}$ is the field due to the single source by itself (with no barrier). This expression describes the true physical situation that gives rise to the observed interference pattern. Now let us plug up the holes with pieces of barrier cut to size. Let $E_{holes}$ be the radiated field due to induced current in the "plugs" that are filling up the holes. For a plugged up barrier

Then
$$E_{net} = E_{in} + E_{barrier} + E_{holes} = 0 \quad \text{(to the right)}$$

$$-E_{holes} = E_{in} + E_{barrier}$$
$$|E_{holes}|^2 = |E_{in} + E_{barrier}|^2$$

**Figure 22-16.** *Water waves from single vibrating blade strike barrier with two holes.*

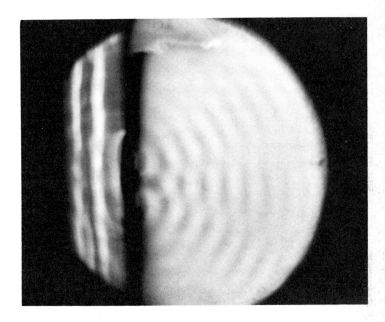

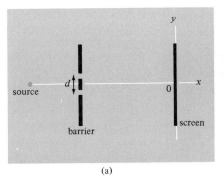

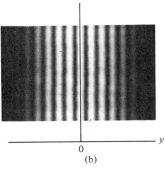

**Figure 22-17.** (a) *Setup for double slit interference of light. The source could be a laser or a single slit illuminated by a monochromatic lamp. The barrier has two open slits a distance d apart.* (b) *The intensity pattern obtained by placing film at screen position. From Educational Development Center.*

SEC. 22-5/HUYGENS' PRINCIPLE 485

The right-hand side is the true physical situation (Eq. 22-10), and we see from the left-hand side that the true situation is mathematically equivalent to the radiation intensity pattern produced by current sources sitting in the holes taken all by themselves. We have demonstrated that if every part of the wavefront that passes thru a barrier is considered as a new point source of radiation, then the radiation pattern will be just that due to the barrier plus the single source. There is a small correction due to edge effects, which we shall ignore. (Our proof assumed the induced currents could cross over the edges of the holes from the plugs to the barrier.)

Our "derivation" of Huygens' principle was for the case where plane waves impinged on a flat barrier. Huygens' principle may also be applied to an entire wavefront of arbitrary shape with no barrier. Then each point on the wavefront is taken as the source of a new wavelet. This kind of application gives correct results for the shape of successive wavefronts, but it does not always give accurate quantitative results for the magnitude of the wave amplitude.

## 22-6 Single Slit Diffraction

Parallel monochromatic light striking a single slit of width $a$ gives an interference pattern on a distant screen as shown in Figure 22-18. This kind of interference effect, which deals with a single slit or single edge of a barrier, is called diffraction. The rays of light are "diffracted" from their original direction.

Using Figure 22-19, we can easily determine the condition for the angle $\theta$ at which the first minimum is observed. According to Huygens' principle we can consider the slit as a series of new sources $S_1, S_2, \ldots, S_N$. If the waves from $S_1$ and the center of the slit are 180° out of phase, they will cancel out [rays (a) and (b) in Figure 22-19]. So will the waves from $S_2$ and the next source below ray (b) cancel each other. And so on for $S_3$ down to the bottom of the slit the rays will cancel out in pairs. The path difference between rays (a) and (b) is $(a/2) \sin \theta$. For these to be 180° out of phase this path difference must be $\lambda/2$.

$$\frac{a}{2} \sin \theta_1 = \frac{\lambda}{2}$$

$$\sin \theta_1 = \frac{\lambda}{a} \quad \text{(single slit intensity minimum)} \quad (22\text{-}11)$$

is the angle for the first intensity minimum.

The intensity at any arbitrary angle $\theta$ is obtained by adding up the phasors from each infinitesimal source as shown in Figure 22-20. They will form the arc of a circle where the total phase difference is

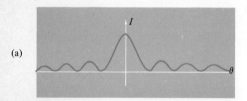

**Figure 22-18.** (a) *Intensity distribution for the diffraction pattern formed from a single slit on a very distant screen.* (b) *Photograph made by placing photographic film in the plane of the screen. The light striking the slit is from a neon–helium laser.*

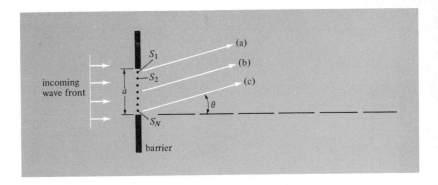

**Figure 22-19.** *Radiation "originating" from a single slit. Rays (a) and (c) are from the edges and (b) is from the center.*

$\Phi = ka \sin \theta$   [the phase difference between rays (a) and (c)]

The resultant amplitude $A$ can be obtained from the right triangle.

$$\sin \frac{\Phi}{2} = \frac{A/2}{R}$$

$$A = 2R \sin \frac{\Phi}{2} \qquad (22\text{-}12)$$

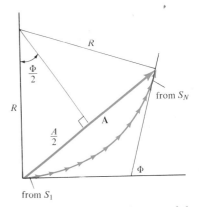

**Figure 22-20.** *Vector* **A** *is the sum of the N phasors shown in Figure 22-10.* $\Phi$ *is the phase difference between the first and last phasors.*

The length of the arc formed by the phasors is $A_0$, the net amplitude seen at 0°. This arc equals $R$ times the angle $\Phi$ in radians:

$$R\Phi = A_0$$

$$R = \frac{A_0}{\Phi}$$

Substituting this into Eq. 22-12 gives

$$A = A_0 \frac{\sin \Phi/2}{\Phi/2}$$

Squaring both sides gives

$$I = I_0 \left( \frac{\sin \Phi/2}{\Phi/2} \right)^2 \quad \text{where} \quad \Phi = ka \sin \theta \qquad (22\text{-}13)$$

This function is plotted in Figure 22-18. The successive minima will occur when $\Phi/2 = n\pi$, or when

$$\frac{ka \sin \theta}{2} = n\pi \quad \text{or} \quad \sin \theta_{\min} = n\frac{\lambda}{a} \quad \text{for } n \geq 1$$

This checks with our earlier result, Eq. 22-11. Note that the central maximum is twice as wide as the secondary maxima.

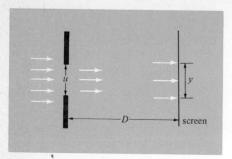

**Figure 22-21.** *Parallel light passes through slit of width u giving band of light of width y on screen.*

**Example 7.** If parallel light strikes a wide slit of width $u$ (Figure 22-21), there will be a band of light of width $y = u$ on the screen. As the slit is gradually closed, the band of light gets narrower until diffraction effects take over and then the band gets wider again. What slit width $u_0$ gives the narrowest band of light on the screen?

ANSWER: The additional width due to diffraction is

$$y_{\text{diff}} \approx \theta_{\text{min}} D \approx \left(\frac{\lambda}{u}\right) D$$

The total width is

$$y \approx u + \frac{\lambda D}{u}$$

When $y$ is at a minimum, $dy/du = 0$ or

$$\left(1 - \frac{\lambda D}{u_0^2}\right) = 0$$

$$u_0 \approx \sqrt{\lambda D}$$

## 22-7 Coherence and Incoherence

Up to now we have studied interference effects due to sources that are in phase with each other (or at some fixed phase relationship). Such sources are called coherent sources. It is possible to obtain separate coherent sources of radio waves and microwaves by feeding two or more antennas from the same oscillator. As we saw in Section 22-5, coherent sources of light can be obtained by illuminating two or more slits or openings from the same point source of light. Also two coherent beams of light can be obtained using a half-silvered mirror as in the Michelson interferometer (see Figure 8-4).

However, if the two arms of the interferometer are made of unequal length, the interference pattern fades out if the path difference is greater than a certain amount $\Delta L_0$ corresponding to a time difference $\Delta t_0 = \Delta L_0/c$. $\Delta L_0$ is called the coherence length and $\Delta t_0$ the coherence time. If the light source of the interferometer (whether it be a laser or a heated gas) is viewed in a high-resolution spectrometer, one sees a sharp line with a natural frequency width $\Delta f$. The measured width is related to the measured coherence time $\Delta t_0$ by

$$2\pi \Delta f \Delta t_0 \approx 1$$

This is the same as Eq. 21-16 and has the same explanation. A good source of monochromatic light, whether it be a laser or an atomic source, behaves as if it is an oscillator of average frequency $f_0$ randomly wandering around in frequency from $f_0 - \Delta f$ to $f_0 + \Delta f$. Two pure sine waves of frequency $\Delta f$ apart will stay in phase for a time $\Delta t \approx 1/(2\pi \Delta f)$ as given by Eq. 21-16. The sharpest lines in atomic spectra have $\Delta t_0 \sim 10^{-8}$ s. A typical laser will drift in frequency by a lesser amount, and thus have a longer coherence time.

From the quantum theory we know that light is emitted photon by photon with each photon coming from a different atom. In lasers these photons are in phase (see Section 26-6). In other sources of light the photons have random phase and are said to be incoherent. However, any two such photons will keep the same phase relation to each other for a time up to $\Delta t_0 \approx 1/(2\pi \Delta f)$, where $\Delta f$ is the measured line width. Also such photons behave as if they are wave packets of length $\Delta L_0 = c\,\Delta t_0 \approx c/(2\pi \Delta f)$.

### Intensity Interferometry

It should be possible to obtain an interference pattern from two independent and incoherent sources of light if the pattern can be measured in a time less than $\Delta t_0 = 1/(2\pi \Delta f)$. Most intensities are too weak to make such a measurement; however, there is a method developed by R. Hanbury Brown and R. Q. Twiss in 1956 that effectively averages over many experiments with a time resolution less than $1/(2\pi \Delta f)$. The trick is to use two separate detectors at the "screen" where the interference pattern is expected. The intensities $I_1$ and $I_2$ from these two detectors are multiplied together instantaneously (this can be done electronically using photomultiplier tubes as the detectors). In Figure 22-22 keep detector 1 at $\theta = 0$. Detector 2 is used to make a series of measurements at neighboring values of $\theta$. Sources $S_1$ and $S_2$ are independent and incoherent sources. If they are an unknown distance $d$ apart, we can determine $d$ by measuring $\overline{I_1 I_2}$ averaged over all values of phase difference $\phi$ between $S_1$ and $S_2$. This can be seen by using Eq. 22-3 for the intensity at each detector. If during a small interval of time less than $\Delta t_0$ the two sources have a phase difference of $\phi$, detector 1 measures

$$I_1 = 2I_0(1 + \cos\phi)$$

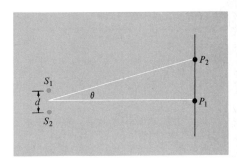

**Figure 22-22.** $S_1$ and $S_2$ are independent sources. Detectors simultaneously measure intensities at $P_1$ and $P_2$.

For detector 2 there is an additional phase difference $\phi_0 = kd \sin \theta$:

$$I_2 = 2I_0[1 + \cos(\phi + \phi_0)]$$

We can average over time by averaging over all possible values of $\phi$:

$$\overline{I_1 I_2} = \frac{\int_0^{2\pi} I_1(\phi) I_2(\phi) \, d\phi}{\int_0^{2\pi} d\phi} = \frac{4I_0^2}{2\pi} \int_0^{2\pi} (1 + \cos \phi)[1 + \cos(\phi + \phi_0)] \, d\phi$$

$$= \frac{2I_0^2}{\pi} \int_0^{2\pi} [1 + \cos \phi + \cos(\phi + \phi_0) + \cos \phi \cos(\phi + \phi_0)] \, d\phi$$

Since $\cos \phi$ and $\cos(\phi + \phi_0)$ are as often positive as negative, their integrals are zero. So

$$\overline{I_1 I_2} = \frac{2I_0^2}{\pi}\left[2\pi + \int_0^{2\pi} \cos \phi (\cos \phi \cos \phi_0 - \sin \phi \sin \phi_0) \, d\phi\right]$$

$$= \frac{2I_0^2}{\pi}[2\pi + \pi \cos \phi_0]$$

$$= 2I_0^2[2 + \cos(kd \sin \theta)] \tag{22-14}$$

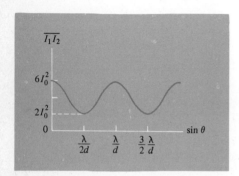

**Figure 22-23.** *Intensity product of two detectors as a function of the angle between them.*

This is plotted versus $\sin \theta$ in Figure 22-23. We see that two incoherent sources can give an interference pattern! In spite of this calculation, in 1956 some physicists at first didn't believe that interference could be observed from incoherent sources.* (Even today some textbooks say this.) But then experiments were performed using radiotelescopes that were able to determine the distance between the two members of a double star and even the diameters of some of the closer stars. These experiments are difficult because the photomultiplier noise becomes comparable to the signal when such short time resolutions are needed.

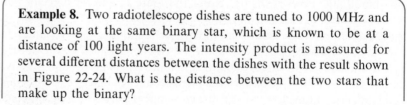

**Example 8.** Two radiotelescope dishes are tuned to 1000 MHz and are looking at the same binary star, which is known to be at a distance of 100 light years. The intensity product is measured for several different distances between the dishes with the result shown in Figure 22-24. What is the distance between the two stars that make up the binary?

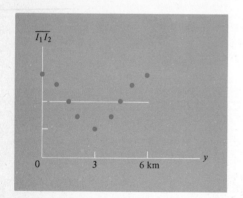

**Figure 22-24**

---

*In defense of these physicists, the explanation becomes more subtle when the light is quantized. It turns out that correct application of quantum mechanics to such a two-photon situation also gives the result stated in Eq. 22-14.

ANSWER: The angle subtended by the two detectors when observing the first minimum is

$$\theta = \frac{3 \text{ km}}{100 \text{ lyr}} = \frac{3 \text{ km}}{9.42 \times 10^{14} \text{ km}} = 3.18 \times 10^{-15}$$

According to Eq. 22-14 the first minimum occurs when

$$\theta = \frac{\lambda}{2d}$$

$$d = \frac{\lambda}{2\theta} = \frac{0.30 \text{ m}}{2(3.18 \times 10^{-15})} = 4.72 \times 10^{10} \text{ km} = 5.0 \times 10^{-3} \text{ lyr}$$

**Figure 22-25.** (a) A portion of the world's largest radiotelescope at Arecibo, Puerto Rico. The 1000-ft diameter reflecting "dish" fits into a natural mountain basin. The reflected signal is focussed onto a movable antenna that is supported 500 ft above the dish by three towers (each equivalent to the Washington Monument). (b) A Cornell student with special shoes for smoothing out bumps in the reflecting surface. Only a small part of the dish is shown. The Arecibo Observatory is part of the National Astronomy and Ionosphere Center, which is operated by Cornell University under contract with the National Science Foundation.

(a)

(b)

In the past 20 years or so the field of optics has experienced a rapid expansion with the advent of radiotelescopes, microwaves, infrared detectors, lasers, quantum detectors, holograms, high speed electronics, computers, and so on. These new developments have made possible what is called modern optics. In the next chapter we shall start out with one more example from modern optics—holography.

## Summary

A traveling wave reflected from the end of a string (or from a mirror if the wave is electromagnetic) interferes with the reflected wave to give a wave amplitude of the form $y = (y_0 \sin \omega t) \sin kx$. This has nodes at positions $x_n = n(\lambda/2)$. If a string is of fixed length $L$ with nodes at either end, the possible standing waves have wavelengths $\lambda_n = 2L/n$.

Two point sources each of intensity $I_0$ oscillating in phase give an interference pattern $I = 2I_0[1 + \cos k(r_2 - r_1)]$ where $r_2 - r_1$ is the path difference. At large distances $r_2 - r_1 = d \sin \theta$, where $d$ is the distance between them. Then $\sin \theta = n\lambda/d$ is the condition for a maximum.

If there are $N$ point sources in a row

$$I = I_0 \frac{\sin^2(N\phi/2)}{\sin^2(\phi/2)} \quad \text{where } \phi = kd \sin \theta$$

For a diffraction grating of $N$ "slits," $\sin \theta = n\lambda/d$ gives the angular positions of spectral lines of wavelength $\lambda$. Two lines separated by $\Delta\lambda$ can be resolved if

$$\frac{\Delta\lambda}{\lambda} \geq \frac{1}{N}$$

Huygens' principle states that each point on a wave front can be taken as the source of a new wavelet.

The diffraction pattern from a single slit of width $a$ is

$$I = I_0 \frac{\sin^2(ka \sin \theta/2)}{(ka \sin \theta/2)^2}$$

This has minima when

$$\sin \theta = \frac{n\lambda}{a} \quad \text{for integral values of } n.$$

If the natural width of a spectral line is $\Delta f$, the coherence time is

$\Delta t_0 = 1/(2\pi \Delta f)$ and the coherence length is $\Delta L_0 = c\,\Delta t_0$. If the intensity product of two incoherent sources is measured over time intervals less than $\Delta t_0$, the average of such measurements will be

$$\overline{I_1 I_2} = 2I_0^2[2 + \cos(kd \sin\theta)]$$

## Exercises

1. In Figure 22-1 what are the directions of $E'$ and $\mathcal{B}'$ just to the right of the reflecting surface?
2. Express Eq. 22-2 in terms of $k_n$ rather than $\lambda_n$.
3. In Figure 22-5 suppose that $E_1 = E_0 \cos(kr_1 - \omega t)$ and $E_2 = 2E_0 \cos(kr_2 - \omega t)$. What is $I$ as a function of $I_0$ and $\theta$?
4. What is the phase difference at the distant screen between the waves from $S_1$ and $S_3$ in Figure 22-14. Give the answer in terms of $\lambda$, $d$, and $\theta$.
5. The wave velocity of a string fixed at both ends is 2 m/s. The string contains standing waves with nodes 3.0 cm apart.
   (a) What is the frequency of vibration?
   (b) How many times per second is the string in a straight line containing no visible waves?
6. $S_1$ and $S_2$ are two sine wave sources of sound. If they are in phase and 3 m apart,
   (a) List three different wavelengths that will give a destructive interference at point $P$.
   (b) List three different wavelengths that will give a constructive interference at $P$.
   (c) What is the lowest frequency that will give a destructive interference at $P$? The velocity of sound is 330 m/s.
7. In the accompanying figure, what are the conditions on $D_1 - D_2$ for an interference maximum and for an interference minimum to occur at point $P$? Assume no change in phase at the reflection.
8. In Exercise 7, suppose the wave is reflected with a 180° change in phase at the reflection. Now what is the condition for an intensity minimum?
9. Light of wavelength 5000 Å is diffracted by a grating of 2000 lines/cm. The screen is 3 m from the grating. What is the distance on the screen between the zeroth and the first-order image?
10. Consider a reflection diffraction grating (lines ruled on a mirror). The incoming beam is perpendicular to the grating and the diffracted beam is at an angle $\theta$ from the mirror.
    (a) What is the path difference in terms of $\theta$ and $d$?
    (b) At what value of $\theta$ would there be an intensity maximum?
11. How many rulings on a diffraction grating are necessary in order to see the natural line width of a typical atomic transition? (See page 461.)
12. In Example 6 what is $\Delta\theta_0$, the artificial line width for each of the sodium D lines? Give answer in degrees.

Exercise 6

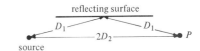

Exercise 7

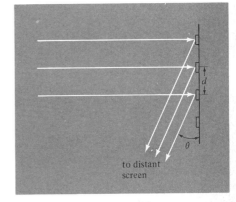

Exercise 10

13. How many rulings on a diffraction grating are needed to just barely separate the sodium D lines?
14. Show that
$$\frac{\sin^2 N(\phi/2)}{\sin^2 \phi/2} = 2(1 + \cos \phi) \quad \text{for } N = 2$$

## Problems

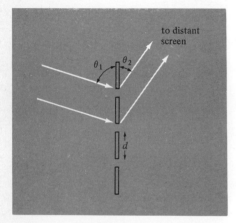

Problem 21

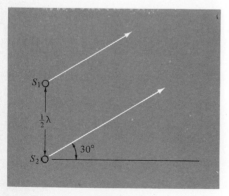

Problem 24

15. Suppose the incoming wave in Figure 22-1 is $E_{\text{in}} = E_0 \sin(\omega t - kx)$. Now what is the expression for the standing wave $E = E_{\text{in}} + E'_{\text{left}}$?
16. Suppose the two dipoles of Figure 22-5 are 180° out of phase. What is $I$ as a function of $\theta$?
17. In Problem 16 show that $I \approx I_0 k^2 d^2 \sin^2 \theta$ when $d \ll \lambda$.
18. Repeat Problem 16 for two dipoles of dipole strength $p_0$ oscillating 180° out of phase in the $y$ direction; $d \ll \lambda$. Define $\theta$ as the angle between $r$ and the $y$ axis. Show that
$$E = \frac{k_0 p_0 d \omega^3}{2c^3 r} \sin 2\theta \cos \omega \left(t - \frac{r}{c}\right)$$

19. In Figure 22-6 suppose that $E_1 \neq E_2$. Express $E'$ in terms of $E_1$, $E_2$, and $\theta$.
20. The second- and third-order visible spectra of a diffraction grating will partially overlap. What $N = 3$ wavelength will appear at the $\lambda = 7000$ Å position of the $N = 2$ spectrum?
21. Consider a diffraction grating where the incident light of wavelength $\lambda$ is not at right angles to the grating but at an angle $\theta_1$ as shown. What is the condition for a maximum in terms of $\theta_1$, $\theta_2$, $\lambda$, and $d$?
22. Repeat Example 3, but for middle C (262 Hz). Give an exact solution where no approximations are made. (You can use trial and error to get a numerical answer.)
23. A certain atomic transition has a mean life $\tau = 2 \times 10^{-4}$ s. The emitted wavelength is 4000 Å. What is the natural line width $\Delta\lambda$, and how many rulings are needed on a diffraction grating in order to measure the natural width?
24. A radio transmitter feeds the same signal to two antennas that are a half wavelength apart. It is important that the radiation at 30° be at a minimum. What must be the phase $\phi$ between the two signals fed to the two antennas in order to give cancellation at 30°?
25. Show that the three-slit interference pattern is
$$I = I_0(1 + 2 \cos \phi)^2 \quad \text{where } \phi = kd \sin \theta$$

26. Add a third incoherent source to Figure 22-22 as shown. Let $\phi_1$ be the random phase between $S_1$ and $S_2$. Let $\phi_2$ be the random phase between $S_2$ and $S_3$. Let $\phi_0 = kd \sin \theta$.
    (a) Show that $I_1 = I_0[3 + 2 \cos \phi_1 + 2 \cos \phi_2 + 2 \cos(\phi_2 - \phi_1)]$.
    (b) What is $I_2$?
    (c) Show that $\overline{I_1 I_2} = I_0^2[6 + (1 + 2 \cos \phi_0)^2]$; that is, the three-slit coherent interference pattern plus a constant term.

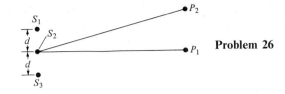

Problem 26

27. A stereo phono pickup gives just two independent signals. How then can these two signals be used to give quadraphonic directionality into four speakers? We shall see that some degree of directionality can be achieved by combining the four original signals in different phases in the two signals on the record, and then recombining the two signals off the record in different phases. Let $L_f$ and $R_f$ be the original signals from the left and right front microphones, respectively. Let $L_b$ and $R_b$ be the signals from the left and right rear microphones. In the SQ matrix system these signals are combined such that the amplitude on the left track of the record is

$$L_t = L_f - \frac{j}{\sqrt{2}} L_b + \frac{R_b}{\sqrt{2}}$$

We have used the notation that $\pm j$ stands for a $\pm 90°$ phase shift of whatever sine wave is being recorded; that is, if the $L_b$ voltage signal is $V_0 \cos \omega t$, it is added as $(1/\sqrt{2})V_0 \cos(\omega t - 90°)$ in forming $L_t$. The amplitude on the right track is

$$R_t = R_f + \frac{j}{\sqrt{2}} R_b - \frac{1}{\sqrt{2}} L_b$$

Then in the stereo amplifier $L_t$ is fed to the left front speaker and $R_t$ to the right front speaker. The voltage $L_b' = (j/\sqrt{2})L_t - (1/\sqrt{2})R_t$ is fed to the left rear speaker and $R_b' = (1/\sqrt{2})L_t - (j/\sqrt{2})R_t$ is fed to the right rear speaker.
   (a) If there is a single sine wave source $L_f = V_0 \cos \omega t$, what is the ratio of the intensities coming out of the four speakers? What are $(L_f')^2$, $(R_f')^2$, $(L_b')^2$, $(R_b')^2$ when $L_f = V_0 \cos \omega t$ and $R_f = L_b = R_b = 0$?
   (b) Repeat for the single source coming from the $R_f$ microphone.
   (c) Repeat for the single source coming from the $L_b$ microphone.
   (d) Repeat for the single source coming from the $R_b$ microphone.
28. A certain atomic transition emits light of coherence length $\Delta L_0$.
   (a) What is the natural line width $\Delta \lambda$ in terms of $\lambda$ and $\Delta L_0$?
   (b) Light from the source illuminates $y_0$ cm of a diffraction grating. What is the minimum value of $\mathcal{N}$ (rulings per centimeter) in order to see the natural line width? Give answers in terms of $\lambda$, $\tau$, and $y_0$. $\tau$ is the lifetime of the excited state.

# Optics

In this chapter we shall deal with various phenomena of electromagnetic waves which are most common in the frequency range corresponding to visible light. Of course, for anything discussed in this chapter there will be an equivalent microwave phenomenon as well as infrared and ultraviolet. We shall even present an application in Section 23-5 where the wavelength is $\sim 10^{-10}$ times that of light.

In more general terms, any study of wave phenomena is also a study of optics—or what is called physical optics as opposed to geometrical optics. In this sense Chapters 21 and 22 could have been called physical optics as well as those parts of Chapter 20 that deal with wave motion.

In this book modern optics (the new developments of the last decade or two) is presented alongside classical optics.

## 23-1 Holography

Holography is one of the many applications of the laser, which was developed in the 1960s. A hologram to ordinary appearance looks like a photographic negative. However, such a negative has the following remarkable properties. It is strictly a two-dimensional surface, but, when viewed with light of one color, a full three-dimensional image is seen floating in space either in front or in back of the negative. The hologram image has the advantage over a stereo photograph in that as one moves to either side, one sees that side of the image come into view just as with a real object. Another remarkable feature is that the image on the negative is not at all that which is seen floating in space. The negative looks like a mess of overlapping fingerprints as shown in Figure 23-1. In spite of the fact that no part of the negative repeats itself, if any piece of the negative is cut out and placed in a beam of monochromatic light, the same floating three-dimensional image is still there, but with reduced optical resolution. No lenses are used either in the exposure of the negative or in the formation of the three-dimensional image.

**Figure 23-1.** *Greatly enlarged part of a hologram negative.*

The principle of the hologram is a graphic illustration of the wave nature of light and the distinction between coherent and incoherent light. A hologram negative of an object is made by shining a beam of coherent light (usually from a laser) on the object and letting the reflected light strike the film. Also some of the original beam strikes the film via a mirror

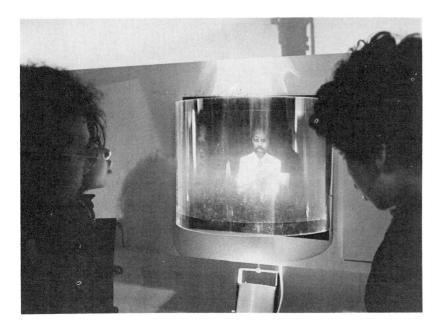

**Figure 23-2.** *Students viewing a holographic negative illuminated by monochromatic light. Students (and camera) see a full three-dimensional image floating in space. [Courtesy Museum of Holography, New York]*

SEC. 23-1/HOLOGRAPHY   497

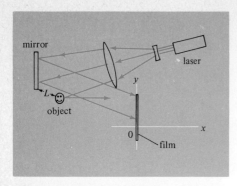

**Figure 23-3.** *How a hologram is made. Both reflected laser light from the object and a reference beam from the mirror strike the film.*

as is shown in Figure 23-3. If the light is coherent, there will be a definite wave interference pattern on the film; however, if the light is incoherent, the film will turn out uniformly gray. In this case the necessary coherence length is 2L, where L is the distance between mirror and object in Figure 23-3.

After the film is developed and a similar beam of light is passed through the film, the original wave front, which was at the position of the film during exposure, is reconstructed. Because of Huygens' principle the reconstructed wave front moves out toward the eye in the same way as the original wave front would have done. But how can a film reconstruct a wave front of correct amplitude and *phase* over its surface when film is only sensitive to light *intensity*? This can be seen in the following simplified mathematical description of holography.

Assume the film is in the $yz$ plane. The amplitude of the reflected light from the object as it reaches the $yz$ plane can be expressed as

$$E = a(y, z) \cos[\omega t + \phi(y, z)] \quad (23\text{-}1)$$

If, after development of the film, we can reproduce this amplitude function, the eye will think it is seeing the original object. Now, in addition to the preceding amplitude on the film, we shine a plane wave from the same laser. Then the net electric field over the plane of the film is

$$E_{\text{net}} = E_0 \cos \omega t + a \cos(\omega t + \phi) \quad \text{where } a = a(y, z) \text{ and } \phi = \phi(y, z)$$

The intensity over the film is proportional to the square of the preceding equation.

$$I = I_0 \cos^2 \omega t + 2E_0 a \cos \omega t \cos(\omega t + \phi) + a^2 \cos^2(\omega t + \phi)$$

The time average of any cosine squared term is $\frac{1}{2}$, hence

$$\bar{I} = \tfrac{1}{2} I_0 + E_0 a \overline{[\cos \phi + \cos(2\omega t + \phi)]} + \tfrac{1}{2} a^2$$

where we have used the trigonometric identity: $\cos A \cos B = \tfrac{1}{2} \cos(B - A) + \tfrac{1}{2} \cos(B + A)$. Since the average of the term $\cos(2\omega t + \phi)$ is zero, we have

$$\bar{I} = K_1 + E_0 a(y, z) \cos \phi(y, z) \quad \text{where } K_1 \equiv \tfrac{1}{2} I_0 + \tfrac{1}{2} a^2 \quad (23\text{-}2)$$

We note that by using light with a coherence length greater than $2L$ we have managed to store the phase information $\phi(y, z)$ on the film. The blackness on the film is proportional to $\bar{I}$. If the negative is illuminated by a similar laser beam of intensity $I' \cos^2 \omega t$, the intensity just after leaving the negative will be

$$I = I' \cos^2 \omega t [1 - K_2(K_1 + E_0 a \cos \phi)]$$

The corresponding electric field is proportional to the square root:

$$E = E' \cos \omega t [1 - (K_2 K_1 + K_2 E_0 a \cos \phi)]^{1/2}$$

$$E \approx K_3 \cos \omega t + 2K_4 a \cos \omega t \cos \phi$$

where

$$K_3 \equiv 1 - \frac{K_2 K_1}{2} \quad \text{and} \quad K_4 \equiv -\frac{K_2 E_0}{4}$$

Now we again use the relation $2 \cos A \cos B = \cos(A - B) + \cos(A + B)$:

$$E \approx K_3 \cos \omega t \ + K_4 a(y, z) \cos[\omega t + \phi(y, z)] + K_4 a(y, z) \cos[\omega t - \phi(y, z)]$$

$$= \text{(direct laser beam)} + K_4 \begin{bmatrix} \text{original light} \\ \text{from object} \end{bmatrix} + K_4 \begin{bmatrix} \text{light from object} \\ \text{with reversed phase} \end{bmatrix}$$

The eye sees the first term as the direct laser beam, and the second term as the reflected light from the object as if it were actually there (see Eq. 23-1); the third term appears as a superfluous secondary real image.

---

**Example 1.** A hologram is made of a very small sphere that is at a distance $x_0 = 50$ cm from the film (see Figure 23-4). The reflected light from the pointlike object has amplitude $E = E_1 \cos(\omega t - kr)$ and the reference beam has amplitude $E_0 \cos(\omega t - kx_0)$. If $\lambda = 6400$ Å, what is the intensity pattern on the film?

ANSWER: Intensity maxima will occur when the phase difference between the reflected and reference beams is $2\pi$ times an integer, or when

$$kr - kx_0 = 2\pi n$$
$$r - x_0 = n\lambda$$
$$\sqrt{y_n^2 + x_0^2} = n\lambda + x_0$$
$$y_n = \sqrt{2n\lambda x_0 + n^2 \lambda^2} \approx 0.08 \sqrt{n} \text{ cm}$$

The intensity pattern consists of concentric circles of radii 0, 0.8 mm, 1.13 mm, 1.39 mm, 1.6 mm, and so on.

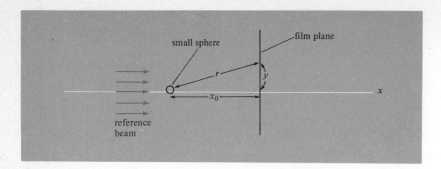

**Figure 23-4.** *Reflected laser light from small sphere and reference beam strike film.*

Holography is a field in itself that, although invented in 1949, became popular in the early 1960s with the development of the laser. It is a major field of current research in optics. Much of the 1960s was spent in understanding various aspects of the theory and developing new techniques. Holography is still too new a field for us to foresee the ultimate practical applications. Research and development is underway for medical applications, such as optical holograms of the eye that show the lens and retina in the same three-dimensional image, or acoustical holograms of the body that may have important advantages over two-dimensional x rays. Other research and development in progress involves video tape cassettes, computer memory storage, and nondestructive structural testing.

## 23-2 Polarization of Light

At any given instant of time a light wave or an electromagnetic wave consists of mutually perpendicular **E** and **ℬ** fields at each point in space. The polarization direction is defined as the direction of **E**. The plane of polarization is defined as the plane containing both **E** and the beam direction. Hence the direction of **ℬ** is perpendicular to the plane of polarization. Electromagnetic radiation in which the direction of **E** stays fixed is called plane-polarized radiation. For a beam from an incoherent source the electric field direction wanders around randomly while staying perpendicular to the beam direction. Such a beam is said to be unpolarized.

### Circular Polarization

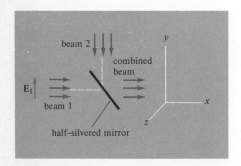

**Figure 23-5.** *Two polarized beams are combined by a half-silvered mirror. The direction of $E_2$ is out of the paper parallel to z axis.*

Suppose two coherent beams of light are superimposed by a half-silvered mirror as shown in Figure 23-5. Beam 1 is vertically polarized ($E_1$ in $xy$ plane) and beam 2 is horizontally polarized ($E_2$ in $xz$ plane). If the phase difference between the two beams is zero, and $E_1 = E_2$ in magnitude, is the resultant beam unpolarized? If the phase difference is $\phi = \pi/2$, is the resultant beam unpolarized?

In answer to the first question let us describe the waves as $E_1 = E_{1_0} \cos(\omega t - kx)$ and $E_2 = E_{2_0} \cos(\omega t - kx)$, where $E_{1_0}$ and $E_{2_0}$ are at right angles, as shown in Figure 23-6. We see that the resultant electric field will always be in a plane at an angle $\alpha$ to the vertical, where $\tan \alpha = E_{2_0}/E_{1_0}$. For $E_{1_0} = E_{2_0}$, $\alpha = 45°$. Hence the resultant beam of light is plane-polarized with the plane of polarization at an angle of 45° from the vertical.

If the two beams are 90° out of phase, we have

$$E_y = E_{1_0} \cos \omega t \quad \text{and} \quad E_z = E_{2_0} \cos\left(\omega t - \frac{\pi}{2}\right)$$

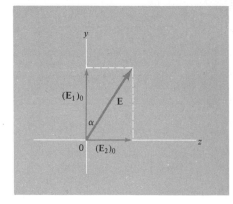

**Figure 23-6.** *View in yz plane of beams from Figure 23-4.*

at $x = 0$. Successive views of the components $E_1$ and $E_2$ are shown in Figure 23-7. It is seen that the resultant vector $E$ stays constant in magnitude and rotates clockwise around the $x$ axis once each period $T$. This is called left-circular polarization. If the direction of $E$ rotates counterclockwise while looking down the beam direction, it is called right-circular polarization.

In some respects circular polarization behaves as an unpolarized beam. However, if a left- and a right-circular polarized beam of equal strength are superimposed, the resultant will be a plane-polarized beam. Clearly this cannot be done with unpolarized beams. Circular polarization does occur in nature. It turns out that individual photons are circularly polarized.

**Figure 23-7.** *Successive views of the electric field in the yz plane when beam 2 is lagging beam 1 by a phase angle of $\pi/2$. The resultant field E, which is shown above each view, rotates clockwise. $E_y$ and $E_z$ are of equal amplitude.*

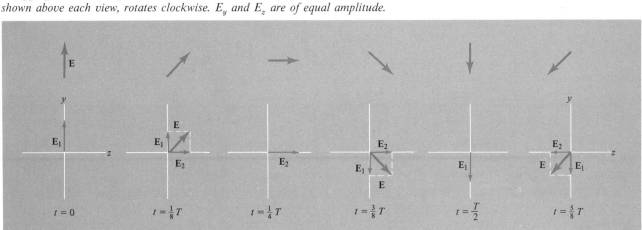

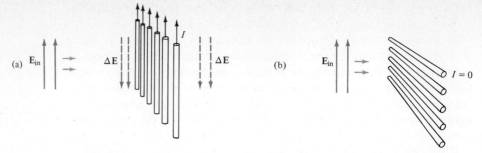

**Figure 23-8.** *Vertically polarized electromagnetic wave impinging on a sheet of parallel wires. (a) The wires are vertical, and the wave is reflected back. (b) The wires are horizontal, and the wave is not reflected back; it passes through the wires unattenuated.*

## Polarizers

Most light sources emit incoherent, unpolarized light. A beam of unpolarized light can be polarized by sending it through what is called a polarizer. A screen of thin parallel wires works fine as a polarizer for microwaves as is seen in Figure 23-8. If the microwave beam is vertically polarized and the wires run vertically as in Figure 23-8(a) there will be an induced current $I$ in each wire. As explained on page 448, the induced current will be such that its radiated field $\Delta \mathbf{E} = -\mathbf{E}_{\text{in}}$. The resultant field to the right of the polarizer will be $\mathbf{E} = \mathbf{E}_{\text{in}} + \Delta \mathbf{E} = 0$. Thus the polarizer in this orientation behaves as a perfect reflector and does not let the beam pass through. If the wires are perpendicular to $E_{\text{in}}$, there is no "room" for a vertical induced current. Hence there is no additional radiation and the incoming wave is undisturbed.

Let us define the axis of the polarizer in Figure 23-8 as perpendicular to the wires. We see in Figure 23-9 that if a polarizer has its axis at an angle $\alpha$ to $\mathbf{E}_{\text{in}}$, it will radiate a field $\Delta E$ at right angles to the axis. Since $\Delta E$ cancels out the component of $\mathbf{E}_{\text{in}}$ in this direction, the resultant field $\mathbf{E}'$ is the component of $\mathbf{E}_{\text{in}}$ parallel to the axis and thus of magnitude

$$E' = E_{\text{in}} \cos \alpha$$

or

$$I' = I_{\text{in}} \cos^2 \alpha \tag{23-3}$$

For an ideal polarizer, the intensity is multiplied by a factor of $\cos^2 \alpha$ where $\alpha$ is the angle between the plane of polarization and the axis of the polarizer. When the axis of a polarizer is aligned with the plane of

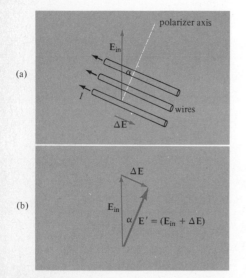

**Figure 23-9.** (a) *View looking down beam (beam is into page). Vertically polarized beam is at angle $\alpha$ to parallel wire polarizer which radiates $\Delta \mathbf{E}$.* (b) $\mathbf{E}'$, *the resultant field behind the wires.*

polarization, the maximum intensity is transmitted. Any radiation leaving a polarizer will be plane-polarized along the axis of the polarizer.

A polaroid filter for light works on the same principle. Polaroid is made out of stretched plastic that is mainly long parallel chains of molecules. These chains are coated so as to conduct current along their length. Thus polaroid is a microscopic version of the parallel wire polarizer. The axis of a polaroid filter is of course perpendicular to the molecular chains. In unpolarized light the components of **E** parallel to the molecular chains are absorbed out of the beam. Only **E** field parallel to the polaroid axis survives after passing through a polaroid filter. If a second polaroid is placed with its axis at right angles to the first as in Figure 23-10(a), the beam is completely absorbed and essentially no light gets through the second polaroid.

Now if a third polaroid is placed between the two crossed polaroids as in view 23-10(b), there will be light where there was none before! How can an additional polaroid "create" light? In order to understand this we let $I_0$ be the intensity of the polarized light entering the center polaroid ($I_0 = \frac{1}{2} I_{in}$). Then the light leaving it has intensity $I' = I_0 \cos^2 \alpha$ and is polarized at the angle $\alpha$. The last polaroid has its axis at an angle $(\pi/2) - \alpha$ to this plane of polarization. Hence

$$I'' = I' \cos^2\left(\frac{\pi}{2} - \alpha\right)$$

$$= (I_0 \cos^2 \alpha) \cos^2\left(\frac{\pi}{2} - \alpha\right)$$

$$= \tfrac{1}{4} I_0 \sin^2 2\alpha$$

This is maximum at $\alpha = 45°$ with a final intensity that is one eighth of $I_{in}$, assuming ideal polaroids.

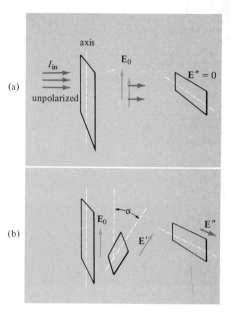

**Figure 23-10.** (a) *Two perpendicular polaroids cut out all the light.* (b) *When a third polaroid is inserted between them, some light gets through.*

---

*** Example 2.** Polarized monochromatic light strikes a double slit as shown in Figure 23-11. The axis of polarization is halfway between the $y$ and $z$ axes (in the direction of the vector $(\mathbf{j} + \mathbf{k})$). When both slits are open (without polaroids), there is a standard double-slit interference pattern at the screen. However, in this problem a polaroid with a horizontal axis is placed over slit 1 and a polaroid with a vertical axis is placed over slit 2. Let $I_0$ be the intensity of light striking the screen when there is just slit 1 or slit 2 by itself (with its polaroid). For points $a$, $b$, $c$, $d$ the path differences are such that $k(r_1 - r_2)$ is $0$, $\pi/2$, $\pi$, and $2\pi$, respectively. What are the intensities and states of polarization at points $a$, $b$, $c$, and $d$?

ANSWER: The field from slit 1 is $\mathbf{j}E_0 \cos(\omega t - kr_1)$ and from slit 2 is $\mathbf{k}E_0 \cos(\omega t - kr_2)$; hence, at point $a$ the electric field is

$$\mathbf{E} = \mathbf{j}E_0 \cos(\omega t - kr_1) + \mathbf{k}E_0 \cos(\omega t - kr_1) \qquad \text{using } r_1 = r_2$$

$E = \sqrt{2}E_0 \cos(\omega t - kr_1)$, pointing at 45° from the vertical. The direction of polarization is $(\mathbf{j} + \mathbf{k})$. We shall call it +45°. The intensity is $E^2 = 2E_0^2 \cos^2(\omega t - kr)$, which is twice that of a single slit.

At point $d$ the two cosine functions are in phase, and we have the same result as at point $a$.

At point $c$ the two cosine functions are 180° out of phase and

$$\mathbf{E} = \mathbf{j}E_0 \cos \omega t - \mathbf{k}E_0 \cos \omega t$$

$E = \sqrt{2}E_0 \cos \omega t$ with the direction of polarization along $(\mathbf{j} - \mathbf{k})$. We shall call it $-45°$. (We have defined $t = 0$ as occurring when the amplitude from slit 1 is at a maximum.) This result also has twice the intensity of a single slit, but the plane of polarization is at right angles to that at points $a$ and $d$.

At point $b$ the two waves are 90° out of phase, and we have the same situation as shown in Figure 23-7. This is a circularly polarized wave where $\overline{E^2} = E_0^2$. Since a single slit gives a time average $\overline{E^2} = \frac{1}{2}E_0^2$, the average intensity is also twice that of a single slit.

We see in this problem that the intensity is uniform over the entire screen but that the state of polarization changes from +45°, to circular, to $-45°$, to circular, to +45°, and so on, as one moves across the screen.

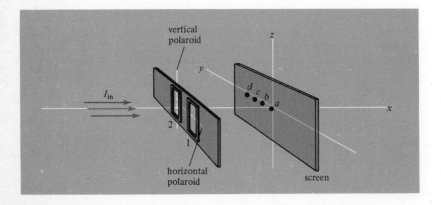

**Figure 23-11.** *Setup for double slit interference. Slits 1 and 2 are covered with horizontal and vertical polaroids, respectively.*

> **\*Example 3.** (a) Suppose in Example 2 that an additional horizontal polaroid is placed just in front of the screen. Now what is the intensity pattern on the screen?
>
> (b) Suppose in Example 2 that a polaroid with axis at $+45°$ (halfway between $y$ and $z$ axes) is placed just in front of the screen. Now what is the intensity pattern?
>
> ANSWER: (a) At any point on the screen a distance $r$ from slit 1, $\mathbf{E} = \mathbf{j}E_0 \cos(\omega t - kr)$ and $E^2 = E_0^2 \cos^2(\omega t - kr)$. This is the intensity pattern from slit 1 alone (with no polaroid at the screen). Light from slit 2 is blocked by the polaroid and cannot reach the screen. If any light reaches the screen we know it came from slit 1.
>
> (b) The radiation at points $a$ and $d$ on the screen have their plane of polarization parallel to the polaroid, whereas at point $c$ the radiation is perpendicular to the polaroid. The intensity is $2I_0$ at points $a$ and $d$, zero at point $c$, and $I_0$ at point $b$. This is the classic double-slit interference pattern corresponding to each slit giving an intensity of $\frac{1}{2}I_0$ by itself.

We note in Examples 2 and 3 that it is possible to tell which slit the light went through depending on the orientation of the polaroid in front of the screen. If it is oriented horizontal or vertical, we can tell whether the light came from slit 1 or slit 2. If it is oriented at $+45°$ or $-45°$, the slit polarization information is lost and it is no longer possible to tell which slit the light came from. In the $\pm 45°$ case we have the classic double-slit interference pattern, but in the case of horizontal or vertical orientation the interference pattern is lost.

## Polarization by Reflection

Almost everyone has observed that polaroid glasses reduce the reflection of sunlight by water, sand, glass, road surfaces, and so on. Apparently unpolarized sunlight after reflection is polarized. Then if the axis of the polaroid in the glasses is perpendicular to the plane of polarization of the reflected light, the reflected light will be cut out. It is shown in Appendix 23-1 that if the angle of incidence $\theta_1$ of the beam of light to the reflecting surface is related to the index of refraction $n$ by $\tan \theta_1 = n$, then the reflected light will be 100% polarized and the electric field will be perpendicular to the plane containing the incident and reflected rays.

**Example 4.** If the index of refraction of water is 1.33, at what angle will the surface reflection completely disappear using polaroid glasses? Should the polaroid axis in the glasses be horizontal or vertical?

ANSWER: For this special angle we have $\tan\theta_1 = 1.33$. Then $\theta_1 = 53°$. Since the reflected light will be horizontally polarized, the polaroid axis in the glasses should be vertical. Then if the sun is 37° above the horizon, its reflection in a smooth surface of water would be cut out by the glasses.

## 23-3 Diffraction by a Circular Aperture

Instead of a single slit of width $a$, let us now consider a circular hole of diameter $a$. According to Huygens' principle we must add up the radiation or wavelets originating from all points inside the open circle in order to get the resultant amplitude. Each wavelet contributes $\cos kr\, dA$ where, according to the Fraunhofer approximation, $r$ is the distance shown in Figure 23-12 and $dA$ is an element of area inside the circle. The amplitude at an angle $\theta$ is the integral

$$\int_{\text{hole}} \cos[kr(\theta)]\, dA$$

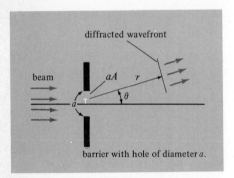

**Figure 23-12.** *Diffraction by a hole of diameter a. Each element of area dA contributes an amplitude in the direction $\theta$.*

This is a fairly complicated integral that results in a function quite familiar to physicists, engineers, and mathematians called the first-order Bessel function $J_1(x)$. This function is plotted in Figure 23-13 and has the form

$$J_1(x) \approx \sqrt{\frac{2}{\pi x}} \sin\left(x - \frac{\pi}{4}\right) \quad \text{for } x > 5$$

The final result for intensity at an angle $\theta$, which is obtained by integrating over the hole and squaring, is

$$I = I_0 \left[\frac{J_1(\Phi/2)}{\Phi/4}\right]^2 \quad \text{where } \Phi = ka \sin\theta \quad (23\text{-}4)$$

Note the similarity to the single-slit result of Eq. 22-10, which is

$$I = I_0 \left[\frac{\sin\Phi/2}{\Phi/2}\right]^2$$

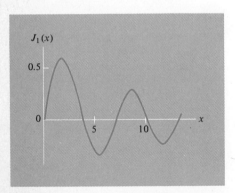

**Figure 23-13.** *Plot of the first order Bessel function.*

For the single slit the first minimum is at $\sin\theta_{\min} = \lambda/a$. For the circular aperture, the first minimum occurs when $J_1(x) = 0$. As seen from

**Figure 23-14.** (*a*) *A photograph of the diffraction pattern produced by monochromatic light striking a circular aperture.* (*b*) *A plot of the intensity pattern given by Eq. 23-4.*

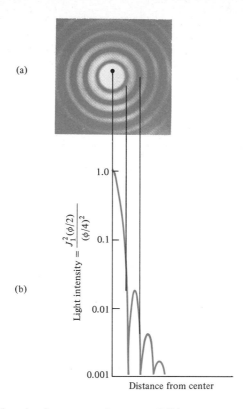

Figure 23-13, this Bessel function has its first zero when $x = 3.84$

$$\left(\frac{\Phi_{min}}{2}\right) = 3.84$$

or

$$\frac{ka \sin \theta_{min}}{2} = 3.84$$

$$\sin \theta_{min} = 1.22 \frac{\lambda}{a} \qquad (23\text{-}5)$$

Equation 23-5 differs from the single-slit result by only 22%. A photograph of the diffraction pattern from a circular aperture is shown in Figure 23-14(a). Its calculated form (Eq. 23-4) is plotted in Figure 23-14(b). The pattern consists of a central disk of light surrounded by faint rings.

We shall next apply the results of this section in order to obtain the resolving powers of the telescope and the microscope. Then we shall apply these results to the supermicroscopes of the particle physicists—the modern high-energy accelerators. We shall show how they are used to determine the sizes and shapes of nuclear particles.

## 23-4 Optical Instruments and Resolution

We have already studied the resolution of a spectrometer on page 483. It is a measure of how close together in wavelength two spectral lines can be and still be seen as two separate lines. The optical resolution of a telescope or a microscope is similar. It is how close together two point sources of light can be and still be seen as two separate sources. In the case of the spectrometer, if two lines are too close they will appear as one line because of the artificial width introduced by the instrument (which is $\Delta\lambda = \lambda/N$ for a diffraction grating). Similarly, in a telescope or microscope, if two point sources are too close they will appear as one because the optical instrument makes a point source look like a small disk or spot of light with circular diffraction rings. In fact, it will be the same diffraction pattern shown in Figure 23-14.

### The Telescope

In order to understand why a telescope cannot give a point image of a point source, we must first briefly explain how a telescope works. Consider the telescope in Figure 23-15, which is viewing two point sources, A and B, that are a great distance away. The telescope consists of an objective lens (or concave mirror) of focal length $F$. A lens or concave mirror has the property that parallel rays are focused to a point at a distance $F$ from the lens where $F$ is called the focal length. We see from Figure 23-15 that if the angle between A and B is $\alpha$, the distance between their images is $y_0 = F\alpha$. In astronomical telescopes these images are recorded directly onto film placed at the focal plane. The images could also be viewed by an eyepiece (another lens) of focal length $f$ placed a distance $f$ to the right of the focal plane. Then the apparent angle between A and B would be

$$\alpha' = \frac{y_0}{f} = \frac{F\alpha}{f}$$

The magnification $M$ is defined as

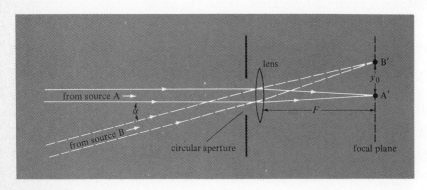

**Figure 23-15.** *Two point sources, A and B, are viewed by this telescope. Their images, A' and B', are in the focal plane.*

$$M \equiv \frac{\alpha'}{\alpha} = \frac{F}{f} \quad \text{(telescope magnification)} \qquad (23\text{-}6)$$

Even though the magnification can be made as large as one desires by choosing appropriate focal lengths, the real limit is resolving power. Resolving power is expressed as $\alpha_{min}$, which is the minimum angle between two point sources that allows the images to be resolved as two spots of light rather than one.

The radius $R_{disk}$ of the central disk of light from a single point source can be obtained using Figure 23-16. According to Eq. 23-5, the first minimum of the diffraction pattern from the circular aperture is

$$\sin\theta_{min} \approx \theta_{min} \approx 1.22\frac{\lambda}{a}$$

The radius of the corresponding disk is

$$R_{disk} = F\theta_{min} = F\left(1.22\frac{\lambda}{a}\right)$$

If the images of two point sources are separated by this amount or more, they can be distinguished as two rather than one disk. This condition is that $y_0$ (or $F\alpha$) be greater than $R_{disk}$:

$$F\alpha > 1.22 F\frac{\lambda}{a}$$

$$\alpha > 1.22\frac{\lambda}{a}$$

The resolution is

$$\alpha_{min} = 1.22\frac{\lambda}{a} \qquad (23\text{-}7)$$

**Figure 23-16.** *Circular aperture in front of lens produces diffraction pattern that is focused in focal plane.*

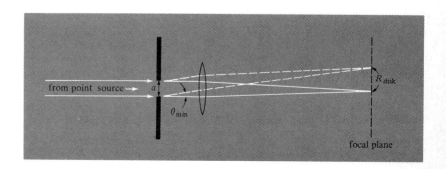

**Example 5.** Two stars are separated by an angle of $10^{-6}$ radian. (a) Can they be resolved by a 100-in. optical telescope? (b) Assuming that they also emit detectable radio signals at 400 MHz, can they be resolved by the Cornell Aricebo radio telescope with its 1000-ft aperture?

ANSWER: A 100-in. telescope has a circular aperture yielding a diffraction angle $\theta_{min} = 1.22\lambda/a$, where $a = 2.54$ m. For light at $\lambda = 5 \times 10^{-7}$ m

$$\theta_{min} = 1.22 \frac{5 \times 10^{-7}}{2.54} = 2.4 \times 10^{-7} \text{ rad}$$

The separation between the two stars is four times this, so they should be resolved (assuming no atmospheric turbulence).

For Aricebo, $a = 305$ m and $\lambda = c/f = (3.00 \times 10^8 \text{ m/s})/(4 \times 10^8 \text{ s}) = 0.75$ m. Then

$$\theta_{min} = 1.22 \frac{0.75}{305} = 3 \times 10^{-3} \text{ rad}$$

We would need 3000 times more resolution to separate these two stars. (In actual practice this sort of thing can be done by electrically connecting together radiotelescopes separated by more than 600 miles.)

## The Microscope

A high-powered optical microscope uses an objective lens with a short focal length $F$. The object to be magnified is placed a distance $\sim F$ from the lens, as shown in Figure 23-17. The image plane is at a much larger distance $D$. Since rays passing through the center of the lens are undeflected, the magnification $M = A'B'/AB$ in Figure 23-17(b). Thus $M = D/F$. As with the telescope, either a photographic plate can be placed at the focal plane or the image there can be examined with an eyepiece.

Light from source A leaving the circular aperture at an angle $\theta_{min} \approx 1.22\lambda/a$, when focused on the focal plane, is at the first intensity minimum. The radius of the disk of light would be this angle times the distance $D$:

**Figure 23-17.** *In the microscope a lens of short focal length focuses image on an image plane a distance D away.*

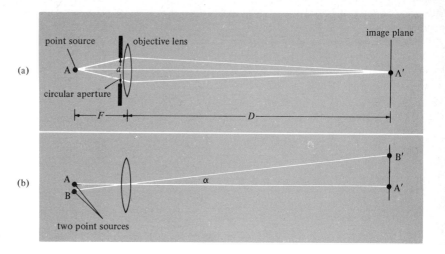

$$R_{\text{disk}} \approx \theta_{\min} D \approx 1.22 \frac{\lambda}{a} D$$

If $\alpha$ is the angle subtended by two point sources at the microscope objective, the distance between their images is $y = D\alpha$ and the condition for minimum resolvable $\alpha$ is

$$y_{\min} = R_{\text{disk}}$$

$$D\alpha_{\min} = 1.22 \frac{\lambda}{a} D$$

$$\alpha_{\min} = 1.22 \frac{\lambda}{a}$$

This is the same as for a telescope (Eq. 23-7).

In a microscope it is of more interest to estimate the minimum resolvable distance between two point sources. Let this distance be $d_{\min}$. Since $d_{\min} = \alpha_{\min} F$, we have

$$d_{\min} = 1.22\lambda \frac{F}{a}$$

The ratio $a/F$ is defined as the numerical aperture. For a good oil-immersion objective it is typically about 1.2. Then

$$d_{\min} \approx \lambda$$

We conclude that a microscope cannot resolve small objects if they are closer than a wavelength of light from each other. Resolution can be improved somewhat by using blue light in a microscope. It can be further improved by using ultraviolet or x rays. In the next section we shall see how the ultimate resolution can be achieved using high energy accelerators. The best resolution obtained so far is about one tenth the radius of the proton or $10^{-16}$ m.

### Resolution of Laser Beam

If parallel light is focused on a screen by a lens of focal length $F$ and aperture $a$, there will be a diffraction disk size $R_{disk} = 1.22\lambda F/a$ as explained on page 509. Since the largest practical value for $a$ is $a \approx F$, we have $R_{disk} \approx \lambda$. This tells us that if we have a perfect source of parallel light, it can be focused down to a spot size of about one wavelength of light. Because a laser gives out a beam of parallel light, all the energy output of the laser can be concentrated down into an area of about $(6 \times 10^{-6} \text{ m})^2$ or $\sim 4 \times 10^{-11}$ m². This is an extremely high energy density capable of producing extremely high temperatures, and is a very important advantage of lasers over other sources of light. Previous to the invention of the laser all sources of light were extended sources, such as an arc or filament of a millimeter or so in diameter. If a lens is close enough to such a source as to collect a significant fraction of the energy, the lens must give an image size comparable to the source size. So there was no way to produce an energy density greater than that of the light source until the invention of the laser.

---

**Example 6.** The light leaving a laser is parallel and of circular cross section with a 5-cm diameter. The wavelength is 6328 Å. If this laser beam is pointed at the dark part of the moon, how large a disk on the moon's surface would be illuminated? The distance is $D = 3.84 \times 10^8$ m.

ANSWER:

$$R_{disk} = 1.22 \frac{\lambda D}{a} = 1.22 \frac{(6328 \times 10^{-10} \text{ m})(3.84 \times 10^8 \text{ m})}{0.05 \text{ m}}$$

$$= 5.9 \text{ km}$$

Shortly after the invention of the laser, such laser spots on the moon were detected using large telescopes.

One could use the equation $\sin\theta_{min} = 1.22\lambda/a$ to determine the unknown diameter of a small pinhole. If a screen is put behind a pinhole, one will see the diffraction pattern of Figure 23-12. Let $R_{disk}$ be the distance from the center of the pattern to the first minimum and let $D$ be the distance to the screen. Then

## 23-5 Diffraction Scattering

$$\theta_{min} = \frac{R_{disk}}{D} = 1.22\frac{\lambda}{a}$$

Solving for $a$ gives

$$a = \frac{1.22\,\lambda D}{R_{disk}}$$

A simple measurement of $R_{disk}$ yields the desired pinhole diameter.

This same method for measuring tiny holes can also be used to measure the diameter of a tiny sphere or black disk. It is easy to see that the diffraction pattern from a black disk will be the same as that from a circular hole. In order to calculate the circular hole we had to invoke Huygens' principle and calculate as if the hole itself was an extended source. In the case of the black disk we have a ready-made extended source due to the induced current in it. The diffraction pattern from a black disk will be the same as in Figure 23-14 except that there will also be a bright spot at the center corresponding to the incoming beam at 0°.

**Example 7.** A laser beam of wavelength $\lambda = 6328$ Å shines on a glass slide covered with lycopodium powder (spherical spores from the lycopodium moss). On a screen at a distance of 5 m one sees a red disk with the first black ring at a distance of 19 cm from the disk center. What is the diameter of a lycopodium spore?

ANSWER:

$$\theta_{min} = \frac{19\text{ cm}}{500\text{ cm}} = 38 \times 10^{-3}\text{ rad} = \frac{1.22\lambda}{a}$$

$$a = \frac{1.22\lambda}{38 \times 10^{-3}} = \frac{1.22 \times 6328 \times 10^{-10}\text{ m}}{38 \times 10^{-3}} = 2.03 \times 10^{-5}\text{ m}$$

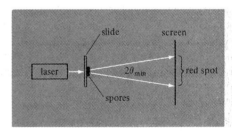

**Figure 23-18.** *Black disks on glass slide give diffraction pattern on screen.*

How can we measure the diameter of a sphere if it is much smaller than a wavelength of light? One could try using "light" of smaller wavelength such as x rays or gamma rays. In place of "light" one could use any other

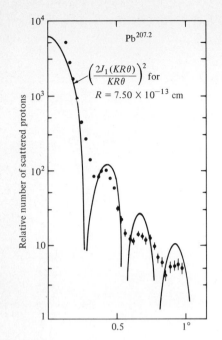

**Figure 23-19.** *Results of scattering 19-BeV protons on lead. Number of scattered protons is plotted versus scattering angle. The smooth curve is the classical diffraction pattern corresponding to light of the same wavelength shining on a disk of radius $7.5 \times 10^{-15}$ m.*

kind of plane waves as long as $\lambda < a$ and the disk (or sphere) is strongly absorbing. Any plane wave with a circle absorbed out of it must give the $[J_1(\Phi/2)/(\Phi/4)]^2$ diffraction pattern.

At this point we must invoke what is the most shocking and revolutionary idea in all of physics. It has been proved beyond any doubt that any moving particle can be represented as a plane wave of wavelength $\lambda = h/p$ where $h = 6.63 \times 10^{-34}$ J s and $p$ is the particle momentum. Such a wild idea was proposed in 1924 and first demonstrated experimentally in 1927. This wave nature of matter will be our central theme starting with the next chapter. For now, let us take it for granted (the evidence for it will be presented in the next chapters). Then a beam of protons of energy 19 GeV ($19 \times 10^9$ eV) could be represented as plane waves of wavelength $\lambda = 0.65 \times 10^{-16}$ m. This is significantly smaller than the diameters of atomic nuclei, which are $\sim 10^{-14}$ m. So if the equation $\lambda = h/p$ is true, we should be able to see a diffraction pattern by "shining" a 19-GeV proton beam on a foil of lead. Such an experiment is easy to perform using a proton beam from a high-energy accelerator. The results of such an experiment are shown in Figure 23-19. This result could be used as one more experimental proof of the wave nature of matter; or, accepting the wave nature of matter, we can use the result to measure the size and edge thickness of the lead nucleus.

Whether the particles in the beam are photons (particles of light) or protons, the solution to the wave equation is the same as that of the classical problem of light waves diffracted by a black disk of radius $R$; that is, Eq. 23-5 must apply to high-energy protons as well. All we need do is to replace $\lambda$ in Eq. 23-5 with $h/p$ to obtain

$$\sin\theta_{min} = 1.22 \frac{h}{pa}$$

Since $a = 2R$ we have

$$\sin\theta_{min} = 0.61 \frac{h}{pR} \qquad (23\text{-}8)$$

The proton diffraction pattern is observed by measuring the number of elastically scattered protons as a function of $\theta$. The points in Figure 23-19 show the results of measurements made at 30 different angles ranging from 0.11° to 1.03°. The smooth curve is the standard intensity pattern of light diffracted by a black disk, assuming light of the same wavelength as the protons. The departures of the experimental points from the smooth curve are not caused by any violation of the wave nature of matter, but because the lead nucleus does not behave as a simple black disk with sharp edges. A black disk with partially transparent edges can give a smooth curve that would agree with all the experimental points. Actually, the experimental results shown in Figure 23-19 can be used to determine the thickness of the edge region as well as the radius of the lead nucleus. The radius is obtained by solving Eq. 23-8 for $R$:

$$R = 0.61 \frac{h}{p \sin \theta_{min}} \qquad (23\text{-}9)$$

The results in Figure 23-19 show that $\theta_{min}$ occurs at 0.3°, which, when substituted into Eq. 23-9 gives $R = 7.5 \times 10^{-15}$ m for the radius of the lead nucleus.

Even scattering of protons against protons gives a diffraction pattern, but in this case the pattern departs even more from that of the black disk with sharp edges because the proton is increasingly transparent from its center to its edge. The effective radius of the proton can be obtained by measuring the width of the central maximum. The result is $1.1 \times 10^{-15}$ m for the "optical" radius when protons are diffracted by protons. But what about possible structure of the proton? Does it have a hard, "black" core as does the atom? A small black disk or core superimposed on a large semitransparent disk will give a complex diffraction pattern: the central maximum will be a superposition of two central maxima of the two corresponding widths. So far no such effect has been clearly seen. An upper limit on the size of a possible core can be obtained from Eq. 23-9 by inserting the highest value of $p \sin \theta$ observed so far.

Proton-proton scattering has been done at the Brookhaven AGS utilizing the full energy of the accelerator. A target containing hydrogen was bombarded by a beam of 32-GeV protons. The scattered and recoil protons each went off at 15° from the beam direction, each with half the beam energy. The product $pc \sin \theta$ is then 4 GeV or $p \sin \theta = 2.13 \times 10^{-18}$ kg m/s which, when inserted into Eq. 23-9, gives a value of $0.19 \times 10^{-15}$ m. Actually, a core of about half this size should give enough of its central maximum to be detectable. Hence we conclude that the proton has no core as large as $0.1 \times 10^{-15}$ m. We see that the AGS and other high-energy accelerators can correctly be thought of as super microscopes that can measure distances down to a fraction of a fermi. (The fermi is a unit of length equal to $10^{-15}$ m.) This is a billion times more resolution than the best optical microscope. In order to see structure in the proton with a resolution better than $0.1 \times 10^{-15}$ m, higher-energy accelerators must be used.

## 23-6 Geometrical Optics

The wavelength of light is so small compared to most optical instruments that interference effects usually do not show up. A wave train or sequence of light waves progresses forward in a straight line perpendicular to the wave front. Any such straight line indicating the direction of motion of light waves is called a light ray. As we shall see, light rays obey the law of reflection (from mirrors) and the law of refraction (for transparent media such as lenses). An entire mathematics or geometry can be developed using these two laws along with the usual rules of Euclidian geometry. This mathematics of rays, a subject in itself, is called geometrical optics. Since the only new physics principles involved here are the laws of reflection and refraction, we shall concentrate on these two laws and treat the rest of the subject rather lightly.

### Law of Reflection

The law of reflection states that when a light ray strikes a reflecting surface, the angle of incidence equals the angle of reflection. The angle of incidence is defined as the angle the incoming beam makes with the perpendicular to the reflecting surface. Hence the angle between incident wavefront and reflecting surface is also the angle of incidence $\theta_{in}$.

In Figure 23-20(a) we see that the incoming electric field induces a surface current $\mathcal{J}(y, t)$ that varies sinusoidally with $y$. This surface current in the conductor adjusts itself so that the field inside the conductor is always zero. This means that the field it radiates to the right must exactly cancel out $E_{in}$. Hence $E'_R = -E_{in}$ and $\theta'_R = \theta_{in}$, as shown in Figure 23-20(b). Symmetry conditions require that $E'_L = E'_R$ and $\theta'_L = \theta'_R$. [If Figure 23-20(b) is turned around left for right, it is the same physical situation.] So we have just proved that the angle of incidence equals the angle of reflection ($\theta_{in} = \theta'_L$) and that, for a conducting surface, the reflected wave is of the same magnitude but with its component along the surface reversed in direction at the reflecting surface.

As an application of the law of reflection, we shall show how a concave mirror behaves as a focusing lens. It is well known that a simple lens or magnifying glass will focus parallel rays to a common point called the focus. This is also true of a concave mirror. For example, a concave shaving mirror can be used to burn a hole in a piece of paper by pointing it toward the sun and holding the paper at the focus. As shown in Figure 23-21(a), the focal distance will be one half the radius of curvature of the mirror. In this figure an arbitrary ray $AP$ is selected out of a bundle of parallel rays. Let $\theta$ be the angle between this ray and the normal ($CP$) to the mirror. Note that $CP$ is the radius of curvature of the mirror. Accord-

(a)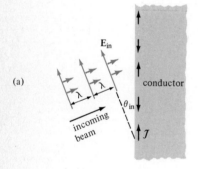

(b)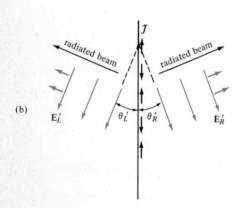

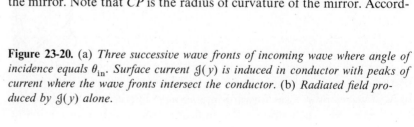

**Figure 23-20.** (a) Three successive wave fronts of incoming wave where angle of incidence equals $\theta_{in}$. Surface current $\mathcal{J}(y)$ is induced in conductor with peaks of current where the wave fronts intersect the conductor. (b) Radiated field produced by $\mathcal{J}(y)$ alone.

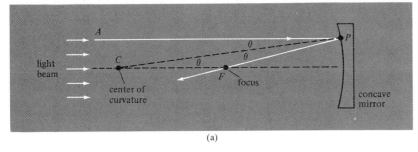

**Figure 23-21.** (a) *Parallel rays of light striking a concave mirror of radius CP.* (b) *If a human eye is placed at the center of curvature of a group of plane mirrors, it will see itself "everywhere." Rays from the object return back to the object.*

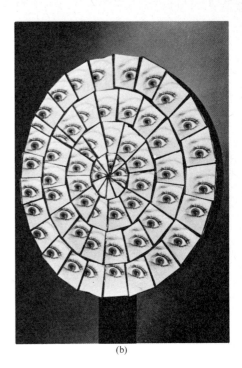

ing to the law of reflection, angle $APC$ must equal angle $FPC$ and the triangle $FPC$ must then be isosceles. Hence the two sides $CF$ and $FP$ are equal and each is very close to half the distance from $C$ to $P$, or radius of curvature.

Figure 23-22 shows how to obtain the image of an object (an arrow) graphically if the focal position $F$ is known. Draw ray (1) from the arrowhead parallel to the axis of the mirror. Draw ray (2) from the arrowhead to the center of the mirror. Where the two rays intersect is the image point of the arrowhead. All other rays from the arrowhead will also pass (or almost pass) through this same image point. A concave mirror can be used to form the image of a distant object. The image can be further magnified with a short focal length lens or eyepiece. This arrangement is a telescope and has a magnification equal to the ratio of the focal lengths as shown in Eq. 23-6. In astronomical telescopes photographic plates are placed directly at the focus of a large concave mirror. This common type of astronomical telescope, which was developed by Isaac Newton, is called a reflecting telescope.

**Figure 23-22.** *Image formation by a concave mirror. The rays* (1) *and* (2) *show how to obtain the image position graphically.*

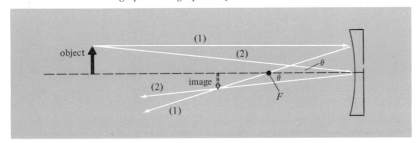

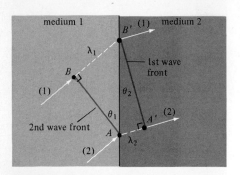

**Figure 23-23.** *Two successive wave fronts crossing a glass–air surface.*

## Law of Refraction (Snell's Law)

The law of refraction states that a light ray changes direction as it passes from one transparent medium to another. For example, if a light ray in air enters water or glass, it is bent toward the perpendicular. We can use Figure 23-23 to see why the direction of wave propagation is altered at an interface. The figure shows a portion of two successive wave fronts $AB$ and $A'B'$. Let $\lambda_1$ be the wavelength in medium 1 and $\lambda_2$ the wavelength in medium 2.

$$\lambda_1 = \frac{u_1}{f} \quad \text{and} \quad \lambda_2 = \frac{u_2}{f} \qquad (23\text{-}10)$$

In the right triangle $ABB'$

$$\sin \theta_1 = \frac{\lambda_1}{AB'}$$

In the right triangle $A'AB'$

$$\sin \theta_2 = \frac{\lambda_2}{AB'}$$

Dividing the first of these two equations by the other gives

$$\frac{\sin \theta_1}{\sin \theta_2} = \frac{\lambda_1}{\lambda_2}$$

Now eliminate $\lambda_1$ and $\lambda_2$ using Eq. 23-10:

$$\frac{\sin \theta_1}{\sin \theta_2} = \frac{u_1}{u_2} = \frac{c/u_2}{c/u_1}$$

We saw on page 450 that $c/u$ is by definition the index of refraction of the medium; hence

$$\frac{\sin \theta_1}{\sin \theta_2} = \frac{n_2}{n_1} \quad \text{(Snell's law)} \qquad (23\text{-}11)$$

where $n_1$ and $n_2$ are the indices of refraction of mediums 1 and 2, respectively.

## Lenses

From this basic equation, which is also called Snell's law, the optical properties of lenses can be calculated. Just as with a concave mirror, a converging lens has the property of bending parallel rays so that they are focused at a distance $F$ from the lens. The distance $F$ is called the focal length.

In Figure 23-24 the image position can be located graphically. First draw ray (1) parallel to the horizontal axis. It is then bent by the lens and passes through the focus $F$. Then draw ray (2), passing directly through the center of the lens. Where the two rays cross is the image point. We can derive a quantitative relation between the object distance $s$ and the image distance $s'$ shown in Figure 23-24. Triangle $ABO$ is similar to triangle $A'B'O$. From these two similar triangles, the ratio

$$\frac{A'B'}{AB} = \frac{s'}{s} \tag{23-12}$$

Triangle $POF$ is similar to triangle $A'B'F$ and the ratio

$$\frac{A'B'}{PO} = \frac{s' - f}{f} \tag{23-13}$$

Since $PO = AB$, the right-hand sides of Eqs. 23-12 and 23-13 are equal. Equating these two right-hand sides gives

$$\frac{s'}{s} = \frac{s' - f}{f} \quad \text{or} \quad \frac{1}{f} = \frac{1}{s} + \frac{1}{s'}$$

This relationship between image and object distance is referred to as the thin lens formula. There is a sign convention to tell whether $s$, $s'$, or $f$ is positive or negative. According to this convention, the problem under study must be oriented so that the light passes through the lens from left to right. Then $s'$ will be positive if the image is to the right of the lens, and

**Figure 23-24.** Object AB is distance s from lens of focal length f. The image A'B' is at distance s'.

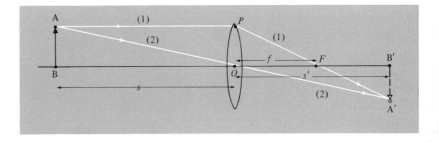

negative if to the left of the lens. The quantity $f$ is negative if the lens is a diverging lens. If the rays passing through the lens are converging on a virtual object (it could be the image produced by a preceding lens to the left), then $s$ will be negative.

## Summary

A hologram is a photographic recording of the interference pattern between reflected laser light from an object and a reference beam. The original wave front of the reflected light is reconstructed by passing monochromatic light through the hologram.

The plane of polarization of an electromagnetic wave (or light) contains the **E** vector. A polarizer consisting of parallel wires (or molecular chains) will allow only the component of **E** perpendicular to the "wires" to pass through. Light reflected by a material of index of refraction $n$ at an angle $\theta_1$ where $\tan \theta_1 = n$ will be 100% polarized. This is called Brewster's law.

Monochromatic light passing through a circular aperture of diameter $a$ will have its first minimum (or dark diffraction ring) at $\sin \theta_{\min} = 1.22 \lambda / a$.

The magnification of a telescope is $M = F/f$, the ratio of objective to eyepiece focal length. The angular resolution of a telescope (also of a microscope) is $\alpha_{\min} = 1.22 \lambda / a$, where $a$ is the diameter of the objective. For the microscope, the minimum separation which can be resolved is $D\alpha_{\min} \approx \lambda$. Laser light can be focused to a spot of this size, but other light sources cannot.

A beam of high-energy protons can be diffraction scattered by a nucleus of radius $R$. The first minimum of the diffraction pattern will be at $\sin \theta_{\min} = 0.61 h/(pR)$, where $h$ is Planck's constant and $p$ is the momentum of a beam proton.

Geometrical optics is based on the law of reflection (angle of reflection equals angle of incidence) and the law of refraction (also called Snell's law). In Snell's law

$$\frac{\sin \theta_1}{\sin \theta_2} = \frac{n_2}{n_1}$$

the inverse ratio of the indices of refraction.

A concave mirror or a converging lens will bring parallel rays to a focal point. The distance from the mirror (or lens) to the focal point is the focal length $F$. If there is an object at a distance $s$ from the lens, there will be an inverted image formed at a distance $s'$ where

$$\frac{1}{s} + \frac{1}{s'} = \frac{1}{F}$$

## Appendix 23-1  Brewster's Law

When a light beam strikes a nonconductor at an angle of incidence $\theta_1$, part of it will be reflected at the same angle, and part will be refracted at an angle $\theta_2$ according to Snell's law:

$$\frac{\sin \theta_1}{\sin \theta_2} = n$$

The source of the reflected ray can only be oscillating atomic electrons in the nonconductor. Because of the factor $\sin \theta$ in Eq. 21-13, these electrons cannot emit any radiation in the direction of their motion. If the incoming ray is polarized as shown in Figure 23-25, the oscillating electrons will be moving in the direction $E'$. Since the would-be reflected ray is in the direction of this motion, there can be no reflected ray. However, if the incoming ray is polarized perpendicular to the page, then a reflected ray is permitted. In Figure 23-25, $\theta_1 + \theta_2 = \pi/2$. If we insert $\theta_2 = \pi/2 - \theta_1$ into Snell's law, we obtain

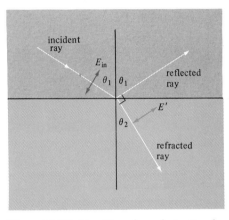

**Figure 23-25.** *Condition for polarization by reflection. Reflected and refracted rays are perpendicular, or $E'$ is in direction of the reflected ray.*

$$\frac{\sin \theta_1}{\sin(\pi/2 - \theta_1)} = n$$

$$\tan \theta_1 = n \quad \text{(Brewster's law)} \quad (23\text{-}12)$$

We see that if unpolarized light enters at this angle, the reflected light must have $E$ perpendicular to the page. This condition for total polarization by reflection is called Brewster's law.

### Exercises

1. Make a plot similar to Figure 23-7 for the case where $E_y = E_0 \cos \omega t$ and $E_z = 2E_0 \cos(\omega t - \pi/2)$. This is elliptical polarization. (The tip of the $E$ vector moves in an ellipse where the ratio of major to minor axis is 2:1.)
2. A right circularly polarized wave is superimposed with a left circularly polarized wave, both of the same amplitude and frequency and traveling together along the $x$ axis. Whenever wave 1 has $E$ pointing along the $+y$ axis, so does wave 2. Describe the polarization state of the resultant wave.
3. Repeat Exercise 2 for the case where the two waves have $E$ pointing in the opposite directions whenever $E$ is lined up with the $y$ axis.

4. The nearest star is 4 light years away. How large must its diameter be in order to measure it with the 200-in. telescope?
5. Suppose a star at a distance of 10 light years has a very large planet in orbit around it and that this planet by itself would be visible in a 200-in. telescope. What must be the distance of this planet from the star in order to resolve it from the star?
6. A 1-cm diameter parallel laser beam passes through a focusing lens of 10 cm focal length. What will be the spot size? Use $\lambda = 6400$ Å.
7. Repeat Exercise 6 for a 1-mm diameter parallel laser beam.
8. It has recently been discovered that the pattern for high-energy proton-proton elastic scattering has a minimum where the transverse momentum of the scattered proton is 1.1 GeV/c; that is, $pc \sin \theta = 1.1$ GeV. A black disk of what radius would give a first minimum at this position?
9. Consider two polaroids in a row. Polaroid A is vertical (vertically polarized light is unabsorbed). Polaroid B is oriented at 45°.
    (a) If vertically polarized light enters from the left, what is the intensity $I$?
    (b) If vertically polarized light enters from the right, what is the intensity $I'$?
10. An object is located between the focus and the center of curvature of a concave mirror. Will the image be inverted? Will it be larger than the object?
11. The distance of an object from a concave mirror is less than the focal length. If you looked in the mirror, would you see an image? If so, would it look larger or smaller than the original object? Would the image be inverted? (Is this the same as looking into a concave shaving mirror?)
12. A light ray enters a flat slab of glass at an angle of 60° from the normal. If $n = 1.5$, at what angle does the ray leave the slab on the other side? The faces of the slab are parallel.

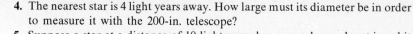

Exercise 9

## Problems

13. The hologram in Example 1 is viewed using light having $\lambda = 4800$ Å. What will be the apparent distance behind the film plane of the image?
14. Repeat the derivation of the reconstructed wave front on page 499 using a photographic positive rather than a negative for the hologram. Then the intensity leaving the film will be

$$I = I' \cos^2 \omega t \left(1 + \frac{E_0}{K_1} a \cos \phi \right)$$

15. Vertically polarized light of intensity $I_0$ passes through nine ideal polaroids. The first polaroid has its axis 10° from the vertical, the second is rotated an additional 10°, and so on, so that the ninth polaroid is rotated by 90°. What is the final intensity?
16. Repeat Problem 15 for 90 polaroids each inclined 1° from its neighbor.
17. A vertically polarized laser beam strikes a screen with two horizontal slits. Whenever one slit is closed, the intensity everywhere on the screen is $I_0$.
    (a) What is the intensity at the screen where $r_2 - r_1 = 0$?

(b) What is the intensity where $r_2 - r_1 = \tfrac{1}{2}\lambda$?
Now cover slit 1 with polaroid at $+45°$ and slit 2 with polaroid at $-45°$.
(c) What is the intensity where $r_2 - r_1 = 0$?
(d) Where it is $\tfrac{1}{4}\lambda$?
(e) Where it is $\tfrac{1}{2}\lambda$?
Now place a vertical polaroid in front of the screen in addition to the polaroids at the slits.
(f) What is the intensity at the screen (after passing through this vertical polaroid) where $r_2 - r_1 = 0$?
(g) Where it equals $\tfrac{1}{2}\lambda$?

18. Suppose the diameter of a star is such that it subtends an angle of $10^{-7}$ rad.
    (a) Can the diameter be measured using an optical telescope?
    (b) Can the diameter be measured using intensity interferometry? Assume the distance to the star is known. (See Example 8 of Chapter 22.)
19. What is the smallest distance that could be resolved using an electron microscope which has a beam of electrons of kinetic energy equal to 1000 eV?
20. There is another method to obtain the average radius of an elementary particle which works even if the first minimum is not observed. The method is to measure the curvature or rate of decrease of the central maximum. At small angles the proton-proton scattering pattern is observed to drop off as $(1 - Ap_\perp^2)$, where $A = (10 \text{ GeV}/c)^{-2}$. For small argument the Bessel function goes as

$$J_1(x) \approx \frac{x}{2} - \frac{x^3}{16}$$

Then

$$\left[\frac{J_1(\Phi/2)}{\Phi/4}\right] \approx 1 - \frac{\Phi^2}{32} \quad \text{where} \quad \frac{\Phi}{2} = kR\sin\theta \quad \text{and} \quad p_\perp = p\sin\theta$$

What value of $R$ in meters gives $A = 10(\text{GeV}/c)^{-2}$?

21. Assume the proton has a hard core of radius $10^{-14}$ cm. Suppose a beam of 200-GeV protons interacted mainly with the core.
    (a) At what angle would the first diffraction minimum occur?
    (b) If the proton had no core and acted like an absorbing disk of $R = 10^{-13}$ cm, at what angle would the first diffraction minimum occur?
22. A pinhole camera consists of a 10-cm long black box with a pinhole for a lens. What must be the pinhole diameter to give the sharpest image?
23. Two lenses of focal lengths $F$ and $-F$, respectively, are a distance $D$ apart where $D < F$. If parallel light enters from the left, will it be brought to a focus, and if so, where?
24. Repeat Problem 23 for parallel light entering from the right rather than from the left.
25. Prove that two thin lenses of focal lengths $F_1$ and $F_2$, if placed close together, behave as a single lens of focal length

$$F = \frac{F_1 F_2}{F_1 + F_2}$$

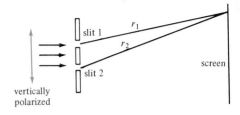

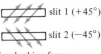

View looking from screen toward slits

Problem 17

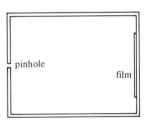

Problem 22

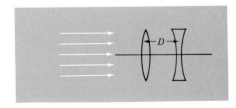

Problem 23

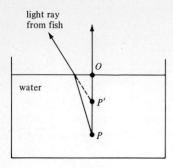

**Problem 26**

26. A fish is at a depth $OP$. What is its apparent depth $OP'$? For water $n = 1.33$.
27. A sheet of stretched plastic has index of refraction $n_\parallel$ for light polarized along the direction of stretch, and $n_\perp$ for $E$ perpendicular to the direction of stretch.
    (a) If outer electrons prefer to oscillate along the direction of stretch, which is greater, $n_\parallel$ or $n_\perp$?
    (b) If polarized light with $E$ at 45° to the direction of stretch passes through such a sheet, its plane of polarization can be rotated by 90°. Such a sheet is called a half-wave plate. What must be the thickness $x_0$ of the sheet in order to be a half-wave plate? Give answer in terms of $f$, $c$, $n_\perp$, and $n_\parallel$.

# The Wave Nature of Matter

Physical laws and the phenomena they describe are commonly classified into two categories: classical and modern. Modern physics makes use of the wave nature of matter, which was not discovered until 1927. Modern physics necessarily involves a new fundamental constant of nature discovered by Max Planck in 1900. (In order to explain the frequency spectrum of radiation from a hot body, Planck proposed that oscillators can emit light only in units of energy $E = hf$ where $f$ is the oscillator frequency and $h$ is called Planck's constant.)

By this definition of modern physics, what we have been discussing up until now is classical physics (except for Section 23-5 on diffraction scattering). However, as we shall see in this chapter, all particles have an intrinsic wave nature that significantly affects their behavior especially at small distances. It would be impossible to understand atoms, molecules, elementary particles, nuclear physics, astrophysics, and solid state physics without first understanding the wave nature of matter. The rules and mathematical formalism that follow from the wave nature of matter are called quantum mechanics. We shall develop some of the basic ideas of quantum mechanics in this and the next chapter. In the remaining chapters we shall discuss important applications, which not only will help us understand the important phenomena under study but will also help us better understand quantum mechanics itself.

Before introducing the radically new concepts of quantum theory, let us stand back and look at the accomplishments and possible shortcomings of classical physics. With Newton's laws we explained falling bodies, projectiles, earth satellites, the motions of the planets, and other macroscopic motion. Newtonian mechanics also gave us the conservation of energy, momentum, and angular momentum.

From nineteenth-century chemistry it was learned that matter is made up of molecules and atoms. This knowledge combined with Newton's laws

## 24-1 Classical Versus Modern Physics

explained the big mystery of heat in what we call the kinetic theory of heat. The richness of electrical and magnetic phenomena was explained by the concept of charge and the basic laws of electricity over a century ago by Maxwell. The crowning achievement of classical physics was reached about 1870 when Maxwell derived the theory of light (and the science of optics) as a mathematical consequence of Maxwell's equations. Historically this led to the difficulty in explaining the ether and the null results of the Michelson–Morley experiment, which led to special relativity in 1905. Then, with hindsight, Maxwell's equations were understood as a necessary relativistic consequence of Coulomb's law (they could be derived from Coulomb's law).

Einstein's revision of our concepts of space and time was shocking enough (how can two differently moving observers looking at the same pulse of light get the same value for its velocity?). However, as we shall see, the wave nature of matter or wave-particle duality and its consequences seem even more shocking and contrary to common sense than Einstein's postulate that all observers get the same value for the speed of light.

Qualitatively, the principle of the wave nature of matter says that all particles have a wave nature, and conversely, all waves have a particle nature. Our first example of the particle nature of waves is the photoelectric effect, the theory for which was also worked out by Einstein in 1905.

## 24-2 The Photoelectric Effect

Shortly after the discovery of the electron in the late nineteenth century, it was observed that, when light strikes certain metal surfaces, electrons are emitted (see Figure 24-1). From the time of Young's double slit experiments, scientists were convinced of the wave nature of light. In this context the photoelectric effect was understandable. We saw in Eq. 21-6 that a free electron in the presence of an oscillating electric field, $E = E_0 \cos \omega t$, has an amplitude of oscillation,

$$A = \frac{eE_0}{m\omega^2}$$

So one would expect electrons near the surface of the metal to escape if $A$ were greater than a certain critical value. The wave theory of light makes these predictions: (1) no electrons are emitted until $E_0$ exceeds a certain critical value, (2) the energy of the emitted electrons should increase with $E_0^2$, and (3) if $E_0$ and thus the light intensity is held constant but $\omega$ is increased, the number of electrons emitted should decrease.

However, the experimental observations were contrary to these predictions. (1) There was no threshold intensity. The number of electrons emitted was strictly proportional to $E_0^2$ no matter how small the intensity. (2) The energy of the electrons was independent of $E_0$. (3) However, the energy did depend on the frequency. There was a threshold frequency, $f_0$, and the electron energies increased linearly with frequency above the threshold frequency. Actually the electrons emitted have kinetic energies ranging from zero to some specific value $K_{max}$ with no electrons above $K_{max}$. The observed relation between $K_{max}$ and $f$ is plotted in Figure 24-2.

Although not all this information was available to Einstein in 1905, he hit upon what turned out to be the correct explanation. He made what in 1905 was a very daring speculation. He proposed that light consists of "light quanta" that have energy $E = hf$ where $h$ is Planck's constant. Einstein proposed that these light quanta (now called photons) behave the same as material particles and that when a photon collides with an electron in a metal it can be absorbed and its energy transferred to the electron. Even Planck himself thought this was ridiculous. How could light, which was known to obey all the wave interference predictions, at the same time consist of particles? In the double slit experiment a *particle* of light would have to go through one slit *or* the other, and then it would not be able to interfere with itself.

But Einstein's theory did explain the peculiar experimental observations. Suppose it takes an energy $W_0$ to pull a surface electron out of a metal. Then after absorbing a photon of energy $hf$, the electron would have a remaining energy of $hf - W_0$ after having left the surface. This would be the maximum kinetic energy possible; that is

$$K_{max} = hf - W_0 \quad \text{(photoelectric effect)} \quad (24\text{-}1)$$

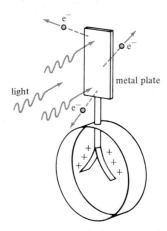

**Figure 24-1.** *A neutral electroscope is connected to a metal plate. When light shines on the plate, photoelectrons are ejected and the leaves then become positive.*

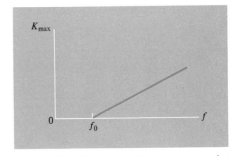

**Figure 24-2.** *Maximum kinetic energy of ejected electron plotted as a function of the frequency of the light.*

This equation agrees with the experimental plot shown in Figure 24-2. We note that Einstein was predicting that the slope of the line should be Planck's constant $h$. Einstein's theory of the photoelectric effect passed this acid test. The slope was

$$h = 6.63 \times 10^{-34} \text{ J s} \quad \text{(Planck's constant)}$$

The units are joule seconds or kg m²s⁻¹.

The energy $W_0$ is called the work function, and it depends upon the particular metal. When a free electron is brought near a metal surface from the outside, it feels an attractive force. If it is initially at rest, it will obtain a kinetic energy $U_0$ after entering the metal. In other words, we are saying that a metal-electron system can be represented as the potential well of depth $U_0$ shown in Figure 24-3. As we shall see in Chapter 28, the outer electrons in a metal are free (not attached to a particular atom) and can have kinetic energies ranging from zero to $K_f$ (called the Fermi

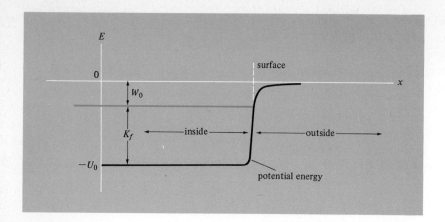

**Figure 24-3.** *Potential well as seen by an electron entering a metal surface. $K_f$ is maximum kinetic energy of the "free" electrons within the metal.*

energy) while inside the metal. If we give additional energy $W_0$ to an electron that has the Fermi energy, it will have $K = (K_f + W_0)$ and just barely get out; that is, when outside it will have $K = 0$. We see from Figure 24-3 that $(W_0 + K_f) = U_0$ or

$$W_0 = U_0 - K_f$$

The photoelectric effect is represented in Figure 24-4. An electron is initially sitting with kinetic energy equal to $K_f$ (the dashed colored line). It absorbs a photon of energy $hf$, thus raising its energy level to the solid colored line. It then has energy $hf - W_0$ on the outside. This is the maximum possible kinetic energy any ejected electron can have. Hence, $K_{max} = hf - W_0$. When an electron below the dashed line absorbs a photon of the same energy, it has energy less than $K_{max}$ when outside the metal.

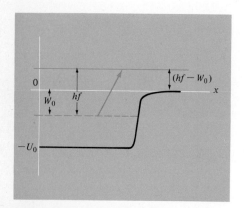

**Figure 24-4.** *Electron of energy $-W_0$ absorbs photon and makes transition to higher energy level.*

**Example 1.** Cesium metal has a work function of 1.8 eV. What is the maximum wavelength light that can eject 2-eV electrons from cesium?

ANSWER: Solving Eq. 24-1 for the frequency:

$$f = \frac{K_m + W_0}{h}$$

$$\lambda = \frac{c}{f} = \frac{hc}{K_m + W_0} = \frac{(6.63 \times 10^{-34} \text{ J s})(3 \times 10^8 \text{ m/s})}{(3.8 \text{ eV})(1.6 \times 10^{-19} \text{ J/eV})}$$

$$= 3.27 \times 10^{-7} \text{ m}$$

In Chapter 22, we saw, using classical electromagnetic theory, that light of energy $E$ must have momentum $p = E/c$. Then a light quantum of energy $E = hf$ must have momentum $p = hf/c$. If we substitute $1/\lambda$ for $f/c$, we obtain

$$p = \frac{h}{\lambda} \qquad (24\text{-}2)$$

Einstein predicted that his light quanta or photons must behave as particles with momentum $p = h/\lambda$. In the photoelectric effect this small amount of momentum must be transferred to the entire piece of metal plus ejected electron. The momentum transfer to the metal is too small to measure; however, if we let the photon collide with a free electron, the momentum transfer could be measured. This process—a photon scattering off of a free electron—is called the Compton effect. It was first verified experimentally by A. Compton in 1923. It is like a billiard-ball collision.

We shall now calculate the wavelength of the scattered photon in terms of the initial wavelength and scattering angle. Initially we have a photon of momentum $\mathbf{p}$ and energy $pc$ collide with a stationary electron of momentum zero and rest energy $mc^2$. After the collision the photon has momentum $\mathbf{p}'$ at an angle $\theta$, as shown in Figure 24-5. The recoil electron has momentum $\mathbf{p}'_e$ and total relativistic energy $E'_e$. Since the electron may attain velocities near the speed of light, we use relativistic mechanics. According to the law of conservation of energy, the total initial energy equals the total final energy:

$$pc + mc^2 = p'c + E'_e$$

$$(p - p' + mc)^2 = \left(\frac{E'_e}{c}\right)^2 \qquad (24\text{-}3)$$

Conservation of momentum gives

$$\mathbf{p} - \mathbf{p}' = \mathbf{p}'_e$$

Squaring both sides gives

$$p^2 - 2\mathbf{p}\cdot\mathbf{p}' + p'^2 = p'^2_e$$

Subtracting this from Eq. 24-3 gives

$$m^2c^2 - 2pp' + 2pmc - 2p'mc + 2pp'\cos\theta = \frac{E'^2_e}{c^2} - p'^2_e$$

## 24-3 The Compton Effect

Before

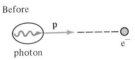

After

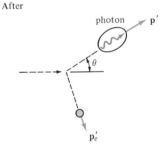

Figure 24-5. *Collision of photon with free electron. The Compton effect.*

Now, Eq. 9-11 says that we can replace the right-hand side with $m^2c^2$:

$$m^2c^2 - 2p'(p + mc - p\cos\theta) + 2pmc = m^2c^2$$

$$p' = \frac{p}{1 + \dfrac{p}{mc}(1 - \cos\theta)}$$

Using $p = h/\lambda$ gives

$$\frac{1}{\lambda'} = \frac{1}{\lambda + \dfrac{h}{mc}(1 - \cos\theta)}$$

$$\lambda' - \lambda = \frac{h}{mc}(1 - \cos\theta) \quad \text{(Compton effect)} \quad (24\text{-}4)$$

Compton used x-ray photons of known wavelength and found that the scattered photons had longer wavelengths consistent with Eq. 24-4.

The Compton effect, the photoelectric effect, and many other atomic-scale experiments with light show that light does behave as if it is made up of particles of energy $hf$ and momentum $h/\lambda$.

## 24-4 Wave-Particle Duality

If the first experiments ever to be done with light had been the Compton effect and the photoelectric effect, one would have been convinced that light is made up of beams of photons that behave like bona fide particles. Then if these particles were made to hit a double slit, one would have been astounded to observe a double slit interference pattern. How could particles behave as classical waves? A particle can go through only one or the other slit. We can rule out the possibility of two photons interfering with each other by reducing the light intensity so that the average time between photons is much longer than the travel time from the light source to the screen. If the screen is 3 m from the source, the travel time is $t = L/c = 10^{-8}$ s. So we use a light intensity $\sim 10^{-11}$ W, which corresponds to less than $10^8$ photons/s.

If we close slit B, we obtain the intensity pattern corresponding to the single slit A shown in Figure 24-6. We can use photomultiplier tubes to count the photons one at a time as they hit the screen. With slit A closed and B open we obtain a similar single slit pattern but displaced slightly, as shown. But, as we know from Chapter 23, the pattern where both slits are open is not the sum of slit A pattern plus slit B pattern; instead it is Young's double slit interference pattern. We are running into a paradox—how can light exhibit both particle properties and wave properties at the same time?

**Figure 24-6.** *Photons passing through slit A give pattern on screen.*

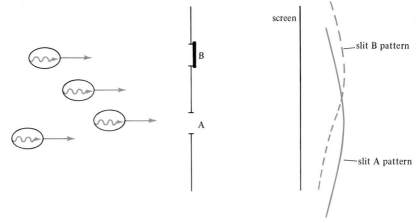

In 1927 the paradox grew even deeper with the discovery that electrons exhibit a similar kind of wave interference property! Actually 3 years before the electron diffraction experiments, Louis de Broglie in his PhD thesis proposed that not just photons but all particles obey Eq. 24-2:

$$p = \frac{h}{\lambda} \quad \text{and} \quad E = hf \quad \text{for all particles} \quad \text{(de Broglie relation)} \quad (24\text{-}5)$$

He was proposing that a beam of any kind of particle when directed at an appropriate double slit will exhibit Young's double slit interference pattern! At the time this was regarded as a wild speculation almost not worthy of a PhD degree. Then only 3 years later, physics received its biggest shock of all time—de Broglie's speculations were confirmed by experiment. It came as such a shock because it almost seemed impossible for particles such as electrons to be both waves and particles at the same time. In the following section we shall describe how present observations appear to be gross violations of common sense and everyday human experience.

## 24-5 The Great Paradox

A possible solution to the paradox is that perhaps a single photon can split in two and then interfere with itself after passing through slits A and B. The paradox becomes sharper if we use an electron beam rather than a photon beam. Never in nature has half an electron or part of an electron been observed. Always a complete electron is observed whether or not a detector is placed at slit A or slit B. This is called the principle of indivisibility and it applies to all elementary particles, including photons. Because of the principle of indivisibility, we would conclude that a single electron can go through only one of the two slits in Figure 24-7. Hence the pattern on the screen must be the direct sum of the two single slit patterns.

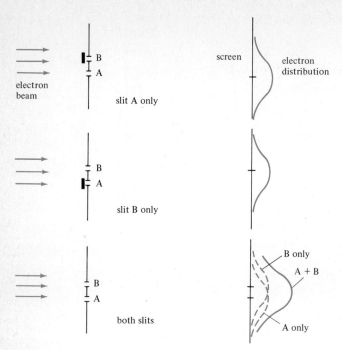

**Figure 24-7.** *Electron distribution according to classical physics.*

Although the logic seems irrefutable, the pattern (A + B) is not observed! Instead, the standard double slit interference pattern shown in Figure 24-8 is observed.

Has pure logic broken down? It is as if 100 + 100 can equal zero. Suppose there is a single geiger counter at point $P_1$ in Figure 24-8 that counts 100 electrons/s when either slit A or slit B is open. Then when both slits are open together, the geiger counter will count zero. Point $P_1$ is at an interference minimum ($r_2 - r_1 = \frac{1}{2}\lambda$). If we open only slit A and then gradually open slit B, we would expect, according to common sense, the counting rate to increase from 100 to 200 counts/s. But instead the rate will decrease from 100 to zero. How can the act of opening slit B possibly influence those electrons that would have gone through slit A? Moreover,

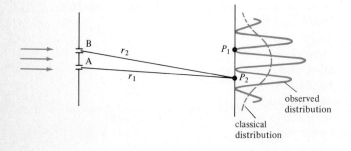

**Figure 24-8.** *Electron distribution according to quantum theory.*

532 CH. 24/THE WAVE NATURE OF MATTER

if the geiger counter is placed at point $P_2$, the rate would increase from 100 to 400 electrons/s as slit B is opened. Then $100 + 100 = 400$.

The only way to resolve the paradox is to invent a mathematical formalism that, while preserving the principle of indivisibility, always predicts the observed interference pattern. One must take care that such a formalism is always self-consistent.

## The Wave Function

The mathematical formalism that resolves the paradox is to represent each particle as a probability amplitude $\psi(x, y, z, t)$ that is a function of position and time. ($\psi$ is pronounced "psi.") **The probability of finding the particle at any value of $x$, $y$, $z$, $t$ is proportional to the intensity** $|\psi(x, y, z, t)|^2$. We take the square modulus or square of the absolute value because $\psi$ can be a complex number. $\psi$ has the mathematical properties of a wave and is therefore often called the **wave function.**

When an event can occur in several alternative ways (such as one path through slit A and another through slit B), the probability amplitude for the event is the sum of the separate probability amplitudes:

$$\psi = \psi_1 + \psi_2 \quad \text{(principle of superposition)}$$

This is the same as the rule for adding wave amplitudes in classical optics. In the above double slit problem $\psi_1$ is a traveling wave passing through slit A and $\psi_2$ is a traveling wave passing through slit B. In the region of the screen both wave functions overlap and give the classical double slit interference pattern, where the angular spacing between successive maxima is $\sin \theta_n = n\lambda/d$ (see Eq. 22-3).

The preceding formalism, which is the basis of wave mechanics or quantum mechanics, may leave the reader with an uneasy feeling that he is missing a deeper understanding. But this is it. There is nothing deeper. It may help the reader to quote from Richard Feynman who received the Nobel Prize for physics in 1965 for applications of quantum mechanics to electrodynamics. In Volume III of his *Lectures on Physics** Feynman says:

One might still like to ask: "How does it work? What is the machinery behind the law?" No one has found any machinery behind the law. No one can "explain" any more than we have just "explained." No one will give you any deeper representation of the situation. We have no ideas about a more basic mechanism from which these results can be deduced.

*Published by Addison-Wesley, Reading, Mass., 1965, p. 1-10.

**Figure 24-9**

> **Example 2.** There is a geiger counter at point P in Figure 24-9. The part of the wave function coming via slit A and reaching P is $\psi_A = 2$ units, and via slit B is $\psi_B = 6$ units. When only slit A is open, 100 electrons per second are observed at P.
> (a) How many electrons per second are observed when only slit B is open?
> (b) Assuming a constructive interference, how many electrons per second are observed when both slits are open?
> (c) Assuming a destructive interference, how many electrons per second are observed when both slits are open?
>
> ANSWER: The ratio of wave intensities $\psi_B^2/\psi_A^2 = 36/4 = 9$. Hence the beam intensity from B is 9 times that from A or 900 electrons/s.
> For part (b) the total wave amplitude is $\psi = \psi_A + \psi_B$, or $\psi = 8$. Since $\psi^2$ is 16 times $\psi_A^2$, there will be 1600 electrons/s.
> For part (c) $\psi_A$ and $\psi_B$ must be of opposite sign to give a destructive interference. Hence $\psi = 2 - 6 = -4$. Now $\psi^2 = 16$, which is four times $\psi_A^2$. This corresponds to 400 electrons/s.

> **Example 3.** What will be the intensity pattern of a double slit interference experiment if four times as many electrons can get through slit B as slit A?
>
> ANSWER: We have $\psi_B^2 = 4\psi_A^2$, or $\psi_B = 2\psi_A$. The total intensity at a maximum is proportional to $(\psi_A + \psi_B)^2$, or
>
> $$I_{max} = (\psi_A + 2\psi_A)^2 = 9\psi_A^2$$
>
> At an intensity minimum
>
> $$I_{min} = (\psi_A - 2\psi_A)^2 = \psi_A^2$$
>
> The ratio $I_{max}/I_{min} = 9$. The intensity pattern would be $I = I_A[5 + 4 \cos k(r_B - r_A)]$, where $r_A$ and $r_B$ are the distances to slits A and B, respectively.

The preceding formalism raises puzzling questions that require some further physical interpretation. Let us shoot only one electron at a time. Then according to this wave picture, each electron is represented by a wave train or wave packet that splits equally between the two slits. But we can put a geiger counter, cloud chamber, or other particle detector at slit A and observe that there is never half of an electron going through the slit. This principle of indivisibility is consistent with the hypothesis that the wave intensity at slit A is the probability of finding one whole electron at that position. Furthermore, if a detector is placed at slit A, the interference pattern smooths out and the classical result is then observed. The presence of a detector changes the result from the interference pattern of Figure 24-8 to the classical result of Figure 24-7. Actually, many physicists, including Einstein, have tried to contrive an experiment that would reveal the slit used by individual particles without destroying the interference, but all such efforts have failed. One such experiment is illustrated by Example 3 of Chapter 23. The slit from which the photons came can be revealed merely by changing the orientation of a polaroid filter, which is placed near the screen. But whenever the polaroid is so altered, the double slit interference pattern is transformed into a single slit pattern.

Just what is it that "waves" in an electron wave? We must give the same kind of answer we gave for photons. Electromagnetic waves travel freely through pure vacuum. In contrast to mechanical waves, no material of any kind is waving. The wave function $\psi$ has no direct physical meaning, and in this sense nothing is waving. It is just that quantum mechanical problems are solved mathematically in the same way that water wave or other kinds of classical wave problems are solved. Classical waves and particle waves both obey the same kind of mathematical wave equation. However, in the case of classical waves, the wave amplitude is directly observable, whereas $\psi$ is not.

## 24-6 Electron Diffraction

The double slit experiment with electrons is difficult to perform because, according to Eq. 24-5, typical electrons have wavelengths much smaller than light. However, C. Jönsson in 1961 was able to obtain a genuine double slit interference pattern of electrons on a photographic screen. The experimental layout is depicted in Figure 24-10, and the results are shown in Figure 24-11(a).

Each electron produces a black spot at the position where it hits the film. This photograph using a double slit source of electrons is the result of a very large number of electron impacts. For comparison, Figure 24-11(b) shows a typical double slit interference pattern using light. We can simulate a low intensity exposure of Figure 24-11(a) by computer generating random positions according to a $\sin^2 x$ probability distribution. Figure 24-12(a) simulates an exposure to 27 electrons. Figures 24-12(b) and (c) correspond to 70 and 735 electrons, respectively.

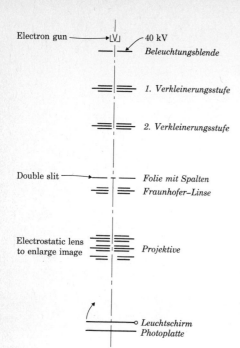

**Figure 24-10.** *C. Jönsson's experimental arrangement for obtaining double slit interference pattern of electrons.* [C. Jönsson, Z. Phys., **161** (1961)]

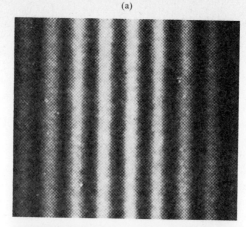

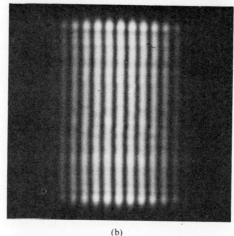

**Figure 24-11.** (a) *A double slit interference pattern of electrons. Each grain in the photographic negative is produced by a single electron. For comparison, (b) is the double slit interference pattern of light shown in Figure 22-17. In this photo, each grain in the negative is produced by a single photon.* [*Photo (a) by Professor C. Jönsson, University of Tubingen*]

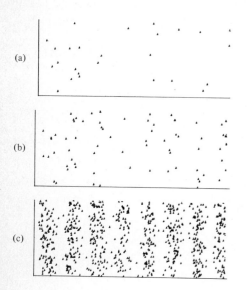

De Broglie's hypothesis was first verified by the observation of a different type of electron diffraction in 1927 by two American physicists, C. J. Davisson and L. H. Germer. It is interesting that in this experiment, as in some others that were of extreme importance to physics, the great discovery was "accidental." Davisson and Germer were not looking for electron diffraction. In fact, in the early stages of their experiment, they had never even heard of electron diffraction. In 1926, Davisson took some of his preliminary data to an international conference in Oxford, England. European physicists suggested to him that his results might be interpreted as electron diffraction rather than the classical electron scattering that he had been studying. Just a few months later, Davisson and Germer obtained data that conclusively demonstrated the wave nature of electrons and gave Planck's constant to an accuracy of about 1%. They scattered

**Figure 24-12.** *Simulated low exposures of double slit electron pattern representing (a) 27 electrons, (b) 70 electrons, and (c) 735 electrons.*

low-energy electrons off the surface of a single metallic crystal. The regular rows of atoms at the surface act as the lines of a very fine diffraction grating. The electron wavelength is determined by knowing the atomic spacing.

An arrangement for observing electron diffraction from a crystal surface is shown in Figure 24-13. The detector could be a fluorescent screen, as in the movie film *Matter Waves* produced by Educational Services, Inc. Planck's constant can be determined by observing the direction $\theta$ at which an intensity maximum appears. As can be seen in Figure 24-13(b), the path difference $\Delta D = d \sin \theta$ and this will be equal to the wavelength $h/p$ at the first intensity maximum. Hence

$$\frac{h}{p} = d \sin \theta$$

and

$$h = pd \sin \theta$$

Shortly after de Broglie's proposal in 1924, the English physicist G. P. Thompson set out in a systematic way to observe electron diffraction. His

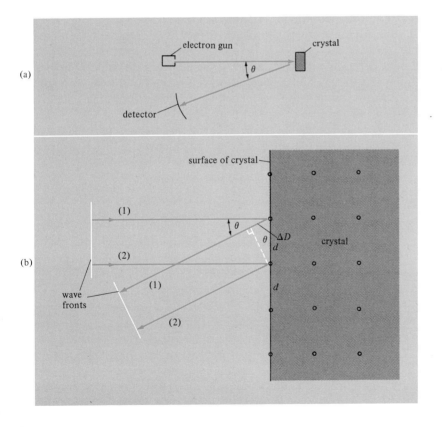

**Figure 24-13.** (a) *Apparatus to observe electron diffraction from a crystal surface.* (b) *Greatly enlarged view of crystal surface.*

method of approach was to make higher-energy electrons pass through a thin metal foil. Since x rays have about the same wavelength, Thompson hoped to get an electron diffraction picture similar in appearance to previously obtained x-ray diffraction pictures. By 1928 Thompson obtained electron diffraction pictures looking almost the same as x-ray diffraction pictures. It is interesting that in this case the "accidental" approach was quicker than the carefully studied, deliberate approach. This may not fit the reader's conception of the scientific method, but it is true, living science. The experiment of Davisson and Germer is a good example of a valid scientific method. If an experimenter observes a strange effect, even by accident, that he does not understand, he should carefully pursue it until it is understood.

By now interference patterns of neutrons, protons, and even whole atoms as well as electrons have been observed. For example, Figure 23-19 is an interference pattern of high-energy protons. In many different ways, the wave nature of matter has been very well established. No violations of the theory have been found.

## Summary

Planck proposed that light be quantized into photons of energy $E = hf$. Einstein explained the photoelectric effect by proposing that the maximum kinetic energy of an electron ejected from a metal surface by a photon is $K_{max} = hf - W_0$, where $W_0$ is the work function of the metal. The work function of a particular metal is related to the depth $U_0$ of the potential well and the maximum kinetic energy $K_f$ of the conduction electrons: $W_0 = U_0 - K_f$.

Photons have momentum $p = h/\lambda$, and when a photon collides with a free electron it transfers some of its energy and momentum to the electron. If the photon wavelength after collision is $\lambda'$, energy and momentum conservation gives the Compton effect relation

$$\lambda' - \lambda = \frac{h}{mc}(1 - \cos\theta)$$

Not only photons but all particles have a wavelength $\lambda = h/p$. The wavelength associated with an electron beam of momentum $p$ reveals itself in a double slit interference pattern when the beam is passed through two slits. If an electron beam strikes a crystalline surface at right angles, there will be a diffracted beam at an angle $\sin\theta = h/(pd)$, where $d$ is the spacing between rows of atoms.

The wave properties of a particle can be described using a wave function $\psi$. The probability of finding the particle at any value of $x, y, z$, and $t$ is proportional to the wave intensity $|\psi(x, y, z, t)|^2$. If an event can occur in several alternative ways, the probability amplitude for the event is the sum of the separate probability amplitudes:

$$\psi = \psi_1 + \psi_2$$

### Exercises

1. What is the wavelength of a 1-MeV photon in angstroms?
2. What is the inertial or relativistic mass of a photon in terms of $h$, $\lambda$, and $c$?
3. If the relativistic mass of a photon is $10^{-15}$ g, what will be its momentum in S.I. units? What will be its wavelength?
4. A handy formula relating wavelength and energy of a photon is $\lambda = K/E$. Find the numerical value of the constant $K$ if $\lambda$ is in angstroms and $E$ is in electron volts.
5. A photon and an electron both have kinetic energies of 1 eV. What are the corresponding wavelengths?
6. Write a formula for the kinetic energy of a nonrelativistic electron in terms of its mass, wavelength, and $h$.
7. For each metal there is a photoelectric threshold, $\lambda_0$. If the work function for copper is 4.4 eV, what is $\lambda_0$?
8. The human eye can barely detect a yellow light that delivers $1.7 \times 10^{-18}$ W to the retina. How many photons per second is this?
9. Express the wavelength of a free electron in terms of $E$, $h$, $m$, and $c$, where $E$ is the total relativistic energy.
10. Express the kinetic energy of a relativistic electron in terms of its wavelength.
11. Suppose in the double slit experiment, the wave amplitudes at a point on the screen coming from slit A and slit B are +3 units and +5 units, respectively, and the counting rate from slit A alone is 60 counts per second.
    (a) What is the counting rate from slit B alone?
    (b) What is the counting rate when both slits are open?
    (c) What would be the counting rate from both slits if the two amplitudes were opposite in sign?
12. Suppose three identical slits are used for the electron diffraction experiment and that the electron counter is at a position where all three waves are in phase.
    (a) If the single slit counting rate is 100 counts per second for each of the slits by themselves, what is the rate when all three slits are open.
    (b) If the intensity of the beam from the electron gun is doubled, by what factor would the above answer be increased?
13. If both the amplitude and the frequency of a plane electromagnetic wave are doubled, by what factor will the number of photons/(s m²) be changed?

### Problems

14. An electron is attracted to an electrically neutral piece of metal. We can represent this attractive force as a potential well of depth $-U_0$. Consider the most energetic electron inside the metal (at the Fermi level). It has $K = 6$ eV while inside the metal. It absorbs a 7-eV photon and after leaving the metal has $K = 5$ eV. Assume it has no collisions.
    (a) What is the work function of the metal in electron volts?

(b) What is $U_0$ in electron volts?
(c) What is the ratio of the electron wavelength after leaving the metal to what it was before it absorbed the photon?

15. A photon of wavelength $\lambda$ collides with a stationary free electron. The photon bounces back (180° scattering angle) with wavelength $\lambda'$.
    (a) What is the electron momentum after the collision in terms of $\lambda$ and $\lambda'$?
    (b) What is the kinetic energy of the electron after the collision? Give a relativistically correct answer in terms of $\lambda$ and $\lambda'$.

16. A photon of energy $E \gg m_e c^2$ backscatters off of a stationary electron. What is the energy of the photon in MeV after scattering?

17. A photon of energy $E \gg M_p c^2$ backscatters off of a stationary proton. What is the energy in MeV of the photon after scattering?

18. In the Compton effect, a 100-keV photon is scattered by 90°.
    (a) What is the energy after scattering?
    (b) What is the electron recoil kinetic energy?
    (c) What is the direction of the recoil electron?

19. Let $E_0$ be the energy of the incoming photon in the Compton effect. Prove that the kinetic energy of the recoil electron is

$$K = \frac{(1 - \cos\theta)E_0}{1 - \cos\theta + mc^2/E_0}$$

20. In the Compton effect let $\phi$ be the angle between the incoming photon and the recoil electron. Express $\phi$ in terms of $\lambda$ and $\lambda'$.

21. Thermal neutrons are in temperature equilibrium with particles at room temperature. At room temperature $kT = 0.025$ eV. What is the average kinetic energy of a thermal neutron? What is the corresponding wavelength?

22. Two very thin slits are separated by 10 $\mu$m. A 1-eV electron beam impinges on these slits. A screen is 10 m behind the slits. What is the separation between successive minima at the screen?

23. In Figure 24-13 the surface of the crystal consists of a rectangular array of atoms spaced 1.5 Å in $y$ and 2.0 Å in $z$. The electron beam has 90 eV kinetic energy and the screen is 10 cm from the small crystal target. A pattern of spots appears on the fluorescent screen as shown. What is the separation in $y$ and $z$ between the spots? In what directions do the spots move when the beam energy is increased?

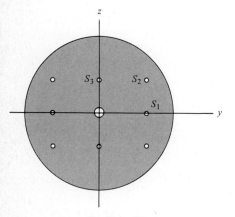

**Problem 23**

# Quantum Mechanics

## 25-1 Wave Packets

The basis of quantum mechanics is the de Broglie relation $p = h/\lambda$. It is convenient to express the momentum in terms of the wave number $k$ ($k \equiv 2\pi/\lambda$) instead of $\lambda$. Then

$$p = \frac{h}{2\pi}\frac{2\pi}{\lambda} = \frac{h}{2\pi}k$$

The quantity $h/2\pi$ occurs so often it is given a symbol all for itself, $\hbar$ (called h-bar).

$$\hbar \equiv \frac{h}{2\pi}$$

$$p = \hbar k \quad \text{(the de Broglie relation)} \quad (25\text{-}1)$$

Now let us consider a particle that has an exact wavelength $\lambda_0$ traveling along the $x$ axis. Its wave number is $k_0 = 2\pi/\lambda_0$. Can we use $\psi = A\cos(k_0 x - \omega t)$ for the wave function? Then the probability distribution would be $|\psi|^2 \propto \cos^2(k_0 x - \omega t)$ and at any time $t$ there would be $x$ positions where the particle could not be found, whereas in actual fact it is equally probable to find the particle at any $x$ position. For this reason it is necessary to use $\psi = Ae^{i(k_0 x - \omega t)}$ for the wave function. Although $\psi$ is still a wave traveling in the positive $x$ direction,

$$|\psi|^2 = \psi^*\psi = (Ae^{-i(k_0 x - \omega t)})(Ae^{+i(k_0 x - \omega t)}) = A^2$$

We see that use of a complex wave function solves the problem and gives us a uniform probability in $x$. Because of Euler's relation

$$e^{i\phi} = \cos\phi + i\sin\phi \quad \text{(Euler's relation)}$$

the real part and the imaginary part of $\psi$ are still sine waves:

$$R_0(\psi) = A \cos(k_0 x - \omega t) \quad \text{and} \quad \text{Im}(\psi) = A \sin(k_0 x - \omega t)$$

In the preceding paragraph we saw that if the momentum of a particle has an exact value, its position in space can be at any point with equal probability. If we have exact knowledge of the momentum, we have no knowledge of the position. However, in most physics applications the position of a particle is known to be in a certain region of space. Consider, for example, the following wave function at $t = 0$:

$$\psi(x, 0) = A \exp\left(-\frac{x^2}{4\sigma_x^2}\right) e^{ik_0 x} \tag{25-2}$$

The real part of this wave function is plotted in Figure 25-1(a). The corresponding probability distribution is shown in Figure 25-1(b). It is

$$|\psi|^2 = A^2 \exp\left(-\frac{x^2}{2\sigma_x^2}\right)$$

Note that the probability is greater than 50% for finding the particle between $x = -\sigma_x$ and $x = +\sigma_x$. The function $e^{-x^2/2\sigma_x^2}$ is the famous Gaussian "error" function; $\sigma_x$ is the standard deviation. We shall be calling it $\Delta x$ or the uncertainty in $x$. Such a localized wave is called a wave packet. Although Figure 25-1(a) is not a pure sine wave, we saw in Appendix 21-2 that such a wave packet can be expressed as the sum of pure sine waves. To illustrate this we shall consider the situation at $t = 0$ and take an appropriate sum of pure sine waves of the form $e^{ikx}$. We want to find the coefficients $B_n$ where

$$\psi = \exp\left(\frac{x^2}{4\sigma_x^2}\right) e^{ik_0 x} = \sum B_n e^{ik_n x}$$

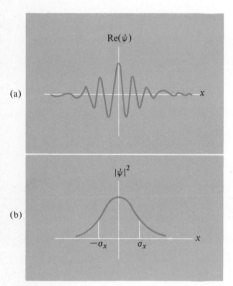

**Figure 25-1.** *Gaussian wave packet. (a) The real part versus x. (b) The square modulus or probability density versus x.*

Actually, it will take an infinite number of sine waves, so we transform the summation into an integral:

$$\psi = \exp\left(\frac{-x^2}{4\sigma_x^2}\right) e^{ik_0 x} = \int B(k) e^{ikx} \, dk \tag{25-3}$$

We make use of the following mathematical identity:

$$\exp\left(\frac{-x^2}{4\sigma_x^2}\right) e^{ik_0 x} = \frac{\sigma_x}{\sqrt{\pi}} \int e^{-\sigma_x^2 (k - k_0)^2} e^{ikx} \, dk \tag{25-4}$$

(The integral on the right-hand side is covered in a typical course in integral calculus.) Comparing Eqs. 25-3 and 25-4 gives

$$B(k) = \frac{\sigma_x}{\sqrt{\pi}} e^{-\sigma_x^2(k-k_0)^2}$$

We use Eq. 25-1 to replace $k$ with $p/\hbar$:

$$B(p) = \frac{\sigma_x}{\sqrt{\pi}} \exp\left[-\frac{(p-p_0)^2}{(\hbar/\sigma_x)^2}\right]$$

## 25-2 The Uncertainty Principle

This momentum distribution function is plotted in Figure 25-2 for wave packets of different widths. Note the narrower the wave packet, the wider is the momentum distribution function. Since the probability of finding the particle with a wave function $B(k)e^{ikx}$ is proportional to the square of this amplitude, the momentum probability function is

$$|B(p)|^2 = \frac{\sigma_x^2}{\pi} \exp\left[-\frac{(p-p_0)^2}{2(\hbar/2\sigma_x)^2}\right] \qquad (25\text{-}5)$$

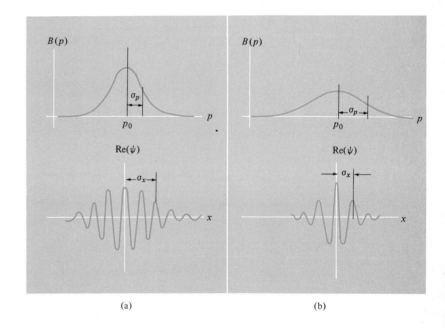

**Figure 25-2.** Momentum distribution function $B(p)$ (above) and its corresponding wave packet (below). The width of the wave packet in (a) is twice that of (b). Note $\sigma_x$ times $\sigma_p$ is the same for (a) and (b).

$|B(p)|^2$ is also a Gaussian distribution in $p$ and can be written as

$$|B(p)|^2 = \frac{\sigma_x^2}{\pi} \exp\left[-\frac{(p-p_0)^2}{2\sigma_p^2}\right] \quad (25\text{-}6)$$

where $\sigma_p$ is the standard deviation or "uncertainty" in $p$. Comparison of Eqs. 25-5 and 25-6 yields

$$\sigma_p = \frac{\hbar}{2\sigma_x}$$

$$\sigma_x \sigma_p = \frac{\hbar}{2} \quad (25\text{-}7)$$

So, for a Gaussian wave function (Eq. 25-2), the width of the wave packet times the width of the momentum probability distribution is $\hbar/2$. For other shapes of wave packets the product can be greater than $\hbar/2$ but never less than $\hbar/2$. The general relation is

$$\Delta x \, \Delta p \geqslant \frac{\hbar}{2} \quad \text{(the uncertainty principle)} \quad (25\text{-}8)$$

This says that if a particle is localized in space with a standard deviation $\Delta x$, it does not have a definite momentum but instead a momentum distribution $|B(p)|^2$ whose "width" is $\Delta p$. Physically it is impossible to specify the exact position and exact momentum of a particle simultaneously.

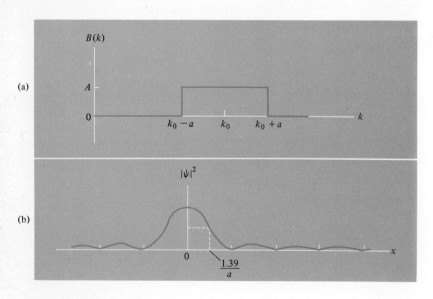

**Figure 25-3.** *View (b) shows the position distribution or wave packet corresponding to the momentum distribution shown in (a).*

**Example 1.** The momentum distribution of a wave packet is known to be a square wave as shown in Figure 25-3(a):

$$B(k) = \begin{cases} 0 & \text{for } k < (k_0 - a) \\ A & \text{for } (k_0 - a) < k < (k_0 + a) \\ 0 & \text{for } k > (k_0 + a) \end{cases}$$

What is $\psi(x)$ and what is $\Delta x\, \Delta p$, where $\Delta x$ and $\Delta p$ are the half widths at half maximum of the position and momentum probability distributions?

ANSWER: According to Eq. 25-3

$$\psi(x) = A \int_{k_0-a}^{k_0+a} e^{ikx}\, dk = \left[\frac{1}{ix} e^{ixk}\right]_{k_0-a}^{k_0+a}$$

$$= \frac{A}{ix} \{\exp[ix(k_0 + a)] - \exp[ix(k_0 - a)]\}$$

$$= 2A \frac{\sin ax}{x} e^{ik_0 x}$$

$$|\psi|^2 = 4A^2 \frac{\sin^2 ax}{x^2}$$

This drops to half its value when $ax = 1.39$. So

$$a\, \Delta x = 1.39$$

Since $a = \Delta k = \Delta p/\hbar$, we have

$$\left(\frac{\Delta p}{\hbar}\right) \Delta x = 1.39$$

$$\Delta x\, \Delta p = 1.39\, \hbar$$

A particle known to be at rest would have $\Delta p = 0$. One might then think that a microscope could be used to determine the position and then the uncertainty principle would be violated. With a microscope, the position of a particle can at best be located to within about one wavelength of the light being used (see page 512). Hence $\Delta x \approx \lambda$. Now, since $\Delta p = 0$, the product $\Delta x\, \Delta p$ would be zero and we have a violation of the uncertainty principle! Or do we? Let us look for some quantum mechanical effect. We used light and the quantum theory tells us that light is quantized into photons of momentum $p = h/\lambda$. In order to detect the particle, at least one of the photons in the converging beam of light from the condensing lens (see Figure 25-4) must be either scattered or absorbed by the particle. Hence the momentum given to the particle would be at least $h/\lambda$. The particle then has an uncertainty in its momentum $\Delta p \geqslant h/\lambda$ at the time it is observed at a position $\Delta x \approx \lambda$. If we multiply these two uncertainties together, we get

$$\Delta x \, \Delta p \geqslant (\lambda)\left(\frac{h}{\lambda}\right) = h$$

which checks with Eq. 25-8. In this example we see that quantum mechanics is self-consistent. Physicists and mathematicians have searched hard for inconsistencies, but none has been found.

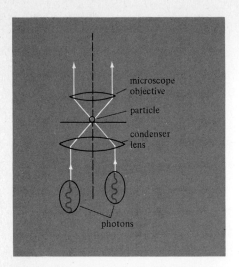

**Figure 25-4.** *Photons from microscope light source interacting with particle.*

### Velocity of Wave Packets

In Chapter 21 we saw that a wave packet does not move along with the wave velocity $u = \omega/k$, but instead it moves with the group velocity $v_g = d\omega/dk$ (see Eq. 21-17). According to de Broglie, $\hbar\omega = E$ and $\hbar k = p$ for all particles. Starting with $E = p^2/2m$ and substituting $\hbar\omega$ for $E$ and $\hbar k$ for $p$, we have

$$\hbar\omega = \frac{(\hbar k)^2}{2m}$$

Now differentiate with respect to $k$:

$$\hbar\frac{d\omega}{dk} = \frac{\hbar^2 k}{m}$$

$$\frac{d\omega}{dk} = \frac{\hbar k}{m} = \frac{p}{m} = v$$

$$v_g = v$$

Hence our wave packet representation of a localized particle does give the classically correct answer. The wave function does move along in space with the particle (as it should).

## Spreading of Wave Packets

Consider two particles with velocities $v_g$ and $(v_g + \Delta v_g)$. If at time $t = 0$ they were together, after a time $t$ they would be separated by

$$\Delta x = (\Delta v_g) t \tag{25-9}$$

We shall now show that a single wave packet contains an intrinsic spread in group velocity $\Delta v_g$ that must contribute to a spreading in width $\Delta x$ given by Eq. 25-9. Let us now estimate $\Delta v_g$. It will be

$$\Delta v_g = \frac{dv_g}{dp} \Delta p$$

We use the final equation in the preceding subsection to substitute $v$ for $v_g$:

$$\Delta v_g = \frac{dv}{dp} \Delta p$$

$$\approx \frac{1}{m} \Delta p \tag{25-10}$$

The initial value of $\Delta p$ is limited by the uncertainty principle to at least $\hbar/\Delta x_0$, where $\Delta x_0$ is the uncertainty in the initial position or the original width of the wave packet. Substituting into Eq. 25-10 gives

$$\Delta v_g = \frac{1}{m}\left(\frac{\hbar}{\Delta x_0}\right)$$

Substituting this into Eq. 25-9 gives

$$\Delta x \approx \frac{\hbar}{m \Delta x_0} t$$

This spread, which is proportional to $t$, is in addition to the initial spread $\Delta x_0$. This is an intrinsic spreading that can only be avoided if the particle is bound in a potential well, as we shall soon see. The appearance of a wave packet as a function of time is shown in Figure 25-5. The behavior of a free-particle wave packet striking a potential barrier is shown in Figure 25-6.

In order to give a quantitative idea of the rate of spreading of a free particle wave packet, consider a free electron that is initially known to be

**Figure 25-5.** *Two successive views of gaussian wave packet. The packet is moving to the right with group velocity the same as particle velocity.*

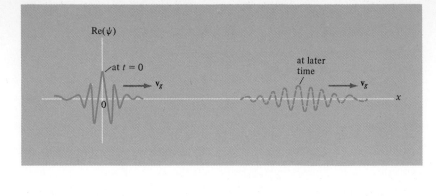

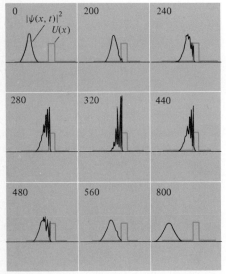

**Figure 25-6.** *Gaussian wave packet colliding with a rectangular potential barrier. The probability density and the potential barrier are plotted on the same axis for convenience (the height of the packet relative to the magnitude of the potential has no significance whatsoever). The potential energy curve is in color. The average energy is one half the barrier height. Note that width of packet increases with time (frame number).* [*From A. Goldberg, H. M. Schey, and J. L. Schwartz, Amer. J. Phys.* 177, *March 1967*]

localized in a region $\Delta x_0 = 10^{-10}$ m (typical size of an atom). After one second

$$\Delta x = \frac{\hbar}{m \, \Delta x_0} t$$

$$= \frac{10^{-34}}{9 \times 10^{-31} \times 10^{-10}} = 1.1 \times 10^6 \text{ m} = 1100 \text{ km}$$

We see that after one second this electron cloud becomes larger than the state of Texas. Even though quantum theory permits an exact determination of the wave function for all future times, once the initial wave function is known, this is of little help in predicting the future because the wave function rapidly spreads out over all space.

Quantum mechanics offers a possible way out of a philosophical dilemma posed by classical physics. In the days of classical physics it was pointed out that if one knew the exact positions and velocities of all the particles in the universe at a time $t_0$, one would in principle be able to calculate the future (and past) course of the universe from the exact laws of physics. The universe was thought of as one giant machine. Using such reasoning, philosophers could conclude that all human actions (even human beings are made up of protons, neutrons, and electrons) would be completely predetermined. Of course, it was realized that such calculations of the future or past would forever be impossible because of the enormous number of particles in the universe. But still, such reasoning was bothersome to believers in free will.

As we see from the uncertainty principle, there is a more fundamental obstacle to the carrying out of such calculations, thus classical determinism is no longer "forced" upon the physicist. This does not mean that we can invoke quantum mechanics as a proof of free will.

We have encountered examples other than the uncertainty principle that equally well refute the necessity of classical determinism. For instance, according to the generally accepted interpretation of quantum theory, there is no way of determining which electron will absorb a photon in the photoelectric effect. All we can do is calculate the probability that the photon will be absorbed by a given electron. The same holds for the position on the screen in Figure 24-8 where a single electron will show up. The wave interference pattern only tells us the probability of finding a given electron for each point on the screen. The same type of phenomenon holds for the decay of a radioactive nucleus such as uranium. There is no way to tell when an individual uranium nucleus will decay. According to the quantum theory all we can ever know is the probability that it may decay in a given interval of time. The predicted probabilities can then be compared with averages of a large number of observations.

We see that in the realm of interactions and structure of small particles, the quantum theory is radically different from the classical theory. If the quantum theory is correct, as we believe, there is no hope in studying elementary phenomena and the elementary structure of matter using classical physics.

## 25-3 Particle in a Box

We shall consider a particle trapped in a one-dimensional box with distance $L$ between perfectly reflecting walls. At the $x = 0$ wall in Figure 25-7 we have the superposition of a wave traveling to the right with a wave traveling to the left. Then

$$\psi(x, t) = Be^{ikx-i\omega t} - Be^{-ikx-i\omega t}$$
$$= B(e^{ikx} - e^{-ikx})e^{-i\omega t}$$

We use a minus sign because $\psi$ must go to zero at $x = 0$. (It is zero outside the wall, and $\psi$ must be continuous.) Using

$$\sin kx = \frac{e^{ikx} - e^{-ikx}}{2i}$$

we have

$$\psi(x, t) = 2iB \sin kx (ie^{-i\omega t})$$

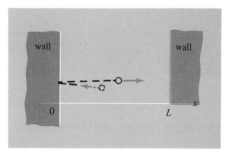

Figure 25-7. *Particle trapped in box of length L being reflected off of left wall.*

Let $A = 2Bi$ and $\psi(x)$ be the spatial part of the wave function, then

$$\psi(x) = A \sin kx \qquad (25\text{-}11)$$

For the same reason that $\psi(x)$ goes to zero at $x = 0$, it must also go to zero at $x = L$. Putting $x = L$ into Eq. 25-11 gives

$$0 = \sin(kL)$$

This is satisfied when $kL = n\pi$ where $n$ is an integer. We see that only the wave numbers $k_n$ are permitted where

$$k_n = n\frac{\pi}{L} \qquad (25\text{-}12)$$

We are requiring that there be an integral number of half wavelengths in the box—the same as the requirement for standing waves on a string:

$$L = n(\tfrac{1}{2}\lambda)$$

The wave functions $\psi_n(x) = A \sin(n\pi/L)x$ for $n = 1$ to 4 are shown in Figure 25-8. The corresponding momenta are

$$p_n = \hbar k_n$$

Using Eq. 25-12 this becomes

$$p_n = n\frac{\pi \hbar}{L} \qquad (25\text{-}13)$$

The corresponding kinetic energies are

$$E_n = \frac{p_n^2}{2m}$$

$$E_n = n^2 \frac{\pi^2 \hbar^2}{2mL^2} \qquad (25\text{-}14)$$

We note that the lowest possible energy is $\pi^2\hbar^2/(2mL^2)$ when $n = 1$ and the wave function is exactly one half of a sine wave. This is called the zero point energy. In classical physics the particle could have zero energy, but in quantum mechanics we see that, because $\psi$ must be nonzero in the box, the particle cannot have energy lower than $\pi^2\hbar^2/(2mL^2)$.

To give a quantitative idea of how large these energies are, consider an electron confined to a box the size of a typical atom of diameter $10^{-10}$ m (1 Å). Then

$$E_n = n^2 \frac{\pi^2 \hbar^2}{2mL^2} = n^2 \frac{(3.14)^2 (1.05 \times 10^{-34})^2}{2(9.11 \times 10^{-31})10^{-20}} = (5.97 \times 10^{-18} \text{ J})n^2$$

$$= (37.3n^2) \text{ eV}$$

The four lowest energies are plotted as an energy level diagram in Figure 25-9. The energy $E_1$ is comparable to the kinetic energy of an electron in a hydrogen atom.

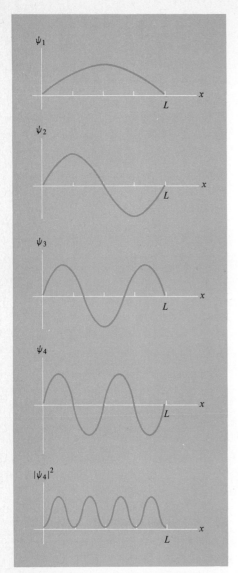

**Figure 25-8.** *The first four standing waves for a particle in a box and (bottom) the probability density for a particle in the $N = 4$ state.*

**Example 2.** Assume an electron in a box $L = 10^{-10}$ m is in the $n = 2$ state and that it can drop down to the lowest energy state by emitting a photon. What would be the wavelength of the photon?

ANSWER: By conservation of energy the photon energy is

$$hf = E_2 - E_1 = 4E_1 - E_1 = 1.79 \times 10^{-17} \text{ J}$$

$$f = \frac{(1.79 \times 10^{-17})}{h} = 2.70 \times 10^{16} \text{ s}^{-1}$$

$$\lambda = \frac{c}{f} = \frac{3 \times 10^8}{2.7 \times 10^{16}} = 1.11 \times 10^{-8} \text{ m} = 111 \text{ Å}$$

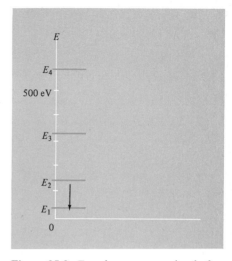

**Figure 25-9.** *Four lowest energy levels for electron in a box of length $10^{-10}$ m. Arrow shows energy transition for Example 2.*

The photon emitted in Example 2 is in the ultraviolet region of the electromagnetic spectrum just as are the most energetic lines in the hydrogen spectrum. Since our electron in the box can have only certain discrete energies, it can only emit photons of discrete energies or wavelengths. Such "light" as viewed by a spectroscope would appear as a line spectrum just as the light emitted from atoms. In fact, an electron confined to a box is a crude approximation to the hydrogen atom. The box is represented by a square well potential, whereas in the hydrogen atom the electron is trapped in a coulomb potential well shown in Figure 25-10. But the qualitative physics is the same. Because the electron must be represented as a standing wave, only certain wave functions $\psi_n$ are possible with their corresponding energies $E_n$.

**Example 3.** Suppose the box is so small that the kinetic energies of the particle are comparable to $mc^2$. Give the exact relativistic expression for the particle energies $E_n$.

ANSWER: Now we must use the relativistic relation

$$E_n = c(p_n^2 + m^2c^2)^{1/2}$$

where $E_n$ is the total energy including the rest energy $mc^2$. Since the de Broglie relation holds for relativistic particles, we can substitute the right-hand side of Eq. 25-13 into the preceding equation:

$$E_n = c\left(n^2 \frac{\pi^2 \hbar^2}{L^2} + m^2c^2\right)^{1/2}$$

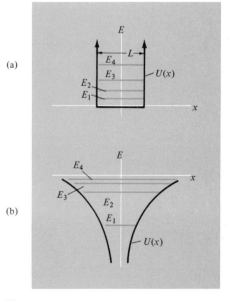

**Figure 25-10.** *(a) Infinite potential well showing the four lowest energy levels. (b) Potential well for attractive electrostatic force showing the four lowest energy levels.*

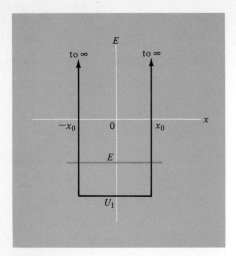

**Figure 25-11.** *Infinite potential well similar to that in Figure 25-10(a) but with energy scale displaced.*

**Example 4.** Suppose a particle of energy $E$ is trapped in the negative potential well shown in Figure 25-11. This is the same potential well as in Figure 25-10(a) except the energy scale has been shifted by an amount $U_1$. Thus inside the well the potential energy is $U_1$. (In this problem $U_1$ is a negative quantity.) (a) What is $d^2\psi/dx^2$ of the particle inside the well? (b) What are the energies $E_n$?

ANSWER: (a) The general solution inside the well will be of the form

$$\psi = A \sin(kx + \phi) \qquad \text{where } k = \frac{p}{\hbar}$$

We can get $p$ from the relation $E = p^2/2m + U_1$:

$$p = \sqrt{2m(E - U_1)}$$

$$k = \sqrt{\frac{2m}{\hbar^2}(E - U_1)}$$

The second derivative of $\psi$ is

$$\frac{d^2\psi}{dx^2} = -k^2 A \sin(kx + \phi) = -k^2\psi$$

$$\frac{d^2\psi}{dx^2} = -\frac{2m}{\hbar^2}(E - U_1)\psi \qquad (25\text{-}15)$$

(b) The kinetic energies are still given by Eq. 25-14 by replacing $L$ with $(2x_0)$ and using $(E_n - U_1)$ for the kinetic energy $K_n$:

$$K_n = n^2 \frac{\pi\hbar^2}{2mL^2}$$

$$(E_n - U_1) = n^2 \frac{\pi\hbar^2}{2m(2x_0)^2}$$

$$E_n = n^2 \frac{\pi^2\hbar^2}{8mx_0^2} + U_1$$

## 25-4 The Schrödinger Equation

Up to now we have been dealing with free particles of fixed momentum and hence fixed wavelength. In the more general situation the particle can be under the influence of an external force, which can be represented by a potential energy function $U(x)$. Then, since

$$E = \frac{p^2}{2m} + U(x) \qquad (25\text{-}16)$$

is constant, if $U$ increases with $x$, $p$ will decrease with $x$ and the corresponding wavelength will increase. The wave function must have a variable wavelength. A wavefunction whose wavelength is increasing with $x$ is illustrated in Figure 25-12(b). The exact form for a $\psi(x)$ with variable wavelength is found by solving a differential equation called the Schrödinger equation.

We shall now obtain the Schrödinger equation by approximating $U(x)$ with a series of steps shown in Figure 25-12(c). We know from Example 4 that $\psi(x)$ in the region of $U_1$ obeys Eq. 25-15:

$$\frac{d^2\psi}{dx^2} = -\frac{2m}{\hbar^2}(E - U_1)\psi$$

Since this equation holds for $U_2, U_3, \ldots, U_j$, and since any $U(x)$ can be broken up into small steps of $U_j$, we can replace $U_1$ with $U(x)$ to obtain

$$\frac{d^2\psi}{dx^2} = -\frac{2m}{\hbar^2}[E - U(x)]\psi \quad \begin{array}{l}\text{(the Schrödinger time-}\\ \text{independent equation)}\end{array} \quad (25\text{-}17)$$

This is the famous Schrödinger time-independent equation in one dimension. It applies to nonrelativistic bodies where the probability distribution is not changing with time; that is, for standing waves. (There is also a Schrödinger time-dependent equation for those problems where the wave packet is changing with time.) We shall now consider several examples of particles bound in potential wells, applying the time-independent equation to find stationary states (standing waves).

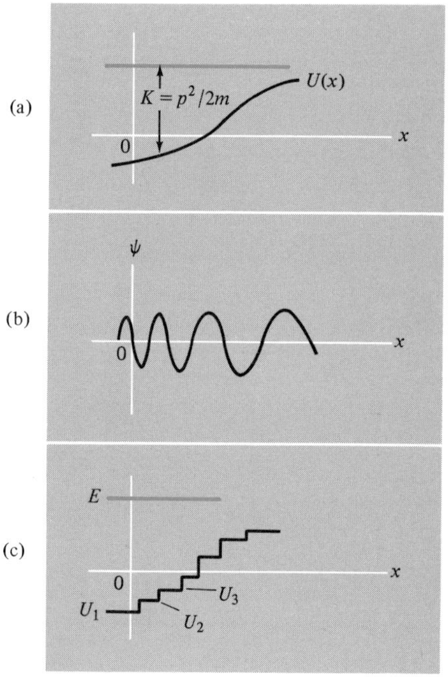

**Figure 25-12.** (a) *Potential energy $U(x)$ is increasing with $x$ showing how $K$ decreases with $x$.* (b) *Corresponding $\psi(x)$ whose wavelength increases with $x$.* (c) *Approximating $U(x)$ in (a) with a series of steps.*

## Boundary Conditions

If the particle is trapped in a potential well, there is zero probability of finding it far away from the well; hence, we have the boundary condition that the probability for finding the particle at large values of $|x|$ is zero. As we shall see in the next section only certain values of $E$ (call them $E_n$) with corresponding $\psi_n$ will satisfy this boundary condition. These values of $E_n$ are called eigenvalues, and the corresponding wave functions are called eigenfunctions.

## 25-5 Finite Potential Wells

The particle in a box of Section 25-3 was in a potential well where the walls were infinitely high. In this section we use the Schrödinger equation to solve the problem of a particle in a potential well where the walls are of finite height $U_0$ as shown in Figure 25-13. The problem is to find those wave functions $\psi_n$ and their energies $E_n$ that satisfy the boundary condition that $\psi(x) \to 0$ for large $|x|$. At first sight one might think that $\psi(x)$ must be zero for $|x| > x_0$ because in that region $(E - U)$ is negative, and this would correspond to a negative value of kinetic energy which is forbidden classically ($K = E - U$). However, the Schrödinger equation in region II of Figure 25-13 is of the form

$$\frac{d^2\psi}{dx^2} = \frac{2m}{\hbar^2}(U_0 - E)\psi$$

and this has the solutions

$$\psi_{II} = e^{-\kappa x} \quad \text{and} \quad e^{+\kappa x}$$

where

$$\kappa \equiv \sqrt{\frac{2m(U_0 - E)}{\hbar^2}} \quad (25\text{-}18)$$

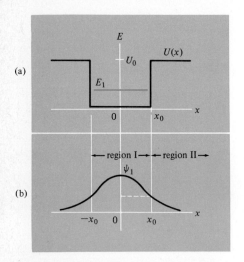

**Figure 25-13.** (a) *Potential well of depth $U_0$ showing the first energy level.* (b) *Corresponding wave function.*

We note that whenever the kinetic energy is negative, $d^2\psi/dx^2$ is of the same sign as $\psi$ and that it must curve away from the $x$ axis; whereas when the kinetic energy is positive (as in region I), $\psi(x)$ will curve toward the $x$ axis as does a sine wave. Curve $b$ in Figure 25-14 shows how $\psi$ behaves if the correct value of $E$ is chosen. If we choose the energy $E$ a little too low, $\psi(x)$ will curve more slowly in region I (see curve $a$.) If $E$ is a little too high, $\psi(x)$ is given by curve $c$. The correct curve $b$ has the form

$$\psi_{II} = Ae^{-\kappa x} \quad \text{(in region II)}$$

It has the form

$$\psi_I = B \cos kx \quad \text{(in region I)}$$

where

$$k = \sqrt{\frac{2mE}{\hbar^2}}$$

At $x = x_0$:

$$\psi_I(x_0) = \psi_{II}(x_0) \quad \text{or} \quad B \cos kx_0 = Ae^{-\kappa x_0}$$

Also at $x_0$ their slopes are the same; so

$$-kB \sin kx_0 = -\kappa Ae^{-\kappa x_0}$$

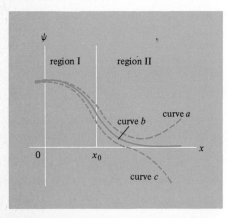

**Figure 25-14.** *Curve b is the same as in Figure 25-13(b). Curve a is for E slightly less than $E_1$, and curve c corresponds to E slightly larger than $E_1$.*

Dividing this equation by the preceding equation gives

$$k \tan kx_0 = \kappa$$

This is a transcendental equation that can be solved for the energy level $E_1$. It can be put into a simpler form using Eq. 25-18

$$\tan kx_0 = \frac{\kappa}{k} = \sqrt{\frac{U_0}{E} - 1}$$

Let

$$y_0 \equiv \sqrt{\frac{2mU_0}{\hbar^2}} x_0 \quad \text{and} \quad y \equiv kx_0$$

Then

$$\tan y = \sqrt{\frac{y_0^2}{y^2} - 1} \qquad (25\text{-}19)$$

Equation 25-19 may have several roots depending on the size of $y_0$. It is interesting to compare the finite potential well with an infinite potential well of length $10^{-10}$ m or $x_0 = 0.5 \times 10^{-10}$ m. Let us now repeat the calculation for an electron in a finite well of 1 Å length where the depth of the well is 800 eV rather than having walls of infinite height. The latter calculation was done on page 550. For $U_0 = 800$ eV we have

$$y_0 = \frac{\sqrt{2mU_0}}{\hbar} x_0$$

$$= \sqrt{2(9.1 \times 10^{-31})(800 \times 1.6 \times 10^{-19})} \frac{0.5 \times 10^{-10}}{1.05 \times 10^{-34}} = 7.27$$

Then there are just three positive roots to Eq. 25-19, which we shall call $y_1$, $y_3$, and $y_5$. They can be found graphically, by iteration, or by "trial and error." They are $y_1 = 1.38$, $y_3 = 4.11$, and $y_5 = 6.69$. Since $k = y/x_0$ and $E = (\hbar k)^2/2m$, we have

$$E_1 = 28.8 \text{ eV} \qquad E_3 = 256 \text{ eV} \quad \text{and} \quad E_5 = 678 \text{ eV}$$

We note that for $n = 2$ and 4, the wave function has the form $\psi_I = B \sin kx$ in region I. Then the matching conditions at $x = x_0$ give

$$B \sin kx_0 = Ae^{-\kappa x_0}$$
$$kB \cos kx_0 = -\kappa Ae^{-\kappa x_0}$$

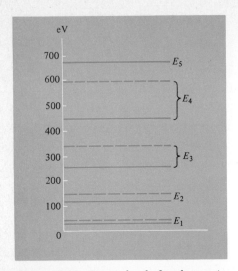

**Figure 25-15.** *Energy levels for electron in a box of length $10^{-10}$ m. Solid lines are for potential well depth of 800 eV. Dashed lines are for infinite potential well (same as in Figure 25-9).*

Dividing gives

$$\cot kx_0 = -\frac{\kappa}{k} \quad (25\text{-}20)$$

$$\cot\left(\sqrt{\frac{2mE}{\hbar^2}}x_0\right) = -\sqrt{\frac{U_0}{E} - 1}$$

or

$$\cot y = -\sqrt{\frac{y_0^2}{y^2} - 1}$$

For $y_0 = 7.27$ there are two positive roots: $y_2 = 2.75$ and $y_4 = 5.44$. The corresponding energies are $E_2 = 115$ eV and $E_4 = 447$ eV for the described electron in a finite potential well. All these energy levels are plotted in Figure 25-15 and the first three wave functions shown in Figure 25-16.

We shall see in Chapter 29 when we study nuclear physics that this general solution to the particle in a finite square well is of practical value. We will be able to approximate the short-range nuclear force between a proton and a neutron as such a square well. Then, by using Eq. 25-20, we will calculate the binding energy of the deuteron and its size (its wave function).

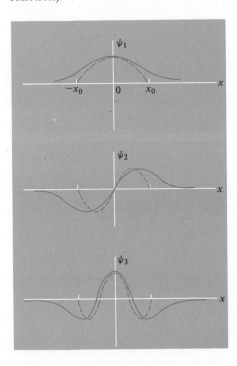

**Figure 25-16.** *The three lowest order standing waves corresponding to $E_1$, $E_2$, and $E_3$ of the previous figure. The corresponding wave functions for an infinite potential well of the same length are shown as dashed lines.*

**Example 5.** Suppose $\psi_1$ in Figure 25-13(b) drops a factor of 2 where $x = x_0$; that is,

$$\frac{\psi_1(x_0)}{\psi_1(0)} = \frac{1}{2}$$

What is $E_1$ in terms of $\hbar$, $m$, and $x_0$?

ANSWER: Since $\psi_1$ is of the form $B \cos kx$.

$$\cos kx_0 = \frac{1}{2}$$

$$kx_0 = \frac{\pi}{3}$$

$$k = \frac{\pi}{3x_0}$$

$$\sqrt{\frac{2mE_1}{\hbar^2}} = \frac{\pi}{3x_0}$$

$$E_1 = \frac{\pi^2 \hbar^2}{18 m x_0^2}$$

## 25-6 The Harmonic Oscillator

Our next application of the Schrödinger equation will be to a particle of mass $m$ under the influence of a harmonic force $F = -kx$. (Here $k$ is the force constant, not the wave number.) As we learned in the chapter on simple harmonic motion, the potential energy is $U(x) = \frac{1}{2}kx^2 = \frac{1}{2}m\omega_c^2 x^2$. In classical physics such a particle executes simple harmonic motion with angular frequency $\omega_c = \sqrt{k/m}$ and it can have any value of energy including zero. The subscript "c" stands for "classical." In quantum mechanics we shall see that because of the boundary condition at large $|x|$, only the energies $E_n = (n - \frac{1}{2})\hbar\omega$ are permitted where $n$ is a positive integer.

Before using the Schrödinger equation, let us make an approximate calculation that will illustrate to us the basic quantum mechanical principles in operation here. Let us try to solve this problem as we did the particle in a box. We see in Figure 25-17 that the box has equivalent length $L = 2x_0$ where $x_0$ is the maximum displacement or amplitude of

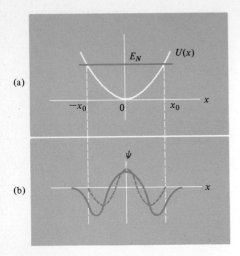

**Figure 25-17.** (a) *Potential energy curve for harmonic oscillator.* (b) *Third order standing wave. Solid line is exact solution; dashed line is approximate solution.*

oscillation. The nth-order standing wave would have $n$ half wavelengths in a distance $L$:

$$n\frac{\bar{\lambda}_n}{2} = 2x_0$$

or

$$\bar{\lambda}_n = \frac{4x_0}{n}$$

The average momentum would be

$$\bar{p} = \frac{h}{\bar{\lambda}} = \frac{h}{4x_0/n} = n\frac{h}{4x_0}$$

The average kinetic energy is

$$\bar{K} = \frac{\bar{p}^2}{2m} = \frac{n^2 h^2}{32 m x_0^2}$$

The total energy $E$ is twice this:

$$E = \frac{n^2 h^2}{16 m x_0^2}$$

$E$ is also equal to the maximum potential energy:

$$E = \tfrac{1}{2} m \omega_c^2 x_0^2$$

Multiplying these two equations together gives

$$E^2 = \frac{n^2 h^2 \omega_c^2}{32}$$

or

$$E = \frac{\pi}{2\sqrt{2}} n \hbar \omega_c$$

To the accuracy of this "calculation" $\pi/(2\sqrt{2}) \approx 1$ and then

$$E_n \approx n \hbar \omega_c \qquad \text{for } n = 1, 2, 3, \ldots$$

As we shall see, the exact calculation gives

$$E_n = (n - \tfrac{1}{2}) \hbar \omega_c \qquad \text{for } n = 1, 2, 3, \ldots$$

Now we shall do the exact calculation using the Schrödinger equation. Putting $U = \frac{1}{2}m\omega_c^2 x^2$ into the Schrödinger equation gives:

$$\frac{d^2\psi}{dx^2} = -\frac{2m}{\hbar^2}(E - \tfrac{1}{2}m\omega_c^2 x^2)\psi \qquad (25\text{-}21)$$

We need a solution with a variable wavelength, so let us try a Gaussian. A legitimate and common procedure for solving differential equations is to "guess" the solution and then see if it works. We try

$$\psi(x) = e^{-ax^2}$$

In order to see if this fits Eq. 25-21, we must take the second derivative and substitute into the left-hand side. The second derivative is

$$\frac{d^2\psi}{dx^2} = -2ae^{-ax^2} + 4a^2 x^2 e^{-ax^2}$$

Substitution into Eq. 25-21 gives

$$(-2a + 4a^2 x^2)e^{-ax^2} = -\frac{2m}{\hbar^2}(E - \tfrac{1}{2}m\omega_c^2 x^2)e^{-ax^2}$$

$$-2a + (4a^2)x^2 = -\frac{2mE}{\hbar^2} + \left(\frac{m^2\omega_c^2}{\hbar^2}\right)x^2$$

Equating the coefficients of $x^2$ gives

$$(4a^2) = \left(\frac{m^2\omega_c^2}{\hbar^2}\right)$$

$$a = \frac{m\omega_c}{2\hbar}$$

Equating the constant terms gives

$$-2a = -\frac{2mE}{\hbar^2}$$

$$E = \frac{\hbar^2 a}{m} = \frac{\hbar^2}{m}\left(\frac{m\omega_c}{2\hbar}\right) = \tfrac{1}{2}\hbar\omega_c$$

We see that our Gaussian guess does work, but only when

$$E = \tfrac{1}{2}\hbar\omega_c$$

and then
$$\psi_1(x) = e^{-(m\omega_c/2\hbar)x^2}$$

The reader can verify by substitution into Eq. 25-21 that the next order standing wave solution is
$$\psi_2(x) = xe^{-(m\omega_c/2\hbar)x^2}$$

This solution only works when $E = \frac{3}{2}\hbar\omega_c$. Note that the energy difference between these two adjacent levels is $E_2 - E_1 = \hbar\omega_c$. This is also true of the higher order energy levels which turn out to be $E_n = (n - \frac{1}{2})\hbar\omega_c$. We see that a harmonic oscillator will emit photons whose frequency is equal to the classical frequency of the oscillator for transitions between adjacent levels.

## Summary

A free particle of momentum exactly equal to $p_0$ has a wave function $\psi(x,t) = A \exp[i(k_0 x - \omega t)]$, where $k_0 = p_0/\hbar$.

If the particle is localized over a Gaussian region of space, $\psi(x,0) = \exp[-x^2/4\sigma_x^2]e^{ik_0 x}$. Then it will have a momentum distribution $B(k) = (\sigma_x/\sqrt{\pi}) \exp[-\sigma_x^2(k - k_0)^2]$ where $\psi(x,0) = \int B(k)e^{ikx}\,dk$.

The standard deviation of the spatial probability distribution is $\sigma_x$ and that of the momentum probability distribution is $\sigma_p = \hbar/2\sigma_x$. For any wave packet $\sigma_x \sigma_p \geq \hbar/2$ (the uncertainty principle).

The group velocity of any quantum mechanical wave packet is the classical particle velocity: $v_g = d\omega/dk = p/m = v$. A wave packet of initial width $\Delta x_0$ gets wider at the rate $\Delta x \approx (\hbar/m\,\Delta x_0)t$.

A particle confined to a box of length $L$ can only have standing waves $\lambda_n = 2L/n$ or kinetic energies
$$E_n = n^2 \frac{\pi^2 \hbar^2}{2mL^2}$$

If the "box" is of nonuniform depth represented by $U = U(x)$, the wave function obeys the time-independent Schrödinger equation:
$$\frac{d^2\psi}{dx^2} = -\frac{2m}{\hbar^2}[E - U(x)]\psi$$

Only certain eigenfunctions $\psi_n(x)$ and eigenvalues $E_n$ will satisfy the boundary condition that $\psi \to 0$ as $x \to \pm\infty$.

In the case of a finite square well $\psi = B\cos(kx + \phi)$ inside the well and is of the form $(A_1 e^{-\kappa x} + A_2 e^{\kappa x})$ with $\kappa \equiv \sqrt{2m(U_0 - E)/\hbar^2}$ outside the well. Conditions for $E$ are obtained by matching $\psi$ and $d\psi/dx$ at the boundaries of the well.

If $U(x) = \frac{1}{2}m\omega_c^2 x^2$, the harmonic oscillator potential, the eigenvalues are $E_n = (n - \frac{1}{2})\hbar\omega_c$, where $n$ is any positive integer. The first two eigenfunctions are

$$\psi_1(x) = \exp\left[-\frac{m\omega_c}{2\hbar}x^2\right]$$

and

$$\psi_2(x) = x\exp\left[-\frac{m\omega_c}{2\hbar}x^2\right]$$

## Exercises

1. If $\psi = \exp[i(kx - \omega t)] + \exp[i(kx + \omega t)]$, what is $\psi^*\psi$?
2. What is $\bar{x} = \int_{-\infty}^{\infty} x\exp(-x^2/2\sigma^2)\,dx$?
3. If $\psi = \exp[i(k_0 + \Delta k)x] + \exp[i(k_0 - \Delta k)x]$
   (a) What is $\psi^*\psi$?
   (b) Let $\Delta x$ be the half width of the central wave packet at half maximum. What is the product $(\Delta x\, \Delta p)$ where $\Delta p = \hbar\, \Delta k$?
4. If $E = \hbar\omega = mc^2 + p^2/(2m)$
   (a) what would be the wave velocity of the corresponding wave packet? Give answer in terms of $c$ and the classical velocity $v$.
   (b) What would be the group velocity?
5. At $t = 0$ an electron wave packet has $\Delta x_0 = 1\,\mu\text{m}$. What is the width of the wave packet at $t = 1$ s?
6. An electron wave packet has a width $\Delta x_0$ such that for each subsequent second its width increases by the same $\Delta x_0$. What is $\Delta x_0$?
7. An electron is confined to a one-dimensional box of length $L = 4 \times 10^{-10}$ m
   (a) What is $E_2$ in eV?
   (b) What wavelength photon would be emitted in the transition $E_4 \to E_2$?
8. If in Example 4, $U_1 = -20$ eV and $x_0 = 5 \times 10^{-11}$ m, what is $E_1$ and what is $K_1$ for an electron?
9. What is the relativistically correct answer to part (b) of Example 4?
10. Suppose $\psi_1$ in Figure 25-13(b) drops by a factor of 3 where $x = x_0$; that is,

    $$\frac{\psi_1(x_0)}{\psi_1(0)} = \frac{1}{3}$$

    What is $E_1$ in terms of $\hbar$, $m$, and $x_0$?
11. At $t = 0$, $|\psi(x)|^2$ of a wave packet is a Gaussian with a standard deviation $\sigma_0$. It is moving along the $x$ axis with velocity $v_0$.
    (a) The envelope of the wave function is also a Gaussian. What is its standard deviation?
    (b) What is $B(k)$ where $\psi(x) = \int B(k)\exp[ikx]\,dk$? Give answer in terms of $m$, $\hbar$, $\sigma_0$, $k$, and $v_0$.

## Problems

Problem 14

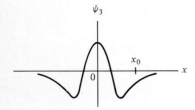

Problem 22

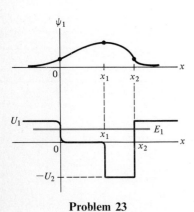

Problem 23

12. If $\psi = \exp\{i[(k_0 + \Delta k)x - (\omega_0 + \Delta\omega)t]\} + \exp\{i[(k_0 - \Delta k)x - (\omega_0 - \Delta\omega)t]\}$, what is $\psi^*\psi$?

13. Express $\psi = A_1 e^{i\phi_1} + A_2 e^{i\phi_2}$ as $\psi = A e^{i\phi}$. What are $A$ and $\phi$ in terms of $A_1$, $A_2$, $\phi_1$, and $\phi_2$?

14. If $B(k)$ is as shown in the figure
    (a) What is the probability that $p = \hbar(k_0 + \tfrac{1}{2}a)$ compared to $p = \hbar k_0$?
    (b) What is $\psi(x)$?
    (c) What is $(\Delta x\, \Delta p)$? $\Delta x$ and $\Delta p$ are half-widths at half maximum.

15. Prove that if $x$ is Gaussian-distributed, $\overline{x^2} = \sigma^2$; that is, prove that

$$\frac{\int_0^\infty x^2 \exp(-x^2/2\sigma^2)\, dx}{\int_0^\infty \exp(-x^2/2\sigma^2)\, dx} = \sigma^2$$

16. If $B(k)$ is as in the figure to Problem 14, what is the probability of finding the particle with a momentum $\hbar(k_0 + \tfrac{1}{2}a) > p > \hbar(k_0 - \tfrac{1}{2}a)$?

17. If $E = \sqrt{p^2 c^2 + m^2 c^4} = \hbar\omega$, what is $v_g = d\omega/dk$?

18. The spreading of a wave packet is given by

$$\Delta x = \sqrt{\left(\frac{\hbar t}{m\, \Delta x_0}\right)^2 + (\Delta x_0)^2}$$

where $\Delta x_0$ is the spread at $t = 0$ when all the component sine waves are in phase. If electron 1 has $\Delta x_0 = 10^{-10}$ m and electron 2 has $\Delta x_0 = 10^{-9}$ m at $t = 0$, at what time $t$ will the two packets have the same width?

19. Suppose an electron is trapped in the well in Example 4 and $x_0 = 10^{-14}$ m and $U_1 = 0$.
    (a) Calculate $E_1$ and $E_2$ nonrelativistically.
    (b) Calculate $E_1$ and $E_2$ relativistically.

20. Suppose an electron is trapped in an atomic nucleus of radius $2 \times 10^{-15}$ m. What would be its zero point energy? Use a one-dimensional square well approximation.

21. Using a "trial and error" method calculate $E_1$ and $E_2$ for an electron in a finite potential well of depth $U_0 = 200$ eV and of length $2x_0 = 10^{-10}$ m.

22. In the figure suppose $\psi_3(x_0)/\psi_3(0) = -\tfrac{1}{4}$. What is $E_3$ in terms of $\hbar$, $m$, and $x_0$? $\psi_3$ is a sine wave in region I.

23. A particle of mass $m$ is bound in a one-dimensional well as shown. If the part of $\psi_1$ between $x = 0$ and $x = x_1$ covers 60° worth of a sine wave, what is $E_1$?

24. Substitute $\psi(x) = xe^{-\alpha x^2}$ into the Schrödinger equation where $U(x) = \tfrac{1}{2}kx^2$.
    (a) What is $\alpha$?
    (b) What is the corresponding energy or eigenvalue?

25. The Schrödinger equation for a particle of mass $m$ is

$$\frac{d^2\psi}{dx^2} = -\frac{2m}{\hbar^2}\left(E - \tfrac{1}{2}kx^2\right)\psi$$

(a) What is the lowest-order eigenfunction $\psi_1(x)$?

(b) What is the corresponding eigenvalue $E_1$? Express it in terms of $\hbar$, $m$, and $k$.

26. The potential that binds a proton and neutron together in a deuteron can be approximated by a square well as shown. The depth is 29 MeV and the radius is $r_0 = 2.3 \times 10^{-15}$ m. We let $u(r)$ be the wave function. Inside the well it is

$$u(r) = \sin kr \qquad \text{where} \quad \frac{(\hbar k)^2}{2M} = E - U$$

Use $M = 8.36 \times 10^{-28}$ kg for the reduced mass of the proton-neutron system. Find a numerical value for $E$ in MeV. This will be the binding energy of the deuteron. (*Hint:* Eq. 25-20 may be of use.) The solution to a transcendental equation can be found by "trial and error" using a pocket calculator.

27. Substitute the wave function $\psi = x \exp[-m\omega x^2/(2\hbar)]$ into the equation

$$\frac{d^2\psi}{dx^2} = -\frac{2m}{\hbar^2}(E - U)\psi \qquad \text{where } U = \tfrac{1}{2}m\omega^2 x^2$$

Solve for $E$.

28. A position wave packet $\psi(x, t)$ is made up of a distribution of pure "sine waves":

$$\psi(x, t) = \int A(k) e^{i(kx - \omega t)}\, dk \qquad \text{where } A(k) = \exp\left[-\frac{(k - k_0)^2}{2a^2}\right]$$

At $t = 0$ what is the half width of the probability distribution in $x$?

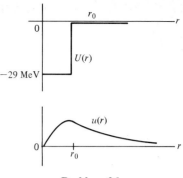

Problem 26

# The Hydrogen Atom

## 26-1 Approximate Solution to Hydrogen Atom

In the previous section we solved the problem of a particle of mass $m$ bound in a potential well of the form $U = \frac{1}{2}kx^2$. Such a potential, which has the form of a parabola, gives rise to simple harmonic motion. Now we shall discuss a potential well in which the potential energy varies inversely with the distance. This kind of a potential occurs when there is a gravitational or electrostatic force between two bodies. In the case of the hydrogen atom the potential energy between the proton and electron is $U = -k_0 e^2/r$, where $r$ is the distance between them.

We shall first approximate this potential well with a square well in order to get an approximate solution, just as we did with the harmonic oscillator. Our approach of first making a simple approximate calculation should help give a more direct understanding of the hydrogen atom while avoiding distracting mathematical details. Our goal is to understand why the hydrogen standing waves are the size they are and why they depend on the fundamental constants $h$, $m$, and $e$ the way they do. We hope to understand why only certain energy levels are permitted and how they depend on fundamental constants and the quantum number $n$.

In this approximate calculation we have the potential well shown in Figure 26-1. If the electron energy level is $E$, its maximum distance according to classical physics will be $R_0$, as seen in Figure 26-1. We take half of this for its average distance $\bar{R}$. For the depth of the equivalent square well we use the value of $U$ when $x = \bar{R}$; that is, $U_0 = k_0 e^2/\bar{R}$. The $n$th-order standing wave in this dashed-line square well has $n$ half wavelengths over the length of the box; that is,

$$n\frac{\lambda_n}{2} = 2R_0$$

or

$$\lambda_n = \frac{4R_0}{n}$$

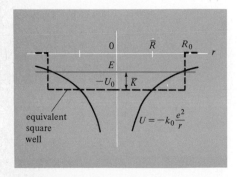

**Figure 26-1.** *Potential well for electron in hydrogen atom. Equivalent square well shown here is used in approximate calculation.*

We make the approximation that the corresponding de Broglie momentum is the average electron momentum:

$$\bar{p}_n \approx \frac{h}{\lambda_n} = \frac{hn}{4R_0}$$

The average kinetic energy is

$$\bar{K} \approx \frac{\bar{p}_n^2}{2m} = \frac{h^2 n^2}{32 m R_0^2}$$

From Figure 26-1 we see that

$$\bar{K} = -E = \frac{k_0 e^2}{R_0} \qquad (26\text{-}1)$$

Equating this to the above expression gives

$$\frac{k_0 e^2}{R_0} \approx \frac{h^2 n^2}{32 m R_0^2}$$

or

$$R_0 \approx \frac{h^2 n^2}{32 k_0 m e^2} = \frac{4\pi^2}{32} \frac{\hbar^2}{k_0 m e^2} n^2 \qquad (26\text{-}2)$$

As we shall see in Section 26-3, the "radius" of the $n$th hydrogen wave function is close: it is

$$R_n = \frac{\hbar^2}{k_0 m e^2} n^2$$

We can obtain approximate solutions for the energy levels by substituting Eq. 26-2 into the right-hand side of Eq. 26-1:

$$E_n = -\frac{16}{\pi^2} k_0^2 \frac{m e^4}{2\hbar^2} \frac{1}{n^2}$$

The exact answer is

$$E_n = -\frac{k_0^2 m e^4}{2\hbar^2 n^2}$$

which is 38% smaller than the preceding result given by our crude approximation. However, the preceding approximation does give the correct dependence on $m$, $e$, $\hbar$, and the quantum number $n$ and it gives us insight into what is going on.

## 26-2 The Three-Dimensional Schrödinger Equation

Two-dimensional standing waves can be produced using a ripple tank with rectangular walls. If the table is tapped or vibrated, standing waves will be induced with lines of nodes in both directions as shown in Figure 26-2. The wave amplitude will be

$$\psi = A(\sin k_x x)(\sin k_y y)$$

where

$$k_x = \frac{\pi}{L_x} n_x \quad \text{and} \quad k_y = \frac{\pi}{L_y} n_y \tag{26-3}$$

satisfy the boundary condition that $\psi = 0$ when $x = L_x$ or $y = L_y$.

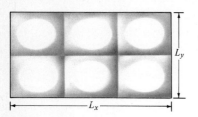

**Figure 26-2.** *Artist's drawing of two-dimensional ripple tank of length $L_x$ and width $L_y$. A possible standing wave pattern is shown with $n_x = 3$ and $n_y = 2$.*

**Example 1.** Derive an expression for the energy levels of a particle of mass $m$ in a three-dimensional box of dimensions $L_x$, $L_y$, $L_z$.

ANSWER: The kinetic energy is

$$K = \frac{p_x^2 + p_y^2 + p_z^2}{2m}$$

According to the de Broglie relation

$$p_x = \hbar k_x \qquad p_y = \hbar k_y \qquad p_z = \hbar k_z$$

Then

$$K = \frac{\hbar^2}{2m}(k_x^2 + k_y^2 + k_z^2)$$

Using Eq. 26-3:

$$K = E_{n_x, n_y, n_z} = \frac{\hbar^2 \pi^2}{2m}\left(\frac{n_x^2}{L_x^2} + \frac{n_y^2}{L_y^2} + \frac{n_z^2}{L_z^2}\right)$$

We see that in three-dimensional space each energy eigenvalue and eigenfunction is dependent upon three quantum numbers.

**Example 2.** An electron is confined to a cubical box the size of an atomic nucleus ($L = 4 \times 10^{-15}$ m). What is its minimum possible kinetic energy?

ANSWER: Using Eq. 26-3, we have

$$k_x = \frac{\pi}{L} n_x \qquad k_y = \frac{\pi}{L} n_y \qquad k_z = \frac{\pi}{L} n_z$$

The lowest energy state has $n_x = n_y = n_z = 1$, so

$$k_x^2 = k_y^2 = k_z^2 = \frac{\pi^2}{L^2}$$

$$p^2 = \hbar^2 k^2 = \hbar^2(k_x^2 + k_y^2 + k_z^2) = 3\hbar^2\left(\frac{\pi^2}{L^2}\right)$$

Just to play safe, we use the formula for relativistic kinetic energy:

$$K = \sqrt{c^2 p^2 + m^2 c^4} - mc^2 = 4.28 \times 10^{-11}\,\text{J} = 267\,\text{MeV}$$

Since this is over 500 times the rest energy of the electron, we were correct in using the relativistic formula.

This problem illustrates why an electron cannot exist inside an atomic nucleus. There is no known force strong enough to bind such an energetic electron to a group of protons and neutrons. The electrostatic potential energy of an electron at the surface of an atomic nucleus is about 10 MeV, not 270 MeV.

In the Examples 1 and 2 the wave function is

$$\psi = A \sin k_x x \sin k_y y \sin k_z z$$

If we take partial derivatives with respect to $x$ we obtain:

$$\frac{\partial \psi}{\partial x} = k_x A \cos k_x x \sin k_y y \sin k_z z$$

(The meaning of the symbol $\partial/\partial x$ is to treat $y$ and $z$ as constants.) Then

$$\frac{\partial^2 \psi}{\partial x^2} = -k_x^2 A \sin k_x x \sin k_y y \sin k_z z = -k_x^2 \psi$$

And

$$\frac{\partial^2 \psi}{\partial x^2} + \frac{\partial^2 \psi}{\partial y^2} + \frac{\partial^2 \psi}{\partial z^2} = -k^2 \psi$$

Note that

$$k^2 = \frac{2m}{\hbar^2}(E - U_0)$$

if there is a fixed potential energy $U_0$. As with the one-dimensional Schrödinger equation, we generalize to a variable potential energy

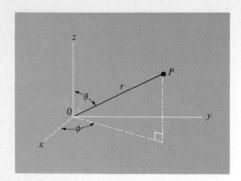

**Figure 26-3.** Point P has Cartesian coordinates x, y, z and spherical coordinates r, θ, φ.

$U(x, y, z)$ by writing

$$\frac{\partial^2 \psi}{\partial x^2} + \frac{\partial^2 \psi}{\partial y^2} + \frac{\partial^2 \psi}{\partial z^2} = -\frac{2m}{\hbar^2}(E - U)\psi$$

(the time-independent Schrödinger equation in three dimensions) (26-4)

In many applications the potential energy depends only on the distance $r = \sqrt{x^2 + y^2 + z^2}$ from the origin. Then it will be more convenient to express the Schrödinger equation in spherical coordinates where

$$x = r \sin\theta \cos\phi$$
$$y = r \sin\theta \sin\phi$$
$$z = r \cos\theta$$

as shown in Figure 26-3. It is a straightforward, but tedious process to transform Eq. 26-4 from the coordinates x, y, z to the coordinates r, θ, φ. The result is

$$\frac{1}{r^2}\frac{\partial}{\partial r}\left(r^2 \frac{\partial \psi}{\partial r}\right) + \frac{1}{r^2 \sin\theta}\frac{\partial}{\partial \theta}\left(\sin\theta \frac{\partial \psi}{\partial \theta}\right) + \frac{1}{r^2 \sin^2\theta}\frac{\partial^2 \psi}{\partial \phi^2}$$

$$= -\frac{2m}{\hbar^2}(E - U)\psi \quad (26\text{-}5)$$

## 26-3 Exact Solutions for Hydrogen Atom

The Schrödinger equation for the hydrogen atom is obtained by putting $U = -k_0 e^2/r$ into Eq. 26-5. Let us first try a simple exponential $\psi = e^{-r/a}$ for a solution. We substitute this into Eq. 26-5, noting that the partial derivatives $\partial \psi/\partial \theta$ and $\partial \psi/\partial \phi$ are zero:

$$\frac{1}{r^2}\frac{\partial}{\partial r}\left(r^2 \frac{\partial(e^{-r/a})}{\partial r}\right) = -\frac{2m}{\hbar^2}\left(E + \frac{k_0 e^2}{r}\right)e^{-r/a}$$

$$\frac{1}{r^2}\frac{\partial}{\partial r}\left(-\frac{r^2}{a}e^{-r/a}\right) = -\frac{2m}{\hbar^2}\left(E + \frac{k_0 e^2}{r}\right)e^{-r/a}$$

$$\frac{1}{r^2}\left(\frac{r^2}{a^2} - \frac{2r}{a}\right) = -\frac{2m}{\hbar^2}\left(E + \frac{k_0 e^2}{r}\right)$$

$$\left[\frac{1}{a^2}\right] - \left(\frac{2}{a}\right)\frac{1}{r} = \left[-\frac{2mE}{\hbar^2}\right] - \left(\frac{2mk_0 e^2}{\hbar^2}\right)\frac{1}{r}$$

Equating the $1/r$ terms gives:

$$\left(\frac{2}{a}\right) = \left(\frac{2mk_0e^2}{\hbar^2}\right)$$

$$a = \frac{\hbar^2}{k_0me^2} \qquad (26\text{-}6)$$

Equating the constant terms gives:

$$\left[\frac{1}{a^2}\right] = \left[-\frac{2mE}{\hbar^2}\right]$$

$$E = -\frac{\hbar^2}{2ma^2}$$

The final result is obtained by using Eq. 26-6 for $a$:

$$E = -k_0^2 \frac{me^4}{2\hbar^2} \qquad (26\text{-}7)$$

We see that a simple exponential is a solution when $a$ and $E$ have the values given by Eqs. 26-6 and 26-7. Putting in the numerical values for $m$, $k_0$, $e$, and $\hbar$ gives

$$E = -21.8 \times 10^{-19} \text{ J} = -13.6 \text{ eV}$$

This is the minimum amount of energy required to remove the electron from a hydrogen atom. By definition this is called the binding energy.

$$\text{(binding energy)} = -E = \text{(energy to remove electron)}$$

According to this definition the binding energy of an electron in a metal is the work function $W_0$ since the highest energy level in Figure 24-3 is $E = -W_0$.

Note that the wave amplitude drops to $1/e$ of its peak value when $r = a$. (Then $e^{-r/a} = e^{-1}$.) This is taken as the radius $R$ of the hydrogen atom. Using Eq. 26-6 we have

$$R = \frac{\hbar^2}{k_0me^2} = 5.3 \times 10^{-11} \text{ m} \qquad \text{(radius of hydrogen atom)} \quad (26\text{-}8)$$

The form $\psi = e^{-r/a}$ has no nodes and is the lowest-order standing wave;

hence, $E = -k_0^2 me^4/2\hbar^2$ is the lowest energy level. We shall call it $E_1$. The wave functions corresponding to the next two energy levels are

$$\psi_2 = \left(1 - \frac{r}{2a}\right)e^{-r/2a}$$

and

$$\psi_3 = \left(1 - \frac{2r}{3a} + \frac{2r^2}{27a^2}\right)e^{-r/3a}$$

These are plotted in Figure 26-4. These solutions can be checked by substitution into Eq. 26-5. They satisfy Eq. 26-5 when $E_2 = \tfrac{1}{4}E_1$ and $E_3 = \tfrac{1}{9}E_1$ (see Example 3).

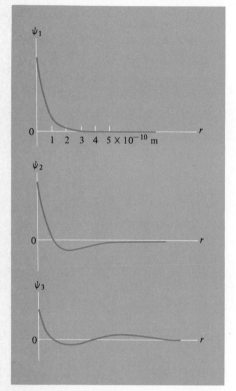

**Figure 26-4.** *Hydrogen wave functions for $n = 1, 2, 3$ and $l = 0$.*

**Example 3.** Verify that $\psi_2 = (1 - r/2a)e^{-r/2a}$ is a solution of the Schrödinger equation applied to the hydrogen atom and find the corresponding energy level $E_2$.

ANSWER: Taking the first derivative:

$$\frac{\partial \psi}{\partial r} = -\frac{1}{2a}\left(2 - \frac{r}{2a}\right)e^{-r/2a}$$

Then

$$\frac{\partial}{\partial r}\left(r^2 \frac{\partial \psi}{\partial r}\right) = -\frac{1}{2a}\frac{\partial}{\partial r}\left[\left(2r^2 - \frac{r^3}{2a}\right)e^{-r/2a}\right]$$

$$\frac{1}{r^2}\frac{\partial}{\partial r}\left(r^2 \frac{\partial \psi}{\partial r}\right) = \left(-\frac{2}{ar} + \frac{5}{4a^2} - \frac{r}{8a^3}\right)e^{-r/2a}$$

According to Eq. 26-5 this is also equal to $-(2m/\hbar^2)(E + k_0e^2/r)\psi$. Hence

$$-\frac{2}{ar} + \frac{5}{4a^2} - \frac{r}{8a^3} = -\frac{2mk_0e^2}{\hbar^2 r} - \frac{2m}{\hbar^2}\left(E - \frac{k_0e^2}{2a}\right) + \frac{mE}{a\hbar^2}r$$

Equating the $1/r$ terms gives

$$-\frac{2}{a} = -\frac{2mk_0e^2}{\hbar^2}$$

or

$$a = \frac{\hbar^2}{k_0 me^2}$$

as before. Equating the $r$ terms gives

$$-\frac{1}{8a^3} = \frac{mE}{a\hbar^2}$$

$$E = -\frac{\hbar^2}{8ma^2}$$

Substituting $\hbar^2/k_0 me^2$ for $a$ gives

$$E = -\frac{k_0^2}{4}\frac{me^4}{2\hbar^2} = -\tfrac{1}{4}(13.6 \text{ eV})$$

It can be shown that there are standing wave solutions corresponding to energy levels

$$E_n = -\frac{1}{n^2}k_0^2\frac{me^4}{2\hbar^2} \quad \text{(hydrogen energy levels)} \quad (26\text{-}9)$$

for $n$ equal to any positive integer; $n$ is called the principal quantum number. However, in order to describe completely a standing wave in three-dimensional space, two other quantum numbers are needed. These are associated with angular momentum and are introduced in the next section.

## 26-4 Orbital Angular Momentum

Suppose a wave packet of wave number $k$ in Figure 26-5 is moving in a circle of radius $R$. It will have angular momentum about the $z$ axis $L_z = Rp = R(\hbar k)$. Over the arc $s$ in Figure 26-5 its wave function will be

$$\psi \propto e^{i(ks-\omega t)} = e^{i(kR\phi-\omega t)}$$

Since $\psi(\phi = 0)$ and $\psi(\phi = 2\pi)$ are measured at the same position in space, we have

$$e^{ikR(0)} = e^{ikR(2\pi)}$$

or

$$1 = e^{i2\pi kR}$$

This can only be true if $(kR)$ is an integer (it is usually called $m_l$).

$$kR = m_l$$

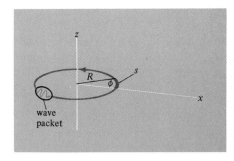

Figure 26-5. *Wave packet moving in circle of radius $R$. The arc $s = R\phi$.*

Now multiply both sides by $\hbar$:

$$\hbar k R = m_l \hbar$$

or

$$L_z = m_l \hbar$$

The corresponding wave function is

$$e^{i(kR)\phi - i\omega t} = e^{im_l \phi - i\omega t}$$

We have just presented a plausibility argument to show that if the factor $e^{im_l \phi}$ is contained in the wave function of a particle, the particle will have a component of angular momentum $L_z = m_l \hbar$ about the $z$ axis. A more rigorous proof exists but is somewhat beyond the scope of this book. We see that angular momentum is quantized in units of $\hbar$; that is, $L_z$ can be 0, $\pm\hbar$, $\pm 2\hbar$, $\pm 3\hbar$, and so on.

So far the solutions to Eq. 26-5 that we have presented have only been functions of $r$. However, there are also other solutions, which are functions of $\theta$ and $\phi$ as well. The most general solution is of the form

$$\psi_{n,l,m_l}(r,\theta,\phi) = R_n(r)\Theta_{l,m_l}(\theta)\Phi_{m_l}(\phi)$$

where

$$\Phi_{m_l}(\phi) = e^{im_l \phi}$$

According to the previous paragraph $\psi_{n,l,m_l}$ must then have $z$ component of angular momentum $= m_l \hbar$. For $n = 2$ we have already seen the solution

$$\psi_{2,0,0} = \left(1 - \frac{r}{2a}\right) e^{-r/2a}$$

but in addition there are the solutions

$$\psi_{2,1,1} = r e^{-r/2a} \sin\theta e^{i\phi}$$
$$\psi_{2,1,0} = r e^{-r/2a} \cos\theta$$
$$\psi_{2,1,-1} = r e^{-r/2a} \sin\theta e^{-i\phi}$$

These solutions can be checked by direct substitution into Eq. 26-5. Percentage contour maps of these $n = 2$ functions are shown in Figure 26-6. We note that for $n = 2$. the quantum number $l$ can be 0 or 1 and for $l = 1$, the quantum number $m_l$ can be 1, 0, or $-1$. According to Eq. 26-9 these four wave functions have exactly the same energy level $E_2 = \frac{1}{4}(13.6 \text{ eV})$. In general each wave function or eigenfunction has its own unique energy level or eigenvalue. However, the coulomb potential

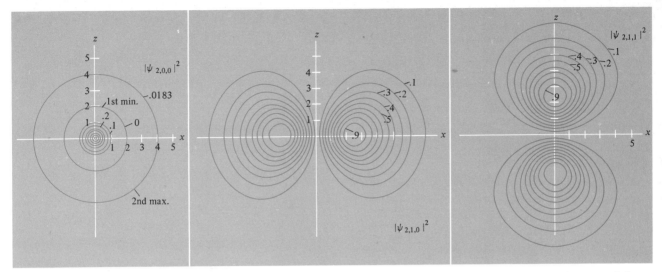

**Figure 26-6.** *Probability densities of electron clouds in the xz plane in the hydrogen atom for the $n = 2$ wave functions. Contour lines of equal probability are shown. The maximum probability is chosen to be 1. Scale units are in angstroms ($10^{-10}$ m).*

has the special property that all eigenfunctions with the same quantum number $n$ have the same energy eigenvalue. It can be shown that the coulomb potential has the property that the quantum number $l$ can never exceed $(n-1)$ and the quantum number $m_l$ runs from $-l$ to $+l$.

Now we shall take $n = 3$ as an example to see which combinations of the three quantum numbers are possible. We see from Table 26-1 that nine different combinations are possible. Thus there are nine different wave functions of $r$, $\theta$, and $\phi$ all having the same energy $E_3 = -\frac{1}{9}(13.6 \text{ eV})$.

---

**Example 4.** How many different wave functions are there for $n = 4$?

ANSWER: The possible $l$ values are 0, 1, 2, 3. In Table 26-1 we saw that $l = 0$, 1, and 2 give rise to nine wave functions. The additional $l = 3$ has seven possibilities for $m_l$ ranging from $-3$ to $+3$. The total number of possibilities is then $9 + 7 = 16$.

---

We note that for fixed $l$, the z component of angular momentum ranges from $-l\hbar$ to $+l\hbar$. If an experiment is performed to measure angular momentum about any axis, the largest value that can be measured is $l\hbar$. The physical interpretation is that an electron wave function with quantum number $l$ has total angular momentum equal to $l\hbar$ and that the z component describes the orientation of the orbital or wave function. All this has been verified by experiment. The ratio of $m_l$ to $l$ determines the angle between the z axis and the angular momentum axis.

The kinetic energy of a spinning rigid body is $K = L^2/(2I)$, where $I$ is

**Table 26-1.** *All possible combinations of n, l, and $m_l$ for $n = 3$*

| n | l | $m_l$ |
|---|---|---|
| 3 | 0 | 0 |
| 3 | 1 | 1 |
| 3 | 1 | 0 |
| 3 | 1 | −1 |
| 3 | 2 | 2 |
| 3 | 2 | 1 |
| 3 | 2 | 0 |
| 3 | 2 | −1 |
| 3 | 2 | −2 |

**Table 26-2.** *Unnormalized radial and angular functions for use in constructing the complete hydrogen wave functions $\psi_{n,l,m_l}$*

| $R_{n,l}(\rho)$ where $\rho \equiv \dfrac{r}{a}$ | | $\Theta_{l,m_l} = \Theta_{l,-m_l}$ | |
|---|---|---|---|
| $R_{10}$ | $e^{-\rho}$ | $\Theta_{00}$ | 1 |
| $R_{20}$ | $\left(1 - \dfrac{\rho}{2}\right)e^{-\rho/2}$ | $\Theta_{10}$ | $\cos\theta$ |
| $R_{21}$ | $\rho e^{-\rho/2}$ | $\Theta_{11}$ | $\sin\theta$ |
| $R_{30}$ | $\left(1 - \dfrac{2\rho}{3} + \dfrac{2e^2}{27}\right)e^{-\rho/3}$ | $\Theta_{20}$ | $(3\cos^2\theta - 1)$ |
| $R_{31}$ | $\rho\left(1 - \dfrac{\rho}{6}\right)e^{-\rho/3}$ | $\Theta_{21}$ | $\sin\theta\cos\theta$ |
| $R_{32}$ | $\rho^2 e^{-\rho/3}$ | $\Theta_{22}$ | $\sin^2\theta$ |
| $R_{n,n-1}$ | $\rho^{(n-1)}e^{-\rho/n}$ | $\Theta_{n,n}$ | $\sin^n\theta$ |

the moment of inertia. When such an energy is measured, the quantum theory predicts (and experiment confirms) that

$$K = \frac{l(l+1)\hbar^2}{2I}$$

suggesting that $L = \sqrt{l(l+1)}\hbar$. So in measurements related to energy one must use $L = \sqrt{l(l+1)}\hbar$; whereas in measurements of angular momentum directly one uses $L = l\hbar$.

The various wave functions can be constructed by use of Table 26-2 using

$$\psi_{n,l,m_l}(r,\theta,\phi) = R_{n,l}(r)\Theta_{l,m_l}(\theta)e^{im_l\phi}$$

Contour plots of $|\psi|^2$ in the $xz$ plane are shown in Figure 26-7. Note that for large $n$, $|\psi|^2$ for $l = (n-1)$ is concentrated in a circle around the $z$ axis with radius $= n^2 a$. It becomes the same orbit as predicted by Bohr, except in the modern theory the electron is smeared out uniformly around the circular orbit.

### Normalization

In this book we specify $|\psi|^2\,dx\,dy\,dz$ as the relative probability for finding the particle in the volume element $dx\,dy\,dz$. It is also possible to specify the same quantity as the absolute probability providing that

$$\iiint\limits_{\text{all space}} |\psi|^2\,dx\,dy\,dz = 1$$

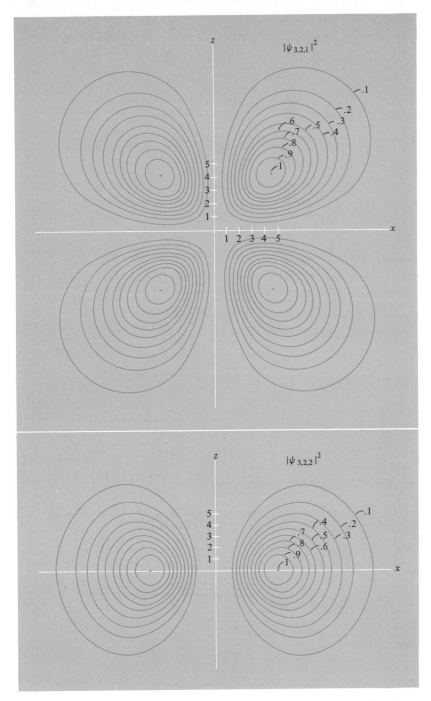

**Figure 26-7.** *Similar to Figure 26-6 except that two of the $n = 3$ standing waves are shown.*

This normalization condition uniquely determines the proportionality constant that multiplies the wave function. Since we have only been concerned with relative probabilities, we have not bothered to normalize our wave functions.

**Expectation Value**

Suppose a particle is known to have a wave function that is a mixture of several eigenfunctions. Let

$$\psi(x) = \sum_j a_j \psi_j(x)$$

be this wave function where each $\psi_j$ is a normalized eigenfunction that has an energy eigenvalue $E_j$. According to the basic postulate in Chapter 24 the probability of finding the particle in the $\psi_j$ state is the square of the amplitude $a_j$. Thus the probability of finding the particle with energy $E_j$ is $|a_j|^2$. Then the expectation value of the energy is

$$\langle E \rangle = \sum_j |a_j|^2 E_j$$

A series of repeated measurements, each starting with the same $\psi(x)$, would give the preceding average measurement of energy. The same holds true for any other physical quantity such as angular momentum; that is,

$$\langle L_z \rangle = \sum_j |a_j|^2 L_j$$

---

*__Example 5.__ An $n = 2$, $l = 1$ hydrogen atom has $L_z = \hbar$. Its wave function must be $\psi(r, \theta, \phi) = R_{2,1} \Theta_{1,1}(\theta) e^{i\phi}$. Suppose Mr. Prime is in a primed frame of reference where the $z'$ axis is tilted at an angle $\alpha$ to the $z$ axis. What value would Mr. Prime get for $\langle L_z' \rangle$? *Hint:* A few steps of trigonometry give the result:

$$\Theta_{11}(\theta) e^{i\phi} = \tfrac{1}{2}(\cos \alpha + 1)\Theta_{11}(\theta') e^{i\phi'} - \frac{\sin \alpha}{\sqrt{2}} \Theta_{10}(\theta') + \tfrac{1}{2}(\cos \alpha - 1)\Theta_{1,-1}(\theta') e^{-i\phi'}$$

(see Problem 28)

ANSWER: In the primed system the original wave function is of the form

$$\psi(r,\theta,\phi) = a_1\psi_1(r,\theta',\phi') + a_2\psi_2(r,\theta',\phi') + a_3\psi_3(r,\theta',\phi')$$

where

$$a_1 = \frac{\cos\alpha + 1}{2} \qquad a_2 = -\frac{\sin\alpha}{\sqrt{2}} \qquad a_3 = \frac{\cos\alpha - 1}{2}$$

$\psi_1$ contains $\Theta_{11}(\theta')e^{i\phi'}$ which has $L_z' = \hbar$; $\psi_2$ has $L_z' = 0$, $\psi_3$ has $L_z' = -\hbar$. Then

$$\langle L_z' \rangle = a_1^2(+\hbar) + a_2^2(0) + a_3^2(-\hbar)$$
$$= \frac{(\cos\alpha + 1)^2}{4}(+\hbar) + \frac{(\cos\alpha - 1)^2}{4}(-\hbar) = (\cos\alpha)\hbar$$

Quantum physics gives the same answer as classical physics.

## 26-5 Photon Emission

After the discovery of the Schrödinger equation, quantum mechanics was applied to the electromagnetic interaction resulting in what is called quantum electrodynamics. The details of quantum electrodynamics are well beyond the scope of this book. All we need be concerned with is that quantum electrodynamics predicts that charged particles can radiate or absorb photons one at a time, and that the probability amplitude for photon emission or absorption can be precisely calculated from the theory.

### Spontaneous Emission

Consider a charged particle attached to a spring with a spring constant such that the natural frequency of oscillation is $\omega_0 = \sqrt{k/m}$. In classical physics the charged particle will radiate at the frequency $\omega_0$ if the spring is stretched and then released. In quantum mechanical language, this corresponds to raising the particle to an energy level above the ground state. The quantum mechanical analog is that an electron at an energy level above the ground state has a certain probability amplitude to emit a photon and end up at a lower energy level. This is called spontaneous emission. If the energy difference is a few electron volts between two hydrogen atom energy levels, the probability amplitude for a transition from one to the other is such that the process of photon emission has a typical value for a half-life on the order of $10^{-8}$ s (see pages 461 and 586). If a photon is emitted in a transition from $E_{n'}$ to $E_n$, the photon energy is $hf = E_{n'} - E_n$. The photon frequency is $f = (E_{n'} - E_n)/h$.

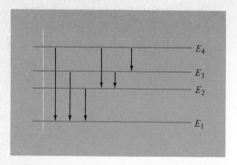

**Figure 26-8.** *All six possible transitions between four energy levels.*

Thus if an atom has four different energy levels, as shown in Figure 26-8, there are six different transitions from higher levels to lower levels available. The "light" emitted by such an atom would contain just the six corresponding frequencies. When analyzed by a spectroscope six spectral lines would be seen. Normally the atoms would be in the ground state and no light would be emitted. However, if a sample of atoms is heated or subjected to an electric discharge, some of the atoms absorb energy in collisions that excite them to higher energy levels. It is these relatively few atoms in the higher energy states that emit the photons.

In the modern theory photons are thought of as elementary particles with spin 1 ($L = \hbar$). Thus when a photon is emitted the $l$ quantum number of the atom will change by at least one unit.

### The Hydrogen Spectrum

Now that we have a quantitative formula for the hydrogen energy levels, we can calculate the entire hydrogen atomic spectrum. Let the higher or excited energy level be

**Figure 26-9.** (a) *All possible lines in the hydrogen spectrum up to* $\lambda = 7000$ Å. (b) *Lyman series (dashed lines) and Balmer series (solid lines) transitions.*

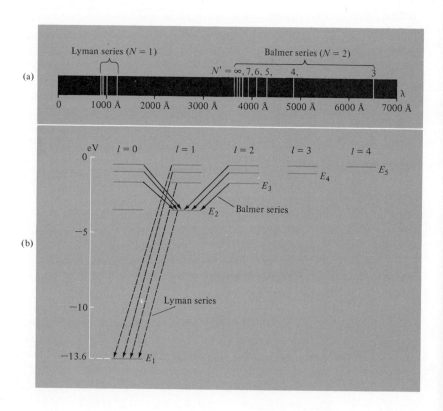

578 CH. 26/THE HYDROGEN ATOM

$$E_{n'} = -k_0^2 \frac{me^4}{2\hbar^2} \frac{1}{n'^2}$$

and the lower level be

$$E_n = -\frac{k_0^2 me^4}{2\hbar^2} \frac{1}{n^2}$$

Then the spectral line frequencies will be

$$f = \frac{k_0^2 me^4}{4\pi\hbar^3} \left(\frac{1}{n^2} - \frac{1}{n'^2}\right) \qquad (26\text{-}10)$$

For $n = 1$ this is a series of spectral lines called the Lyman series. These lines are all in the ultraviolet region of the electromagnetic spectrum. For $n = 2$ there is another series of lines called the Balmer series. The hydrogen spectrum is shown in Figure 26-9(a). Some of the corresponding energy transitions are shown in Figure 26-9(b). In such an energy level diagram it is customary to plot states of increasing angular momentum horizontally.

**Example 6.** What are the wavelengths of the lines in the Balmer series? How many of the lines are visible?

ANSWER: The photon energies are $hf = 13.6 \text{ eV}(1/4 - 1/n'^2)$, where $n' = 3, 4, 5, \ldots$. The first four of these are:

$$hf_1 = 13.6 \left(\frac{1}{4} - \frac{1}{9}\right) = 1.89 \text{ eV}$$

$$hf_2 = 13.6 \left(\frac{1}{4} - \frac{1}{16}\right) = 2.55 \text{ eV}$$

$$hf_3 = 13.6 \left(\frac{1}{4} - \frac{1}{25}\right) = 2.86 \text{ eV}$$

$$hf_4 = 13.6 \left(\frac{1}{4} - \frac{1}{36}\right) = 3.02 \text{ eV}$$

For very large $n$:

$$hf_\infty = 13.6 \left(\frac{1}{4} - 0\right) = 3.40 \text{ eV}$$

$$\lambda = \frac{c}{f} = \frac{hc}{hf} = \frac{(6.63 \times 10^{-34})(3.00 \times 10^8)}{hf}$$

$$= \frac{19.9 \times 10^{-26} \text{ J m}}{hf} = \frac{12.4 \times 10^{-7} \text{ eV m}}{hf}$$

$\lambda_1 = 656$ nm  $\quad \lambda_2 = 4.86$ nm  $\quad \lambda_3 = 433$ nm  $\quad \lambda_4 = 411$ nm

$\lambda_\infty = 365$ nm

There will be an infinite number of lines converging on the wavelength $\lambda_\infty = 3650$ Å, which is in the near ultraviolet. $\lambda_1$ at 6560 Å is in the red part of the visible spectrum, $\lambda_2$ is blue, and $\lambda_3$ is violet. $\lambda_4$ is at the borderline between visible and ultraviolet.

### Absorption

If a continuous spectrum (as is emitted by a red-hot body) is passed through a cold gas, atoms in the ground state will absorb photons of just the right energies to make transitions to the higher energy states. If the cold gas is hydrogen, photons of frequencies corresponding to the Lyman series will be absorbed. If one views the continuous spectrum after passing through the gas, there will be missing photons of energies $(E_2 - E_1)$, $(E_3 - E_1)$, $(E_4 - E_1)$, and so on. These missing photons show up on a spectrogram as dark lines. The process of exiting atoms into higher energy levels by shining light on the sample of atoms is called optical pumping.

## 26-6 Stimulated Emission

Quantum electrodynamics tells us that, in addition to the processes of spontaneous emission and absorption, there is a process called stimulated emission. Suppose an atom is already in an excited state $E_{n'}$ and would like to emit a photon of energy $(E_{n'} - E_n)$. Then if this atom is put in the presence of external radiation already containing photons of energy $(E_{n'} - E_n)$, the probability for radiation by the atom will be increased. This effect of speeding up the atomic transitions by shining "light" on the excited atoms is called stimulated emission. But this is not all; a photon resulting from stimulated emission will be exactly in phase and pointing in the same direction as the external photon that induced the atom to radiate.

We can again consider a charged particle attached to a spring as a classical analog. Suppose the spring constant is such that the natural frequency of oscillation is $\omega_0$. Then if an external field $E = E_0 \cos \omega t$ is applied, the particle will oscillate with frequency $\omega$ and radiate. The closer $\omega$ is to $\omega_0$, the larger will be the amplitude of oscillation (see Eq. 21-6). Note that the radiated frequency is that of the external field and not $\omega_0$.

### The Laser

Suppose we had a sample of atoms (or molecules) with many of them already in an excited state $E_{n'}$. The sample could have been prepared in many ways: by optical pumping, by collisions with an electron current, collisions with other excited atoms, as a result of spontaneous emission from an even higher energy state, and so on. Now consider the first photon that happens to be emitted. As it moves along through the gas it will stimulate neighboring atoms to emit their photons by stimulated emission. If there are mirrors at both ends, there will be a chain reaction until all the atoms have radiated (see Figure 26-10). Now all the photons will be in phase with each other. If one of the mirrors is partially reflecting, the escaping beam of light will be coherent. It will be one continuous sine wave of electromagnetic radiation just as is electromagnetic radiation from a radio transmitter. In order for stimulated emission to dominate over absorption, it is necessary for more atoms to be in the higher energy state. Such a scheme for the helium–neon gas laser as shown in Figure 26-11 has more atoms in state $E_{n'}$ than in $E_n$.

Until the development of the laser in 1960 all radiation available to mankind in the infrared, visible, and ultraviolet was incoherent—the individual photons were of random phase with respect to each other. As explained on page 512, coherent light is capable of giving much smaller spot sizes over long distances. Since the first laser, the technology has grown so that now both pulsed and continuous lasers are available using gases, solids, and liquids with wavelengths in the infrared, visible, and ultraviolet regions.

(a)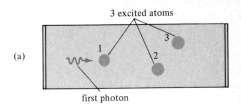

(b)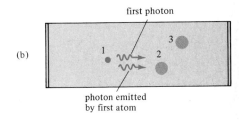

(c)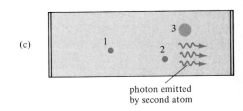

(d)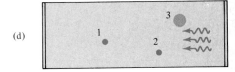

**Figure 26-10.** *Stimulated emission in a laser. (a) The first photon is about to "hit" an excited atom. (b) The atom has emitted its photon, which is about to "hit" a second excited atom. (d) The three photons have been reflected and are about to "collect" a fourth photon.*

**Figure 26-11.** *Three energy levels for the neon atom. In the helium–neon laser neon atoms are excited to $E_{n'}$ by collisions with excited helium atoms. The transition to $E_n$ is by stimulated emission. Then the neon atoms quickly drop to the ground state by collisions with the walls.*

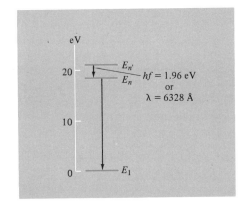

**Figure 26-12.** *Niels Bohr skiing at Los Alamos. [Courtesy of Mrs. Laura Fermi]*

## 26-7 The Bohr Model

The present-day model of the atom as presented in this chapter was developed in 1926 shortly after the introduction of the Schrödinger equation. However, 13 years earlier Niels Bohr proposed a theory that very accurately explained the entire hydrogen spectrum and also provided a physical model for a stable atomic structure.

Bohr's theory, although now known to be wrong, is so simple and of such great historical importance that we present it here. Symbols of Bohr's outmoded theory are still used today (see Figure 26-13). Bohr thought of the possible electron orbits as classical circular planetary orbits and he looked for some rule that would allow only certain energies or orbit radii. The rule he invented was that the angular momentum be an integral multiple of $\hbar$:

$$mvR = n\hbar \qquad (26\text{-}11)$$

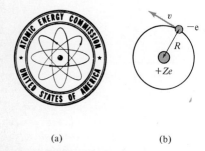

(a)  (b)

**Figure 26-13.** (a) *Obsolete atomic symbol.* (b) *Bohr model of hydrogen-like atom.*

Note that the Bohr postulate differs from our present knowledge of the hydrogen atom in two ways. First, we now know that classical orbits do not apply and that the electron must be expressed as a standing wave. Second, we know that the angular momentum is not $n\hbar$, but it is $l\hbar$, which is always less than Bohr's value. It is a lucky accident that the Bohr theory leads to the correct hydrogen energy levels. It is not uncommon to get the right answer for the wrong reason. This can even happen to great men.

We shall now go through Bohr's derivation of the energy levels of an electron in the field of a nucleus of charge $Ze$. The Bohr postulate (Eq. 26-11) gives for the radius of the $n$th orbit

$$R_n = n\frac{\hbar}{mv} \qquad (26\text{-}12)$$

582   CH. 26/THE HYDROGEN ATOM

Equating the centripetal force to the electrostatic force:

$$\frac{mv^2}{R_n} = \frac{k_0 Z e^2}{R_n^2}$$

$$mv^2 = \frac{k_0 Z e^2}{R_n} = -U \quad \text{(the potential energy)} \quad (26\text{-}13)$$

$$v^2 = \frac{k_0 Z e^2}{m R_n} \quad (26\text{-}14)$$

Now substitute the right hand side of Eq. 26-12 into Eq. 26-14:

$$v^2 = \frac{k_0 Z e^2}{m \left( n \dfrac{\hbar}{mv} \right)}$$

$$v = \frac{k_0 Z e^2}{n \hbar} \quad (26\text{-}15)$$

The energy level is defined as

$$E_n = \tfrac{1}{2} mv^2 + U$$

Using Eq. 26-13 for $U$:

$$E_n = \tfrac{1}{2} mv^2 + (-mv^2) = -\tfrac{1}{2} mv^2$$

The final result is obtained by squaring the right-hand side of Eq. 26-15 and substituting into the above equation:

$$E_n = -k_0^2 \frac{Z^2 m e^4}{2\hbar^2} \frac{1}{n^2} = -13.6 \left( \frac{Z^2}{n^2} \right) \text{eV} \quad (26\text{-}16)$$

The energy can also be expressed in terms of $R$ by substituting Eq. 26-14 into $E = -\tfrac{1}{2} mv^2$. This gives

$$E_n = -\frac{k_0 e^2}{2 R_n} \quad (26\text{-}17)$$

Equation 26-16 agrees with the present-day theory (see Eq. 26-7 and 26-9). The Bohr model also gives a simple answer for the size of atoms. The formula for $R_n$ is obtained by substituting Eq. 26-15 into Eq. 26-12. Then

$$R_n = n^2 \frac{\hbar^2}{k_0 Z m e^2} \quad (26\text{-}18)$$

This is in agreement with the average radius of the electron cloud (Eq. 26-8).

However, a serious failure of the Bohr model is that it is unable to explain the spectra of atoms starting with helium (a nucleus of $Z = 2$ surrounded by two electrons). Although complicated by three mutually interacting particles, the helium energy levels can be calculated using quantum mechanics. Using the modern theory and high-speed computers, the helium spectrum has now been calculated to great accuracy and found to check with experiment. Physicists and chemists are confident that present-day quantum mechanics can in principle explain all atomic spectra and chemical properties.

---

**\*Example 7.** Prove that the Bohr result for the radius ($R_n = n^2 a$) agrees with the hydrogen wave functions for $l = n - 1$. Define the "average" radius as that distance where the probability distribution in $r$ is at its maximum.

ANSWER: The probability distribution is $|\psi_{n,n-1}|^2 \, dV$ where $dV = 4\pi r^2 \, dr$ is a volume element. Hence the probability distribution in $r$ is proportional to

$$r^2 (R_{n,n-1}(r))^2 \, dr$$

According to Table 26-2

$$(R_{n,n-1})^2 \propto r^{2n-2} e^{-2r/na}$$

Let $P(r)$ be the probability distribution:

$$P(r) = r^{2n} e^{-2r/na}$$

This has its maximum value when $dP/dr = 0$.

$$\frac{dP}{dr} = 2n r^{2n-1} e^{-2r/na} - \frac{2}{na} r^{2n} e^{-2r/na}$$

Setting this equal to zero and dividing out the exponentials gives:

$$2n r^{2n-1} - \frac{2}{na} r^{2n} = 0$$

$$n^2 a r^{2n-1} = r^{2n}$$

$$r = n^2 a$$

Note that this is true for all values of $n$ including the ground state wave function.

## Orbit Stability

We can use the Bohr theory plus the formula for radiation from a point charge to calculate the transition time from a circle of radius $R_{n'}$ to radius $R_n$. According to Eq. 21-14 the rate of energy loss is

$$\frac{dE}{dt} = \frac{2}{3} k_0 \frac{e^2}{c^3} a^2 \quad \text{where } a = F/m = k_0 \frac{e^2}{mR^2}$$

$$\frac{dE}{dt} = -\frac{2}{3} k_0^3 \frac{e^6}{m^2 c^3 R^4} \quad (26\text{-}19)$$

We shall need $dE/dR$, which can be obtained from the relation

$$E = -k_0 \frac{e^2}{2R}$$

$$\frac{dE}{dR} = k_0 \frac{e^2}{2R^2} \quad (26\text{-}20)$$

Now we can get $dt/dR$, which is

$$\frac{dt}{dR} = \left(\frac{dt}{dE}\right)\left(\frac{dE}{dR}\right)$$

Substituting from Eqs. 26-19 and 26-20 gives

$$\frac{dt}{dR} = \left(-\frac{2}{3} k_0^3 \frac{e^6}{m^2 c^3 R^4}\right)^{-1} \left(k_0 \frac{e^2}{2R^2}\right) = -\frac{3}{4} \frac{m^2 c^3 R^2}{k_0^2 e^4}$$

The transition time is

$$t = \int_{R_{n'}}^{R_n} \frac{dt}{dR} dR = \int_{R_{n'}}^{R_n} \left(\frac{3}{4} \frac{m^2 c^3 R^2}{k_0^2 e^4}\right) dR$$

$$t = \frac{m^2 c^3}{4 k_0^2 e^4} (R_{n'}^3 - R_n^3) \quad (26\text{-}21)$$

For the first Balmer line $R_{n'} = R_3 = 0.477$ nm and $R_n = R_2 = 0.212$ nm. Then

$$t = \frac{(9.11 \times 10^{-31} \text{ kg})^2 (3 \times 10^8 \text{ m/s})^3}{4(9 \times 10^9 \text{ N m}^2/\text{C}^2)^2 (1.6 \times 10^{-19} \text{ C})^4}$$

$$\times [(4.77 \times 10^{-10} \text{ m})^3 - (2.12 \times 10^{-10} \text{ m})^3]$$

$$= 1.04 \times 10^{-8} \text{ s}$$

As with other aspects of the Bohr model, this agrees reasonably well with a correct quantum mechanical calculation which makes no use of classical physics.

> **Example 8.** Use Eq. 26-21 to calculate how long it would take the electron in the ground state of hydrogen in the Bohr model to spiral in and hit the proton. Assume the proton has a radius $R = 10^{-15}$ m.
>
> ANSWER: We can use Eq. 26-21 if we put $R_{n'} = R_1 = 0.53 \times 10^{-10}$ m for the initial radius and $r = 10^{-15}$ m for $R_n$, the final radius. Then
>
> $$t = \frac{m^2 c^3}{4 k_0^2 e^4}[(0.53 \times 10^{-10} \text{ m})^3 - (10^{-15} \text{ m})^3] = 1.57 \times 10^{-11} \text{ s}$$
>
> This tells us that if the electron in the ground state Bohr orbit should radiate classically, hydrogen atoms would all collapse in $10^{-11}$ s. Bohr's answer to this serious problem was to postulate that electrons in the ground state do not radiate.

## Summary

The Schrödinger equation in three-dimensional space is

$$\frac{\partial^2 \psi}{\partial x^2} + \frac{\partial^2 \psi}{\partial y^2} + \frac{\partial^2 \psi}{\partial z^2} = -\frac{2m}{\hbar^2}(E - U)\psi$$

If $\psi(x, y, z)$ is a function of $r$ alone, this becomes

$$\frac{1}{r^2}\frac{d}{dr}\left(r^2 \frac{d\psi}{dr}\right) = -\frac{2m}{\hbar^2}[E - U(r)]\psi$$

In the hydrogen atom $U = -k_0(e^2/r)$ and the lowest energy solution is $\psi_1 = \exp(-r/a)$ where $a = \hbar/(k_0 m e^2)$ is the radius and $E_1 = -k_0(me^4/2\hbar^2) = -13.6$ eV is the energy. Higher-order solutions are of the form $E_n = (-13.6/n^2)$ eV. These higher-order solutions can also be functions of the polar angles $\theta$ and $\phi$. The $\phi$ dependence is of the form $\exp(im_l\phi)$, where $L_z = m_l \hbar$ is the $z$ component of angular momentum. The $\theta$ dependence involves a third quantum number $l$ where $L = l\hbar$ is the highest value $L_z$ can attain; that is, $m_l$ can be any integer from $-l$ to $+l$. The quantum number $l$ can have an integral value from 0 to $(n-1)$.

If $n' > n$, there can be spontaneous emission of a photon from a state $E_{n'}$ to $E_n$. The photon energy is $hf = E_{n'} - E_n$. If light of that frequency collides with an atom in the state $E_n$, a photon can be absorbed by raising the atom from energy $E_n$ to $E_{n'}$. If hydrogen gas is heated so that some of the atoms are in higher energy states, the possible photon energies in the emission spectrum will be

$$hf = 13.6 \left(\frac{1}{n^2} - \frac{1}{n'^2}\right) \text{eV}$$

A photon can stimulate excited atoms to emit photons of the same frequency and phase. Thus, one can obtain coherent light from a sample of atoms in an appropriate excited state. This is the principle of the laser.

The Bohr model, which gives the correct hydrogen energy levels and orbital radii, starts from the hypothesis that the electron is in a classical circular orbit with $mvR = n\hbar$.

## Exercises

1. Repeat the derivation of Eq. 26-2 for an electron bound to a nucleus of charge $Ze$.
2. An electron is at rest near the surface of a uranium nucleus ($Z = 92$). If the nuclear radius is $5.5 \times 10^{-15}$ m, how much energy is needed to move the electron off to infinity?
3. A cubical box contains an electron. If $L$ is increased by 1% the area of the box increases by 2% and the volume by 3%. What will be the percentage change in the standing wave energy levels?
4. Put $U = -k_0(Ze^2/r)$ into Eq. 26-5 and try $\psi_1 = e^{-r/a}$. What do you get for $a$ and for $E$?
5. How many different wave functions are there for $n = 5$?
6. What is the minimum energy which can be absorbed by a hydrogen atom when in its ground state?
7. What is the ground state Bohr radius for $He^+$?
8. In the ground state of the Bohr model of the hydrogen atom
   (a) What is the kinetic energy in electron volts?
   (b) What is the potential energy in electron volts?
   (c) What is the binding energy in electron volts?
   (d) If the electron were at rest, but still at a distance equal to the Bohr radius from the proton, how much energy would be needed to remove the electron?
9. A sample of hydrogen gas is excited to $n = 5$. What is the total number of lines that can appear in the emission spectrum of this gas?

10. The following three lines are observed in the absorption spectrum of element X: $f_1 = 2.2 \times 10^{15}$ Hz, $f_2 = 3.0 \times 10^{15}$ Hz, and $f_3 = 3.5 \times 10^{15}$ Hz.
    (a) Will the three lines also appear in the emission spectrum?
    (b) List the frequencies of three other lines appearing in the emission spectrum.
11. Find a line in the He$^+$ spectrum having the same wavelength as a line in the hydrogen spectrum. What is this wavelength?
12. A mu-mesic atom consists of a nucleus of charge Z with a captured muon (a particle 207 times as heavy as an electron) in the ground state.
    (a) What is the binding energy of a $\mu^-$ captured by a proton?
    (b) What is the corresponding $n = 1$ Bohr radius?
    (c) What energy photon is emitted when the above $\mu^-$ jumps from $n = 2$ to the ground state?
13. Write out explicitly the hydrogen wave function $\psi_{3,2,-2}(r, \theta, \phi)$.
14. Normalize the $n = 1$ wavefunction; that is, let $\psi = Ce^{-r/a}$. Then

$$\int \psi^* \psi \, dV = 1 \quad \text{or} \quad C^2 \int e^{-2r/a} \, dV = 1$$

15. Use Eq. 26-21 to estimate the transition time from $n = 4$ to $n = 3$.
16. Give the wavelengths of the first four lines in the absorption spectrum of hydrogen.

## Problems

17. Consider an electron in a three-dimensional cubical box of length $L = 10^{-10}$ m. What are the four lowest energy levels in electron volts? How many states correspond to each energy level? Write out the wave function of each such state.
18. A box has $L_y = L_z = 2L_x$. What are the three lowest eigenvalues in terms of $n_x, n_y, n_z, m, \hbar$, and $L_x$? Write out the corresponding eigenfunctions. Note that two different eigenfunctions correspond to the second energy level.
19. Show by substitution into Eq. 26-5 that $\psi(r, \theta, \phi) = (re^{-r/2a})(\sin \theta)(e^{i\phi})$ is a hydrogen atom solution. What do you get for $a$ and $E$?
20. Show by substitution into Eq. 26-5 that

$$\psi = \left(1 - \frac{2r}{3a} + \frac{2r^2}{27a^2}\right)e^{-r/3a}$$

is a hydrogen atom solution.
21. What is the average value of $1/r$ in the ground state of hydrogen?

$$\frac{1}{r} = \frac{\int \frac{1}{r} \psi^* \psi \, dV}{\int \psi^* \psi \, dV}$$

22. Use Eq. 26-21 to estimate the transition time from $n = 4$ to $n = 3$. What would be $\Delta\lambda/\lambda$ of this spectral line? What must be the line spacing in a 5 cm wide diffraction grating in order to see this natural linewidth?

23. Use the Bohr model to calculate the radius of an electron orbit around a neutron.
24. A light particle of mass $m$ and positive charge $q$ is in a circular orbit of radius $R$ about a heavy particle of mass $M$ and charge $-Q$. The only significant force between them is the electrostatic force.
    (a) What is the potential energy of $q$ in terms of its mass $m$ and its velocity $v$?
    (b) What is the total energy $E$ in terms of $m$ and $v$?
    (c) Using the Bohr postulate to quantize the angular momentum, give an expression for $v$ in terms of $R$, $m$, $\hbar$, and $n$.
    (d) What are the quantized $R$ and $E$ in terms of $m$, $\hbar$, $Q$, $q$, and $n$?
25. Consider element $Q$, a hypothetical atom of valence $+1$. The binding energy of the outer electron is 3.2 eV. It is also known that the energy levels for three excited states of the outer electron are $-1.0$ eV, $-1.4$ eV, and $-2.0$ eV.
    (a) What is the energy level of the ground state in eV?
    (b) List all the lines that should appear in the emission spectrum of element $Q$. Give photon energies in eV.
26. (a) What is the probability that the electron in the hydrogen atom is at a position $r > a$ where $a$ is the Bohr radius?
    (b) What is the probability that the electron is at $r < 10^{-15}$ m? (Assume the proton is a point charge. This still gives the correct answer for the probability of finding the electron inside the proton.)
27. Suppose $\psi$ is only a function of $r$. Let $u(r) \equiv r\psi$. Express the radial three-dimensional Schrödinger equation in terms of $u$ rather than $\psi$.
28. In Example 5 the transformation equations from the primed to unprimed system are

$$\begin{cases} x = x'\cos\alpha - z'\sin\alpha \\ y = y' \\ z = x'\sin\alpha + z'\cos\alpha \end{cases}$$

The normalized angular functions are $\Theta_{11} = \sqrt{\tfrac{3}{4}}\sin\theta$ and $\Theta_{10} = \sqrt{\tfrac{3}{2}}\cos\theta$.
(a) Express $\Theta_{11}(\theta)e^{i\phi}$ as a function of $x$, $y$, and $z$.
(b) Now express it as a function of $x'$, $y'$ and $z'$.
(c) What is $\Theta_{11}(\theta)e^{i\phi}$ as a function of $\theta'$ and $\phi'$?
(d) What are the coefficients $a_1$, $a_2$, and $a_3$ in

$$\Theta_{11}(\theta)e^{i\phi} = a_1\Theta_{11}(\theta')e^{i\phi'} + a_2\Theta_{10}(\theta') + a_3\Theta_{1-1}(\theta')e^{-i\phi'}$$

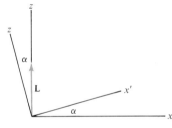

Problem 28

# Atomic Physics

## 27-1 The Pauli Exclusion Principle

As seen in the periodic table of the elements (see Table 27-2) and in Figure 27-1, the elements repeat their chemical and physical properties in groups of 2, 8, 8, 18, 18, 32. In 1925 Wolfgang Pauli proposed a simple rule that automatically generated groups of 2, 8, 18, 32. Pauli proposed that no more than two electrons can occupy the same electron orbital or standing wave. Then the $n = 1$ orbital could hold two electrons. For $n = 2$ there are 4 orbitals: $(n, l, m_l) = (2, 0, 0)$ or $(2, 1, 1)$ or $(2, 1, 0)$ or $(2, 1, -1)$. Hence up to 8 electrons could have $n = 2$. So far we have generated the numbers 2 and 8. The number 18 can be obtained by adding the 5 orbitals of $l = 2$ to the 4 orbitals of $l = 0$ and 1. 18 electrons will fit into these 9 orbitals. We see that the periods 2, 8, and 18 are a direct consequence of the exclusion principle and the quantum mechanical result that

**Figure 27-1.** (a) *Ionization energy of the elements as a function of atomic number.* (b) *Atomic volume as a function of Z.*

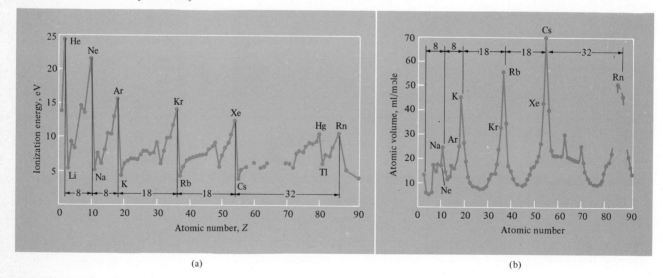

(a)

(b)

**Figure 27-2.** *Wolfgang Pauli, 1900–1958.* [Courtesy American Institute of Physics]

$-1 \leq m_l \leq +l$ and $0 \leq l \leq n-1$. The repetition of the "shells" of 8 and 18 will be explained in Section 27-3. At the time it was proposed, Pauli's exclusion principle was a new ad hoc rule that could not be derived.

## *Electron Spin*

However, a year later it was discovered that all electrons have an intrinsic angular momentum or spin, $L_{\text{intrinsic}} = \frac{1}{2}\hbar$. Note that this is one half the natural unit of orbital angular momentum. It is as if the electron is a sphere spinning around an axis of rotation with a fixed angular momentum of $\frac{1}{2}\hbar$. This intrinsic spin can never be increased or decreased. It is unique for each type of elementary particle.

Shortly after the discovery of electron spin, P. M. Dirac and Pauli developed a relativistic theory of spin-$\frac{1}{2}$ particles and they were pleased to learn that the condition of relativistic invariance gave them electron wave functions that automatically obeyed the Pauli exclusion principle. The exclusion principle was not an ad hoc assumption after all! A particle of spin $\frac{1}{2}$ can have components of spin along the $z$ axis of only $+\frac{1}{2}\hbar$ or $-\frac{1}{2}\hbar$. This spin component is needed in order to specify completely an electron state. When electron states are so specified, the exclusion principle says that no more than one electron can occupy a given state. Since two spin orientations are possible for a given orbital state, this is just another way of stating Pauli's original rule that no more than two electrons can occupy a given orbital state.

## 27-2 Multielectron Atoms

Armed with the Schrödinger equation and the Pauli principle, it is now possible to calculate properties of the elements without resorting to chemical experiments. In fact, in 1929 P. M. Dirac said: "The underlying physical laws necessary for the mathematical theory of a large part of physics and the whole of chemistry are completely known." With modern high speed computers the electron densities and binding energies of atoms heavier than hydrogen have been calculated. Even chemical reactions and molecular binding can be studied as illustrated in Figure 27-3. However, for most problems in chemistry, the computer approach is impractical; it is far easier to make the measurement. (Some typical problems would take years of computer time to calculate.)

Using the Pauli exclusion principle, we can specify where each electron goes in an atom. For example, consider a bare neon nucleus ($Z = 10$). If it is given just one electron, the electron will quickly drop down to the $n = 1$ orbital. The same is true for a second electron. These two electrons completely fill up the $n = 1$ orbital wave function. Now if the eight remaining electrons are also added, they will completely fill up the four possible $n = 2$ orbitals $(l, m_l) = (0, 0)$ or $(1, 1)$ or $(1, 0)$ or $(1, -1)$. We shall now describe some of the atomic structures predicted by the quantum theory, starting with hydrogen. Without resorting to detailed calculations, we shall be able to estimate chemical valences and give crude estimates of ionization potentials as well as predict sizes and shapes.

### $Z = 1$ (Hydrogen)

The structure of this atom was worked out in the previous chapter. The single electron is in the $n = 1$ state, which has an energy of $-13.6$ eV. Thus the binding energy or ionization energy is 13.6 eV. Electrons accelerated by a potential difference of 13.6 V are just barely capable of ionizing hydrogen atoms. This minimum voltage needed for ionization is called the ionization potential. Thus the ionization potential of hydrogen is 13.6 V. The ionization potentials of the elements are plotted in Figure 27-1(a). Note the periodicity pattern 2, 8, 8, 18, 18, 32.

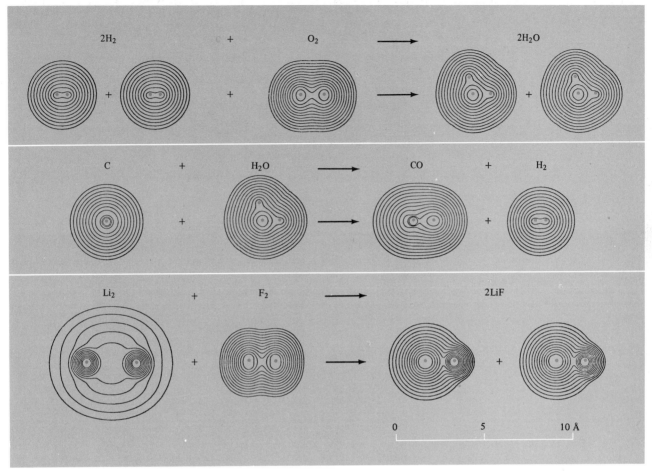

**Figure 27-3.** *Three different chemical reactions. Initial and final electron densities computer calculated by Arnold C. Wahl. With each successive contour the electron density decreases by a factor of 2. For intermediate stages of a chemical reaction, see Figure 27-6. [Adapted from Arnold C. Wahl: Sci. Am. (April 1970)]*

## $Z = 2$ (Helium)

First let us consider ionized helium, He$^+$, which consists of a helium nucleus and a single electron. Any nucleus of atomic number $Z$ and a single electron is the same as the hydrogen atom except that the strength of the electrostatic force has been scaled up by a factor of $Z$. Standing waves of the same shape occur, but with the proportionality constant $k_0$ scaled up by a factor of $Z$. The result will be the same as in Eq. 26-16:

$$E_n = -\frac{(k_0 Z)^2 m e^4}{2\hbar^2 n^2} = -13.6 \frac{Z^2}{n^2} \text{eV} \qquad (27\text{-}1)$$

and the constant $a$ appearing in $R_{nl}^{(r)}$ will be $a = \hbar/(k_0 Z m e^2)$ rather than

$\hbar/(k_0 me^2)$. Because of the factor $Z^2$, the ionization potential of He$^+$ would be $4 \times 13.6$ or 54.4 V. This checks with experiment.

If a second electron is brought near He$^+$ it will first see an object that appears to have a charge of $(Z - 1)$. But when this second electron gets down to the $n = 1$ shell, half the time it is closer to the nucleus than the original electron and then it sees a nuclear charge of $Z$. The straight average of these two values is $(Z - \frac{1}{2})$. Thus we predict that the effective nuclear charge will be $Z_{\text{eff}} = 1.5e$ for an electron in helium. We can generalize Eq. 27-1 to read

$$E_{n,l} = -13.6 \frac{Z_{\text{eff}}^2}{n^2} \text{ eV}$$

where $Z_{\text{eff}}$ depends both on $n$ and $l$. From this estimate of $Z_{\text{eff}}$, we would expect the ionization potential of helium to be about $(1.5)^2 \times 13.6$, or 30 V. Actually, because of the positive potential energy of repulsion of the two electrons we would expect the binding to be not quite so large. Experimentally the ionization potential of helium is 24.6 V. This is the highest ionization potential of any element. Helium is very inert chemically because of its large ionization potential and because there is no more room for a third electron in the $n = 1$ shell. No chemical forces are strong enough to supply 24.6 eV in order to form the positive ion He$^+$. If we tried to form the negative ion He$^-$, the extra electron must be in a $n = 2$ standing wave, which is well outside both the nucleus of charge $+2e$ and the two $n = 1$ electrons of negative charge. Hence the net charge at the center of the $n = 2$ wave is zero and there is no attractive potential available to hold an $n = 2$ wave; that is, for $n = 2$, $Z_{\text{eff}} \approx 0$. Consequently helium does not form molecules with any element. It and the other closed shell atoms are called the noble gases. Several of the heavier noble gases can form certain special compounds.

> **Example 1.** Given that the ionization potential of helium is 24.6 V, what is the total binding energy of helium; that is, how much energy is needed to disassociate a helium atom into two free electrons?
>
> ANSWER: It takes 24.6 eV to remove the first electron and 13.6 $Z^2$ or 54.4 eV to remove the final electron. The total binding energy is the sum of these or 79 eV.

## Z = 3 (Lithium)

Doubly ionized lithium Li$^{2+}$ will have a hydrogenlike spectrum with the energy levels $(3)^2$ or 9 times that of hydrogen. Singly ionized lithium has a helium-type spectrum with a $Z_{\text{eff}}$ of about $(3 - \frac{1}{2})$ instead of the $(2 - \frac{1}{2})$ for helium. Because of the exclusion principle, neutral lithium must have its third electron in the $n = 2$ shell. For this electron $Z_{\text{eff}}$ should be somewhat larger than one. Thus we expect the ionization potential of lithium to be somewhat more than $13.6/n^2 = 13.6/2^2 = 3.4$ V. The experimental value is 5.4 V which corresponds to a $Z_{\text{eff}}$ of 1.25. The second ionization potential (for removing the second electron) is 75.6 V. Thus lithium should always appear in compounds with a valence of $+1$ (gives up one electron), never with $+2$ (gives up two electrons).

What is the $l$ quantum number of the outer electron in lithium? According to all that we have learned so far, the $(n = 2, l = 0)$ and $(n = 2, l = 1)$ states should have the same energy. However, it can be seen by referring to Figure 27-4 that the $l = 0$ state should be more strongly bound than the $l = 1$ state. This is because the lower angular momentum state ($l = 0$) has more of its electron wave near the nucleus than higher angular momentum states. In fact, all electron waves having $l$ greater than zero have $\psi = 0$ when $r = 0$.

That part of an electron wave near the nucleus sees a $Z_{\text{eff}}$ almost as large as $Z$, whereas that part of the wave far from the nucleus sees a $Z_{\text{eff}}$ of about 1. Hence a $l = 0$ wave sees overall a higher $Z_{\text{eff}}$ than an $l = 1$ wave. This is why $Z_{\text{eff}}$ is a function of $l$ as well as $n$. This effect can cause an appreciable energy difference between the $l = 0$ and $l = 1$ or 2 "subshells." In fact, for $Z = 19$ (potassium) the effect is so strong that the $(n = 4, l = 0)$ energy level is lower than the $(n = 3, l = 2)$ level. Table 27-1 shows the order in which energy levels occur. Another way to think of this effect is that a higher angular momentum orbit is more circular and consequently farther away from the nucleus than a lower angular momentum orbit. Hence the lower "$l$" states are more strongly bound.

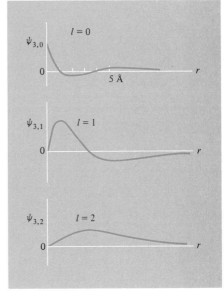

**Figure 27-4.** $n = 3$ hydrogen wave functions for $l = 0, 1,$ and $2$. Note that only the $l = 0$ wave is nonzero at the origin.

## Z = 4 (Beryllium)

According to the exclusion principle there is room for two electrons in the $(n = 2, l = 0)$ state. Because $Z_{\text{eff}}$ for that part of the electron wave near the nucleus is now greater than for that of lithium, the ionization potential should be larger. The experimental value is 9.32 V as compared to the 5.39 V for lithium. However, for beryllium the second ionization potential is not much larger since this electron also comes from an $n = 2$ state. Thus beryllium has a valence of $+2$ in compounds.

## Z = 5 (Boron), Z = 6 (Carbon), Z = 7 (Nitrogen), Z = 8 (Oxygen), Z = 9 (Fluorine), and Z = 10 (Neon)

These atoms are formed by filling up the $l = 1$ states of the $n = 2$ shell. Since $l = 1$ has three different values of $m_l$, the $(n = 2, l = 1)$ subshell can take up to six electrons. Boron, carbon, and nitrogen would have three, four, and five electrons, respectively, in the $n = 2$ states with corresponding valences of +3, +4, and +5. There is a reason why all three of the $n = 2$ electrons of boron (or all four of carbon) play identical roles when compounds are formed. The explanation involves a process called hybridization, which is discussed in Section 27-6.

For oxygen and fluorine there is a new phenomenon called electron affinity. A single fluorine atom can take up an extra electron and become stable Fl$^-$. This extra electron, which sees a large $Z_{eff}$ over part of its wave, is bound with an energy of 3.6 eV. Thus the valence of fluorine is $-1$. The electron affinity for forming O$^-$ is 2.2 V. Oxygen and nitrogen commonly appear in chemical compounds with valences of $-2$ and $-3$, respectively. In neon the $n = 2$ states are all filled up and we have a closed shell. Because part of the $n = 2$ electron waves get fairly close to the nucleus (where now $Z_{eff}$ goes up to 10), the ionization potential is high, 21.6 V. Thus we would expect neon, like helium, to be quite inert chemically.

## 27-3 The Periodic Table of the Elements

If we continue the previous element by element presentation we shall immediately run into elements that have very similar properties and shapes as the elements discussed above.

### $Z = 11$ (Sodium) to $Z = 18$ (Argon)

In sodium the exclusion principle forces the eleventh electron to go into the $n = 3$ wave with a $Z_{eff} \approx 1$ which is a much larger diameter wave than the $n = 2$ of the preceding element neon. Hence, the theory predicts that every time the outer electron is in a higher $n$ orbital, the atomic size will be significantly larger. These abrupt increases in size are observed to occur for $Z = 3, 11, 19, \ldots$ as shown in Figure 27-1(b). In the series of eight elements—sodium to argon—the $(n = 3, l = 0)$ and $(n = 3, l = 1)$ states are filled up in exactly the same way as the preceding eight elements. The chemical properties are thus quite similar to the corresponding eight preceding elements. This is the explanation of the "periodic system" of chemistry. So far we have explained the periodicity 2, 8, 8. Now we shall see why the next period must be 18.

### $Z = 19$ (Potassium) and Up

We might expect that the next element would have its outer electron in the $(n = 3, l = 2)$ state. However, as mentioned in the discussion of $Z = 3$

(lithium), the $(n = 4, l = 0)$ wave sees a considerably larger $Z_{eff}$ than the $(n = 3, l = 2)$ wave because the $l = 0$ wave is most concentrated at $r = 0$ where the effective charge is the greatest. For this $(n = 4, l = 0)$ wave, $Z_{eff} = 2.26$ and the binding energy is $13.6 \, Z_{eff}^2/4^2 = 4.34$ eV; whereas for $(n = 3, l = 2)$, $Z_{eff}$ is somewhat less than 1.7 corresponding to a binding energy less than 4.34 eV. If the nineteenth electron were put into the $(n = 3, l = 2)$ state, it would quickly drop down to the lower energy state $(n = 4, l = 0)$ emitting a photon equal to the energy difference.

When we get to $Z = 21$ (scandium), the $(n = 4, l = 0)$ wave is all filled up and now the $(n = 4, l = 1)$ competes with the $(n = 3, l = 2)$ for the twenty-first electron. Now, as we would expect, the $(n = 3)$ state is the

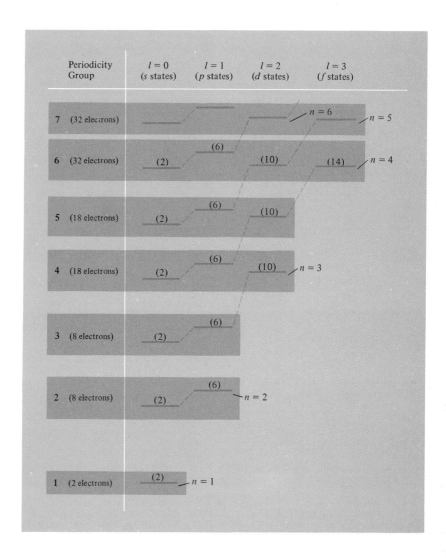

**Figure 27-5.** *Relative spacings of electron state energy levels in atoms of high Z (not drawn to scale). States of common principal quantum number n are connected by dashed lines. Note that energy levels increase with increasing l. The states group themselves into periodicity groups of 2, 8, 8, 18, 18, 32, 32.*

lower one, so at scandium the ten $l = 2$ states of $n = 3$ start filling up. The six $(n = 4, l = 1)$ states follow next. So we have a total of $(2 + 10 + 6) = 18$ electron states all with similar energies. The 18 states for $Z = 19$ to 36 are illustrated in Group 4 of Figure 27-5. This figure shows the relative spacings of atomic energy levels according to the quantum number $n$ and $l$. These energy levels could be measured by taking a bare nucleus of large atomic number and adding electrons one by one. One could measure the energy released when each electron drops down to the lowest possible level. By this means one could map out the relative spacing of the energy levels of all the inner electrons.

We expect that for any given $n$ value, the energy levels increase in successive steps for increasing $l$ values as shown in Figure 27-5. This is what is observed experimentally. Note also the especially large energy gaps following electrons 2, 10, 18, 36, 54, and 86. Hence elements of atomic number $Z = 2, 10, 18, 36, 54$, and 86 have closed shells where the outer electrons are especially tightly bound. These are the noble gases He, Ne, Ar, Kr, Xe, and Rn. One can predict in advance the quantum numbers and energy level of each electron in each element. Such predictions of electron configurations are listed in Table 27-1. The corresponding periodic chart of the elements is shown in Table 27-2. Elements in the same column have the same valence and similar chemical properties.

This concludes our "derivation" of chemistry as far as atoms are concerned. To calculate precisely the ionization potentials and electron affinities by computer requires an enormous amount of calculation; but we have the theory and it can be done in principle, thus explaining all of chemistry with one simple theory, the quantum mechanics of the spin-$\frac{1}{2}$ electron. The computer approach of how atoms can combine to form molecules will be dealt with briefly in Sections 27-5 and 27-6.

---

**Example 2.** What $n$ and $l$ values would you expect for the outermost electrons of the newly discovered element 106? What would be the electronic configuration of this element?

ANSWER: The seventh period starts with $Z = 86$. According to Figure 27-5 the eighty-seventh and eighty-eighth electrons are $7s^2$ ($n = 7, l = 0$). Then almost at the same energy come 14 more electrons as $5f^{14}$ ($n = 6, l = 3$). The remaining 4 electrons are $6d^4$ ($n = 6, l = 2$). In Figure 27-4 all these levels are very close to each other and it is not completely clear which should come first. Table 27-1, which is more quantitative, shows that these levels do switch around with $Z$. From this table we expect a filled Rn core followed by $5f^{14}6d^47s^2$ for the electronic configuration.

**Table 27-1.** *Electronic configuration of the atoms.*

| Z | Symbol | Ground Configuration | Ionization Energy, eV |
|---|--------|---------------------|----------------------|
| 1 | H | $1s$ | 13.595 |
| 2 | He | $1s^2$ | 24.581 |
| 3 | Li | [He] $2s$ | 5.390 |
| 4 | Be | $2s^2$ | 9.320 |
| 5 | B | $2s^2 2p$ | 8.296 |
| 6 | C | $2s^2 2p^2$ | 11.256 |
| 7 | N | $2s^2 2p^3$ | 14.545 |
| 8 | O | $2s^2 2p^4$ | 13.614 |
| 9 | F | $2s^2 2p^5$ | 17.418 |
| 10 | Ne | $2s^2 2p^6$ | 21.559 |
| 11 | Na | [Ne] $3s$ | 5.138 |
| 12 | Mg | $3s^2$ | 7.644 |
| 13 | Al | $3s^2 3p$ | 5.984 |
| 14 | Si | $3s^2 3p^2$ | 8.149 |
| 15 | P | $3s^2 3p^3$ | 10.484 |
| 16 | S | $3s^2 3p^4$ | 10.357 |
| 17 | Cl | $3s^2 3p^5$ | 13.01 |
| 18 | Ar | $3s^2 3p^6$ | 15.755 |
| 19 | K | [Ar] $4s$ | 4.339 |
| 20 | Ca | $4s^2$ | 6.111 |
| 21 | Sc | $3d 4s^2$ | 6.54 |
| 22 | Ti | $3d^2 4s^2$ | 6.83 |
| 23 | V | $3d^3 4s^2$ | 6.74 |
| 24 | Cr | $3d^5 4s$ | 6.76 |
| 25 | Mn | $3d^5 4s^2$ | 7.432 |
| 26 | Fe | $3d^6 4s^2$ | 7.87 |
| 27 | Co | $3d^7 4s^2$ | 7.86 |
| 28 | Ni | $3d^8 4s^2$ | 7.633 |
| 29 | Cu | $3d^{10} 4s$ | 7.724 |
| 30 | Zn | $3d^{10} 4s^2$ | 9.391 |
| 31 | Ga | $3d^{10} 4s^2 4p$ | 6.00 |
| 31 | Ge | $3d^{10} 4s^2 4p^2$ | 7.88 |
| 33 | As | $3d^{10} 4s^2 4p^3$ | 9.81 |
| 34 | Se | $3d^{10} 4s^2 4p^4$ | 9.75 |
| 35 | Br | $3d^{10} 4s^2 4p^5$ | 11.84 |
| 36 | Kr | $3d^{10} 4s^2 4p^6$ | 13.996 |
| 37 | Rb | [Kr] $5s$ | 4.176 |
| 38 | Sr | $5s^2$ | 5.692 |
| 39 | Y | $4d 5s^2$ | 6.377 |
| 40 | Zr | $4d^2 5s^2$ | 6.835 |
| 41 | Nb | $4d^4 5s$ | 6.881 |
| 42 | Mo | $4d^5 5s$ | 7.10 |
| 43 | Tc | $4d^5 5s^2$ | 7.228 |
| 44 | Ru | $4d^7 5s$ | 7.365 |
| 45 | Rh | $4d^8 5s$ | 7.461 |
| 46 | Pd | $4d^{10}$ | 8.33 |
| 47 | Ag | $4d^{10} 5s$ | 7.574 |
| 48 | Cd | $4d^{10} 5s^2$ | 8.991 |
| 49 | In | $4d^{10} 5s^2 5p$ | 5.785 |
| 50 | Sn | $4d^{10} 5s^2 5p^2$ | 7.342 |
| 51 | Sb | $4d^{10} 5s^2 5p^3$ | 8.639 |
| 52 | Te | $4d^{10} 5s^2 5p^4$ | 9.01 |
| 53 | I | $4d^{10} 5s^2 5p^5$ | 10.454 |
| 54 | Xe | $4d^{10} 5s^2 5p^6$ | 12.127 |
| 55 | Cs | [Xe] $6s$ | 3.893 |
| 56 | Ba | $6s^2$ | 5.210 |
| 57 | La | $5d 6s^2$ | 5.61 |
| 58 | Ce | $4f 5d 6s^2$ | 6.54 |
| 59 | Pr | $4f^3 6s^2$ | 5.48 |
| 60 | Nd | $4f^4 6s^2$ | 5.51 |
| 61 | Pm | $4f^5 6s^2$ | |
| 62 | Fm | $4f^6 6s^2$ | 5.6 |
| 63 | Eu | $4f^7 6s^2$ | 5.67 |
| 64 | Gd | $4f^7 5d 6s^2$ | 6.16 |
| 65 | Tb | $4f^9 6s^2$ | 6.74 |
| 66 | Dy | $4f^{10} 6s^2$ | 6.82 |
| 67 | Ho | $4f^{11} 6s^2$ | |
| 68 | Er | $4f^{12} 6s^2$ | |
| 69 | Tm | $4f^{13} 6s^2$ | |
| 70 | Yb | $4f^{14} 6s^2$ | 6.22 |
| 71 | Lu | $4f^{14} 5d 6s^2$ | 6.15 |
| 72 | Hf | $4f^{14} 5d^2 6s^2$ | 7.0 |
| 73 | Ta | $4f^{14} 5d^3 6s^2$ | 7.88 |
| 74 | W | $4f^{14} 5d^4 6s^2$ | 7.98 |
| 75 | Re | $4f^{14} 5d^5 6s^2$ | 7.87 |
| 76 | Os | $4f^{14} 5d^6 6s^2$ | 8.7 |
| 77 | Ir | $4f^{14} 5d^7 6s^2$ | 9.2 |
| 78 | Pt | $4f^{14} 5d^8 6s^2$ | 8.88 |
| 79 | Au | [Xe, $4f^{14} 5d^{10}$] $6s$ | 9.22 |
| 80 | Hg | $6s^2$ | 10.434 |
| 81 | Tl | $6s^2 6p$ | 6.106 |
| 82 | Pb | $6s^2 6p^2$ | 7.415 |
| 83 | Bi | $6s^2 6p^3$ | 7.287 |
| 84 | Po | $6s^2 6p^4$ | 8.43 |
| 85 | At | $6s^2 6p^5$ | |
| 86 | Rn | $6s^2 6p^6$ | 10.745 |
| 87 | Fr | [Rn] $7s$ | |
| 88 | Ra | $7s^2$ | 5.277 |
| 89 | Ac | $6d 7s^2$ | 6.9 |
| 90 | Th | $6d^2 7s^2$ | |
| 91 | Pa | $5f^2 6d 7s^2$ | |
| 92 | U | $5f^3 6d 7s^2$ | 4.0 |
| 93 | Np | $5f^4 6d 7s^2$ | |
| 94 | Pu | $5f^6 7s^2$ | |
| 95 | Am | $5f^7 7s^2$ | |
| 96 | Cm | $5f^7 6d 7s^2$ | |
| 97 | Bk | $5f^8 6d 7s^2$ | |
| 98 | Cf | $5f^{10} 7s^2$ | |
| 99 | Es | $5f^{11} 7s^2$ | |
| 100 | Fm | $5f^{12} 7s$ | |
| 101 | Mv | $5f^{13} 7s^2$ | |
| 102 | No | $5f^{14} 7s^2$ | |
| 103 | Lw | $5f^{14} 6d 7s^2$ | |
| 104 | Ku | $5f^{14} 6d^2 7s^2$ | |

The notation $s, p, d, f$ stands for $l = 0, 1, 2, 3$, respectively. For example the configuration listed for oxygen ($Z = 8$) is [He] $2s^2 2p^4$. The square bracket means the inner shells are filled the same as in He. The $2s^2 2p^4$ means there are two electrons in the $2s$ state and four electrons in the $2p$ state. The $2p$ state is ($n = 2, l = 1$).

**Table 27-2.** *Periodic Table of the Elements*

| Outer Electron States | IA | IIA | IIIB | IVB | VB | VIB | VIIB | VIII | | | IB | IIB | IIIA | IVA | VA | VIA | VIIA | 0 |
|---|---|---|---|---|---|---|---|---|---|---|---|---|---|---|---|---|---|---|
| 1s | | | | | | | | | | | | | | | | | | 2 He 4.003 |
| | 1 H 1.008 | | | | | | | | | | | | | | | | | |
| 2s, 2p | 3 Li 6.941 | 4 Be 9.012 | | | | | | | | | | | 5 B 10.81 | 6 C 12.011 | 7 N 14.007 | 8 O 15.9994 | 9 F 18.998 | 10 Ne 20.18 |
| 3s, 3p | 11 Na 22.990 | 12 Mg 24.31 | | | | | | | | | | | 13 Al 26.98 | 14 Si 28.09 | 15 P 30.97 | 16 S 32.06 | 17 Cl 35.453 | 18 Ar 39.95 |
| 4s, 4p, 3d | 19 K 39.10 | 20 Ca 40.08 | 21 Sc 44.96 | 22 Ti 47.90 | 23 V 50.94 | 24 Cr 52.00 | 25 Mn 54.94 | 26 Fe 55.85 | 27 Co 58.93 | 28 Ni 58.71 | 29 Cu 63.55 | 30 Zn 65.37 | 31 Ga 69.72 | 32 Ge 72.59 | 33 As 74.92 | 34 Se 78.96 | 35 Br 79.90 | 36 Kr 83.80 |
| 5s, 5p, 4d | 37 Rb 85.47 | 38 Sr 87.62 | 39 Y 88.91 | 40 Zr 91.22 | 41 Nb 92.91 | 42 Mo 95.94 | 43 Tc 98.91 | 44 Ru 101.07 | 45 Rh 102.91 | 46 Pd 106.4 | 47 Ag 107.87 | 48 Cd 112.40 | 49 In 114.82 | 50 Sn 118.69 | 51 Sb 121.75 | 52 Te 127.60 | 53 I 126.90 | 54 Xe 131.30 |
| 6s, 6p, 5d | 55 Cs 132.91 | 56 Ba 137.34 | 57* La 138.91 | 72 Hf 178.49 | 73 Ta 180.95 | 74 W 183.85 | 75 Re 186.2 | 76 Os 190.2 | 77 Ir 192.22 | 78 Pt 195.09 | 79 Au 196.97 | 80 Hg 200.59 | 81 Tl 204.37 | 82 Pb 207.2 | 83 Bi 208.98 | 84 Po (210) | 85 At (210) | 86 Rn (222) |
| 7s, 7p, 6d | 87 Fr (223) | 88 Ra 226.03 | 89† Ac (227) | 104 (Ku) (261) | 105 (Ha) (260) | 106 | | | | | | | | | | | | |

| 4f *Lanthanide series: | 58 Ce 140.12 | 59 Pr 140.91 | 60 Nd 144.24 | 61 Pm (147) | 62 Sm 150.4 | 63 Eu 151.96 | 64 Gd 157.25 | 65 Tb 158.93 | 66 Dy 162.50 | 67 Ho 164.93 | 68 Er 167.26 | 69 Tm 168.93 | 70 Yb 173.04 | 71 Lu 174.97 |
|---|---|---|---|---|---|---|---|---|---|---|---|---|---|---|
| 5f †Actinide series: | 90 Th 232.04 | 91 Pa 231.04 | 92 U 238.03 | 93 Np 237.05 | 94 Pu (244) | 95 Am (243) | 96 Cm (245) | 97 Bk (247) | 98 Cf (249) | 99 Es (249) | 100 Fm (255) | 101 Md (256) | 102 No (254) | 103 Lr (257) |

## 27-4 X Rays

As we have seen in the previous section, the first two electrons in all atoms make up a heliumlike core. They have binding energy of 13.6 $Z_{eff}^2$ in electron volts. In atoms of high atomic number $Z_{eff} \approx (Z - 2)$ because of the shielding contributions of the several $l = 0$ electrons. For atoms of large atomic number the binding energies of the two electrons in the $n = 1$ shell are well above 10,000 eV compared to the binding energies of the outer electrons which are of a few electron volts. When an outer electron is removed by placing a sample of the element in an electric arc or discharge tube, the available quantum jumps will be on the order of a few electron volts and the characteristic spectrum will consist of lines of corresponding wavelength. Since a 1-eV photon has $\lambda = 12{,}390$ Å, the characteristic spectrum of a typical element has lines in the infrared, visible, and ultraviolet.

But suppose an inner electron could be removed. Then an outer electron could "jump down" to replace the missing inner electron and a photon would be emitted having thousands of times the usual amount of energy. Its wavelength would be hundreds or thousands of times shorter than that of visible light. Such photons where $0.1 \text{ Å} < \lambda < 100 \text{ Å}$ are called X rays. But how can an inner electron be removed? It is easy. Just bombard a sample with a beam of electrons whose kinetic energy exceeds the binding energy of the inner electrons. In the following example we shall see that an electron beam of 1650 eV or greater can remove the $n = 1$ electrons from aluminum. In x ray terminology the $n = 1$ electrons are called $K$-shell electrons.

**Example 3.** What are the highest energy x rays that can be emitted by aluminum and lead? Give the photon energies and wavelengths.

ANSWER: The highest energy x rays occur when a free electron (zero energy) jumps all the way down to a vacancy in the $K$ shell. For aluminum, $Z = 13$ and $hf = 13.6(13 - 2)^2 \text{ eV} = 1.65 \text{ keV}$. The corresponding wavelength is $\lambda = 7.5 \text{ Å}$.
For lead, $Z = 82$ and $hf = 13.6(80)^2 \text{ eV} = 87 \text{ keV}$ with $\lambda = 0.14 \text{ Å}$.

Example 3 illustrates a very reliable method that has been used to determine the $Z$ of newly discovered elements. One merely determines the wavelengths of high-energy x rays emitted from the unknown sample when it is bombarded with electrons.

Since x rays have wavelengths comparable to the interatomic spacings in solids, they are also a very useful tool for the determination of the structure of solids. As mentioned in previous chapters, the periodically repeating planes of atoms in a crystal behave as the rulings of a diffraction grating. Thus if the x-ray wavelength is known, the interatomic spacings can be determined by measuring the x-ray diffraction angles (see page 537).

## 27-5 Molecular Binding

We shall discuss two different mechanisms that cause atoms to bind together into molecules: ionic binding and covalent binding (also called bonding).

### Ionic Binding

In ionic binding, as the two neutral atoms are slowly brought closer together, they reach a point where the outer electron from one atom

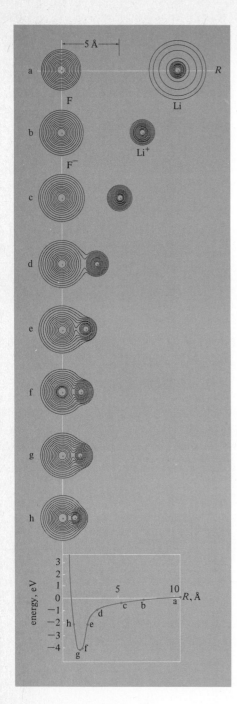

**Figure 27-6.** *Computer calculation of total electron density in lithium and fluorine atoms as the distance R between them is decreased. Net energy as a function of R is plotted below. Electron density decreases by a factor of two for each successive contour (it has dropped to 3% at the fifth contour). Note that at position b the lithium outer electron "jumps over" to the fluorine atom.* [*Calculations by Arnold C. Wahl, published in the April 1970* Scientific American]

prefers to be attached to the other. The atom with the missing electron behaves as a positive particle of charge $+e$ and the atom with the extra electron is a negative particle of charge $-e$. The electrostatic potential energy between them is $U = -k_0 e^2/R$ where $R$ is the separation between the centers of the two ions.

As an example, we shall consider the formation of the lithium fluoride (LiF) molecule. Since the ionization energy of lithium is 5.4 eV and the electron affinity of fluorine is 3.6 eV, the net energy needed to remove the electron from lithium and attach it to fluorine is (5.4 eV − 3.6 eV) = 1.8 eV. When the two atoms are at a distance $R = 8$ Å the quantity $k_0 e^2/R = 1.8$ eV. Hence, at distances of $R$ less than 8 Å, the outer electron of lithium can reach a lower energy state by "jumping" over to the fluorine atom. This is seen in Figure 27-6, which is a computer calculation of the total electron densities in both atoms. The net energy of the system is shown at the bottom. Note that when the atoms reach a distance of $R = 8$ Å, as in view (b), the lithium outer electron moves over to the fluorine atom. In view (h) the energy starts increasing because of the positive repulsive energy between the inner electrons of both atoms. The minimum energy occurs at a distance $R = 1.5$ Å where the net binding energy is 4.3 eV [see view (g)].

### Covalent Binding

Another very prevalent mechanism for molecular binding that occurs in most organic molecules is called covalent binding. A covalent bond is sharing of electrons by two atoms. The simplest example of covalent binding is the hydrogen molecule. First we shall consider the ionized hydrogen molecule $H_2^+$. This consists of two protons surrounded by an electron cloud. The binding energy of the electron in the presence of two protons is of course larger than in the presence of one proton only. On the other hand, the electrostatic repulsion of the two protons tends to oppose the binding. However, since the electron wave gets in close and tends to concentrate between the two protons, the effect of the electrostatic attraction of the electron to the two protons dominates. The binding energy of the hydrogen atom to the hydrogen ion in $H_2^+$ is 2.65 eV; that is,

**Figure 27-7.** *Covalent binding. Two hydrogen atoms treated as described for Figure 27-6. Note how the electrons like to fill up each other's unfilled shell. [Calculations by Arnold C. Wahl, published in the April 1970 Scientific American]*

$(2.65 + 13.6)$ eV is required to dissociate completely an $H_2^+$ ion into two protons and one electron.

According to the Pauli exclusion principle there is room for a second electron to fit in the same electron wave as the first electron. This system of two electrons and protons is the neutral hydrogen molecule. Here the electron wave function is somewhat more spread out than that of the single electron in $H_2^+$ because of the electrostatic repulsion between the two electrons. The binding energy of the two hydrogen atoms in the neutral $H_2$ molecule is 4.48 eV. Computer calculations of electron density contours and binding energy for two H atoms as a function of distance are displayed in Figure 27-7.

Carbon atoms usually form covalent bonds. The carbon atom has a tendency to share with four other electrons in an attempt to fill up its $n = 2$, $l = 1$ shell. The simplest such case is $CH_4$ (methane) shown in Figure 27-8. As with the hydrogen molecule, the electron waves tend to concentrate between the positive charges where they will make the greatest contribution to the binding energy. Since these four electron clouds are mutually repulsive, the lowest energy configuration is achieved when they are the farthest from each other as in Figure 27-8. The shapes of electron waves in molecules are determined by the Schrödinger equation along with the condition that the energy levels be as low as possible.

**Figure 27-8.** *(a) The relative orientations of the carbon nucleus and four hydrogen nuclei (protons) in the $CH_4$ molecule. (b) The four electron clouds or lobes extending out from the carbon nucleus and enveloping the four protons. Each lobe contains two electrons in a covalent bond.*

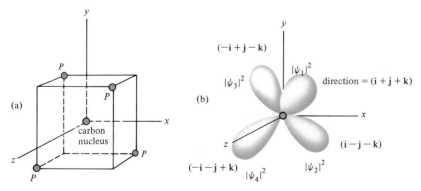

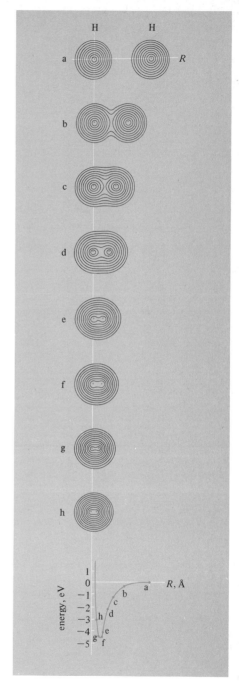

SEC. 27-5/MOLECULAR BINDING 603

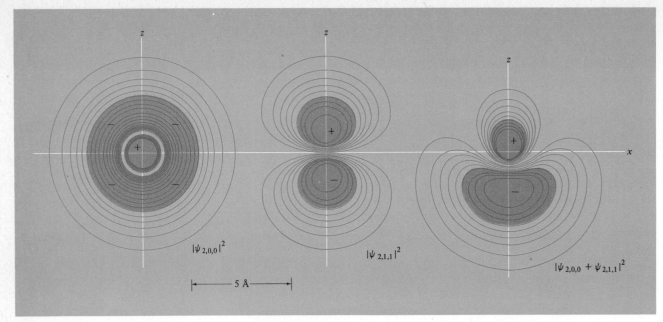

**Figure 27-9.** $\psi_{2,0,0} + \psi_{2,1,1} = \psi_{sp}$. *Percentage contours of electron density. The highest 40% is shaded. The plus and minus signs indicate the polarity of the wave function before it is squared.* [*With permission of Professor George H. Gerhold*]

## 27-6 Hybridization

We see that when forming molecules, the four $n = 2$ electrons of carbon can extend out and play equal roles giving carbon a valence of 4. But how could this be, when the first two $n = 2$ carbon electrons are more strongly bound and in an $l = 0$ closed subshell? Similarly, why should boron ($Z = 5$) with its one $l = 1$ electron not have a valence of 1 rather than 3? The answer is that atoms in molecules are not the same as when they are alone. In the pure atom, there is a distinction between the $l = 0$ and $l = 1$ electrons. As emphasized in Section 27-2, the $l = 0$ electrons are indeed more tightly bound than the $l = 1$ electrons. But, molecules are not pure atoms. The $CH_4$ molecule, for example, has in addition to the carbon nucleus, four other centers of positive charge. With a system of five atomic nuclei, the solution to Schrödinger's equation for standing waves is now more complicated, and gives the result for the ground state shown in Figure 27-8.

Even though the standing waves for the four outer electrons of the carbon atom look vastly different from the usual atomic standing waves, they can be expressed mathematically as a mixture of hydrogenlike wave functions (see Example 5 of Chapter 26). In fact any standing wave can be expanded in terms of hydrogenlike wave functions. In the case of the carbon atom the binding energies of the four outer electrons are the same and correspond to the $n = 2$ binding energy; hence, we would expect the electron standing waves to be mixtures of mainly $\psi_{2,0,0}$, $\psi_{2,1,1}$, $\psi_{2,1,0}$, and $\psi_{2,1,-1}$. The mixtures which give the result shown in Figure 27-8 are

$$\psi_1 = \frac{1}{2}(s + p_x + p_y + p_z) \qquad \text{where } s = -\psi_{200}$$

$$\psi_2 = \frac{1}{2}(s + p_x - p_y - p_z) \qquad p_x = \frac{1}{\sqrt{2}}(\psi_{211} + \psi_{21-1})$$

$$\psi_3 = \frac{1}{2}(s - p_x + p_y - p_z) \qquad p_y = \frac{1}{\sqrt{2}}(\psi_{211} - \psi_{21-1})$$

$$\psi_4 = \frac{1}{2}(s - p_x - p_y + p_z) \qquad p_z = \psi_{210}$$

Figure 27-9 illustrates how a mixture of $\psi_{2,0,0}$ and $\psi_{2,1,1}$ can give an electron cloud which sticks out mainly in one direction. Here $\psi_{sp} = (\psi_{2,0,0} + \psi_{2,1,1})$. Note that except for the small central region, $\psi_{2,0,0}$ is negative in all directions; whereas $\psi_{2,1,1}$ is positive in the positive $z$ direction and negative in the negative $z$ direction. Hence $\psi_{2,0,0}$ and $\psi_{2,1,1}$ are of opposite sign in the positive $z$ direction, giving rise to a destructive interference, whereas they interfere constructively in the negative $z$ direction. The addition of single-atom wave functions to form lobes that stick out is called hybridization. The electron cloud in Figure 27-9 is called an *sp* hybrid and the four clouds in Figure 27-8 are called *sp*³ hybrids.

---

*Example 4.* From Table 26-2 and Figure 26-7 one sees that $\psi_{2p_z} \equiv \psi_{2,1,0}$ is shaped like a dumbbell with its axis in the $z$ direction. Show that

$$\psi_{2p_x} \equiv \frac{1}{\sqrt{2}}(\psi_{2,1,1} + \psi_{2,1,-1})$$

has the same dumbbell shape, but with its axis in the $x$ direction.

ANSWER: Using Table 26-2:

$$\psi_{2p_x} = \frac{r}{\sqrt{2}a} e^{-r/a} \sin\theta \, (e^{i\phi} + e^{-i\phi}).$$

Since $\cos\theta = (e^{i\phi} + e^{-i\phi})/2$, we have

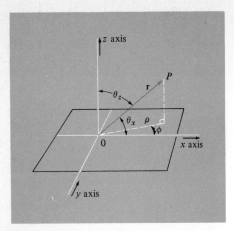

**Figure 27-10.** *The vector* **r** *to point P is at an angle $\theta_z$ to the z axis and $\theta_x$ to the x axis. $\rho$ is the projection of* **r** *in the xy plane.*

$$\psi_{2p_x} = \frac{2r}{\sqrt{2}a} e^{-r/a} \sin\theta \cos\phi$$

In Figure 27-10, $\sin\theta = \rho/r$ and $\cos\phi = x/\rho$. The product is $x/r$, hence

$$\psi_{2p_x} = \frac{2r}{\sqrt{2}a} e^{-r/a} \left(\frac{x}{r}\right)$$

$$\psi_{2p_x} = \frac{2r}{\sqrt{2}a} e^{-r/a} \cos\theta_x$$

where $\theta_x$ is the angle between **r** and the x axis. This is the same mathematical form as $\psi_{2,1,0}$, which is $(r/a) e^{-r/2a} \cos\theta_z$ where $\theta_z$ is the usual polar angle $\theta$. We have also shown that $\psi_{2p_x}$ is symmetric with respect to the x axis, whereas $\psi_{2p_z}$ is symmetric with respect to the z axis. Hence $\psi_{2p_x}$ is a dumbbell pointing along the x axis.

## Summary

The chemical and physical properties of the elements reveal a striking similarity of properties in groups of 2, 8, 8, 18, 18, 32, 32. The groups 2, 8, 18, 32 can be obtained from the hydrogen standing waves or orbitals by noting that because of the Pauli exclusion principle no more than one electron can be in the same state. Since electrons have intrinsic angular momentum of $\frac{1}{2}\hbar$, there are two possible states to each orbital.

Properties of the elements such as valence and ionization potential can be explained semiquantitatively by using hydrogenlike standing waves that have binding energies of $(13.6/n^2) Z_{eff}^2$ in eV. $Z_{eff}$ is the "average" charge of the core seen by the electron in question.

For the same $n$ value, electron waves with higher $l$ are farther out and see smaller $Z_{eff}$, so they are less strongly bound. The result is that energy levels of higher $l$ form groups with other levels of higher $n$ having lower $l$ in such a way that the groups sizes are 2, 8, 8, 18, 18, 32, 32.

X rays are photons over 100 eV and are emitted from heavy atoms that have missing inner electrons. Outer electrons quickly cascade down to fill a missing electron state. The K-shell energy is about $13.6(Z-2)^2$ in eV, which is the maximum energy x ray that can be emitted.

Atoms bind together into molecules by ionic and covalent binding. In ionic binding the potential energy between the positive and negative ion is $U \approx -k_0(e^2/R)$ where $R$ is the separation between ion centers. In covalent binding two or more atomic centers share the same outer electrons.

## Exercises

1. Use Figure 27-1 to determine the spacing between atoms in liquid sodium. Repeat for potassium.
2. In a large atom like uranium, how many electrons are there having principal quantum number $n = 4$? How many are there with $n = 5$?
3. Which lines of the $Li^{2+}$ spectrum could be seen by eye?
4. What is $Z_{eff}$ seen by an outer electron in $Fl^-$? (Use the measured value for electron affinity.)
5. Some of the $He^+$ spectral lines are visible. What are their wavelengths?
6. What is $Z_{eff}$ for either electron in the ion $Li^+$? (Use the measured value for electron binding.)
7. What is the highest energy x ray emitted by uranium?
8. The highest energy x rays emitted by an unknown sample have a wavelength of 2.16 Å. What is the highest $Z$ element in the sample?
9. How much energy is needed to dissociate a $H_2$ molecule into two free protons plus two free electrons?
10. What electron configuration would you expect for element 107?

## Problems

11. Extend Figure 27-5 to periodicity group 8. How many elements would be in this group?
12. If the net binding of a LiF molecule is 4.3 eV, what is the heat of formation per mole of LiF?
13. In x ray terminology the eight $n = 2$ electrons are in the $L$ shell.
    (a) Estimate $Z_{eff}$ for a $2p$ electron in the $L$ shell. (Assume the $2s$ electrons are inside.)
    (b) The transition of an electron from the $2p$ to the $K$ shell gives the $K_\alpha$ line in x ray terminology. Estimate wavelengths of the $K_\alpha$ lines for aluminum and lead.
14. A hydrogen atom has $n = 10$, $l = 9$. For fixed $r$, what is the ratio of finding the electron at 70° compared to 90°? You may use Table 26-2. Make a qualitative sketch of the electron cloud. The angles are measured with respect to the angular momentum axis.
15. Show that $\psi_{2p_y} \equiv (1/\sqrt{2})(\psi_{2,1,1} - \psi_{2,1,-1})$ has the same dumbbell shape as $\psi_{2,1,0}$, but that the axis is in the $y$ rather than $z$ direction.
16. (a) Express $\psi_{2,1,1}$ in terms of $\psi_{2p_x}$ and $\psi_{2p_y}$.
    (b) Express $\psi_1, \psi_2, \psi_3,$ and $\psi_4$ on page 605 in terms of $\psi_{2,0,0}, \psi_{2,1,1}, \psi_{2,1,0},$ and $\psi_{2,1,-1}$.
17. The $K_\alpha$ lines in Al and Pb are measured to be 8.3 Å and 0.17 Å. What then is $Z_{eff}$ for $2p$ electrons in Al and in Pb?
18. The distance between the two protons in $H_2^+$ is 1.06 Å. What would be this distance if they are bound together by a $\mu^-$ meson (207 times the electron mass)?
19. Elements repeat their properties in groups of 2, 8, 8, 18, 18, 32, 32, . . . . How many elements would be in the next group starting with $Z = 119$? (Assume stable atomic nuclei.)

# Condensed Matter

When an element or compound in the gaseous state or liquid state is cooled sufficiently it condenses into the solid state where the relative positions of the atoms stay approximately fixed. The study of the properties and phenomena of solids and liquids is called condensed matter or solid state physics and is at present one of the major fields of physics research. Not only does solid state physics aim at a fundamental understanding of the phenomena, but the fundamental understanding is used to predict new phenomena that otherwise might never have been seen.

Most elements and compounds when examined under a microscope reveal a crystalline structure. Grains of table salt (NaCl) under a microscope appear as perfect cubes. Many other solids, although they do not appear as obvious crystals, actually are made up of many tiny crystals (this is called polycrystalline structure). The same mechanisms that bind atoms together into molecules can bind them into an unlimited solid periodic structure which can be thought of as a super molecule. Just as we had ionic and covalent molecules, we can have ionic and covalent solids. Some solids are held together by a third type of binding called metallic binding, which has no diatomic molecule counterpart. We shall study all three types of binding with most emphasis on metals and semiconductors. We shall also discuss other large-scale quantum mechanical phenomena such as superconductivity, superfluidity, and field emission. In Section 28-5 we shall try to give some idea of the impact semiconductor physics has had on our technology and civilization. In addition to the basic theory we shall discuss some of the applications such as solid state diodes, transistors, light emitting diodes, solid state lasers, photocells, solar cells, thermistors, integrated circuits, and so on. These and other valuable applications would not have been possible without a basic understanding of solid state theory.

**Figure 28-1.** *The crystal structure of NaCl. The small circles indicate the positions of the centers of the Na and Cl atoms. (a) Centers of atom. (b) Entire atoms.*

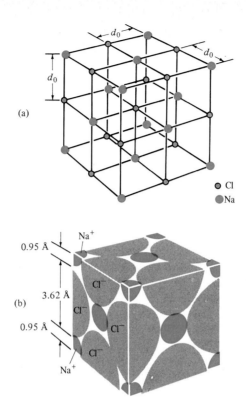

## 28-1 Types of Binding

### Ionic Binding

Just as there are ionic and covalent molecules, there are also ionic and covalent crystals. Figure 28-1 shows the structure of the NaCl ionic crystal. Notice that each Na⁺ ion has six Cl⁻ ions for its nearest neighbors. This type of spatial configuration of Na⁺ and Cl⁻ ions has the lowest energy (it gives off the most heat during formation) of all other possible configurations. This explains the tendency for NaCl and many other substances to form pure crystals as they are cooled below their freezing points. As the temperature of a crystal is raised, the thermal kinetic energy eventually becomes large enough to overcome the binding into a regular crystal, and the crystal melts.

### Covalent Binding

We saw in the last chapter that carbon, which has a valence of 4, likes to form four covalent bonds with other atomic electrons. The $CH_4$ molecule

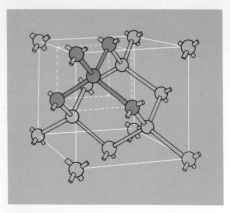

**Figure 28-2.** *The diamond crystal structure. Note that each atom has four nearest neighbors.*

in Figure 27-8 is an example of this. The most tightly bound form of carbon in the solid state is diamond (see Figure 28-2). If the four hydrogen atoms of Figure 27-8 are replaced with carbon atoms, we then have the basic cell of the diamond structure shown in Figure 28-2. Other valence 4 elements such as germanium, silicon, and tin crystallize into the same diamond-type of structure. Contours of equal charge density have been calculated for the valence electrons in a germanium crystal. As seen in Figure 28-3(a) the electron clouds are concentrated in covalent bonds lying midway between each pair of germanium cores.

## Metallic Binding

Now suppose one of the valence electrons was excited to the next highest energy level. We would expect its wave function to be more spread out. But then it would be closer to its nearest neighbor atomic cores. Since these positively charged cores attract the electron, they tend to spread out the wave function even more so that it overlaps the next nearest neighbors. There is a snowballing effect with the wavefunction being pulled out uniformly over the entire crystal (with some concentration in the region of each attracting core). Charge density contours of the next highest set of energy states are shown in Figure 28-3(b). Note that the charge density for the excited state varies only by about a factor of three with minimum charge halfway between any two cores; whereas, in Figure 28-3(a) the charge is quite localized between each pair of cores and varies by a factor of 28. This latter is typical of covalent binding. We are tempted to conclude that whenever a valence electron in a germanium crystal is excited to the next highest level, it behaves as a free electron (a conduction electron). In germanium the energy required to raise a valence electron to the next highest level is 0.72 eV. Another way of saying this is that there is a 0.72-eV energy gap between the covalent state and the conduction state. In Section 28-4 we shall return to the study of germanium and other semiconductors. They are called semiconductors because normally they have only a few electrons in the conduction state.

The charge density plot for crystals of valence-1 atoms (such as Li, Na, K) is similar to Figure 28-3(b). Crystals held together in this way are called metals. Metallic binding occurs when the atoms are pushed together closer than the size of the electron cloud of the outer electrons. Because of the Pauli exclusion principle, such a configuration will tend to raise the outer electrons to higher energies. However, in the case of the metals this configuration still has a lower energy than if the atoms were held farther apart.

If the atoms are crowded together so that their inner closed shells are touching, the neighboring nuclei will be inside what was the outer electron cloud for the free atom. In such a case the outer electron is attracted by the

**Figure 28-3.** (a) *Contours of electron charge density averaged over valence electron states in germanium crystal.* (b) *Same as* (a) *but for the next excited state.* [*Courtesy Prof. M. L. Cohen.*]

neighboring nuclei, which both increases its binding energy and spreads out its size even more. This permits it to be near even more remote neighbors, which in turn "pull out" the electron cloud even more. The end result is that each outer electron wave function gets uniformly spread out over the entire crystal!

We can begin to see that quantum theory provides a reasonable explanation of why metals conduct electricity and why other substances do not (or almost do not). That metals contain at least one "free" electron per atom is due in part to the wave nature of the electrons. These "free" or conduction electrons are not bound to any particular atom and are free to flow anywhere in the metal.

In ionic and covalent crystals the outer electrons are bound or localized; hence these crystals generally do not conduct electricity. They are called insulators. The fact that pure metallic crystals can have free electrons should be considered as a large-scale quantum mechanical phenomenon. Classically each electron would belong to its own atom.

## 28-2 Free Electron Theory of Metals

To a first approximation, the attractive forces of the nuclei on an outer electron can be averaged out to correspond to a uniform attractive potential energy, the magnitude of which we shall call $U_0$. A plot of this averaged potential energy is the potential well shown in Figure 28-4. Each outer electron is described by a standing wave confined to this potential well. We now see that the hypothetical example in Chapter 25 of an electron trapped in a box is not so hypothetical after all. In this approximation we can consider a metal of volume $V$ as a box of volume $V$ holding $n$ free electrons. Because of the Pauli exclusion principle only two of these electrons are permitted to occupy each of the states specified by Eq. 25-1. All $n$ of these electrons try to crowd into the lowest energy states, forming what is known as a Fermi gas. Such a gas has certain interesting, nonclassical properties first pointed out by Enrico Fermi. The $n$ electrons will fill up all the energy states from the lowest state to a state of kinetic energy $K_f$ called the Fermi level. The value of $K_f$ should depend on $n$ and the volume $V$. As we shall now show, it depends only on the ratio $\mathfrak{N} \equiv n/V$, the number of conduction electrons per unit volume. The kinetic energy of a particle in a box is called its Fermi energy.

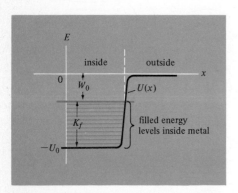

**Figure 28-4.** *Solid curve is the approximate potential energy of an outer electron as it crosses the surface of a metal. Fermi energy $K_f$ and work function $W_0$ are shown.*

The calculation is made easier by choosing a cubical box of volume $L^3$. If a single electron is put into this box it will quickly radiate down to the lowest energy level. A second electron would also drop down to this lowest energy level. Because of the Pauli exclusion principle, a third electron must occupy the next highest or second energy level. A fifth electron will occupy the third energy level and so on. If the box contains a total of $n$ electrons, the Fermi level has the kinetic energy of the $(n/2)$-th energy level. We saw on page 566 that the energy levels in a three-dimensional box depend on three quantum numbers $n_x$, $n_y$, $n_z$ where

$$n_x = \frac{2LP_x}{h} \quad n_y = \frac{2LP_y}{h} \quad n_z = \frac{2LP_z}{h} \quad P^2 = \frac{h^2}{4L^2}(n_x^2 + n_y^2 + n_z^2).$$

Let $P_f$ stand for the electron momentum at the Fermi level. Then $P_f$ is the maximum possible value of $P_x$ for any of the $n$ electrons in the box. The highest value of $n_x$ still corresponding to a filled state is then $(n_x)_f = 2LP_f/h$. The same is true for $n_y$ and $n_z$. To obtain the total number of states that are filled we have to count up all possible combinations of three integers $n_x$, $n_y$, and $n_z$ up to the above limit of $2LP_f/h$. This counting would be extremely tedious without the aid of Figure 28-5. In this figure, for convenience, we number at 1-cm intervals $n_x$ along the x axis, $n_y$ along the y axis, and $n_z$ along the z axis. Then each possible state (or combination of the three integers) is represented as a unique point in space. These points make up a lattice of 1-cm cubes. Note that there are the same number of 1-cm cubes as of these points. Now use the fact that the filled states will be inside a radius $R = 2LP_f/h$. The total number of filled states will then numerically be the number of 1-cm cubes enclosed by the spherical surface shown in Figure 28-5. Since this volume is one-eighth of an entire sphere,

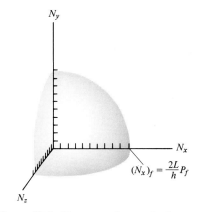

**Figure 28-5.** *Upper quadrant of sphere of radius $R = (2L/h)P_f$.*

$$\text{total number of states} = \frac{1}{8} \times \frac{4}{3}\pi R^3 = \frac{1}{6}\pi \left(\frac{2LP_f}{h}\right)^3 = \frac{4\pi L^3 P_f^3}{3h^3}$$

Since there are two electrons per state, the total number of electrons is

$$n = \frac{8\pi V P_f^3}{3h^3} \qquad (28\text{-}1)$$

where $V = L^3$ is the volume of the box.

Solving for $P_f$ gives

$$P_f = \left(\frac{3h^3}{8\pi}\frac{n}{V}\right)^{1/3} \qquad \text{(maximum Fermi momentum)} \qquad (28\text{-}2)$$

Substitution into $K_f = P_f^2/2m$ gives

$$K_f = \frac{h^2}{8m}\left(\frac{3}{\pi}\mathfrak{N}\right)^{2/3} \qquad \text{(Fermi level)} \qquad (28\text{-}3)$$

The result is independent of the particular shape or volume of the piece of metal. It depends only on how tightly the free electrons are crowded together.

SEC. 28-2/FREE ELECTRON THEORY OF METALS

**Example 1.** The density of lithium is 0.534 g/cm³. What is the maximum Fermi energy of a conduction electron in lithium in electron volts?

ANSWER: Lithium has only one outer electron, hence $\mathfrak{N}$ is the number of atoms per cubic centimeter. Since the atomic weight of lithium is 6.94

$$\mathfrak{N} = (0.534 \text{ g/cm}^3)(6.02 \times 10^{23} \text{ atoms/mole})\left(\frac{1}{6.94 \text{ g/mole}}\right)$$

$$= 4.63 \times 10^{22} \frac{\text{atoms}}{\text{cm}^3} = 4.63 \times 10^{28} \frac{\text{electrons}}{\text{m}^3}$$

Using Eq. 28-3 we have

$$K_f = \frac{h^2}{8m}\left(\frac{3}{\pi} \times 4.63 \times 10^{28}\right)^{2/3} = 7.55 \times 10^{-19} \text{ J}$$

$$= 4.7 \text{ eV}$$

***Example 2.** What is the average kinetic energy $\bar{K}$ in terms of $K_f$? The total Fermi energy in a sample containing $n$ conduction electrons would be $n\bar{K}$.

ANSWER: If $dn$ is the number of electrons in the momentum interval $dp$, by definition the average kinetic energy is

$$\bar{K} = \frac{\int K(p) \, dn}{\int dn}$$

Differentiation of Eq. 28-1 gives

$$dn = \frac{8\pi V p^2 dp}{h^3}$$

$$\bar{K} = \frac{\int K p^2 dp}{\int p^2 dp} = \frac{\int \left(\frac{p^2}{2m}\right) p^2 dp}{\left[\frac{p^3}{3}\right]_0^{p_f}} = \frac{1}{2m} \frac{\left[\frac{p^5}{5}\right]_0^{p_f}}{\frac{1}{3}p_f^3} = \frac{3}{5} K_f$$

*__Example 3.__ If the electron density is extremely high (as in some astronomical bodies), then nearly all the electrons will be extremely relativistic; that is, $E = K = pc$. Then what will be $K_f$ in terms of $\mathfrak{N}$ and what is $\bar{K}$?

ANSWER: The relativistic Fermi energy is obtained by multiplying both sides of Eq. 28-2 by $c$:

$$K_f = cP_f = c\left(\frac{3h^3}{8\pi}\mathfrak{N}\right)^{1/3}$$

$$\bar{K} = \frac{\int K\, dn}{\int dn} = \frac{\int (cp)p^2\, dp}{\int p^2\, dp} = c\frac{p_f^4/4}{p_f^3/3} = \frac{3}{4}K_f$$

The approximate potential energy of an electron in the edge region of a metal is shown in Figure 28-4. We define zero energy as the energy of a free electron at rest outside the metal. The energy levels of the electron Fermi gas are indicated as fine horizontal lines starting at $-U_0$ and going up an energy interval $K_f$ from the bottom of the potential well. The minimum energy required to remove an electron from the metal is then $U_0 - K_f$. By definition this is the work function $W_0$ defined in the section on photoelectric effect,

$$W_0 = U_0 - K_f$$

Although an electron at the Fermi level has kinetic energy equal to $K_f$, its total energy is

$$E_f = K_f + U = K_f + (-U_0) = -(U_0 - K_f)$$

$$E = -W_0$$

Hence the Fermi level is at $-W_0$ on the energy scale. Actually, one has a sharply defined work function only at absolute zero. At a temperature $T$ Kelvin, the electrons are in thermal equilibrium so that they will have some thermal energy in addition to their Fermi energy. As shown in Chapter 13 the average thermal energy per particle in a classical gas is $\frac{3}{2}kT$. In a Fermi gas, only the particles with kinetic energies close to $K_f$ can have thermal energy. Thus there will be some electrons with kinetic energies a little higher than $K_f$; at room temperature $kT$ is 0.025 eV, whereas $K_f$ and $W_0$ are on the order of several electron volts.

### Contact Potential

Whenever two dissimilar metals are joined, a potential difference called the contact potential appears between them. Now that we know about Fermi levels, we can explain this phenomenon with the use of potential well diagrams. Consider two different metals A and B as in Figure 28-6a. The Fermi levels of A and B are at $-2$ and $-3$ eV respectively, and the electron potential energies inside the two metals are $-4$ and $-6$ eV respectively—all with respect to the energy of an electron at rest just outside the metal. Figure 28-6b shows the situation just as the two metals are first brought into contact. Now electrons in A are free to move to B, where lower energy states are available. But as electrons move into B, it rapidly acquires a negative charge with respect to A. Now more work must be done to bring a negative electron to the negatively charged metal B; in other words, the entire potential energy diagram of B is raised with respect to A. This process continues until the Fermi levels meet as shown in Figure 28-6c. This equilibrium situation is achieved after a very small fraction of the conduction electrons have moved from A to B. As seen in Figure 28-6c, the potential energy difference is $\Delta U$ which must be equal to the initial energy difference of the Fermi levels; that is, if the Fermi levels of two metals differ by 1 V, then, when brought together, there will be a potential difference of 1 V between the two metals.

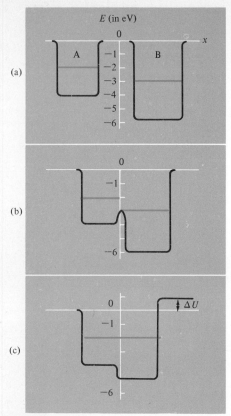

**Figure 28-6.** *When two dissimilar metals are joined, electrons will flow until the Fermi levels line up as in (c).*

## 28-3 Electrical Conductivity

We have seen that in a perfect metallic crystal lattice the outer electrons behave as free electrons in a box. Since these electrons can carry electrical current, we would expect the electrical resistance of a perfect metal to be zero. However, real metals have impurities and lattice imperfections. A free electron can interact and lose energy to imperfections and impurities. The electrical resistance in ohms depends on the mean free path for electron collisions with the imperfections and impurities. From this theory of electrical resistance, we can easily derive Ohm's law as was done in Chapter 17 on page 339. Ohm's law states that the resistance is completely independent of the value of the current and only dependent on the temperature. According to Eq. 17-5, the resistivity

$$\rho = \frac{m\bar{u}}{e^2 \mathfrak{N} L}$$

The increase of electrical resistance with temperature is easily seen according to this theory of electrical conduction. An intrinsic source of lattice imperfection is the vibrational motion of the atoms due to the fact that they are not at absolute zero. Hence the collision mean free path $L$ should decrease with temperature and we have predicted that the resistance of a pure metal should increase with increased thermal motion of the atoms. The theory predicts the resistance of a pure crystal should approach zero as the temperature approaches absolute zero. This prediction agrees with experiment.

## *Superconductivity*

The fact that a pure metal can have zero resistance or infinite conductivity at absolute zero should not be confused with a different quantum phenomenon called superconductivity. Superconductivity is infinite conductivity at temperatures several degrees above absolute zero. Actually quite a few metals and metallic alloys have this strange property of superconductivity. For each superconductor there is a critical temperature above which the metal is a normal conductor and below which it is a superconductor. The highest known critical temperature is about 20 K.

Once a circular current has been started up in a superconductor, it should keep going by itself until the cooling system breaks down. Such currents have kept going by themselves for years in the laboratory. The quantum mechanical explanation of superconductivity is one of the current problems in theoretical solid state physics. Recently, considerable progress has been made in the understanding of this most remarkable phenomenon.

A brief qualitative description of the theory goes as follows. Below a certain temperature, the disturbance of the lattice by a conduction electron is greater than the thermal motion of the lattice disturbing the electron. The disturbance of the lattice by electron A will show an effect on the motion of electron B. The net effect is an effective attractive force between electrons A and B, which in some materials is greater than the electrostatic repulsive force. Hence, if both electrons are set in motion in the same direction (a net current), this will be the lowest energy state available for the electrons and they must stay in that state because there is no lower state available for them; consequently, there will be a permanent net electron current in their direction of motion.

There are several practical applications of superconductivity presently under development. We shall discuss three of them: (1) strong-field magnets, (2) low-loss power transmission, and (3) high-speed mass transit systems.

Many electromagnets have been built using superconducting coils.

They can give fields up to ~100,000 gauss (10 tesla) with zero electrical power loss in the coils. The only power needed is for the refrigeration system.

As shown on page 392, a major power loss is the $I^2R$ loss in transmission lines. Superconducting power lines would have no electrical power loss for dc current because $R = 0$. A superconducting cable designed to transmit $4 \times 10^9$ W of ac power is shown in Figure 28-7. Such cable should be less expensive than conventional underground cable and have less power loss. If ever a desert region of the United States is to produce power for the entire country using solar energy, superconducting power lines would be a necessity for such long distance transmission. When the cost of solar power becomes competitive with coal and nuclear power, a total system using solar power with superconducting transmission may be the ultimate answer. There are associated economic questions, such as the large quantity of liquid helium needed for cooling. At present, helium gas is being exhausted into the atmosphere and it ultimately escapes from the earth (its molecular velocity is comparable to the velocity of escape). The earth's supply of helium is quite limited and at present there is no adequate helium conservation program. If government planners would look far enough into the future they probably would be conserving helium now.

Last, but not least, we discuss the application of superconductivity to mass transportation. High speed ground transportation requires floatation

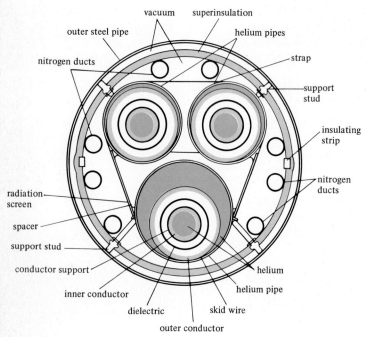

**Figure 28-7.** *Cross section of a semiflexible superconducting ac cable. This system has an outer diameter of 4.7 cm, and it is rated at 275 kV. The design is representative of those being developed by several laboratories. Such a single cable could transmit up to $4 \times 10^9$ W of electrical power.*

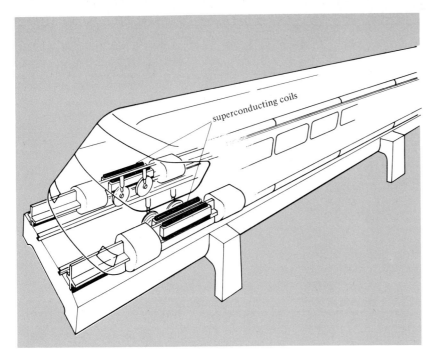

**Figure 28-8.** *Artist's concept of the AEG–BBC–Siemens superconducting train, showing the location of the superconducting coils for levitation and guidance as well as the coils for the linear induction motor. Note the use of a continuous track.* [*Courtesy of Siemens Research Laboratories, Erlangen, Germany*]

of passenger cars above the tracks. Then, speeds up to 500 km/h are possible. Suppose each passenger car had superconducting coils near a conducting "track." Then when the car reached a certain speed the induced current in the track would repel the superconducting coil with enough force to "float" the car. Such systems are presently under development, mainly in Japan and Germany (see Figure 28-8). Efficient, high-speed mass transportation systems are needed to replace unnecessary waste of the limited world supply of petroleum by private automobiles.

## 28-4 Band Theory of Solids

So far in our description of electron states in a metal we have ignored the periodic variation of the electron potential energy due to the atomic cores. We have been assuming the floor of the potential well in Figure 28-4 is flat rather than having periodic dips. If one takes into account such a periodic variation, one finds that the possible energies are limited to certain energy bands. The reason for this is illustrated in the following examples.

For our first example consider two one-dimensional square wells of width $x_0$ and depth $U_0$. This simulates the effect of two atomic cores in a row. We shall see that what was originally the ground state energy level for each well by itself becomes a "band" of two energies. And when there are $n$ square wells in a row we get a band of $n$ energy levels corresponding to

**Figure 28-9.** (a) *The first two energy levels of a square well of width $2x_0$ made up of two contiguous wells of width $x_0$ each. The corresponding standing waves are shown. $U_0 = 50\hbar^2/mx_0^2$. (b) The two single wells in (a) are split apart by a distance $\frac{1}{10}x_0$. Note that the two energy levels become closer together. (c) When the two single wells are far enough apart the two energy levels become almost the same and the wave function in the region of each well is almost the same as the ground state wave function for the single well.*

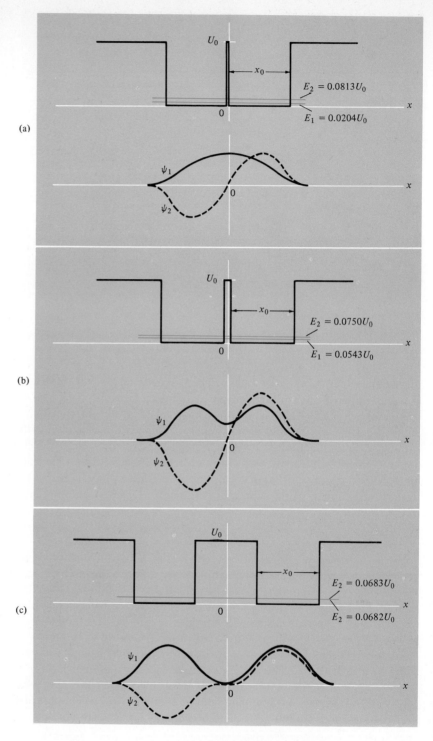

the ground state. According to Eq. 25-14 the energy levels in a single square well are approximately

$$E_n \approx \frac{n^2}{L^2}\frac{h^2}{8m} \quad \text{where } L \text{ is the width}$$

Now in Figure 28-9(a) the two wells are joined into one well of double width $L = 2x_0$ and

$$E_n \approx \frac{n^2}{4x_0^2}\frac{h^2}{8m}$$

$$E_1 \approx \frac{1}{4}\frac{h^2}{8mx_0^2} \quad \text{and} \quad E_2 \approx \frac{h^2}{8mx_0^2}$$

In Figure 28-9(b) the "double width" well is split into two nearby single-width wells. Note that the energies corresponding to $\psi_1$ and $\psi_2$ have come closer together. In Figure 28-9(c) the two single wells are so far apart that the standing wave shown in each well is essentially the same as the lowest order wave in one well by itself; hence

$$E_1 \approx E_2 \approx \frac{h^2}{8mx_0^2}$$

We have just shown that as the two wells in view (c) are moved closer together, the two energy levels corresponding to the $\psi_1$ and $\psi_2$ shown in Figure 28-9 move farther apart. What was originally the lowest energy state in a single well results in two energy states when there are two wells and, the closer the two wells, the wider the spacing between the two energy levels.

Next we repeat the preceding analysis for four wells in a row, as shown in Figure 28-10. View (a) can be treated as a single well of fourfold width $L = 4x_0$. Then the energy levels are

$$E_n = \frac{n^2}{16}\frac{h^2}{8mx_0^2}$$

In view (b), where the wells are now slightly separated, the band of four energies is less wide. In view (c) the wells are so far apart that the four corresponding energies are almost the same as the single-well ground state energy. It is easy to generalize to $n$ wells in a row. Then there would be a band of $n$ energy levels in the neighborhood of the original level. Increasing $n$ does not change the width of the band as long as the spacing between wells is kept the same. In a typical solid $n$ can be on the order of $10^{23}$, so the band of energy levels can be treated as a continuum.

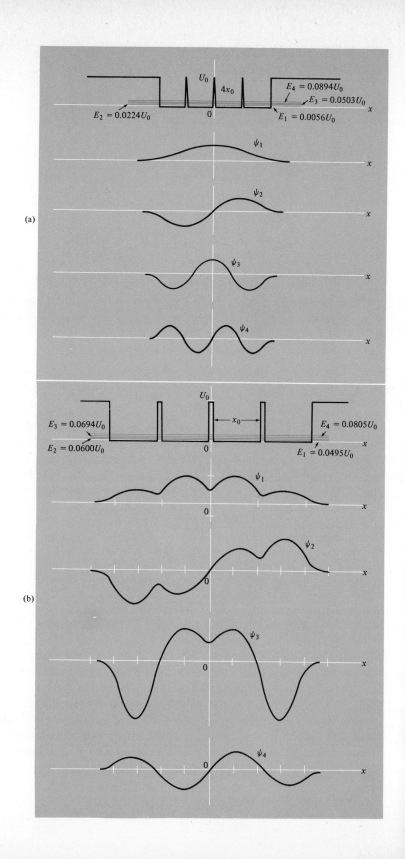

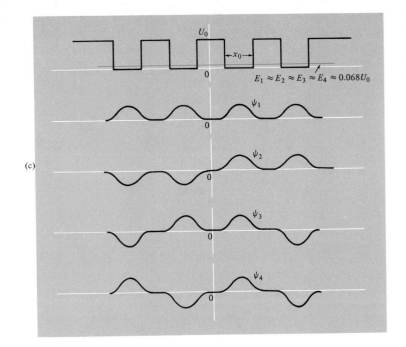

**Figure 28-10.** (a) *Four contiguous single wells making a well of width $4x_0$. The first four energy levels are shown.* (b) *The separation between each well is $x_0/10$.* (c) *The separation between each well is $x_0$. Now the four energy levels are almost the same.*

In a real solid the one-dimensional square well must be replaced with the three-dimensional atomic potential well. Then each atomic state results in a band of $n$ corresponding energies as $n$ atoms are pushed together. The closer they are pushed together, the wider the bands, as illustrated in Figure 28-11.

---

**Example 4.** Suppose there are 100 square wells in a row of width $x_0 = 1$ Å and depth $U_0 = 378$ eV. (a) If the spacing between the wells is also 1 Å, what will be the kinetic energies of the first 100 energy levels? (b) If the separation between wells is 0.1 Å, what will be the first, fiftieth, and one-hundredth energy levels?

Answer: The depth of the well is the same as in Figure 28-10 where $U_0 = 50\hbar^2/mx_0^2 = 378$ eV. Part (a) of this problem has the same spacing as in Figure 28-10(c) where the ground state band (the first 100 energy levels) is at $K \approx 0.068\ U_0 = 25.7$ eV. Part (b) corresponds to Figure 28-10(b). Now the band runs from $0.0495\ U_0$ to $0.0805\ U_0$ or from 18.7 to 30.4 eV. Hence $K_1 = 18.7$ eV and $K_{100} = 30.4$ eV. $K_{50}$ will be about halfway between them, so $K_{50} \approx 24.5$ eV.

**Figure 28-11.** *Energy bands in sodium as functions of the internuclear spacing R. The 2s band is at $-63.4$ eV, and the 1s band is at $-1041$ eV. Both of these bands are even narrower than the 2p band.* [*From J. C. Slater,* Phys. Rev., **45**:794 *(1934)*]

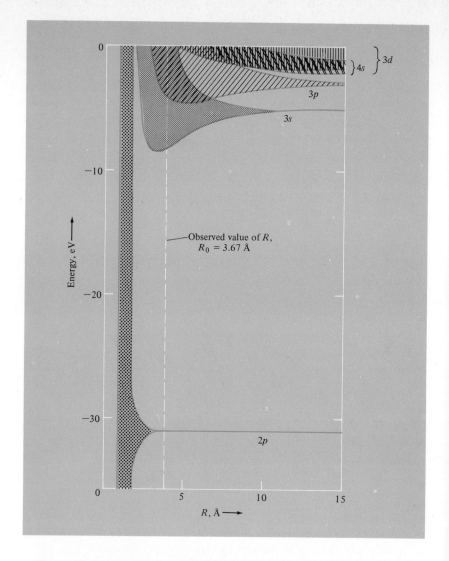

The wells are so close to each other in metallic binding that the outer electron is in an energy band which overlaps with other bands. This is the case for the sodium metal energy bands shown in Figure 28-11. Then there are unlimited empty energy states available to the outer electron as in our free electron model of metals.

In a covalent solid like germanium or silicon the atomic well spacing is such that the valence electron energy bands do not overlap. It turns out that for silicon there is an energy gap of 1.09 eV between the filled band containing the valence electrons and the next highest unfilled band. In the case of germanium the corresponding energy gap is 0.72 eV. In order for silicon or germanium to conduct an electric current, there must be some

**Figure 28-12.** *Schematic comparison of valence and next excited state bands of (a) conductor, (b) semiconductor, and (c) insulator. In (c) $E_{gap} \gg kT$.*

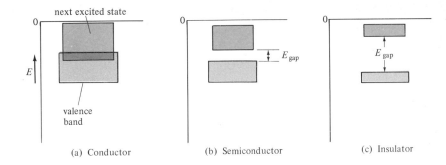

(a) Conductor  (b) Semiconductor  (c) Insulator

electrons in the unfilled energy band. At room temperature there will be a relatively small number of thermally excited electrons in the unfilled conduction band. The conductivity of silicon and germanium is thus much less than that of a normal metal. Hence they are called semiconductors. If the energy gap of a solid is too large to reach by thermal excitation, it is called an insulator as illustrated in Figure 28-12. At absolute zero a pure crystal of an insulator or even of a semiconductor would have infinite resistance.

The square of the average wave function corresponding to the valence band of germanium is shown in Figure 28-3(a). The square of the average wave function corresponding to the conduction band is shown in Figure 28-3(b).

## 28-5 Semiconductor Physics

In this section, not only shall we discuss the basic nature of semiconductors, but we shall try to give some idea of the immense practical applications which have revolutionized electronics and vast areas of present-day technology.

When an electron in the valence band of germanium is thermally excited, it leaves an empty state in the valence band and creates a filled state in the conduction band. The empty state is called a **hole**. Such an excited electron and its resulting hole are symbolized in Figure 28-13. In the presence of an external electric field a neighboring electron in the valence band will move into the hole, leaving a new hole, which is then filled by the next neighboring electron, and so on. Thus the hole moves along in the opposite direction to the electrons and behaves as a positive carrier of charge. Each hole can be considered as a positive conduction electron.

If one goes deeper into statistical mechanics than this book, it can be shown that the probability of an electron near the top of the valence band being thermally excited into the conduction band is proportional to

$$e^{-E_{gap}/kT} \qquad (28\text{-}4)$$

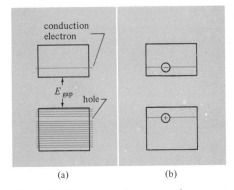

**Figure 28-13.** (a) *An electron in the valence band has been excited to the conduction band, leaving a hole.* (b) *An alternate representation of the situation in (a).*

At room temperature $kT = 0.026$ eV and $E_{gap}/kT \approx 29$ for germanium. In spite of the small number of conduction electrons at room temperature, germanium and silicon are much better conductors than an insulator such as diamond which has $E_{gap} \approx 7$ eV. By the same token, germanium and silicon will be insulators at $\sim 30$ K.

A thermally excited electron in a semiconductor will eventually collide with a hole and drop back down into the valence band, so the rate of loss of conduction electrons is proportional to $(N^-N^+)$, where $N^-$ is the number of thermally excited electrons and $N^+$ is the number of holes. As stated in the preceding paragraph, the rate of pair creation is proportional to $\exp[-E_{gap}/kT]$. When in equilibrium these two rates must be equal, so

$$(N^-N^+) \propto \exp\left[-\frac{E_{gap}}{kT}\right]$$

For a pure semiconductor $N^- = N^+$, and we have

$$N^- \propto \exp\left[-\frac{E_{gap}}{2kT}\right]$$

Since the conductivity $\sigma$ is proportional to $N^-$, there will be a very rapid increase in conductivity with temperature.

**Example 5.** A sample of silicon is raised from 0°C to 10°C. By what factor will the conductivity increase?

ANSWER: The ratio

$$\frac{N^{-\prime}}{N^-} = \frac{\exp[-E_{gap}/2kT']}{\exp[-E_{gap}/2kT]} = \exp\left[\frac{E_{gap}}{2k}\left(\frac{1}{T} - \frac{1}{T'}\right)\right]$$

Since $E_{gap} = 1.1$ eV $= 1.76 \times 10^{-19}$ J,

$$\frac{N^{-\prime}}{N^-} = \exp\left[\frac{1.76 \times 10^{-19}}{2(1.38 \times 10^{-23})}\left(\frac{1}{273} - \frac{1}{283}\right)\right] = 2.28$$

### The Thermistor

We see that for a sample of pure silicon the resistance decreases more than a factor of 2 for each 10 degree temperature increase. Hence pure silicon can be used as a very sensitive electronic temperature sensing device. Such a device (consisting of pure semiconductor) is called a thermistor.

## Doping of Semiconductors

If a small amount of arsenic (valence 5) is added to molten germanium as a crystal is being grown, the arsenic will fit into the lattice structure using four of its five valence electrons to form the necessary four covalent bonds. The left-over fifth electron will be in an energy state just below the conduction band and a small amount of thermal agitation will raise it to the conduction band. Hence there will be almost as many conduction electrons from this source as there are arsenic atoms. Typically this is a much larger number of conduction electrons than are supplied by thermal excitation from the valence band. Such a semiconductor is called *n*-type, where *n* stands for negative carriers.

Germanium can also be doped with gallium (valence 3). In this case the typical gallium atom in the lattice will capture a neighboring electron in order to complete the four valence bonds. Hence the typical gallium atom creates a hole and we have a *p*-type semiconductor where *p* stands for positive carriers.

## p–n Junctions

If a *p*-type and an *n*-type semiconductor are placed in intimate contact, some electrons will flow from the *n* to the *p* type and holes from the *p* to the *n* type until the Fermi levels are equalized as explained on page 616. The *p* type after receiving the extra electrons will become more negative and the *n* type more positive. There will be a contact potential equal to the original difference in Fermi levels which is close to $E_{\text{gap}}$. (As shown in Figure 28-14 the Fermi levels are close to the highest energy electron states which are occupied at absolute zero.) If we call the contact potential

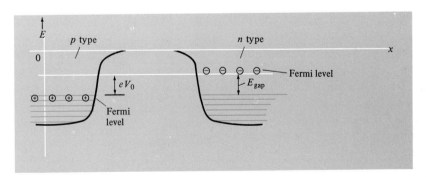

**Figure 28-14.** *Potential energy diagram of p-type and n-type semiconductors before being put into contact. $V_0$ is the difference in potential between the Fermi levels.*

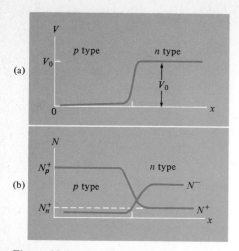

**Figure 28-15.** (a) *Electric potential of the two samples in the previous figure when in close contact as a function of x. A contact potential $V_0$ has developed.* (b) *The x distributions of both positive and negative carriers.*

$V_0$, the electric potential $V$ (not potential energy) will be as shown in Figure 28-15(a) after the two samples are in close contact. Figure 28-15(b) shows the corresponding densities of positive and negative carriers in the two samples. In the next paragraph we shall discuss possible currents due to holes only. The contribution due to negative carriers is discussed later.

Actually there will be a small current $I_0$ of holes flowing in both directions across the surface of contact. The rate of holes flowing to the left in Figure 28-15(b) will be proportional to $N_n^+$, the density of holes in the n-type sample. As they cross the surface, they will be accelerated by the potential difference $V_0$. So $I_{\text{left}} = CN_n^+$ where $C$ is a proportionality constant. Similarly, the current of holes to the right will be proportional to $N_p^+$ times the fraction of holes that can climb the potential hill. As stated in Eq. 28-4, this fraction comes from statistical mechanics and is $\exp[-eV_0/kT]$. So

$$I_{\text{right}} = CN_p^+ \exp\left[-\frac{eV_0}{kT}\right]$$

Since $I_{\text{left}} = I_{\text{right}} = I_0$, we have

$$N_n^+ = N_p^+ \exp\left[-\frac{eV_0}{kT}\right]$$

and

$$I_0 = CN_p^+ \exp\left[-\frac{eV_0}{kT}\right]$$

Now we shall apply an external potential $V$ across the p–n junction as shown in Figure 28-16. Then the n-type will be at a potential $V_0 - V$ with respect to the p type and the current to the right will be

$$I_{\text{right}} = CN_p^+ \exp\left[-\frac{e(V_0 - V)}{kT}\right] = \left\{CN_p^+ \exp\left[-\frac{eV_0}{kT}\right]\right\} \exp\left[\frac{eV}{kT}\right]$$

$$= I_0 e^{eV/kT}$$

The current to the left is still $I_0$ because all the $N_n^+$ still get through; hence the net current is

$$I = I_{\text{right}} - I_{\text{left}} = I_0(e^{eV/kT} - 1) \quad \text{(current in p–n junction)} \quad (28\text{-}5)$$

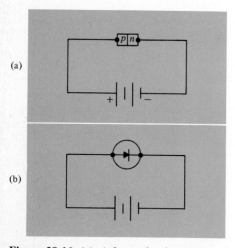

**Figure 28-16.** (a) *A forward voltage V is applied to a p-n junction.* (b) *The more conventional representation of a p-n junction.*

This net current of holes flows into the n-type sample where the holes are eventually annihilated by the conduction electrons. The electrons that are lost in this annihilation will be made up by a current of electrons from the external voltage source shown in Figure 28-16. Equation 28-5 is plotted in

Figure 28-17. We note that for a positive voltage the current is typically many times $I_0$, whereas for a negative voltage (also called a back voltage) the maximum current is $I_0$. A device having this kind of nonlinear response is called a junction diode. The heart of an AM radio receiver is the diode detector, which is usually a $p$–$n$ junction. This application is described in Appendix 28-1.

We have not bothered to discuss the current of conduction electrons due to $N_p^-$ and $N_n^-$ because the analysis is exactly the same and gives the same result (Eq. 28-5). If the negative carriers are also taken into account, $I_0$ is the maximum current that can flow due to both positive and negative carriers in the presence of a back voltage.

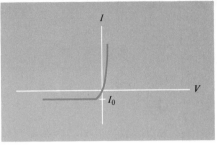

Figure 28-17. *Plot of I versus V from Eq. 28-5. To a first approximation, positive V gives large current and negative V gives almost no current.*

---

**Example 6.** A $p$–$n$ junction is at $0°C$. Its resistance is 10 ohms when a forward voltage of 0.1 V is applied. What is its resistance if the voltage polarity is reversed?

ANSWER: According to Eq. 28-5 the resistance is proportional to $|\exp(eV/kT) - 1|$. Hence the ratio of back to forward resistance for an applied voltage $V_1$ is

$$\frac{R_{back}}{R_{forward}} = \frac{|\exp(eV_1/kT) - 1|}{|\exp(-eV_1/kT) - 1|}$$

where $eV_1 = 0.1$ eV and $kT = (1.38 \times 10^{-23})(273)$ J $= 0.0236$ eV. Then

$$\frac{eV_1}{kT} = 4.237 \quad \text{and} \quad \frac{R_{back}}{R_{forward}} = \frac{|\exp(4.237) - 1|}{|\exp(-4.237) - 1|} = 69.2$$

$$R_{back} = 692 \text{ ohms}$$

---

## Solar Cells

If light shines on the transition region between the $p$- and $n$-type samples, photons will be absorbed by electrons in the valence band and be promoted to the conduction band. Each such photon produces an electron-hole pair. The newly formed holes are pushed by the electric field in Figure 28-15(a) into the $p$-type region and the electrons into the $n$-type region. These additional current carriers can travel around a closed external circuit. Thus light can be directly converted into electrical power. A silicon solar cell behaves as a $\sim\frac{1}{2}$-V battery and can convert sunlight into electrical energy with up to 15% efficiency. Solar cells are used as a source of power for space capsules. They could also be used on earth to generate electricity from solar energy if their cost could be made competitive with conventional power plants (see Example 1 of Chapter 1). For this reason it is important to develop rapid, low-cost fabrication of solar cell semiconductors and there are now some serious efforts in that direction.

## Photodiodes

If a back voltage is applied to a solar cell, the normally weak current $I_0$ will be increased manyfold when light is used to produce extra current carriers. The photocurrent will be proportional to the rate of photons incident on the photocell. Such a device is very sensitive to the intensity of light reaching it and is used in various applications to detect a change in light intensity.

## Light-Emitting Diodes

The pure red spots of light used in digital readout of pocket calculators, wristwatches, and so on, are called LED's or light-emitting diodes. They are merely small junction diodes with a strong enough forward voltage applied so that the conduction electrons in their typical collisions create electron-hole pairs. Then each time an electron recombines with a hole, a photon of energy $E_{\text{gap}}$ is emitted. In order to get red light $hf \approx E_{\text{gap}}$ must be about 2 eV. This is possible to achieve by using gallium arsenide crystals. If operated in a sufficiently high current region, LED's are close to 100% efficient in conversion of electrical energy to visible light. A closely related device is the solid state laser.

## The Transistor

A *pnp* transistor is illustrated in Figure 28-18. It can be thought of as the same *p–n* junction shown in Figure 28-16 with an extra *p*-type region added on the right-hand side (called the collector). In Figure 28-18 there is a forward voltage $V_b$ applied to the "diode part," which causes a large current of holes from the left-hand *p*-type region (called the emitter) to the *n*-type region (called the base). The "trick" to the transistor is to make the base so thin that most of the holes diffuse across the thin base into the collector region. The collector is made more negative than the base in order to encourage the positive carriers (holes) to go to the collector. In a typical transistor only ~1% of the emitter current leaves the base terminal

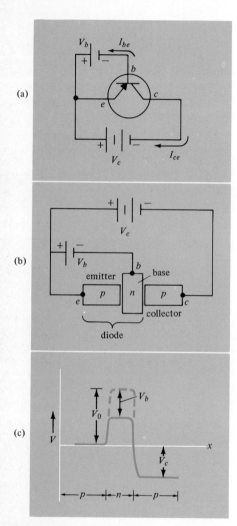

**Figure 28-18.** (a) *A conventional schematic representation of a transistor.* (b) *The connections to the n-type and p-type semiconductors.* (c) *A plot of the potential inside the transistor. If $V_b$ were removed, the contact potential between base and emitter would be $V_0$.*

*b*. The other 99% leave the collector terminal *c*. The ratio of collector current to base current is called β, the current amplification factor;

$$\beta \equiv \frac{I_c}{I_b} \quad \text{(current amplification factor)}$$

For a typical transistor with $\beta = 100$, a small input current at the base can control a 100 times larger output current at the collector. Transistors can be used to amplify small signals into large signals.

For example, $I_{be}$ in Figure 28-18 could be a weak current picked up by a radio antenna. Then the current $I_{ce}$ would have the same time variations but amplified by a factor of 100. *npn* transistors have the same characteristics as *pnp*. The main difference is that the major charge carriers in *npn* transistors are electrons rather than holes.

### Integrated Circuits

Not only has electronics been revolutionized by the host of new, inexpensive solid state devices, but there has been a second revolution caused by the integrated circuit (IC). We would go too far afield to discuss the various ingenious and sophisticated methods for construction of junction diodes and junction transistors. However, the concept of integrated circuits is so exciting and almost unbelievable that it is worth describing briefly. A pocket electronic calculator requires literally thousands of transistors connected together in complicated circuits. These transistors and all their interconnections are contained on a thin wafer of about 1 cm² area. The record so far is 20,000 solid state devices crowded into an area of about 0.4 cm². (This is comparable to the density of neurons in the human brain.)

The "wiring" is so small that a microscope would be needed to see it. Thin layers of crystal are grown on the wafer using optical masks (the circuits are photoreduced and projected onto a photosensitive mask). If such a large scale integrated circuit contains 10,000 transistors and sells for $10, the cost per transistor is one tenth of a cent ($0.001). This is about three orders of magnitude cheaper than the vacuum tube technology it has superceded. It is also an improvement of several orders of magnitude in weight, size, reliability, and power consumption.

### Other Devices

There are a host of other solid state devices that have valuable practical applications. This book, which deals with basic principles, is not the proper place to discuss such detailed devices. All we can do here is list some of them: the tunnel diode, the silicon controlled rectifier (SCR), the zener diode, the field effect transistor (FET), the solid state laser, thermoelectric devices, and so on.

## 28-6 Superfluidity

Another strange quantum mechanical phenomenon that occurs near absolute zero is the superfluidity of liquid helium. As helium gas is cooled down it liquefies at 4.2 K. As the liquid is cooled down further, it suddenly changes its properties at 2.2 K. Then large-scale phenomena occur that are completely contrary to common experience. For example, a partially filled vessel that is open at the top will quickly empty itself of this strange form of liquid helium (called superfluid helium II). The explanation is that the liquid crawls up the inside surface of the vessel (no matter how tall it is) over the rim and down the outside. For the same reason the reverse phenomenon also occurs (see Figure 28-19). If an empty glass is partially immersed, the helium will quickly creep up the glass as shown until the beaker is filled to the same level. As the temperature is decreased below 2.2 K, the amount of superfluid increases. At absolute zero it would be all superfluid.

Another strange property of the pure superfluid is that it cannot exert forces on anything. A high pressure firehose shooting a stream of this liquid could not even knock over a coin balanced on edge. The liquid would freely flow around the coin without exerting any net force on the coin. The superfluid phase of liquid helium has zero viscosity. But why is the viscosity zero? Like superconductivity, the peculiar properties of liquid helium are actively under study. Considerable progress in the theoretical understanding has been made. Helium atoms do not obey the Pauli exclusion principle because they have zero spin. Any number of spin zero particles can be in the same state and obey what is called Bose-Einstein statistics. All the helium atoms in the superfluid state are in the ground state. In order to have nonzero viscosity they would have to make transitions to other states and the energy is not available.

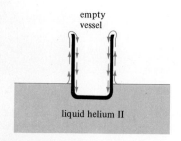

**Figure 28-19.** *Arrows represent surface film creep of liquid helium II into empty vessel. The surface acts as a siphon.*

## 28-7 Barrier Penetration

### Thermionic Emission

If an electric field is applied that tends to pull electrons away from a metal surface, it is found that a steady current of electrons will actually leave the metal. The potential energy due to a uniform field $E$ is $U = -eEx$. The combined potential energy curve seen by a conduction electron is shown in Figure 28-20(a). We can see that those few electrons that have thermal energy greater than $W'$ can escape from their metallic prison. This phenomenon is called thermionic emission. Since the probability that an electron will have thermal energy $W'$ is proportional to

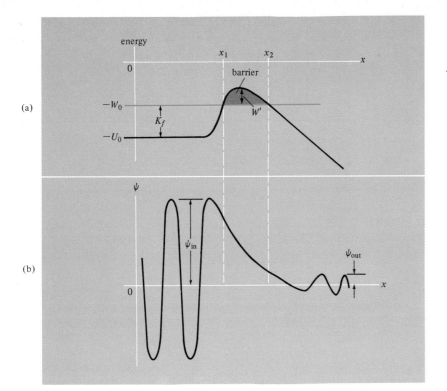

**Figure 28-20.** (a) *The same potential energy as in Figure 28-4, except for external electric field.* (b) *The corresponding wave function of an inside electron at the Fermi level.*

$\exp(-W'/kT)$ we would expect that a small increase in temperature would cause a large increase in electron emission. This is why the cathodes in electronic vacuum tubes are heated. But even if the cathodes are cooled to absolute zero, still some electrons are emitted! This is a quite different phenomenon and is discussed in the next paragraph.

## Field Emission

It is observed that a smaller electron current is still emitted from a cathode even at very low temperatures where no electrons can have thermal energies as high as $W'$. This phenomenon is called field emission and is an example of an important quantum mechanical phenomenon that blatantly violates classical physics. The striking phenomenon referred to is the quantum mechanical penetration of a potential barrier. We first encountered barrier penetration when we studied the square well with finite walls in Chapter 25. There we saw that the wave function penetrated exponentially into the classically forbidden region. In the present example we have a potential barrier of height $W'$ (see Figure 28-20). Classically an electron of kinetic energy $K_f$ inside the metal would have zero kinetic

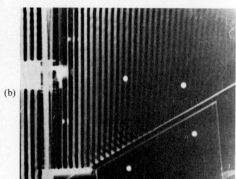

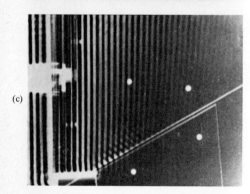

**Figure 28-22.** *Barrier penetration in a ripple tank.* (a) *Waves are totally reflected from a gap of deeper water. As this gap is narrowed in* (b) *and* (c), *a transmitted wave appears. The transmitted wave increases in intensity as the gap decreases.* [*Courtesy Educational Development Center.*]

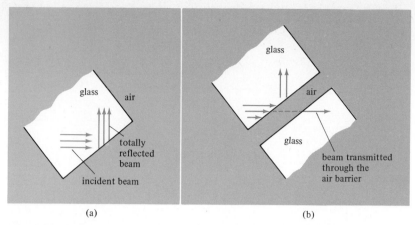

**Figure 28-21.** *Barrier penetration of light. When second plate of glass is brought near the first, some of the light can then escape from the first into the second.*

energy at the position $x_1$. The attractive potential of the metal would then pull it back in. Classically no electron could ever penetrate the slightest amount into the barrier. In the region between $x_1$ and $x_2$ such an electron would have negative kinetic energy, which is impossible. However, we know from the Schrödinger equation that the electron wave could still exist in this region. It must curve away from the $x$ axis as shown in Figure 28-20(b). Note that in this figure there is some probability of finding the electron outside the metal. According to quantum mechanics the chance that a given electron gets through the barrier must be $\psi_{out}^2/\psi_{in}^2$ for each time the electron collides with the barrier. A classical example of a potential energy barrier would be a marble rolling inside a bowl. If the marble is released inside the bowl with its center at the height of the rim, it will roll back and forth and never get out. However, according to quantum mechanics there is an extremely small probability (about one chance in $10^{(10^{29})}$) that the marble will escape.

Actually there is an example of barrier penetration in classical physics. Since the Schrödinger equation is of the same form as the wave equation for light or water waves, we might expect to find an example of barrier penetration in optics. The surface of a piece of glass presents a barrier to light inside the glass trying to get out. If the light strikes the surface at an angle of incidence greater than the critical angle (angles forbidden by Snell's law), then the light beam cannot get past the barrier and is consequently totally reflected back into the glass as shown in Figure 28-21(a). However, if another piece of glass is brought near (within a wavelength or two) as in Figure 28-21(b), some light will penetrate the barrier and continue on in the second piece of glass. The ripple tank analogy of this is shown in Figure 28-22.

## Summary

In covalent binding of a solid, the valence electron clouds are concentrated in "lobes" connecting the nearest neighbor nuclei. In metallic binding there is no such concentration of the valence clouds—it is as if the valence electrons are spread out over the entire piece of metal. The maximum kinetic energy of such conduction electrons is

$$K_f = \frac{h^2}{8m}\left(\frac{3}{\pi}\mathcal{N}\right)^{2/3}$$

where $\mathcal{N}$ is the number of atoms per unit volume. $K_f$ is called the Fermi level. These conduction electrons are trapped in a potential well of average depth $U_0$. The work function is $W_0 = |U_0| - K_f$.

When dissimilar metals are put into contact, there is a small, quick flow of conduction electrons until the Fermi levels line up. Then there is a contact potential equal to the difference in work functions.

The electrical conductivity is proportional to the collision mean free path for conduction electrons. This mean free path increases as the temperature decreases, and for some conductors (called superconductors) it becomes infinite before the temperature reaches absolute zero.

When there are $n$ identical atoms in a row equally spaced, each energy level of the single atom becomes split into a band of $n$ closely spaced energy levels. In the case of metallic binding, the energy band of the ground state of the valence electrons is either partially filled, or else it overlaps the energy band of a higher energy level.

In the case of semiconductors, the energy band of the valence electrons is filled and separated by an energy gap from the next highest energy band (called the conduction band). The number of conduction electrons thermally excited into the conduction band is proportional to $\exp(-E_{gap}/2kT)$. A semiconductor can be doped to contain extra conduction electrons ($n$ type) or holes ($p$ type). A $p$-$n$ junction consists of $p$ and $n$ types in contact. Then if an external voltage $V$ is applied, the current will be

$$I_0\left[\exp\left(\frac{eV}{kT}\right) - 1\right]$$

A *pnp* transistor can be thought of as a $p$-$n$ junction (emitter-base) with a $p$-type collector fastened to the base. Then a small base current $I_b$ can control a large collector current $I_c$. The current amplification factor $\beta = I_c/I_b$ can be $\sim 100$.

Conduction electrons can be "pulled out" of the surface of a cold piece of metal by applying a strong external electric field. This process of field emission through a potential barrier can be explained in terms of the exponentially decreasing wave function inside the barrier.

# Appendix 28-1  Applications of the *p-n* Junction (Radio and TV)

The heart of a radio or television set is the diode detector. As shown in Figure 28-23 the usual diode detector consists of a resistor $R$ in series with a *p-n* junction diode. As was explained on page 629 the diode's resistance is strongly dependent on the direction of the applied potential difference. In one direction the resistance is almost zero, and in the other direction it is very large. Any circuit element having this property is called a diode. In Figure 28-23 we see that when a positive voltage is applied across the diode resistor combination, the diode is essentially a short circuit and the positive voltage appears across the resistor $R$. But when a negative voltage is applied, the diode resistance is much greater than $R$ and most of the negative voltage appears across the diode rather than the resistor. Although the voltage on $A$ is negative 50% of the time, the voltage on the output $B$ is never negative. The voltage on $B$ can be used to charge up a capacitor. This application of the diode, called a diode rectifier (see Figure 28-24), can convert an ac voltage into a dc voltage. All ac operated radio and television sets have diode rectifiers to convert the ac into dc.

**Figure 28-23.** *The diode rectifier. An ac voltage is applied to the input terminal* **A.** *The output voltage at* **B** *has the negative peaks "clipped" off.*

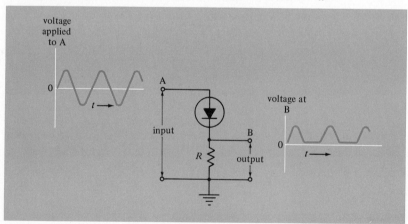

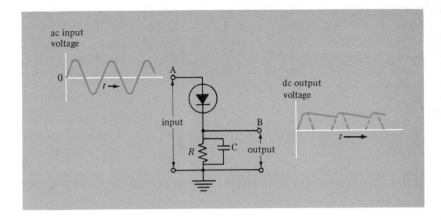

**Figure 28-24.** *Diode rectifier with capacitor C. Without C the output voltage would be the dashed curve.*

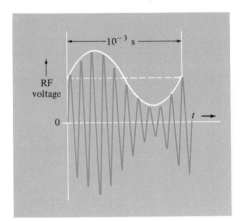

**Figure 28-25.** *The color curve is a plot of RF (radio frequency) voltage as a function of time. In this figure the RF or high frequency electromagnetic wave is amplitude modulated with a pure sine wave or musical note of 1000 Hz (white curve). For the sake of clarity, the RF frequency is greatly reduced.*

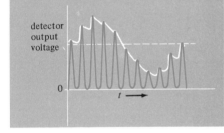

**Figure 28-26.** *White curve is output voltage from diode detector. Note that output voltage is original audio signal with small RF ripple. The color curve represents the RF signal with the negative peaks clipped off.*

## Radio

An AM radio station transmits an electromagnetic wave at a fixed frequency somewhere in the broadcast band. The broadcast band includes frequencies from 0.5 to 1.6 MHz. The amplitude of the electromagnetic wave is modulated or varied with the audio signal that is being transmitted. As an example Figure 28-25 is a plot of an electromagnetic wave modulated with a pure musical note of 1000 Hz. In a radio receiver, this weak signal is picked up by an antenna, amplified by an RF (radio-frequency) amplifier, and then fed to a diode detector. The output voltage from the diode is the original audio signal of 1000 Hz, as shown in Figure 28-26. This audio signal is amplified by an audio amplifier and fed into a loudspeaker. The block diagram of an AM radio receiver is shown in Figure 28-27.

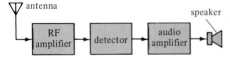

**Figure 28-27.** *Block diagram of AM radio receiver.*

**Figure 28-28.** *The video signal of the letter "N" using nine scan lines.* (a) *The image on an oscilloscope of picture tube using the video signal of* (b) *to control the electron beam intensity.*

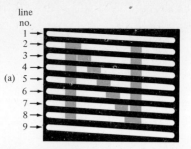

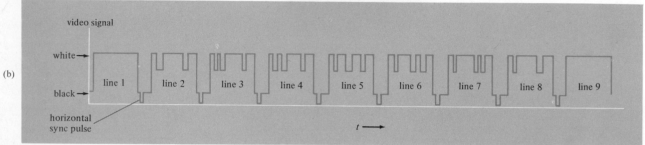

## Television

In television the video signal is also amplitude modulated onto a RF carrier. Thus a television receiver is similar to an AM receiver except that the final video signal is used to modulate the intensity of the electron beam that strikes the screen of the television picture tube. In the final video signal, the voltage is directly proportional to the brightness of the image scanned. In the time intervals between the scan lines of the video signal are voltage pulses that tell the picture tube sweep-circuit to start sweeping another line. The scanning sequence of the letter "N" is shown in Figure 28-28(a), and the corresponding video signal is shown in Figure 28-28(b). A block diagram of a television receiver, including the sweep circuits, is shown in Figure 28-29. Commercial television in the United States uses 525 scan lines to the frame and 30 complete frames per second.

**Figure 28-29.** *Block diagram of a television receiver.*

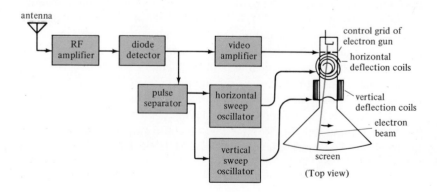

## Exercises

1. In a metal the depth of the potential well is 11 eV. The work function is 4 eV.
   (a) What is the total mechanical energy of a conduction electron at the Fermi level?
   (b) When an electron enters the surface of this metal, how much kinetic energy does it gain?
2. How many conduction electrons are there per gram of sodium? How many conduction electrons are there per gram of germanium seeded with 3 parts per million of arsenic?
3. An electron with 3 eV of kinetic energy strikes a metal. As it enters the surface its kinetic energy increases to 8 eV. What is the depth of the potential well?
4. If two metals have work functions of 2.8 eV and 3.2 eV respectively, what will be the contact potential after they are brought into contact? Which one will have the higher potential?
5. Metal A has $U_0 = 4$ eV and $K_f = 3$ eV. Metal B has $U_0 = 3.5$ eV and $K_f = 2$ eV. When connected together, what is the contact potential and which metal is at the higher potential?
6. Solar radiation at the earth's surface is 2 cal min$^{-1}$ cm$^{-2}$. What must be the area of a 15% efficient solar battery in order to generate 100 W?
7. Consider a hypothetical substance, metal A. When photons of 7 eV are absorbed by this metal, photoelectrons having energies up to 3 eV are emitted. The density of conduction electrons in metal A is such that the conduction electrons have kinetic energies up to 5 eV inside the metal.
   (a) What is the Fermi level?
   (b) What is the work function?
   (c) What is the depth of the potential well?
   (d) How much kinetic energy does an electron lose as it leaves the surface?
   (e) What is the photoelectric threshold in eV?
8. A 1-cm diameter vessel contains a 5-cm column of liquid helium II. How long after the top is removed will it take before all the liquid creeps out of the vessel? The creep velocity is 50 cm/s and the film thickness is $10^{-5}$ cm.
9. A certain metal has an attractive potential $U_0$ for electrons. A certain electron outside the metal has a wavelength of 10 Å. As it penetrates the metal, its wavelength decreases to 4 Å. What is $U_0$ in eV?
10. Repeat Example 5 for germanium.
11. Repeat Example 6 for a voltage of 1 V.
12. Repeat Example 6 for a temperature of 100°C.
13. What must be the semiconductor energy gap in a green LED? $\lambda = 500$ nm.
14. Assume there are $\sim 10^{10}$ neurons in the human brain. What must be the approximate thickness of an IC chip in order to have the same density of solid state devices as neurons in the brain?
15. What is the horizontal sweep frequency in American TV?

## Problems

16. If each atom in a crystal lattice is held in position by a force proportional to the displacement, how would average displacement increase with temperature? If the force constant is $\kappa$, derive a formula for the root mean squared displacement in terms of $\kappa$, $T$, and any other constants that are necessary.

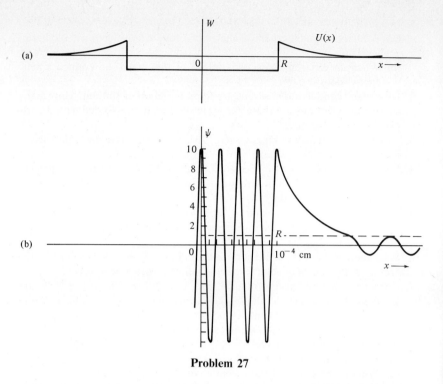

(a)

(b)

**Problem 27**

17. What is the density of states in phase space where an element of phase space is defined as $(dx\, dy\, dz\, dp_x\, dp_y\, dp_z)$. We want $\rho$ where $dn = \rho\, dx\, dy\, dz\, dp_x\, dp_y\, dp_z$. (Hint: $\int_0^{P_f} dp_x\, dp_y\, dp_z = \int_0^{P_f} 4\pi p^2\, dp = \frac{4}{3}\pi P_f^3$.)

18. If $P_f = mc$ for free electrons in a white dwarf star, what is the ratio $\bar{K}/K_f$? It is necessary to use the exact expression $K = (c^2 p^3 + m^2 c^4)^{1/2} - mc^2$.

19. Using Figure 28-11, at what value of interatomic spacing would sodium become a conductor? What would be the corresponding density in grams per cubic centimeter.

20. The sun has a mass $2.0 \times 10^{30}$ kg. Suppose it suddenly contracted into a sphere of radius 5 km. Assume uniform density.
    (a) Assuming equal numbers of electrons, protons, and neutrons, what would be the maximum electron kinetic energy and the maximum neutron kinetic energy in electron volts? What would be the total kinetic energy of all the particles in joules?
    (b) If all the protons and electrons converted into neutrons, what would be the total kinetic energy?

21. The work function of lithium is 2.36 eV. The density is 0.534 g/cm³. What is $U_0$?

22. If $U_0$ were infinite in Figure 28-9(a), what would be $E_1$ and $E_2$ in terms of $\hbar^2/mx_0^2$? Compare with the value in the figure.

23. Use the method of Chapter 25 for solving for energy levels in a finite well to get $E_1$ in Figure 28-9(c).
24. If a cylinder of pure germanium is cooled from 300 K to 30 K, by what factor does its resistance increase?
25. What fraction of arsenic doping is needed to double the conductivity of germanium at room temperature? Assume the fraction of conduction electrons in pure Ge is $\exp(-E_{gap}/2kT)$.
26. Repeat Example 4 for 200 square wells in a row rather than 100. Keep all the other numbers the same. What will be the energy of the 200th level as well as the first, fiftieth, and one hundredth?
27. Assume the potential energy curve between an electron and a negatively charged speck of dust of $10^{-4}$ cm radius would be as in Figure (a). Consider an electron inside the sphere which has a wave function as in Figure (b).
    (a) What is the electron wavelength inside the sphere?
    (b) What is the electron velocity inside the sphere?
    (c) What is the probability per collision for the electron to escape from the sphere?
    (d) Assuming the electron loses no energy, what is the mean-life for this electron to escape from the sphere?
28. A small, cold cathode has $10^{18}$ conduction electrons per second colliding with the high field surface. The electron wave function is reduced by a factor of $10^3$ in passing through the barrier. What is the field emission current in amperes?
29. An ac voltage is applied to point $A$ in the accompanying figure. Plot a curve of the voltage appearing at $B$ as a function of time.
30. Consider the circuit of three resistors and two ideal diodes.
    (a) What is the current in the 2-ohm resistor?
    (b) What is the current in the 3-ohm resistor?
    (c) What is the current in the 4-ohm resistor?
    (d) If the battery voltage is reversed, what will be the current in the 4-ohm resistor?
31. In a Fermi gas the density of states is $\mathfrak{N} = \frac{8}{3}\pi(P_f^3/h^3)$. What is the average magnitude of the momentum $P$ in terms of the maximum momentum $P_f$?
32. Derive a formula for $K_f$ in two-dimensional space. Let $\mathfrak{N}$ be the number of particles per unit area.
33. Repeat the above problem for extremely relativistic particles.

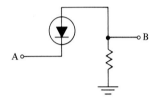

**Problem 29**

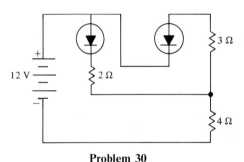

**Problem 30**

PROBLEMS 641

# Nuclear Physics

This chapter is mainly devoted to the understanding of the atomic nucleus, its structure, and properties. Also applications of nuclear interactions such as radioisotopes, fission and fusion reactors, and bombs will be discussed. Our presentation of nuclear physics will bear a close analogy to our presentation of atomic physics. Starting with the basic force law we shall try to predict the sizes, binding energies, angular momenta, energy levels, and so on of the various possible nuclei. Some of the practical applications involve multibillion dollar industries and the very survival of human civilization.

## *Terminology*

Any atomic nucleus consists of protons and neutrons bound together by the nuclear force (or strong interaction). Because neutrons and protons have almost the same mass and very similar properties, the word nucleon is used for either a proton or a neutron. The word nuclide is used for any nucleus larger than a nucleon. (The words "nucleus" and "nuclide" can be used interchangeably.) Nuclides having the same number of protons but different numbers of neutrons are called isotopes. Small and medium-sized nuclides have approximately the same number of protons and neutrons. The letter $A$ is used for the mass number; that is, the number of protons plus the number of neutrons. The number of neutrons is $(A - Z)$ where $Z$ is the atomic number or the number of protons. The $A$ value of an atomic nucleus is very close to the atomic weight of the corresponding atom. In order to identify a particular nuclide the mass number is used as a superscript preceding the atomic symbol. For example $^{14}C$ is that isotope of carbon that has 6 protons and 8 neutrons. The nuclide $^{12}C$ happens to be defined as having an atomic weight exactly equal to 12. The scale of atomic weights is based on $^{12}C$.

Perhaps the most direct way to measure the size and mass distribution of a nucleus is to "look" at it optically by using a beam of wavelength significantly smaller than the size of the nucleus. This technique, called diffraction scattering, has been explained in Section 23-5. Beams of high energy protons or neutrons can be used as the short wavelength probe. The result of this and other techniques is that all but the very smallest nuclei have a mean radius of

## 29-1 Nuclear Size

$$R \approx (1.2 \times 10^{-15} \text{ m}) A^{1/3} \qquad (29\text{-}1)$$

In nuclear and particle physics the unit $10^{-15}$ m occurs so often it is given the special name of the fermi. 1 fermi = 1 fm = $10^{-15}$ m. Diffraction scattering reveals not only the size of a given nucleus but also the distribution of nuclear matter within the nucleus.

---

**Example 1.** What are the mass density and particle density of nuclear matter?

ANSWER: For a nucleus of volume $\frac{4}{3}\pi R^3$ the number of particles per unit volume is

$$\mathcal{N} = \frac{A}{\frac{4}{3}\pi R^3} = \frac{A}{\frac{4}{3}\pi[(1.2 \times 10^{-15} \text{ m})A^{1/3}]^3}$$

$$= 1.38 \times 10^{44} \text{ nucleons/m}^3 \qquad (29\text{-}2)$$

The mass density is this times the nucleon mass:

$$\rho = \mathcal{N} M_p = (1.38 \times 10^{44})(1.67 \times 10^{-27}) \text{ kg/m}^3$$

$$= 2.3 \times 10^{17} \text{ kg/m}^3$$

---

Hence 1 cm³ of nuclear matter would contain 230 million tons. Note that the density of nuclear matter is independent of size because the volume is proportional to $A$.

### High Energy Electron Scattering

A second method for determining the distribution of nuclear matter uses high energy electrons. Since the force on the electrons passing by a nucleus is due to the nuclear charge, electron scattering can determine the distribution of electric charge or distribution of protons within the nucleus. The charge distributions inside the carbon and gold nuclei as determined

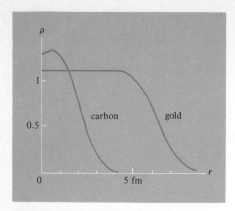

**Figure 29-1.** *Nuclear charge densities of carbon and gold in units of $10^{19}$ C/cm³ as determined by electron scattering.* [After R. Hofstadter, Ann. Rev. Nucl. Sci., **7**:231 (1957)]

by electron scattering are shown in Figure 29-1. As we shall now see, if a nucleus has its charge concentrated near the center, it can deflect electrons by larger angles than if the same charge is spread out over the entire sphere.

If a high energy electron passes by a nucleus of charge $Ze$ at a distance $b$, it will be deflected toward the nucleus by an angle $\theta$ (see Figure 29-2). Because of Coulomb's law, the closer the approach to the nucleus, the stronger will be the force, and the greater will be the deflection. An equation can be derived giving $\theta$ in terms of $b$. However, if the electron gets so close that it penetrates inside the nucleus, it no longer sees a strong Coulomb force (see Figure 16-3). For a charged spherical shell the maximum angle of deflection occurs when $b = R$, the radius. This is because inside such a shell the field is zero. In such a case the radius could be determined by measuring $\theta_{\max}$. All we need is the formula relating $b$ to $\theta$. An approximate derivation goes as follows.

Let $\Delta p_y$ be the change in momentum of the electron due to the Coulomb force. From Newton's definition of force we have

$$\Delta p_y = \int F_y \, dt = e \int E_y \, dt = e \int E_y \left(\frac{ds}{v}\right)$$

For the case of a small angle deflection $ds \approx dx$, and we have

$$\Delta p_y \approx \frac{e}{v} \int E_y \, dx = \frac{e}{2\pi bv} \int E_y \, 2\pi b \, dx$$

Note that $2\pi b \, dx$ is a ring of cylindrical area $dA$. Then

$$\Delta p_y \approx \frac{e}{2\pi bv} \oint \mathbf{E} \cdot d\mathbf{A}$$

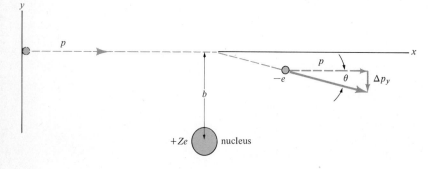

**Figure 29-2.** *Electron of momentum $p$ scattered by atomic nucleus of charge $Ze$.*

where the integral is taken over the area of a cylinder of radius $b$ with the nucleus sitting on the axis of the cylinder. We can use Gauss' law (Eq. 15-7, which says $\oint \mathbf{E} \cdot d\mathbf{A} = 4\pi k_0 Q_{\text{in}}$) to evaluate the integral. Then, since the net charge enclosed by the cylinder is $Ze$, we have

$$\Delta p_y \approx \frac{e}{2\pi bv}(4\pi k_0 Ze)$$

$$= \frac{2k_0 Ze^2}{bv}$$

From Figure 29-2 we see that $\Delta p_y = p \tan \theta$. Substituting this into the preceding equation gives

$$(p \tan \theta) \approx \frac{2k_0 Ze^2}{bv}$$

$$b \approx \frac{k_0 Ze^2}{pv \tan \theta /2}$$

With more mathematical effort it can be shown that the exact formula for all values of $\theta$ is

$$b = \frac{k_0 Ze^2}{pv \tan(\theta /2)} \tag{29-3}$$

If the nuclear charge were all concentrated on the surface of a sphere of radius $R$, the nuclear radius could be obtained by setting $\theta = \theta_{\text{max}}$ in Eq. 29-3. Then

$$R = \frac{k_0 Ze^2}{pv \tan(\theta_{\text{max}}/2)}$$

So far we have been using classical mechanics. When the wave nature of the electron is taken into account, there can be a small amount of scattering beyond $\theta_{\text{max}}$. For a more uniformly charged sphere, electrons penetrating inside the nucleus will experience some force due to charge inside the nucleus, and hence be deflected somewhat more than $\theta_{\text{max}}$. So for this additional reason $\theta_{\text{max}}$ should be thought of as a fuzzy limit. But if one measures the shape and extent of this "fuzziness" in the scattering distribution, one can calculate back and determine the charge distribution inside the nucleus. This can be done for electron-proton scattering as well as for electron-nucleus scattering with the result that no charged core to the proton has been found.

**Example 2.** A 10-GeV electron is "aimed" at the edge of a lead nucleus ($R = 7.5$ fm). What would be the classical angle of deflection?

ANSWER: Solving Eq. 29-3 for $\tan(\theta/2)$ and using $v = c$ gives

$$\tan\frac{\theta}{2} = \frac{k_0 Z e^2}{pcb} \tag{29-4}$$

The energy is $pc = 10 \text{ GeV} = 1.6 \times 10^{-10}$ J, and $b = 7.5 \times 10^{-15}$ m.

$$\tan\frac{\theta}{2} = \frac{(9 \times 10^9)(82)(1.6 \times 10^{-19})^2}{(1.6 \times 10^{-10})(7.5 \times 10^{-15})} = 0.0157$$

$$\theta = 1.80°$$

From Example 2 we see that if all the charge of the lead nucleus was on the surface, no 10-GeV electrons would scatter more than $\theta_{max} = 1.8°$ according to classical physics.

**Example 3.** Repeat Example 2 for a 10-GeV electron beam bombarding a hydrogen target. If the proton radius is $R = 1$ fermi, what is the corresponding scattering angle?

ANSWER: Putting $Z = 1$ and $b = 10^{-15}$ m into Eq. 29-4 gives

$$\tan\frac{\theta}{2} = \frac{k_0 e^2}{pcR} = 0.00144$$

$$\theta = 0.17°$$

We see from Example 3 that if the angular distribution in 10-GeV electron-proton scattering starts fading away at angles beyond 0.17°, then the radius of the proton must be about 1 fm. In the 1960s such experiments were performed yielding a value of 0.8 fm for the root mean square charge radius of the proton.

**Example 4.** If one third of the proton's charge were concentrated in a core region, $R_{core} \sim 0.1$ fm, what would be the corresponding $\theta_{max}$ for 10-GeV electrons?

ANSWER: Now the charge of the target particle ($Ze$) in Eq. 29-4 must be replaced with $e/3$. Using $b = 10^{-16}$ m gives $(\theta_{max})_{core} = 0.55°$. Note that this is $\sim 3$ times as large as for Example 3.

Example 4 shows that if the proton had a charged core, scatterings would be seen at larger angles than is observed. In Chapter 31 we shall learn that the proton might be made up of three subnuclear particles called quarks. However, none of these quarks is expected to sit quietly at the center and behave as a core to the proton.

### Neutron Absorption Cross Sections

The area, and hence the radius, of an atomic nucleus can be determined in a third way by bombarding a plate made up of the atoms in question with high energy neutrons (see Figure 29-3). A neutron will travel in a straight line through the plate until it has a direct collision with a nucleus. Hence the number of neutrons lost from the beam will be proportional to the area of the nucleus. As long as the neutron wavelength is much smaller than the size of the nucleus the effective cross-sectional area presented by a nucleus is $\sigma = \pi R^2$ where $R$ is the nuclear radius.

The probability of losing a neutron is the ratio of the total nuclear area to the area of the plate:

$$\text{probability} = \frac{N_a \sigma}{A}$$

where $N_a$ is the total number of atoms in the plate, $\sigma$ is the effective area of each atomic nucleus, and $A$ is the area of the plate. Let

$$\mathcal{N} = \frac{N_a}{Ax}$$

be the number of atoms per unit volume and $x$ be the plate thickness. Then the probability of losing a neutron is $\mathcal{N}\sigma\, dx$ for a plate of thickness $dx$. If $N$ is the number of neutrons in the beam after penetrating a distance $x$, the change in $N$ equals the number of particles in the beam times the probability $\mathcal{N}\sigma\, dx$:

$$dN = -N\mathcal{N}\sigma\, dx$$

$$\frac{dN}{N} = -\mathcal{N}\sigma\, dx$$

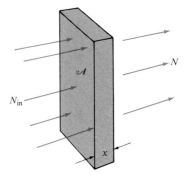

**Figure 29-3.** *Beam of neutrons striking plate of area A.*

Integrating both sides gives

$$\ln N = -\mathcal{N}\sigma x + \text{constant}$$

Taking the antilog of both sides gives

$$N = N_{in} \exp(-\mathcal{N}\sigma x)$$

where $N_{in}$ is the number of incoming neutrons. ($N = N_{in}$ at $x = 0$.) The cross section is

$$\sigma = \frac{1}{\mathcal{N}x} \ln \frac{N_{in}}{N} \qquad (29\text{-}5)$$

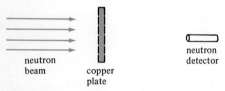

**Figure 29-4.** *When copper plate is inserted, the counting rate drops from $N$ to $N'$.*

**Example 5.** Assume that 10,000 neutrons per minute are measured by the neutron detector shown in Figure 29-4. When a copper plate of 2-cm thickness is interposed, the counting rate drops to 8950 counts per minute. What is the neutron cross section for copper, and what is the corresponding radius?

ANSWER: We can obtain the numerical density from the mass density $\rho = 9.0$ g/cm³ and the atomic weight of 63.5:

$$\mathcal{N} = \frac{9.0 \text{ g/cm}^3}{63.5 \text{ g/mole}} \left( 6.02 \times 10^{23} \frac{\text{atoms}}{\text{mole}} \right) = 8.53 \times 10^{22} \frac{\text{atoms}}{\text{cm}^3}$$

Putting this and $x = 2$ cm and $N_{in}/N = 1.117$ into Eq. 29-5 gives

$$\sigma = 0.65 \times 10^{-23} \text{ cm}^2$$

The unit $10^{-24}$ cm² is a typical nuclear cross section. It occurs so often in nuclear physics that it is given a special name—the barn ($10^{-24}$ cm² = 1 barn). If we set the preceding result equal to $\pi R^2$ we obtain

$$\pi R^2 = 0.65 \times 10^{-24} \text{ cm}^2$$

$$R = 4.55 \text{ fm}$$

## 29-2 The Basic Nucleon-Nucleon Force

The main goal of physics is to explain all physical phenomena by means of a small number of simple, fundamental principles. Since material objects are made up of electrons and nuclei, our approach so far has been to study the fundamental interactions of electrons, nuclei, and photons.

We saw in the last chapter that this approach has met with great success. It has given a complete (although difficult to calculate) explanation of the structure and interactions of matter. In fact, the modern theory of quantum electrodynamics is so good that when applied to atomic physics it predicts results more accurately than can be measured. So far no discrepancy between experiment and theory has been found despite the fact that some of the experimental results are more accurate than one part in ten million.

On the other hand, quantum electrodynamics cannot explain the structure of the atomic nucleus that we now know to be made up of protons and neutrons. A new, fundamental force law is needed to explain what holds the protons so tightly together within the nucleus. This force must be even stronger than the electrostatic force in order to overcome the electrostatic repulsion of the protons. This new force is called the nuclear force or strong interaction. As seen in Figure 29-5 its potential well is an order of magnitude deeper than the potential energy due to the electrostatic repulsion between two protons.

Except for this rather minor electrostatic repulsion, the strong proton-proton, proton-neutron, and neutron-neutron nuclear forces are all the same and are called the nucleon-nucleon force. Although the detailed form of this force is still not known, a crude plot of the potential energy of two nucleons is shown in Figure 29-5. The $k_0 e^2/r$ electric potential energy

**Figure 29-5.** *Potential energy diagram of the elementary nucleon-nucleon force.*

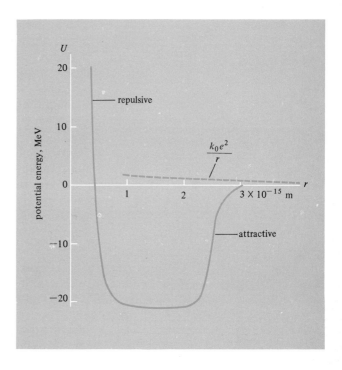

of two protons is shown for comparison (dashed curve). This nuclear force also has the peculiar feature that it looks like Figure 25-5 if the nuclear spins are parallel; on the other hand, when the spins are antiparallel, the nuclear force is about a factor of 2 weaker. As we shall see, the depth of the potential well shown in Figure 29-5 can be determined from the binding energy of the deuteron ($^2$H, the nucleus of heavy hydrogen). More detailed information on the shape of this potential comes from proton-proton and neutron-proton scattering experiments.

In atomic physics, the only atom we were able to analyze easily was the one containing just one proton and one electron, the hydrogen atom. A similar situation holds in nuclear physics: the easiest nucleus to analyze is made up of just one proton and one neutron, the deuteron. In the deuteron, the neutron and the proton are bound together with an energy of 2.22 MeV. This is determined by measuring the masses or rest energies of the proton, neutron, and deuteron, which are 938.21 MeV, 939.50 MeV, and 1875.49 MeV, respectively. The binding energy is the sum of the masses of the individual nucleons minus the mass of the nucleus. In the case of the deuteron the binding energy is $(m_p + m_n) - m_d = 2.22$ MeV.

As with the hydrogen atom, which has a binding energy of 13.6 eV, one should be able to calculate the binding energy of the deuteron once the force between the two particles is known. The problem is to find the lowest order wave function corresponding to the potential energy curve of Figure 29-5. To first approximation the potential energy can be drawn as a "square well" of radius $r_0 = 2.3 \times 10^{-15}$ m. This is shown in Figure 29-6(a). The colored horizontal line is the energy $E$ corresponding to the lowest order standing wave which is shown in Figure 29-6(b).

We are now in position to make a quantitative calculation. We have the choice of calculating the lowest energy level $E$ knowing the depth of the well, or else of calculating the depth of the well knowing $|E|$ (the measured binding energy). The latter is what was actually done historically, and is thus what we choose to do here. We will start with the knowledge of the binding energy and the range $r_0$ of the nuclear force. Our goal is to obtain from this information the depth of the potential well which is a measure of the strength of the nuclear force.

As with the hydrogen atom, one must use Schrödinger's equation in three dimensions. For the lowest order wave function, which is independent of $\theta$ and $\phi$, the three-dimensional Schrödinger equation is of the form given by Eq. 26-5:

$$\frac{1}{r^2}\frac{d}{dr}\left(r^2 \frac{d\psi}{dr}\right) = -\frac{2\mu}{\hbar^2}(E - U)\psi \qquad (29\text{-}6)$$

Here we use $r$ as the relative displacement (distance between the proton and neutron). Then as suggested on page 220 one must use the reduced

**Figure 29-6.** (a) *The approximate potential well for the neutron-proton force.* (b) *The lowest energy wave function with energy level of* $E = -2.2$ *MeV.*

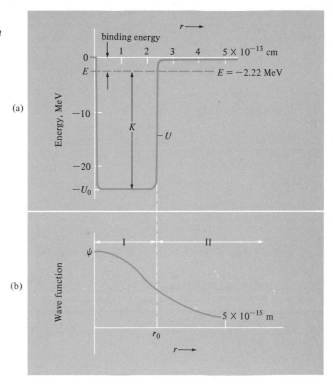

mass of the two particles for $\mu$. Since the proton and neutron have almost equal masses $M$, $\mu = M/2$. Equation 29-6 can be simplified by using $u(r) \equiv r\psi(r)$. Then

$$\frac{d\psi}{dr} = \frac{1}{r}\frac{du}{dr} - \frac{u}{r^2}$$

and Eq. 29-6 becomes

$$\frac{1}{r^2}\frac{d}{dr}\left(r\frac{du}{dr} - u\right) = -\frac{M}{\hbar^2}(E - U)\frac{u}{r}$$

or

$$\frac{d^2u}{dr^2} + \frac{M}{\hbar^2}(E - U)u = 0 \qquad (29\text{-}7)$$

In region I the quantity $(E - U)$ is a positive constant and the solution of Eq. 29-7 is a sine wave. Then

$$\psi = A\frac{\sin kr}{r} + B\frac{\cos kr}{r} \qquad \text{where } k = \sqrt{\frac{M}{\hbar^2}(E - U)}$$

SEC. 29-2/THE BASIC NUCLEON-NUCLEON FORCE  651

Since $\psi$ at $r = 0$ must not be infinite, $B = 0$ and

$$u_I = A \sin kr \quad \text{where } k \equiv \sqrt{\frac{M}{\hbar^2}(E - U)} = \sqrt{\frac{M}{\hbar^2}(U_0 - |E|)}$$

In region II Eq. 29-7 becomes

$$\frac{d^2 U_{II}}{dr^2} - \frac{M}{\hbar^2}|E|U_{II} = 0$$

$$u_{II} = Be^{-Kr} \quad \text{where } K \equiv \sqrt{\frac{M}{\hbar^2}|E|}$$

Note that this is exactly the same mathematical situation as for $\psi_2$ in Figure 25-16. The solution for $E$ is obtained by matching $u_I$ and $u_{II}$ and their slopes at the boundary between regions I and II, with the result given by Eq. 25-20:

$$\cot(kr_0) = -\frac{K}{k} \tag{29-8}$$

or

$$\tan x = -\frac{x}{Kr_0} \quad \text{where } x \equiv kr_0$$

Putting in the numbers $r_0 = 2.3 \times 10^{-15}$ m and $|E| = 2.22$ MeV gives

$$Kr_0 = \frac{r_0}{\hbar}\sqrt{M|E|} = 0.53$$

so

$$\tan x = -\frac{x}{0.53}$$

This equation can be solved graphically by plotting $\tan x$ (the left-hand side) and also $-x/0.53$ (the right-hand side) to find where these two quantities are equal. The result is $x = 1.85$. It can also be solved by trial and error using a pocket calculator. According to the definition of $x$ we have

$$kr_0 = 1.85 \tag{29-9}$$

$$k = \frac{1.85}{2.3 \times 10^{-15} \text{ m}} = 8.04 \times 10^{14} \text{ m}^{-1}$$

Now we replace $k$ with its definition and square both sides:

$$\frac{M}{\hbar^2}(U_0 - |E|) = (8.04 \times 10^{14})^2 \text{ m}^{-2}$$

$$(U_0 - |E|) = \frac{8.04^2 \times 10^{28} \times (1.05 \times 10^{-34})^2}{1.67 \times 10^{-27}} \text{ J}$$

$$= 4.27 \times 10^{-12} \text{ J} = 26.7 \text{ MeV}$$

$$U_0 = |E| + 26.7 \text{ MeV} = 28.9 \text{ MeV}$$

We see that measurements of the deuteron binding energy tell us that the depth of the neutron-proton potential well should be about 29 MeV.

**Example 6.** In the preceding model of the deuteron, the probability density is greatest at $r = 0$. By what factor is the probability density (or mass density) reduced at $r = r_0$?

ANSWER: The solution in region I is $u(r) = r\psi(r) = A \sin kr$.

$$\psi(r) = A \frac{\sin kr}{r} = kA \frac{\sin kr}{kr}$$

$$\frac{\psi(r_0)}{\psi(0)} = \frac{kA(\sin kr_0)/kr_0}{kA} = \frac{\sin kr_0}{kr_0}$$

Using Eq. 29-9, we have

$$\frac{\psi(r_0)}{\psi(0)} = \frac{\sin 1.85}{1.85} = 0.52$$

$$\frac{|\psi(r_0)|^2}{|\psi(0)|^2} = 0.27$$

We see that the 25% contour for the deuteron is outside the range of the nuclear force.

**Example 7.** Suppose the nuclear force were such that the neutron and proton in the deuteron were just barely bound; that is, the binding energy $|E|$ would be close to zero. Then what would be the depth of the potential well?

ANSWER: As $E \to 0$, so does $K$ in Eq. 29-8:

$$\cot kr_0 = -\frac{K}{k} = 0$$

Then
$$kr_0 = \frac{\pi}{2} \quad \text{or} \quad k = \frac{\pi}{2r_0}$$

$$\frac{M}{\hbar^2}(U_0 - |E|) = \left(\frac{\pi}{2r_0}\right)^2$$

$$U_0 = \frac{1}{M}\left(\frac{\hbar\pi}{2r_0}\right)^2 = 3.08 \times 10^{-12} \text{ J} = 19.2 \text{ MeV}$$

## 29-3 The Structure of Heavy Nuclei

The reason for the high density of nucleons in heavy nuclei can be clarified by the following explanation. Start with a large number of free nucleons with a distance $s$ between adjacent particles. Now slowly push them together (decrease $s$). Suddenly, when $s$ becomes less than $2.5 \times 10^{-13}$ cm, the nucleons feel the strong attractive force of their neighbors, and their binding energy increases correspondingly. On the other hand, we saw in Chapter 28 that when free electrons are crowded closely together, their average kinetic energy must increase because of the Pauli exclusion principle (see Eq. 28-3). Since protons and neutrons are also spin-$\frac{1}{2}$ particles, they must also obey the Pauli principle. Thus the effect of the exclusion principle is to decrease the binding energy as $s$ is decreased. Fortunately, the nucleon-nucleon attractive force is just barely strong enough to allow a "happy medium" between these two effects; that is, there exists a value of $s$ where the binding energy is maximum (if the nucleon-nucleon force had been 30% weaker, the effect of the exclusion principle would dominate and no nuclei could ever exist). The value of $s$ that gives the maximum binding energy determines the size of the nucleus. The experimental result is $s = 1.9 \times 10^{-13}$ cm as determined from Eq. 29-2.

Let us now consider a single neutron inside of a heavy nucleus. The neutron sees an attractive force averaged over all the other nucleons in the nucleus. In Chapter 28 an analogous situation was the potential seen by a free electron inside of a metal. The average potential energy curve seen by our neutron is shown in Figure 29-7. It is about 42 MeV deep for all medium and large nuclei. Adding more nucleons will not increase the strength of the net force on a given nucleon because, as pointed out before, a nucleon is only attracted by its nearest neighbors. Actually about $A/2$ neutrons are crowded into this potential well. Because of the Pauli

**Figure 29-7.** *Average nuclear potential seen by a neutron in a nucleus of radius R. Occupied states are shown as solid lines and unoccupied or excited states as dashed lines. The neutron binding energy (energy required to remove a neutron) is shown.*

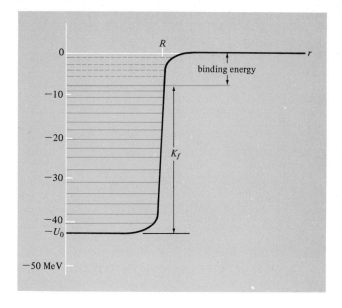

principle they will occupy different states or energy levels up to the Fermi level. The height of the Fermi level can be calculated using Eq. 28-3, which is

$$K_f = \frac{h^2}{8M}\left(\frac{3}{\pi}\mathcal{N}\right)^{2/3} \qquad (29\text{-}10)$$

where $\mathcal{N}$ is the number of neutrons per unit volume. This would be approximately one half the nuclear particle density, which, according to Eq. 29-2, comes to $0.7 \times 10^{44}$ neutrons/m³. Then

$$K_f = \frac{(6.62 \times 10^{-34})^2}{8(1.67 \times 10^{-27})}\left[\frac{3}{\pi} \times 0.7 \times 10^{44}\right]^{2/3} \text{ J} = 5.4 \times 10^{-12} \text{ J}$$

$$= 33.8 \text{ MeV}$$

We see that the highest occupied neutron energy level is about 34 MeV above the bottom of the potential well, or about 8 MeV below zero energy. Thus a minimum of 8 MeV is needed to remove a single neutron from a typical nucleus. The remaining nucleus will have the same well depth and Fermi energy, and thus it would take another 8 MeV to remove a second neutron. The end result for a medium-sized nucleus is that the binding energy per nucleon is about 8 MeV. Smaller nuclei would not have quite as deep a potential well because the average number of nearest neighbors would be reduced; hence the binding energy per nucleon would be less than 8 MeV. For large nuclei the electrostatic potential energy which goes

**Figure 29-8.** *Experimental values of binding energy per nucleon as a function of the mass number A.*

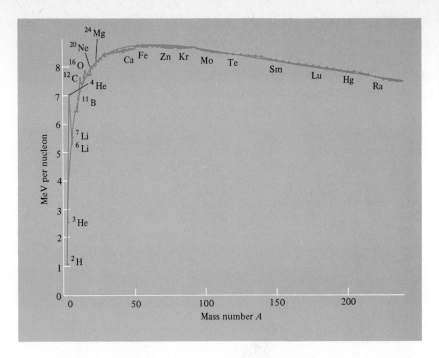

as $Z^2/R$ is an additional contribution to the total energy of the nucleus; hence the binding energy per nucleon for large nuclei should also be less than 8 MeV per nucleon. The experimental results are shown in Figure 29-8 and agree with such predictions. We have only discussed neutrons; however, the protons in a nucleus experience the same nuclear force corresponding to the potential well depth $U_0 = 42$ MeV. But, in the case of protons, this well depth is raised a few MeV by the electrostatic contribution to the potential energy which will be $\sim k_0(Ze^2/R)$.

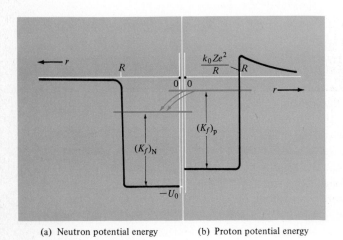

(a) Neutron potential energy    (b) Proton potential energy

**Figure 29-9.** *Neutron and proton potential energy diagrams for nucleus of radius R. If the nucleus has the same number of neutrons and protons, the Fermi energies $(K_f)_N = (K_f)_p$ and the highest energy protons will convert into neutrons by emitting positron-neutrino pairs.*

Suppose a nucleus of radius $R$ initially had the same number of neutrons and protons. Then the neutron and proton potential energy diagrams would appear as in Figure 29-9 and the highest energy proton would be at a few MeV above the highest energy neutron.

We shall see in Section 29-5 that whenever this is the case such a proton would change into a neutron by emitting a positron and neutrino. Thus our potential well model has predicted that the larger nuclei should ultimately have more neutrons than protons. For example, the $^{238}$U nucleus has 146 neutrons and 92 protons.

The following quantitative example gives an idea why the ratio of neutrons to protons is greater than 1, and why this ratio increases with increasing $A$.

---

*Example 8.** According to the potential well model of the nucleus, what would be the number of protons $Z$ for a nucleus of mass number $A = 51$?

ANSWER: Let $U_0 = 42$ MeV be the neutron potential well depth and $[U_0 - k_0(Z-1)e^2/R]$ be the proton well depth. Let $E_n$ be the energy level of the highest energy neutron and $E_p$ be that for the highest energy proton. Then

$$E_n = E_p$$

or

$$U_0 - (K_f)_n = \left(U_0 - k_0 \frac{(Z-1)e^2}{R}\right) - (K_f)_p$$

$$U_0 - \frac{h^2}{8M}\left[\frac{3}{\pi}\frac{(A-Z)}{\frac{4}{3}\pi R^3}\right]^{2/3}$$

$$= U_0 - k_0\frac{e^2(Z-1)}{R} - \frac{h^2}{8M}\left(\frac{3}{\pi}\frac{Z}{\frac{4}{3}\pi R^3}\right)^{2/3}$$

We use $R = 4.45 \times 10^{-15}$ m obtained from Eq. 29-1. This gives us a numerical equation for $Z$, which in MeV is

$$42 - 3.87(51-Z)^{2/3} = 42 - 0.324(Z-1) - 3.87 Z^{2/3}$$

or

$$0.324(Z-1) - 3.87[(51-Z)^{2/3} - Z^{2/3}] = 0$$

The solution can be found in a few minutes by trial and error using a pocket calculator. It is $Z = 21.7$. This is close to the element vanadium, which has $A = 51$ and $Z = 23$. There is also a radioisotope of titanium which has $A = 51$ and $Z = 22$.

We see that this simplified square well model of nuclear structure gives a fairly good prediction of the ratio of protons to neutrons in a nucleus as well as binding energies. One can make other predictions, such as the energies of the excited states (the dashed lines in Figure 29-7). We can estimate the spacing between energy levels by taking the logarithm of both sides of Eq. 29-10 and differentiating:

$$\frac{\Delta E}{E} = \frac{2}{3}\frac{\Delta \mathcal{N}}{\mathcal{N}}$$

For a nucleus having 20 neutrons, when one more neutron is added $\Delta \mathcal{N}/\mathcal{N} = 0.05$ and $\Delta E/E \approx 0.03$ which gives $\Delta E \approx 0.03 \times 34$ MeV = 1.1 MeV. This is a typical spacing between low-lying energy levels for a small to medium-sized nucleus. For larger nuclei $\Delta \mathcal{N}/\mathcal{N}$ is smaller and so is the spacing between energy levels.

Not only can the energy values of excited states be crudely predicted but so can the angular momenta. This is discussed in the next paragraph where an important refinement will be made to the potential well model.

### The Shell Model

Each of the neutron energy levels shown in Figure 29-7 corresponds to a standing wave or orbital of definite energy and angular momentum. This is true for both the occupied levels and the higher levels or excited states. Such nucleon orbitals would be "stable" and have a definite energy only if the mean free path for a nucleon in nuclear matter is greater than the size of the nucleus. Actually the mean free path is significantly greater than $10^{-15}$ m because of the Pauli principle. What would normally be considered a typical collision cannot take place because there is no place for the scattered nucleon to go. The momentum state where it would like to go is already occupied.

Figure 29-10 shows the calculated energy levels (and corresponding angular momenta) of the possible standing waves corresponding to the potential well shown in Figure 29-7. The energy levels have been modified to take into account the observation that the force on a nucleon is greater

| N | l | j | Number of Nucleons per Energy Shell | Accumulated Number of States |
|---|---|---|---|---|
| 4 | 0 | 1/2 | 2 | |
| 5 | 2 | 3/2 | 4 | |
| 6 | 4 | 7/2 | 8 | |
| 5 | 2 | 5/2 | 6 | |
| 7 | 6 | 11/2 | 12 | |
| 6 | 4 | 9/2 | 10 | |
| 5 | 3 | 5/2 | 6 | 126 ← |
| 4 | 1 | 1/2 | 2 | 120 |
| 7 | 6 | 13/2 | 14 | 118 |
| 4 | 1 | 3/2 | 4 | 104 |
| 5 | 3 | 7/2 | 8 | 100 |
| 6 | 5 | 9/2 | 10 | 92 |
| 3 | 0 | 1/2 | 2 | 82 ← |
| 6 | 5 | 11/2 | 12 | 80 |
| 4 | 2 | 3/2 | 4 | 68 |
| 4 | 2 | 5/2 | 6 | 64 |
| 5 | 4 | 7/2 | 8 | 58 |
| 5 | 4 | 9/2 | 10 | 50 ← |
| 3 | 1 | 1/2 | 2 | 40 |
| 3 | 1 | 3/2 | 4 | 38 |
| 4 | 3 | 5/2 | 6 | 34 |
| 4 | 3 | 7/2 | 8 | 28 ← |
| 2 | 0 | 1/2 | 2 | 20 ← |
| 3 | 2 | 3/2 | 4 | 18 |
| 3 | 2 | 5/2 | 6 | 14 |
| 2 | 1 | 1/2 | 2 | 8 ← |
| 2 | 1 | 3/2 | 4 | 6 |
| 1 | 0 | 1/2 | 2 | 2 ← |

**Figure 29-10.** *Relative spacing of nuclear energy levels taking into account spin-orbit interaction (force is larger if spin and orbital angular momentum are parallel). As with the hydrogen atom, each energy level or "shell" contains subshells corresponding to the quantum number $m_l$. The total number of nucleons required to fill these subshells is given in the right-hand column. Magic numbers are indicated by arrows.*

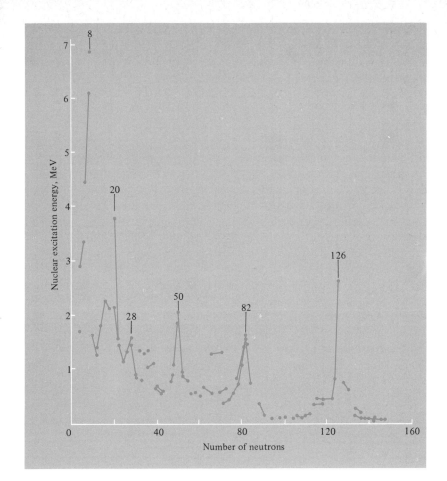

**Figure 29-11.** *Measured nuclear excitation energies for nuclei with even numbers of protons and neutrons. Solid lines connect nuclei having the same proton number.*

if its spin and orbital angular momentum are in the same direction. The sum of the spin and orbital angular momentum is the total angular momentum $j$. We see that a nucleus that has the $n = 5$, $l = 4$, $j = \frac{9}{2}$ shell completely filled has a total of 50 neutrons (or 50 protons). We also note from Figure 29-10 that there is a large energy gap between this energy shell and the next highest one. Thus we would expect that nuclei having 50 neutrons ($A - Z = 50$) or 50 protons ($Z = 50$) would be strongly bound and particularly stable. This is dramatically illustrated in Figure 29-11, which shows that over 2 MeV is required to excite a nuclide having 50 neutrons to its next highest state, whereas neighboring nuclei require less excitation energy. Another example of how nuclei with magic number 50 are more tightly bound is the fact that tin ($Z = 50$) has 10 stable isotopes, which is more than any other element. Also the natural abundance of nuclei with 50 neutrons or 50 protons is significantly higher than those with 51 neutrons or protons.

The experimental results shown in Figure 29-11 agree with the theoretical predictions of Figure 29-10 in that nuclides having 2, 8, 20, 28, 50, 82, or 126 neutrons (or protons) are more tightly bound. These are called magic numbers and are analogous to the closed shells 2, 10, 18, 36, 54, and 86 in atomic physics (see Figure 27-1).

## 29-4 Alpha Decay

For historical reasons the $^4$He nucleus is called an alpha particle. Many of the larger nuclides above $Z = 82$ (lead) are observed to undergo radioactive decay by emitting an alpha particle. Since the binding energy per nucleon in an alpha particle is greater than the binding energy per nucleon in such a larger nucleus, alpha decay is energetically possible. For example, a sample of $^{238}$U is observed to emit alpha particles with a half-life of $4.5 \times 10^9$ years. The spontaneous nuclear reaction is

$$^{238}\text{U} \rightarrow {}^{234}\text{Th} + {}^4\text{He} + 4.2 \text{ MeV}$$

After $4.5 \times 10^9$ years half of the $^{238}$U nuclei will have decayed. The mass difference between $^{238}$U and its decay products is 4.2 MeV.

We can obtain an understanding of why alpha decay occurs by study of the potential energy diagram of the alpha particle and residual nucleus as shown in Figure 29-12. In this figure $E_\alpha$ is the kinetic energy of the emitted

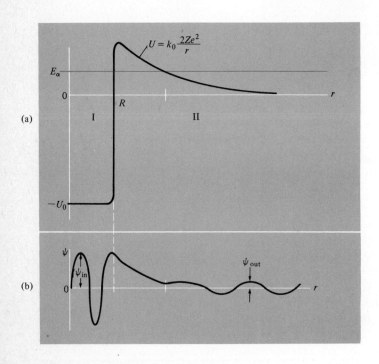

**Figure 29-12.** (a) *The potential energy between an $\alpha$ particle and a residual nucleus of charge Ze. $E_\alpha$ is the kinetic energy of the $\alpha$ particle at a large distance.* (b) *The corresponding wave function.*

alpha particle. Initially the alpha particle is in region I and can be represented as a standing wave of amplitude $\psi_{in}$. However, because of barrier penetration as explained on page 633, there will be a small "tail" to the wave function which has an amplitude $\psi_{out}$ in the region far away from the nucleus. Hence, the probability that the alpha particle escapes each time it collides with the barrier is

$$\frac{\text{prob}}{\text{coll.}} \approx \frac{|\psi_{out}|^2}{|\psi_{in}|^2}$$

The number of collisions per second is roughly

$$\frac{\text{coll.}}{\text{sec}} \approx \frac{v}{2R}$$

where $v$ is the velocity of the alpha particle in region I. If we multiply the two preceding equations together, we obtain

$$\frac{\text{prob.}}{\text{sec}} \equiv \frac{dp_r}{dt} \approx \left(\frac{v}{2R}\right) \times \frac{|\psi_{out}|^2}{|\psi_{in}|^2} \qquad (29\text{-}11)$$

Let the symbol $\tau$ (tau) stand for the quantity $(2R/v)(|\psi_{in}|^2/|\psi_{out}|^2)$:

$$\tau \equiv \frac{2R}{v} \frac{|\psi_{in}|^2}{|\psi_{out}|^2} \qquad (29\text{-}12)$$

Combining Eqs. 29-11 and 29-12 gives

$$\frac{dp_r}{dt} = \frac{1}{\tau}$$

For a sample of $n$ nuclei the number of decays per second (or the rate of decrease of $n$) is

$$\frac{dn}{dt} = -n\frac{dp_r}{dt} = -n\left(\frac{1}{\tau}\right)$$

$$\frac{dn}{n} = -\frac{1}{\tau} dt$$

$$\int \frac{dn}{n} = -\int \frac{1}{\tau} dt$$

$$\ln n = -\frac{t}{\tau} + \text{constant}$$

Taking the antilog of both sides gives

$$n = (\text{constant}) \times e^{-t/\tau}$$

If we let $n_0$ equal the number of nuclei at $t = 0$, then

$$n = n_0 e^{-t/\tau} \tag{29-13}$$

This is called the radioactive decay curve and is plotted in Figure 29-13.

The half-life $T$ is defined as that value of $t$ where $n = \tfrac{1}{2} n_0$. Putting these values into Eq. 29-13 gives

$$(\tfrac{1}{2} n_0) = n_0 e^{-T/\tau}$$

$$2 = e^{T/\tau}$$

Taking the natural log of both sides gives

$$\ln 2 = \frac{T}{\tau}$$

$$T = 0.693\tau$$

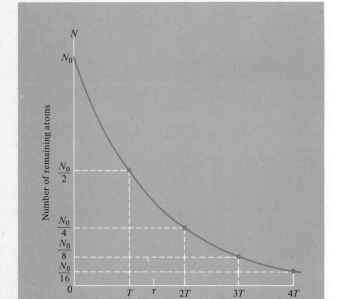

**Figure 29-13.** *Radioactive decay curve. The number of atoms remaining as a function of time over a time interval of four half-lives.*

Eq. 29-13 can be reexpressed in terms of powers of 2 by replacing $\tau$ with $T$:

$$n = n_0 e^{-0.693t/T} = n_0 (e^{-0.693})^{t/T} = n_0 (\tfrac{1}{2})^{t/T}$$

If we replace $\tau$ with $(1.4T)$ in Eq. 29-12, we obtain a formula for the half-life:

$$T \approx \frac{1.4R}{v} \frac{|\psi_{in}|^2}{|\psi_{out}|^2} \qquad (29\text{-}14)$$

**Example 9.** What is the mean life in terms of $\tau$? (Mean life is the average length of time a nucleus survives before decay.)

ANSWER: If we start with a sample of $n_0$ radioactive nuclei, we must form an average of $t$ with $n$ as the weighting factor:

$$t_{mean} = \frac{\int tn\, dt}{\int n\, dt} = \frac{\int_0^\infty t \exp(-t/\tau)\, dt}{\int_0^\infty \exp(-t/\tau)\, dt} = \frac{[\tau^2 e^{-t/\tau}(-t/\tau - 1)]_0^\infty}{[-\tau e^{-t/\tau}]_0^\infty} = \tau$$

We see that $\tau$, which was defined as the reciprocal of the decay rate for an individual nucleus, is the mean life.

**Example 10.** In $^{238}$U an alpha particle collides with the potential barrier $5 \times 10^{20}$ times per second and $\psi_{out}/\psi_{in} = 10^{-19}$. (a) What is the probability that this nucleus decays in one second? (b) What would be the mean life for this nucleus?

ANSWER: Using Eq. 29-11, the decay rate is

$$\frac{dp_r}{dt} \approx \left(\frac{\text{coll.}}{\text{sec}}\right) \times \left(\frac{\psi_{out}}{\psi_{in}}\right)^2 = (5 \times 10^{20})(10^{-38}) = 5 \times 10^{-18}\ \text{s}^{-1}$$

$$\Delta p_r \approx 5 \times 10^{-18}\, \Delta t$$

For $\Delta t = 1$ s,

$$\Delta p_r \approx 5 \times 10^{-18}$$

The mean life can be obtained by inverting the equation $dp_r/dt = 1/\tau$:

$$\tau = \frac{1}{\dfrac{dp_r}{dt}} = \frac{1}{5 \times 10^{-18} \text{ s}^{-1}} = 2 \times 10^{17} \text{ s} = 6.5 \times 10^9 \text{ year}$$

The half-life is $T = 0.693\tau = 4.5 \times 10^9$ year.

Equation 29-14 demonstrates how quantum mechanics can be used to explain radioactivity. Present quantum theory gives as full an explanation of alpha decay and other forms of radioactive decay as it does of anything else. It is worth recalling that the nature of probability is such that, if a given nucleus has by rare coincidence managed to survive many half-lives without decaying, this previous history will in no way influence its future chance of decay. The same is true for coin tossing. If you happen to toss five heads in a row, the probability is is still one half that the sixth toss will also be a head. We can never predict when a given nucleus will decay. All nuclei of the same kind always have exactly the same probability for decay no matter how long they have lived. For example, half the nuclei of a radioactive isotope of 1 year half-life will decay in the first year, but an individual nucleus that survived the first year will still have a 50-50 chance of decaying in the second year. If it survives the first two years, its probability of decay during the third year is still one half.

## 29-5 Gamma and Beta Decay

### Gamma Decay

If a nuclide is excited to a higher energy level (one of the dashed lines in Figure 29-7), it can undergo spontaneous emission of a photon to a lower energy level as explained on page 577. Since the spacing of nuclear energy levels is on the order of an MeV, the photons emitted by nuclei are on the order of hundreds or thousands of times more energetic than those emitted by atoms. These higher energy photons emitted by nuclei are often called gamma rays.

A nucleus in an excited state can easily be obtained using low energy neutrons. For example, as a slow neutron passes through a piece of $^{238}$U it will feel the attractive nuclear force whenever it is within range of the atomic nucleus. Then it is quite probable that the neutron will be captured, thus forming $^{239}$U* in an excited state. (The star is used to indicate an excited state.) Such an excited nuclide will cascade down to the ground state by emitting one or more gammas. These processes are indicated by the following nuclear reaction notation:

$$n + {}^{238}U \rightarrow {}^{239}U^*$$

$$^{239}U^* \rightarrow {}^{239}U + \gamma$$

**Beta Decay**

Shortly after the discovery of radioactivity in 1896 the properties of the radiation were studied and it was concluded that there existed three different kinds of radiation. They were named alpha, beta, and gamma. After more years of study the alpha radiation was found to consist of helium nuclei and the gamma rays were found to be high energy photons. The beta radiation was found to consist of electrons or positrons. Certain nuclides are observed to emit electrons when they decay. Also certain nuclides are observed to emit positrons. More careful measurements reveal that such an electron or positron is always accompanied by a neutrino or antineutrino. A neutrino is an elementary particle of zero charge and rest mass but with the same spin as the electron.

**Example 11.** Use the uncertainty principle to demonstrate that an electron cannot be contained in an atomic nucleus.

ANSWER: The uncertainty in the electron momentum would be at least

$$\Delta p \approx \frac{\hbar}{\Delta x} \approx \frac{1.05 \times 10^{-34}}{10^{-15}} \text{ kg m s}^{-1}$$

$$= 1.05 \times 10^{-19} \text{ kg m s}^{-1} = 197 \text{ MeV}/c$$

According to Eq. 9-11 an electron of this momentum would have a total energy

$$E = \sqrt{(m_e c^2)^2 + (cp)^2} = \sqrt{(0.51)^2 + (197)^2} \text{ MeV} = 197 \text{ MeV}$$

using $m_e c^2 = 0.51$ MeV. In order for an electron with 197 MeV of kinetic energy to be bound, the electrostatic binding energy must be more than this. However, the electrostatic binding energy is $k_0 Ze^2/R$, which is less than 10 MeV for all nuclei.

The modern theory of beta decay is based on a theory advanced by Enrico Fermi in 1931. He proposed that a proton or neutron can emit an electron-neutrino pair by essentially the same mechanism that a charged particle can emit a photon. The electron-neutrino pair is created by the weak interaction just as the photon is created by the electromagnetic interaction. Before the beta decay there was no electron or neutrino inside the nucleus.

The simplest example of beta decay is the decay of a free neutron into a proton with a half-life of 12 min:

$$n \rightarrow p + (e^- + \bar{\nu})$$

where the symbol $\bar{\nu}$ stands for an antineutrino. (The distinction between $\nu$ and $\bar{\nu}$ will be covered in Chapter 31.) The rest mass of the neutron happens to be 1.3 MeV greater than that of the proton, hence the total energy of the emitted electron-neutrino pair will be 1.3 MeV. Taking away the 0.5 MeV of electron rest mass leaves 0.8 MeV of kinetic energy to be shared by the electron and neutrino.

We saw in Figure 29-9 that in a typical nucleus the highest energy neutron has almost the same energy level as the highest energy proton. In such a nucleus all the neutrons are prevented by the law of conservation of energy from decaying into protons. If a neutron were added to such a nucleus and if its energy level were more than 0.5 MeV above the highest proton energy level, then creation of an electron-neutrino pair would be possible and beta decay could take place. As an example consider the bombardment of $^{238}$U by neutrons. We saw on page 665 that neutron capture results in formation of $^{239}$U. This isotope of uranium has its highest neutron energy level 1.8 MeV above the highest proton. Hence an electron-neutrino pair is emitted with kinetic energy of 1.3 MeV. The observed half-life is 24 min:

$$^{239}U \rightarrow \,^{239}Np + e^- + \bar{\nu}$$

It turns out that the neptunium is also unstable to beta decay:

$$^{239}Np \rightarrow \,^{239}Pu + e^- + \bar{\nu}$$

with a half-life of 2.35 days. In the preceding two-stage process, the two highest neutron energy levels have been vacated and two proton energy levels filled with the resulting $^{239}$Pu having its highest proton and neutron energy levels close to each other. Hence beta decay is forbidden for $^{239}$Pu (plutonium). However it can alpha decay with a half-life of 24,000 years. As we shall see in the next section an even more important property of $^{239}$Pu is that it has a short lifetime for fission when stimulated by a neutron.

As we saw in Figure 29-8, the binding energy per nucleon increases with $A$ up to $A \approx 50$. This effect can be explained as the result of addition of forces; a single nucleon is more strongly bound if attracted by several others rather than by just one or two. However, above $A \approx 50$ the binding energy per nucleon gradually decreases with increasing $A$. This suggests that the attractive nuclear force is of short range (about the diameter of a single nucleon). Past this range the electrostatic repulsive force dominates; that is, when two protons are more than $\sim 2.5 \times 10^{-15}$ m apart, the force is repulsive rather than attractive.

Two consequences of this behavior of the binding energy as a function of $A$ are nuclear fusion and nuclear fission. First consider what happens when an electron and proton are brought together. Then 13.6 eV of energy is released with the result that the mass of the hydrogen atom is 13.6 eV less than the mass of a free electron plus proton. Similarly, two light nuclei will have more mass or rest energy than will their sum. If they can be brought together they will "fuse" into their sum with a release of energy corresponding to the mass difference. This process is called nuclear fusion. In Section 29-7 we will see that this mass difference can be larger than one-half of 1%. On the other hand, if a *heavy* nucleus is split into two smaller nuclei, the two pieces will have less mass than does their parent by as much as one tenth of 1%. Thus there is a tendency for a heavy nucleus to fission into two smaller nuclei with a release of energy. The energy of an A-bomb and of a nuclear reactor is the energy released in nuclear fission. The energy of an H-bomb is energy released in fusion.

Alpha decay (see Section 29-4) can be thought of as a lopsided fission where the parent nucleus $M$ splits into a small alpha particle and a large residual nucleus $M'$. Alpha decay is possible only if in the reaction

$$M \rightarrow M' + \alpha$$

the mass $M$ is greater than $M'$ plus the alpha particle mass. Then the nucleus will be radioactive and can alpha decay. It turns out that $M > (M' + M_\alpha)$ for all nuclei with $Z > 82$ (lead). Above $Z = 92$ (uranium) the alpha decay half-lives become significantly shorter than the age of the Earth. This is why no elements of atomic number greater than 92 occur naturally on the Earth. Such elements, however, can be produced artificially by man. For example, plutonium ($Z = 94$) can be produced from uranium in nuclear reactors; this process has now become sufficiently common so that the price is about $15 per gram. So far elements up to $Z = 106$ have been produced, but at much higher prices and usually in very minute quantities. It is expected that radiochemists will eventually succeed in producing extremely minute quantities of new elements even beyond $Z = 106$.

If one could slice a big uranium nucleus in two, the resulting two groups of nucleons left over would reorder themselves into more tightly bound

## 29-6 Nuclear Fission

nuclei—releasing energy while in the process of reordering. We see that spontaneous fission is permitted by the law of conservation of energy. However, in the case of the naturally occurring nuclei, the potential barrier is so great that the probability for spontaneous fission is even smaller than that for alpha decay. For example, the half-life of $^{238}$U due to spontaneous fission alone is $8 \times 10^{15}$ years. This is more than one million times the age of the Earth. On the other hand, if such a nucleus is hit by a neutron, it can be excited to a higher energy level closer to the top of the electrostatic potential barrier and thus would have higher probability for fission. Also, such an excited state can have high angular momentum and be egg shaped. The far ends find it much easier to penetrate the barrier because they have already partially penetrated it, hence the barrier is reduced even more when the nucleus is egg shaped. When $^{235}$U or $^{239}$Pu captures a slow neutron, states are formed that have extremely short lifetimes for fission. The mass difference between the uranium nucleus and typical fission products is such that 200 MeV of energy is released in the average uranium fission. Since the rest mass of a uranium nucleus is $2.2 \times 10^5$ MeV, the fraction of rest mass converted to energy is 200 MeV divided by this or about 0.1%.

Since one gram of any substance has $mc^2 = 9 \times 10^{13}$ J, the fission of one gram of uranium releases almost $9 \times 10^{10}$ J. This is about 3 million times more than the $2.9 \times 10^4$ J released by the burning of one gram of coal. On the other hand, one gram of uranium is more expensive than one gram of coal. However, the cost per joule is 400 times more for coal than for uranium fuel. Present experience by United States utilities using both coal and nuclear reactors is a cost of 1.7 ¢/(kW h) for coal plants and 1.05 ¢/(kW h) for nuclear plants (1974 figures). It is expected that nuclear power will compete favorably with fossil fuels in the northeastern part of

**Figure 29-14.** (a) *Each mousetrap has been set with two Ping-Pong balls. A mousetrap simulates a $^{235}$U nucleus, and the Ping-Pong balls simulate the neutrons that can be released in fission.* (b) *Snapshot taken a few seconds after one extra ball (neutron) has been dropped on the mousetraps. The chain reaction lasts for several seconds.*

(a)        (b)

the United States. However, power production using coal is expected to be more economical than nuclear in coal-producing areas.

Nuclear fission can be made self-sustaining by a chain reaction process. Each fission releases two or three neutrons. Then, if one of these neutrons manages to induce fission in another uranium nucleus, the process is self-sustaining (see Figure 29-14). An assembly of fissionable material that meets this criterion is called a critical assembly. The first of these, called a nuclear pile, was constructed by Enrico Fermi in a squash court at the University of Chicago. A bronze plaque is mounted at the site that reads "On December 2, 1942, man achieved here the first self-sustaining chain reaction and thereby initiated the controlled release of nuclear energy." Along with the plaque is the sculpture shown in Figure 29-15.

A mass of $^{235}$U or $^{239}$Pu can also be made supercritical. Here the neutrons from one fission induce more than one secondary fission. Since neutrons travel with velocities greater than $10^8$ cm/s, a supercritical assembly can be all used up (or blown apart) in much less than a thousandth of a second. This device is called an A-bomb. The standard method for making a sphere of plutonium supercritical is the implosion technique. A subcritical sphere of plutonium is surrounded by chemical explosives. When these go off, the plutonium sphere is momentarily compressed. Because its density is then significantly increased, it will absorb neutrons at a rate faster than the rate it loses neutrons to the outside. This is the condition for supercriticality. Apparently the explosion of an A-bomb can be made reasonably efficient (most of the plutonium is consumed rather than blown apart). Chemical energies are such that 1 ton of TNT releases $10^9$ cal, or $4 \times 10^9$ J. An A-bomb consuming 1 kg of plutonium or $^{235}$U has an energy release of $\sim 8 \times 10^{13}$ J or about 20,000 times as much as 1 ton of TNT. This is called a 20-kiloton bomb. Present-day megaton bombs are roughly a million times more powerful than conventional TNT "blockbusters." Not only is the energy release a million times greater but also each gram of plutonium or uranium consumed must result in almost a gram of fission products, which are all initially radioactive. This is an extremely large amount of radioactivity.

Whether or not uranium fuel is "burnt" quickly as in a bomb, or slowly as in a nuclear reactor, the resultant radioactivity is a serious problem. Much of the cost of nuclear reactors is due to the elaborate safety precautions taken to prevent accidents as well as handling of the radioactive wastes. In the case of a rare accident such as loss of cooling water or malfunction of the control rods, a nuclear reactor would not explode like an A-bomb; however, it could melt down and possibly leak some radioactivity to the surrounding neighborhood. For the purpose of safety, commercial nuclear reactors are provided with emergency core cooling systems and high pressure containment tanks. It is difficult to make reliable estimates of the probabilities of reactor accidents and the resulting amount of life lost. Many observers feel that more damage is done to the

**Figure 29-15.** *The Atomic Age. Sculpture by Henry Moore, commissioned by The University of Chicago, to commemorate the attainment of nuclear power.*

human race by pollution from conventional coal burning power plants than by possible radiation hazards of nuclear power plants. There is also the problem of long-term storage of the radioactive fission products (nuclear waste).

A more indirect hazard of nuclear power not usually included in such assessments might be its proliferation to smaller nations leading to further proliferation of nuclear weapons leading to a possible nuclear war. If the world could have given up nuclear power, perhaps there would be less chance of nuclear war. Nuclear power is expected to help meet the world energy needs at a time when oil production will no longer be able to keep up with demand. The world oil shortage is expected to become severe around 1990 with depletion occurring perhaps 20 years later. The $^{235}$U enriched uranium fuel needed for the United States light water reactors is expected to run out at about the same time. However, conversion to heavy water reactors fed by fuel produced in a smaller number of breeder reactors can extend the availability of nuclear fission fuel indefinitely.

## 29-7 Nuclear Fusion

Figure 29-8 shows that two light nuclei have more mass or rest energy than their sum. If they can be brought together, the resultant nucleus will be of less mass and there will be a release of energy corresponding to the mass difference. For example, if two deuterons can be fused together in order to form a helium nucleus, the helium nucleus is about 24 MeV less mass than two deuterons and 24 MeV of fusion energy would be released. Thus when two deuterons are joined together into helium, six tenths of 1% of the original rest mass is converted into energy. We see that if this fusion process could be used for energy production, it would be about six times more efficient than fission of uranium. Furthermore, there is an unlimited and inexpensive supply of deuterium in the water of the lakes and oceans, which is not so for other fuels. The world's supply of gas and oil will be depleted in a few decades. Even the supply of coal and uranium will last only a few centuries at most. The big stumbling block to obtaining unlimited energy from "sea water" is Coulomb's law. The electrostatic repulsion between two deuterons at room temperature does not permit them to get within the range of each other's short-range, attractive nuclear force.

**Example 12.** Assume two deuterons must get as close as $10^{-14}$ m in order for the nuclear force to overcome the repulsive electrostatic force. What is the height in MeV of the electrostatic potential barrier?

ANSWER:

$$U = k_0 \frac{e^2}{r} = (9 \times 10^9) \frac{(1.6 \times 10^{-19})^2}{10^{-14}} \text{ J}$$

$$= 2.3 \times 10^{-14} \text{ J} = 0.14 \text{ MeV}$$

**Example 13.** Suppose each deuteron has $\frac{3}{2}kT$, what temperature is needed to overcome the potential barrier?

ANSWER: From Example 12, the two deuterons must have 0.14 MeV or 0.07 MeV per deuteron:

$$\tfrac{3}{2}kT = 0.07 \text{ MeV} = 1.15 \times 10^{-14} \text{ J}$$

$$T = 5.6 \times 10^8 \text{ K}$$

From this example we see that if deuterium could be heated to $\sim 5 \times 10^8$ K, fusion would take place. Because of barrier penetration, the temperature need not be this high. A temperature of $\sim 5 \times 10^7$ K is adequate for controlled thermonuclear power and also will serve to ignite the thermonuclear explosion of an H-bomb. Nuclear reactions that require temperatures on the order of millions of degrees are called thermonuclear reactions. The temperatures momentarily obtained in an A-bomb explosion are high enough to ignite the thermonuclear fuel.

Rather than using liquid deuterium for this, the compound LiH is used where only the isotope $^6$Li is used for the lithium and $^2$H (deuterium) is used for the hydrogen. $^6$Li absorbs neutrons emitted by the reaction

$$^2\text{H} + {}^2\text{H} \rightarrow {}^3\text{He} + \text{n}$$

that is,

$$\text{n} + {}^6\text{Li} \rightarrow {}^3\text{H} + {}^4\text{He}$$

Then the tritium ($^3$H) is used in the reaction

$$^2\text{H} + {}^3\text{H} \rightarrow {}^4\text{He} + \text{n}$$

The net effect is the burning of inexpensive lithium-6 deuteride ($^6$Li$^2$H) to form $^3$He, $^4$He, and neutrons. Once the thermonuclear reactions are started, the extra energy they release can sustain the high temperature until much of the material is quickly "burned." Then we have what is called an H-bomb. The thermonuclear fuel (lithium-6 deuteride) for an H-bomb is very inexpensive, and there is no limit to how much can be used in an individual bomb. Bombs as large as 60 megatons ($6 \times 10^7$ tons of TNT equivalent) have been tested.

The energy release of a purely thermonuclear H-bomb can almost be doubled at very little extra cost by using $^{238}$U for the bomb casing. Then some of the neutrons from the thermonuclear reactions cause fission in the $^{238}$U, which supplies even more neutrons to burn the $^6$Li, and so on. Hence most H-bombs derive almost as much energy from fission as from fusion with all the dangerous radioactive fallout that accompanies fission.

### Controlled Fusion

In order to obtain usable power from fusion, we must have some control over the thermonuclear reactions. We must have some way of generating and sustaining temperatures of many millions of degrees. One of the technical problems is that of confining the high temperature gas or plasma in such a way that the container walls will not melt. A large amount of effort has gone into solving this technological problem. Strong magnetic fields are used in an attempt to keep the plasma from the walls. The problem is to keep the plasma confined for a long enough time to generate more power than what is used to run the thermonuclear reactor. So far the best confinement times are about a factor of 50 too low. Some experts in the field predict success before the end of this century. A very preliminary design is shown in Figure 29-16.

A thermonuclear power plant is much less radioactive than a nuclear reactor because of the absence of fission products. However, many neutrons are released and when captured, radioactive isotopes are usually produced. It is proposed to surround the plasma chamber with a lithium "blanket." Then the neutrons produce tritium ($^3$H with a 12-year half-life) which can be recycled as fuel. Another possibility is to use a uranium or thorium blanket in order to breed fission fuel for nuclear reactors. If the latter possibility can produce power at overall lower cost, thermonuclear power plants might ultimately result in the same kind of radioactivity as is now the case with fission power. A closely related device is called the hybrid reactor. Here neutrons resulting from fusion produce fission power in $^{238}$U. The power output comes from both fusion and fission at the same time. So far all suggested design concepts for fusion "reactors" show little promise of being economically competitive with fossil fuel energy production.

Another possible approach to thermonuclear power is to find a practical way of tapping the energy release of H-bombs. There have been studies of obtaining energy from small H-bombs exploded deep underground in reusable cavities. Such an approach could possibly prove economically competitive with the magnetic confinement approach.

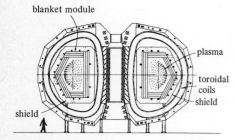

**Figure 29-16.** *Preliminary design of a thermonuclear reactor using magnetic confinement. Superconducting coils supply the magnetic field. Such a reactor would produce ~750 MW. This design is the result of the Oak Ridge National Laboratory Fusion Power Demonstration Study. [Courtesy Don Steiner of Oak Ridge National Laboratory.]*

A smaller scale version of this is to use much smaller (well under 1-cm diameter) "H-bombs." A small pellet containing thermonuclear fuel can be quickly compressed to high temperature by delivering a group of high intensity laser pulses to the pellet. Electron or heavy ion beams could be used in place of the laser beams. Several pellets per minute would be imploded in order to achieve appropriate power output.

There is in nature another method of confinement that permits continuous thermonuclear power generation. This is gravitational confinement. However, a mass of about the size of the sun is needed to provide a strong enough gravitational field. As we shall see in the next chapter, the energy source of the stars is indeed thermonuclear.

## Summary

Nuclear size can be determined by high energy diffraction scattering, electron scattering, or neutron absorption. The result is that nuclei have a radius $R \approx (1.2 \times 10^{-15} \text{ m})A^{1/3}$.

If an electron is "pointed" at the edge of a nucleus of charge $Ze$ and radius $R$, its classical angle of deflection is given by

$$\tan \frac{\theta}{2} = k_0 \frac{Ze^2}{R\, pv}$$

The basic nucleon-nucleon force can be approximated by a square well of radius $r_0 = 2.3 \times 10^{-15}$ m and depth of 29 MeV (when the two spins are parallel). The lowest energy nucleon standing wave in such a potential has the energy level $E = -2.22$ MeV, which is the observed binding energy of the deuteron.

In nuclei with many nucleons the effective well depth is increased to about 42 MeV. Then putting the known density of neutrons into the formula for Fermi energy gives $K_f = 34$ MeV. The corresponding energy level is $E = K_f + U = (34 - 42)$ MeV $= -8$ MeV. The highest energy protons in a nucleus will be at just about the same energy level and the energy required to remove a proton or a neutron from a typical medium-weight nucleus will be 8 MeV. If the observed spin-orbit interaction is added to this potential well model, we have the shell model which results in the magic numbers of 2, 8, 20, 50, 82, 126 for the more tightly bound nuclei.

Alpha decay can be calculated using quantum mechanical penetration of the alpha particle through the coulomb barrier. The mean life for alpha decay is

$$\tau \approx \frac{2R}{v} \frac{|\psi_{\text{in}}|^2}{|\psi_{\text{out}}|^2}$$

If we start with $n_0$ radioactive nuclei, after a time $t$ there will be $n = n_0 \exp(-t/\tau)$.

A nucleus in an excited energy state can emit a photon by spontaneous emission just as an excited atom. This is called gamma decay. Beta decay is a somewhat similar quantum mechanical mechanism except that an electron-neutrino pair is emitted rather than a photon.

Plutonium can be produced by irradiating $^{238}$U with neutrons producing $^{239}$U, which beta decays into $^{239}$Np, which then beta decays into $^{239}$Pu. Both $^{235}$U and $^{239}$Pu undergo nuclear fission quickly after absorption of a low energy neutron. Because the fission products are more tightly bound (by about 1 MeV per nucleon), about 200 MeV is released in fission.

Energy can also be released by fusing together two small nuclei such as $^2$H, $^3$H, or $^6$Li. Very high temperatures ($\sim 10^8$ K) are needed in order to bring two positively charged nuclei close enough to feel the attractive nuclear force that can bind them together into a more tightly bound larger nucleus.

## Exercises

1. Suppose the neutron consists of a core of radius $R_{core} = 0.1$ fm with charge $+e$ surrounded by a larger charged cloud of radius 1 fm and charge $-e$. For 10-GeV high energy electron scattering, what would be $\theta_{max}$ corresponding to the core?
2. What is the electrostatic potential energy of two protons separated by 1 fm?
3. If there are $1.38 \times 10^{44}$ nucleons/m$^3$, what is the distance between adjacent nucleons? (Assume the nucleons are distributed on a cubic lattice.)
4. What is the electrostatic potential energy of a proton at the surface of a copper nucleus?
5. Use Figure 29-10 to get the next magic number above 126.
6. In Example 10, what value of $v$ is used for the average velocity of the alpha particle inside the nucleus?
7. If the half-life of radium is 1600 years, what fraction of a sample of radium will have decayed after 3200 years?
8. Which is longer, 3 half-lives or 2 mean lives?
9. What percent of a radioactive sample decays during one mean life? During two mean lives?
10. Consider a sample of 1000 radioactive nuclei with a half-life $T$. Approximately how many will be left after a time $T/2$?
11. The sun has a mass of $2 \times 10^{30}$ kg and an average density of $1.4 \times 10^3$ kg/m$^3$. What would be the diameter of the sun if it had the same mass but the density of nuclear matter?
12. In a thermonuclear bomb, 18 kg of explosive can give an energy release equivalent to 1 million tons of TNT. One ton of TNT releases $10^9$ cal. How many grams of this thermonuclear explosive get converted to energy?
13. A big nucleus X contains 204 nucleons and has a binding energy per nucleon equal to 8 MeV. Assume that the rest energy of one free proton or one free neutron is 940 MeV.
    (a) Find the rest energy of nucleus X.

(b) Nucleus X emits an alpha particle (28 MeV binding energy) and changes into nucleus Y, which has 8.1 MeV binding energy per nucleon. Find the kinetic energy released in this process.

14. The rest energies of tritium and helium-3 are $M(^3H) = 2805.205$ MeV and $M(^3He) = 2804.676$ MeV. The electron rest energy is 0.511 MeV. Will tritium beta decay, and if so, what will be the beta-ray energy?

## Problems

15. If a star has a radius $R = 2GM/c^2$, light cannot leave it—it is a black hole. Consider a star whose radius is given by this equation and which has a density equal to that of a typical atomic nucleus. What is the radius and what is the mass? Compare it with the solar mass.

16. Show that the neutron collision cross section is

$$\sigma = \frac{N_{in} - N}{N_{in}} \frac{A}{N_a}$$

when the plate thickness $x$ is small. $A$ is the plate area and $N_a$ is the number of atoms in the plate.

17. If the cross section of copper for neutrons is 0.65 barns, for what thickness will a neutron beam be reduced by a factor of $1/e$?

18. Suppose we know $U_0 = 25$ MeV and the binding energy of the deuteron is 2.2 MeV. Use Eq. 29-8 to solve for $r_0$, the range of the effective square well.

19. In a certain heavy nucleus an alpha particle collides with the potential barrier $10^{22}$ times per second and $\psi_{in}/\psi_{out} = 10^{14}$.
    (a) What is the probability that this nucleus would decay in one second?
    (b) What is the mean life and the half-life?

20. A sample of radioactive material contains $10^{12}$ radioactive atoms. If the half-life is 1 hour, how many of these atoms will decay in 1 second?

21. Consider the following photonuclear reaction:

$$\gamma + {}^{63}Cu \rightarrow {}^{63}Cu^*$$
$$^{63}Cu^* \rightarrow {}^{62}Ni + p$$

Assume the photon is absorbed by a proton at the Fermi level. If the photon is 13 MeV,
(a) What will be the energy of the emitted proton?
(b) Give a crude estimate of the number of collisions per second this proton makes against the potential barrier.
(c) If $\psi_{out}/\psi_{in} \approx 10^{-8}$, give a crude estimate of the lifetime of the intermediate state ${}^{63}Cu^*$.
(d) Suppose it had been a 12-MeV photon. Would the lifetime be longer or shorter?
(e) What is the energy threshold for this photonuclear reaction? (What is the lowest possible energy photon which could give rise to an emitted proton?)

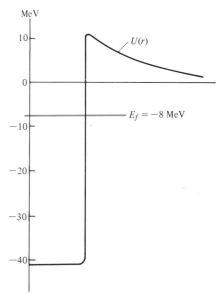

Problem 21. *Potential energy curve for proton in* ${}^{63}Cu$.

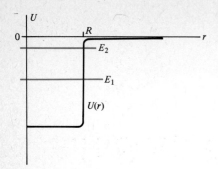

**Problem 22**

22. Assume that in three-dimensional space the potential energy is a function only of $r$ as shown. Plot the function $u(r) = r\psi(r)$ for the lowest two energy levels in this well.
23. Plot $u(r)$ and $\psi(r)$ for Example 7.
24. What fraction of the deuteron is inside the range $r_0$? [*Hint:* This fraction is $(\int_0^{r_0}\psi^2 r^2\, dr)/(\int_0^\infty \psi^2 r^2\, dr)$.]
25. Repeat Example 8 for $A = 60$. What is the proton to neutron ratio?
26. Repeat Example 8 for $A = 40$. What is the proton to neutron ratio?
27. How many grams per day of uranium must be fissioned to produce 1000 MW of electric power? Assume 30% conversion efficiency.
28. United States electric power consumption is about $2 \times 10^{12}$ kW h per year. This is about 30% efficiency in converting heat energy into electric energy.
    (a) How many tons of coal (or oil) are needed per year? How many tons of uranium would be needed if all the electric power plants were nuclear?
    (b) If all the electric power were supplied by exploding 1 megaton H-bombs underground, how many bombs per day would be needed?
29. The present number of United States nuclear bombs is about 33,000. Assume each bomb contains ~2 kg of $^{239}$Pu. This plutonium was produced in military nuclear reactors where each fission gives one neutron for the reaction $n + {}^{238}U \rightarrow {}^{239}U$.
    (a) How many kilograms of fission products were formed in these reactors?
    (b) Assume that commercial nuclear power production in the United States reaches a level of 10% of present electrical power production for 30 years. How many kilograms of fission products would be formed in these commercial reactors?

# 30

# Astrophysics

In this chapter we shall try to understand how stars and planets might have been formed, where the enormous energy radiated by stars comes from, and what happens to a star when its energy source runs out. We shall see that a star could be formed by the gravitational collapse of a cold hydrogen cloud of sufficient mass, and that most of a star's radiated energy is supplied by conversion of the hydrogen into helium via thermonuclear reactions.

---

**Example 1.** How long could the sun live using chemical combustion? How long could it live by converting hydrogen into helium? Assume that its mass is $2 \times 10^{30}$ kg and that it radiates $4 \times 10^{23}$ kW of energy.

ANSWER: A typical energy release by chemical reactions is the $7.9 \times 10^6$ J/kg of $CO_2$ when carbon is burned. If we multiply this by the mass of the sun we obtain $1.6 \times 10^{37}$ J. Since power is $P = E/t$, we have

$$t = \frac{E}{P} = \frac{1.6 \times 10^{37} \text{ J}}{4 \times 10^{26} \text{ W}} = 4 \times 10^{10} \text{ s} = 1250 \text{ years}$$

Hence, if the sun were a mixture of carbon and oxygen and its energy were obtained by burning the carbon, it would burn out after 1250 years.

The energy release by converting hydrogen to helium is 0.7% of the rest energy; hence $E = (7 \times 10^{-3})Mc^2 = 1.3 \times 10^{45}$ J. Then

$$t = \frac{E}{P} = \frac{1.3 \times 10^{45}}{4 \times 10^{26}} \text{ s} = 10^{11} \text{ yr}$$

which is ~20 times longer than the present age of the sun.

When the nuclear fuel is exhausted the star can collapse further into a white dwarf, a neutron star, or a black hole. In this chapter we shall describe white dwarfs, neutron stars, and black holes and give approximate derivations for their masses and radii. These derivations will give us an understanding of the basic properties of such large-scale states of matter. We shall see that the laws of physics which are based on microscopic phenomena seem to work well when applied to astronomically large systems. We shall encounter striking examples of the unity of science such as the interpretation of a neutron star as a giant "atomic" nucleus $\sim 10^{56}$ times larger than an ordinary atomic nucleus.

## 30-1 The Energy Source of Stars

Most cosmological theories start with mainly hydrogen gas as the progenitor of stars and planets. As we shall see, heavier elements up to iron are formed inside stars. The heavier elements beyond iron are believed to be formed during supernova explosions. In the formation of planets like the earth, heavy elements are preferentially selected because the gravitational field is not strong enough to hold the light elements in the gaseous state and they tend to boil off. Star formation starts with a mass of cold hydrogen gas shrinking in size due to gravitational attraction. As the hydrogen atoms "fall" toward each other they pick up kinetic energy or temperature. This heated mass of gas has pressure that resists the gravitational collapse. However, as heat energy of the gas is lost via electromagnetic radiation, the collapse continues to a point where a new source of heat takes over. This new source of energy is thermonuclear, and, as we shall see, in a star it starts up at a temperature of about $10^7$ K.

Now we have enough information to "calculate" the radius of the sun in terms of its mass ($M_s = 2 \times 10^{30}$ kg). In order to simplify the calculation we shall assume the sun is of uniform density. Although the core region of a star is more dense than the surface region, this assumption gives answers correct to within an order of magnitude or better. Gravitational collapse of $2 \times 10^{30}$ kg of hydrogen gas would continue until the pressure due to the thermonuclear heating offsets the gravitational pressure. We saw in Example 3 of Chapter 12 that the gravitational pressure at the center of a uniform sphere of radius $R$ is $P = \frac{1}{2}\rho g R$ where $g = GM_s/R^2$ is the gravitational acceleration at the surface. So

$$P = \tfrac{1}{2}\rho G \frac{M_s}{R}$$

According to Eq. 12-10 the thermal pressure is

$$P = \frac{\rho k T}{M_p} \qquad \text{where } M_p \text{ is the proton mass}$$

Equating these two pressures gives

$$\frac{kT}{M_p} = \frac{1}{2}\frac{GM_s}{R}$$

or

$$R = \frac{GM_s M_p}{2kT} \qquad (30\text{-}1)$$

Using $T \sim 10^7$ K gives $kT \sim 1.38 \times 10^{-16}$ J or

$$R \approx \frac{(6.67 \times 10^{-11}\text{ N m kg}^{-2})(2 \times 10^{30}\text{ kg})(1.67 \times 10^{-27}\text{ kg})}{2(1.38 \times 10^{-16}\text{ J})}$$
$$= 8 \times 10^8 \text{ m}$$

The measured value is $7 \times 10^8$ m. For such an order of magnitude calculation, this is good agreement.

If the starting mass is small, the collapse procedes until the atoms are touching each other. The final result is a planet such as the earth. If the starting mass is larger so that the pressures and densities achieved are high enough to cause overlap of the atomic wave functions, the result is a plasma where the final radius is given by Eq. 30-8. The planet Jupiter is large enough to fit into this category. Here the gravitational pressure is offset by the quantum mechanical pressure of the plasma. (Quantum mechanical pressure is discussed in Section 30-4.)

It can be shown that if the starting mass is greater than 0.08 solar masses the temperatures will be high enough to initiate the following thermonuclear reactions:

$$p + p \rightarrow D + e^+ + \nu \qquad \text{where D = deuterium (}^2\text{H)}$$
$$p + D \rightarrow {}^3\text{He} + \gamma$$
$$^3\text{He} + {}^3\text{He} \rightarrow {}^4\text{He} + p + p$$

This set of three thermonuclear reactions is shown in Figure 30-1 and is known as the proton-proton cycle. It is the main mechanism for producing energy in the sun and other hydrogen-rich stars. The net effect is that 4 protons are used up to form an alpha particle, 2 positrons, 2 neutrinos and 2 photons with total kinetic energies of about 26 MeV. Note that the first reaction, which involves emission of a positron-neutrino pair, is an example of the weak interaction. It is the same as the conversion of a proton in a radioactive nucleus to a neutron plus positron-neutrino pair (a form of beta decay).

**Figure 30-1.** *Schematic representation of the proton-proton cycle. Six protons are used to form* $^4$He *plus 2 protons plus 2* $e^+$ *plus* $2\nu$ *plus* $2\gamma$.

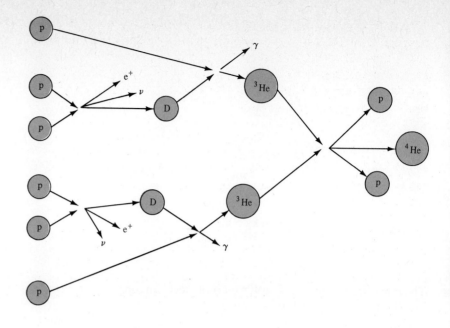

**Example 2.** What temperature is needed to bring two protons to within $5 \times 10^{-15}$ m of each other?

ANSWER: If each proton has $\tfrac{3}{2}kT$, the total kinetic energy of the two is $3kT$, which we equate to $k_0 e^2/R$:

$$T = \frac{k_0 e^2}{3kR} = 1.1 \times 10^9 \text{ K}$$

Even at temperatures one or two orders of magnitude below this, in the center of a star there will be an adequate number of protons above the average energy to keep the reaction going.

Example 2 suggests that a collapsing mass of hydrogen must generate a temperature greater than $10^7$ K due to gravitational collapse before it starts on the proton-proton cycle.

## 30-2 Star Death

We see from Example 1 that the sun, which was formed about $4.5 \times 10^9$ years ago, has so far lived only a small part of its life. Can there be other stars that have used up all their nuclear fuel? Stars of higher masses will have higher temperatures and will burn their hydrogen at faster rates. There are many stars in the sky that, due to their higher mass, have higher intrinsic brightness along with a more bluish color due to the higher temperature. Such stars can use up all their hydrogen in less than $10^{10}$ years (the "age of the universe"). When the supply of hydrogen is exhausted, such a star keeps on radiating and starts to contract; half of the gravitational energy released feeds the radiation and the other half heats the star's interior to a higher temperature $T$ as the star's radius decreases. This contraction plus heating is halted when the temperature is great enough for helium to undergo further thermonuclear reactions (conversion into carbon, oxygen, and neon). Given enough time this process would continue until most of the material in the star's interior has been converted into $^{56}$Fe. This iron isotope has the most stable nucleus of all, and any further nuclear reactions can only use up energy instead of releasing it.

When the thermonuclear fuel has run out in the star's iron core, one obstacle (thermal pressure) to continuous further contraction is removed. A further obstacle for stars of low mass is quantum mechanical pressure, which will be discussed in Section 30-4. However, for sufficiently massive stars the contraction may not only continue, but accelerate into a catastrophic collapse. Various rival mechanisms that would lead to such acceleration have been suggested; which of them actually causes a supernova catastrophe is still controversial. However, we can derive conditions on mass and on radius.

## 30-3 Black Holes

Assuming there are no other kinds of repulsive forces or pressure, the star will keep collapsing. However, there will be a limiting radius $R_0$ called the Schwartzschild radius or event horizon, beyond which we can no longer see the star. If $R < R_0$, then any particles (even photons) emitted by the star will fall back into it. The star would no longer be able to communicate with the outside world. This is illustrated in Figure 30-2(c), which shows all photons falling back into the star. Such photons can travel some

**Figure 30-2.** *Equal mass stars with three different radii. (a) and (b) The radii are greater than $R_0$ where $R_0 = 2GM/c^2$. (c) The radius is less than $R_0$ and the gravitational field is so strong that no photons can escape. In each case photons are emitted at 0°, 30°, and 60° to the star's surface.*

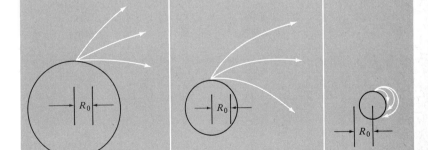

distance away from the star, but they cannot reach infinite distance unless $R > R_0$.

We can estimate $R_0$ by evaluating the gravitational potential energy of a photon emitted from the surface. It is $U = -GMm/R$ where $m = \varepsilon/c^2$ is the gravitational mass of a photon of energy $\varepsilon$ and $M$ is the star's mass. Because of energy conservation, the photon energy at large $r$ will be reduced by $GMm/R$. So if $\varepsilon < GMm/R$ the photon will not get all the way out to infinity. The limiting condition on $R$ is

$$\varepsilon = \frac{GMm}{R_0} = \frac{GM}{R_0}\left(\frac{\varepsilon}{c^2}\right) \quad \text{or} \quad R_0 = \frac{GM}{c^2}$$

**Figure 30-3.** *Artist's rendition of a black hole, from the front cover of* The New York Times Magazine, *July 14, 1974, where a lengthy article on the subject appeared.* [*Copyright 1974 by The New York Times Company. Reprinted by permission.*]

In this calculation we have ignored general relativity. General relativity must be used when gravitational energies are comparable to total energies. A correct calculation gives the exact answer:

$$R_0 = \frac{2GM}{c^2} \quad \text{(event horizon)} \quad (30\text{-}2)$$

For $M = M_s$, the mass of the sun ($2 \times 10^{30}$ kg), Eq. 30-2 gives $R_0 = 3$ km. If a star collapses until its radius is less than or equal to $2GM/c^2$, no particles or light emitted by it could reach the earth (or any other distant observer). However, its gravitational field could still be felt. In fact, its gravitational mass is essentially undiminished and particles (and light) could fall in towards the star—hence the name "black hole." Although the "hole" itself is "black," any gas clouds falling towards it would get compressed and heated and could radiate before reaching the radius $2GM/c^2$. Figure 30-3 shows an artist's rendition of a black hole.

## 30-4 Quantum Mechanical Pressure

We have to consider next an important effect, akin to a repulsive force, which contributes kinetic energy and a strong pressure at high densities (even at low temperature). According to the Pauli exclusion principle, as particles of spin $\frac{1}{2}$ (in particular electrons, protons, and neutrons) are crowded together in a box, they are not permitted to occupy the same momentum states. Thus, the particles cannot be at rest even at zero temperature, but occupy states up to the so-called Fermi momentum, $p_f$. If $n$ of these particles occupy a volume $V$, the spacing between particles is a length of order $(V/n)^{1/3}$. If each particle were confined to such a length, the Heisenberg uncertainty momentum would be $\sim h/(V/n)^{1/3}$. The exact expression for $p_f$ is given by Eq. 28-2:

$$p_f = h\left(\frac{3}{8\pi}\frac{n}{V}\right)^{1/3} \quad (30\text{-}4)$$

A particle of momentum $p_f$ has a kinetic energy $K_f$, the so-called Fermi energy, which increases with increasing density $n/V$ and contributes a pressure. How the energy $K_f$ varies with $p_f$ depends on whether the particle of rest mass $m$ is relativistic or not. The exact expression is

$$K_f = (c^2 p_f^2 + m^2 c^4)^{1/2} - mc^2$$

For $p_f \ll mc$ the particle is nonrelativistic and

$$K_f = \frac{p_f^2}{2m} = \frac{h^2}{8m}\left(\frac{3}{\pi}\frac{n}{V}\right)^{2/3} \quad (30\text{-}5)$$

According to Eq. 12-7 the pressure times volume is

$$PV = n\left(\frac{\overline{mv^2}}{3}\right) = \frac{2}{3}n\overline{K}$$

It is shown in Example 2 of Chapter 28 that $\overline{K} = \frac{3}{5}K_f$ is the average particle kinetic energy. Then

$$PV = \frac{2}{3}n\left(\frac{3}{5}K_f\right)$$

$$P = \frac{2}{5}\frac{nK_f}{V} = \frac{h^2}{20m}\left(\frac{3}{\pi}\right)^{2/3}\left(\frac{n}{V}\right)^{5/3}$$

This is the quantum mechanical pressure of a nonrelativistic Fermi gas.

In the case where the Fermi energy is extremely relativistic, $p_f \gg mc$ and the Fermi energy is

$$K_f = p_f c = \frac{1}{2}hc\left(\frac{3}{\pi}\frac{n}{V}\right)^{1/3} \qquad (30\text{-}6)$$

Then according to Example 3 of Chapter 28, the total kinetic energy of all the particles would be $\frac{3}{4}nK_f$.

## 30-5 White Dwarf Stars

We shall calculate the total energy content of a cold, burned-out star and find that it is a function $E(R)$ of the radius $R$. Such an object will keep radiating and contracting even after exhausting its nuclear fuel until it reaches the lowest possible value of $E(R)$. The total energy consists of the total Fermi energy of its particles ($n\overline{K} = \frac{3}{5}nK_f$) plus the gravitational potential energy $U$. We saw in Example 9 of Chapter 16 that the potential energy of a spherical shell of mass $M$ would be $U = -GM^2/2R$. For a uniform solid sphere it is $U = -\frac{3}{5}GM^2/R$.

We shall consider a star made up of atoms of mass number $A$. Let $n$ be the total number of nucleons. Then there are $n/A$ nuclei each with $Z$ protons and $(A - Z)$ neutrons. We let $x \equiv Z/A$ which in the case of these atomic nuclei will be $\sim\frac{1}{2}$. The total mass of the star is $M \approx nM_p$ where $M_p$ is the proton mass. The total number of electrons (and protons) is $n_e = xn$. We shall see that all electrons (and nuclei) remain nonrelativistic if a star of sufficiently small mass contracts after it has exhausted its thermonuclear fuel. Equation 30-5 then applies to the quantum mechanical kinetic energy for both electrons $(K_f)_e$ and nuclei, but the contribution from the nuclei is negligible (compared to that from the electrons) because of the much larger mass occurring in the denominator. If the star is sufficiently cool that the thermal energy can be neglected compared with the quantum-mechanical energy, the total energy of the star is

$$E(R) = n_e \bar{K}_e + U$$
$$= \tfrac{3}{5} n_e (K_f)_e - \tfrac{3}{5} \frac{GM^2}{R}$$

assuming uniform density. Using Eq. 30-5 with $m_e$ for $m$, $xn$ for $n$, and $V = \tfrac{4}{3}\pi R^3$, we get explicitly

$$E(R) = \tfrac{3}{5} \frac{xn}{R^2} \frac{h^2}{8m_e} \left( \frac{9}{4\pi^2} xn \right)^{2/3} - \tfrac{3}{5} G \frac{n^2 M_p^2}{R} \qquad (30\text{-}7)$$

This is plotted in Figure 30-4.

A star keeps radiating and contracting even after it has exhausted its nuclear fuel, so its total energy $E(R)$ must continue to decrease. For a

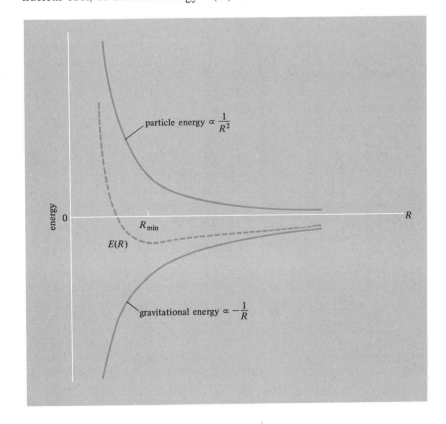

**Figure 30-4.** *The two contributions to the energy of a white dwarf. The upper curve is the total kinetic energy of the electrons, and the lower curve is the gravitational potential energy. The sum of the two is the dashed curve, which has a minimum at $R = R_{\min}$.*

given radius $R$, the lowest value of $E(R)$ is reached if the star has cooled off so that thermal energy is negligible and Eq. 30-7 applies. The lowest value of $E(R)$ can be found by plotting $E$ versus $R$ as in Figure 30-4 or by solving the equation $dE/dR = 0$. This solution occurs for

$$R = \frac{xh^2}{4m_e}\left(\frac{9}{4\pi^2}xn\right)^{2/3}\frac{1}{GnM_p^2} \tag{30-8}$$

If this result is substituted into Eq. 30-7, we see that the gravitational binding energy $(-U)$ is twice that of the total kinetic energy (the first term). White dwarf stars are stars that are still cooling off but have almost contracted down to the radius given by Eq. 30-8. For a typical white dwarf of mass about 15% less than the sun ($n = 10^{57}$ and $x = \frac{1}{2}$), Eq. 30-8 gives $R \approx 8000$ km, which corresponds to a density $\rho \approx 3 \times 10^6$ g/cm$^3$. We see that white dwarfs are about the same size as the earth, but a million times heavier. Even the planet Jupiter can be considered as a white dwarf. This is because Eq. 30-8 holds whether or not the mass is large enough to have thermonuclear reactions. However, the mass must be large enough so that the electrons are in a common Fermi sea (a plasma). Equation 30-8 is plotted in Figure 30-5 and it is seen that it agrees well with the values for the mass and radius of Jupiter.

If Eq. 30-7 is correct, the star cannot shrink below the radius given by Eq. 30-8. Then $R \propto n^{-1/3}$, and the Fermi energy per electron increases as $n^{4/3}$. As $n$ becomes larger, Eq. 30-7 becomes incorrect due to the fact that the energy per electron becomes comparable to its rest-mass energy $m_ec^2$. When the exact relativistic expression for $(K_f)_e$ is used in Eq. 30-7, one

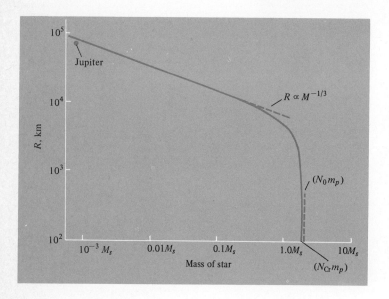

**Figure 30-5.** *The value of $R$ that minimizes the sum of the electron Fermi energy plus the gravitational energy is plotted as a function of the star's mass $M$. Uniform density is assumed and the exact relativistic expression is used for Fermi energy. For small masses this exact result approaches the nonrelativistic Eq. 30-8. Note the limiting mass given by Eq. 30-10 when the electrons become extremely relativistic.*

obtains the result shown in Figure 30-5 for $R$ as a function of mass. In the extreme relativistic limit Eq. 30-6 should be used instead of Eq. 30-5 for the "Fermi energy" of the electrons and then Eq. 30-7 should be replaced by

$$E(R) = \tfrac{3}{4} \frac{xn}{R} \frac{hc}{2} \left( \frac{9}{4\pi^2} xn \right)^{1/3} - \tfrac{3}{5} G \frac{n^2 M_p^2}{R} \qquad (30\text{-}9)$$

Note that now both terms have the same dependence on $R$ and that the second term dominates if the mass $nM_p$ is sufficiently large. If $n$ exceeds a critical number $n_{cr}$, the energy $E(R)$ decreases continuously with decreasing radius. There is no stable configuration at any radius without thermal energy and thermal pressure. This leaves open the possibility for those stars that happen to be heavy enough of continued contraction as the star continues to radiate (for the eventual fate of the star see the following sections).

Equation 30-9 is not strictly true unless the electrons are extremely relativistic and is based on the approximation of uniform density (in reality the density is larger at the center of a star than further out). Nevertheless we can get an estimate of the critical number $n_{cr}$ by finding the value of $n$ for which the two terms in Eq. 30-9 are equal. Then

$$\tfrac{3}{4} \frac{xn_{cr}}{R} \frac{hc}{2} \left( \frac{9}{4\pi^2} xn_{cr} \right)^{1/3} \approx \tfrac{3}{5} G \frac{n_{cr}^2 M_p^2}{R}$$

or

$$n_{cr} \approx \frac{(125\pi)^{1/2}}{4} x^2 \left( \frac{\hbar c}{GM_p^2} \right)^{3/2}$$

Note the occurrence in this expression of a dimensionless quantity $n_0$,

$$n_0 \equiv \left( \frac{\hbar c}{GM_p^2} \right)^{3/2} = 2.4 \times 10^{57} \qquad (30\text{-}10)$$

An exact calculation of $n_{cr}$ requires numerical integrations and gives $n_{cr} = 0.7(x/0.5)^2 n_0$. The corresponding critical mass, $n_{cr} M_p$, is called the "Chandrasekhar limiting mass" and is the largest mass a white dwarf star can have and still cool off to a stable cold state of finite radius and density. This mass is only about 40% heavier than that of the sun, $M_s = 0.49 n_0 M_p$.

Note that the quantity $n_0$ does not depend on the value of the electron mass $m_e$, even though electron kinetic energy dominates in white dwarfs. This is due to the fact that the extreme relativistic form for energy in Eq. 30-6 does not depend on mass. On the other hand, the density $\rho_r$, at which the Fermi momentum $p_f = m_e c$, does depend on $m_e$ as seen in Example 3.

> **Example 3.** What is the density $\rho_r$ of a white dwarf when the electron Fermi momentum $p_f = m_e c$?
>
> ANSWER: Solving Eq. 30-4 for $(n/V)$ gives
>
> $$\frac{n_e}{V} = \frac{8\pi}{3}\left(\frac{p_f}{h}\right)^3$$
>
> Replace $n_e$ with $(xn)$ and $p_f$ with $(m_e c)$ and multiply by $M_p$:
>
> $$\frac{(xn)M_p}{V} = \frac{8\pi}{3}\left(\frac{m_e c}{h}\right)^3 M_p$$
>
> $$\rho_r = \frac{8\pi M_p}{3x}\left(\frac{m_e c}{h}\right)^3 \quad \text{when } p_f = m_e c \quad (30\text{-}11)$$
>
> $$= \frac{0.97 \times 10^6}{x} \text{ g/cm}^3$$
>
> For $x = \frac{1}{2}$, such a star has a density of about $2 \times 10^6$ g/cm³.

## 30-6 Neutron Stars

Suppose a white dwarf of mass near the sun's mass could be compressed by external forces. We shall see that a much lower value for the total energy is possible at a much lower value of $R$ if the value of $x$ could be reduced from $\sim\frac{1}{2}$ to near zero. According to Eq. 30-7 such fictitious external forces must at first do work. However, beyond a certain point the electron energies would be so high that protons would start converting into neutrons via the weak interaction:

$$e^- + p \rightarrow n + \nu$$

This process is called inverse beta decay and requires electrons of large kinetic energy, but high energies are available if the density is sufficiently large (see Eq. 30-6). The detailed conversion of ordinary matter into neutron-rich material is complicated, but calculations show that neutrons are much more abundant than protons at densities $\rho > 10^{11}$ g/cm³. We are then dealing with a "neutron star," but we still have to find the equilibrium values of radius $R$ and of $x$ for a given total number of nucleons with $n_p = n_e = xn$. Since we are now dealing with densities much greater than $(\rho_r)$ in Eq. 30-11, the electrons will be extremely relativistic for any neutron star and Eq. 30-6 applies for the electrons. On the other hand, neutrons with their much larger mass $M_p$ remain nonrelativistic as long as the density is less than the neutron equivalent of Eq. 30-11:

$$(\rho_r)_n = \frac{8\pi M_p}{3(1-x)} \left(\frac{M_p c}{h}\right)^3 = \frac{6 \times 10^{15}}{1-x} \text{ g/cm}^3 \quad (30\text{-}12)$$

In the case where the star density is less than that of Eq. 30-12, Eq. 30-5 can be used for the neutrons and the energy of the protons can be neglected entirely because they make up only a small fraction of the nucleons. So in order to obtain $E(R, x)$, we merely add the total neutron kinetic energy to Eq. 30-9:

$E(R, x)$ = neutron energy + electron energy + gravitational energy

$$= \tfrac{3}{5} \frac{(1-x)n}{R^2} \frac{h^2}{8M_p} \left[\frac{9}{4\pi^2}(1-x)n\right]^{2/3}$$

$$+ \tfrac{3}{4} \frac{xn}{R} \frac{hc}{2} \left(\frac{9}{4\pi^2} xn\right)^{1/3} - \tfrac{3}{5} G \frac{n^2 M_p^2}{R} \quad (30\text{-}13)$$

The values of $R$ and $x$ that minimize this expression for a given value of the stellar mass $M = nM_p$ could be found graphically or by solving the two simultaneous equations

$$\frac{\partial E}{\partial R} = 0 \quad \text{and} \quad \frac{\partial E}{\partial x} = 0$$

Because $x$ is very small, one can get a good approximation to the solution by a succession of two simpler procedures: first use $x = 0$ in the equation $\partial E/\partial R = 0$ to get $R$. Then use this value of $R$ in the equation $\partial E/\partial x = 0$ to get $x$. Putting $x = 0$ into the equation $\partial E/\partial R = 0$ gives

$$R = \frac{h^2}{4M_p} \left(\frac{9}{4\pi^2} n\right)^{2/3} \frac{1}{GnM_p^2} \quad (30\text{-}14)$$

This expression is very similar to Eq. 30-8, but numerically smaller because the nucleon mass rather than the electron mass appears in the denominator. Hence the neutron star radius will be about 1000 times smaller than the white dwarf radius. For a star of one solar mass, $n = 1.2 \times 10^{57}$ and Eq. 30-14 gives $R = 12.6$ km and a density of $2.4 \times 10^{14}$ g/cm$^3$. If a star of the mass of the sun is collapsed down to a 15-mile diameter sphere, it has a density that happens to be almost the same as that inside atomic nuclei. Finally, in order to determine $x$, the fraction of protons in this giant "atomic nucleus," we put $R = 12.6$ km into the equation $\partial E/\partial x = 0$ and solve for $x$. The result is $x = 0.005$. We see that about 99.5% of the nucleons in such a giant "atomic nucleus" would be neutrons. Ordinary atomic nuclei contain no electrons; however, in neutron stars about 0.5% of the particles will be electrons.

In our equation for $E(R, x)$ we have neglected the effect of nuclear forces. Since the densities inside the star are comparable with those inside atomic nuclei, the nuclear forces are, in fact, appreciable. However, for masses between about half a solar mass to one solar mass the attractive and the repulsive components of the nuclear forces tend to cancel each other and accurate calculations for $R$ and $x$ give rather similar numerical values to our approximate ones.

---

**Example 4.** If the sun could contract down to $R = 20$ km without losing any angular momentum, what would be the frequency of its rotation? The present period of rotation is 27 days, which corresponds to $f = 4.3 \times 10^{-7}\,\text{s}^{-1}$. The present radius is $R = 1.4 \times 10^6$ km.

ANSWER: If $I$ is the moment of inertia, the angular momentum is

$$L = I\omega = I'\omega'$$

$$\omega' = \left(\frac{I}{I'}\right)\omega$$

$$f' = \left(\frac{R}{R'}\right)^2 f = \left(\frac{1.4 \times 10^6\,\text{km}}{20\,\text{km}}\right)^2 (4.3 \times 10^{-7}\,\text{s}^{-1}) = 2.1 \times 10^3\,\text{rev/s}$$

---

The collapse of a burnt-out star down to a neutron star is a catastropic event (called a supernova). Large gas clouds which contain much of the angular momentum are ejected. However, we might expect from Example 4 that a newly formed neutron star could be making dozens of revolutions per second. If the surface of such a neutron star had a disturbance akin to sunspots, the radiation eminating from it would be like a giant searchlight beacon. For example, the star at the center of the Crab Nebula is such a neutron star rotating at 30 revolutions per second. Figure 30-6 shows the gas cloud remains of the supernova explosion that was observed in the year A.D. 1054. The explosion was so spectacular that it could be seen during the daylight hours. The star at the center is observed both by radiotelescopes and optical telescopes to flash on and off 30 times per

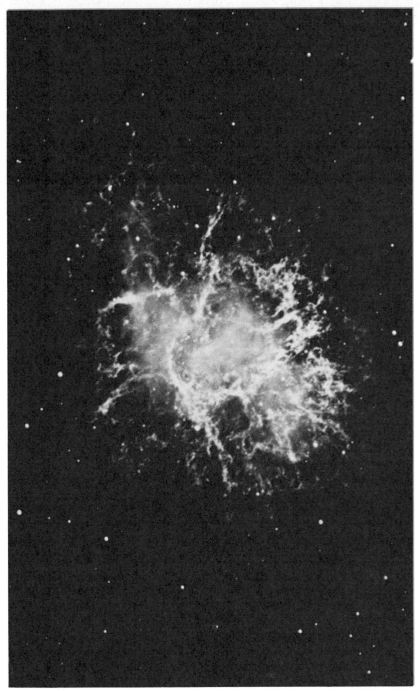

**Figure 30-6.** *Photo of the Crab Nebula showing present distribution of the ejected gas cloud. [Courtesy California Institute of Technology.]*

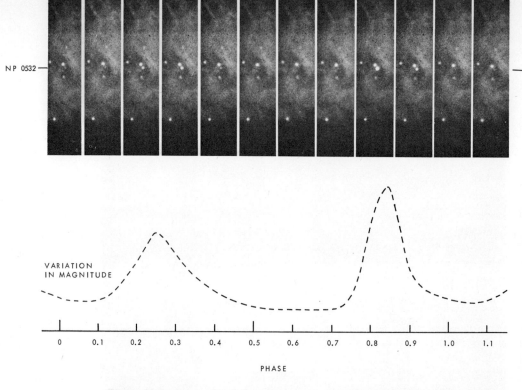

**Figure 30-7.** *Accumulated short time exposures of central region of the Crab Nebula. The parent neutron star is level with the tick marks. Its large brightness charges are plotted below. The phase from 0 to 0.1 corresponds to $\frac{1}{300}$ s. [Kitt Peak photograph.]*

second as shown in Figure 30-7. Such flashing neutron stars are called pulsars. When the first pulsar was discovered in 1967 by a radiotelescope, it was not clear what mechanism could give such accurately timed radio pulses at such a fast rate. It was a completely new and unanticipated astronomical phenomenon. One of the early hypotheses was the LGM theory where LGM stands for "little green men." This theory of outside intelligence trying to communicate with us was abandoned the following year when a second pulsar was discovered.

A key test of the neutron star hypothesis is that there should be a reduction of rotational energy equal to the energy that is radiated away. Recent measurements do detect a slowing down in agreement with the prediction.

Even without nuclear forces, we saw that for a neutron star of about one solar mass the density should be similar to ordinary nuclear matter where the typical Fermi energy per neutron is ~40 MeV. For a more massive star the density would be greater and the neutron kinetic energy could approach its rest energy of 939 MeV. For densities exceeding $(\rho_r)_n$ of Eq. 30-12, one must use the relativistic form (Eq. 30-6) for neutron Fermi energy. Just as we did for white dwarfs, we will get an upper limit to the mass at which neutron stars can be stable; that is, a critical mass above which the star must become a black hole (unless it loses mass first).

For the crudest kind of estimate for this critical mass we put $x = 0$, neglect nuclear forces and general relativity, and consider only special relativity and pure neutrons. For extremely relativistic neutrons we have for $E(R)$ the expression in Eq. 30-9, but with the factor $x$ replaced by 1 (since all the particles are neutrons instead of the $x$ fraction of electrons). This estimate for the critical mass is only larger by $1/x^2$ or a factor 4 from what we had before for the white dwarf critical mass. It should be the same order of magnitude as $n_0 M_p$, where $n_0$ is given by Eq. 30-10. More accurate calculations have been carried out with general relativity and nuclear forces included. The effects of general relativity are only moderately large and well understood. The effects of nuclear forces are also only moderate, but the details of nuclear forces are not sufficiently well known at the moment and results of different calculations for the critical mass vary from about 0.7 to 4 times the mass of the sun.

## 30-7 Black Hole Critical Mass

In the previous sections we gave mainly theoretical discussions of white dwarf stars, neutron stars, and black holes as three different possibilities for the end result of evolution of stars. Which possibilities happen under which conditions cannot be predicted with any reliability from purely theoretical grounds at the moment. But we know that ultimately each star in the sky must end up as a white dwarf, a neutron star, or a black hole.

One can estimate reasonably well the rate at which stars in our galaxy should die, and this rate is about one or a few stars per year. (A very crude estimate would be the total number of stars divided by the age of the galaxy, which agrees within a factor 5 of the given estimate.) White dwarf stars are one category of stars that is quite noncontroversial. They are very numerous (about 10% of all stars in our galaxy) and at least the nearer ones can be seen and studied in detail. Analysis of statistical data indicates that white dwarfs are cooling off as expected theoretically and that most star deaths proceed in a fairly unspectacular manner through the white dwarf stage.

Supernova events are very spectacular and rare (about one per hundred star deaths in total). That they occur and that a shell of gas is ejected is

## 30-8 Summary of Experimental Evidence

uncontroversial, but what happens to the core of the original star is controversial. The two theoretical possibilities are a neutron star and a black hole. For a neutron star there is now a lot of recently obtained evidence that is indirect, but nevertheless quite compelling. Most of this evidence comes from pulsars (discovered in 1967), which are sources emitting radio noise in pulsed bursts at very regularly repeated time intervals. The period of this pulsing is so short (down to about 0.03 sec) that it is hard to think of any visible object other than a neutron star. Fairly detailed theories for pulsars have been put forward; the details of the radiation mechanism are still very controversial, but all agree that a rotating neutron star must underlie the emitting region and that its rotational kinetic energy supplies the energy source of radiation. One pulsar is seen in a known supernova remnant (the Crab Nebula, explosion observed A.D. 1054) and is also seen to pulse in visible light. So far about 70 pulsars have been discovered.

Even more recently, a number of compact x-ray sources have been discovered from detectors in rockets and in satellites. These sources all show short time variations, which all indicate an object as small in diameter as a neutron star or else matter in the vicinity of a black hole (for larger objects, the difference in light travel time from different parts of the object would smear out short time variations). Most of these time variations are erratic, but in a few cases they are regular enough so they can be studied in detail. In six or seven cases these studies indicate a binary system with the compact x-ray source and a more or less normal star orbiting about each other. In two cases (Hercules X-1 and Centaurus X-3) the orbital velocity of the x-ray source has been measured from Doppler shifts and the x-ray sources appear to be pulsars; in a third case (Cygnus X-1), Doppler shift measurements gave the velocity of the ordinary or optical star. Unfortunately, simultaneous velocity measurements for both partners in a pair have not yet been achieved.

Nevertheless, in each of these three cases one can make some estimates of the mass of the x-ray source (unfortunately only by making a few assumptions, however). The most clear-cut case is Hercules X-1 where the x-ray source probably has a mass of the order of $0.6 \pm 0.2$ solar mass, a reasonable value for a neutron star. In this case a lot of observational data on the x-ray source and the optical companion is available and the data are all compatible with the x-rays coming from the surface of a neutron star which is bombarded by matter expelled from the ordinary star and falling in onto the neutron star. For Cygnus X-1 the mass estimates indicate about $5M_s$ for the x-ray source. Since this value exceeds the critical mass for a stable neutron star, it is likely that this x-ray source is—or rather surrounds—a black hole (if the assumptions that went into the estimate turn out to be correct). The hole itself is "black," but matter expelled from the ordinary star is accelerated by gravity and compressed as it falls toward the black hole. In such a model the compressed matter also heats up and radiates x rays during its infall, but only *before* it passes the critical radius $R_0$, given by Eq. 30-2. In 1978 a second binary star with characteristics similar to Cygnus X-1 was found in the constellation Scorpio. This black hole is called Scorpii V-861. Also there is evidence that the distant galaxy M87 has a large black hole at its center.

Gravitational collapse into a neutron star or black hole should be catastrophic with the radius $R$ decreasing at a rate comparable to the speed of light in the final stages. Such acceleration of mass according to general relativity must emit a pulse of gravitational radiation analogous to electromagnetic radiation produced by an accelerating charge. According to the theory it should be possible to construct gravity-wave detectors here on earth with enough sensitivity to detect such a collapse (or supernova). Preliminary versions of such gravity-wave detectors have come into operation recently and others with improved sensitivity are still under construction. With them it might be possible to detect supernovas or collisions with black holes in the center of our galaxy that otherwise would be undetectable.

## Summary

In a star of mass $M$ the gravitational pressure is offset by the thermal pressure, which gives rise to the relation

$$kT \approx \frac{GMM_p}{2R}$$

Heat to supply the thermal pressure comes from decrease of gravitational potential energy during collapse, but once a temperature greater than $10^7$ K is reached, the proton-proton thermonuclear cycle starts up and the slow collapse is arrested.

When the thermonuclear fuel is used up, there will then be further gravitational collapse until the quantum mechanical pressure offsets the gravitational pressure. In a white dwarf the quantum mechanical pressure comes from the electron Fermi energy (the pressure is $P = \frac{2}{5}nK_f/V$). In some cases this final collapse is so catastrophic that most of the electrons and protons convert into neutrons, giving a neutron star where the quantum mechanical pressure is supplied by the neutrons as well as the few remaining electrons. Then the radius is $\sim 10$ km and the density is about the same as that inside an atomic nucleus. For a white dwarf, the radius is $\sim 10^4$ km and density about $10^6$ times that of the earth. The white dwarf radius goes as $M^{-1/3}$ up to the point where the electrons become relativistic giving a limiting mass $\sim 1.4 M_s$.

Neutron stars have a limiting mass $\sim 3 M_s$ for the same reason (the neutrons become relativistic). A collapsing star of mass greater than $\sim 3 M_s$ should form a black hole providing it all holds together during collapse.

Once the radius $R_0 = 2GM/c^2$ is reached, no photons or other particles can leave the region of the "star." Signals cannot be received from a black hole, although its gravitational field can be felt.

Because neutron stars (also called pulsars) retain the angular momentum of their parents, they rotate several revolutions per second and any radiation they emit will have this fast periodicity or pulse rate. Some neutron stars and probably some black holes have an ordinary star as a close partner. Some of the outer gas of the ordinary star can be sucked into a black hole partner emitting characteristic x rays. This is the best experimental evidence so far for a black hole.

## Exercises

1. The solar flux received at the Earth is 1.4 kW/m². What then is the total power output of the sun? The earth–sun distance is $1.5 \times 10^{11}$ m.
2. Assume Eq. 30-1 applied to Jupiter while it was collapsing during formation. What then would be the temperature of Jupiter's core? Its mass and radius are $6 \times 10^{24}$ kg and $7.2 \times 10^4$ km, respectively.
3. In the proton-proton cycle each proton liberates about 6 MeV of kinetic energy. If the sun's output is $4 \times 10^{23}$ kW, how many protons per second are undergoing thermonuclear reactions?
4. In order for a material particle to escape from a black hole at a distance equal to $R_0$, the event horizon, it must have an escape velocity equal to the speed of light. Use the nonrelativistic formula for escape velocity to solve for $R_0$.
5. Consider the entire visible universe as one giant black hole of radius $R_u = 10^{10}$ light years. What must be the average density $\bar{\rho}$ in order for $R_u$ to equal $2GM_u/c^2$? Give a numerical value. (Actually the observed density is uncertain, but it could equal this value.)
6. Use Eq. 30-7 to obtain $dE/dR$. Set this equal to zero and solve for $R$.
7. If the radius of a white dwarf is decreased by a factor of 2, by what factor does the Fermi energy per electron increase?
8. What is the ratio of the "critical" density for a neutron star to the density of an atomic nucleus? (See Eq. 30-12.)
9. Use Eq. 30-13 to obtain $\partial E/\partial R$. Set this equal to zero and solve for $R$ when $x = 0$.

## Problems

10. Assume that little energy is radiated away as a large hydrogen cloud of mass $M$ collapses; that is, assume that the decrease in potential energy equals the increase in kinetic energy. If the kinetic energy is $\frac{3}{2}nkT$ and $U = -\frac{3}{5}GM^2/R$, solve for $R$ in terms of $T$, $M$, and $M_p$.
11. Derive a formula for the critical angular velocity $\omega_c$ of a uniform sphere of mass $M$ and radius $R$. (If $\omega > \omega_c$, the sphere will fly apart.) (Consider a mass $\Delta m$ on the equator.) ANSWER: $\omega_c = (GM/R^3)^{1/2}$.
12. Use conservation of angular momentum to calculate the smallest radius the sun can reach in gravitational collapse, assuming no angular momentum is lost. Use the result of Problem 11.

13. Suppose the universe is a finite sphere with radius $R$ and average density $\bar{\rho} = 10^{-28}$ g/cm$^3$. What is the velocity of escape as a function of $R$? At what value of $R$ would it equal the velocity of light? (It is interesting to note that the answer to this problem turns out to be close to the observed velocity of recession of galaxies at a distance $R$. Also note that the relation between $\bar{\rho}$, $G$, and $R_u$ is the same as for Exercise 5.)
14. Repeat the derivation of Eq. 12-7 to show that for relativistic particles $PV = \frac{1}{3}n\overline{pv}$. (Hint: $\bar{F}_x = \Delta p_x/\Delta t$ and $\overline{p_x v_x} = \overline{p_y v_y} = \overline{p_z v_z} = \frac{1}{3}\overline{pv}$.)
15. Using the result of Problem 14, show that for extreme relativistic particles the quantum mechanical pressure due to Fermi energy is

$$P = \frac{hc}{8}\left(\frac{3}{\pi}\right)^{1/3} \mathcal{N}^{4/3}$$

16. What white dwarf density is needed to give electrons up to 1 MeV of kinetic energy? Assume $x = \frac{1}{2}$.
17. A box contains $\mathcal{N}$ photons per unit volume of frequency $f_0$. What is the pressure?
18. Use Eq. 30-8 to obtain the density of a white dwarf as a function of the mass. Equate this to $\rho$ in Eq. 30-11 and solve for $n$. (This is another way to get an estimate for the limiting mass for a white dwarf.)
19. If all the protons in the sun are at $T = 10^7$ K, how many of them will have an energy at least as large as $\frac{1}{2}(k_0 e^2/R)$ where $R = 10$ fm? Use $\exp(-E/kT)$ for the relative probability as a function of energy.

# Particle Physics

The main goal of physics is to explain all physical phenomena by means of a small number of simple, fundamental principles. Since all matter is made up of elementary particles, the countless number of physical phenomena and properties of matter can be explained in principle by the few simple properties of a small number of elementary particles.

In man's search for the elementary particles, he first found that compounds were made up of "elementary" molecules. Then it was learned that molecules are made up of "elementary" atoms. Centuries later these "elementary" atoms were discovered to be made up of "elementary" nuclei and orbital electrons. These successive probings to learn what is truly elementary are like peeling away successive layers of an onion. A later stage in this peeling down of ordinary matter was the discovery that all nuclei are made of protons and neutrons.

Have we finally reached the core of the onion? Are the proton and neutron made up of even smaller elementary particles? Up until recently the proton and neutron were thought of as elementary with no internal structure. Assuming that the proton, neutron, electron, and photon are the basic building blocks of ordinary matter as explained in Chapters 28 and 29, this would seem like an appropriate place to end this book. However, since a central purpose of this book is to cover fundamental topics of physical reality, before closing we must ask: Are there any other physically real elementary particles that happen not to appear in ordinary matter? The answer is an emphatic yes! Since 1933 physicists have discovered over 200 other elementary particles. And the evidence is getting stronger that most of these 200 particles can be explained as being made up of four basic subparticles called **quarks.** So far, however, no particle has been split apart into its constituent quarks. For a reference on quarks, see Figure 31-1.

The next few sections will introduce some of the new particles that have

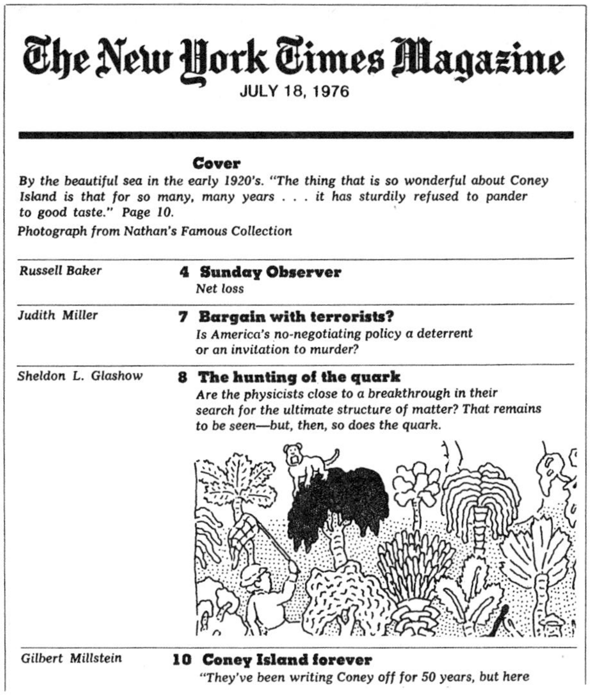

The New York Times Magazine
JULY 18, 1976

**Cover**
*By the beautiful sea in the early 1920's. "The thing that is so wonderful about Coney Island is that for so many, many years . . . it has sturdily refused to pander to good taste." Page 10.*
*Photograph from Nathan's Famous Collection*

*Russell Baker* — **4 Sunday Observer**
*Net loss*

*Judith Miller* — **7 Bargain with terrorists?**
*Is America's no-negotiating policy a deterrent or an invitation to murder?*

*Sheldon L. Glashow* — **8 The hunting of the quark**
*Are the physicists close to a breakthrough in their search for the ultimate structure of matter? That remains to be seen—but, then, so does the quark.*

*Gilbert Millstein* — **10 Coney Island forever**
*"They've been writing Coney off for 50 years, but here*

**Figure 31-1.** *The elusive quark makes the* Times. [*Copyright 1976 by The New York Times Company. Reprinted by permission.*]

been observed. Then Section 31-6 will outline the quark theory. This is followed by a summary of the known conservation laws.

## 31-1 The Weak Interaction

Most of the new elementary particles are unstable—they decay or transform into other elementary particles of smaller mass. Before discussing the new particles themselves, it is necessary to learn more about beta decay and what is called the weak interaction. The weak interaction can be thought of as a universal disease attacking all particles with the same strength. It tries to transform elementary particles into electrons and neutrinos as a final product. Historically these decay electrons are called beta rays.

Since beta decay always involves neutrinos, a description of the neutrino is essential. A neutrino is an elementary particle that has no charge and no rest mass. Furthermore, the interaction between a neutrino and anything else is so weak as to be almost unobservable. If a beam of $10^{12}$ neutrinos were shot at the earth, all but one would pass through the earth completely unaffected. So far this particle sounds like a figment of the imagination. However, the neutrino is not a hoax dreamed up by theoretical physicists. The evidence, both experimental and theoretical, has become so convincing that no competent scientist questions the existence of the neutrino.

In 1958 great progress was made in understanding the weak interaction. A specific interaction that can transform particles into electrons and neutrinos was proposed. This specific interaction called the universal Fermi interaction was developed by R. Feynman and others. For example, the new theory gives an accurate prediction of the muon lifetime.

One example of a weak interaction is the beta decay of a free neutron by the process

$$n \rightarrow p + e^- + \bar{\nu}_e$$

with a half-life of 12 min and energy release of 0.8 MeV. The neutron-proton mass difference is 1.3 MeV. Since the rest mass of the electron is 0.5 MeV there is 0.8 MeV left over, which goes into the kinetic energy of the electron plus the kinetic energy of the antineutrino. (The symbol $\bar{\nu}_e$ is used for the antineutrino that accompanies an electron.) The difference between the neutrino and the antineutrino is explained in Section 31-3.

Note that the decay electron and antineutrino can share the 0.8 MeV of kinetic energy in any way they choose. Thus, in a large group of neutron decays, electrons may be found having values of kinetic energy anywhere from zero to 0.8 MeV. If the neutron decayed into a proton and electron only, the electron would always have the full kinetic energy of 0.8 MeV. Experimentally the electron energy can be determined by measuring its radius of curvature in a magnetic field (see Figure 31-3). The experimental result is that beta rays rarely have their maximum permissible kinetic energy. Historically, this "puzzle" of what happened to the "missing" kinetic energy was the reason the neutrino was "invented" in the first place.

In order to preserve conservation of energy, W. Pauli proposed in 1930 that the missing energy may have been carried off by an undetectable, light, neutral particle. Shortly thereafter E. Fermi named this particle the neutrino (little neutral one) and worked out the theory of beta decay that is similar to the modern, more universal theory of weak interactions, called the universal Fermi interaction.

The evidence for neutrinos as decay products was quite satisfying. However, the physicists wanted to be more satisfied by observing a direct interaction starting with a beam of neutrinos. According to the Fermi theory the following reaction should take place for antineutrinos which have sufficient energy to make up the mass difference:

$$\bar{\nu}_e + p \rightarrow n + e^+$$

where $e^+$ is a positron (positive electron). However, the predicted occurrence of this direct neutrino interaction was close to zero. The only

**Figure 31-2.** *R. P. Feynman on a visit to Cornell University. Professor Feynman received the Nobel Prize for his contributions to the modern theory of quantum electrodynamics and the universal Fermi interaction.* [Photo by H. Y. Chiu]

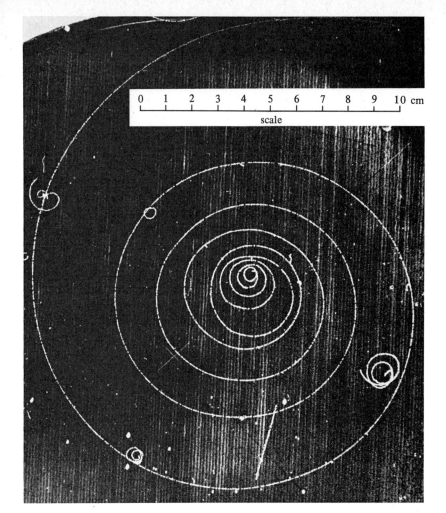

**Figure 31-3.** *Electron track in liquid hydrogen bubble chamber. Track is curved because of uniform magnetic field pointing out of the page. Radius of curvature decreases as electron is slowed down by liquid hydrogen. [Photograph courtesy Alvarez group, Lawrence Radiation Laboratory, University of California.]*

possibility of observing this reaction would be to obtain an extremely intense beam of antineutrinos. The development of the nuclear reactor made this possible. Each fission in a nuclear reactor leads to several beta decays, and hence several antineutrinos. Such an experiment would require a high-power nuclear reactor and an elaborate large-sized detector. A group from Los Alamos undertook this experiment using the detector shown in Figure 31-4, with the result that the elusive neutrino was first trapped by man in 1956.

At high energy accelerators it is possible to produce beams of neutrinos of much higher energy, but less intensity, than those produced in nuclear reactors. Fortunately the cross section for a neutrino interaction increases linearly with energy, so that in the last few years thousands of high energy neutrino interactions have been obtained using these high energy beams.

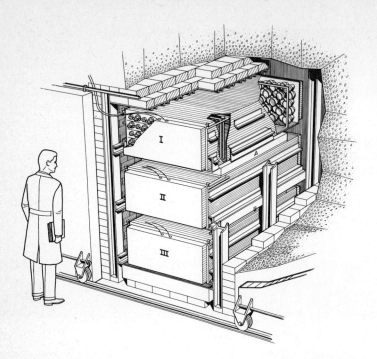

**Figure 31-4.** *Cutaway drawing of neutrino detector in the wall of a nuclear reactor. Each of the three tanks (I, II, and III) contains 370 gal of liquid scintillator and is monitored by 110 photomultiplier tubes. The positron in the reaction $\bar{\nu} + P \to N + e^+$ will produce distinctive light pulses in the liquid scintillator.* [Courtesy Los Alamos Scientific Laboratory and Dr. F. Reines.]

### Law of Conservation of Baryons

If the free neutron can decay by weak interaction, one might think that the free proton should also. However, careful measurements have shown that free protons are stable. This observation plus other interactions among strongly interacting particles have led to the formulation of the law of conservation of baryons. The proton, neutron, and about half of the strongly interacting particles are baryons. The law of conservation of baryons states that in a closed system the total number of baryons minus antibaryons stays the same. So, since there is no baryon lighter than the proton, there is nothing available for the proton to decay into. Baryons have half-integral intrinsic spin; that is, $\frac{1}{2}\hbar$, $\frac{3}{2}\hbar$, $\frac{5}{2}\hbar$, and so on.

## 31-2 High Energy Accelerators

Without some discussion of how they are produced and detected, the new elementary particles might seem the product of a superhuman imagination.

We shall define high energy accelerators as those that give proton or electron beams at energies above 1 GeV ($10^9$ eV). Except for one linear accelerator, all the world's high energy accelerators are synchrotrons. In a synchrotron the particles circulate in an evacuated hollow donut between the poles of a ring-shaped magnet (see Figure 31-5). The particles are

**Figure 31-5.** *Early Cornell electron synchrotron before insertion of the donut-shaped vacuum tank. Electric field for acceleration is in the cavity at the right.*

accelerated by passing through regions of electric field generated by a radio frequency oscillator. The radius of curvature can be obtained by equating the centripetal force $mv^2/R$ to the magnetic force $ev\mathcal{B}$:

$$\frac{mv^2}{R} = ev\mathcal{B}$$

$$\frac{p}{R} = e\mathcal{B}$$

$$R = \frac{1}{e\mathcal{B}}p \qquad (31\text{-}1)$$

Thus if $\mathcal{B}$ is kept proportional to the momentum $p$, the particles will keep the same radius of curvature and stay in the center of the donut-shaped vacuum tank.

---

**Example 1.** A synchrotron of radius $R$ has four straight sections of length $L$ each. If the period of the RF oscillator corresponds to the time of one revolution, what is the magnetic field in terms of $R$, $L$, and the oscillator frequency $f$?

ANSWER: The oscillator period is

$$T = \frac{\text{(path length)}}{v} = \frac{2\pi R + 4L}{v} = \frac{1}{f}$$

$$v = (2\pi R + 4L)f$$

Now substitute this into Eq. 31-1:

$$\mathcal{B} = \frac{M\gamma v}{eR} = \frac{M}{eR}(2\pi R + 4L)f\left[1 - \frac{(2\pi R + 4L)^2 f^2}{c^2}\right]^{-1/2}$$

In synchrotrons a computer is used to maintain this relation between magnetic field and oscillator frequency.

---

Much of the early work on the new particles was done using a 3-GeV proton synchrotron at Brookhaven National Laboratory and a 6.2-GeV proton synchrotron at Berkeley called the Bevatron. A 10-GeV machine similar to the Bevatron is located at Dubna near Moscow.

In the early 1960s two 30-GeV proton synchrotrons came into operation, one at Brookhaven and the other at the international CERN laboratory just outside Geneva, Switzerland. The Soviet Union then built a similar machine, but of twice the energy at Serpukhov near Moscow.

At present the world's highest energy accelerator is at the Fermi National Accelerator laboratory near Chicago, Illinois. It is a 500-GeV proton synchrotron (see Figure 31-6). This photo shows a 200-MeV proton linear accelerator that feeds into an 8-GeV booster synchrotron, that then feeds into the main ring, which is over 6 km in circumference. After being transferred into the main ring, the protons are accelerated from 8 to 500 GeV. A similar 400-GeV machine was completed at the CERN laboratory in 1976. It is so large that it spills over from Switzerland into France.

**Example 2.** What is the peak energy of the Fermi National Accelerator when the peak magnetic field is at 20 kilogauss? $R = 800$ m.

ANSWER: Solving Eq. 31-1 for $p$ gives

$$p = e\mathcal{B}R$$
$$pc = e\mathcal{B}Rc = (1.6 \times 10^{-19})(2.0)(800)(3 \times 10^8) \text{ J}$$
$$= 4.8 \times 10^{11} \text{ eV} = 480 \text{ GeV}$$

Because the energy is so much greater than the rest energy of 0.94 GeV, the relativistic energy is almost equal to $pc$ (see Eq. 9-11).

---

The highest energy electron synchrotron is at Cornell University, which delivers an electron beam at 12 GeV. Electrons of 20 GeV are obtained at the Stanford Linear Accelerator Center using a linear accelerator of about 2 km length. In the electron synchrotron the electrons travel about 2000 km while being accelerated and thus require much less electric field for acceleration than in a linear accelerator.

Another approach that achieves very high effective energies is to use

**Figure 31-6.** *Aerial photograph of the 500-GeV accelerator complex at the Fermi National Accelerator Laboratory 30 miles west of Chicago.* [Photo Courtesy Photography Department, Fermilab.]

colliding beams. The CERN 28-GeV proton synchrotron can feed alternately into two intersecting rings where the beam is stored. Interactions equivalent to a conventional 1500-GeV machine take place in the colliding regions, but at a low rate. At Brookhaven National Laboratory a 400-GeV proton colliding-beam facility is under construction.

There are also electron-positron colliding-beam storage rings in Novosibirsk, Rome, Hamburg, and Stanford; 18-GeV storage rings are under construction in Hamburg and Stanford, and 8-GeV storage rings are under construction at Cornell University.

Without these accelerators, it would be impossible to study the fundamental properties and interactions of the elementary particles of matter. Most of the elementary particles were first discovered with the help of high-energy accelerators. A new high-energy accelerator may cost $100 million, but it is necessary if we are to advance the knowledge upon which all of physical science is ultimately based.

## 31-3 Antimatter

The relativistic quantum theory of spin-$\frac{1}{2}$ particles not only gives us the exclusion principle but it also predicts the existence of what is called an antiparticle. The antiparticle of a given particle should have exactly the same mass but opposite charge. Also an antiparticle can be annihilated by its corresponding particle. Then the two rest masses are directly converted into energy, in the form of other particles such as photons. The first antiparticle known to man was the positron which was discovered in 1933 in a cloud chamber exposed to cosmic rays. The first positron was found by accident even though its existence had already been predicted by the relativistic quantum theory. The positron or positive electron has the same mass, but opposite charge, as that of an electron. When a positron comes to rest in matter, it is quickly annihilated by an electron usually into two photons:

$$e^+ + e^- \rightarrow 2\gamma$$

Then each photon must have an energy of 0.51 MeV, which is the rest mass of an electron. Positrons are easily produced by a process called pair production (a high-energy photon strikes a nucleus and is completely converted into an electron-positron pair).

$$\gamma \rightarrow e^+ + e^-$$

The preceding reaction is one of many examples of direct conversion of energy into rest mass. As mentioned in Section 31-1, both electrons and positrons can be products of beta decay. In beta decay a positron is always produced along with a neutrino.

The antiparticle of a proton is called the antiproton or negative proton $\bar{p}$. The standard notation for an antiparticle is a bar over the symbol. Hence $\bar{p}$ stands for antiproton and $\bar{n}$ for antineutron. The antielectron (the positron) would be $\overline{e^-}$, but for this the usual convention is $e^+$. After the discovery of the positron in 1933, many physicists felt that there must also be an antiproton. According to theory, antiprotons could be produced by bombarding nuclei with protons of 6 billion eV kinetic energy. One of the production reactions should be then

$$p + p \rightarrow p + p + \bar{p} + p$$

The 6 GeV is directly converted into the rest mass energy of a proton-antiproton pair in addition to the kinetic energy of the final particles.

The possibility of discovering the antiproton was one of the main considerations that led the United States AEC to build the Bevatron, a high-energy proton synchrotron located in Berkeley, California. The Bevatron accelerates protons to a kinetic energy of 6.2 GeV, which is just barely enough to produce nucleon-antinucleon pairs. In 1955, the second year of Bevatron operation, the antiproton was discovered. The antineutron $\bar{n}$ was discovered a year later. Since the neutron is neutral in charge, so must the antineutron be neutral. However, the antineutron is quickly annihilated by either a neutron or a proton. With antinucleon annihilation, the annihilation products are usually pions (see Section 31-5). In Figure 31-7 an antiproton enters a liquid hydrogen bubble chamber, slows

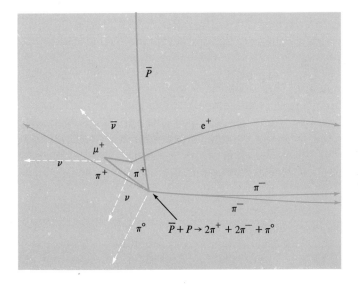
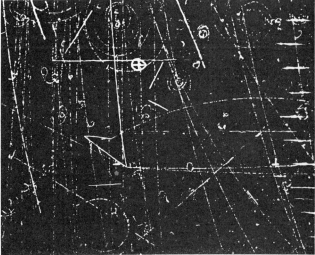

**Figure 31-7.** *Antiproton coming to rest in a liquid hydrogen bubble chamber. The antiproton is annihilated by a proton. The annihilation products are five pions: two positive, two negative, and one neutral. One of the positive pions also comes to rest and then decays into a $\mu^+$. The $\mu^+$ comes to rest and decays into a positron. [Courtesy liquid hydrogen bubble chamber group of the Lawrence Radiation Laboratory.]*

down to rest, and is annihilated by one of the hydrogen nuclei. In this particular photograph, the rest mass of the antiproton and proton is converted into five pions:

$$\bar{p} + p \rightarrow \pi^+ + \pi^+ + \pi^- + \pi^- + \pi^0$$

The question arises why all hydrogen atoms are made up of positive protons and negative electrons in preference to negative protons (antiprotons) and positive electrons (positrons). Such a "reversed" hydrogen atom is called an antihydrogen atom. Matter made of antinucleons and orbital positrons is called antimatter. By general symmetry considerations one would expect half the atoms in the universe to be antimatter. It is difficult to understand why there should be a preference to positive charge over negative charge. On the other hand, if there were any antimatter on earth, or even in our galaxy, it would not last very long. It would quickly annihilate away releasing energy with ∼1000 times the efficiency of an H-bomb. At present there is speculation that some galaxies may be made of antimatter, but so far there is no sufficient evidence.

As already mentioned, the antiparticle of the neutrino is the antineutrino $\bar{\nu}$. According to the theory, the photon must be its own antiparticle. Counting the antiparticles, our list of elementary particles covered thus far has grown to nine: $\gamma$, $\nu$, $\bar{\nu}$, $e^-$, $e^+$, $p$, $\bar{p}$, $n$, $\bar{n}$.

### Antiparticle Symmetry

Suppose during the course of any physical experiment that all the particles were suddenly changed to their corresponding antiparticles. Would the experiment proceed to give the same results? Up until 1957 physicists had assumed that antiparticles should obey exactly the same laws of physics as their counterparts. In principle there should be no way to tell whether a certain physical system is built out of antimatter or ordinary matter. We shall call this fundamental symmetry principle antiparticle symmetry. Theoretical physicists usually call it the law of charge conjugation invariance. Charge conjugation is a mathematical operation that changes every particle to its antiparticle while leaving everything else the same. The charge conjugate of a hydrogen atom is antihydrogen. The principle of antiparticle symmetry predicts, for example, that the spectrum emitted by antihydrogen gas should be exactly the same as that from ordinary hydrogen. Since antiparticles are hard to produce (antihydrogen has never been produced), some aspects of the principle of antiparticle symmetry are difficult to check by experiment.

Actually, in 1957 physicists were suddenly shocked to learn that the principle of antiparticle symmetry is violated by the weak interaction. The nature of this violation is discussed in Section 31-7.

Not only does the weak interaction cause a heavy particle to emit an electron and neutrino, but in some cases a muon and neutrino are emitted instead. The muon is an elementary particle that is the same as the electron in all respects except for its rest mass. It happens to have a rest mass 207 times that of the electron. The muon should be thought of as a heavy electron. In 1975 some evidence began to accumulate for a second heavy electron with a mass about 20 times that of the muon. Just why there should be two or three kinds of "electrons" nobody knows.

Muons are not uncommon. They were discovered in 1936 in cosmic rays. Actually at sea level cosmic rays consist mainly of muons and electrons in the ratio of 4 to 1 muons to electrons. Just as the $e^+$ is the antiparticle of the $e^-$, so is the $\mu^+$ the antiparticle of the $\mu^-$. The reason why ordinary matter is not made of muons as well as electrons is because muons can decay by the weak interaction into electrons with a half-life of $1.5 \times 10^{-6}$ s:

$$\mu^- \rightarrow \nu_\mu + e^- + \bar{\nu}_e$$

## 31-4 Conservation of Leptons

Note that we are giving subscripts e and $\mu$ to the decay neutrinos. This is because in 1963 a group using the Brookhaven synchrotron discovered that there are two distinct kinds of neutrinos: the $\nu_\mu$ which is associated with the muon, and the $\nu_e$, which is associated with the electron. The $\nu_\mu$ should be thought of as the neutral muon and the $\nu_e$ as the neutral electron. We could have used the notation $\mu°$ for $\nu_\mu$ and $e°$ for $\nu_e$. As a class, muons, electrons, and their corresponding neutrinos are called leptons. Leptons are involved in weak interactions. It is observed that whenever leptons are created, they are created in pairs. This observation has led to a new conservation law called the conservation of leptons. Actually there are two independent conservation of lepton laws, one for electrons and one for muons. The electron $e^-$ and its neutrino $\nu_e$, are given electron lepton number $+1$ and their antiparticles, $e^+$ and $\bar{\nu}_e$, are given electron lepton number $-1$. In any closed system the total electron lepton number must be conserved. For example, in neutron decay

$$n \rightarrow p + e^- + \bar{\nu}_e$$
Lepton no.: $\quad (0) \rightarrow (0) + (1) + (-1)$

the total lepton number before decay is zero. Hence the total lepton number on the right-hand side must also be zero. We see that it is $(1) + (-1) = 0$. Similarly in a beta decay, where a positron is emitted, the neutrino must be of opposite lepton number. Since the positron lepton number is $-1$, the emitted neutrino must have lepton number $+1$; hence it must be $\nu_e$ rather than $\bar{\nu}_e$.

Our final example of conservation of leptons is the decay of the muon:

$$\mu^- \to \nu_\mu + e^- + \bar{\nu}_e$$

Muon lepton no.:  (1) → (1) + (0) + (0)
Electron lepton no.:  (0) → (0) + (1) + (−1)

The initial muon lepton number is +1. This value of +1 is conserved if the muon decays into a neutral muon $\nu_\mu$. Electric charge is conserved by emitting an electron, but then to conserve the electron lepton number, an antineutrino, $\bar{\nu}_e$, must also be emitted.

> **Example 3.** When high energy positive muons collide with atomic electrons they occasionally produce two neutrinos: $\mu^+ + e^- \to 2\nu$. What kind of neutrinos are these?
>
> ANSWER: One neutrino must have muon lepton number of −1, which is the same as the $\mu^+$. Hence one neutrino is $\bar{\nu}_\mu$. The other neutrino is a $\nu_e$, which has the same electron lepton number as an atomic electron.

In summary there are eight well-established leptons: ($\mu^-, \nu_\mu, e^-, \nu_e$) and their antiparticles ($\mu^+, \bar{\nu}_\mu, e^+, \bar{\nu}_e$) and these particles occur in pairs in weak interactions. In addition there is a heavy lepton with rest mass of 1.8 GeV and its neutrino of zero rest mass: ($\tau^-, \nu_\tau$) and antiparticles ($\tau^+, \bar{\nu}_\tau$).

## 31-5 The Hadrons

Just as the leptons are the particles of the weak interaction, the hadrons are the particles of the strong or nuclear interaction. The electromagnetic interaction cuts across these boundaries and interacts with any charged particle as well as the photon. It interacts with charged leptons as well as charged hadrons.

The most familiar hadrons are our friends the proton and neutron. But since 1947 many other unstable hadrons have been discovered; 21 of them (see Figure 31-10) have long enough half-lives so that their tracks can be seen in nuclear emulsions or bubble chambers. Most of these 21 have half-lives of about $10^{-10}$ s. A particle of velocity $v = c$ will travel 3 cm in $10^{-10}$ s. In spite of the fact that certain unifying properties and groupings among these particles will become evident, the amount of data presented here may seem overwhelming, making us wonder whether there is something more elementary and simpler than the elementary particles. This indeed is the central question physics is now facing. In this book we have now reached the outer frontier of human knowledge and understanding of the physical world. The remaining sections of the book attempt to give some feeling of the present-day "pioneer" exploration of the physical world.

There are two types of hadrons called mesons and baryons. The mesons have spin 0, 1, 2, or some whole integer. The baryons have half-integer spin; namely, $\frac{1}{2}, \frac{3}{2}, \frac{5}{2}$, and so on. The baryons obey the law of conservation of baryons and always have a proton or antiproton as a final decay product. All baryons such as the proton and neutron have baryon number $+1$ and all antibaryons such as $\bar{p}$ and $\bar{n}$ have baryon number $-1$. For a closed system, the total baryon number must remain constant. Mesons have zero baryon number.

## The Mesons

The two longest-lived mesons are the pion and the K meson. The pion mass is about one seventh that of a proton and the K meson is about one-half a proton in mass.

Pions have spin zero and occur with negative, positive, and neutral charge ($\pi^-$, $\pi^+$, $\pi^0$). The $\pi^-$ is the antiparticle of the $\pi^+$. As with the photon, the $\pi^0$ is its own antiparticle. The "reason" for the existence of pions is easier to understand theoretically than that for muons. In fact, in 1936, eleven years before its discovery, the pion was predicted by H. Yukawa. Yukawa tried to explain the strong nuclear force in the same way that quantum electrodynamics explains the electric force. In quantum electrodynamics the electric force is explained in terms of an electric charge continually emitting and reabsorbing virtual quanta (photons). Yukawa invented a new type of virtual quantum to explain the strong, short-range nuclear force. The quantum theory specifies the mass of this new type of particle (quantum) in terms of the range of the nuclear force. This can be seen crudely using the uncertainty principle. If the range of these virtual quanta is $R$, the uncertainty principle says

$$R \Delta p \approx \hbar$$

where $\Delta p$ is the uncertainty in momentum, which will be on the order of $m_\pi v$. Hence

$$(M_\pi v)R \approx \hbar$$

or

$$M_\pi \approx \frac{\hbar}{Rv}$$

The smallest value this can have is when $v = c$. Then the above equation says

$$M_\pi \approx \frac{\hbar}{Rc}$$

and predicts a mass that agrees well with the measured mass of the pion. In addition to predicting the correct mass, Yukawa also predicted that pions would interact strongly with nucleons. For example, pions are easily produced by collisions of nucleons. In this case kinetic energy of the nucleons is directly converted into rest mass. Some pion production reactions are

$$p + p \rightarrow p + n + \pi^+$$
$$p + n \rightarrow p + p + \pi^-$$
$$\gamma + p \rightarrow n + \pi^+$$
$$\gamma + p \rightarrow p + \pi^0$$

Protons of several hundred million electron volts are needed to produce pions. Such proton beams are provided by an older type of proton accelerator called the synchrocyclotron.

In 1947 the pion was first discovered in cosmic rays by examining the tracks they made in nuclear emulsions (see Figure 31-8). A year later the first man-made pions were detected at the Berkeley synchrocyclotron. The charged pions usually decay by the weak interaction as follows

$$\pi^+ \rightarrow \mu^+ + \nu_\mu$$
$$\pi^- \rightarrow \mu^- + \bar{\nu}_\mu$$

with a half-life of $1.8 \times 10^{-8}$ s. The $\pi^0$ decays very much faster into two photons by the electromagnetic rather than the weak interaction. The half-life of the $\pi^0$ is about $10^{-16}$ s. The decay of a $\pi^+$ and also the decay of a $\mu^+$ can be seen in the bubble chamber picture of Figure 31-7. One of the two $\pi^+$ tracks actually comes to rest in the liquid hydrogen. Then it decays into a visible $\mu^+$ track and invisible neutrino. After traveling 1.1 cm, the $\mu^+$ comes to rest also and decays into a visible $e^+$ and two invisible neutrinos.

Actually the muon was discovered almost at the same time that Yukawa had predicted the pion. Since the muon mass is close to that of the pion, physicists thought for many years that the muon was indeed Yukawa's particle. Just before the pion was finally discovered, most physicists had given up on the muon (or else on Yukawa) because it never interacted strongly as Yukawa had predicted.

**The $\rho$ meson.** There is a sister to the pion called the $\rho$ meson. It also occurs in all three charge states, but it has spin 1 compared to spin 0 for

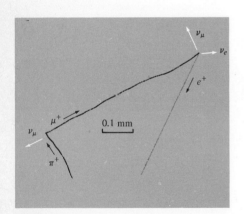

**Figure 31-8.** *Drawing of track left by stopping pion in nuclear emulsion. The decay chain $\pi^+ \rightarrow \mu^+ + \nu$ followed by $\mu^+ \rightarrow e^+ + \bar{\nu} + \nu$ is seen (except for the neutrinos). Grains of a photographic emulsion can be sensitized by collisions with a charged particle as well as by light.*

the pion. Also the ρ meson is about five times heavier than the pion. Since the ρ is a strongly interacting particle, it can decay via the strong interaction as well as be produced by the strong interaction. It is easily produced in proton-proton collisions or in pion-proton collisions. When a high-energy pion or proton travels through an atomic nucleus, on the average it travels about $3 \times 10^{-15}$ m before interacting. This is a travel time of

$$t \approx \frac{D}{v} \approx \frac{3 \times 10^{-15} \text{ m}}{3 \times 10^8 \text{ m/s}} = 10^{-23} \text{ s}$$

Hence it takes about $10^{-23}$ seconds for the strong interaction to act. Decay by the strong interaction takes the same length of time; that is, the half-life for the decay $\rho^\circ \to \pi^+ + \pi^-$ is $\sim 10^{-23}$ s.

**The K meson.** At first sight the K meson seems similar to the ρ meson. There are $K^+$, $K^0$, and $K^-$ just as there are $\rho^+$, $\rho^0$, and $\rho^-$. K mesons are easily produced via the strong interaction. The K meson is of comparable mass to the ρ meson and can also decay into two pions: $K^0 \to \pi^+ + \pi^-$. However, the half-life is $\sim 10^{-10}$ s instead of the expected $10^{-23}$ s for a strongly interacting particle! How can it be that a strongly interacting particle refuses to decay by the strong interaction even though its decay products are also strongly interacting particles? In 1950, at the time of its discovery, this aspect of the K meson seemed strange to say the least. So physicists called it a strange particle, and then proposed a new conservation law called the conservation of strangeness. The K mesons were assigned a new quantum number, $S$, called strangeness. The $K^+$ and $K^0$ have $S = +1$. The law of conservation of strangeness says that in a closed system the total strangeness number is conserved in all strong and electromagnetic interactions, but not in weak interactions. Thus, in the $K^0$

**Table 31-1.** *Categories of elementary particles*

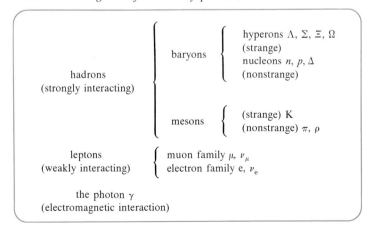

decay strangeness is not conserved:

$$K^0 \rightarrow \pi^+ + \pi^-$$
Strangeness no.: $(+1) \rightarrow (0) + (0)$

The original strangeness is $+1$ and the final strangeness is zero. Thus this decay mode must be via the weak interaction, which is $\sim 10^{13}$ times weaker than the strong interaction; that is, the expected lifetime of $10^{-23}$ s should be increased by a factor of $10^{13}$. This agrees with the measured lifetime of $\sim 10^{-10}$ s.

### The Hyperons (Strange Baryons)

Now we must ask how K mesons can be produced by strong interactions in the first place. In 1950 there were also discovered some baryons heavier than the proton that also have lifetimes $\sim 10^{-10}$ s. For example, the $\Lambda^0$ (lambda) hyperon has a mass of 1116 MeV and decays into a proton and a pion with a half-life of $2 \times 10^{-10}$ s:

$$\Lambda^0 \rightarrow p + \pi^-$$

The $\Lambda^0$ is assigned a strangeness number $S = -1$, which explains its long lifetime. But note that a $\Lambda^0$ plus a $K^0$, which have a total strangeness of zero, could be produced together by the strong interaction in a pion-proton collision:

$$\pi^- + p \rightarrow \Lambda^0 + K^0$$
Strangeness no.: $(0) + (0) \rightarrow (-1) + (+1)$

This "trick" of associated production in the context of conservation of strangeness explains how two strange particles can be easily produced by the strong interaction, but, once produced, each strange particle can only decay by weak interaction. A liquid hydrogen bubble chamber photograph of the reaction $\pi^- + p \rightarrow \Lambda^0 + K^0$ is displayed in Figure 31-9.

In addition to the lambda there are three other kinds of hyperons (strange baryons) that have lifetimes $\sim 10^{-10}$ s. They are the $\Sigma$ (sigma with $S = -1$), the $\Xi$ (xi with $S = -2$), and the $\Omega$ (omega with $S = -3$). Since the antiparticle of a particle must have all its quantum numbers reversed in sign, an antilambda ($\overline{\Lambda}^0$) must have $S = +1$. Then strangeness is conserved in the following reaction;

$$\bar{p} + p \rightarrow \overline{\Lambda}^0 + \Lambda^0$$

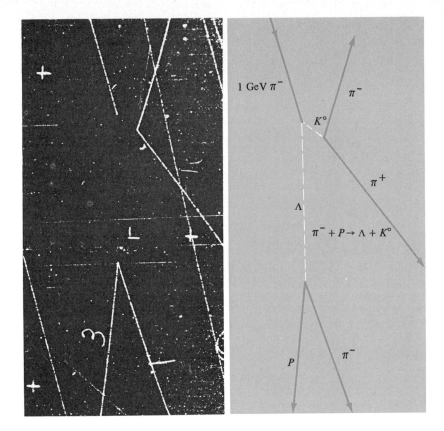

**Figure 31-9.** *Associated production of a lambda hyperon and K meson in a liquid hydrogen bubble chamber. A 1-GeV $\pi^-$ from the Bevatron enters the chamber, hits a proton, and produces the $\Lambda$ and $K^0$. [Courtesy liquid hydrogen bubble chamber group of the Lawrence Radiation Laboratory.]*

### The "Resonances"

In the past few years quite a few new mesons and baryons have been discovered that decay directly by strong interaction. The first example, a baryon, was discovered by Fermi and coworkers in the early 1950s. It is called the $\Delta$ and is 160 MeV heavier than a proton plus a pion. It decays into a proton and a pion the same as the $\Lambda$, but with a lifetime of about $10^{-23}$ s rather than the $10^{-10}$ s of the $\Lambda$. Thus the $\Delta$ must have $S = 0$.

In the past few years dozens of new hadrons have been discovered that have lifetimes ranging from $10^{-22}$ s to $10^{-23}$ s. These all decay by the strong interaction and for historical reasons are called resonances.

Most of these newly discovered particles fit into simple groups of nine or ten based on their quantum numbers (see next section). It is now felt that the proton or pion is no more elementary than any other hadron. Each hadron is just another state or energy level of strong-interaction matter. Over 200 such elementary particles have been found thus far.

Since the weak interaction is about $10^{14}$ times weaker than the strong interaction, the lifetimes of particles that decay by weak interaction will be

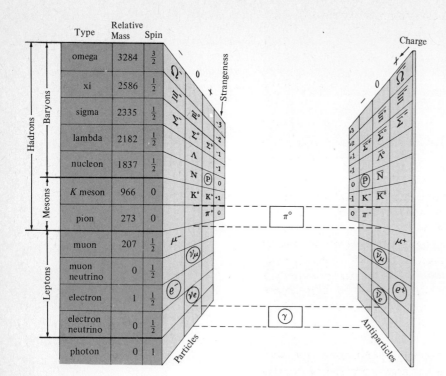

**Figure 31-10.** Table of the "long-lived" elementary particles. The particles are tabulated on the left. The "reflections" on the right are the corresponding antiparticles. The stable particles are circled.

about $10^{14}$ times $10^{-23}$ s or about $10^{-9}$ s. All of the charged particles listed in Figure 31-10 decay by weak interaction or else are stable. As we have pointed out, some hadrons decay by weak interaction rather than by strong interaction because conservation laws such as the conservation of strangeness forbid their decay via the strong interaction.

## 31-6 Quarks

In the introduction to this chapter it was mentioned that many of the present elementary particles may be made up of subparticles. In the 1960s the quark theory of hadrons was developed. In the simplest form of this theory there are just three basic particles called quarks, and their three antiparticles or antiquarks. All the baryons are explained as combinations of three quarks in appropriate angular momentum states. The mesons are explained as quark-antiquark combinations. The daring part of the theory was to propose basic particles of fractional charge and fractional baryon number. The postulated quantum numbers are given in Table 31-2. There is nothing in theoretical physics that would forbid fractional charge, however.

We note that if a quark and antiquark are to be in the same angular momentum state, there are just the nine different combinations shown in Table 31-3(a). Thus the theory would predict a group of nine spin-zero

**Table 31-2.** *Quantum numbers of the three quarks (the three antiquarks have the same values but with opposite sign)*

| Quark Symbol | Charge | Spin | Baryon Number | Strangeness |
|---|---|---|---|---|
| $u$ | $\frac{2}{3}$ | $\frac{1}{2}$ | $\frac{1}{3}$ | 0 |
| $d$ | $-\frac{1}{3}$ | $\frac{1}{2}$ | $\frac{1}{3}$ | 0 |
| $s$ | $-\frac{1}{3}$ | $\frac{1}{2}$ | $\frac{1}{3}$ | $-1$ |

mesons with predetermined values of charge and strangeness. In spite of the large number of mesons that have been discovered, there are only nine of spin zero and they happen to have the very same quantum numbers predicted by the theory. We note that the theory predicts that mesons can only have strangeness numbers contributed by an $s$ or $\bar{s}$ quark; that is, mesons can only have strangeness equal to $-1, 0,$ or $1$. It also predicts that baryons can only have $S = 0, -1, -2,$ or $-3$ because the $\bar{s}$ quark cannot be a constituent of a baryon. Since there are only the ten possible three-fold combinations of $u$, $d$, and $s$ shown in Table 31-3(b), the theory predicts families of ten baryons each.

Not only does the theory predict the observed groups of baryons and mesons, but it gives crude estimates for masses, reaction rates, and magnetic moments of hadrons. Essentially all these predictions agree with experiment.

There are other rather stringent predictions involving collisions of high energy electrons and neutrinos with protons and neutrons that seem to check out. One big puzzle, however, is that fractionally charged particles have never been observed. When a composite particle like an atomic nucleus is hit with sufficient energy, it breaks up into its constituents. But when a proton is hit even with hundreds of times its rest energy, it is never observed to break up into quarks. Some reasonable explanations have been given that have the property that free quarks could never be produced.

**Table 31-3.** (a) *The nine quark-antiquark combinations with spin 0 and spin 1 particle assignments.* (b) *The ten three-quark combinations with spin $\frac{3}{2}$ particle assignments. All the particles in this table have been discovered and are well measured.*

| | | | | | | | | | | |
|---|---|---|---|---|---|---|---|---|---|---|
| (a) mesons | $q\bar{q}$ combinations: | $u\bar{d}$ | $u\bar{s}$ | $d\bar{u}$ | $d\bar{s}$ | $s\bar{d}$ | $s\bar{u}$ | $d\bar{d}$ | $u\bar{u}$ | $s\bar{s}$ |
| | ↑↓ (spin zero) | $\pi^+$ | $K^+$ | $\pi^-$ | $K^0$ | $\bar{K}^0$ | $K^-$ | $\pi^0$ | $\eta$ | $\eta'$ |
| | ↑↑ (spin one) | $\rho^+$ | $K^{+*}$ | $\rho^-$ | $K^{0*}$ | $\bar{K}^{0*}$ | $K^{-*}$ | $\rho^0$ | $\omega$ | $\phi$ |
| (b) baryons | $qqq$ combinations: | $uuu$ | $uud$ | $uus$ | $ddd$ | $ddu$ | $dds$ | $sss$ | $ssu$ | $ssd$ | $uds$ |
| | ↑↑↑ (spin $\frac{3}{2}$) | $\Delta^{++}$ | $\Delta^+$ | $\Sigma^+$ | $\Delta^-$ | $\Delta^0$ | $\Sigma^-$ | $\Omega^-$ | $\Xi^0$ | $\Xi^-$ | $\Sigma^0$ |

### Charmed Quarks

In order to better explain the behavior of hadrons undergoing the weak interaction, a fourth quark with a new quantum number called "charm" was postulated in 1963. The quantum number $C$ for charm would play a similar role to that played by the quantum number $S$ (strangeness). Both $C$ and $S$ are conserved in strong interactions but are not conserved by the weak interaction. Up until 1974 the charm theory was considered a rather far-out type of prediction. If it were true, there should be a whole new set of mesons and baryons as illustrated in Example 4. The charmed quark would have $C = 1$, $S = 0$, $Q = \frac{2}{3}$, and $B = \frac{1}{3}$ as well as spin $\frac{1}{2}$.

---

**Example 4.** Use the quark theory to predict how many different spin 0 charmed mesons ($C = +1$) there should be. What would be their charges and other quantum numbers ($c$ is the symbol for charmed quark)?

ANSWER: The possible $c\bar{q}$ combinations are $c\bar{u}$, $c\bar{d}$, and $c\bar{s}$. Since the $c$ quark has charge $\frac{2}{3}$ and the $\bar{u}$ has $-\frac{2}{3}$, the $c\bar{u}$ meson has charge $Q = \frac{2}{3} + (-\frac{2}{3}) = 0$. It has $S = 0$ and is called the $D^0$. The $c\bar{d}$ has $Q = \frac{2}{3} + \frac{1}{3} = +1$ and $S = 0$. It is called the $D^+$. The $c\bar{s}$ has $Q = \frac{2}{3} + \frac{1}{3} = +1$ and $S = -1$. It is called the $F^+$.

---

Finally in 1974 and 1975 several new mesons of over 3-GeV rest mass were discovered which were most easily explained as being $c\bar{c}$ combinations ($c$ is the symbol for the charmed quark). These new heavy mesons, like all other particles up to that time would have $C = 0$, but it would be difficult for them to decay into mesons made up of "ordinary" quarks and thus they had lifetimes somewhat longer than that of other nonstrange hadrons. But the real proof of the charm theory did not come until 1976 when some of the long-predicted charmed mesons and baryons were finally discovered. All the charmed particles are over 1.8 GeV in mass because the $c$ quark is ~1.5 GeV in mass and the other three quarks are ~0.5 GeV in mass.

In 1977–78 the above history repeated itself. A long-lived meson of much higher mass (9.46 GeV) was discovered. Information obtained in 1978 indicated that it is a $b\bar{b}$ system of a new quark called the $b$ quark ($b$ for beauty). This fifth quark has charge $-\frac{1}{3}$ and must lead to another new family of particles above 5 GeV in mass that possess a new quantum number, tentatively called "beauty."

*"How would you like to live in a Looking-glass, Kitty?
I wonder if they'd give you milk in there? Perhaps
Looking-glass milk isn't good to drink." Alice*

## 31-7 Nonconservation of Parity

In this section we will learn that Lewis Carroll was right—looking-glass milk is indigestible if not poisonous. Also we shall see that not only is the law of conservation of parity violated by the weak interaction but so is the principle of antiparticle symmetry, which is discussed in Section 31-3. The conservation of parity is the mathematical formalism of the symmetry principle called reflection invariance. The principle of reflection invariance states that the mirror image of any physical phenomenon itself is just as true a physical phenomenon. According to conservation of parity, if someone observed any physics experiment in a mirror and was not told he was looking in a mirror, there would be no way he could tell from the results whether he was looking in a mirror. Another way of saying it is that all the fundamental laws of physics should have the same mathematical form whether one is using a left-handed or a right-handed coordinate system. One consequence of the conservation of parity is that there is no way for an absent-minded professor to determine which is his right hand by performing experiments. It would be cheating for the professor to determine which side of his body his heart lies on. That would be equivalent to handing him a glove labeled "left."

Actually, the molecules of his body and, for that matter, all life on earth are equivalent to labeled gloves. Biologically produced protein molecules are built up from amino acids—all of which are of a left-handed screw-type structure (except for antibiotics, such as penicillin, which contain a certain percentage of right-handed amino acids. This is thought to make these molds poisonous to bacteria and accounts for their use as antibiotics). On the other hand, right-handed proteins can be synthesized by chemists; and, as would be expected from conservation of parity, they have exactly the same chemical properties as the natural varieties. The only difference is that one is the mirror image of the other (see Figure 31-11). The fact that biology on the earth always produces molecules of one definite mirror symmetry, whereas the same molecules when synthesized by chemists are always produced as a fifty-fifty mixture of right- and left-handed varieties, may seem puzzling. An explanation may be that early forms of life on earth started out both left- and right-handed. Then the animals and plants of one form would be indigestible and probably poisonous to those of the other form. Finally one of the two forms of life would win out over the other in the battle for survival.

For reasons beyond the scope of this book, the conservation of parity forbids the K meson to have both two-pion and three-pion decay modes. The K meson is permitted to do one or the other, but not both. Both modes were observed experimentally. This led T. D. Lee and C. N. Yang

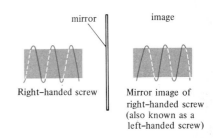

**Figure 31-11.** *A right-handed screw and its mirror image.*

in 1956 to question seriously the "self-evident truth" that nature should have no preference of right over left or vice versa. Lee and Yang proposed that the weak interactions do indeed violate the "sacred" principle of conservation of parity. They also proposed some specific experiments to test their hypothesis. We shall now discuss one of these experiments in detail. Our example shall be the decay of the $\pi^+$, which is a consequence of the weak interactions.

$$\pi^+ \rightarrow \mu^+ + \nu$$

Lee and Yang suggested that the decay muons and neutrinos might have a preferred spin direction along their direction of motion.

We shall now proceed to show that, if this is the case, reflection invariance would be violated. We shall represent spin schematically as the motion in the equatorial plane of the spinning particle. The particles will be drawn as spinning spheres. Lee and Yang proposed that the $\pi^+$ decay should always look as shown in Figure 31-12(a). The $\mu^+$ and $\nu$ must be spinning in opposite directions as shown because their spins must add up to the initial pion spin, which is zero. Figure 31-12(b) shows the mirror image of the decay particles of Figure 31-12(a). Note that in the mirror the spheres will appear as if they are spinning in the opposite sense. A "piece" of the equator of the $\mu^+$ or $\nu$ would trace out a left-handed screw in the original view (a) and a right-handed screw in the mirror image (b). The situation shown in (a) was first observed in 1957 using cyclotron produced muons. The mirror image experiment shown in (b) never occurs in nature. Thus, conservation of parity is clearly violated. The neutrinos always come off as left-handed screws. Nonconservation of parity was first observed in a beta decay experiment performed by C. S. Wu of Columbia University and a group at the National Bureau of Standards in Washington. It is now known that neutrinos are always spinning as left-handed screws, whereas antineutrinos are always spinning as right-handed screws.

It is still hard to believe that space has a built-in preference for left over right, but the evidence is so simple and clear that everyone almost immediately accepted it as true. Thus, all an absent-minded professor need do to determine his left hand is to look at any neutrino or at a $\mu^+$ coming from a $\pi^+$ decay. Here is a clear case of a law of nature that is not symmetrical. One reason for all the excitement is that this is the first time that violation of a basic symmetry principle has been found.

### Violation of Antiparticle Symmetry

If we charge conjugate Figure 31-12(a), we obtain the decay as shown in Figure 31-13(a). Note that this would make the antineutrino left-handed. But we now know from experiment that the antineutrino is always

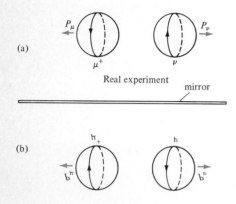

**Figure 31-12.** (a) *Decay of the $\pi^+$.* (b) *The mirror image of the decay of the $\pi^+$.*

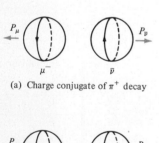

**Figure 31-13.** (a) *Charge conjugate of $\pi^+$ decay, obtained from the $\pi^+$ decay by replacing particles with their antiparticles.* (b) *The observed experimental result.*

right-handed, as shown in Figure 31-13(b). Figure 31-13(b) is what the $\pi^-$ decay looks like experimentally. Thus here is a case where changing particles to their corresponding antiparticles predicts results that do not occur in nature. This is a clear violation of antiparticle symmetry.

Note that if we reflect $\pi^+$ decay (Figure 31-12a) in a mirror and also change particles to their antiparticles, we arrive at the correct answer for $\pi^-$ decay (Figure 31-13b). Thus an overall symmetry is preserved. There still is no way to tell an absent-minded professor in a remote galaxy which is his right hand because we do not know whether his galaxy is made of antimatter or not, and we would have no way to distinguish a $\mu^+$ from a $\mu^-$ for him. Conversely, there is no way to tell him whether his atoms are made of orbital electrons or positrons unless he knows right from left. This overall symmetry is called CP invariance by the theoretical physicists. Is the Professor T. D. Lee in Figure 31-14 the real T. D. Lee or a mirror image?

Then, in 1965, a group working at the Brookhaven synchrotron found a small violation of CP invariance by carefully measuring decay modes in the decay of the $K^0$. Of the various observed decay modes are

$$(K^0 \text{ or } \bar{K}^0) \begin{cases} e^+ + \nu + \pi^- \\ e^- + \bar{\nu} + \pi^+ \end{cases}$$

**Figure 31-14.** *T. D. Lee, whose work in collaboration with C. N. Yang led to the downfall of parity conservation. Close examination of the buttons on Professor Lee's jacket establishes that this is the true image and not the mirror image.*

It turns out that the $e^+$ decay mode is slightly prefered over the $e^-$. And this is the same whether the original parent is a $K^0$ or $\bar{K}^0$. So now there is an absolute way to tell a professor in a remote galaxy which is his right hand. Tell him to build a high energy accelerator and to produce a beam of $K^0$ mesons. If his planet is made of antimatter his beam will be $\bar{K}^0$ instead, but that is no worry, because we tell him to look at the two above decay modes and to measure which one is more abundant. We tell him the electron in the more abundant mode is positive in charge. Then he can look at pion decay of the same charge in order to determine which is his left hand. So far the only violation of CP invariance that has been found is in $K^0$ decay. Just why there is such a weak violation of CP invariance is a current puzzle to theoretical physicists.

## 31-8 Summary of the Conservation Laws

The basic conservation laws of physics are in a sense stronger than other "laws." This is because anything that can happen does happen, unless it is forbidden by a conservation law. We cannot tell how often something will happen from the conservation laws—although the relative strengths of the four basic interactions are a good guideline. We recall that electromagnetic interactions are about $10^{-2}$ as probable as the strong interaction, and weak interactions are about $10^{-14}$ as probable as the strong interaction. The important point is that any reaction or decay mode that you can think of will occur in nature, provided it is not prohibited by any of the conservation laws.

Some of the 14 conservation laws are stated as symmetry principles. In quantum mechanics it is always possible to find a mathematically equivalent conservation law corresponding to each symmetry principle. The sudden and recent loss of conservation of parity and antiparticle symmetry serves as an additional warning to scientists and philosophers that other sacred laws of physics may also not be correct. For example, one can never prove in a foolproof way that the law of conservation of energy is true. However, if a single clear-cut violation of this law were ever found, this would be an absolute proof that the law of conservation of energy as now stated is incorrect. With this warning we list our final collection of conservation laws. For completeness, some of the laws not covered in the text are also listed.

1. Conservation of total energy. Rest mass must be included.
2. Conservation of total linear momentum.
3. Conservation of total angular momentum.
4. Conservation of charge.
5. Conservation of baryons. The nucleons and hyperons have baryon number $+1$. Their antiparticles have baryon number $-1$. This law says that the total baryon number must remain unchanged.

6. Conservation of leptons. This law may be thought of as the light-particle counterpart of law 5. The leptons $v_e$ and $e^-$ have electron lepton number $+1$ and their antiparticles have lepton number $-1$. According to this law the total electron lepton number must remain unchanged before and after any interaction. There is a corresponding, but independent, law of conservation of muon leptons (and of tau leptons).
7. Charge independence (also called conservation of isotopic spin). This law holds only for the strong interactions. Because of the electromagnetic interactions its predictions are only accurate to within about 1%. Charge independence makes such predictions as that the proton-proton force should be the same as the proton-neutron force.
8. Conservation of strangeness (associated production of strange particles). This laws holds for all strong and electromagnetic interactions but not for the weak interactions. It is because of this law that hyperons and K mesons decay slowly enough for us to see their tracks.
9. Antiparticle symmetry. This law holds also for all strong and electromagnetic interactions but not for the weak interactions.
10. Conservation of parity. This law also holds for all strong and electromagnetic interactions but not for the weak interactions.
11. Overall antiparticle-parity symmetry (CP invariance). This law says that if all the particles in the mirror image of any experiment are changed to their corresponding antiparticles, this new experiment is also a legitimate experiment. This law appears to hold for strong and electromagnetic interactions, but a small violation has been observed in the decay of the neutral K meson.
12. CPT invariance. Overall antiparticle-parity-time reversal symmetry. This law says that if all the particles in the mirror image of any experiment are changed to their corresponding antiparticles, and all velocities and rotations are reversed, the new experiment so obtained is a legitimate experiment. This law is believed to hold for all interactions.
13. Time reversal invariance. This law says that if the velocities and rotations of all the particles of any experiment are reversed, this new experiment is also a legitimate experiment. It appears to hold for strong and electromagnetic interactions, but must be slightly violated for the weak interaction because CP is slightly violated and CPT is not.
14. Conservation of charm. This law is similar to the law of conservation of strangeness. Just as some hadrons can have a quantum number called strangeness, some other hadrons have a quantum number called charm. This law is postulated to hold for the strong and electromagnetic interactions, but not for the weak interactions. There is a corresponding conservation of "beauty" for those hadrons that have the $b$ quark as a constituent.

## 31-9 Problems for the Future

The use of the term "elementary" begins to look ridiculous when we start talking of over 200 elementary particles. It might be reasonable to think in terms of just four different kinds of particles: the photon, electron leptons, muon leptons, and hadrons. A hopeful prediction for the future is that there may be a reduction in the number of truly elementary particles. The fact that all these elementary particles can transform into one another at will (consistent with the conservation laws) raises the hope that there might be one grand field of which these particles are different "quantum states." Such a grand unified theory would have to predict the masses of the existing "elementary" particles. Also with such an ultimate theory we should be able to calculate the charge of the electron and all other physical constants. At present the physical constants such as $c, e, h, m_e, m_p$, and so on, are completely independent. In general, as we get closer to the ultimate truth we should be able to calculate some of these constants in terms of the others. For example, the binding energy of the hydrogen atom can now be calculated from $e$, $h$, and $m_e$. Another example is that the recent theory of the universal Fermi interaction permits us to calculate the muon lifetime in terms of the neutron lifetime.

There has been some recent progress in unifying the weak and the electromagnetic interactions. In this approach, which is not yet adequately tested, the lifetimes of the muon and neutron can be expressed in terms of the charge of the electron.

The ultimate theory should not only give us the way to calculate the charge of the electron (or strength of the electromagnetic interaction), but it also should explain the strong, weak, and gravitational interactions. The gravitational interaction is much weaker even than the weak interaction. There have been unsuccessful attempts to explain gravity in terms of neutrinos. Perhaps someday gravity will be explained in terms of other apparently unrelated phenomena. Other unsolved problems are the origin, size, and evolution of the universe. Are there just as many galaxies made out of antimatter as ordinary matter?

Our understanding of the physical world has come a long way from the days of Aristotle when everything was explained in terms of the four basic elements: Fire, Water, Air, and Earth. We now have a thorough and satisfying explanation of the structure of ordinary matter in terms of quantum electrodynamics. However, we are just scratching the surface in our attempt to understand what we believe to be truly fundamental, the multitude of elementary particles and their interactions.

## Summary

The weak interaction usually involves an electron-$\nu_e$ pair or a muon-$\nu_\mu$ pair. The $e^-$ and $\nu_e$ have electron lepton number $+1$. The $\mu^-$ and $\nu_\mu$ have muon lepton number $+1$. Both electron and muon lepton numbers are strictly conserved as is baryon number. Hadrons (strongly interacting particles) with half-integral spin have baryon number $+1$. Hadrons with integral spin are called mesons and have zero baryon number. All elementary particles have antiparticles with quantum numbers of opposite sign. For example the antiproton has baryon number $B = -1$ and charge $Q = -1$.

Beams of high energy protons and electrons can be obtained by accelerating them in synchrotrons. In a synchrotron the magnetic field is adjusted so that $\mathcal{B} = p/(eR)$ where $p$ is the particle momentum and $R$ is the radius of curvature of the particle orbits in a donut-shaped vacuum tank.

Hadrons can be produced copiously by the strong interaction in $\sim 10^{-23}$ s if sufficient energy is available. A hadron will decay into lighter hadrons in $\sim 10^{-23}$ s if it is energetically permitted and if not forbidden by other conservation laws such as conservation of charge, baryon number, and strangeness. Strange particles take $\sim 10^{13}$ times longer to decay because conservation of strangeness holds except for the weak interaction.

All observed hadrons fit into regular groups that can be explained in terms of four subparticles called quarks. The quarks have charge $Q = -\frac{1}{3}$ or $+\frac{2}{3}$ and baryon number $B = \frac{1}{3}$. The mesons are quark-antiquark combinations ($q\bar{q}$) and the baryons are three quark combinations ($qqq$). "Ordinary" particles such as the nucleon, pion, and $\rho$ meson are made up of $u$ and $d$ quarks. Strange particles have at least one $s$ quark and charmed particles have at least one $c$ quark.

Not only does the weak interaction violate the conservation of strangeness and conservation of charm but it also violates conservation of parity and charge conjugation invariance (antiparticle symmetry).

Elementary particles will interact in all conceivable ways unless prohibited by one or more conservation laws. Fourteen such laws are listed in this chapter.

## Exercises

1. What is the half-life of the antineutron? What are its decay products?
2. The $^{238}$U alpha decay is followed by two successive beta (e$^-$) decays. What are the $Z$ and $A$ of this great-granddaughter of $^{238}$U?
3. Of the four kinds of neutrinos ($\nu_e, \bar{\nu}_e, \nu_\mu, \bar{\nu}_\mu$) which is the correct choice for each of the following reactions?
   (a) $(?) + p \to n + e^+$
   (b) $(?) + n \to p + \mu^-$
   (c) $(?) + n \to p + e^-$
4. Assume that none of the annihilation products in Figure 31-5 has further interactions but that they all decay. Assume the charged pions decay into muons, and then the muons decay. The final products will all be $e^-$, $e^+$, $\nu$, $\bar{\nu}$, and photons. How many of each will there be?
5. In $\mu^-$ capture by the proton the two can convert by the universal Fermi interaction to a neutron plus another particle. What is this other particle?

6. The following decay modes are forbidden. For each decay mode list the conservation laws that would be violated.
   (a) $\Lambda \rightarrow \pi^+ + \pi^-$
   (b) $K^+ \rightarrow \pi^+ + \pi^- + \pi^0$
   (c) $\bar{n} \rightarrow e^- + p + \bar{\nu}$
   (d) $p \rightarrow n + e^+ + \nu$
   (e) $n \rightarrow e^- + e^+ + \nu$

7. For each of the following reactions state whether or not it is forbidden. If it is forbidden, state a conservation law that is violated.
   (a) $\Lambda \rightarrow p + \pi^0$
   (b) $\bar{p} + p \rightarrow \mu^+ + e^-$
   (c) $n \rightarrow p + e^- + \nu_e$
   (d) $p \rightarrow n + e^+ + \nu_e$
   (e) $\Sigma^+ \rightarrow \Lambda^0 + \pi^+$

8. Draw a sketch of what a bubble chamber picture of $K^- + p \rightarrow \Sigma^- + \pi^+$ would look like. Assume the $K^-$ comes to rest and the $\Sigma^-$ decays in flight. Assume the magnetic field is pointing into the page.

9. The image of a right-handed screw is projected onto a frosted glass screen and appears as a right-handed screw. When the screen is viewed from the other side, is the image a right- or left-handed screw?

10. A right-handed screw is being screwed into a threaded hole. As viewed from the hole, does it appear as a right-handed or left-handed screw?

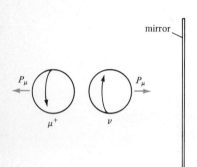

Exercise 11

11. In the accompanying figure the spinning spheres are viewed in the mirror. Draw a picture of what the mirror image would look like. Is the $\mu^+$ in the mirror image right- or left-handed?

12. When $^{60}$Co nuclei are lined up with their spins pointing up, more beta rays are observed in the down direction than the up direction. Draw the mirror image of this.
    (a) When the mirror is horizontal, what is the spin direction of the reflected $^{60}$Co? Always use the right-hand rule convention for spin direction.
    (b) When the mirror is vertical, what is the spin direction if the $^{60}$Co image?
    (c) In each of the above mirror images state whether the electron shown is moving parallel or antiparallel to the spin direction. Is the law of conservation of parity violated by this decay?

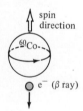

Exercise 12

13. Experimentally when the left-handed positive muons come to rest, their decay positrons come off predominantly in the backward direction. Consider the case of a positron coming off exactly backward and the $\nu_e$ and $\bar{\nu}_\mu$ going forward. Is the positron left-handed or right-handed?

## Problems

14. Consider a beam particle of total energy $E_b$ striking a target particle at rest. Both particles have the same rest mass $m$. In Problem 8 of Chapter 9 we saw that

$$E' = \left(\frac{mc^2}{2}(E_b + mc^2)\right)^{1/2}$$

is the energy of either particle in the center-of-mass system.

(a) Consider two colliding proton beams of 1000 GeV each. What is the equivalent beam energy $E_b$ for a single beam striking a fixed proton target? (There are plans to use the Fermi National Accelerator to achieve such colliding beams.)

(b) Show that for electron-positron colliding beams the equivalent $e^+$ beam energy (assuming a fixed target of electrons) is $E_b = 3914 E'^2$ GeV, when $E'$ is in GeV.

15. Suppose a heavy lepton $\tau^-$ and its neutrino $\nu_\tau$ exist. Sometimes it will decay into three muons and two neutrinos. Specify the two neutrinos.

16. A $^{239}$Pu ($Z = 94$) nucleus alpha decays. The daughter nucleus beta decays (emits $e^-$) into the granddaughter nucleus, which also beta decays (emits $e^-$). Then this great-granddaughter is bombarded by neutrons and absorbs four neutrons. What is the final product?

17. In the decay $\pi^+ \to \mu^+ + \nu$ the neutrino energy $cp$ plus the muon kinetic energy must be supplied by the mass difference. Find $p$ and the muon kinetic energy in MeV. Use classical mechanics for the muon.

18. Repeat Problem 17 using relativistic mechanics for the muon.

19. What is the maximum electron energy in the decay $\mu^- \to e^- + \nu + \bar{\nu}$? Assume the electron is so relativistic that its momentum $p = E/c$.

20. How many spin-$\frac{3}{2}$ charmed baryons should there be? What are their charges and other quantum numbers?

21. The range of the nucleon-nucleon force is $\sim 2$ fm. What mass virtual particle is needed to give a force of this range?

# Appendix A

## Physical Constants

| | | |
|---|---|---|
| speed of light | $c$ | $2.998 \times 10^8$ m/s |
| acceleration due to gravity | $g$ | $9.807$ m/s$^2$ |
| gravitational constant | $G$ | $6.672 \times 10^{-11}$ N m$^2$/kg$^2$ |
| elementary charge | $e$ | $1.602 \times 10^{-19}$ C |
| Boltzmann constant | $k$ | $1.381 \times 10^{-23}$ J/K |
| Avogadro's number | $N_0$ | $6.022 \times 10^{23}$/mole |
| gas constant | $R = N_0 k$ | $8.314$ J/(K mole) |
| electron mass | $m_e$ | $9.110 \times 10^{-31}$ kg |
| proton mass | $m_p$ | $1.673 \times 10^{-27}$ kg |
| permeability constant | $\mu_0$ | $4\pi \times 10^{-7}$ H/m |
| permittivity constant | $\epsilon_0$ | $8.854 \times 10^{-12}$ F/m |
| coulomb constant | $k_0 = \dfrac{1}{4\pi\epsilon_0}$ | $8.988 \times 10^9$ N m$^2$/C$^2$ |
| | $\dfrac{k_0}{c^2} = \dfrac{\mu_0}{4\pi}$ | $10^{-7}$ N s$^2$/C$^2$ |
| Bohr radius | $a$ | $5.292 \times 10^{-11}$ m |
| Planck's constant | $h$ | $6.626 \times 10^{-34}$ J s |
| | $\hbar = \dfrac{h}{2\pi}$ | $1.055 \times 10^{-34}$ J s |
| one atmosphere of pressure | $P_0$ | $1.013 \times 10^5$ N/m$^2$ |
| density of water | | $1.00 \times 10^3$ kg/m$^3$ |
| density of air | | $1.293$ kg/m$^3$ |
| absolute zero | | $-273.16\,°$C |

## Astronomical Constants

|  | Mass, kg | Diameter, km | Distance from Sun, km | Surface Gravity ($\times g$) |
|---|---|---|---|---|
| sun | $1.99 \times 10^{30}$ | $1.39 \times 10^6$ |  | 28.0 |
| earth | $5.977 \times 10^{24}$ | $1.27 \times 10^4$ | $1.49 \times 10^8$ | 1.00 |
| moon | $7.36 \times 10^{22}$ | $3.48 \times 10^3$ |  | 0.17 |
| Mercury | $3.28 \times 10^{23}$ | $5.14 \times 10^3$ | $5.8 \times 10^7$ | 0.40 |
| Venus | $4.82 \times 10^{24}$ | $1.26 \times 10^4$ | $1.08 \times 10^8$ | 0.90 |
| Mars | $6.4 \times 10^{23}$ | $6.86 \times 10^3$ | $2.28 \times 10^8$ | 0.40 |
| Jupiter | $1.90 \times 10^{27}$ | $1.44 \times 10^5$ | $7.78 \times 10^8$ | 2.70 |
| Saturn | $5.7 \times 10^{26}$ | $1.21 \times 10^5$ | $1.43 \times 10^9$ | 1.20 |

Earth–moon distance = $3.80 \times 10^5$ km

# Appendix B

## Conversion of Units

| Unit | mks | cgs | British | Other |
|---|---|---|---|---|
| meter | 1 m | 100 cm | 39.37 in. | $10^{10}$ Å |
| kilogram | 1 kg | $10^3$ g | 2.205 lb | |
| second | 1 s | 1 s | 1 s | |
| joule | 1 J | $10^7$ erg | $9.48 \times 10^{-4}$ Btu | 0.2389 cal; $6.24 \times 10^{18}$ eV |
| electron volt | $1.602 \times 10^{-19}$ J | $1.602 \times 10^{-12}$ erg | | |
| newton | 1 N | $10^5$ dyne | 0.2248 lbf | |
| watt | 1 W | $10^7$ erg/s | 3.413 Btu/h | $1.341 \times 10^{-3}$ hp |
| Kelvin | 1 K | 1 K | 1.8°F | |
| N/m² (pressure) | 1 N/m² | 10 dyne/cm² | $1.45 \times 10^{-4}$ psi | $9.869 \times 10^{-6}$ atm |
| m/s (speed) | 1 m/s | 100 cm/s | 3.281 ft/s | 2.237 mph |

## Electrical Units

| Quantity | | mks | cgs or gaussian |
|---|---|---|---|
| charge | $Q$ | 1 C (coulomb) | $2.998 \times 10^9$ statcoul |
| current | $I$ | 1 A (ampere) | $2.998 \times 10^9$ statamp |
| voltage | $V$ | 1 V (volt) | $3.336 \times 10^{-3}$ statvolt |
| magnetic field | $\mathcal{B}$ | 1 T (tesla) | $10^4$ gauss |
| electric field | $E$ | 1 V/m | $3.336 \times 10^{-5}$ statvolt/cm |

To convert equations in mks to gaussian form replace $\mathcal{B}$ with $B/c$, replace $\epsilon_0$ with $1/4\pi$, replace $\mu_0$ with $4\pi/c^2$, and set $k_0$ equal to 1.

# Appendix C

# Mathematical Appendix

## Geometry

Area of circle $= \pi r^2$

Area of sphere $= 4\pi r^2$

Volume of sphere $= \frac{4}{3}\pi r^3$

Volume of spherical shell of thickness $dr$ is $4\pi r^2 dr$

## Trigonometry

$\sin \theta = \dfrac{y}{r}$ $\qquad$ $\sin(-\theta) = -\sin \theta$

$\cos \theta = \dfrac{x}{r}$ $\qquad$ $\cos(-\theta) = \cos \theta$

$\tan \theta = \dfrac{y}{x}$ $\qquad$ $\sin^2 \theta + \cos^2 \theta = 1$

$\sin\left(\dfrac{\pi}{2} - \theta\right) = \cos \theta$ $\qquad$ $\sin 2\theta = 2 \sin \theta \cos \theta$

$\cos 2\theta = 2 \cos^2 \theta - 1 = 1 - 2 \sin^2 \theta$

$e^{\pm i\theta} = \cos \theta \pm i \sin \theta$

$\sin(\alpha \pm \beta) = \sin \alpha \cos \beta \pm \cos \alpha \sin \beta$

$\cos(\alpha \pm \beta) = \cos \alpha \cos \beta \mp \sin \alpha \sin \beta$

$\sin \alpha \pm \sin \beta = 2 \sin \tfrac{1}{2}(\alpha \pm \beta) \cos \tfrac{1}{2}(\alpha \mp \beta)$

$a^2 = b^2 + c^2 - 2bc \cos A$ $\qquad$ $\dfrac{\sin A}{a} = \dfrac{\sin B}{b} = \dfrac{\sin C}{c}$

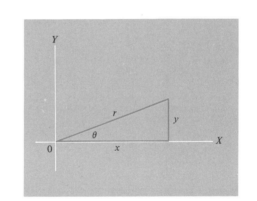

## Binomial Expansion

$$(1 + x)^n = 1 + \frac{nx}{1!} + \frac{n(n-1)x^2}{2!} + \cdots$$

## Quadratic Equation

If $ax^2 + bx + c = 0$,

then $$x = \frac{-b \pm \sqrt{b^2 - 4ac}}{2a}$$

## Some Derivatives

$$\frac{d}{dx}(au) = a\frac{du}{dx}$$

$$\frac{d}{dx}x^n = nx^{n-1}$$

$$\frac{d}{dx}\ln x = \frac{1}{x}$$

$$\frac{d}{dx}(uv) = u\frac{dv}{dx} + v\frac{du}{dx}$$

$$\frac{d}{dx}e^{ax} = ae^{ax}$$

$$\frac{d}{dx}\sin ax = a\cos ax$$

$$\frac{d}{dx}\cos ax = -a\sin ax$$

## Some Indefinite Integrals
(add an arbitrary constant)

$$\int du = u$$

$$\int x^n dx = \frac{x^{n+1}}{n+1} \quad n \neq -1$$

$$\int x^{-1} dx = \ln x$$

$$\int e^{ax} dx = \frac{1}{a}e^{ax}$$

$$\int \sin ax \, dx = -\frac{1}{a}\cos ax$$

$$\int \cos ax \, dx = \frac{1}{a}\sin ax$$

## Vector Products

$\mathbf{A} \cdot \mathbf{B} = |A| \, |B| \cos \alpha$

$\mathbf{A} \times \mathbf{B} = \hat{\mathbf{n}} |A| \, |B| \sin \alpha$

$\mathbf{A} \cdot \mathbf{B} = \mathbf{B} \cdot \mathbf{A}$

$\mathbf{A} \cdot (\mathbf{B} \times \mathbf{C}) = (\mathbf{A} \times \mathbf{B}) \cdot \mathbf{C}$

where $\alpha$ is angle between $\mathbf{A}$ and $\mathbf{B}$ and $\hat{\mathbf{n}}$ is unit vector perpendicular to plane containing $\mathbf{A}$ and $\mathbf{B}$

## The Greek Alphabet

| | | | | | | | | | | | | |
|---|---|---|---|---|---|---|---|---|---|---|---|---|
| alpha | A | $\alpha$ | eta | H | $\eta$ | nu | N | $\nu$ | tau | T | $\tau$ |
| beta | B | $\beta$ | theta | $\Theta$ | $\theta$ | xi | $\Xi$ | $\xi$ | upsilon | $\Upsilon$ | $\upsilon$ |
| gamma | $\Gamma$ | $\gamma$ | iota | I | $\iota$ | omicron | O | $o$ | phi | $\Phi$ | $\phi, \varphi$ |
| delta | $\Delta$ | $\delta$ | kappa | K | $\kappa$ | pi | $\Pi$ | $\pi$ | chi | X | $\chi$ |
| epsilon | E | $\epsilon$ | lambda | $\Lambda$ | $\lambda$ | rho | P | $\rho$ | psi | $\Psi$ | $\psi$ |
| zeta | Z | $\zeta$ | mu | M | $\mu$ | sigma | $\Sigma$ | $\sigma$ | omega | $\Omega$ | $\omega$ |

# Answers to Odd-numbered Exercises and Problems

## CHAPTER 1

**1.** 235 mi  **3.** $K = P^2/2M$  **5.** $\dfrac{1-\beta}{\sqrt{1-\beta^2}} = \sqrt{\dfrac{1-\beta}{1+\beta}}$
**7.** $V_2/V_1 = 1.5^2 = 3.375$  **9.** $m_H = 1.66 \times 10^{-24}$ g/atom
**11.** $\exp\left(-\ln\dfrac{1}{x}\right) = \exp(\ln x) = x$
**13.** $t_{1/2} = (\ln 2)\tau = 0.693\tau$  **15.** $y = 1.745$ m
**17.** $E = -e^2/(2R)$  **19.** $E = \sqrt{M_e^2 c^2 + c^2 p^2}$  **21.** $v = \sqrt{Y/\rho}$
**23.** (a) 3.87 g/s  (b) This is $116 \times 10^3$ W or 116 kW. A typical home uses several kilowatts, the car uses about 30 times more energy.
**25.** $v = c - \dfrac{c}{2\gamma^2} - \dfrac{c}{8\gamma^4} + \cdots$
**27.** (a) Accuracy of mean for watch $A = 1.58 \times 10^{-2}$ s
(b) Accuracy of mean for watch $B = 1.58 \times 10^{-2}$ s  (c) Accuracy of mean of above two measurements $= 1.12 \times 10^{-2}$ s
**31.** $\sin 2A = 2\sin A \sqrt{1 - \sin^2 A}$  **33.** 30%; $C = 2.0 \times 10^1$

## CHAPTER 2

**1.** 1 km/h = 0.911 ft/s  **3.** $\bar{v} = \tfrac{1}{2}(v_1 + v_2)$  **5.** $\bar{v} = \dfrac{x_1 v_1 + x_2 v_2}{x_1 + x_2}$

**7.** **9.**

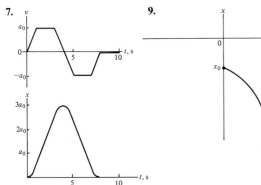

**11.** Total time = 237 s = 3.94 min
**13.** Total time = 2828 s = 47 min  **15.** 1.225 m
**17.** $a = 387.2$ ft/s$^2$ = 12.1 g  **19.** $h = 154.3$ m
**21.** $v = \dfrac{2x}{T} - v_0$
**23.** Ball takes 2 s to rise and 2 s to fall. Total time is 4 s. Thus the runner was 0.5 s late.  **25.** (a) $a = -2.5$ m/s$^2$  (b) $T = 4$ s
**27.** $v = 2.8$ km/s  $T = 286$ s = 4.76 min
**29.** Classical: $v = 1.54 \times 10^8$ m/s;  relativistic: $v = 1.37 \times 10^8$ m/s
**31.** $T = 2.15 \times 10^8$ s = 6.84 yr
**33.** It fell 19.4 m before reaching top of the window; hence it started from the tenth floor.  **35.** $\bar{v} = \tfrac{2}{3}v_1$  **37.** $v = \tfrac{1}{2}At^2$

## CHAPTER 3

**1.** $\mathbf{C} = \mathbf{B} - \mathbf{A}$;  $\mathbf{Z} = \mathbf{X} - \mathbf{Y}$  **3.** $B \cos 30° = 1.732$ m
**5.** $F_\perp = F_g \cos \alpha$  **7.** $\theta = 15°$ or 75°
**9.** When $\theta' = \theta'' = 45°$, $E'' = \tfrac{1}{2}E$. When $\theta' = 30°$, $E' = 0.866E$. When $\theta'' = 60°$, $E'' = \tfrac{1}{2}E' = 0.433E$
**11.** Bearing is 56.44° W of N. Speed with respect to shore is 3.32 km/h.
**13.** $y = 3.82x - 0.098x^2$

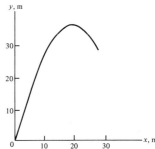

**15.** $\alpha = \tfrac{1}{2}\theta$  **17.** $h = \dfrac{v_0^2 \sin^2 \theta}{2g}$  **19.** 490 m

**21.** (a) 4 h  (b) 4.167 h  (c) 4.0825 h
**23.** (a) $v = 7$ m/s  (b) 22.3 rev/s    **25.** 27.5 days

## CHAPTER 4

**1.** $F_{net} = 0$
**3.** $T_1 = \dfrac{M_1 F}{M_1 + M_2 + M_3}$   $T_2 = \dfrac{(M_1 + M_2)F}{M_1 + M_2 + M_3}$
Net forces: $(F_1)_{net} = \dfrac{M_1 F}{M_1 + M_2 + M_3}$   $(F_2)_{net} = \dfrac{M_2 F}{M_1 + M_2 + M_3}$
$(F_3)_{net} = \dfrac{M_3 F}{M_1 + M_2 + M_3}$
**5.** 1 lbf = 32 poundals  1 N = 7.226 poundals  1 slug = 14.528 kg
**7.** $F = -940.8$ N
**9.** The net force on the plow is zero. Under this condition the plow can move with constant velocity.
**11.** $7.08 \times 10^{15}$ rev/s  $v = 2.22 \times 10^6$ m/s
**13.** (a) $T = 1.9456$ s  (b) ratio of periods = 0.9694
(c) ratio of periods = 0.841
**15.** (a) $F_{net} = 100$ N at an angle of 0°
(b) $F_{net} = 500.5$ N at an angle of 37°
**17.** (a) $a = F/(4M)$  (b) $F_2 = 0.75F$
**21.** (a) $F = 58.8$ N  (b) $F = 70.8$ N   $T = 47.2$ N
**23.** (a) $a = 0.333g$  (b) $T = 0.667Mg$   **25.** $\Delta P = Ft_0$
**27.** (a) $M_2 = 2M_1$  (b) $a_2 = g/8$ pointing up
**29.** period = 1.0035 s   $\theta = 60°$
**31.** $v = 4.11334$ km/s  This is less than the orbital velocity of 8 km/s
**33.** $v_{father} = 0.0833$ m/s  $v_{daughter} = 0.238$ m/s
**35.** $\alpha = \tan^{-1}\left(\dfrac{v^2}{gR}\right)$   **37.** $v = 16.6$ m/s
**39.** (a) $a = 0.327g$  (b) $v = 0.189$ m/s

## CHAPTER 5

**1.** 490 N   **3.** 1.87 yr or 684 days   **5.** No; no; yes
**7.** The same   **9.** (a) The top  (b) 490 m
**11.** $3.46 \times 10^8$ m from the center of the earth. Passengers would be weightless whenever the rocket motors are turned off, which would be for nearly all of the trip. (This is for a nonrotating spaceship.)
**13.** $2.006 \times 10^{30}$ kg   **15.** $5.95 \times 10^{-3}$ m/s²   **17.** $\tfrac{4}{9}\pi^2 G \rho^2 R^4$
**19.** $\sqrt{2}$ times the driver's normal weight
**21.** (a) 264.6 N pointing down  (b) 264.6 N pointing up
**23.** The orbit radius would be $\tfrac{1}{4}$ the present radius. The period of revolution would be $\tfrac{1}{8}$ the present period.
**25.** $F = 4\pi^2 \dfrac{mR}{T^2}$   For part (a), $R = 3.84 \times 10^8$ m and $T = 27$ days $= 2.33 \times 10^6$ s, so $F = 2.79 \times 10^{-3}$ N. For part (b) $R = 1.50 \times 10^{11}$ m and $T = 1$ yr, so $F = 4.58 \times 10^{-2}$ N.
**27.** $T = 7071$ s = 1.96 h
**29.** $4.73 \times 10^{-5}$ radians = $2.70 \times 10^{-3}$ degrees

**31.** $T = 2\pi \sqrt{\dfrac{R^3}{GM}}$   **33.** $v_1/v_2 = (R_1/R_2)^{-1/2}$
**35.** $v_2/v_1 = \dfrac{a - \sqrt{a^2 - b^2}}{b}$   **37.** Apparent weight = 3 mg
**39.** $T = 2\pi \sqrt{\dfrac{M}{M'} \dfrac{L}{g}}$   $T_c/T_{Pb} = 1.0005$   **41.** 2.0 yr

## CHAPTER 6

**1.** The ratio = $6.27 \times 10^{-4}$
**3.** (a) Zero  (b) Negative work of magnitude $8.14 \times 10^6$ J is done on the satellite.   **5.** $2.34 \times 10^5$ MW   **7.** $F = -2Ax$
**9.** 5% of the room's oxygen will be used up. Ventilation is not needed to replenish the oxygen supply; however, it might be helpful to reduce the $CO_2$ concentration or to remove unpleasant odors.
**11.** A $3 \times 10^{-5}$ J increase in potential energy
**13.** $v_0 = 0.5$ m/s   **15.** $\Delta U = -kx_1^3/3$
**17.** (a) 0
(b) $U(r) = \dfrac{GMm}{R_2^3 - R_1^3}\left(\dfrac{r^2}{2} + \dfrac{R_1^3}{r} - \dfrac{3}{2}R_1^2\right)$
(c) $U(r) = \dfrac{GMm}{R_2^3 - R_1^3}\left(\dfrac{R_2^2}{2} + \dfrac{R_1^3}{R_2} - \dfrac{3}{2}R_1^2\right) + GMm\left(\dfrac{1}{R_2} - \dfrac{1}{r}\right)$
**19.** The mass must drop 10.2 m independent of its initial velocity. If starting from rest, $t = 1.44$ s. This time interval depends on the initial velocity.   **21.** (a) $-2000$ N  (b) $6 \times 10^4$ W  (c) 15 km
**23.** (a) $U = -GM_2 m/r$  (b) $U = -GM_2 m/R_2$

## CHAPTER 7

**1.** $v = \sqrt{\dfrac{3}{4}\dfrac{k}{m}} x_0$   **3.** (a) 0.045 s  (b) $66.7 \times 10^3$ m/s² = $6800g$
**5.** In the collision 8/9 of $M_1$'s energy is transferred to $M_2$.
**7.** $v_1 = 0.983 v_0$   **9.** $K/K_{in} = 0.331$   **11.** 1.512 kW
**13.** $v = \sqrt{gL(1 - \cos\theta)}$   **15.** $F_B = 18Mg$   **19.** $\theta = 90°$
**21.** (a) $v = 9.68 \times 10^{-34}$ m/s  (b) $R = 1.19 \times 10^{29}$ m
**23.** (a) $V = \sqrt{\tfrac{1}{2}gR}$  (b) $h = \tfrac{1}{4}R$   **25.** $9.09 \times 10^{-2}$
**27.** (a) 0.049 J  (b) 0.07 J  (c) 2.0 m/s   **29.** 319 N
**31.** $y_{max} = 12.25$ m   $a_{max} = 17.33g$
**33.** (a) $Mgh$  (b) $v = \sqrt{gh}$  (c) $a = v^2/(2L)$   **35.** 2.22 gal/h  74 kW
**37.** 4.52 m to the right; a rise of 0.204 m   **39.** 316 m/s

## CHAPTER 8

**1.** (a) 0.1222 s  (b) 0.012273 s  (c) $6.67 \times 10^{-3}$ s
**3.** 94.9 km/s   **5.** $v = c\sqrt{1 - L'/L}$
**7.** (a) 22.37  (b) $4.026 \times 10^{-7}$ s  (c) 120.66 m
**9.** $(20 - 3.44 \times 10^{-13})$ m
**11.** A would be 100.08 yr old; B would be 23.58 yr old.
**13.** $\dfrac{df}{f} = \dfrac{d\beta}{1 - \beta^2}$   **15.** $\rho$ increases with increasing velocity.

**19.** $\Delta t_A - \Delta t_B = 0$   **21.** $x = \gamma x' - \gamma v t'$   $t = \gamma t' - \gamma(v/c^2)x'$
**23.** $x_2 - x_1 = \sqrt{\dfrac{1+\beta}{1-\beta}} L_0$   **25.** $f'/f = 1 - \beta$
**27.** (a) $-4 \times 10^{-7}$ s  (b) $5.333 \times 10^{-7}$ s  (c) $1.11 \times 10^{-6}$ s
**29.** (a) $1.484 \times 10^{-12}$ s  (b) $1 + 7.42 \times 10^{-13}$
(c) $7.42 \times 10^{-13}$; $2.47 \times 10^{-15}$

## CHAPTER 9

**1.** $u_x = \dfrac{u'_x - v}{1 - u'_x v/c^2}$   **3.** $0.0654$ g   **5.** $5.01\%$   **7.** $35.1\ \mu g$
**9.** (a) 117 MeV  (b) 145 MeV   **11.** $0.75c$   **13.** 1.25
**15.** $P^2/2M_0$ is greatest.
**17.** (a) $E = M_p c^2 + K_0$  (b) $p = (1/c)\sqrt{K_0(2M_p c^2 + K_0)}$
(c) $v = pc^2/E$   **19.** $u_x = \dfrac{u'_x - v}{1 - vu'_x/c^2}$
**23.** (a) 93.8 MeV  (b) $2.4 \times 10^8$ m/s  (c) 1.67  (d) 1.337
**25.** $p = (1/c)\sqrt{K(2Mc^2 + K)}$   **27.** $E_b = 1918$ GeV
**31.** For small $t$, $v \to a_0 t$; for large $t$, $v \to c$.
**33.** (a) $F = ma(1 - u^2/c^2)^{-1/2} + ma(u^2/c^2)(1 - u^2/c^2)^{-3/2}$
(b) $F'_y = (dp_y/dt)(\gamma + \gamma(v/c^2)u_x)$   **35.** (a) $E/c^2$;  (b) $gh/c^2$
**37.** (b) $c/\sqrt{2}$

## CHAPTER 10

**1.** $\omega = b + 2ct_0$   $\alpha = 2c$.
**3.** (a) $\bar{\omega} = \omega_0 + \tfrac{1}{2}\alpha_0 t$  (b) $\bar{\omega} = \dfrac{\theta - \theta_0}{t}$
**5.** $f_2 = 36f_1$  The final angular velocity is 36 times the initial angular velocity. The final kinetic energy is 36 times the initial kinetic energy.
**7.** It will take 10.72 s to reach 20 mph.   **9.** $P = I\alpha\omega$
**11.** $F = (R_2 R_4/R_1 R_3)Mg$
**13.** 3.33 in favor of the fused quartz flywheel   **15.** 66.67 cm
**17.** 49 N
**21.** $\mathbf{A} \times \mathbf{B} = \mathbf{i}(A_y B_z - A_z B_y) + \mathbf{j}(A_z B_x - A_x B_z) + \mathbf{k}(A_x B_y - A_y B_x)$
$\sin\alpha = \left[\dfrac{(A_y B_z - A_z B_y)^2 + (A_z B_x - A_x B_z)^2 + (A_x B_y - A_y B_x)^2}{(A_x^2 + A_y^2 + A_z^2)(B_x^2 + B_y^2 + B_z^2)}\right]$
**23.** (a) $L = 2M_0 R_0 v_0$  (b) $F = G(M_0^2/4R_0^2)$  (c) $F = M_0 v_0^2/R_0$
(d) $R_0 = \hbar^2/GM_0^3$   **25.** $\tfrac{3}{4}Mv^2$
**27.** Ratio of rotational kinetic energy to total kinetic energy of the dumbbell is $\dfrac{4 - (4y_0/y) + (y_0^2/y^2)}{4 - (4y_0/y) + 2(y_0^2/y^2)}$
**29.** (a) $3.59 \times 10^{34}$ kg m²/s  (b) $6.0 \times 10^8$ m  (c) 53.3 days
**31.** $I = M[(\tfrac{1}{2}L - x)^2 + L^2/12]$
**33.** (a) $v_A/2$  (b) $L = \tfrac{1}{2}mr_0^2\omega_A$  (c) $\tfrac{1}{6}\omega_A$  (d) $\tfrac{1}{4}mr_0^2\omega_A^2 + \tfrac{1}{2}mv_A^2$
(e) $\tfrac{5}{24}mr_0^2\omega_A^2 + \tfrac{1}{4}mv_A^2$
**37.** The man can climb a distance $x = 0.741L$ before the ladder slips.
**41.** $T = (2m + 3M)g$

## CHAPTER 11

**1.** Amplitude $= 1.5$ cm  period $= 4$ s
maximum velocity $= 0.75\pi$ cm/s
maximum acceleration $= 0.375\pi^2$ cm/s²   **3.** $2\pi/A$
**5.** 0.248 m   **7.** $0.4\pi$ s   **9.** $f = 9.3 \times 10^{11}$ s$^{-1}$   **11.** 6 db
**13.** (a) $x = x_0 \cos\sqrt{\dfrac{k}{M}}(t - t_1)$  (b) $v = -\sqrt{\dfrac{k}{M}}x_0 \sin\sqrt{\dfrac{k}{M}}(t - t_1)$
**15.** (a) $x = -x_0 \sin\sqrt{\dfrac{k}{M}}t$  (b) $v = -x_0\sqrt{\dfrac{k}{M}}\cos\sqrt{\dfrac{k}{M}}t$
**17.** (a) $x = R_0 \cos\omega(t + \tfrac{3}{4}\pi)$  $y = R_0 \cos\omega(t + \tfrac{1}{4}\pi)$  $r = R_0$
(b) $v = \omega R_0$  (c) a circle
**19.** (a) $I = \tfrac{1}{3}ML^2$;  (b) $k = \dfrac{F_0 L}{\theta_0}$  (c) $f = \dfrac{1}{2\pi}\sqrt{\dfrac{2F_0}{ML\theta_0}}$
**21.** $\dfrac{d^2 x_1}{dt^2} = -k\left(\dfrac{M_1 + M_2}{M_1 M_2}\right)x_1$  $T = 2\pi\sqrt{\dfrac{\mu}{k}}$
**23.** $\dfrac{d^2(\Delta x_2)}{dt^2} = -3\dfrac{k_0}{M}\Delta x_2$   **25.** $r_0 = \left(\dfrac{5a}{b}\right)^{1/4}$
**27.** $f = 27.57$ Hz   **29.** db $= 20\log(V_1 - V_2)$

## CHAPTER 12

**1.** $6.24 \times 10^7$ N/m²   **3.** 8.245 km
**5.** 10% of the volume is above the water line. This is independent of the shape.   **7.** $\Delta p = mv$
**9.** volume $= 10.2$ liters  density $= 3000$ kg/m³
**11.** (a) 28.8 g  (b) 28.8 g  (c) 12 cm²   **13.** 2.42 g lost per week
**15.** 0.36°C   **17.** $\Delta T = 0.468$°C
**19.** (a) $N\varepsilon$  (b) $\dfrac{2\varepsilon}{3k}$  (c) $\dfrac{2N\varepsilon}{3V}$  (d) the temperature remains the same and the pressure drops by a factor of two.   **21.** (a) one  (b) $\tfrac{1}{2}$
**23.** $5 \times 10^{-5}$ m   **25.** $d = RP/2S$
**27.** (b) $h = h_0 \ln 2 = 5.50$ km  (c) $3.33 \times 10^4$ N/m²
**31.** $y = 2.16 \times 10^{10}/R^3$ when $R$ is in meters

## CHAPTER 13

**1.** $-12.3$ J   **3.** (a) $N_0/2.016$  (b) $N_0/18$  (c) $N_0/180$
**5.** $3.68 \times 10^3$ molecules/cm³
**7.** 0.75 cal/(g K) for He; 2.5 cal/(g K) for H₂; 5/28 cal/(g K) for N₂
**9.** $C_v = 10$ cal/(mole K)   **11.** $T_2/T_1 = 0.488$
**13.** $T_{\text{monatomic}}/T_{\text{diatomic}} = (V_1/V_2)^{0.267}$
**15.** (a) 9 cal/K  (b) 6 cal/K, assuming no vibrational degrees of freedom   **17.** $RT\ln\left[\dfrac{V_2 - V_0}{V_1 - V_0}\right]$
**19.** (a) $P_2 = 0.379$ atm  $T_2 = 207$ K  (b) $W_{12} = 1.37 = 10^3$ J
(c) $W_{23} = -1.19 \times 10^3$ J  (d) 175 J  (e) 207 K  (f) $1.37 \times 10^3$ J
**21.** (a) 6  (b) 1  (c) $3R$  (d) $4R$  (e) $T = mv_0^2/3k$

**23.** (a) $T_2 = T_1(V_2/V_1)$  (b) $P_1(V_2 - V_1)$  (c) $\frac{5}{2}R(T_2 - T_1)$
(d) $V_3 = V_2(T_2/T_3)^{1/(\gamma-1)}$   **25.** 13.2 hp
**27.** (a) $W_{AB} = 3.37 \times 10^3$ J  $W_{BC} = 7.00 \times 10^3$ J  $W_{CD} = 0$
(b) $Q_{AB} = 8417$ J absorbed  $Q_{BC} = 7000$ J rejected  $Q_{CD} = 5047$ J rejected  $Q_{DA} = 7397$ J absorbed  (c) $\varepsilon = \dfrac{Q_{AB} + Q_{BC} - Q_{DA}}{Q_{AB} + Q_{BC}} = 0.287$

**13.** $2k_0 Q \dfrac{r^2 + L^2/4}{(r^2 - L^2/4)^2}$

**15.** The net electrostatic force would be $1.86 \times 10^{-64}$ N. This is the same as the gravitational force. This hypothesis predicts that neutral elementary particles such as the neutron would experience zero gravitational force. Also it predicts that the gravitational mass of $^{238}$U would be 92 times that of the hydrogen atom rather than 238 times.

**17.** $E = k_0 Q \left\{ \dfrac{x + R/2}{[(x + R/2)^2 + R^2]^{3/2}} + \dfrac{x - R/2}{[(x - R/2)^2 + R^2]^{3/2}} \right\}$

**19.** $2\pi k_0 \sigma$   **21.** $2k_0 \lambda / y_0$   **23.** $E = 3k_0 QL^2 / x^4$
**25.** Both derivatives are equal to zero.

## CHAPTER 14

**1.** $\Delta W_1$, $\Delta Q_1$, and $\Delta Q_2$ are all negative. The relation $\Delta W = \Delta Q_1 - \Delta Q_2$ still holds.   **3.** 767 kW   **5.** 74 J
**7.** 10 J   **9.** $\Delta S = 0.10$ cal/K   **11.** 1/32   **17.** $3.22 \times 10^4$ J 32.2 W
**19.** $(T_1 - T_3)/T_1$   **21.** 1443 cal  $\Delta W = 1.75 \times 10^5$ J   **23.** 35%
**25.** 66.7°C   **27.** (a) Specific heat is 0.214 cal/(g K).
(b) (i) $\Delta S_A = -8.90$ cal/K  (ii) $\Delta S_w = 10.1$ cal/K  (iii) 1.2 cal/K

## CHAPTER 16

**1.** $E_{\text{I}} = 0$  $\quad V_{\text{I}} = k_0 \left( \dfrac{Q_1}{R_1} + \dfrac{Q_2}{R_2} + \dfrac{Q_3}{R_3} \right)$

$E_{\text{II}} = k_0 \dfrac{Q_1}{r^2}$  $\quad V_{\text{II}} = k_0 \left( \dfrac{Q_1}{r} + \dfrac{Q_2}{R_2} + \dfrac{Q_3}{R_3} \right)$

## CHAPTER 15

**1.** $-4.16 \times 10^{42}$   **3.** $\dfrac{2k_0 Q^2 r}{(r^2 + L^2/4)^{3/2}}$   **5.** decrease

$E_{\text{III}} = k_0 \dfrac{Q_1 + Q_2}{r^2}$  $\quad V_{\text{III}} = k_0 \left( \dfrac{Q_1 + Q_2}{r} + \dfrac{Q_3}{R_3} \right)$

$E_{\text{IV}} = k_0 \dfrac{Q_1 + Q_2 + Q_3}{r^2}$  $\quad V_{\text{IV}} = k_0 \dfrac{Q_1 + Q_2 + Q_3}{r}$

**7.** (a)

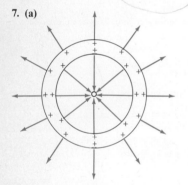

**3.** 0.167 m radius  $Q = 9.26 \times 10^{-6}$ C
**5.** Let left edge of left slab be at $x = 0$: $E_{\text{I}} = 0$; $E_{\text{II}} = 4\pi k_0 \rho x$;
$E_{\text{III}} = 4\pi k_0 \rho x_0$; $E_{\text{IV}} = 4\pi k_0 \rho (2x_0 + d - x)$; $E_{\text{V}} = 0$
**7.** $1.13 \times 10^7$ m²   **9.** $C = \left[ \dfrac{1}{C_1} + \dfrac{1}{C_2} + \dfrac{1}{C_3} \right]^{-1}$
**11.** (a) zero  (b) zero  (c) $E = 1.8 \times 10^5$ V/m
**13.** $3.09 \times 10^6$ m/s   **15.** 400 V
**17.** (a) $4.28 \times 10^{-2}$ C  (b) $E = 1.54 \times 10^{13}$ V/m  $V = 7.7 \times 10^{10}$ V
**19.** 1153 V   **21.** 1.28
**23.** $E_{\text{I}} = 0$  $E_{\text{II}} = k_0 \dfrac{Q(r^3 - R_2^3)}{(R_1^3 - R_2^3) r^2}$  $E_{\text{III}} = k_0 \dfrac{Q}{r^2}$

**25.** $\mathbf{E} = \frac{4}{3} \pi k_0 \rho \left\{ \dfrac{R_2^3 x_0 \mathbf{i}}{(x_0^2 + y^2)^{3/2}} + y \left[ 1 - \dfrac{R_2^3}{(x_0^2 + y^2)^{3/2}} \right] \mathbf{j} \right\}$

(b) $1.70 \times 10^5$ N m²/C  (c) $2.5 \times 10^{-5}$ C   **9.** $-1.24 \times 10^{-36}$

**11.**

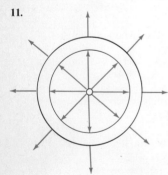

**27.** $V_{\text{I}} = k_0 \dfrac{Q}{R_1} + k_0 \dfrac{Q}{R_1^3 - R_2^3} \left[ \dfrac{R_1^2 - R_2^2}{2} + \dfrac{R_2^3}{R_1}(R_2 - R_1) \right]$

$V_{\text{II}} = k_0 \dfrac{Q}{R_1} + k_0 \dfrac{Q}{R_1^3 - R_2^3} \left[ \dfrac{R_1^2 - r^2}{2} + R_2^3 \left( \dfrac{1}{R_1} - \dfrac{1}{r} \right) \right]$

$V_{\text{III}} = k_0 \dfrac{Q}{r}$

**29.** (a) $3.39 \times 10^4$ V/m  (b) $1.13 \times 10^4$ V/m
(c) $-3.39 \times 10^4$ V/m  (d) 226 V
**33.** (a) 0.5 MeV  (b) 0.5 MeV  (c) $2.88 \times 10^{-15}$ m
**35.** (a) 12 MeV  (b) $-10$ MeV  (c) 4 MeV  (d) 6 MeV

**37.** $C' = \dfrac{\varepsilon_0 A}{\frac{x_1}{\kappa} - x_1 + x_0}$

**39.** $Q_1 = V_0 R_1/k_0 \quad Q_2 = V_0 R_2/k_0$

## CHAPTER 17

**1.** $I = 2.69$ A  **3.** $\rho = \dfrac{2mu}{\mathcal{N}e^2 L}$

**5. (a)** $I = 80$ A  **(b)** $P = 640$ W, 6.4 times that of a 100-W light bulb

**7. (a)** $\mathcal{B} = 0$  **(b)** $\mathcal{B} = 2 \times 10^{-5}$ tesla

**9. (a)** $a = eE/m$ pointing up  **(b)** $t = \sqrt{\dfrac{2m(d-h)}{eE}}$

**(c)** $F = -e\mathbf{v}_0 \times \mathcal{B}$ pointing out of the page

**11.** 1.05 mA  **13.** $I = Q/T = 4\pi R^2 \sigma/T$

**15.** $R_1 = \frac{1}{2}(R_{12} + R_{13} - R_{23}) \quad R_2 = \frac{1}{2}(R_{12} - R_{13} + R_{23})$
$R_3 = \frac{1}{2}(R_{23} - R_{12} + R_{13})$  **17. (a)** 100 ohms  **(b)** 300 ohms

**19. (a)** $\mathcal{B}$ is vertical pointing up.  **(b)** 0.4 tesla or 4000 gauss

**21.** $m = 58.5 m_e$

**23. (a)** $1.13 \times 10^6$ N/C  **(b)** $1.41 \times 10^6$ N/C
**(c)** $2.82 \times 10^{-3}$ tesla = 28.2 gauss pointing into the page

**25. (a)** $4 \times 10^{-4}$ tesla pointing into the page  **(b)** $5 \times 10^{-4}$ tesla
**(c)** $3 \times 10^{-4}$ V/m pointing down  **27.** $\varepsilon = 10$ V  $r = 20$ ohms

**29. (a)** $\varepsilon_1 \dfrac{rR_2}{rR_1 + R_1 R_2 + rR_2}$  **(a)** $\varepsilon_1 \dfrac{R_2}{R_1 + R_2}$

## CHAPTER 18

**1.** $\mathcal{B}_\text{I} = \mathcal{B}_\text{III} = 4\pi k_0 \mathcal{J}/c^2 \quad \mathcal{B}_\text{II} = 0 \quad \mathcal{B}_\text{I}$ points out of the page; $\mathcal{B}_\text{III}$ points into the page.

**3.** $\mathcal{B} = 4\pi \dfrac{k_0}{c^2}(\mathcal{J}_1 + \mathcal{J}_2)$ pointing into the page,

when $r < R_1$; $\mathcal{B} = 4\pi \dfrac{k_0}{c^2}\mathcal{J}_2$ pointing into the page, when $R_1 < r < R_2$; $\mathcal{B} = 0$, when $r > R_2$.

**5.** It travels with constant velocity down the axis.

**7. (a)** $4 \times 10^{-6}$ T  **(b)** $1.2 \times 10^{-4}$ T  **9.** 0.02 dyne/cm

**11.** $\mathcal{B} = 0$, when $r < R_1$; $\mathcal{B} = 2 \dfrac{k_0}{c^2} I \dfrac{r^2 - R_1^2}{(R_2^2 - R_1^2)r}$,

when $R_1 < r < R_2$; $\mathcal{B} = 2 \dfrac{k_0}{c^2}\dfrac{I}{r}$, when $r > R_2$.

**13.** $\mathcal{B} = 2\dfrac{k_0}{c^2}\dfrac{I}{R_2^2 - R_1^2}\left[\mathbf{i}\left(y - \dfrac{R_1^2 y}{x_0^2 + y^2}\right) - \mathbf{j}\dfrac{R_1^2 x_0}{x_0^2 + y^2}\right]$, for $y < R_2$.

**17.** $\mathcal{B} = 2\pi \dfrac{k_0}{c^2} IR^2 \left\{\left[\left(x + \dfrac{R}{2}\right)^2 + R^2\right]^{-3/2} + \left[\left(\dfrac{R}{2} - x\right)^2 + R^2\right]^{-3/2}\right\}$

**19.** $2\pi \sqrt{\dfrac{I_0}{IA\mathcal{B}}}$  **21.** 0.467 T

**23. (a)** $L = \frac{1}{2}M\omega R^2$  **(b)** $\mu = \frac{1}{4}Q\omega R^2$  **(c)** $\mu/L = Q/2M$

**25.** Same answers as for Exercise 2 except that $\mathcal{B} = 4\pi \dfrac{k_0}{c^2} jx_0$ pointing out of the page in the region between the slabs.

## CHAPTER 19

**1.** $4\pi R^2 \mathcal{B}$  **3.** counterclockwise

**5.** $\mathcal{E} = 2\pi^2 f \mathcal{B} R^2 \sin \omega t \quad I = \mathcal{E}/R_m$  **7.** $10^{-2}$ W  **9.** $f = R/(2\pi L)$

**11.** $6 \times 10^{-5}$ tesla = 0.6 gauss  **13.** $4.58 \times 10^{-7}$ gauss

**15.** $4\mathcal{B} l_1 l_2$  **17.** $P_2$  **19.** 3.16 kV

**23.** $C = \dfrac{1}{\omega_0 QR} \quad L = \dfrac{QR}{\omega_0} \quad Z = R\sqrt{1 + Q^2\left(\dfrac{\omega}{\omega_0} - \dfrac{\omega_0}{\omega}\right)^2}$

**25.** $\mathcal{B} = 2\dfrac{k_0}{c^2}\dfrac{NI}{r} \quad L = 2\dfrac{k_0}{c^2}N^2(R_2 - R_1)\ln\dfrac{R_2}{R_1}$  **27.** $3 \times 10^{-3}$ C

**29. (a)** side b  **(b)** side a

**31. (a)** $E = \dfrac{r}{2}\dfrac{\partial \mathcal{B}}{\partial t}$  **(b)** $E = 0.2$ V/m  **33.** 12.5 m/s

**37.** Let us choose $L = 10^{-3}$ henrys. Then $C = 25.33$ pF and $R = 36.3$ ohms.

**39. (a)** $I = 0.4237$ A  **(b)** $\bar{P} = 17.95$ W  **(c)** The maximum instantaneous power across the capacitor is 47.6 W.

**41.** 0.693 $RC$  **43.** $I = \dfrac{\mathcal{E}}{R}(1 - e^{-Rt/L})$

## CHAPTER 20

**1.** $\Phi_E = 4\pi k_0 Q_\text{in} \quad \oint \mathbf{E} \cdot d\mathbf{s} = -\dfrac{d\Phi_B}{dt} \quad \Phi_B = 0$

$\oint \mathcal{B} \cdot d\mathbf{s} = 4\pi \dfrac{k_0}{c^2} I_\text{in} + \dfrac{1}{c^2}\dfrac{d\Phi_E}{dt}$

**3.** $-ab\omega E_0 \sin \omega t$  **5.**

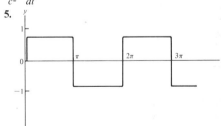

**7.** Along the positive direction  **9.** $T = 36$ N  **11.** $T = 329.6$ N

**13. (a)** $\mathcal{J} = 2.65$ A/m  **(b)** $\mathcal{B} = 1.67 \times 10^{-6}$ tesla

**15.** 400 to 700 nm  **17. (a)** $E = \dfrac{4k_0 I_c \Delta t}{R^2}$  **(b)** $\mathcal{B} = \dfrac{r}{2c^2}\dfrac{E}{\Delta t}$

**19.** $L = NK_1/K_2$  **21.** $u = c$  **23. (a)** $\mathcal{B} = 0$  **(b)** $\mathcal{B} = 0$

**25.** $E_\text{rad} = 150.8 \cos \omega t$ in V/m  The ratio is 0.377.

**27. (a)** $\mathcal{J}_0 = \sqrt{A_1^2 + A_2^2}$  **(b)** $\phi = \tan^{-1}\left(\dfrac{A_2}{A_1}\right)$

**29.** (a) $\dfrac{\overline{P}}{y_0 z_0} = \dfrac{x_0 V_0^2}{2\rho y_0^2}$ (b) $\dfrac{\overline{P}_{rad}}{y_0 z_0} = \dfrac{\pi k_0}{c}\dfrac{z_0^2 V_0^2}{\rho^2 y_0^2}$

## CHAPTER 21

**1.** $S = \dfrac{c^3 \mathcal{B}^2}{4\pi k_0} = \dfrac{E\mathcal{B}}{\mu_0}$

**3.** Thickness $= 0.75\ \mu$m just cancels out the gravitational attraction. The sail must be thinner than this in order to have a net repulsive force from the sun.

**5.** (a) $E_{max} = 3 \times 10^{-7}$ V/m (b) $S_{max} = 2.39 \times 10^{-16}$ W/m² (c) $\overline{P} = 1.00 \times 10^{-9}$ W

**7.** (a) $v = 2.79 \times 10^6$ m/s (b) $v = 27.9$ m/s   **9.** $n = 1.00025$

**11.** (a) approximately $2 \times 10^7$ J energy increase per orbit (b) approximately 1 yr

**13.** (a) $F = 6.67 \times 10^{-6}$ N (b) The rms induced current is 36.8 A/m. (c) $4.244 \times 10^{-4}$ J/m²

**15.** (a) $y = \dfrac{F_0}{k - m\omega^2}\sin \omega t$ (b) $y = \dfrac{F_0}{k - m\omega^2}\cos \omega t$
(c) $\omega_0 = \sqrt{k/m}$ (d) out of phase

**17.** $\mathcal{N} = 0.025$ electron/cm³

**19.** (a) $E_{trans} = 0.98 E_{in}$   $E_{refl} = 10^{-4} E_{in}$
(b) $E_{trans} = 0.96 E_{in}$   $E_{refl} = 1.98 \times 10^{-4} E_{in}$

**21.** (a) $\Delta U = 4.39 \times 10^{-5}$ eV (e) $t = 1.57 \times 10^{-11}$ s

**25.** $E = E_1 + E_2 = \dfrac{k_0 Q \omega^3}{2c^3 r}\sin 2\theta$

**27.** $E_{rad} = \dfrac{k_0 Q q a}{c^2 r}\sin^2 \theta$   The direction is along $-z$.

## CHAPTER 22

**1.** $E'$ is along the negative $y$-axis. $\mathcal{B}'$ is along the negative $z$-axis.
**3.** $I = 5I_0 + 4I_0 \cos (kd \sin \theta)$
**5.** (a) $f = 33.3$ Hz (b) 66.7 times per second
**7.** $D_1 - D_2 = (n + \tfrac{1}{2})\lambda$ is the condition for a minimum.
**9.** 30 cm   **11.** $N = 4.25 \times 10^7$   **13.** $N = 982$
**15.** $E = -2E_0 \sin kx \cos \omega t$
**19.** $E'^2 = E_1^2 + E_2^2 + 2E_1 E_2 \cos (kd \sin \theta)$
**21.** $\sin \theta_1 + \sin \theta_2 = n\lambda/d$
**23.** $\Delta \lambda = 4.24 \times 10^{-19}$ m; $9.4 \times 10^{11}$ rulings would be needed.
**27.** The entries in the following table should be multiplied by $V_0^2$.

|     | $L_f'$ | $R_f'$ | $L_b'$ | $R_b'$ |
|-----|--------|--------|--------|--------|
| (a) | 1      | 0      | $\tfrac{1}{2}$ | $\tfrac{1}{2}$ |
| (b) | 0      | 1      | $\tfrac{1}{2}$ | $\tfrac{1}{2}$ |
| (c) | $\tfrac{1}{2}$ | $\tfrac{1}{2}$ | 1 | 0 |
| (d) | $\tfrac{1}{2}$ | $\tfrac{1}{2}$ | 0 | 1 |

## CHAPTER 23

**1.**
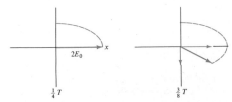
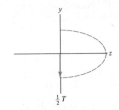

**3.** The resultant is plane-polarized in the $z$-direction.
**5.** $10^{-6}$ lightyear $= 9.5 \times 10^9$ m
**7.** spot size $= 7.72 \times 10^{-5}$ m    **9.** (a) $I = \tfrac{1}{2}I_0$ (b) $I' = \tfrac{1}{4}I_0$
**11.** Larger, but not inverted; same image as produced by a concave shaving mirror.    **13.** 66.7 cm    **15.** $I' = 0.759 I_0$
**17.** (a) $4I_0$ (b) zero (c) $I_0$ vertically polarized
(d) circularly polarized with same average energy as in part (c)
(e) $I_0$ horizontally polarized (f) $I_0$ (g) zero    **19.** 0.389 nm
**21.** (a) $3.79 \times 10^{-2}$ rad (b) $3.79 \times 10^{-3}$ rad
**23.** $F(F - D)/D$ to the right of the concave lens
**27.** (a) $n_\| > n_\perp$ (b) $x_0 = \dfrac{c}{2f(n_\| - n_\perp)}$

## CHAPTER 24

**1.** 0.0124 nm    **3.** $3.0 \times 10^{-10}$ kg m/s    $\lambda = 2.21 \times 10^{-24}$ m
**5.** $\lambda_{photon} = 1.24 \times 10^{-6}$ m   $\lambda_{electron} = 1.23 \times 10^{-9}$ m
**7.** $\lambda_0 = 2816$ nm    **9.** $\lambda = \dfrac{hc}{\sqrt{E^2 - m^2 c^4}}$
**11.** (a) 166.7 counts/s (b) 426.7 counts/s (c) 26.7 counts/s
**13.** by a factor of 2
**15.** (a) $p_e = h\left(\dfrac{1}{\lambda} - \dfrac{1}{\lambda'}\right)$ (b) $K = \left[c^2 h^2 \left(\dfrac{1}{\lambda} - \dfrac{1}{\lambda'}\right)^2 + m^2 c^4\right]^{1/2} - mc^2$
**17.** 469 MeV    **21.** $K = 0.0375$ eV    $\lambda = 6.35 \times 10^{-9}$ m

**23.**

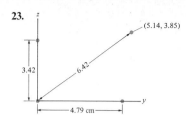

The spots move closer together as the beam energy is increased.

## CHAPTER 25

**1.** $\psi^*\psi = 2(1 + \cos 2\omega t) = 4\cos^2 \omega t$
**3.** $\psi^*\psi = 4\cos^2(\Delta kx)$   $\Delta p\,\Delta x = \dfrac{\pi}{4}\hbar$   **5.** $\Delta x = 115\,m$
**7. (a)** $E_2 = 9.42 \times 10^{-2}$ eV  **(b)** $\lambda = 438$ nm
**9.** $K = c\left(\dfrac{N^2\pi^2\hbar^2}{4x_0^2} + m^2c^2\right)^{1/2} - mc^2$
**11. (a)** $\sqrt{2}\sigma_0$  **(b)** $B(k) = \dfrac{\sigma_0}{\sqrt{\pi}}\exp[-\sigma_0^2(k - mv_0/\hbar)^2]$
**13.** $\phi = \tan^{-1}\left[\dfrac{A_1\sin\phi_1 + A_2\sin\phi_2}{A_1\cos\phi_1 + A_2\cos\phi_2}\right]$
$A = \sqrt{A_1^2 + A_2^2 + 2A_1A_2\cos(\phi_1 - \phi_2)}$   **17.** $\dfrac{d\omega}{dk} = \dfrac{pc^2}{E} = v$
**19. (a)** $E_1 = 942.4$ MeV  $E_2 = 3770$ MeV  **(b)** $E_1 = 31.1$ MeV
$E_2 = 62.2$ MeV  **21.** $E_1 = 22.76$ eV  $E_2 = 88.23$ eV
**23.** $E_1 = \dfrac{\hbar^2}{72\,mx_1^2}$   **25. (a)** $\psi_1(x) = \exp\left(-\dfrac{\sqrt{mk}}{2\hbar}x^2\right)$
**(b)** $E_1 = \tfrac{1}{2}\hbar\sqrt{k/m}$   **27.** $E = \tfrac{3}{2}\hbar\omega$

## CHAPTER 26

**1.** $R_0 = \dfrac{\pi^2}{8}\dfrac{\hbar^2}{k_0mze^2}n^2$
**3.** 2% energy decrease for nonrelativistic particles
**5.** 25 different standing waves
**7.** one half that of the hydrogen atom or $R_{He^+} = 0.265$ nm   **9.** ten
**11.** The $n = 4$ to 2 transition in $He^+$ will give the same energy photon as the $n = 2$ to 1 in H. The wavelength is 1215 nm.
**13.** $\psi_{3,2,-2} = \dfrac{r^2}{a^2}e^{-r/(3a)}\sin^2\theta\,e^{-2i\phi}$   **15.** $t = 4.7 \times 10^{-8}$ s
**17.** All wave functions are of the form
$$\psi_{n_x,n_y,n_z} = \sin\dfrac{n_x\pi}{L}\sin\dfrac{n_y\pi}{L}\sin\dfrac{n_z\pi}{L}.$$
$E_1 = 112$ eV with $\psi = \psi_{1,1,1}$
$E_2 = 224$ eV with $\psi = \psi_{1,1,2};\ \psi_{1,2,1};$ and $\psi_{2,1,1}$
$E_3 = 336$ eV with $\psi = \psi_{1,2,2};\ \psi_{2,1,2};$ and $\psi_{2,2,1}$
$E_4 = 410$ eV with $\psi = \psi_{1,1,3};\ \psi_{1,3,1};$ and $\psi_{3,1,1}$

**19.** $a = \dfrac{\hbar^2}{k_0me^2}$   **21.** $\left\langle\dfrac{1}{r}\right\rangle = \dfrac{1}{a} = \dfrac{k_0me^2}{\hbar^2}$
**23.** $R = \dfrac{\hbar^2n^2}{GM_nm_e^2} = 1.2 \times 10^{30}$ m for $n = 1$
**25. (a)** $E_1 = -3.2$ eV  **(b)** Three transitions to the ground state giving $hf = 1.2,\ 1.8,$ and $2.2$ eV; two transitions to the first excited state giving $hf = 0.6$ and $1.0$ eV; one transition to the second excited state giving $hf = 0.4$ eV.

## CHAPTER 27

**1.** $3.40 \times 10^{-10}$ m for sodium; $4.24 \times 10^{-10}$ m for potassium
**3.** The transitions from $n = 9, 10, 11, 12, 13,$ or 14 to $n = 6$; also $n = 5$ to 4; also $n = 7$ or 8 to 5; also $n = 9$ through 23 to 6; also $n = 12$ or more to 7; also $n = 21$ or more to 8.
**5.** 6560, 5411, 4859, 4541, and 4338 Å   **7.** 110 keV   **9.** 31.7 eV
**11.** 50 elements for Group 8.
**13. (a)** $Z_{eff} = Z - 5$  **(b)** $hf = 13.6\left[(z-2)^2 - \dfrac{(z-5)^2}{4}\right]$ gives
8.68 Å for Al and 0.185 Å for Pb.
**17.** $z_{eff} = 6.7$ for $2p$ in Al $= 6.7$   $z_{eff} = 64.5$ for $2p$ in Pb   **19.** fifty

## CHAPTER 28

**1. (a)** $E = K + U = -4$ eV  **(b)** $K_{in} = 11$ eV   **3.** $U_0 = 5$ eV
**5.** contact potential $= 0.5$ V   Metal A is at a higher potential after contact.
**7. (a)** $E = -4$ eV  **(b)** $W_0 = 4$ eV  **(c)** $U_0 = 9$ eV  **(d)** 9 eV  **(e)** 4 eV
**9.** $U_0 = 7.91$ eV   **11.** $R_{back}/R_{forward} = 2.5 \times 10^{18}$
**13.** $E_{gap} = 2.48$ eV   **15.** 15,750 Hz   **17.** $\rho = 1/h^3$
**19.** 6.77 Å   $\rho = 124$ kg/m$^3$   **21.** $U_0 = 7.06$ eV   **23.** $E_1 = 0.06833\,U_0$

## CHAPTER 29

**1.** $\theta_{max} = 1.7°$   **3.** $2.0 \times 10^{-15}$ m   **5.** 148   **7.** 3/4
**9.** 63.2% decays in 1 mean life; 86.5% decays in 2 mean lives
**11.** The diameter would be reduced by a factor of $5.47 \times 10^4$. It would be 25.4 km.   **13. (a)** 190.128 GeV  **(b)** $K = 16$ MeV
**15.** $R = 26.5$ km   $M = 1.78 \times 10^{31}$ kg, or 8.9 times the solar mass
**17.** 18.0 cm   **19. (a)** $10^{-6}$ s$^{-1}$  **(b)** $\tau = 10^6$ s   $T = 6.93 \times 10^5$ s
**21. (a)** 5 MeV  **(b)** $\dfrac{coll.}{sec} = 8.4 \times 10^{21}$ s$^{-1}$  **(c)** $\tau = 1.2\,\mu$s
**(d)** longer  **(e)** 8 MeV

**23.**

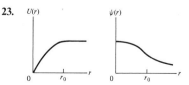

**25.** $z = 25.3$    **27.** about 3 kg per day
**29. (a)** ~66 tons of fission products **(b)** about 350 tons of fission products over 30 yr

## CHAPTER 30

**1.** $3.96 \times 10^{26}$ W    **3.** $4.17 \times 10^{38}$ protons per second
**5.** $\bar{\rho} = 1.8 \times 10^{-26}$ kg/m³
**7.** A factor of 4 if nonrelativistic and 2 if relativistic
**9.** $R = \dfrac{h^2}{4GM_p^3 n}\left(\dfrac{9n}{4\pi^2}\right)^{2/3}$
**13.** $v = \sqrt{\tfrac{8}{3}\pi G\rho}\, R = (7.47 \times 10^{-18}\text{ s}^{-1})R$
$R = c/\sqrt{\tfrac{8}{3}\pi G\rho} = 4 \times 10^{25}$ m $= 4.26 \times 10^9$ lightyears
**17.** $P = \tfrac{1}{3}\mathcal{N}hf_0$    **19.** $6.5 \times 10^{20}$

## CHAPTER 31

**1.** $T = 12$ min    $\bar{n} \to \bar{p} + e^+ + \nu_e$    **3. (a)** $\bar{\nu}_e$ **(b)** $\nu_\mu$ **(c)** $\nu_e$
**5.** $\mu^- + p \to n + \nu_\mu$

**7.** The following conservation laws would be violated: **(a)** charge **(b)** muon lepton number and electron lepton number **(c)** electron lepton number **(d)** energy **(e)** energy    **9.** left-handed
**11.** The mirror image of the $\mu^+$ is right-handed.

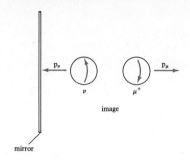

**13.** The decay positron is right-handed.
**15.** $\tau^- \to \mu^+ + \mu^- + \mu^- + \bar{\nu}_\mu + \nu_\tau$    **17.** $K_\mu = 4.18$ MeV
**19.** $K_e = 52.6$ MeV    **21.** $Mc^2 \sim 100$ MeV

# Index

A-bomb, 667, 669
Ac series circuit, 401
Ac voltage, 387, 392, 398
Ac meter, 404
Ac generator, 384–87
Ac current, 395
Absolute temperature, 241–44
Absorption, 580, 581
Acceleration, 27–34, 42
   angular, 191
   centripetal, 47–49, 70
   gravitational, 28, 39
   instantaneous, 28
   in simple harmonic motion, 216, 217
   uniform, 30, 31
Accelerators,
   high energy, 702–706
Adiabatic expansion, 260–63
Air, 303
   breakdown field, 319
Air bag, 140
Air conditioners, 274, 275, 288
Air resistance, 9, 33, 61, 135, 136
Alpha decay, 660–63, 667
American Physical Society, 271
Ampere, 335
Ampere's law, 365–68, 417
Amperian current, 376
Ångstrom, 440, 482
Angular acceleration, 191
Angular momentum, 86
   conservation of, 194–99
   definition, 193
   molecular, 257
   in quantum mechanics, 571, 572
   of a star, 690
Angular velocity, 190–92
Annihilation, 177, 291, 292, 707
Antihydrogen, 708
Antimatter, 706–708, 722
Antineutrino, 666, 700

Antiparticle symmetry, 708
Antiparticles, 706–708
Antiproton, 707
Apollo lunar module, 81
Archimedes' principle, 238, 239
Aricebo radiotelescope, 491, 510
Aristotle, 58, 84
Arsenic, 627
Artificial gravity, 89
Associated production, 714
Astrology, 2
Astrophysics, 677–95
Atmospheric pressure, 236
Atomic clocks, 4, 162, 163, 182
Atomic number, 642
Atomic physics, 590–600
Atomic radiation, 460–62
Atomic transition, 460, 461
Atomic volumes, 590
Atomic weight, 254, 642
Atwood's machine, 69, 70, 77
Audio amplifier, 637
Automobile, 26, 30, 37, 58, 65, 134–37
   flywheel engine, 209, 210
Available energy, 282
Average velocity, 24, 25, 26
Avogadro's hypothesis, 255
Avogadro's number, 254, 255

Back emf, 392
Back voltage, 629
Ballistic pendulum, 132
Balloon, 238, 239
Balmer series, 578
Band theory of solids, 615–25
Bar magnet, 371, 376–78, 391
Barn, 648
Barometer, 236, 237
Barrier penetration, 168, 633, 634, 661
Baryons, 178, 716, 717, 722
   conservation of, 702

Basal metabolic rate, 133
Basic interactions, 1, 2
Beryllium, 595
Beta decay, 176, 665, 666, 699, 709
Bevatron, 363, 704, 707
Bicycle, 199, 200
Billiards, 123
Binding energy, 569
   nuclear, 655, 656
Binomial expansion, 14
Biological energy, 133, 134
Biot-Savart law, 356, 372
Black hole, 182, 675, 681–83, 693, 694
Black disk, 513–15
Blue sky, 470
Body fat, 247, 249
Bohr, Niels, 582
Bohr model, 126, 322, 582–86
   radiation, 461, 462
Bohr postulate, 582
Boltzmann constant, 241
Bombs, 176
   nuclear, 16, 17
Bose–Einstein statistics, 632
Boundary conditions, 553
Boyle's law, 240
Breeder reactor, 670
Brewster's law, 521
Brookhaven AGS, 515
Brookhaven National Laboratory, 704, 706
Brown and Twiss, 489
Bruno, Giordano, 84
Btu/hour, 275
Bubble chamber, 178, 707, 710, 714, 715
Bullet, 132
Buoyant force, 238, 239

Calculus, 13, 23, 24, 26
Calorie, 134, 247
Capacitance, 323–25, 327, 328, 330
Capacitive reactance, 399

745

Capacitor, 323–25
  RC circuit, 405, 406
Car, 9, 26, 30, 37, 112, 113, 121
Carbon, 603
Carnot cycle (or engine), 270–75
  efficiency, 271–73
Cavendish, Henry, 82, 83
Celsius scale, 242, 243
Center of mass, 200–202
  velocity, 200, 210
Central force, 196
Centrifugal force, 48
Centripetal acceleration, 47–49, 70, 117
CERN, 188, 704, 706
Cgs system, 5, 292
Chain reaction, 581, 668, 669
Chandrasekhar limiting mass, 687
Charge, 291, 292
  accelerating, 421
  conservation of, 291, 292
  induced, 304, 327–29
  moving point, 347, 348
  point, 299, 317
Charge conjugation, 708
Charge density, 294, 348
Charged parallel plates, 314
Charged plane, 313, 314
Charged slab, 315
Charged sphere, 303, 308–11
Charged wire, 312, 313, 355
Charm, 718, 723
Chemical energy, 276
Chemical reactions, 592
Chemistry, 592, 598
Circuit
  ac series, 401
  parallel, 342
  series, 341, 342
Circuit theory, 341–46
Circular motion, 190, 191
Climber, 118, 119
Clock, 220, 222
  atomic, 4
  light, 150–52
Clock paradox, 161–64
Closed system, 72, 119
Coal, 176, 668
Coaxial cable, 313, 321, 381
  capacitance, 324
Coefficient of friction, 64, 135
Coefficient of performance, 274
Coherence, 488–90
Coherence length, 488, 489, 498
Coherent light, 497, 498, 581
Collisions, 120–24
  elastic, 122–24
  inelastic, 130, 131
Common sense, 526, 531
Commutator ring, 414
Compton effect, 529, 530

Condensed matter, 608
Conduction electrons, 336–39, 348, 612–17
Conductivity, 339, 626
Conductors, 301–302
Conical pendulum, 70, 71
Conservation laws, list of, 722–23
Conservation of angular momentum, 86, 194–99
Conservation of energy, 115–20, 129, 130, 139
Conservation of momentum, 71–74
  relativistic, 172
Conservative force, 107, 108, 115, 129
Constants (tables)
  astronomical, 730
  physical, 729
Contact force, 60, 68
Contact potential, 616
Continuous spectrum, 580
Conversion of units (table), 731
Copernicus, 84
Cornell electron synchrotron, 703, 705
Corona discharge, 319
Cosines, law of, 102
Cosmic rays, 709, 712
Cosmology, 183
Coulomb, 291, 292, 303
Coulomb potential in quantum mechanics, 564
Coulomb's law, 292, 293, 349
Covalent binding, 602, 603
CP invariance, 721–23
Crab nebula, 690, 691
Cross product, 192, 193
Cross section, 647, 648
Crystal, specific heat, 258
Crystal imperfections, 616, 617
Crystalline structure, 608–10
Current, 335, 336
  amperian, 376
  mks unit, 347
  transformation of, 359, 360
Current amplification factor, 631
Current density, 335, 336, 367–69
Current distributions, 368, 369
Current element, 372, 374
Current loop, 372, 373
Current in $p$–$n$ junction, 628
Current sheets, 369, 370
Cygnus X-1, 694

Damped oscillations, 226
Davison and Germer experiment, 536
Dc circuit theory, 341–46
De Broglie, Louis, 531
De Broglie relation, 531, 536, 541
Decibel scale, 230
Degree(s) of freedom, 246
  rotational, 256, 257
Density, 2, 5, 61
  ideal gas, 243

Derivative, 23, 24
Determinism, 548
Deuterium, 670–72
Deuteron, 563, 650–53
  binding energy, 650, 653
Diamond, 610
Dicke, R., 91
Dielectric constant, 327–29
Differential equations, 216
Diffraction
  by circular aperture, 506–11
  second order, 482
  single slit, 486–88
Diffraction grating, 481, 482, 537
Diffraction scattering, 513–15, 643
Dimensional analysis, 7
Dimensions, 3
Diode, light-emitting, 630
Diode detector, 637
Diode rectifier, 636
Dipole, 294, 295, 307, 320
  magnetic, 373, 376–78
Dipole antenna, 472
Dipole moment, 311, 312
Dipole radiation, 459
Dirac, P. M., 591, 592
Dispersion, 454
Displacement current, 419
Doping of semiconductors, 627
Doppler effect, 159, 160
Double slit experiment, 526
  with electrons, 535, 536
Double slit interference, 486, 487, 503, 504
  pattern, 530–32
Drag, 136
Drift velocity, 336–39
Dulong and Petit law, 258, 259, 283
Dynamics, 55
Dyne, 61, 62

Ear, 440
Eardrum, 229, 230, 233
Earth, 81, 83, 108, 109, 218, 219
  magnetic field, 373, 376, 413
  pressure at center, 237
  orbit, 32
Earth satellite, 49–52
Eddy currents, 391
Effective nuclear charge, 594
Efficiency
  gasoline engine, 265, 266
  heat engines, 271–73, 277, 278
Eigenfunction, 553
Eigenvalue, 553
Einstein, Albert, 145, 149, 182, 526, 527, 535
Einstein's addition of velocities, 169–71
Electric charge, 291, 292
Electric current, 335, 336

Electric dipole, 294, 295, 320, 328
  radiation, 459
Electric field, 295, 303, 384, 388–90
Electric fields (table), 316
Electric flux, 296–301
Electric generator, 384–87
Electric heating, 273, 276
Electric motor, 387
Electric potential, 317–22
Electric potential energy, 317
Electric power, 340, 392
Electric power plants, 273
Electric quadrupole, 307
Electric transmission lines, 392
Electrical induction, 304
Electromagnetic field, 355
Electromagnetic field energy, 398
Electromagnetic force, 346
Electromagnetic induction, 357
Electromagnetic pulse, 430
Electromagnetic radiation, 421–27
  energy of, 442–47
Electromagnetic theory, 149
Electromagnetic wave, 398, 425, 500
Electromotive force, 340
Electrostatic force, 290
Electrostatic potential energy of nucleus, 656
Electron, 177, 290, 315
  conduction, 336–39
  inner, 598
  magnetic moment, 378
  radius, 326
  self-energy, 326
Electron affinity, 596
Electron beam, 346, 601
Electron charge density, 592
Electron cloud, 310, 311, 449, 462, 584, 605
Electron configuration of atoms (table), 559
Electron diffraction, 537
Electron hole, 625
Electron microscope, 523
Electron orbital, 590
Electron spin, 591, 592
Electron synchrotron, 363
Electron volt, 99, 322
Elementary particles, 591, 698, 716
Elevator, 87, 88
Ellipse, 84, 85, 95
Emf, 340, 384–90
  induced, 389
  magnetic, 384–86
Emission
  spontaneous, 577, 578, 664
  thermionic, 632, 633
Energy, 97–100
  available, 282
  biological, 133, 134
  of capacitor, 325, 326
  of charge distributions, 326

  chemical, 133, 276
  conservation law, 115–20, 139, 253
  density of electromagnetic fields, 442, 446
  electric potential, 317
  of electromagnetic radiation, 442–47
  equipartition of, 245, 246
  gravitational, 124
  gravitational potential, 108, 109, 125, 126
  heat, 247–49
  in simple harmonic motion, 223, 224
  internal, 129, 253
  ionization, 323
  kinetic, 103–106, 131
  macroscopic, 253
  mechanical, 115–20, 276
  microscopic, 253
  molecular, 128
  nuclear, 133
  potential, 106–10, 118
  relativistic, 173–80
  stored in inductor, 397
  total mechanical, 126
Energy density, magnetic, 327, 398
Energy efficiency ratio, 274, 275
Energy levels, nuclear, 658
Energy loss, 131
Energy transmission, 435
Engine, gasoline, 209, 262–66
  see also Carnot cycle; Otto cycle
Engineering, mechanical, 206
Entropy, 277, 279–84
  of mixing, 280
Eötvös, R., 91
Equal areas, law of, 84, 85
Equilibrium, 128, 129
  thermal, 244, 245
Equipartition of energy, 245, 246
Equipotential surface, 323
Erg, 99
Errors
  random, 11, 21
  systematic, 11
Escape velocity, 124, 127
Escher, M. C., 116
Ether, 146–49
Euler's relation, 541
Event horizon, 681, 683
Exclusion principle, 590–92
Exhaust velocity, of rocket, 74
Expansion
  adiabatic, 260–63
  isothermal, 260
Expectation value, 576
Exponential, 406
Exponents, 15
External force, 60
Eye, 539

Fahrenheit scale, 242, 243
Falling bodies, 30

Farad, 323
Faraday's law, 357, 384, 386, 388–90
Fat, 134
Fermi, 515, 643
Fermi, Enrico, 612, 666, 669, 700, 715
Fermi energy, 527, 528, 612, 614, 615, 683, 684, 687
  relativistic, 615
Fermi gas, 612
Fermi level, 612, 613, 616, 627, 655
Fermi momentum, 613, 683, 687, 688
Fermilab, 177, 179, 188, 352, 353, 704, 705
Ferromagnetism, 379
Feynman, Richard, 533, 700
Field
  electric, 295–303
  gravitational, 91, 92
  magnetic, 347, 349–57, 365–73, 388
  physical reality of, 448
  radiation, 424
Field emission, 633, 634
Field theory, 422
Field transformations, 355–58
Fields, table of, 316
Fission
  nuclear, 178, 186, 667–70
  spontaneous, 668
Fission products, 669, 676
Fluctuations, 284
Fluids, 234–39
Fluorine, 596
Flux, 297–301
  magnetic, 367, 386
Flux linkage, 393
Flywheel, 209–12
Focal length, 517
Food calorie, 134, 247
Force
  buoyant, 238, 239
  centrifugal, 48
  conservative, 107, 108
  contact, 60
  definition, 56
  drag, 135
  electromagnetic, 55, 346
  electrostatic, 294
  external, 60
  frictional, 129
  gravitational, 55, 295
  harmonic, 215
  interaction, 60
  magnetic, 346–50
  net, 59
  from potential energy, 108
  relativistic, 181
  spring, 102, 103, 110
  units of, 62
Fourier analysis, 455
Fourier expansion, 428
Fourier integral, 430

Fourier transform, 467
Fourth dimension, 155
Franklin, Benjamin, 291, 336
Fraunhofer approximation, 477, 506
Free body diagrams, 66, 67
Free electron theory of metals, 612–16
Free expansion of gas, 244, 280, 284
Free will, 548
Frequency, 431
  of oscillation, 218
Friction, 64–66, 99, 107, 134, 135, 208, 209, 247
  kinetic, 64
  static, 64
Furnace, 276
Fusion, nuclear, 667, 670–74

Galaxy, 95, 183, 693, 708
Galileo, 150, 220
Gallium, 627
Gallium arsenide, 630
Galvanometer, 346
Gamma decay, 664, 665
Gas, ideal, 239
Gas compression, 263, 264
Gas constant $R$, 255
Gas law, ideal, 239–44
Gasoline, 137, 265
Gasoline engine, 209, 262–66
Gauss, 353
Gauss' law, 300–303, 308, 645
Gaussian function, 466, 467, 542, 559
Gaussian surface, 299
Gaussian system, 353, 354
Gaussian units, 293
Geiger counter, 532, 533
General relativity, 91, 181–83
Generator
  ac, 384–87
  dc, 414
Germanium, 610, 611, 624–27
Giga-, 4
Gravitation, universal law of, 80–84
Gravitational acceleration, 28, 39, 51
Gravitational collapse, 677, 678, 680, 695
Gravitational constant, 82, 83
Gravitational energy
  conservation of, 124
  of a star, 685
Gravitational field, 91, 92, 309
Gravitational force, 55, 65, 79, 80, 86, 103, 290
Gravitational mass, 90, 91, 96, 180–83, 291
Gravitational potential energy, 108, 114
Gravitational pressure, 678
Gravitational red shift, 182
Gravitational waves, 183, 695
Gravity-wave detectors, 183, 695
Greek alphabet, 735
Ground potential, 343
Ground state, 578
Group velocity, 455–57, 467–69, 546

H-bomb, 667, 671, 672
Hadrons, 710–16
Half-life, 152, 602, 663
Half-wave plate, 524
Hall effect, 415
Handcrank, 207
Harmonic force, 215, 216, 315
Harmonic oscillator, in quantum mechanics, 557–61
Hearing threshold, 229, 230
Heat, 247–49, 276, 340
Heat exchanger, 274
Heat pumps, 275, 276, 288
Heavy lepton, 710
Heavy nucleus, 654–57
Helium, 593, 618, 632
Helium atom, 584
Helium-neon laser, 581
Helium nucleus, 331
Helmholtz coils, 383
Henry, 393
High energy accelerators, 514, 515, 702–706
High energy electron scattering, 643–46
Holes, 625, 628
Hologram, 497–500
Holography, 495–500
Homogeneity of space, 72
Homogeneity of time, 72
Hooke's law, 102, 118, 215
Horsepower, 99, 100, 133, 135
Huggen's principle, 484–86, 498
Hybrid, 605
Hybridization, 596, 604–606
Hydroelectric power, 105
Hydrogen atom, 75, 126, 254, 306, 310, 322, 362, 551, 564, 565, 568–76
  ground state, 568, 569
  radius, 569
  wave functions, 570, 572–74
Hydrogen energy levels, 565, 571
Hydrogen gas, 592, 678
Hydrogen molecule, 232, 257, 602, 603
Hydrogen spectrum, 578, 579
Hydrogen wave functions (table), 574
  contour maps, 572, 573, 575
Hydrostatics, 234–39
Hyperons, 714

Ideal gas, 239
Image, optical, 517, 518
Impedance, 401–403
  of free space, 427
Impulse, 120, 121
Inclined plane, 66, 68
Incoherence, 488–90
Incoherent radiation, 581
Index of refraction, 450, 453–55, 506, 518
Indivisibility principle, 531, 533
Induced current, 391, 425
Induced surface current, 448
Inductance, 392–94
  of solenoid, 393, 394
Induction, electrical, 304
Inductive reactance, 400
Inductor, 395, 396
  LR circuit, 408
Inertial mass, 90, 91, 96, 182
Inertial system, 59
Infrared, 226
Integral, surface, 298, 299
Integral calculus, 81, 102, 103
Integrated circuits, 631
Intensity, 229, 230, 435, 477, 480, 498
Intensity interferometry, 489
Interaction(s)
  basic, 1, 2
  of radiation with matter, 443–55
  strong, 642, 649
Interaction force, 60
Interference
  double slit, 486, 487, 503–505
  two source, 477, 478
  of waves, 473–82
Interferometer, 148
Internal energy, 246, 247
Inverse beta decay, 688
Ionic binding, 601, 602
Ionization energy, 323
  of elements, 590, 599
Ionization potential, 592
Ionosphere, 316, 455, 456, 480
Ions, 602
Iron filings, 351, 371(f)
Irreversible heat flow, 281, 282
Irreversible processes, 284
Isothermal expansion, 260
Isotope, 642
Isotopic spin, 723

Joule, 99
Junction diode, 629
Junction theorem, 345
Jupiter, 83, 251, 686

K meson, 711, 713
$K$ shell, 601
Kelvin scale, 242, 243
Kepler, Johannes, 84–86
Kepler's laws, 84–86
Kilo-, 4
Kinematics, 22–54
Kinetic energy, 103–106
  relativistic, 178–79
  rotational, 201, 212
Kirchhoff's laws, 344

Laser, 488, 489, 497–99, 512, 513, 581, 582, 673
Lead nucleus, 514, 515
Lee, T. D., 719–21
Length, 3
Lens, 519, 520

Lenz's law, 390, 391
Leptons, 709, 710, 723
Lever, 207
Life, 152, 161, 284, 719
Light, 144-49, 421
  blue shift, 159
  Doppler effect, 159, 160, 163
  red shift, 159, 160
  speed of, 28, 29, 145, 149, 292, 350, 455
Light bulb, 340
Light clock, 150-52
Light-emitting diode, 630
Light polarization, 500-502
Light quantum, 527
Light ray, 516-18
Light spectrum, 454
Line charge, 312, 313
Line spectrum, 551
Line width, artificial, 483
Lines of force, 297-301
Lines of magnetic field, 351
Liquid helium, 632
Liter, 255
Lithium, 594, 595, 614
Lithium fluoride, 602
Loop of arbitrary shape, 411
Loop theorem, 345
Lorentz contraction, 156, 348
Lorentz transformation, 155, 173, 184
Loudspeaker, 227, 228
Lycopodium, 513
Lyman series, 578, 580

Mack's principle, 472
Macroscopic approach, 107, 133
Macroscopic quantities, 234
Magic numbers, 659, 660
Magnetic charge, 376, 377, 383
Magnetic dipole, 373, 376-78
Magnetic domains, 379
Magnetic field, 347, 349-57, 365-73, 388
  of moving charged bodies, 369
Magnetic field energy, 397
Magnetic flux, 351, 367, 390, 391
Magnetic force, 346-49, 385
Magnetic pole, 376, 377
Magnetism, 335
Magnetized iron, 377, 378
Magnification of telescope, 508, 509
Mass, 3, 5, 61, 83
  definition, 56
  of electron, 326
  gravitational, 90, 91, 96, 180-83
  inertial, 90, 91, 96, 182
  relativistic, 59, 175-78, 180, 181, 326
  rest, 59
Mass-energy relationship, 176
Mass number, 642
Mass transportation, 618
Mathematical appendix, 733, 734
Maxwell, C., 418, 421

Maxwell's equations, 301, 379, 417, 420-22
  table, 420
Mean, 11
Mean free path, 338, 361, 616
Mean life, 461, 663, 664
Mechanical advantage, 207, 235
Mechanical energy, 115-20, 276
Mechanical equivalent of heat, 248
Mega-, 4
Mercury barometer, 236
Mesons, 711-13, 717
Metabolism, 133
Metal, 336-38, 527, 528
Metallic binding, 608, 610-12
Meter, 4
Methane, 603
Metric system, 3, 4
Metric ton, 9
Michelson-Morley experiment, 146-49
Micro-, 4
Microscope, 510-12, 546
Microscopic approach, 107
Microwaves, 474
Milli-, 4
Mirror, 425
Mks system, 5, 292, 354
Modern optics, 492
Modern physics, 525
Modulation, 455, 637
Mole, 254, 255
Molecular binding, 592, 601-603
Molecular weight, 254
Molecule
  diatomic, 246, 256, 257
  hydrogen, 232, 257, 602, 603
  nonpolar, 329
  polar, 328
  potential energy, 128
  vibrational frequency, 225, 226, 257
Moment arm, 194
Moment of inertia, 203-206
Moments of inertia (table), 205
Momentum, 71, 105, 106, 120-22
  definition, 56
  of photons, 529
  quantum mechanical, 542
  relativistic, 172-75, 184, 185
Momentum distribution function, 543
Momentum of electromagnetic radiation, 446, 447
Moon, 79, 80, 81, 194, 512
Mossbauer clock, 153, 162, 168
Mossbauer effect, 182
Motor, 387
Muon, 177, 709-12
Muscle energy, 133

Nano-, 4
NASA, 81(f)
Neptunium, 666
Net force, 62, 63

Neutrino, 657, 666, 699-701, 709
Neutron, 171, 176, 177, 642, 657-59, 688, 689, 690
Neutron absorption, 647, 648
Neutron capture, 666
Neutron decay, 700
Neutron-proton potential well, 653
Neutron star, 678, 688-95
Newton, Isaac, 50, 58, 79-81, 301, 309, 421, 517
Newton's laws of motion, 57-61
Noble gases, 594, 598
Node, 474, 476
Nonconductors, 301
Nuclear binding energy, 655, 656
Nuclear emulsions, 710
Nuclear energy levels, 658
Nuclear fission, 178, 186, 667-70
Nuclear force, 55, 642, 649, 711
Nuclear fusion, 667, 670-74
Nuclear matter, 643
Nuclear physics, 642-74
Nuclear radius, 645
Nuclear reactor, 20, 667-70, 701
Nuclear waste, 670
Nuclear weapons, 16, 17, 21, 176, 186, 252, 676
Nucleon, 642
Nucleon-nucleon force, 649, 650, 653, 654
Nucleus, 642
  charge distribution, 644
  structure of, 654-57
Nuclide, 642
Numerical aperture, 511

Objective lens, 508, 510
Occam's razor, 1, 13
Ohm, Georg, 337
Ohmmeter, 362, 363
Ohm's law, 337, 339, 616
Optical instruments, 508-12
Optical pumping, 580, 581
Optics, 417, 496-524
  geometrical, 515-24
  modern, 496-500
  physical, 496-515
Orbit, 84, 85
  of earth, 32
Orbit stability in Bohr model, 585, 586
Orbital, 590
Orbital angular momentum, 571, 572
Orbital velocity, 49
Oscillating charge, 458, 459
Oscillation, small, 224-27
Oscillator, 394, 395
Otto cycle (or engine), 264, 265, 272, 273, 288

Parabola, 38, 39
Parallel axis theorem, 213
Parity, 719-23

Particle in a box, 549–52
Particle in a finite box, 554–56
Particle physics, 698–724
Particle waves, 535
Pascal's principle, 235
Path difference, 479
Path integral, 365, 366, 423
Pauli, W., 590, 591, 700
Pauli exclusion principle, 590–92, 654, 658, 683
Pendulum
  ballistic, 132
  conical, 70, 71
  physical, 221, 222
  simple, 71, 220–22, 231, 232
Period, 71
  of oscillation, 218–22
  of revolution, 48
Periodic table of the elements, 600
Periodicity groups, 597
Periodicity of the elements, 590, 596
Permanent magnet, 371, 376–78
Permittivity, 293
Perpetual motion, 115
  of the first kind, 277
  of the second kind, 276, 277
Phase difference, 479
Phase space, 640
Phasor, 463, 464, 486
Philosophers, 548
Photodiode, 630
Photoelectric effect, 526–28, 549
Photon, 448, 489, 527, 546, 560, 600, 681
  emission, 577–80
  gravitational mass, 682
Physical constants (table), 729
Pico-, 4
Pinhole camera, 523
Pion, 152, 153, 178, 187, 711–14, 720
Planck, Max, 525, 527
Planck's constant, 378, 525, 527
Plane polarization, 501
Planets, 84, 85
Plasma, 316, 455, 672, 679
Plasma frequency, 316
Plutonium, 666–69
$p$–$n$ junctions, 627–29
Point charge, 299, 317
  moving, 347, 348
Polar molecule, 328
Polarization
  circular, 501, 502, 504
  of light, 500–502
  by reflection, 505, 506, 521
Polarizer, 502
Polaroid, 503
Polaroid glasses, 506
Pollution, 670
  sound, 230
  thermal, 273

Polycrystal, 608
Polygon rule of addition, 41
Positron, 177, 657, 706
Potential, 317–22
  ground, 343
Potential difference, 318
Potential energy, 106–10, 118
  electric, 317
  gravitational, 108, 109, 125, 126
  of spring, 110
Pound, 61, 62
Poundal, 62
Power, 100, 135
  ac circuits, 403, 404
  electric, 340, 392
  solar, 15, 16
Power factor, 404
Poynting vector, 444–47
Pressure, 234–41
  atmospheric, 236
  quantum mechanical, 679, 681, 683, 697
Principle of equivalence, 90, 91, 181
Principle of relativity, 149, 150, 388
Principle of superposition, 294, 533
Prism, 454
Probability, 279, 284
Probability amplitude, 533
Probability distribution, 542
Projectile, 104, 124
Projectile motion, 45
Proliferation of nuclear weapons, 670
Proper time, 151
Proton, 310, 642, 657
Proton beam, 514
Proton core, 647
Proton structure, 515
Proton synchrotron, 352, 353
Proton-proton cycle, 679, 680
Proton-proton elastic scattering, 515, 522
Ptolemy, 84
Pulley, 69
Pulsar, 456, 457, 692, 694
Pulse of radiation, 464–69

Quadraphonic sound reproduction, 495
Quadrupole, 471
Quantum electrodynamics, 577, 724
Quantum mechanical pressure, 679, 681, 683, 697
Quantum mechanics, 257, 533, 541, 548
Quantum of light, 527
Quantum theory, 449
Quark, 698, 699, 716–18
Quasars, 160, 161, 165

Radian, 191
Radiation
  electromagnetic, 422–27
  by point charge, 457–59
Radiation field, 442, 443
  energy of, 447

Radiation power, 459
Radio, 421, 636, 637
Radio transmitter, 460, 480
Radio waves, 455-57
Radioactive decay, 176, 549
  curve, 662
Radioactivity, 664, 665
Radioastronomy, 490, 491
Radiometer, 447
Radiotelescope, 456, 490, 491, 510, 690
Random error, 21
Range, 45
Ray of light, 516–18
RC circuit, 405
Reactance
  capacitative, 399
  inductive, 400
Reactor
  hybrid, 672
  nuclear, 20, 667–70
Red shift
  gravitational, 182
  light, 159, 160
Reduced mass, 220, 650, 651
Reflection, 425
  of electromagnetic waves, 448, 449
  law of, 516, 517
  polarization by, 505, 506, 521
  of waves, 473
Reflection invariance, 719
Refraction, law of, 518
Refrigerators, 274, 275, 288
Relativistic energy, 173–80
Relativistic kinetic energy, 178–79
Relativistic mass, 59, 175–78, 180, 181, 326
Relativistic mechanics, 529
Relativistic momentum, 172–75, 184, 185
Relativity, 28, 29, 335, 346
  general, 91, 181–83
  principle of, 149, 150, 388
  special theory of, 144, 145, 149–65
Resistance, 337, 339
  air, 33
  total, 341
Resistivity, 339, 616
Resolution
  laser beam, 512
  optical, 483, 484
  telescope, 509, 510
Resolving power, 507–11
Resonance, 402, 403
  particle, 715
Resonant frequency, 396
Rest energy, 176
Rest mass, 59
Reversibility, 263, 264, 273
Rho meson, 712, 713
Rigid body, 190, 203–208
Right-hand rule, 192, 193, 350, 351
  circulation, 390

Ripple tank, 478, 634
Rock climbing, 118, 119
Rocket, 125, 126
  multistage, 77, 78
Rocket propulsion, 73, 74
Rocket ride, 88
Rollercoaster, 67, 68
Root mean square, 404
Rotating coil, 386
Rotational kinetic energy, 201, 202

Sailboat, 43
Satellite, 49–52
  synchronous, 82
Scale, spring, 57
Schrödinger equation, 553, 554, 557, 604
  three-dimensional, 567, 568, 650
Schwartzschild radius, 681
Scientific notation, 15
Seat belt, 121
Second-law efficiency, 271
Second law of thermodynamics, 270–86
Self inductance, 392–94
Semiconductor, 610, 611
Shell model, 658–60
Shoot-the-monkey problem, 46
Short circuit, 343
S.I. system, 5
Significant figures, 11
Silicon, 624, 625
Simple harmonic motion, 216–20
  energy, 223, 224
Simultaneity, 157, 158
Sine wave, 424
Single slit diffraction, 486–88
Sky, color of, 470
SLBM, 45
Slip rings, 387
Slope, 24
Slug, 62
Small oscillations, 224–27
Snell's law, 518
Society, and science, 17, 18
Sodium chloride, crystal structure, 609
Sodium energy bands, 624
Sodium line, 484
Solar cell, 629
Solar energy, 111, 276, 618, 630
Solar power, 15, 16
Solar sailing, 469
Solenoid, 370, 371
Sound, 8, 20, 227–33, 478, 495
Sound intensity, 229, 230
Sound pollution, 230
Sound velocity, 8, 440
Spaceship, 89
Spark, 303
Speaker cone, 228
Special theory of relativity, 144, 145, 149–65
Specific heat, 255–59
  constant pressure, 258, 259
  constant volume, 256–58
  tables, 259
Spectral lines, 578
Spectrometer, 483
Spectrum, 454, 551
Speed, 22
  of light, 28, 29, 145, 149, 292, 350, 455
  of sound, 8, 440
Sphere, gravitational potential energy, 114
Spherical charge distribution, 308–11
Spherical coordinates, 568
Spontaneous emission, 577, 578, 664
Spontaneous fission, 668
Spreading of wave packet, 547
Spring
  force, 110, 116–18, 215, 216, 219, 220
  potential energy, 110
Spring scale, 57
Sputnik, 50
Square modulus, 533
Square well, 551, 552
  finite depth, 554–56
Square wells in a row, 619–23
Standard deviation, 467, 542
Standing wave(s), 473, 475
  quantum mechanical, 550, 553
  on a string, 550
  three-dimensional, 566, 567
  two-dimensional, 566
Stanford Linear Accelerator Center, 705, 706
Star formation, 677
Statcoulomb, 292
Statics, 206–209
Statistical mechanics, 281, 625
Stereo sound, 495
Stereo speakers, 478
Stimulated emission, 580, 581
Strangeness, 713, 714, 723
Strange particles, 713, 714
String
  standing waves on, 474–76
  vibrating, 228
Strong interaction, 642, 649
Sun, 213, 640, 677, 680, 686, 690
  temperature at center, 243
Superconducting loop, 390
Superconducting power cable, 618
Superconductivity, 617–19
Supercritical, 669
Superfluidity, 632
Supernova, 183, 678, 681, 690–94
Superposition, 294
Surface, gaussian, 299
Surface charge, 303
Surface current, 377, 422, 423
Surface integral, 298, 299
Symmetry principles, 72
Synchronous earth satellite, 82

Synchrotron, 352, 353, 702–706
Systematic errors, 11

Tau lepton, 710
Telescope, 508–10
  reflecting, 517
Television, 638
Temperature, 241–44
  thermodynamic, 278, 279
Tempered scale, 440
Tensile strength, 210, 214, 251
Tension, 69, 476
Tera-, 4
Terminal velocity, 35
Tesla, 349, 353
Test charge, 295
Thermal pollution, 273
Thermionic emission, 632, 633
Thermistor, 626, 627
Thermodynamics, 234
  first law, 253, 254, 271
  second law, 270–86
  zeroth law, 244, 245
Thermometer, 242, 243
Thermonuclear energy, 678
Thermonuclear power, 671–73
Thermonuclear reaction, 244, 679
Thin lens formula, 519
Tides, 194
Time, 3, 4, 155, 284
Time constant, 408
Time dilation, 150–53
Time reversal, 284, 723
Time reversal invariance, 286
TNT, 669
Toroid, 382, 383
Torque, 195, 196
  on current loop, 375
Total energy, 126
Total internal reflection, 634
Trajectory, 45
Transformation of charge distributions, 359, 360
Transformation of current, 359, 360
Transformation of electric and magnetic fields, 355–58
Transformer, 391, 392
Transient, 405
Transistor, 630, 631
Transition probability, 577
Transmission lines, 392
Transmitted power, 435
Traveling wave, 228, 431, 432, 467–69
Trigonometric formulas, 733
Tritium, 671
Turbulence, 266
TV picture tube, 346
TV tuner, 396, 402, 403
Twin paradox, 161–64

Ultraviolet, 551
Uncertainty principle, 548, 665, 711
Unified weak and electromagnetic interaction, 724
Unit vectors, 44, 192
Units, 3
  cgs, 5
  conversions, 6
  mks, 5
  prefixes, 4
  S.I., 5
Unity of science, 678
Universal Fermi interaction, 700, 724
Universal law of gravitation, 80–84
Universe, 183, 697
Uranium, 238, 660

Vacuum tubes, 631
Valence, 592
Valence band, 625
Valence electrons, 610
Van de Graaf generator, 318, 319
Vector
  addition, 40
  cross product, 192, 193
Vector analysis, 13
Vectors, 39–44, 101, 102
  dot product, 101, 102
Velikovsky, I., 3
Velocity, 22–27, 41, 42
  addition, 42
  angular, 190–92
  average, 24, 25, 26
  center of mass, 200, 201
  constant, 22
  escape, 124, 127
  group, 467–69
  instantaneous, 23, 24
  orbital, 49
  in simple harmonic motion, 216, 217
  terminal, 35
  wave, 431–33, 450
Vibrational motion, 257
Volt, 317
Voltage regulator, 344, 345
Voltmeter, 364

Water waves, 478
Wave amplitude, 569
Wave equation, 423, 433, 434
Wave front, 484–86
Wave function, 533–35, 553, 554, 574, 576
Wave interference, 473–82
Wave intensity, 229, 230, 435, 447, 534, 535
Wave mechanics, 533
Wave modulation, 455
Wave nature of matter, 2, 514, 525, 526, 538
Wave packet, 464–69, 542–47
  spreading of, 547
  traveling, 228
  width, 467
Wave velocity, 431–33, 450
Wavelength, 424, 431, 432
Wavelet, 486
Wavenumber, 432
Waves, 417
  gravitational, 183
Weak interaction, 286, 699–701, 715, 716, 720
Weber, 353
Weighing the earth, 83
Weight, 86–88
  apparent, 87, 88
  atomic, 254
  definition, 65
Weighted average, 25
Weightlessness, 68, 86, 87
Wheatstone bridge, 344, 346
White dwarf, 640, 678, 684–88, 693
Work, 98, 99, 102
  done by a gas, 262
Work-energy theorem, 104, 129
Work function, 527, 528, 569, 615

X-rays, 530, 538, 600, 601
X-ray stars, 694

Yukawa, H., 711, 712

Zero point energy, 550
Zero law of thermodynamics, 244, 245

# Some Physical Constants

*See Appendix A for a more complete list with four-place accuracy.*

| | | |
|---|---|---|
| Speed of light | $c$ | $3.00 \times 10^8$ m/s |
| Acceleration due to gravity | $g$ | $9.8$ m/s$^2$ |
| Gravitational constant | $G$ | $6.67 \times 10^{-11}$ N m$^2$/kg$^2$ |